Lecture Notes in Computer Science 6487

Commenced Publication in 1973
Founding and Former Series Editors:
Gerhard Goos, Juris Hartmanis, and Jan van Leeuwen

Marc Joye   Atsuko Miyaji   Akira Otsuka (Eds.)

Pairing-Based Cryptography – Pairing 2010

4th International Conference
Yamanaka Hot Spring, Japan, December 13-15, 2010
Proceedings

 Springer

Volume Editors

Marc Joye
Technicolor, Security and Content Protection Labs
35576 Cesson-Sévigné Cedex, France
E-mail: marc.joye@technicolor.com

Atsuko Miyaji
Japan Advanced Institute of Science and Technology (JAIST)
Nomi, Ishikawa 923-1292, Japan
E-mail: miyaji@jaist.ac.jp

Akira Otsuka
National Institute of Advanced Industrial Science and Technoloy (AIST)
Tokyo 101-0021, Japan
E-mail: a-otsuka@aist.go.jp

Library of Congress Control Number: 2010939850

CR Subject Classification (1998): E.3, K.6.5, D.4.6, C.2, E.4, I.1

LNCS Sublibrary: SL 4 – Security and Cryptology

ISSN	0302-9743
ISBN-10	3-642-17454-X Springer Berlin Heidelberg New York
ISBN-13	978-3-642-17454-4 Springer Berlin Heidelberg New York

springer.com

© Springer-Verlag Berlin Heidelberg 2010
Printed in Germany

Typesetting: Camera-ready by author, data conversion by Scientific Publishing Services, Chennai, India
Printed on acid-free paper   06/3180

Preface

The 4th International Conference on Pairing-Based Cryptography (Pairing 2010) was held in Yamanaka Hot Spring, Japan, during December 13-15, 2010. It was jointly co-organized by the National Institute of Advanced Industrial Science and Technology (AIST), Japan, and the Japan Advanced Institute of Science and Technology (JAIST).

The goal of Pairing 2010 was to bring together leading researchers and practitioners from academia and industry, all concerned with problems related to pairing-based cryptography. We hope that this conference enhanced communication among specialists from various research areas and promoted creative interdisciplinary collaboration.

The conference received 64 submissions from 17 countries, out of which 25 papers from 13 countries were accepted for publication in these proceedings. At least three Program Committee (PC) members reviewed each submitted paper, while submissions co-authored by a PC member were submitted to the more stringent evaluation of five PC members. In addition to the PC members, many external reviewers joined the review process in their particular areas of expertise. We were fortunate to have this energetic team of experts, and are deeply grateful to all of them for their hard work, which included a very active discussion phase. The paper submission, review and discussion processes were effectively and efficiently made possible by the Web-based system iChair.

Furthermore, the conference featured three invited speakers: Jens Groth from University College London, Joseph H. Silverman from Brown University, and Gene Tsudik from University of California at Irvine, whose lectures on cutting-edge research areas— "Pairing-Based Non-interactive Zero-Knowledge Proofs," "A Survey of Local and Global Pairings on Elliptic Curves and Abelian Varieties," and "Some Security Topics with Possible Applications for Pairing-Based Cryptography," respectively— contributed in a significant part to the richness of the program.

We are very grateful to our supporters and sponsors. In addition to AIST and JAIST, the event was supported by the Special Interest Group on Computer Security (CSEC), IPSJ, Japan, the Japan Technical Group on Information Security (ISEC), IEICE, Japan, and the Technical Committee on Information and Communication System Security (ICSS), IEICE, Japan, and co-sponsored by the National Institute of Information and Communications Technology (NICT), Japan, Microsoft Research, Voltage Security, Hitachi, Ltd., and NTT Data.

Finally, we thank all the authors who submitted papers to this conference, the Organizing Committee members, colleagues and student helpers for their valuable time and effort, and all the conference attendees who made this event a truly intellectually stimulating one through their active participation.

December 2010

Marc Joye
Atsuko Miyaji
Akira Otsuka

Pairing 2010
The 4th International Conference on
Pairing-Based Cryptography

Jointly organized by

National Institute of Advanced Industrial Science and Technology (AIST)
and
Japan Advanced Institute of Science and Technology (JAIST)

General Chair

Akira Otsuka   AIST, Japan

Program Co-chairs

Marc Joye   Technicolor, France
Atsuko Miyaji   JAIST, Japan

Organizing Committee

Local Arrangements Co-chairs	Shoichi Hirose (University of Fukui, Japan)
	Natsume Matsuzaki (Panasonic, Japan)
	Kazumasa Omote (JAIST, Japan)
	Yuji Suga (IIJ, Japan)
	Tsuyoshi Takagi (Kyushu University, Japan)
Finance Co-chairs	Mitsuhiro Hattori (Mitsubishi Electric, Japan)
	Shoko Yonezawa (AIST, Japan)
Publicity Co-chairs	Tomoyuki Asano (Sony, Japan)
	Tetsutaro Kobayashi (NTT Labs, Japan)
	Ryo Nojima (NICT, Japan)
Liaison Co-chairs	Hiroshi Doi (IISEC, Japan)
	Masaki Inamura (KDDI R&D Labs Inc., Japan)
	Toshihiko Matsuo (NTT Data, Japan)
System Co-chairs	Nuttapong Attrapadung (AIST, Japan)
	Atsuo Inomata (NAIST, Japan)
	Yasuharu Katsuno (IBM Research - Tokyo, Japan)
	Dai Yamamoto (Fujitsu Laboratories, Japan)
	Toshihiro Yamauchi (Okayama University, Japan)
Publication Co-chairs	Goichiro Hanaoka (AIST, Japan)
	Takeshi Okamoto (Tsukuba University of Technology, Japan)

Registration Co-chairs   Hideyuki Miyake (Toshiba, Japan)
                                Katsuyuki Okeya (Hitachi, Japan)

Program Committee

Michel Abdalla	Ecole Normale Supérieure and CNRS, France
Paulo S.L.M. Barreto	University of São Paulo, Brazil
Daniel Bernstein	University of Illinois at Chicago, USA
Jean-Luc Beuchat	University of Tsukuba, Japan
Xavier Boyen	Université de Liège, Belgium
Ee-Chien Chang	National University of Singapore, Singapore
Liqun Chen	HP Labs, UK
Reza Rezaeian Farashahi	Macquarie University, Australia
David Mandell Freeman	Stanford University, USA
Jun Furukawa	NEC Corporation, Japan
Craig Gentry	IBM Research, USA
Juan González Nieto	Queensland University of Technology, Australia
Vipul Goyal	Microsoft Research, India
Shai Halevi	IBM Research, USA
Antoine Joux	University of Versailles and DGA, France
Jonathan Katz	University of Maryland, USA
Kwangjo Kim	KAIST, Korea
Kristin Lauter	Microsoft Research, USA
Pil Joong Lee	Pohang University of Science and Technology, Korea
Reynald Lercier	DGA and Université de Rennes, France
Benoît Libert	Université Catholique de Louvain, Belgium
Mark Manulis	TU Darmstadt, Germany
Giuseppe Persiano	Università di Salerno, Italy
C. Pandu Rangan	IIT Madras, India
Christophe Ritzenthaler	IML, France
Germán Sáez	UPC, Spain
Michael Scott	Dublin City University, Ireland
Alice Silverberg	University of California at Irvine, USA
Katsuyuki Takashima	Mitsubishi Electric, Japan
Keisuke Tanaka	Tokyo Institute of Technology, Japan
Edlyn Teske	University of Waterloo, Canada
Frederik Vercauteren	K.U. Leuven, Belgium
Bogdan Warinschi	University of Bristol, UK
Duncan S. Wong	City University of Hong Kong, China
Bo-Yin Yang	Academia Sinica, Taiwan
Sung-Ming Yen	National Central University, Taiwan
Fangguo Zhang	Sun Yat-sen University, P.R. China
Jianying Zhou	I2R, Singapore

External Reviewers

Joonsang Baek, Angelo De Caro, Wouter Castryck, Emanuele Cesena, Melissa Chase, Kuo-Zhe Chiou, Sherman Chow, Cheng-Kang Chu, Iwen Coisel, Vanesa Daza, Jérémie Detrey, Sungwook Eom, Essam Ghadafi, Goichiro Hanaoka, Javier Herranz, Qiong Huang, Xinyi Huang, Vincenzo Iovino, David Jao, Ezekiel Kachisa, Dalia Khader, Woo Chun Kim, Fabien Laguillaumie, Tanja Lange, Wei-Chih Lien, Hsi-Chung Lin, Georg Lippold, Jerome Milan, Michael Naehrig, Toru Nakanishi, Greg Neven, Daniel Page, Elizabeth Quaglia, Carla Rafols, Francisco Rodríguez-Henríquez, Alexandre Ruiz, Peter Schwabe, Sharmila Deva Selvi, Jae Woo Seo, Hakan Seyalioglu, Andrew Shallue, Igor Shparlinski, Dan Shumow, Kate Stange, Dongdong Sun, Koutarou Suzuki, Jheng-Hong Tu, Sree Vivek, Christian Wachsmann, Jia Xu, Lingling Xu, Greg Zaverucha, Ye Zhang, Xingwen Zhao

Table of Contents

Key Agreement

Invited Talk 2

Applications: Code Generation, Time-Released Encryption, Cloud Computing

Point Encoding and Pairing-Friendly Curves

ID-Based Encryption Schemes

Invited Talk 3

Efficient Hardware, FPGAs, and Algorithms

An Analysis of Affine Coordinates
for Pairing Computation

Kristin Lauter, Peter L. Montgomery, and Michael Naehrig

Microsoft Research, One Microsoft Way, Redmond, WA 98052, USA
{klauter,petmon,mnaehrig}@microsoft.com

Abstract. In this paper we analyze the use of affine coordinates for
pairing computation. We observe that in many practical settings, e. g.
when implementing optimal ate pairings in high security levels, affine
coordinates are faster than using the best currently known formulas for
projective coordinates. This observation relies on two known techniques
for speeding up field inversions which we analyze in the context of pairing
computation. We give detailed performance numbers for a pairing implementation based on these ideas, including timings for base field and extension field arithmetic with relative ratios for inversion-to-multiplication
costs, timings for pairings in both affine and projective coordinates, and
average timings for multiple pairings and products of pairings.

Keywords: Pairing computation, affine coordinates, optimal ate pairing,
finite field inversions, pairing cost, multiple pairings, pairing products.

1 Introduction

Cryptographic pairing computations are required for a wide variety of new cryptographic protocols and applications. All cryptographic pairings currently used
in practice are based on pairings on elliptic curves, requiring both elliptic curve
operations and function computation and evaluation to compute the pairing of
two points on an elliptic curve [36]. For a given security level, it is important to
optimize efficiency of the pairing computation, and much work has been done
on this topic (see for example [6,7,5,30,35,44,42]).

Elliptic curve operations can be implemented using various coordinate systems, such as affine or different variants of projective coordinates (for an overview
see [10]). It has long been the case that many implementers have found affine
coordinates slow for elliptic curve operations because of the relatively high costs
of inversions and the relatively fast modular multiplication that can be achieved
for special moduli such as generalized Mersenne primes. Thus projective coordinates were also suggested for pairing implementations [33,41], and very efficient
explicit formulas were found for various parameter choices [1,16]. So recently
there has been a bias in the literature towards the use of projective coordinates for pairings as well. Nevertheless, researchers had previously concluded
that affine coordinates can be superior in many situations (see [22, Section 5]
and [21, Section IX.14]).

M. Joye, A. Miyaji, and A. Otsuka (Eds.): Pairing 2010, LNCS 6487, pp. 1–20, 2010.
© Springer-Verlag Berlin Heidelberg 2010

In this paper we analyze the use of affine coordinates for pairing computation in different settings. We propose the use of two known techniques for speeding up field inversions and analyze them in the context of pairing computation. Based on these, we find that in many practical settings, for example when implementing one of the optimal pairings [44] based on the ate pairing [30] in high security levels, affine coordinates will be much faster than projective coordinates.

The first technique we investigate is computing inverses in extension fields by using towers of extension fields and successively reducing inverse computation to subfield computations via the norm map. We show that this technique drastically reduces the ratio of the costs of inversions to multiplications in extension fields. Thus when computing the ate pairing, where most computations take place in a potentially large extension field, the advantage of projective coordinates is eventually erased as the degree of the extension gets large. This happens for example when implementing pairings on curves for higher security levels such as 256 bits, or when special high-degree twists can not be used to reduce the size of the extension field.

The second technique we investigate is the use of inversion-sharing for pairing computations. Inversion-sharing is a standard trick whenever several inversions are computed at once. As the number of elements to be inverted grows, the average ratio of inversion-to-multiplication costs approaches 3. Inversion-sharing can be used in a single pairing computation if the binary expansion is read from right-to-left instead of left-to-right. This approach also has the advantage that it can be easily parallelized to take advantage of multi-core processors. Inversion-sharing for pairing computation can also be advantageous for computing multiple pairings or for computing products of pairings, as was suggested by Scott [41] and analyzed by Granger and Smart [25].

Ironically, although the two techniques we investigate can be used simultaneously, it is often not necessary to do so, since either technique alone can reduce the inversion to multiplication ratio. Either technique alone makes affine coordinates faster than projective coordinates in some settings.

To illustrate these techniques, we give detailed performance numbers for a pairing implementation based on these ideas. This includes timings for base field and extension field arithmetic with relative ratios for inversion-to-multiplication costs and timings for pairings in both affine and projective coordinates, as well as average timings for multiple pairings and products of pairings. In our implementation, affine coordinates are faster than projective coordinates even for Barreto-Naehrig curves [8] with a high-degree twist at the lowest security levels. However, we expect that for other implementations, the benefits of affine coordinates would only be realized for higher security levels or for curves without high-degree twists.

The paper is organized as follows: Section 2 provides the necessary background on the ate pairing and discusses the costs of doubling and addition steps in Miller's algorithm. In Section 3, we show how variants of the ate pairing can benefit from using affine coordinates due to the fact that the inversion-to-multiplication ratio in an extension field is much smaller than in the base field.

Section 4 is dedicated to revisiting the well-known inversion-sharing trick and its application in pairing computation. Finally, Section 5 gives benchmarking results for our pairing implementation based on the Microsoft Research bignum library.

2 Pairing Computation

Let $p > 3$ be a prime and $\mathbb{F}_q$ be a finite field of characteristic p. Let E be an elliptic curve defined over $\mathbb{F}_q$, given by $E : y^2 = x^3 + ax + b$, where $a, b \in \mathbb{F}_q$ and $4a^3 + 27b^2 \neq 0$. We denote by $\mathcal{O}$ the point at infinity on E. Let $n = \#E(\mathbb{F}_q) = q + 1 - t$, where t is the trace of Frobenius, which fulfills $|t| \leq 2\sqrt{q}$. We fix a prime r with $r \mid n$. Let k be the embedding degree of E with respect to r, i.e. k is the smallest positive integer with $r \mid q^k - 1$. This means that $\mathbb{F}_{q^k}^*$ contains the group μ_r of r-th roots of unity. The embedding degree of E is an important parameter, since it determines the field extensions over which the groups that are involved in pairing computation are defined.

For $m \in \mathbb{Z}$, let $[m]$ be the multiplication-by-m map. The kernel of $[m]$ is the set of m-torsion points on E; it is denoted by $E[m]$ and we write $E(\mathbb{F}_{q^\ell})[m]$ for the set of $\mathbb{F}_{q^\ell}$-rational m-torsion points ($\ell > 0$). If $k > 1$, which we assume from now on, we have $E[r] \subseteq E(\mathbb{F}_{q^k})$, i.e. all r-torsion points are defined over $\mathbb{F}_{q^k}$.

Most pairings that are suitable for use in practical cryptographic applications are derived from the Tate pairing, which is a map $E(\mathbb{F}_{q^k})[r] \times E(\mathbb{F}_{q^k})/rE(\mathbb{F}_{q^k}) \to \mathbb{F}_{q^k}^*/(\mathbb{F}_{q^k}^*)^r$ (for details see [18,19]). In this paper, we focus on the ate pairing [30], variants of which are often the most efficient choices for implementation.

2.1 The Ate Pairing

Given $m \in \mathbb{Z}$ and $P \in E[r]$, let $f_{m,P}$ be a rational function on E with divisor $(f_{m,P}) = m(P) - ([m]P) - (m-1)(\mathcal{O})$. Let ϕ_q be the q-power Frobenius endomorphism on E. Define two groups of prime order r by $G_1 = E[r] \cap \ker(\phi_q - [1]) = E(\mathbb{F}_q)[r]$ and $G_2 = E[r] \cap \ker(\phi_q - [q]) \subseteq E(\mathbb{F}_{q^k})[r]$. The *ate pairing* is defined as

$$a_T : G_2 \times G_1 \to \mu_r, \quad (Q, P) \mapsto f_{T,Q}(P)^{(q^k - 1)/r}, \tag{1}$$

where $T = t - 1$. The group G_2 has a nice representation by an isomorphic group of points on a twist E' of E, which is a curve that is isomorphic to E. Here, we are interested in those twists which are defined over a subfield of $\mathbb{F}_{q^k}$ such that the twisting isomorphism is defined over $\mathbb{F}_{q^k}$. Such a twist E' of E is given by an equation $E' : y^2 = x^3 + (a/\alpha^4)x + (b/\alpha^6)$ for some $\alpha \in \mathbb{F}_{q^k}$ with isomorphism $\psi : E' \to E$, $(x, y) \mapsto (\alpha^2 x, \alpha^3 y)$. If ψ is minimally defined over $\mathbb{F}_{q^k}$ and E' is minimally defined over $\mathbb{F}_{q^{k/d}}$ for a $d \mid k$, then we say that E' is a twist of degree d. If $a = 0$, let $d_0 = 4$; if $b = 0$, let $d_0 = 6$, and let $d_0 = 2$ otherwise. For $d = \gcd(k, d_0)$ there exists exactly one twist E' of E of degree d for which $r \mid \#E'(\mathbb{F}_{q^{k/d}})$ (see [30]). Define $G_2' = E'(\mathbb{F}_{q^{k/d}})[r]$. Then the map ψ is a group isomorphism $G_2' \to G_2$ and we can represent all elements in G_2 by the corresponding preimages in G_2'. Likewise, all arithmetic that needs to be done in G_2 can be carried out in G_2'. The advantage of this is that points in G_2' are defined over a smaller field than those in G_2. Using G_2', we may now view the ate pairing as a map $G_2' \times G_1 \to \mu_r$, $(Q', P) \mapsto f_{T,\psi(Q')}(P)^{(q^k-1)/r}$.

The computation of $a_T(Q', P)$ is done in two parts: first the evaluation of the function $f_{T,\psi(Q')}$ at P, and second the so-called final exponentiation to the power $(q^k - 1)/r$. The first part is done with Miller's algorithm [36]. We describe it for even embedding degree in Algorithm 1 which shows how to compute $f_{m,\psi(Q')}(P)$ for some integer $m > 0$. We denote the function given by the line through two points R_1 and R_2 on E by l_{R_1,R_2}. If $R_1 = R_2$, then the line is given by the tangent to the curve passing through R_1. Throughout this paper, we assume that k is even so that denominator elimination techniques can be used (see [6,7]).

Algorithm 1. Miller's algorithm for even k and ate-like pairings

---

Input: $Q' \in G_2', P \in G_1, m = (1, m_{l-2}, \ldots, m_0)_2$
Output: $f_{m,\psi(Q')}(P)$ representing a class in $\mathbb{F}_{q^k}^* / (\mathbb{F}_{q^k}^*)^r$
1: $R' \leftarrow Q'$, $f \leftarrow 1$
2: **for** i from $\ell - 1$ downto 0 **do**
3:     $f \leftarrow f^2 \cdot l_{\psi(R'),\psi(R')}(P)$, $R' \leftarrow [2]R'$
4:     **if** $(m_i = 1)$ **then**
5:         $f \leftarrow f \cdot l_{\psi(R'),\psi(Q')}(P)$, $R' \leftarrow R' + Q'$
6:     **end if**
7: **end for**
8: **return** f

Miller's algorithm builds up the function value $f_{m,\psi(Q')}(P)$ in a square-and-multiply-like fashion from line function values along a scalar multiplication computing $[m]Q'$ (which is the value of R' after the Miller loop). Step 3 is called a *doubling step*, it consists of squaring the intermediate value $f \in \mathbb{F}_{q^k}$, multiplying it with the function value given by the tangent to E in $R = \psi(R')$, and doubling the point R'. Similarly, an *addition step* is computed in Step 5 of Algorithm 1.

The final exponentiation in (1) maps classes in $\mathbb{F}_{q^k}^* / (\mathbb{F}_{q^k}^*)^r$ to unique representatives in μ_r. Given the fixed special exponent, there are many techniques that improve its efficiency significantly over a plain exponentiation (see [42,24]).

The most efficient variants of the ate pairing are so-called optimal ate pairings [44]. They are optimal in the sense that they minimize the size of m and with that the number of iterations in Miller's algorithm to $\log(r)/\varphi(k)$, where φ is the Euler totient function. For these minimal values of m, the function $f_{m,\psi(Q')}$ alone usually does not give a bilinear map. To get a pairing, these functions need to be adjusted by multiplying with a small number of line function values; for details we refer to [44].

Secure and efficient implementation of pairings can be done only with a careful choice of the underlying elliptic curve. The curve needs to be pairing-friendly, i.e. the embedding degree k needs to be small, while r should be larger than $\sqrt{q}$.

A survey of methods to construct such curves can be found in [20]. For security reasons, the parameters need to have certain minimal sizes which lead to optimal values for the embedding degree k for specific security levels (see for example the keysize recommendations in [43] and [4]).

Furthermore, it is advantageous to choose curves with twists of degree 4 or 6, so-called high-degree twists, since this results in higher efficiency due to the more compact representation of the group G_2. To achieve security levels of 128 bits or higher, embedding degrees of 12 and larger are optimal. Because the degree of the twist E' is at most 6, this means that when computing ate-like pairings at such security levels, all field arithmetic in the doubling and addition steps in Miller's algorithm takes place over a proper extension field of $\mathbb{F}_q$.

2.2 Costs for Doubling and Addition Steps

In this section, we take a closer look at the costs of the doubling and addition steps in Miller's algorithm. We begin by describing the evaluation of line functions in affine coordinates, i.e. a point P on E, $P \neq \mathcal{O}$, is given by two affine coordinates as $P = (x_P, y_P)$. Let $R_1, R_2, S \in E$ with $R_1 \neq -R_2$ and $R_1, R_2 \neq \mathcal{O}$. Then the function of the line through R_1 and R_2 (tangent to E if $R_1 = R_2$) evaluated at S is given by $l_{R_1,R_2}(S) = y_S - y_{R_1} - \lambda(x_S - x_{R_1})$, where $\lambda = (3x_{R_1}^2 + a)/2y_{R_1}$ if $R_1 = R_2$ and $\lambda = (y_{R_2} - y_{R_1})/(x_{R_2} - x_{R_1})$ otherwise. The value λ is also used to compute $R_3 = R_1 + R_2$ on E by $x_{R_3} = \lambda^2 - x_{R_1} - x_{R_2}$ and $y_{R_3} = \lambda(x_{R_1} - x_{R_3}) - y_{R_1}$. If $R_1 = -R_2$, then we have $x_{R_1} = x_{R_2}$ and $l_{R_1,R_2}(S) = x_S - x_{R_1}$.

Before analyzing the costs for doubling and addition steps, we introduce notations for field arithmetic costs. Let $\mathbb{F}_{q^m}$ be an extension of degree m of $\mathbb{F}_q$ for $m \geq 1$. We denote by $\mathbf{M}_{q^m}$, $\mathbf{S}_{q^m}$, $\mathbf{I}_{q^m}$, $\mathbf{add}_{q^m}$, $\mathbf{sub}_{q^m}$, and $\mathbf{neg}_{q^m}$ the costs for multiplication, squaring, inversion, addition, subtraction, and negation in the field $\mathbb{F}_{q^m}$. When we omit the indices in all of the above, this indicates that we are dealing with arithmetic in a fixed field and field extensions do not play a role. The cost for a multiplication by a constant $\omega \in \mathbb{F}_{q^m}$ is denoted by $\mathbf{M}_{(\omega)}$. We assume the same costs for addition of a constant as for a general addition.

Let notations be as described in Section 2.1. Let $e = k/d$, then $G_2' = E'(\mathbb{F}_{q^e})[r]$. Let $P \in G_1$, $R', Q' \in G_2'$ and let $R = \psi(R')$, $Q = \psi(Q')$. Furthermore, we assume that $\mathbb{F}_{q^k} = \mathbb{F}_{q^e}(\alpha)$ where $\alpha \in \mathbb{F}_{q^k}$ is the same element as the one defining the twist E', and we have $\alpha^d = \omega \in \mathbb{F}_{q^e}$. This means that each element in $\mathbb{F}_{q^k}$ is given by a polynomial of degree $d - 1$ in α with coefficients in $\mathbb{F}_{q^e}$ and the twisting isomorphism ψ maps (x', y') to $(\alpha^2 x', \alpha^3 y')$.

Doubling steps in affine coordinates: We need to compute

$$l_{R,R}(P) = y_P - \alpha^3 y_{R'} - \lambda(x_P - \alpha^2 x_{R'}) = y_P - \alpha\lambda' x_P + \alpha^3(\lambda' x_{R'} - y_{R'})$$

and $R_3' = [2]R'$, where $x_{R_3'} = \lambda'^2 - 2x_{R'}$ and $y_{R_3'} = \lambda'(x_{R'} - x_{R_3'}) - y_{R'}$. We have $\lambda' = (3x_{R'}^2 + a/\alpha^4)/2y_{R'}$ and $\lambda = (3x_R^2 + a)/2y_R = \alpha\lambda'$. Note that $[2]R' \neq \mathcal{O}$ in the pairing computation.

The slope λ' can be computed with $\mathbf{I}_{q^e} + \mathbf{M}_{q^e} + \mathbf{S}_{q^e} + 4\mathbf{add}_{q^e}$, assuming that we compute $3x_{R'}^2$ and $2y_{R'}$ by additions. To compute the double of R' from the slope λ', we need at most $\mathbf{M}_{q^e} + \mathbf{S}_{q^e} + 4\mathbf{sub}_{q^e}$. We obtain the line function value with a cost of $e\mathbf{M}_q$ to compute $\lambda' x_P$ and $\mathbf{M}_{q^e} + \mathbf{sub}_{q^e} + \mathbf{neg}_{q^e}$ for $d \in \{4, 6\}$. When $d = 2$, note that $\alpha^2 = \omega \in \mathbb{F}_{q^e}$ and thus we need $(k/2)\mathbf{M}_q + \mathbf{M}_{q^{k/2}} + \mathbf{M}_{(\omega)} + 2\mathbf{sub}_{q^{k/2}}$ for the line.

We summarize the operation counts in Table 1. We restrict to even embedding degree and $4 \mid k$ for $b = 0$ as well as to $6 \mid k$ for $a = 0$ because these cases allow using the maximal-degree twists and are likely to be used in practice. We compare the affine counts to costs of the fastest known formulas using projective coordinates taken from [31] and [16]; see these papers for details. For an overview of the most efficient explicit formulas known for elliptic-curve operations see the EFD [10]. We transfer the formulas in [31] to the ate pairing using the trick in [16] where the ate pairing is computed entirely on the twist. In this setting we assume field extensions are constructed in a way that favors the representation of line function values. This means that the twist isomorphism can be different from the one described in this paper. Still, in the case $d = 2$, evaluation of the line function can not be done in $k\mathbf{M}_q$; instead 2 multiplications in $\mathbb{F}_{q^{k/2}}$ need to be done (see [16]). Furthermore, we assume that all precomputations are done as described in the above papers and small multiples are computed by additions.

Table 1. Operation counts for the doubling step in the ate-like Miller loop omitting $1\mathbf{S}_{q^k} + 1\mathbf{M}_{q^k}$

DBL	d	coord.	$\mathbf{M}_q$	$\mathbf{I}_{q^e}$	$\mathbf{M}_{q^e}$	$\mathbf{S}_{q^e}$	$\mathbf{M}_{(\cdot)}$	$\mathbf{add}_{q^e}$	$\mathbf{sub}_{q^e}$	$\mathbf{neg}_{q^e}$
$ab \neq 0$	2	affine	$k/2$	1	3	2	$1\mathbf{M}_{(\omega)}$	4	6	—
$2 \mid k$		Jac. [31]	—	—	3	11	$1\mathbf{M}_{(a/\omega^2)}$	6	17	—
$b = 0$	4	affine	$k/4$	1	3	2	—	4	5	1
$4 \mid k$		$W_{(1,2)}$ [16]	$k/2$	—	2	8	$1\mathbf{M}_{(a/\omega)}$	9	10	1
$a = 0$	6	affine	$k/6$	1	3	2	—	4	5	1
$6 \mid k$		proj. [16]	$k/3$	—	2	7	$1\mathbf{M}_{(b/\omega)}$	11	10	1

Addition steps in affine coordinates: The line function value has the same shape as for doubling steps. Note that we can replace R' by Q' in the line and compute

$$l_{R,Q}(P) = y_P - \alpha^3 y_{Q'} - \lambda(x_P - \alpha^2 x_{Q'}) = y_P - \alpha\lambda' x_P + \alpha^3(\lambda' x_{Q'} - y_{Q'})$$

and $R'_3 = R' + Q'$, where $x_{R'_3} = \lambda'^2 - x_{R'} - x_{Q'}$ and $y_{R'_3} = \lambda'(x_{R'} - x_{R'_3}) - y_{R'}$. The slope λ' now is different, namely $\lambda' = (y_{R'} - y_{Q'})/(x_{R'} - x_{Q'})$. Note that $R' = -Q'$ does not occur when computing Miller function values of degree less than r. The cost for doing an addition step is the same as that for a doubling step, except that the cost to compute the slope λ' is now $\mathbf{I}_{q^e} + \mathbf{M}_{q^e} + 2\mathbf{sub}_{q^e}$.

Table 2 compares the costs for affine addition steps to those in projective coordinates. Again, we take these operation counts from the literature (see [1,16,15] for the explicit formulas and details on the computation). Concerning the field

Table 2. Operation counts for the addition step in the ate-like Miller loop omitting $1\mathbf{M}_{q^k}$

ADD	d	coord.	$\mathbf{M}_q$	$\mathbf{I}_{q^e}$	$\mathbf{M}_{q^e}$	$\mathbf{S}_{q^e}$	$\mathbf{add}_{q^e}$	$\mathbf{sub}_{q^e}$	$\mathbf{neg}_{q^e}$
$ab \neq 0$	2	affine	$k/2$	1	3	1	–	8	–
$2 \mid k$		Jacobian [1]	–	–	8	6	6	17	–
$b = 0$	4	affine	$k/4$	1	3	1	–	7	1
$4 \mid k$		$W_{(1,2)}$ [16]	$k/2$	–	9	5	7	8	1
$a = 0$	6	affine	$k/6$	1	3	1	–	7	1
$6 \mid k$		projective [15,16]	$k/3$	–	11	2	1	7	–

and twist representations and line function evaluation, similar remarks as for doubling steps apply here.

The multiplication with ω in the case $d = 2$ can be done as a precomputation, since Q' is fixed throughout the pairing algorithm. Since other formulas do not have multiplications by constants, we omit this column in Table 2.

Affine versus projective: Doubling and addition steps for computing pairings in affine coordinates include one inversion in $\mathbb{F}_{q^e}$ per step. The various projective formulas avoid the inversion, but at the cost of doing more of the other operations. How much higher these costs are exactly, depends on the underlying field implementation and the ratio of the costs for squaring to multiplication.

A rough estimate of the counts in Table 2 shows that for $\mathbf{S}_{q^e} = \mathbf{M}_{q^e}$ or $\mathbf{S}_{q^e} = 0.8\mathbf{M}_{q^e}$ (commonly used values in the literature, see [10]), the cost traded for the inversion in the projective addition formulas is at least $9\mathbf{M}_{q^e}$. For doubling steps, it is smaller, but larger than $3\mathbf{M}_{q^e}$ in all cases. Since doubling steps are much more frequent in the pairing computation (especially when a low Hamming weight for the degree of the used Miller function is chosen), the traded cost in the doubling case is the most relevant to consider.

Example 1. Let $ab \neq 0$, i.e. $d = 2$. The cost that has to be weighed against the inversion cost for a doubling step is $9\mathbf{S}_{q^{k/2}} - (k/2)\mathbf{M}_q + \mathbf{M}_{(a/\omega^2)} - \mathbf{M}_{(\omega)} + 2\mathbf{add}_{q^{k/2}} + 11\mathbf{sub}_{q^{k/2}}$. Clearly, $(k/2)\mathbf{M}_q < \mathbf{S}_{q^{k/2}}$, and we assume $\mathbf{M}_{(\omega)} \approx \mathbf{M}_{(a/\omega^2)}$ and $\mathbf{add}_{q^{k/2}} \approx \mathbf{sub}_{q^{k/2}}$. If $\mathbf{S}_{q^{k/2}} \approx 0.8\mathbf{M}_{q^{k/2}}$, we see that if an inversion costs less than $6.4\mathbf{M}_{q^{k/2}} + 13\mathbf{add}_{q^{k/2}}$, then affine coordinates are better than Jacobian.

Example 2. In the case $a = 0$, $d = 6$, and $\mathbf{S}_{q^{k/6}} \approx 0.8\mathbf{M}_{q^{k/6}}$, similar to the previous example, we deduce that if an inversion in $\mathbb{F}_{q^{k/6}}$ is less than $3\mathbf{M}_{q^{k/6}} + (k/6)\mathbf{M}_q + \mathbf{M}_{(b/\omega)} + 12\mathbf{add}_{q^{k/6}}$, then affine coordinates beat the projective ones.

To compare affine to projective formulas, we need to look at the relative cost of an inversion that is used in the affine formulas versus the cost of the additional operations needed for the projective formulas. Therefore, an important measure that determines whether the affine formulas are competitive with the projective formulas is the ratio of the cost of an inversion to the cost of a multiplication. For a positive integer ℓ, define the inversion-to-multiplication ratio in the field $\mathbb{F}_{q^\ell}$ by $\mathbf{R}_{q^\ell} = \mathbf{I}_{q^\ell}/\mathbf{M}_{q^\ell}$.

In implementations of prime fields, inversions are usually very expensive, i.e. the ratio $\mathbf{R}_q$ is very large. So the costs for inversions are much higher than the above-mentioned costs to avoid them. Thus it does not make sense to use affine coordinates. But it is possible to obtain much smaller ratios, e.g. when computing in extension fields. Since the ate pairing requires inversions only in $\mathbb{F}_{q^e}$ this could be in favor of affine coordinates. Depending on the specific ratio $\mathbf{R}_q$ for a given implementation, affine coordinates might even be faster than projective.

3   Inversions in Extension Fields for the Ate Pairing

In this section, we describe and analyze a way to compute inversions in finite field extensions. It is based on a given, fixed implementation of arithmetic in the underlying prime field and explains that the inversion-to-multiplication ratio $\mathbf{R} = \mathbf{I}/\mathbf{M}$ decreases when moving up in a tower of field extensions.

3.1   Inverses in Field Extensions

The method we suggest for computing the inverse of an element in an extension of some finite field $\mathbb{F}_q$ was originally described by Itoh and Tsujii [32] for binary fields using normal bases. Kobayashi et al. [34] generalize the technique to large-characteristic fields in polynomial basis and use it for elliptic-curve arithmetic. It is a standard way to compute inverses in optimal extension fields (see [2,27] and [17, Sections 11.3.4 and 11.3.6]).

We require $\mathbb{F}_{q^\ell} = \mathbb{F}_q(\alpha)$ where α has minimal polynomial $X^\ell - \omega$ for some $\omega \in \mathbb{F}_q^*$ and assume $\gcd(\ell, q) = 1$. Then, the inverse of $\beta \in \mathbb{F}_{q^\ell}^*$ can be computed as

$$\beta^{-1} = \beta^{v-1} \cdot \beta^{-v},$$

where $v = (q^\ell - 1)/(q - 1) = q^{\ell-1} + \cdots + q + 1$. Note that β^v is the norm of β and thus lies in the base field $\mathbb{F}_q$. So the cost for computing the inverse of β is the cost for computing β^{v-1} and β^v, one inversion in the base field $\mathbb{F}_q$ to obtain β^{-v}, and the multiplication of β^{v-1} with β^{-v}. The powers of β are obtained by using the q-power Frobenius automorphism on $\mathbb{F}_{q^\ell}$.

We give a brief estimate of the cost of the above. A Frobenius computation using a look-up table of $\ell - 1$ pre-computed values in $\mathbb{F}_q$ consisting of powers of ω costs at most $\ell - 1$ multiplications in $\mathbb{F}_q$ (see [34, Section 2.3], note $\gcd(\ell, q) = 1$). According to [29, Section 2.4.3] the computation of β^{v-1} via an addition chain approach, using a look-up table for each needed power of the Frobenius, costs at most $\lfloor \log(\ell - 1) \rfloor + h(\ell - 1)$ Frobenius computations and fewer multiplications in $\mathbb{F}_{q^\ell}$. Here $h(m)$ denotes the Hamming weight of an integer m. Knowing that $\beta^v \in \mathbb{F}_q$, its computation from β^{v-1} and β costs at most ℓ base field multiplications, one multiplication with ω, and $\ell - 1$ base field additions. The final multiplication of β^{-v} with β^{v-1} can be done in ℓ base field multiplications. This leads to an upper bound for the cost of an inversion in $\mathbb{F}_{q^\ell}$ as follows:

$$\mathbf{I}_{q^\ell} \leq \mathbf{I}_q + (\lfloor \log(\ell - 1) \rfloor + h(\ell - 1))(\mathbf{M}_{q^\ell} + (\ell - 1)\mathbf{M}_q)$$
$$+ 2\ell\mathbf{M}_q + \mathbf{M}_{(\omega)} + (\ell - 1)\mathrm{add}_q. \tag{2}$$

Let $M(\ell)$ be the minimal number of multiplications in $\mathbb{F}_q$ needed to multiply two different, non-trivial elements in $\mathbb{F}_{q^\ell}$ not lying in a proper subfield of $\mathbb{F}_{q^\ell}$. Then the following lemma bounds the ratio of inversion to multiplication costs in $\mathbb{F}_{q^\ell}$ from above by $1/M(\ell)$ times the ratio in $\mathbb{F}_q$ plus an explicit constant. Thus the ratio in the extension improves by roughly a factor of $M(\ell)$.

Lemma 1. *Let $\mathbb{F}_q$ be a finite field, $\ell > 1$, $\mathbb{F}_{q^\ell} = \mathbb{F}_q(\alpha)$ with $\alpha^\ell = \omega \in \mathbb{F}_q^*$. Then using the above inversion algorithm in $\mathbb{F}_{q^\ell}$ leads to*

$$\mathbf{R}_{q^\ell} \leq \mathbf{R}_q/M(\ell) + C(\ell),$$

where $C(\ell) = \lfloor \log(\ell-1) \rfloor + h(\ell-1) + \frac{1}{M(\ell)}\left(3\ell + (\ell-1)(\lfloor \log(\ell-1) \rfloor + h(\ell-1))\right)$.

Proof. Since $M(\ell)$ is the minimal number of multiplications in $\mathbb{F}_q$ needed for multiplying two elements in $\mathbb{F}_{q^\ell}$, we can assume that the actual cost for the latter is $\mathbf{M}_{q^\ell} \geq M(\ell)\mathbf{M}_q$. Using inequality (2), we deduce

$$\mathbf{R}_{q^\ell} = \mathbf{I}_{q^\ell}/\mathbf{M}_{q^\ell} \leq \mathbf{I}_q/(M(\ell)\mathbf{M}_q) + \tilde{C}(\ell) = \mathbf{R}_q/M(\ell) + \tilde{C}(\ell),$$

where $\tilde{C}(\ell) = \lfloor \log(\ell-1) \rfloor + h(\ell-1) + (2\ell + (\ell-1)(\lfloor \log(\ell-1) \rfloor + h(\ell-1)))/M(\ell) + (\mathbf{M}_\omega + (\ell-1)\mathbf{add}_q)/(M(\ell)\mathbf{M}_q)$. Since $\mathbf{M}_{(\omega)} \leq \mathbf{M}_q$ and $\mathbf{add}_q \leq \mathbf{M}_q$, we get $\mathbf{M}_\omega + (\ell-1)\mathbf{add}_q \leq \ell\mathbf{M}_q$ and thus $\tilde{C}(\ell) \leq C(\ell)$. $\qquad\square$

In Table 3 we give values for the factor $1/M(\ell)$ and the additive constant $C(\ell)$ that determine the improvements of $\mathbf{R}_{q^\ell}$ over $\mathbf{R}_q$ for several small extension degrees ℓ. We take the numbers for $M(\ell)$ from the formulas given in [38].

Table 3. Constants that determine the improvement of $\mathbf{R}_{q^\ell}$ over $\mathbf{R}_q$

ℓ	2	3	4	5	6	7
$1/M(\ell)$	1/3	1/6	1/9	1/13	1/17	1/22
$C(\ell)$	3.33	4.17	5.33	5.08	6.24	6.05

For small-degree extensions, the inversion method can be easily made explicit. We state and analyze it for quadratic and cubic extensions.

Quadratic extensions: Let $\mathbb{F}_{q^2} = \mathbb{F}_q(\alpha)$ with $\alpha^2 = \omega \in \mathbb{F}_q$. An element $\beta = b_0 + b_1\alpha \neq 0$ can be inverted as

$$\frac{1}{b_0 + b_1\alpha} = \frac{b_0 - b_1\alpha}{b_0^2 - b_1^2\omega} = \frac{b_0}{b_0^2 - b_1^2\omega} - \frac{b_1}{b_0^2 - b_1^2\omega}\alpha.$$

In this case the norm of β is given explicitly by $b_0^2 - b_1^2\omega \in \mathbb{F}_q$. The inverse of β thus can be computed in $\mathbf{I}_{q^2} = \mathbf{I}_q + 2\mathbf{M}_q + 2\mathbf{S}_q + \mathbf{M}_{(\omega)} + \mathbf{sub}_q + \mathbf{neg}_q$.

We assume that we multiply $\mathbb{F}_{q^2}$-elements with Karatsuba multiplication, which costs $\mathbf{M}_{q^2} = 3\mathbf{M}_q + \mathbf{M}_{(\omega)} + 2\mathbf{add}_q + 2\mathbf{sub}_q$. As in the general case above, we assume that the cost for a full multiplication in the quadratic extension is at least $\mathbf{M}_{q^2} \geq 3\mathbf{M}_q$, i.e. we restrict to the average case where both elements have both of their coefficients different from 0. Thus

$$\mathbf{R}_{q^2} = \mathbf{I}_{q^2}/\mathbf{M}_{q^2} \leq (\mathbf{I}_q/3\mathbf{M}_q) + 2 = \mathbf{R}_q/3 + 2,$$

where we roughly assume that $\mathbf{I}_{q^2} \leq \mathbf{I}_q + 6\mathbf{M}_q$. This bound shows that for $\mathbf{R}_q > 3$ the ratio becomes smaller in $\mathbb{F}_{q^2}$. For large ratios in $\mathbb{F}_q$ it becomes roughly $\mathbf{R}_q/3$.

Cubic extensions: Let $\mathbb{F}_{q^3} = \mathbb{F}_q(\alpha)$ with $\alpha^3 = \omega \in \mathbb{F}_q$. Similar to the quadratic case we can invert $\beta = b_0 + b_1\alpha + b_2\alpha^2 \in \mathbb{F}_{q^3}^*$ by

$$\frac{1}{b_0 + b_1\alpha + b_2\alpha^2} = \frac{b_0^2 - \omega b_1 b_2}{N(\beta)} + \frac{\omega b_2^2 - b_0 b_1}{N(\beta)}\alpha + \frac{b_1^2 - b_0 b_2}{N(\beta)}\alpha^2$$

with $N(\beta) = b_0^3 + b_1^3\omega + b_2^3\omega^2 - 3\omega b_0 b_1 b_2$. We start by computing ωb_1 and ωb_2 as well as b_0^2 and b_1^2. The terms in the numerators are obtained by a 2-term Karatsuba multiplication and additions and subtractions via $3\mathbf{M}_q$ computing $b_0 b_2$, $\omega b_1 b_2$ and $(\omega b_2 + b_0)(b_2 + b_1)$. The norm can be computed by 3 more multiplications and 2 additions. Thus the cost for the inversion is $\mathbf{I}_{q^3} = \mathbf{I}_q + 9\mathbf{M}_q + 2\mathbf{S}_q + 2\mathbf{M}_{(\omega)} + 4\mathbf{add}_q + 4\mathbf{sub}_q$. A Karatsuba multiplication can be done in $\mathbf{M}_{q^3} = 6\mathbf{M}_q + 2\mathbf{M}_{(\omega)} + 9\mathbf{add}_q + 6\mathbf{sub}_q$. We use $\mathbf{M}_{q^3} \geq 6\mathbf{M}_q$, assume $\mathbf{I}_{q^3} \leq \mathbf{I}_q + 18\mathbf{M}_q$ and obtain $\mathbf{R}_{q^3} = \mathbf{I}_{q^3}/\mathbf{M}_{q^3} \leq (\mathbf{I}_q/6\mathbf{M}_q) + 3 = \mathbf{R}_q/6 + 3$.

Towers of field extensions: Baktir and Sunar [3] introduce optimal tower fields as an alternative for optimal extension fields, where they build a large field extension as a tower of small extensions instead of one big extension. They describe how to use the above inversion technique recursively by passing down the inversion in the tower, finally arriving at the base field. They show that this method is more efficient than computing the inversion in the corresponding large extension with the Itoh-Tsujii inversion directly.

In pairing-based cryptography it is common to use towers of fields to represent the extension $\mathbb{F}_{q^k}$, where k is the embedding degree. Benger and Scott [9] discuss how to best choose such towers, but do not address inversions.

3.2   Extension-Field Inversions for the Ate Pairing

We have seen in Section 2 that for the ate pairing, the inversions in the doubling and addition steps are inversions in a proper extension field of $\mathbb{F}_q$. We now take a closer look at specific high-security levels to see which degrees these extension fields have. For a pairing-friendly elliptic curve E over $\mathbb{F}_q$ with embedding degree k with respect to a prime divisor $r \mid \#E(\mathbb{F}_q)$, we define the ρ-value of E as $\rho = \log(q)/\log(r)$. This value is a measure of the base field size relative to the size of the prime-order subgroup on the curve.

Table 4 gives the recommendations by NIST [4] and ECRYPT II [43] for equivalent levels of security for the discrete logarithm problems in the elliptic curve subgroup of order r and in a subgroup of $\mathbb{F}_{q^k}^*$. For efficiency reasons, it is

Table 4. NIST [4] and ECRYPT II [43] recommendations for bitsizes of r and q^k providing equivalent levels of security on elliptic-curve point groups and in finite fields

security (bits)	r (bits)	NIST [4] q^k (bits)	ρk	ECRYPT II [43] q^k (bits)	ρk
128	256	3072	12	3248	12.69
192	384	7680	20	7936	20.67
256	512	15360	30	15424	30.13

Table 5. Extension fields for which inversions are needed when computing ate-like pairings for different examples of pairing-friendly curve families suitable for the given security levels

security	construction in [20]	curve	k	ρ	ρk	d	extension
128	Ex. 6.8	$a = 0$	12	1.00	12.00	6	$\mathbb{F}_{p^2}$
	Ex. 6.10	$b = 0$	8	1.50	12.00	4	$\mathbb{F}_{p^2}$
	Section 5.3	$a, b \neq 0$	10	1.00	10.00	2	$\mathbb{F}_{p^5}$
	Constr. 6.7+	$a, b \neq 0$	12	1.75	21.00	2	$\mathbb{F}_{p^6}$
192	Ex. 6.12	$a = 0$	18	1.33	24.00	6	$\mathbb{F}_{p^3}$
	Ex. 6.11	$b = 0$	16	1.25	20.00	4	$\mathbb{F}_{p^4}$
	Constr. 6.3+	$a, b \neq 0$	14	1.50	21.00	2	$\mathbb{F}_{p^7}$
256	Constr. 6.6	$a = 0$	24	1.25	30.00	6	$\mathbb{F}_{p^4}$
	Constr. 6.4	$b = 0$	28	1.33	37.24	4	$\mathbb{F}_{p^7}$
	Constr. 6.24+	$a, b \neq 0$	26	1.17	30.34	2	$\mathbb{F}_{p^{13}}$

desirable to balance the security in both groups. The group sizes are linked by the embedding degree k, which leads to desired values for ρk as given in Table 4.

To implement pairings at a given security level, one needs to find a pairing-friendly elliptic curve with parameters of at least the sizes given in Table 4; for efficiency it is even desirable to obtain ρk as close to the desired value as possible. An overview of construction methods for pairing-friendly elliptic curves is given in [20]. In Table 5, we list suggestions for curve families by their construction in [20] for high-security levels of 128, 192, and 256 bits. The last column in Table 5 shows the field extensions in which inversions are done to compute the line function slopes. We not only give families of curves with twists of degree 4 and 6, but also more generic families such that the curves only have a twist of degree 2. Of course, in the latter case the extension field, in which inversions for the affine ate pairing need to be computed, is larger than when dealing with higher-degree twists. Because curves with twists of degree 4 and 6 are special (they have j-invariants 1728 and 0), there might be reasons to choose the more generic curves. Note that curves from the given constructions are all defined over prime fields. Therefore we use the notation $\mathbb{F}_p$ in Table 5.

Remark 1. The conclusion to underline from the discussion in this section, is that, using the improved inversions in towers of extension fields described here, there are at least two scenarios where *most implementations of the ate pairing would be more efficient using affine coordinates*:

When higher security levels are required, so that k is large. For example 256-bit security with $k = 28$, so that most of the computations for the ate pairing take place in the field extension of degree 7, even using a degree-4 twist (second-to-last line of Table 5). In that case, the **I/M** ratio in the degree-7 extension field would be roughly 22 times less (plus 6) than the ratio in the base field (see the last entry in Table 3). The costs for doubling and addition steps given in the second lines of Tables 1 and 2 for degree-4 twists show that the cost of the inversion avoided in a projective implementation should be compared with roughly $6\mathbf{S}_{q^7} + 5\mathbf{add}_{q^7} + 5\mathbf{sub}_{q^7}$ extra for a doubling (and an extra $6\mathbf{M}_{q^7} + 4\mathbf{S}_{q^7} + 7\mathbf{add}_{q^7} + \mathbf{sub}_{q^7}$ for an addition step). In most implementations of the base field arithmetic, the cost of these 16 or 17 operations in the extension field would outweigh the cost of one improved inversion in the extension field. See for example our sample timings for degree-6 extension fields in Table 6 in Section 5. Note there that even the cost for additions and subtractions is not negligible as is usually assumed.

When special high-degree twists are not being used. In this scenario there are two reasons why affine coordinates will be better under most circumstances:

First, the costs for doubling and addition steps given in the first lines of Tables 1 and 2 for degree-2 twists are not nearly as favorable towards projective coordinates as the formulas in the case of higher degree twists. For degree-2 twists, both the doubling and addition steps require roughly at least 9 extra squarings and 13 or 15 extra field extension additions or subtractions for the projective formulas.

Second, the degree of the extension field where the operations take place is larger. See the bottom row for each security level in Table 5, so we have extension degree 6 for 128-bit security up to extension degree 13 for 256-bit security.

4   Sharing Inversions for Pairing Computation

In this section, we revisit a well-known trick for efficiently computing several inverses at once, asymptotically achieving an **I/M**-ratio of 3. We point out and recall possibilities to improve pairing computation in affine coordinates by using this trick.

4.1   Simultaneous Inversions

The inverses of s field elements $a_1, \ldots, a_s$ can be computed simultaneously with Montgomery's well-known sharing-inversions trick [37, Section 10.3.1.] at the cost of 1 inversion and $3(s-1)$ multiplications. It is based on the following idea: to compute the inverse of two elements a and b, one computes their product ab and its inverse $(ab)^{-1}$. The inverses of a and b are then found by $a^{-1} = b \cdot (ab)^{-1}$ and $b^{-1} = a \cdot (ab)^{-1}$.

In general, for s elements one first computes the products $c_i = a_1 \cdots a_i$ for $2 \leq i \leq s$ with $s - 1$ multiplications and inverts c_s. Then we have

$a_s^{-1} = c_{s-1}c_s^{-1}$. We get a_{s-1}^{-1} by $c_{s-1}^{-1} = c_s^{-1}a_s$ and $a_{s-1}^{-1} = c_{s-2}c_{s-1}^{-1}$ and so forth (see [17, Algorithm 11.15]), where we need $2(s-1)$ more multiplications to get the inverses of all elements.

The cost for s inversions is replaced by $\mathbf{I} + 3(s-1)\mathbf{M}$. Let $\mathbf{R}_{\text{avg},s}$ denote the ratio of the cost of s inversions to the cost of s multiplications. It is bounded above by $\mathbf{R}_{\text{avg},s} = \mathbf{I}/(s\mathbf{M}) + 3(s-1)/s \leq \mathbf{R}/s + 3$, i.e. when the number s of elements to be inverted grows, the ratio $\mathbf{R}_{\text{avg},s}$ gets closer to 3. Note that most of the time, this method improves the efficiency of an implementation whenever applicable. However, as discussed in Section 3, in large field extensions, the $\mathbf{I}/\mathbf{M}$-ratio might already be less than 3 due to the inversion method from Section 3.1, in which case the sharing trick would make the average ratio worse.

4.2   Sharing Inversions in a Single Pairing Computation

Schroeppel and Beaver [40] demonstrate the use of the inversion-sharing trick to speed up a single scalar multiplication on an elliptic curve in affine coordinates. They suggest postponing addition steps in the double-and-add algorithm to exploit the inversion sharing. In order to do that, the double-and-add algorithm must be carried out by going through the binary representation of the scalar from right to left. First, all doublings are carried out and the points that will be used to add up to the final result are stored. When all these points have been collected, several additions can be done at once, sharing the computation of inversions among them.

Miller's algorithm can also be done from right to left. The doubling steps are computed without doing the addition steps. The required field elements and points are stored in lists and addition steps are done in the end. The algorithm is summarized in Algorithm 2. Unfortunately, addition steps cost much more than in the conventional left-to-right algorithm as it is given in Algorithm 1. In the right-to-left version, each addition step in Line 10 needs a general $\mathbb{F}_{q^k}$-multiplication and a multiplication with a line function value. The conventional algorithm only needs a multiplication with a line. These huge costs can not be compensated by using affine coordinates with the inversion-sharing trick.

Parallelizing a single pairing. However, the right-to-left algorithm can be parallelized, and this could lead to more efficient implementations taking advantage of the recent advent of many-core machines. Grabher, Großschädl, and Page [23, Algorithm 2] use a version of Algorithm 2 to compute a single pairing by doing addition steps in parallel on two different cores. They divide the lists with the saved function values and points into two halves and compute two intermediate values which are in the end combined in a single addition step. For their specific implementation, they conclude that this is not faster than the conventional non-parallel algorithm. Still, this idea might be useful for two or more cores, once multiple cores can be used with less overhead. It is straightforward to extend this algorithm to more cores.

So we suggest that the parallelized algorithm can be combined with the shared inversion trick when doing the addition steps in the end. The improvements

---

Algorithm 2. Right-to-left version of Miller's algorithm with postponed addition steps for even k and ate-like pairings

Input: $Q' \in G'_2, P \in G_1, m = (1 = m_{l-1}, m_{l-2}, \ldots, m_0)_2$
Output: $f_{m,\psi(Q')}(P)$ representing a class in $\mathbb{F}^*_{q^k}/(\mathbb{F}^*_{q^k})^r$
1: $R' \leftarrow Q', \quad f \leftarrow 1, \quad j \leftarrow 0$
2: **for** i from 0 to $\ell - 1$ **do**
3:     **if** $(m_i = 1)$ **then**
4:         $A_{R'}[j] \leftarrow R', \quad A_f[j] \leftarrow f, \quad j \leftarrow j + 1$
5:     **end if**
6:     $f \leftarrow f^2 \cdot l_{\psi(R'),\psi(R')}(P), \quad R' \leftarrow [2]R'$
7: **end for**
8: $R' \leftarrow A_{R'}[0], \quad f \leftarrow A_f[0]$
9: **for** $(j \leftarrow 1; \; j \leq h(m) - 1; j + +)$ **do**
10:     $f \leftarrow f \cdot A_f[j] \cdot l_{\psi(R'),\psi(A_{R'}[j])}(P), \quad R' \leftarrow R' + A_{R'}[j]$
11: **end for**
12: **return** f

---

achieved by this approach strongly depend on the Hamming weight of the value m in Miller's algorithm. If it is large, then savings are large, while for very sparse m there is almost no improvement. Therefore, when it is not possible to choose m with low Hamming weight, combining the parallelized right-to-left algorithm for pairings with the shared inversion trick can speed-up the computation. Grabher et al. [23] note that when multiple pairings are computed, it is better to parallelize by performing one pairing on each core.

4.3   Multiple Pairings and Products of Pairings

Many protocols involve the computation of multiple pairings or products of pairings. For example, multiple pairings need to be computed in the searchable encryption scheme of Boneh et al. [13]; and the non-interactive proof systems proposed by Groth and Sahai [26] need to check pairing product equations. In these scenarios, we propose sharing inversions when computing pairings with affine coordinates. In the case of products of pairings, this has already been proposed and investigated by Scott [41, Section 4.3] and Granger and Smart [25].

Multiple pairings. Assume we want to compute s pairings on points Q'_i and P_i, i.e. a priori we have s Miller loops to compute $f_{m,\psi(Q'_i)}(P_i)$. We carry out these loops simultaneously, doing all steps up to the first inversion computation for a line function slope for all of them. Only after that, all slope denominators are inverted simultaneously, and we continue with the computation for all pairings until the next inversion occurs. The s Miller loops are not computed sequentially, but rather sliced at the slope denominator inversions. The costs stay the same, except that now the average inversion-to-multiplication cost ratio is $3 + \mathbf{R}_{q^e}/s$, where $e = k/d$ and d is the twist degree.

So when computing enough pairings such that the average cost of an inversion is small enough, using the sliced-Miller approach with inversion sharing in affine coordinates is faster than using the projective coordinates explicit formulas described in Section 2.2.

Products of pairings. For computing a product of pairings, more optimizations can be applied, including the above inversion-sharing. Scott [41, Section 4.3] suggests using affine coordinates and sharing the inversions for computing the line function slopes as described above for multiple pairings. Furthermore, since the Miller function of the pairing product is the product of the Miller functions of the single pairings, in each doubling and addition step the line functions can already be multiplied together. In this way, we only need one intermediate variable f and only one squaring per iteration of the product Miller loop. Of course in the end, there is only one final exponentiation on the product of the Miller function values. Granger and Smart [25] show that by using these optimizations the cost for introducing an additional ate pairing to the product can be as low as 13% of the cost of a single ate pairing.

5   Example Implementation

The implementation described in this section is an implementation of the optimal ate pairing on a Barreto-Naehrig (BN) curve [8] over a 256-bit prime field, i.e. the curve has a 256-bit prime number n of $\mathbb{F}_p$-rational points and embedding degree $k = 12$ with respect to n.

The implementation is part of the Microsoft Research pairing library. It is specialized to the BN curve family but is not specialized for a specific BN curve. It is based on Microsoft Research's general purpose library for big number arithmetic, which can be compiled under 32-bit or 64-bit Windows. On top of that, we use the tower of field extensions $\mathbb{F}_{p^{12}}/\mathbb{F}_{p^6}/\mathbb{F}_{p^2}/\mathbb{F}_p$ to realize field arithmetic in $\mathbb{F}_{p^{12}}$. In Table 6 we give timings for the required field arithmetic in the fields $\mathbb{F}_p$, $\mathbb{F}_{p^2}$, $\mathbb{F}_{p^6}$, and $\mathbb{F}_{p^{12}}$ for the 32-bit and 64-bit versions, respectively. The 32-bit timings are for a pure software C-implementation, while the 64-bit software makes use of assembly code for base field multiplications, i.e. special code for Montgomery multiplication with a prime modulus of 256 bits, only using the fixed size of the modulus. Note that the timings in cycles and miliseconds stem from two different measurements and thus do not exactly translate.

The last column in Table 6 gives the **I/M**-ratios for the corresponding extension field and demonstrates the effect of using the inversion method for extension field towers described in Section 3.1. The ratios are even smaller than predicted by the theoretical upper bounds in Lemma 1 and Table 3. This is explained by the fact that actual multiplication costs for elements in $\mathbb{F}_{q^\ell}$ are higher than the estimates given there that take into account only multiplications from the base field and neglect all other base field operations.

The pairing implementation uses the usual optimizations. First of all, a twist $E'/\mathbb{F}_{p^2}$ provides the group G'_2 to represent elements in G_2 as described in Section 2.1. The affine doubling and addition steps in Miller's algorithm are computed as shown in Section 2.2. The projective steps use the explicit formulas

Table 6. Field arithmetic timings in a 256-bit prime field, on an Intel Core 2 Duo E8500 @ 3.16 GHz under 32-bit/64-bit Windows 7. Average over 1000 operations in cpucycles (cyc) and microseconds (μs).

		add		sub		M		S		I		R = I/M
		cyc	μs	cyc	μs	cyc	μs	cyc	μs	cyc	μs	
32-bit	$\mathbb{F}_p$	327	0.11	309	0.10	988	0.32	945	0.32	13285	4.18	13.45
	$\mathbb{F}_{p^2}$	588	0.19	585	0.18	4531	1.44	2949	0.91	18687	5.65	4.13
	$\mathbb{F}_{p^6}$	1746	0.54	1641	0.52	38938	12.09	26364	8.44	78847	24.98	2.03
	$\mathbb{F}_{p^{12}}$	3300	1.06	3233	1.03	123386	38.97	88249	27.94	210907	66.90	1.71
64-bit	$\mathbb{F}_p$	189	0.06	163	0.05	414	0.13	414	0.13	9469	2.98	22.87
	$\mathbb{F}_{p^2}$	329	0.10	300	0.10	2122	0.67	1328	0.42	11426	3.65	5.38
	$\mathbb{F}_{p^6}$	931	0.29	834	0.26	18544	5.81	12929	4.05	40201	12.66	2.17
	$\mathbb{F}_{p^{12}}$	1855	0.57	1673	0.51	60967	19.17	43081	13.57	103659	32.88	1.70

from the recent paper of Costello et al. [16]. The final exponentiation is done as described in [42], and uses the special squaring formulas given by Granger and Scott [24].

Table 7 gives benchmarking results for several pairing functions in the library, compiled under 32-bit and 64-bit Windows 7, respectively. All functions compute the optimal ate pairing for BN curves as described for example in [39]. The line entitled "20 at once (per pairing)" gives the average timing for one pairing out of 20 that have been computed at the same time. This function uses the inversion-sharing trick as described in Section 4.3. The function corresponding to the line "product of 20" computes the product of 20 pairings using the optimizations described in Section 4.3. The lines with the attribute "1st arg. fixed" mean functions that compute multiple pairings or a product of pairings, where the first input point is fixed for all pairings, and only the second point varies. In this case, the operations depending only on the first argument are done only once. We list separately the final exponentiation timings. They are included in the pairing timings of the other lines.

Table 7. Optimal ate pairing timings on a 256-bit BN curve, measured on an Intel Core 2 Duo E8500 @ 3.16 GHz under 32-bit/64-bit Windows 7. Average over 20 pairings in cpucycles (cyc) and milliseconds (ms).

optimal ate pairings		32-bit		64-bit	
		cyc	ms	cyc	ms
projective		32,288,630	10.06	15,426,503	4.88
affine	single pairing	30,091,044	9.49	14,837,947	4.64
	20 at once (per pairing)	29,681,288	9.39	14,442,433	4.53
	20 at once, 1st arg. fixed (per pairing)	27,084,852	8.53	13,124,802	4.12
	product of 20 (per pairing)	10,029,724	3.16	4,832,725	1.52
	product of 20, 1st arg. fixed (per pairing)	7,316,501	2.32	3,563,108	1.12
single final exponentiation		15,043,435	4.75	7,266,020	2.28

Implementation Notes

1. For both the 32-bit and 64-bit versions of the library, a single pairing is computed faster with affine coordinates than with projective coordinates. This is due to the relatively low $\mathbf{I}/\mathbf{M}$-ratios in the base field $\mathbb{F}_p$ (13.45 and 22.87 respectively) and in the quadratic extension (ratios 4.13 and 5.38 respectively). These low ratios are due to a relatively efficient inversion implementation in the base field combined with the improved inversion for quadratic extensions given in Section 3.1.
2. At this security level (128-bits) and using the special high-degree-6 twist, the projective implementation is almost on par with the affine implementation, so that even a small improvement in the base field multiplication would tip the balance in favor of a projective implementation.
3. However, as was explained in Remark 1 in Section 3.2, either for higher security levels or for curves without special high degree twists, affine coordinates will be much faster than projective coordinates given our base field and extension field arithmetic. Indeed, our $\mathbf{I}/\mathbf{M}$-ratio in a degree 6 extension is already roughly 2, for both our 32-bit and 64-bit versions. With a ratio of 2, projective coordinates are not a good choice.
4. Because our $\mathbf{I}/\mathbf{M}$-ratios in the quadratic field extension are already so close to 3, there is little improvement expected or observed from using the shared inversion tricks discussed in Section 4.
5. Note that field addition and subtraction costs are not negligible, as one might think from the fact that they are not often included in the operation counts when comparing various methods for elliptic curve operations and pairing implementations. In our base field arithmetic, 1 multiplication costs roughly the same as 3 field additions or subtractions, but the relative cost of additions and subtractions in extension fields is significantly less.
6. Note that the ratio of squarings to multiplications changes in the extension fields as well. A squaring in the quadratic extension is done with only 2 multiplications using the fact that the extension is generated by $\sqrt{-1}$. This improvement carries through to squarings in the higher field extensions.

Comparison to related work. We compare our work with the best results for optimal ate pairing implementations on BN curves that we are aware of.

The software described in [28] needs about $10,000,000$ cycles on an Intel Core 2 for the R-ate pairing. Modular multiplication takes 310 cycles which is about 25% faster than ours and seems to mostly account for the difference in performance with our implementation for a pairing in projective coordinates.

Recently, there has been significant improvement on pairing implementations for BN curves. The paper [39] presents an implementation that computes the optimal ate pairing on a 256-bit BN curve using one core of an Intel Core 2 Quad in about $4,380,000$ cycles. The implementation described in [11] computes the same pairing on a 254-bit BN curve in $2,490,000$ cycles on an Intel Core i7.

Software as described in [39] and [11] is much faster than our implementation for the following reason. The above implementations gain their efficiency by special curve parameter choices combined with a careful instruction scheduling specific to the parameters and certain computer architectures or even processors, in particular resulting in highly efficient multiplications in the base field and the quadratic extension field. Instead, our implementation is based on a general-purpose library for the base field arithmetic which can be compiled on many platforms and works for all BN curves. Thus our implementation is not competitive with specially tailored ones as in [39] and [11]. Nevertheless, the effects implied by the use of affine coordinates that we demonstrated with the help of our implementation also apply to implementations with faster field multiplications. Affine coordinates will then be better only when working with larger extension degrees that occur for higher security levels.

Acknowledgements. We would like to thank Dan Shumow and Tolga Acar for their help with the development environment for our implementation. We thank Steven Galbraith, Diego F. Aranha, and the anonymous referees for their helpful comments to improve the paper.

References

1. Arène, C., Lange, T., Naehrig, M., Ritzenthaler, C.: Faster computation of the Tate pairing. Journal of Number Theory (2010), doi:10.1016/j.jnt.2010.05.013
2. Bailey, D.V., Paar, C.: Efficient arithmetic in finite field extensions with application in elliptic curve cryptography. Journal of Cryptology 14(3), 153–176 (2001)
3. Baktir, S., Sunar, B.: Optimal tower fields. IEEE Transactions on Computers 53(10), 1231–1243 (2004)
4. Barker, E., Barker, W., Burr, W., Polk, W., Smid, M.: Recommendation for key management - part 1: General (revised). Technical report, NIST National Institute of Standards and Technology. Published as NIST Special Publication 800-57 (2007), http://csrc.nist.gov/groups/ST/toolkit/documents/SP800-57Part1_3-8-07.pdf
5. Barreto, P.S.L.M., Galbraith, S.D., Ó' hÉigeartaigh, C., Scott, M.: Efficient pairing computation on supersingular abelian varieties. Designs, Codes and Cryptography 42(3), 239–271 (2007)
6. Barreto, P.S.L.M., Kim, H.Y., Lynn, B., Scott, M.: Efficient algorithms for pairing-based cryptosystems. In: Yung, M. (ed.) CRYPTO 2002. LNCS, vol. 2442, pp. 354–368. Springer, Heidelberg (2002)
7. Barreto, P.S.L.M., Lynn, B., Scott, M.: Efficient implementation of pairing-based cryptosystems. Journal of Cryptology 17(4), 321–334 (2004)
8. Barreto, P.S.L.M., Naehrig, M.: Pairing-friendly elliptic curves of prime order. In: Preneel, B., Tavares, S. (eds.) SAC 2005. LNCS, vol. 3897, pp. 319–331. Springer, Heidelberg (2006)
9. Benger, N., Scott, M.: Constructing tower extensions of finite fields for implementation of pairing-based cryptography. In: Anwar Hasan, M., Helleseth, T. (eds.) WAIFI 2010. LNCS, vol. 6087, pp. 180–195. Springer, Heidelberg (2010)
10. Bernstein, D.J., Lange, T.: Explicit-formulas database, http://www.hyperelliptic.org/EFD

11. Beuchat, J.-L., González Díaz, J.E., Mitsunari, S., Okamoto, E., Rodríguez-Henríquez, F., Teruya, T.: High-speed software implementation of the optimal ate pairing over Barreto-Naehrig curves. IACR ePrint Archive, report 2010/354 (2010), http://eprint.iacr.org/2010/354
12. Blake, I.F., Seroussi, G., Smart, N.P. (eds.): Advances in Elliptic Curve Cryptography. Cambridge University Press, Cambridge (2005)
13. Boneh, D., Di Crescenzo, G., Ostrovsky, R., Persiano, G.: Public key encryption with keyword search. In: Cachin, C., Camenisch, J.L. (eds.) EUROCRYPT 2004. LNCS, vol. 3027, pp. 506–522. Springer, Heidelberg (2004)
14. Cohen, H., Frey, G., Doche, C. (eds.): Handbook of Elliptic and Hyperelliptic Curve Cryptography. Chapman and Hall/CRC, Boca Raton (2005)
15. Costello, C., Hisil, H., Boyd, C., Nieto, J.M.G., Wong, K.K.-H.: Faster pairings on special Weierstrass curves. In: Shacham, H., Waters, B. (eds.) Pairing 2009. LNCS, vol. 5671, pp. 89–101. Springer, Heidelberg (2009)
16. Costello, C., Lange, T., Naehrig, M.: Faster pairing computations on curves with high-degree twists. In: Nguyen, P.Q., Pointcheval, D. (eds.) PKC 2010. LNCS, vol. 6056, pp. 224–242. Springer, Heidelberg (2010)
17. Doche, C.: Finite Field Arithmetic. In: [14], ch. 11, pp. 201–237. CRC Press, Boca Raton (2005)
18. Duquesne, S., Frey, G.: Background on Pairings. In: [14], ch. 6, pp. 115–124. CRC Press, Boca Raton (2005)
19. Duquesne, S., Frey, G.: Implementation of Pairings. In: [14], ch. 16, pp. 389–404. CRC Press, Boca Raton (2005)
20. Freeman, D., Scott, M., Teske, E.: A taxonomy of pairing-friendly elliptic curves. Journal of Cryptology 23(2), 224–280 (2010)
21. Galbraith, S.D.: Pairings. In: [12], ch. IX, pp. 183–213. Cambridge University Press, Cambridge (2005)
22. Galbraith, S.D., Harrison, K., Soldera, D.: Implementing the Tate pairing. In: Fieker, C., Kohel, D.R. (eds.) ANTS 2002. LNCS, vol. 2369, pp. 324–337. Springer, Heidelberg (2002)
23. Grabher, P., Großschädl, J., Page, D.: On software parallel implementation of cryptographic pairings. In: Avanzi, R.M., Keliher, L., Sica, F. (eds.) SAC 2008. LNCS, vol. 5381, pp. 35–50. Springer, Heidelberg (2009)
24. Granger, R., Scott, M.: Faster squaring in the cyclotomic group of sixth degree extensions. In: Nguyen, P.Q., Pointcheval, D. (eds.) PKC 2010. LNCS, vol. 6056, pp. 209–223. Springer, Heidelberg (2010)
25. Granger, R., Smart, N.P.: On computing products of pairings. Cryptology ePrint Archive, Report 2006/172 (2006), http://eprint.iacr.org/2006/172/
26. Groth, J., Sahai, A.: Efficient non-interactive proof systems for bilinear groups. In: Smart, N.P. (ed.) EUROCRYPT 2008. LNCS, vol. 4965, pp. 415–432. Springer, Heidelberg (2008)
27. Guajardo, J., Paar, C.: Itoh-Tsujii inversion in standard basis and its application in cryptography and codes. Designs, Codes and Cryptography 25, 207–216 (2001)
28. Hankerson, D., Menezes, A.J., Scott, M.: Software implementation of pairings. In: Joye, M., Neven, G. (eds.) Identity-Based Cryptography. Cryptology and Information Security Series, vol. 2. IOS Press, Amsterdam (2008)
29. Hankerson, D., Menezes, A.J., Vanstone, S.: Guide to Elliptic Curve Cryptography. Springer, New York (2003)
30. Heß, F., Smart, N.P., Vercauteren, F.: The eta pairing revisited. IEEE Transactions on Information Theory 52, 4595–4602 (2006)

31. Ionica, S., Joux, A.: Another approach to pairing computation in Edwards coordinates. In: Chowdhury, D.R., Rijmen, V., Das, A. (eds.) INDOCRYPT 2008. LNCS, vol. 5365, pp. 400–413. Springer, Heidelberg (2008)
32. Itoh, T., Tsujii, S.: A fast algorithm for computing multiplicative inverses in GF(2^m) using normal bases. Inf. Comput. 78(3), 171–177 (1988)
33. Izu, T., Takagi, T.: Efficient computations of the Tate pairing for the large MOV degrees. In: Lee, P.J., Lim, C.H. (eds.) ICISC 2002. LNCS, vol. 2587, pp. 283–297. Springer, Heidelberg (2003)
34. Kobayashi, T., Morita, H., Kobayashi, K., Hoshino, F.: Fast elliptic curve algorithm combining Frobenius map and table reference to adapt to higher characteristic. In: Stern, J. (ed.) EUROCRYPT 1999. LNCS, vol. 1592, pp. 176–189. Springer, Heidelberg (1999)
35. Lee, E., Lee, H.S., Park, C.-M.: Efficient and generalized pairing computation on Abelian varieties. IEEE Trans. on Information Theory 55(4), 1793–1803 (2009)
36. Miller, V.S.: The Weil pairing and its efficient calculation. Journal of Cryptology 17(4), 235–261 (2004)
37. Montgomery, P.L.: Speeding the Pollard and elliptic curve methods of factorization. Mathematics of Computation 48(177), 243–264 (1987)
38. Montgomery, P.L.: Five, six, and seven-term Karatsuba-like formulae. IEEE Transactions on Computers 54(3), 362–369 (2005)
39. Naehrig, M., Niederhagen, R., Schwabe, P.: New software speed records for cryptographic pairings. In: Abdalla, M. (ed.) LATINCRYPT 2010. LNCS, vol. 6212, pp. 109–123. Springer, Heidelberg (2010), corrected version: http://www.cryptojedi.org/papers/dclxvi-20100714.pdf
40. Schroeppel, R., Beaver, C.: Accelerating elliptic curve calculations with the reciprocal sharing trick. In: Mathematics of Public-Key Cryptography (MPKC), University of Illinois at Chicago (2003)
41. Scott, M.: Computing the Tate pairing. In: Menezes, A. (ed.) CT-RSA 2005. LNCS, vol. 3376, pp. 293–304. Springer, Heidelberg (2005)
42. Scott, M., Benger, N., Charlemagne, M., Dominguez Perez, L.J., Kachisa, E.J.: On the final exponentiation for calculating pairings on ordinary elliptic curves. In: Shacham, H., Waters, B. (eds.) Pairing 2009. LNCS, vol. 5671, pp. 78–88. Springer, Heidelberg (2009)
43. Smart, N. (ed.): ECRYPT II yearly report on algorithms and keysizes (2009-2010). Technical report, ECRYPT II – European Network of Excellence in Cryptology, EU FP7, ICT-2007-216676. Published as deliverable D.SPA.13 (2010), http://www.ecrypt.eu.org/documents/D.SPA.13.pdf
44. Vercauteren, F.: Optimal pairings. IEEE Transactions on Information Theory 56(1), 455–461 (2010)

High-Speed Software Implementation of the Optimal Ate Pairing over Barreto–Naehrig Curves

Jean-Luc Beuchat[1], Jorge E. González-Díaz[2], Shigeo Mitsunari[3],
Eiji Okamoto[1], Francisco Rodríguez-Henríquez[2], and Tadanori Teruya[1]

[1] Graduate School of Systems and Information Engineering, University of Tsukuba,
1-1-1 Tennodai, Tsukuba, Ibaraki, 305-8573, Japan
[2] Computer Science Department, Centro de Investigación y de Estudios Avanzados
del IPN, Av. Instituto Politécnico Nacional No. 2508, 07300 México City, México
[3] Cybozu Labs, Inc., Akasaka Twin Tower East 15F, 2-17-22 Akasaka, Minato-ku,
Tokyo 107-0052

Abstract. This paper describes the design of a fast software library for the computation of the optimal ate pairing on a Barreto–Naehrig elliptic curve. Our library is able to compute the optimal ate pairing over a 254-bit prime field $\mathbb{F}_p$, in just 2.33 million of clock cycles on a single core of an Intel Core i7 2.8GHz processor, which implies that the pairing computation takes 0.832msec. We are able to achieve this performance by a careful implementation of the base field arithmetic through the usage of the customary Montgomery multiplier for prime fields. The prime field is constructed via the Barreto–Naehrig polynomial parametrization of the prime p given as, $p = 36t^4 + 36t^3 + 24t^2 + 6t + 1$, with $t = 2^{62} - 2^{54} + 2^{44}$. This selection of t allows us to obtain important savings for both the Miller loop as well as the final exponentiation steps of the optimal ate pairing.

Keywords: Tate pairing, optimal pairing, Barreto–Naehrig curve, ordinary curve, finite field arithmetic, bilinear pairing software implementation.

1 Introduction

The protocol solutions provided by pairing-based cryptography can only be made practical if one can efficiently compute bilinear pairings at high levels of security. Back in 1986, Victor Miller proposed in [26, 27] an iterative algorithm that can evaluate rational functions from scalar multiplications of divisors, thus allowing to compute bilinear pairings at a linear complexity cost with respect to the size of the input. Since then, several authors have found further algorithmic improvements to decrease the complexity of Miller's algorithm by reducing its loop length [3, 4, 12, 20, 21, 38], and by constructing pairing-friendly elliptic curves [5, 14, 29] and pairing-friendly tower extensions of finite fields [6, 24].

M. Joye, A. Miyaji, and A. Otsuka (Eds.): Pairing 2010, LNCS 6487, pp. 21–39, 2010.
© Springer-Verlag Berlin Heidelberg 2010

Roughly speaking, an asymmetric bilinear pairing can be defined as the non-degenerate bilinear mapping, $\hat{e} : \mathbb{G}_1 \times \mathbb{G}_2 \rightarrow \mathbb{G}_3$, where both $\mathbb{G}_1$, $\mathbb{G}_2$ are finite cyclic additive groups with prime order r, whereas $\mathbb{G}_3$ is a multiplicative cyclic group whose order is also r. Additionally, as it was mentioned above, for cryptographic applications it is desirable that pairings can be computed efficiently. When $\mathbb{G}_1 = \mathbb{G}_2$, we say that the pairing is symmetric, otherwise, if $\mathbb{G}_1 \neq \mathbb{G}_2$, the pairing is asymmetric [15].

Arguably the η_T pairing [3] is the most efficient algorithm for symmetric pairings that are always defined over supersingular curves. In the case of asymmetric pairings, recent breakthroughs include the ate pairing [21], the R-ate pairing [25], and the optimal ate pairing [38].

Several authors have presented software implementations of bilinear pairings targeting the 128-bit security level [1,8,10,16,18,23,31,32]. By taking advantage of the eight cores of a dual quad-core Intel Xeon 45nm, the software library presented in [1] takes 3.02 millions of cycles to compute the η_T pairing on a supersingular curve defined over $\mathbb{F}_{2^{1223}}$. Authors in [8] report 5.42 millions of cycles to compute the η_T pairing on a supersingular curve defined over $\mathbb{F}_{3^{509}}$ on an Intel Core i7 45nm processor using eight cores. The software library presented in [32] takes 4.470 millions of cycles to compute the optimal ate pairing on a 257-bit BN curve using only one core of an Intel Core 2 Quad Q6600 processor.

This paper addresses the efficient software implementation of asymmetric bilinear pairings at high security levels. We present a library,[1] that performs the optimal ate pairing over a 254-bit Barreto–Naehrig (BN) curve in just 2.33 million of clock cycles on a single core of an Intel i7 2.8GHz processor, which implies that the optimal ate pairing is computed in 0.832msec. To the best of our knowledge, this is the first time that a software or a hardware accelerator reports a high security level pairing computation either symmetric or asymmetric, either on one core or on a multi-core platform, in less than one millisecond. After a careful selection of a pairing-friendly elliptic curve and the tower field (Sections 2 and 3), we describe the computational complexity associated to the execution of the optimal ate pairing (Section 4). Then, we describe our approach to implement arithmetic over the underlying field $\mathbb{F}_p$ and to perform tower field arithmetic (Section 5), and we give benchmarking results of our software library (Section 6).

2   Optimal Ate Pairing over Barreto–Naehrig Curves

Barreto and Naehrig [5] described a method to construct pairing-friendly ordinary elliptic curves over a prime field $\mathbb{F}_p$. Barreto–Naehrig curves (or BN curves) are defined by the equation $E : y^2 = x^3 + b$, where $b \neq 0$. Their embedding degree k is equal to 12. Furthermore, the number of $\mathbb{F}_p$-rational points of E, denoted by r in the following, is a prime. The characteristic p of the prime field, the group order r, and the trace of Frobenius t_r of the curve are parametrized as follows [5]:

---

[1] An open source code for benchmarking our software library is available at
http://homepage1.nifty.com/herumi/crypt/ate-pairing.html

$$p(t) = 36t^4 + 36t^3 + 24t^2 + 6t + 1,$$
$$r(t) = 36t^4 + 36t^3 + 18t^2 + 6t + 1, \tag{1}$$
$$t_r(t) = 6t^2 + 1,$$

where $t \in \mathbb{Z}$ is an arbitrary integer such that $p = p(t)$ and $r = r(t)$ are both prime numbers. Additionally, t must be large enough to guarantee an adequate security level. For a security level equivalent to AES-128, we should select t such that $\log_2(r(t)) \geq 256$ and $3000 \leq k \cdot \log_2(p(t)) \leq 5000$ [14]. For this to be possible t should have roughly 64 bits.

Let $E[r]$ denote the r-torsion subgroup of E and π_p be the Frobenius endomorphism $\pi_p : E \to E$ given by $\pi_p(x, y) = (x^p, y^p)$. We define $\mathbb{G}_1 = E[r] \cap \mathrm{Ker}(\pi_p - [1]) = E(\mathbb{F}_p)[r]$, $\mathbb{G}_2 = E[r] \cap \mathrm{Ker}(\pi_p - [p]) \subseteq E(\mathbb{F}_{p^{12}})[r]$, and $\mathbb{G}_3 = \mu_r \subset \mathbb{F}_{p^{12}}^*$ (*i.e.* the group of r-th roots of unity). Since we work with a BN curve, r is a prime and $\mathbb{G}_1 = E(\mathbb{F}_p)[r] = E(\mathbb{F}_p)$. The optimal ate pairing on the BN curve E is a non-degenerate and bilinear pairing given by the map [30, 32, 38]:

$$a_{\mathrm{opt}} : \mathbb{G}_2 \times \mathbb{G}_1 \longrightarrow \mathbb{G}_3$$
$$(Q, P) \longmapsto \left(f_{6t+2,Q}(P) \cdot l_{[6t+2]Q,\pi_p(Q)}(P) \cdot \right.$$
$$\left. l_{[6t+2]Q+\pi_p(Q),-\pi_p^2(Q)}(P) \right)^{\frac{p^{12}-1}{r}},$$

where

- $f_{s,Q}$, for $s \in \mathbb{N}$ and $Q \in \mathbb{G}_2$, is a family of normalized $\mathbb{F}_{p^{12}}$-rational functions with divisor $(f_{s,Q}) = s(Q) - ([s]Q) - (s - 1)(\mathcal{O})$, where $\mathcal{O}$ denotes the point at infinity.
- l_{Q_1,Q_2} is the equation of the line corresponding to the addition of $Q_1 \in \mathbb{G}_2$ with $Q_2 \in \mathbb{G}_2$.

Algorithm 1 shows how we compute the optimal ate pairing in this work. Our approach can be seen as a signed-digit version of the algorithm utilized in [32], where both point additions and point subtractions are allowed. The Miller loop (lines 3–10) calculates the value of the rational function $f_{6t+2,Q}$ at point P. In lines 11–13 the product of the line functions $l_{[6t+2]Q,\pi_p(Q)}(P) \cdot l_{[6t+2]Q+\pi_p(Q),-\pi_p^2(Q)}(P)$ is multiplied by $f_{6t+2,Q}(P)$. The so-called final exponentiation is computed in line 14. A detailed summary of the computational costs associated to Algorithm 1 can be found in Section 4.

The BN curves admit a sextic twist $E'/\mathbb{F}_{p^2} : y^2 = x^3 + b/\xi$ defined over $\mathbb{F}_{p^2}$, where $\xi \in \mathbb{F}_{p^2}$ is an element that is neither a square nor a cube in $\mathbb{F}_{p^2}$, and that has to be carefully selected such that $r | \#E'(\mathbb{F}_{p^2})$ holds. This means that pairing computations can be restricted to points P and Q' that belong to $E(\mathbb{F}_p)$ and $E'(\mathbb{F}_{p^2})$, respectively, since we can represent the points in $\mathbb{G}_2$ by points on the twist [5, 21, 38].

3   Tower Extension Field Arithmetic

Since $k = 12 = 2^2 \cdot 3$, the tower extensions can be created using irreducible binomials only. This is because $x^k - \beta$ is irreducible over $\mathbb{F}_p$ provided that $\beta \in \mathbb{F}_p$

Algorithm 1. Optimal ate pairing over Barreto–Naehrig curves.

Input: $P \in \mathbb{G}_1$ and $Q \in \mathbb{G}_2$.
Output: $a_{\mathrm{opt}}(Q, P)$.
 1. Write $s = 6t + 2$ as $s = \sum_{i=0}^{L-1} s_i 2^i$, where $s_i \in \{-1, 0, 1\}$;
 2. $T \leftarrow Q, f \leftarrow 1$;
 3. **for** $i = L - 2$ to 0 **do**
 4. $f \leftarrow f^2 \cdot l_{T,T}(P); T \leftarrow 2T$;
 5.   **if** $s_i = -1$ **then**
 6. $f \leftarrow f \cdot l_{T,-Q}(P); T \leftarrow T - Q$;
 7.   **else if** $s_i = 1$ **then**
 8. $f \leftarrow f \cdot l_{T,Q}(P); T \leftarrow T + Q$;
 9.   **end if**
10. **end for**
11. $Q_1 \leftarrow \pi_p(Q); Q_2 \leftarrow \pi_{p^2}(Q)$;
12. $f \leftarrow f \cdot l_{T,Q_1}(P); T \leftarrow T + Q_1$;
13. $f \leftarrow f \cdot l_{T,-Q_2}(P); T \leftarrow T - Q_2$;
14. $f \leftarrow f^{(p^{12}-1)/r}$;
15. **return** f;

is neither a square nor a cube in $\mathbb{F}_p$ [24]. Hence, the tower extension can be constructed by simply adjoining a cube or square root of such element β and then the cube or square root of the previous root. This process should be repeated until the desired extension of the tower has been reached.

Accordingly, we decided to represent $\mathbb{F}_{p^{12}}$ using the same tower extension of [18], namely, we first construct a quadratic extension, which is followed by a cubic extension and then by a quadratic one, using the following irreducible binomials:

$$
\begin{aligned}
\mathbb{F}_{p^2} &= \mathbb{F}_p[u]/(u^2 - \beta), \text{ where } \beta = -5, \\
\mathbb{F}_{p^6} &= \mathbb{F}_{p^2}[v]/(v^3 - \xi), \text{ where } \xi = u, \\
\mathbb{F}_{p^{12}} &= \mathbb{F}_{p^6}[w]/(w^2 - v).
\end{aligned}
\tag{2}
$$

We adopted the tower extension of Equation (2), mainly because field elements $f \in \mathbb{F}_{p^{12}}$ can be seen as a quadratic extension of $\mathbb{F}_{p^6}$, and hence they can be represented as $f = g + hw$, with $g, h \in \mathbb{F}_{p^6}$. This towering will help us to exploit the fact that in the hard part of the final exponentiation we will deal with field elements $f \in \mathbb{F}_{p^{12}}$ that become *unitary* [35, 36], *i.e.*, elements that belong to the cyclotomic subgroup $\mathbb{G}_{\Phi_6}(\mathbb{F}_{p^2})$ as defined in [17]. Such elements satisfy, $f^{p^6+1} = 1$, which means that $f^{-1} = f^{p^6} = g - hw$. In other words, inversion of such elements can be accomplished by simple conjugation. This nice feature opens the door for using addition-subtraction chains in the final exponentiation step, which is especially valuable for our binary signed choice of the parameter t. We also stress that our specific t selection permits to use $\xi = u \in \mathbb{F}_p$, which will yield important savings in the arithmetic computational cost as discussed next.

3.1 Computational Costs of the Tower Extension Field Arithmetic

The tower extension arithmetic algorithms used in this work were directly adopted from [18]. Let (a, m, s, i), $(\tilde{a}, \tilde{m}, \tilde{s}, \tilde{i})$, and (A, M, S, I) denote the cost of field addition, multiplication, squaring, and inversion in $\mathbb{F}_p$, $\mathbb{F}_{p^2}$, and $\mathbb{F}_{p^6}$, respectively. From our implementation (see Section 5), we observed experimentally that $m = s = 8a$ and $i = 48.3m$. We summarize the towering arithmetic costs as follows:

- In the field $\mathbb{F}_{p^2}$, we used Karatsuba multiplication and the complex method for squaring, at a cost of 3 and 2 field multiplications in $\mathbb{F}_p$, respectively. Inversion of an element $A = a_0 + a_1 u \in \mathbb{F}_{p^2}$, can be found from the identity, $(a_0 + a_1 u)^{-1} = (a_0 - a_1 u)/(a_0^2 - \beta a_1^2)$. Using once again the Karatsuba method, field multiplication in $\mathbb{F}_{p^6}$ can be computed at a cost of $6\tilde{m}$ plus several addition operations. All these three operations require the multiplication in the base field by the constant coefficient $\beta \in \mathbb{F}_p$ of the irreducible binomial $u^2 - \beta$. We refer to this operation as m_β Additionally, we sometimes need to compute the multiplication of an arbitrary element in $\mathbb{F}_{p^2}$ times the constant $\xi = u \in \mathbb{F}_p$ at a cost of one multiplication by the constant β. We refer to this operation as m_ξ, but it is noticed that the cost of m_ξ is essentially the same of that of m_β.
- Squaring in $\mathbb{F}_{p^6}$ can be computed via the formula derived in [9] at a cost of $2\tilde{m} + 3\tilde{s}$ plus some addition operations. Inversion in the sextic extension can be computed at a cost of $9\tilde{m} + 3\tilde{s} + 4m_\beta + 5\tilde{a} + \tilde{i}$ [34].
- Since our field towering constructed $\mathbb{F}_{p^{12}}$ as a quadratic extension of $\mathbb{F}_{p^6}$, the arithmetic costs of the quadratic extension apply. Hence, a field multiplication, squaring and inversion costs in $\mathbb{F}_{p^{12}}$ are, $3M + 5A$, $2M + 5A$ and $2M + 2S + 2A + I$, respectively. However, if $f \in \mathbb{F}_{p^{12}}$, belongs to the cyclotomic subgroup $\mathbb{G}_{\Phi_6}(\mathbb{F}_{p^2})$, its field squaring f^2 can be reduced to three squarings in $\mathbb{F}_{p^4}$ [17].

Table 1 lists the computational costs of the tower extension field arithmetic in terms of the $\mathbb{F}_{p^2}$ field arithmetic operations, namely, $(\tilde{a}, \tilde{m}, \tilde{s}, \tilde{i})$.

Table 1. Computational costs of the tower extension field arithmetic

Field	Add./Sub.	Mult.	Squaring	Inversion
$\mathbb{F}_{p^2}$	$\tilde{a} = 2a$	$\tilde{m} = 3m + 3a + m_\beta$	$\tilde{s} = 2m + 3a + m_\beta$	$\tilde{i} = 4m + m_\beta$ $+2a + i$
$\mathbb{F}_{p^6}$	$3\tilde{a}$	$6\tilde{m} + 2m_\beta + 15\tilde{a}$	$2\tilde{m} + 3\tilde{s} + 2m_\beta + 8\tilde{a}$	$9\tilde{m} + 3\tilde{s} + 4m_\beta$ $+4\tilde{a} + \tilde{i}$
$\mathbb{F}_{p^{12}}$	$6\tilde{a}$	$18\tilde{m} + 6m_\beta + 60\tilde{a}$	$12\tilde{m} + 4m_\beta + 45\tilde{a}$	$25\tilde{m} + 9\tilde{s} + 12m_\beta$ $+61\tilde{a} + \tilde{i}$
$\mathbb{G}_{\Phi_6}(\mathbb{F}_{p^2})$	$6\tilde{a}$	$18\tilde{m} + 6m_\beta + 60\tilde{a}$	$9\tilde{s} + 4m_\beta$ $+30\tilde{a}$	Conjugation

3.2   Frobenius Operator

Raising an element $f \in \mathbb{F}_{p^{12}} = \mathbb{F}_{p^6}[w]/(w^2 - v)$ to the p-power, is an arithmetic operation needed in the final exponentiation (line 14) of the optimal ate pairing (Algorithm 1). We briefly describe in the following how to compute f^p efficiently.

We first remark that the field extension $\mathbb{F}_{p^{12}}$ can be also represented as a sextic extension of the quadratic field, i.e., $\mathbb{F}_{p^{12}} = \mathbb{F}_{p^2}[W]/(W^6 - u)$, with $W = w$. Hence, we can write $f = g + hw \in \mathbb{F}_{p^{12}}$, with g, $h \in \mathbb{F}_{p^6}$ such that, $g = g_0 + g_1 v + g_2 v^2$, $h = h_0 + h_1 v + h_2 v^2$, where $g_i, h_i \in \mathbb{F}_{p^2}$, for $i = 1, 2, 3$. This means that f can be equivalently written as, $f = g + hw = g_0 + h_0 W + g_1 W^2 + h_1 W^3 + g_2 W^4 + h_2 W^5$.

We note that the p-power of an arbitrary element in the quadratic extension field $\mathbb{F}_{p^2}$ can be computed essentially free of cost as follows. Let $b \in \mathbb{F}_{p^2}$ be an arbitrary element that can be represented as $b = b_0 + b_1 u$. Then, $(b)^{p^{2i}} = b$ and $(b)^{p^{2i-1}} = \bar{b}$, with $\bar{b} = b_0 - b_1 u$, for $i \in \mathbb{N}$.

Let $\bar{g}_i$, $\bar{h}_i$, denote the conjugates of g_i, h_i, for $i = 1, 2, 3$ respectively. Then, using the identity $W^p = u^{(p-1)/6} W$, we can write, $(W^i)^p = \gamma_{1,i} W^i$, with $\gamma_{1,i} = u^{i(p-1)/6}$, for $i = 1, \ldots, 5$. From the definitions given above, we can compute f^p as,

$$
\begin{aligned}
f^p &= \left(g_0 + h_0 W + g_1 W^2 + h_1 W^3 + g_2 W^4 + h_2 W^5\right)^p \\
&= \bar{g}_0 + \bar{h}_0 W^p + \bar{g}_1 W^{2p} + \bar{h}_1 W^{3p} + \bar{g}_2 W^{4p} + \bar{h}_2 W^{5p} \\
&= \bar{g}_0 + \bar{h}_0 \gamma_{1,1} W + \bar{g}_1 \gamma_{1,2} W^2 + \bar{h}_1 \gamma_{1,3} W^3 + \bar{g}_2 \gamma_{1,4} W^4 + \bar{h}_2 \gamma_{1,5} W^5.
\end{aligned}
$$

The equation above has a computational cost of 5 multiplications in $\mathbb{F}_p$ and 5 conjugations in $\mathbb{F}_{p^2}$. We can follow a similar procedure for computing f^{p^2} and f^{p^3}, which are arithmetic operations required in the hard part of the final exponentiation of Algorithm 1. For that, we must pre-compute and store the per-field constants $\gamma_{1,i} = u^{i \cdot (p-1)/6}$, $\gamma_{2,i} = \gamma_{1,i} \cdot \bar{\gamma}_{1,i}$, and $\gamma_{3,i} = \gamma_{1,i} \cdot \gamma_{2,i}$ for $i = 1, \ldots, 5$.

4   Computational Cost of the Optimal Ate Pairing

In this work we considered several choices of the parameter t, required for defining $p(t)$, $r(t)$, and $t_r(t)$ of Equation (1). We found 64-bit values of t with Hamming weight as low as 2 that yield the desired properties for p, r, and t_r. For example, the binomial $t = 2^{63} - 2^{49}$ guarantees that p and r as defined in Equation (1) are both 258-bit prime numbers. However, due to the superior efficiency on its associated base field arithmetic, we decided to use the trinomial $t = 2^{62} - 2^{54} + 2^{44}$, which guarantees that p and r as defined in Equation (1) are 254-bit prime numbers. Since the automorphism group $\text{Aut}(E)$ is a cyclic group of order 6 [30], it is possible to slightly improve Pollard's rho attack and get a speedup of $\sqrt{6}$ [11]. Therefore, we achieve a 126-bit security level with our choice of parameters. The curve equation is $E : Y^2 = X^3 + 5$ and we followed the procedure outlined in [6, 36] in order to find a generator $P = (x_P, y_P) = (1, \sqrt{6})$ for the group $E(\mathbb{F}_p)$, and one generator $Q' = (x_{Q'}, y_{Q'})$ for the group $E'(\mathbb{F}_{p^2})[r]$, given as,

$$x_{Q'} = \texttt{0x19B0BEA4AFE4C330DA93CC3533DA38A9F430B471C6F8A536E81962ED967909B5}$$
$$+ \texttt{0xA1CF585585A61C6E9880B1F2A5C539F7D906FFF238FA6341E1DE1A2E45C3F72}u,$$
$$y_{Q'} = \texttt{0x17ABD366EBBD65333E49C711A80A0CF6D24ADF1B9B3990EEDCC91731384D2627}$$
$$+ \texttt{0xEE97D6DE9902A27D00E952232A78700863BC9AA9BE960C32F5BF9FD0A32D345}u.$$

In this Section, we show that our selection of t yields important savings in the Miller loop and the hard part of the final exponentiation step of Algorithm 1.

4.1 Miller Loop

We remark that the parameter $6t + 2$ of Algorithm 1 has a bitlength $L = 65$, with a Hamming weight of 7. This implies that the execution of the Miller loop requires 64 doubling step computations in line 4, and 6 addition/subtraction steps in lines 6 and 8.

It is noted that the equation of the tangent line at $T \in \mathbb{G}_2$ evaluated at P defines a sparse element in $\mathbb{F}_{p^{12}}$ (half of the coefficients are equal to zero). The same observation holds for the equation of the line through the points T and $\pm Q$ evaluated at P. This sparsity allows us to reduce the number of operations on the underlying field when performing accumulation steps (lines 4, 6, 8, 12, and 13 of Algorithm 1).

We perform an interleaved computation of the tangent line at point T (respectively, the line through the points T and Q) evaluated at the base point P, with a point doubling (respectively, point addition) using the formulae given in [2]. We recall that the field extension $\mathbb{F}_{p^{12}}$ can be also represented as, $\mathbb{F}_{p^{12}} = \mathbb{F}_{p^2}[W]/(W^6 - u)$, with $W = w$.

Doubling step (line 4). We represent the point $T \in E'(\mathbb{F}_{p^2})$ in Jacobian coordinates as $T = (X_T, Y_T, Z_T)$. The formulae for doubling T, *i.e.*, the equations that define the point $R = 2T = (X_R, Y_R, Z_R)$ are,

$$X_R = 9X_T^4 - 8X_TY_T^2, \; Y_R = 3X_T^2(4X_TY_T^2 - X_R) - 8Y_T^4, \; Z_R = 2Y_TZ_T.$$

Let the point $P \in E(\mathbb{F}_p)$ be represented in affine coordinates as $P = (x_P, y_P)$. Then, the tangent line at T evaluated at P can be calculated as [32],

$$l_{T,T}(P) = 2Z_RZ_T^2y_P - (6X_T^2Z_T^2x_P)W + (6X_T^3 - 4Y_T^2)W^2 \in \mathbb{F}_{p^{12}}.$$

Hence, the computational cost of the interleaving computation of the tangent line and the doubling of the point T is, $3\tilde{m} + 8\tilde{s} + 16\tilde{a} + 4m$. Other operations included in line 4 are f^2 and the product $f^2 \cdot l_{T,T}(P)$, which can be computed at a cost of, $12\tilde{m} + 45\tilde{a} + 4m_\beta$ and $13\tilde{m} + 39\tilde{a} + 2m_\beta$, respectively. In summary, the computational cost associated to line 4 of Algorithm 1 is given as, $28\tilde{m} + 8\tilde{s} + 100\tilde{a} + 4m + 6m_\beta$.

Addition step (lines 6 and 8). Let $Q = (X_Q, Y_Q, Z_Q)$ and $T = (X_T, Y_T, Z_T)$ represent the points Q and $T \in E'(\mathbb{F}_{p^2})$ in Jacobian coordinates. Then the point $R = T + Q = (X_R, Y_R, Z_R)$, can be computed as,

$$X_R = (2Y_Q Z_T^3 - 2Y_T)^2 - 4(X_Q Z_T^2 - X_T)^3 - 8(X_Q Z_T^2 - X_T)^2 X_T,$$
$$Y_R = (2Y_Q Z_T^3 - 2Y_T)(4(X_Q Z_T^2 - X_T)^2 X_T - X_R) - 8Y_T(X_Q Z_T^2 - X_T)^3,$$
$$Z_R = 2Z_T(X_Q Z_T^2 - X_T).$$

Once again, let the point $P \in E(\mathbb{F}_p)$ be represented in affine coordinates as $P = (x_P, y_P)$. Then, the line through T and Q evaluated at the point P is given as,

$$l_{T,Q}(P) = 2Z_R y_P - 4x_P(Y_Q Z_T^3 + Y_T)W + (4X_Q(Y_Q Z_T^3 X_Q - Y_T) - 2Y_Q Z_R)W^2 \in \mathbb{F}_{p^{12}}.$$

The combined cost of computing $l_{T,Q}(P)$ and the point addition $R = T + Q$ is, $7\tilde{m} + 7\tilde{s} + 25\tilde{a} + 4m$. Finally we must accumulate the value of $l_{T,Q}(P)$ by performing the product $f \cdot l_{T,Q}(P)$ at a cost of, $13\tilde{m} + 39\tilde{a} + 2m_\beta$.

Therefore, the computational cost associated to line 6 of Algorithm 1 is given as, $20\tilde{m} + 7\tilde{s} + 64\tilde{a} + 4m + 2m_\beta$. This is the same cost of line 8.

Frobenius application and final addition step (lines 11–13). In this step we add to the value accumulated in $f = f_{6t+2,Q}(P)$, the product of the lines through the points $Q_1, -Q_2 \in E'(\mathbb{F}_{p^2})$, namely, $l_{[6t+2]Q,Q_1}(P) \cdot l_{[6t+2]Q+Q_1,-Q_2}(P)$.

The points Q_1, Q_2 can be found by applying the Frobenius operator as, $Q_1 = \pi_p(Q)$, $Q_2 = \pi_p^2(Q)$. The total cost of computing lines 11–13 is given as, $40\tilde{m} + 14\tilde{s} + 128\tilde{a} + 4m + 4m_\beta$.

Let us recall that from our selection of t, $6t + 2$ is a 65-bit number with a low Hamming weight of 7.[2] This implies that the Miller loop of the optimal ate pairing can be computed using only 64 point doubling steps and 6 point addition/subtraction steps. Therefore, the total cost of the Miller loop portion of Algorithm 1 is approximately given as,

$$\begin{aligned}
\text{Cost of Miller loop} = {}& 64 \cdot (28\tilde{m} + 8\tilde{s} + 100\tilde{a} + 4m + 6m_\beta) + \\
& 6 \cdot (20\tilde{m} + 7\tilde{s} + 64\tilde{a} + 4m + 2m_\beta) + \\
& 40\tilde{m} + 14\tilde{s} + 128\tilde{a} + 14m + 4m_\beta \\
= {}& 1952\tilde{m} + 568\tilde{s} + 6912\tilde{a} + 294m + 400m_\beta.
\end{aligned}$$

4.2   Final Exponentiation

Line 14 of Algorithm 1 performs the final exponentiation step, by raising $f \in \mathbb{F}_{p^{12}}$ to the power $e = (p^{12} - 1)/r$. We computed the final exponentiation by following the procedure described by Scott $et\ al.$ in [36], where the exponent e is split into three coefficients as,

$$e = \frac{p^{12} - 1}{r} = (p^6 - 1) \cdot (p^2 + 1) \cdot \frac{p^4 - p^2 + 1}{r}. \tag{3}$$

---

[2] We note that in the binary signed representation with digit set $\{-1, 0, 1\}$, the integers $t = 2^{62} - 2^{54} + 2^{44}$ and $6t + 2 = 2^{64} + 2^{63} - 2^{56} - 2^{55} + 2^{46} + 2^{45} + 2$ have a signed bitlength of 63 and 65, respectively.

As it was discussed in Section 3, we can take advantage of the fact that raising f to the power p^6 is equivalent to one conjugation. Hence, one can compute $f^{(p^6-1)} = \bar{f} \cdot f^{-1}$, which costs one field inversion and one field multiplication in $\mathbb{F}_{p^{12}}$. Moreover, after raising to the power $p^6 - 1$, the resulting field element becomes a member of the cyclotomic subgroup $\mathbb{G}_{\Phi_6}(\mathbb{F}_{p^2})$, which implies that inversion of such elements can be computed by simply conjugation (see Table 1). Furthermore, from the discussion in Section 3.2 raising to the power $p^2 + 1$, can be done with five field multiplications in the base field $\mathbb{F}_p$, plus one field multiplication in $\mathbb{F}_{p^{12}}$. The processing of the third coefficient in Equation (3) is referred as the *hard part* of the final exponentiation, *i.e*, the task of computing $m^{(p^4-p^2+1)/r}$, with $m \in \mathbb{F}_{p^{12}}$. In order to accomplish that, Scott *et al.* described in [36] a clever procedure that requires the calculation of ten temporary values, namely,

$$m^t, \ m^{t^2}, \ m^{t^3}, \ m^p, \ m^{p^2}, \ m^{p^3}, \ m^{(tp)}, \ m^{(t^2p)}, \ m^{(t^3p)}, \ m^{(t^2p^2)},$$

which are the building blocks required for constructing a vectorial addition chain whose evaluation yields the final exponentiation f^e, by performing 13 and 4 field multiplication and squaring operations over $\mathbb{F}_{p^{12}}$, respectively.[3] Taking advantage of the Frobenius operator efficiency, the temporary values m^p, m^{p^2}, m^{p^3}, $m^{(tp)}$, $m^{(t^2p)}$, $m^{(t^3p)}$, and $m^{(t^2p^2)}$ can be computed at a cost of just 35 field multiplications over $\mathbb{F}_p$ (see Section 3.2). Therefore, the most costly computation of the hard part of the final exponentiation is the calculation of $m^t, m^{t^2} = (m^t)^t, m^{t^3} = (m^{t^2})^t$. From our choice, $t = 2^{62} - 2^{54} + 2^{44}$, we can compute these three temporary values at a combined cost of $62 \cdot 3 = 186$ cyclotomic squarings plus $2 \cdot 3 = 6$ field multiplications over $\mathbb{F}_{p^{12}}$. This is cheaper than the t selection used in [32] that requires $4 \cdot 3 = 12$ more field multiplications over $\mathbb{F}_{p^{12}}$.

Consulting Table 1, we can approximately estimate the total computational cost associated to the final exponentiation as,

$$\begin{aligned}
\text{F. Exp. cost} = \ &(25\tilde{m} + 9\tilde{s} + 12m_\beta + 61\tilde{a} + \tilde{i}) + (18\tilde{m} + 6m_\beta + 60\tilde{a}) + \\
&(18\tilde{m} + 6m_\beta + 60\tilde{a}) + 10m + \\
&13 \cdot (18\tilde{m} + 6m_\beta + 60\tilde{a}) + 4 \cdot (9\tilde{s} + 4m_\beta + 30\tilde{a}) + 70m + \\
&186 \cdot (9\tilde{s} + 4m_\beta + 30\tilde{a}) + 6 \cdot (18\tilde{m} + 6m_\beta + 60\tilde{a}) \\
= \ &403\tilde{m} + 1719\tilde{s} + 7021\tilde{a} + 80m + 898m_\beta + \tilde{i}.
\end{aligned}$$

Table 2 presents a comparison of $\mathbb{F}_{p^2}$ arithmetic operations of our work against the reference pairing software libraries [18, 32]. From Table 2, we observe that our approach saves about 39.5% and 13% $\mathbb{F}_{p^2}$ multiplications when compared against [18] and [32], respectively. We recall that in our work, the cost of the operation m_ξ is essentially the same of that of m_β. This is not the case in [18,32], where the operation m_ξ is considerably more costly than m_β.

---

[3] We remark that the cost of the field squaring operations is that of the elements in the cyclotomic subgroup $\mathbb{G}_{\Phi_6}(\mathbb{F}_{p^2})$ listed in the last row of Table 1.

Table 2. A Comparison of arithmetic operations required by the computation of the ate pairing variants

		$\tilde{m}$	$\tilde{s}$	$\tilde{a}$	i	m_ξ
Hankerson *et al.* [18] R-ate pairing	Miller Loop	2277	356	6712	1	412
	Final Exp.	1616	1197	8977	1	1062
	Total	3893	1553	15689	2	1474
Naehrig *et al.* [32] Optimal ate pairing	Miller Loop	2022	590	7140		410
	Final Exp.	678	1719	7921	1	988
	Total	2700	2309	15061	1	1398
This work Optimal ate pairing	Miller Loop	1952	568	6912		400
	Final Exp.	403	1719	7021	1	898
	Total	2355	2287	13933	1	1298

5 Software Implementation of Field Arithmetic

In this work, we target the x86-64 instruction set [22]. Our software library is written in C++ and can be used on several platforms: 64-bit Windows 7 with Visual Studio 2008 Professional, 64-bit Linux 2.6 and Mac OS X 10.5 with gcc 4.4.1 or later, etc. In order to improve the runtime performance of our pairing library, we made an extensive use of Xbyak [28], a x86/x64 just-in-time assembler for the C++ language.

5.1 Implementation of Prime Field Arithmetic

The x86-64 instruction set has a **mul** operation which multiplies two 64-bit unsigned integers and returns a 128-bit unsigned integer. The execution of this operation takes about 3 cycles on Intel Core i7 and AMD Opteron processors. Compared to previous architectures, the gap between multiplication and addition/subtraction in terms of cycles is much smaller. This means that we have to be careful when selecting algorithms to perform prime field arithmetic: the schoolbook method is for instance faster than Karatsuba multiplication in the case of 256-bit operands.

An element $x \in \mathbb{F}_p$ is represented as $x = (x_3, x_2, x_1, x_0)$, where $x_i, 0 \leq i \leq 3$, are 64-bit integers. The addition and the subtraction over $\mathbb{F}_p$ are performed in a straightforward manner, *i.e.*, we add/subtract the operands followed by reduction into $\mathbb{F}_p$. Multiplication and inversion over $\mathbb{F}_p$ are accomplished according to the well-known Montgomery multiplication and Montgomery inversion algorithms, respectively [19].

5.2 Implementation of Quadratic Extension Field Arithmetic

This section describes our optimizations for some operations over $\mathbb{F}_{p^2}$ defined in Equation (2).

Multiplication. We implemented the multiplication over the quadratic extension field $\mathbb{F}_{p^2}$ using a Montgomery multiplication scheme split into two steps:

1. The straightforward multiplication of two 256-bit integers (producing a 512-bit integer), denoted as, **mul256**.
2. The Montgomery reduction from a 512-bit integer to a 256-bit integer. This operation is denoted by **mod512**.

According to our implementation, **mul256** (resp. **mod512**) contains 16 (resp. 20) **mul** operations and its execution takes about 55 (resp. 100) cycles.

Let $P(u) = u^2 + 5$ be the irreducible binomial defining the quadratic extension $\mathbb{F}_{p^2}$. Let $A, B, C \in \mathbb{F}_{p^2}$ such that, $A = a_0 + a_1 u$, $B = b_0 + b_1 u$, and $C = c_0 + c_1 u = A \cdot B$. Then, $c_0 = a_0 b_0 - 5 a_1 b_1$ and $c_1 = (a_0 + a_1)(b_0 + b_1) - a_0 b_0 - a_1 b_1$. Hence, in order to obtain the field multiplication over the quadratic extension field, we must compute three multiplications over $\mathbb{F}_p$, and it may seem that three **mod512** operations are necessary. However, we can keep the results of the products **mul256**(a_0, b_0), **mul256**(a_1, b_1), and **mul256**$(a_0 + a_1, b_0 + b_1)$ in three temporary 512-bit integer values. Then, we can add or subtract them without reduction, followed by a final call to **mod512** in order to get $c_0, c_1 \in \mathbb{F}_p$. This approach yields the saving of one **mod512** operation as shown in Algorithm 2. We stress that the **addNC/subNC** functions in lines 1, 2, 6, and 7 of Algorithm 2, stand for addition/subtraction between 256-bit or 512-bit integers without checking the output carry. We explain next the rationale for using addition/subtraction without output carry check.

The addition $x + y$, and subtraction $x - y$, of two elements x, $y \in \mathbb{F}_p$ include an unpredictable branch check to figure out whether $x + y \geq p$ or $x < y$. This is a costly check that is convenient to avoid as much as possible. Fortunately, our selected prime p satisfies $7p < N$, with $N = 2^{256}$, and the function **mod512** can reduce operands x, whenever, $x < pN$. This implies that we can add up to seven times without performing an output carry check. In line 8, d_0 is equal

---

Algorithm 2. Optimized multiplication over $\mathbb{F}_{p^2}$.

Input: A and $B \in \mathbb{F}_{p^2}$ such that $A = a_0 + a_1 u$ and $B = b_0 + b_1 u$.
Output: $C = A \cdot B \in \mathbb{F}_{p^2}$.

1. $s \leftarrow \textbf{addNC}(a_0, a_1)$;
2. $t \leftarrow \textbf{addNC}(b_0, b_1)$;
3. $d_0 \leftarrow \textbf{mul256}(s, t)$;
4. $d_1 \leftarrow \textbf{mul256}(a_0, b_0)$;
5. $d_2 \leftarrow \textbf{mul256}(a_1, b_1)$;
6. $d_0 \leftarrow \textbf{subNC}(d_0, d_1)$;
7. $d_0 \leftarrow \textbf{subNC}(d_0, d_2)$;
8. $c_1 \leftarrow \textbf{mod512}(d_0)$;
9. $d_2 \leftarrow 5 d_2$;
10. $d_1 \leftarrow d_1 - d_2$;
11. $c_0 \leftarrow \textbf{mod512}(d_1)$;
12. **return** $C \leftarrow c_0 + c_1 u$;

to $(a_0 + a_1)(b_0 + b_1) - a_0 b_0 - a_1 b_1 = a_0 b_1 + a_1 b_0 < 2p^2 < pN$. Hence, we can use **addNC/subNC** for step 1, 2, 6, and 7. In line 9, we multiply d_2 by the constant value 5, which can be computed with no carry operation. By applying these modifications, we manage to reduce the cost of the field multiplication over $\mathbb{F}_{p^2}$ from about 640 cycles (required by a non-optimized procedure) to just 440 cycles.

In line 10, $d_1 = a_0 b_0 - 5a_1 b_1$. We perform this operation as a 512-bit integer subtraction with carry operation followed by a **mod512** reduction. Let x be a 512-bit integer such that $x = a_0 b_0 - 5a_1 b_1$ and let t be a 256-bit integer. The aforementioned carry operation can be accomplished as follows: if $x < 0$, then $t \leftarrow p$, otherwise $t \leftarrow 0$, then $d_1 \leftarrow x + tN$, where this addition operation only uses the 256 most significant bits of x.

Squaring. Algorithm 3 performs field squaring where some carry operations have been reduced, as explained next. Let $A = a_0 + a_1 u \in \mathbb{F}_{p^2}$, $C = A^2 = c_0 + c_1 u$, and let $x = (a_0 + p - a_1)(a_0 + 5a_1)$. Then $c_0 = x - 4a_0 a_1 \bmod p$. However, we observe that $x \leq 2p \cdot 6p = 12p^2 < N^2$ where $N = 2^{256}$. Also we have that,

$$x - 4a_0 a_1 \geq a_0(a_0 + 5a_1) - 4a_0 a_1 = a_0(a_0 + a_1) \geq 0,$$

which implies,

$$\max(x - 4a_0 a_1) = \max(a_0(a_0 + p) + 5a_1(p - a_1))$$
$$< p \cdot 2p + 5(p/2)(p - p/2) < pN.$$

We conclude that we can safely add/subtract the operands in Algorithm 3 without carry check.

Fast reduction for multiplication by small constant values. The procedures of point doubling/addition and line evaluation in Miller loop, and the

---

Algorithm 3. Optimized squaring over $\mathbb{F}_{p^2}$.

Input: $A \in \mathbb{F}_{p^2}$ such that $A = a_0 + a_1 u$.
Output: $C = A^2 \in \mathbb{F}_{p^2}$.
1. $t \leftarrow$ **addNC**(a_1, a_1);
2. $d_1 \leftarrow$ **mul256**(t, a_0);
3. $t \leftarrow$ **addNC**(a_0, p);
4. $t \leftarrow$ **subNC**(t, a_1);
5. $c_1 \leftarrow 5a_1$;
6. $c_1 \leftarrow$ **addNC**(c_1, a_0);
7. $d_0 \leftarrow$ **mul256**(t, c_1);
8. $c_1 \leftarrow$ **mod512**(d_1);
9. $d_1 \leftarrow$ **addNC**(d_1, d_1);
10. $d_0 \leftarrow$ **subNC**(d_0, d_1);
11. $c_0 \leftarrow$ **mod512**(d_0);
12. **return** $C \leftarrow c_0 + c_1 u$;

operations m_ξ, m_β in the tower field arithmetic, involve field multiplications of an arbitrary element $A \in \mathbb{F}_{p^2}$ by small constant values 3, 4, 5, and 8.

We first remark that m_ξ requires the calculation of a field multiplication by the constant u. Given $A = a_0 + a_1 u \in \mathbb{F}_{p^2}$, then $A \cdot u = \beta \cdot a_1 + a_0 u = -5a_1 + a_0 u$. Computing this operation using shift-and-add expressions such as $5n = n + (n \ll 2)$ for $n \in \mathbb{F}_p$ may be tempting as a means to avoid full multiplication calculations. Nevertheless, in our implementation we preferred to compute those multiplication-by-constant operations using the x86-64 **mul** instruction, since the cost in clock cycles of **mul** is almost the same or even a little cheaper than the one associated to the shift-and-add method.

Multiplications by small constant values require the reduction modulo p of an integer x smaller than $8p$. Note that we need five 64-bit registers to store $x = (x_4, x_3, x_2, x_1, x_0)$. However, one can easily see that $x_4 = 0$ or $x_4 = 1$, and then one can prove that x div $2^{253} = (x_4 \ll 3)|(x_3 \gg 61)$. Division by 2^{253} involves only three logical operations and it can be efficiently performed on our target processor. Furthermore, the prime p we selected has the following nice property:

$$(ip) \text{ div } 2^{253} = \begin{cases} i & \text{if } 0 \leq i \leq 9, \\ i+1 & \text{if } 10 \leq i \leq 14. \end{cases}$$

Hence, we built a small look-up table p-Tbl defined as follows:

$$p\text{-Tbl}[i] = \begin{cases} ip & \text{if } 0 \leq i \leq 9, \\ (i-1)p & \text{if } 10 \leq i \leq 14. \end{cases} \tag{4}$$

We then get $|x - p\text{-Tbl}[x \gg 253]| < p$. Algorithm 4 summarizes how we apply this strategy to perform a modulo p reduction.

Algorithm 4. Fast reduction $x \bmod p$.

Input: $x \in \mathbb{Z}$ such that $0 \leq x < 13p$ and represented as $x = (x_4, x_3, x_2, x_1, x_0)$, where $x_i, 0 \leq i \leq 4$, are 64-bit integers. Let p-Tbl be the precomputed look-up table defined in Equation (4).

Output: $z = x \bmod p$.

1. $q \leftarrow (x_4 \ll 3)|(x_3 \gg 61)$; $(q \leftarrow \lfloor x/2^{253} \rfloor)$
2. $z \leftarrow x - p\text{-Tbl}[q]$;
3. **if** $z < 0$ **then**
4. $\quad z \leftarrow z + p$;
5. **end if**
6. **return** z;

---

6   Implementation Results

We list in Table 3 the timings that we achieved on different architectures. Our library is able to evaluate the optimal ate pairing over a 254-bit prime field $\mathbb{F}_p$, in just 2.33 million of clock cycles on a single core of an Intel Core i7 2.8GHz processor, which implies that the pairing computation takes 0.832msec. To our best knowledge, we are the first to compute a cryptographic pairing in less than one millisecond at this level of security on a desktop computer.

According to the second column of Table 3, the costs (in clock cycles) that were measured for the $\mathbb{F}_{p^2}$ arithmetic when implemented in the Core i7 processor are $\tilde{m} = 435$ and $\tilde{s} = 342$. Additionally, we measured $\tilde{a} = 40$, and $\tilde{i} = 7504$. Now, from Table 2, one can see that the predicted computational cost of the optimal ate pairing is given as,

$$\begin{aligned} \text{Opt. ate pairing cost} &= 2355\tilde{m} + 2287\tilde{s} + 13933\tilde{a} + \tilde{i} \\ &= 2355 \cdot 435 + 2287 \cdot 342 + 13933 \cdot 40 + 7504 \\ &= 2{,}371{,}403. \end{aligned}$$

We observe that the experimental results presented in Table 3 have a reasonable match with the computational cost prediction given in Section 4.

For comparison purpose, we also report the performance of the software library for BN curves developed by Naehrig et al. [32], which is the best software implementation that we know of.[4] Naehrig et al. combined several state-of-the art optimization techniques to write a software that is more than twice as fast as the previous reference implementation by Hankerson et al. [18]. Perhaps the most original contribution in [32] is the implementation of the arithmetic over the quadratic extension $\mathbb{F}_{p^2}$ based on a tailored use of SIMD floating point instructions. Working in the case of hardware realizations of pairings, Fan et al. [13] suggested to take advantage of the polynomial form of $p(t)$ and introduced a new hybrid modular multiplication algorithm. The operands a and $b \in \mathbb{F}_p$ are converted to degree-4 polynomials $a(t)$ and $b(t)$, and multiplied according to Montgomery's algorithm in the polynomial ring. Coefficients of the results must be reduced modulo t. Fan et al noticed that, if $t = 2^m + s$, where s is a small constant, this step consists of a multiplication by s instead of a division by t.

Table 4 summarizes the best results published in the open literature since 2007. All the works featured in Table 4, targeted a level of security equivalent to that of AES-128. Aranha et al. [1] and Beuchat et al. [8] considered supersingular elliptic curves in characteristic 2 and 3, respectively. All other authors worked with ordinary curves.

Several authors studied multi-core implementations of a cryptographic pairing [1, 8, 16]. In the light of the results reported in Table 4, it seems that the acceleration achieved by an n-core implementation is always less than the ideal $n\times$ speedup. This is related to the extra arithmetic operations needed to combine the partial results generated by each core, and the dependencies between the different operations involved in the final exponentiation. The question that arises is therefore: how many cores should be utilized to compute a cryptographic pairing? We believe that the best answer is the one provided by Grabher et al.: "if the requirement is for two pairing evaluations, the slightly moronic conclusion

---

[4] The results on the Core 2 Quad processor are reprinted from [32]. We downloaded the library [33] and made our own experiments on an Opteron platform.

Table 3. Cycle counts of multiplication over $\mathbb{F}_{p^2}$, squaring over $\mathbb{F}_{p^2}$, and optimal ate pairing on different machines

	Our results			
	Core i7[a]	Opteron[b]	Core 2 Duo[c]	Athlon 64 X2[d]
Multiplication over $\mathbb{F}_{p^2}$	435	443	558	473
Squaring over $\mathbb{F}_{p^2}$	342	355	445	376
Miller loop	1,330,000	1,360,000	1,680,000	1,480,000
Final exponentiation	1,000,000	1,040,000	1,270,000	1,081,000
Optimal ate pairing	2,330,000	2,400,000	2,950,000	2,561,000

	dclxvi [32, 33]			
	Core i7	Opteron[b]	Core 2 Quad[e]	Athlon 64 X2[d]
Multiplication over $\mathbb{F}_{p^2}$	–	695	693	1714
Squaring over $\mathbb{F}_{p^2}$	–	614	558	1207
Miller loop	–	2,480,000	2,260,000	5,760,000
Final exponentiation	–	2,520,000	2,210,000	5,510,000
Optimal ate pairing	–	5,000,000	4,470,000	11,270,000

[a] Intel Core i7 860 (2.8GHz), Windows 7, Visual Studio 2008 Professional
[b] Quad-Core AMD Opteron 2376 (2.3GHz), Linux 2.6.18, gcc 4.4.1
[c] Intel Core 2 Duo T7100 (1.8GHz), Windows 7, Visual Studio 2008 Professional
[d] Athlon 64 X2 Dual Core 6000+(3GHz), Linux 2.6.23, gcc 4.1.2
[e] Intel Core 2 Quad Q6600 (2394MHz), Linux 2.6.28, gcc 4.3.3

Table 4. A comparison of cycles and timings required by the computation of the ate pairing variants. The frequency is given in GHz and the timings are in milliseconds.

	Algo.	Architecture	Cycles	Freq.	Calc. time
Devegili et al. [10]	ate	Intel Pentium IV	69,600,000	3.0	23.20
Naehrig et al. [31]	ate	Intel Core 2 Duo	29,650,000	2.2	13.50
Grabher et al. [16]	ate	Intel Core 2 Duo (1 core)	23,319,673	2.4	9.72
		Intel Core 2 Duo (2 cores)	14,429,439		6.01
Aranha et al. [1]	η_T	Intel Xeon 45nm (1 core)	17,400,000	2.0	8.70
		Intel Xeon 45nm (8 cores)	3,020,000		1.51
Beuchat et al. [8]	η_T	Intel Core i7 (1 core)	15,138,000	2.9	5.22
		Intel Core i7 (8 cores)	5,423,000		1.87
Hankerson et al. [18]	R-ate	Intel Core 2	10,000,000	2.4	4.10
Naehrig et al. [32]	a_{opt}	Intel Core 2 Quad Q6600	4,470,000	2.4	1.80
This work	a_{opt}	Intel Core i7	2,330,000	2.8	0.83

is that one can perform one pairing on each core [. . .], doubling the performance versus two sequential invocations of any other method that does not already use multi-core parallelism internally" [16].

7   Conclusion

In this paper we have presented a software library that implements the optimal ate pairing over a Barreto–Naehrig curve at the 126-bit security level. To the best of our knowledge, we are the first to have reported the computation of a bilinear pairing at a level of security roughly equivalent to that of AES-128 in less than one millisecond on a single core of an Intel Core i7 2.8GHz processor. The speedup achieved in this work is a combination of two main factors:

- A careful programming of the underlying field arithmetic based on Montgomery multiplication that allowed us to perform a field multiplication over $\mathbb{F}_p$ and $\mathbb{F}_{p^2}$ in just 160 and 435 cycles, respectively, when working in an Opteron-based machine. We remark that in contrast with [32], we did not make use of the 128-bit multimedia arithmetic instructions.
- A binary signed selection of the parameter t that allowed us to obtain significant savings in both the Miller loop and the final exponentiation of the optimal ate pairing.

Our selection of t yields a prime $p = p(t)$ that has a bitlength of just 254 bits. This size is slightly below than what Freeman *et al.* [14] recommend for achieving a high security level. If for certain scenarios, it becomes strictly necessary to meet or exceed the 128-bit level of security, we recommend to select $t = 2^{63} - 2^{49}$ that produces a prime $p = p(t)$ with a bitlength of 258 bits. However, we warn the reader that since a 258-bit prime implies that more than four 64-bit register will be required to store field elements, the performance of the arithmetic library will deteriorate.

Consulting the cycle count costs listed in Table 3, one can see that for our implementation the cost of the final exponentiation step is nearly 25% cheaper than that of the Miller loop.

Authors in [13, 32] proposed to exploit the polynomial parametrization of the prime p as a means to speed up the underlying field arithmetic. We performed extensive experiments trying to apply this idea to our particular selection of t with no success. Instead, the customary Montgomery multiplier algorithm appears to achieve a performance that is very hard to beat by other multiplication schemes, whether integer-based or polynomial-based multipliers.

The software library presented in this work computes a bilinear pairing at a high security level at a speed that is faster than the best hardware accelerators published in the open literature (see for instance [7, 13, 23, 37]). We believe that this situation is unrealistic and therefore we will try to design a hardware architecture that can compute 128-bit security bilinear pairing in shorter timings. Our future work will also include a study of the parallelization possibilities on pairing-based protocols that specify the computation of many bilinear pairing during their execution.

Acknowledgements

We thank Michael Naehrig, Ruben Niederhagen, and Peter Schwabe for making their pairing software library [33] freely available for research purposes.

The authors also want to thank Diego Aranha, Paulo S.L.M. Barreto, Darrel Hankerson, Alfred Menezes, and the anonymous referees for their valuable comments.

References

1. Aranha, D.F., López, J., Hankerson, D.: High-speed parallel software implementation of the η_T pairing. In: Pieprzyk, J. (ed.) CT-RSA 2010. LNCS, vol. 5985, pp. 89–105. Springer, Heidelberg (2010)
2. Arène, C., Lange, T., Naehrig, M., Ritzenthaler, C.: Faster computation of the Tate pairing. Cryptology ePrint Archive, Report 2009/155 (2009), http://eprint.iacr.org/2009/155.pdf
3. Barreto, P.S.L.M., Galbraith, S.D., Ó hÉigeartaigh, C., Scott, M.: Efficient pairing computation on supersingular Abelian varieties. Designs, Codes and Cryptography 42, 239–271 (2007)
4. Barreto, P.S.L.M., Kim, H.Y., Lynn, B., Scott, M.: Efficient algorithms for pairing-based cryptosystems. In: Yung, M. (ed.) CRYPTO 2002. LNCS, vol. 2442, pp. 354–368. Springer, Heidelberg (2002)
5. Barreto, P.S.L.M., Naehrig, M.: Pairing-friendly elliptic curves of prime order. In: Preneel, B., Tavares, S. (eds.) SAC 2005. LNCS, vol. 3897, pp. 319–331. Springer, Heidelberg (2006)
6. Benger, N., Scott, M.: Constructing tower extensions for the implementation of pairing-based cryptography. Cryptology ePrint Archive, Report 2009/556 (2009), http://eprint.iacr.org/2009/556.pdf
7. Beuchat, J.-L., Detrey, J., Estibals, N., Okamoto, E., Rodríguez-Henríquez, F.: Fast architectures for the η_T pairing over small-characteristic supersingular elliptic curves. Cryptology ePrint Archive, Report 2009/398 (2009), http://eprint.iacr.org/2009/398.pdf
8. Beuchat, J.-L., López-Trejo, E., Martínez-Ramos, L., Mitsunari, S., Rodríguez-Henríquez, F.: Multi-core implementation of the Tate pairing over supersingular elliptic curves. In: Garay, J.A., Miyaji, A., Otsuka, A. (eds.) CANS 2009. LNCS, vol. 5888, pp. 413–432. Springer, Heidelberg (2009)
9. Chung, J., Hasan, M.A.: Asymmetric squaring formulae. In: Kornerup, P., Muller, J.-M. (eds.) Proceedings of the 18th IEEE Symposium on Computer Arithmetic, pp. 113–122. IEEE Computer Society, Los Alamitos (2007)
10. Devegili, A.J., Scott, M., Dahab, R.: Implementing cryptographic pairings over Barreto–Naehrig curves. In: Takagi, T., Okamoto, T., Okamoto, E., Okamoto, T. (eds.) Pairing 2007. LNCS, vol. 4575, pp. 197–207. Springer, Heidelberg (2007)
11. Duursma, I., Gaudry, P., Morain, F.: Speeding up the discrete log computation on curves with automorphisms. In: Lam, K.-Y., Okamoto, E., Xing, C. (eds.) ASIACRYPT 1999. LNCS, vol. 1716, pp. 103–121. Springer, Heidelberg (1999)
12. Duursma, I., Lee, H.S.: Tate pairing implementation for hyperelliptic curves $y^2 = x^p - x + d$. In: Laih, C.S. (ed.) ASIACRYPT 2003. LNCS, vol. 2894, pp. 111–123. Springer, Heidelberg (2003)
13. Fan, J., Vercauteren, F., Verbauwhede, I.: Faster $\mathbb{F}_p$-arithmetic for cryptographic pairings on Barreto–Naehrig curves. In: Clavier, C., Gaj, K. (eds.) CHES 2009. LNCS, vol. 5747, pp. 240–253. Springer, Heidelberg (2009)
14. Freeman, D., Scott, M., Teske, E.: A taxonomy of pairing-friendly elliptic curves. Journal of Cryptology 23(2), 224–280 (2010)

15. Galbraith, S., Paterson, K., Smart, N.: Pairings for cryptographers. Discrete Applied Mathematics 156, 3113–3121 (2008)
16. Grabher, P., Großschädl, J., Page, D.: On software parallel implementation of cryptographic pairings. In: Avanzi, R.M., Keliher, L., Sica, F. (eds.) SAC 2008. LNCS, vol. 5381, pp. 34–49. Springer, Heidelberg (2008)
17. Granger, R., Scott, M.: Faster squaring in the cyclotomic subgroup of sixth degree extensions. Cryptology ePrint Archive, Report 2009/565 (2009), http://eprint.iacr.org/2009/565.pdf
18. Hankerson, D., Menezes, A., Scott, M.: Software implementation of pairings. In: Joye, M., Neven, G. (eds.) Identity-based Cryptography. Cryptology and Information Security Series, ch. 12, pp. 188–206. IOS Press, Amsterdam (2009)
19. Hankerson, D., Menezes, A., Vanstone, S.: Guide to Elliptic Curve Cryptography. Springer, New York (2004)
20. Hess, F.: Pairing lattices. In: Galbraith, S.D., Paterson, K.G. (eds.) Pairing 2008. LNCS, vol. 5209, pp. 18–38. Springer, Heidelberg (2008)
21. Hess, F., Smart, N., Vercauteren, F.: The Eta pairing revisited. IEEE Transactions on Information Theory 52(10), 4595–4602 (2006)
22. Intel Corporation. Intel 64 and IA-32 Architectures Software Developer's Manuals, http://www.intel.com/products/processor/manuals/
23. Kammler, D., Zhang, D., Schwabe, P., Scharwaechter, H., Langenberg, M., Auras, D., Ascheid, G., Mathar, R.: Designing an ASIP for cryptographic pairings over Barreto–Naehrig curves. In: Clavier, C., Gaj, K. (eds.) CHES 2009. LNCS, vol. 5747, pp. 254–271. Springer, Heidelberg (2009)
24. Koblitz, N., Menezes, A.: Pairing-based cryptography at high security levels. Cryptology ePrint Archive, Report 2005/076 (2005), http://eprint.iacr.org/2005/076.pdf
25. Lee, E., Lee, H.-S., Park, C.-M.: Efficient and generalized pairing computation on abelian varieties. Cryptology ePrint Archive, Report 2008/040 (2008), http://eprint.iacr.org/2008/040.pdf
26. Miller, V.S.: Short programs for functions on curves (1986), http://crypto.stanford.edu/miller
27. Miller, V.S.: The Weil pairing, and its efficient calculation. Journal of Cryptology 17(4), 235–261 (2004)
28. Mitsunari, S.: Xbyak: JIT assembler for C++, http://homepage1.nifty.com/herumi/soft/xbyak_e.html
29. Miyaji, A., Nakabayashi, M., Takano, S.: New explicit conditions of elliptic curve traces for FR-reduction. IEICE Trans. Fundamentals E84, 1234–1243 (2001)
30. Naehrig, M.: Constructive and Computational Aspects of Cryptographic Pairings. PhD thesis, Technische Universiteit Eindhoven (2009), http://www.cryptojedi.org/users/michael/data/thesis/2009-05-13-diss.pdf
31. Naehrig, M., Barreto, P.S.L.M., Schwabe, P.: On compressible pairings and their computation. In: Vaudenay, S. (ed.) AFRICACRYPT 2008. LNCS, vol. 5023, pp. 371–388. Springer, Heidelberg (2008)
32. Naehrig, M., Niederhagen, R., Schwabe, P.: New software speed records for cryptographic pairings. Cryptology ePrint Archive, Report 2010/186 (2010), http://eprint.iacr.org/2010/186.pdf
33. Schwabe, P.: Software library of "New software speed records for cryptographic pairings", http://cryptojedi.org/crypto/#dclxvi (accessed June 4, 2010)

34. Scott, M.: Implementing cryptographic pairings. In: Takagi, T., Okamoto, T., Okamoto, E., Okamoto, T. (eds.) Pairing 2007. LNCS, vol. 4575, pp. 177–196. Springer, Heidelberg (2007)
35. Scott, M., Barreto, P.S.L.M.: Compressed pairings. In: Franklin, M.K. (ed.) CRYPTO 2004. LNCS, vol. 3152, pp. 140–156. Springer, Heidelberg (2004)
36. Scott, M., Benger, N., Charlemagne, M., Dominguez Perez, L.J., Kachisa, E.J.: On the final exponentiation for calculating pairings on ordinary elliptic curves. Cryptology ePrint Archive, Report 2008/490 (2008), http://eprint.iacr.org/2008/490.pdf
37. Shu, C., Kwon, S., Gaj, K.: Reconfigurable computing approach for Tate pairing cryptosystems over binary fields. IEEE Transactions on Computers 58(9), 1221–1237 (2009)
38. Vercauteren, F.: Optimal pairings. IEEE Transactions on Information Theory 56(1), 455–461 (2010)

Some Security Topics with Possible Applications for Pairing-Based Cryptography

Gene Tsudik

University of California at Irvine, USA

Abstract. Over the last decade, pairing-based cryptography has found a wide range of interesting applications, both in cryptography and in computer/network security. It often yields the most elegant (if not always the most efficient) techniques. This talk overviews several topics to which pairing-based methods either have not been applied, or where they have not reached their potential. The first topic is "privacy-preserving set operations", such as private set intersection (PSI) protocols. Despite lots of prior work, state-of-the-art (in terms of efficiency) PSI is grounded in more mundane non-pairing-based number theoretic settings. This is puzzling, since the same does not hold with closely related secret handshakes and affiliation-hiding key exchange (AH-AKE) techniques. The second topic is more applied: "security in unattended wireless sensor networks" (UWSNs). We discuss certain unique security issues occurring in UWSNs, overview some protection measures, and consider whether pairing-based cryptography has some applications in this context. The third topic is "privacy in mobile ad hoc networks" (MANETs). The central goal is to achieve privacy-preserving (i.e., tracking-resistant) mobility in the presence of malicious insiders, while maintaining security. Since security is based on authentication, which is, in turn, usually based on identities, routing and packet forwarding are very challenging. Pairing-based cryptography might offer some useful techniques for reconciling security and privacy in this context. Finally, we consider the topic of "secure code attestation for embedded devices" where the main challenge is: how an untrusted (and possibly compromised) device can convince a trusted verifier that it runs appropriate code. After discussing current approaches, once again, consider whether pairing techniques can be of use.

M. Joye, A. Miyaji, and A. Otsuka (Eds.): Pairing 2010, LNCS 6487, p. 40, 2010.

A New Construction of Designated Confirmer Signature and Its Application to Optimistic Fair Exchange

(Extended Abstract)

Qiong Huang[1], Duncan S. Wong[1], and Willy Susilo[2]

[1] City University of Hong Kong, China
csqhuang@gmail.com,
duncan@cityu.edu.hk
[2] University of Wollongong, Australia
wsusilo@uow.edu.au

Abstract. Designated confirmer signature (DCS) extends undeniable signature so that a party called confirmer can also confirm/disavow non-self-authenticating signatures on the signer's behalf. Previous DCS constructions, however, can only let the signer confirm her own signatures but not disavow an invalid one. Only confirmer is able to disavow. In this work, we propose a new suite of security models for DCS by adding the formalization that the signer herself can do both confirmation and disavowal. We also propose a new DCS scheme and prove its security in the standard model. The new DCS scheme is efficient. A signature in this new DCS consists of only three group elements (i.e. 60 bytes altogether for 80-bit security). This is much shorter than any of the existing schemes; it is less than 12% in size of the Camenisch-Michels DCS scheme (Eurocrypt 2000); and it also compares favorably with those proven in the random oracle model, for example, it is less than 50% in size of the Wang et al.'s DCS scheme (PKC 2007). This new DCS scheme also possesses a very efficient signature conversion algorithm. In addition, the scheme can be easily extended to support multiple confirmers (and threshold conversion). To include an additional confirmer, the signer needs to add only one group element into the signature.

Due to the highly efficient properties of this new DCS scheme, we are able to build a practical ambiguous optimistic fair exchange (AOFE) scheme which has short partial and full signatures. A partial signature consists of three elements in an elliptic curve group and four in $\mathbb{Z}_p$ (altogether 140 bytes), and a full signature has only three group elements (altogether 60 bytes), which are about 70% and 21% in size when compared with Garay et al.'s scheme (Crypto 1999), respectively.

Keywords: designated confirmer signature, optimistic fair exchange, ambiguity, standard model.

M. Joye, A. Miyaji, and A. Otsuka (Eds.): Pairing 2010, LNCS 6487, pp. 41–61, 2010.

1   Introduction

Digital signature, the digital analogy of handwriting signature, is publicly verifiable but easy to copy. Anyone can easily convince others that a signer's signature is indeed from the signer. This is not desirable in some scenarios, such as software purchase [5,11] and e-payment [6]. Chaum introduced the notion of undeniable signature [13], in which a signer's signature is non-self-authenticating. To verify a signature, a verifier has to interact with the signer so to let the signer confirm or disavow the signature. In convertible undeniable signature [5], the signer can further convert a (valid) signature to a conventional, publicly verifiable one. The signer is responsible for all the confirmation or disavowal of signatures as well as signature conversion. Chaum [12] then introduced the notion of *designated confirmer signature* (DCS), to alleviate the burden of the signer on confirming, disavowing and converting signatures. In a DCS scheme, there is a party called *confirmer*, which can confirm or disavow a signature on the signer's behalf. The confirmer can also convert a DCS signature to a publicly verifiable one so that if the DCS signature is valid, the publicly verifiable one after conversion will also be valid; otherwise, it will not be valid either. As we can see, the motivation of introducing the notion of DCS in [12] is to have a confirmer share the workload with the signer on confirming, disavowing and converting valid/invalid signatures. However, valid signatures can only be generated by the signer.

On the construction of DCS schemes, one common issue in previous DCS schemes [8,14,23,24,32,33,36,39] is that the signer cannot disavow an invalid signature, though it is able to confirm signatures generated by itself (for example, using the randomness of the signature generation). Because of this, existing DCS security models only formalize the confirmation and disavowal capabilities of the confirmer but not the signer. It remains open to build a DCS scheme which allows *both* signer and confirmer to disavow invalid signatures. The ability of disavowing invalid signatures by the signer is important, and is actually the original motivation of DCS [12], that is, to alleviate the burden of the signer on confirming and disavowing signatures rather than removing the capabilities of doing so from the signer. To see the necessity of allowing the signer to disavow, consider the following scenario.

Suppose that Apple releases a software, say Snow Leopard, and designates MacOne as the confirmer, who is retailer that sells the software. In the case where the retailers are bankrupt, for example, it doesn't mean that the signer, which is Apple, is also bankrupt. Then in this case, Apple can confirm their software themselves, and in the meanwhile, it should be able to deny those softwares not released by Apple.

A naive solution to this problem is to let the signer and the confirmer share the same confirmation key. For example, besides its signing key pair, the signer also generates a confirmation key pair and gives it to the confirmer. In this way the signer is able to confirm and disavow signatures just like the confirmer does. However, this approach is not appropriate in practice, because in the case where multiple signers share the same confirmer, the confirmer has to obtain a key pair from each signer, and use different keys to confirm and disavow for different signers. The complexity for the confirmer to provide confirmation and disavowal service is high.

1.1 Our Contributions

In this paper, we re-formalize the notion of DCS to capture the signer's capability of disavowing invalid signatures. Since the introduction of DCS in [12], this problem has ever been discussed by Galbraith and Mao in [20], but a formal definition is still missing. Besides re-formalizing DCS, we also propose a new scheme and prove its security under the new security models we defined without random oracles. The new DCS scheme is efficient, and to the best of our knowledge, has the *shortest* signature, which consists of only three elements of a bilinear group $\mathbb{G}$, i.e. about 60 bytes for 80-bit security. This is much shorter than any existing schemes with security in the standard model, for example, it is less than 12% in size of the scheme due to Camenisch and Michels [8]. It also compares favorably with those proven in the random oracle model, for example, it is less than 50% in size when compared with the one due to Wang et al. [36]. Our scheme also supports very efficient conversion of DCS signatures. Furthermore, it can be easily extended to support multiple confirmers as well as threshold conversion. To add a confirmer, the signer only needs to add one group element into the signature. The scheme also has an additional feature that one can easily sample a signature uniformly at random from a signer's signature space. We will see that this feature is useful when applying DCS to the construction of a practical ambiguous optimistic fair exchange (AOFE).

On the application of DCS, we construct an efficient AOFE using this new DCS scheme. The notion of AOFE was introduced by Garay et al. [21], in which AOFE is called *abuse-free optimistic fair exchange*. It was later called *ambiguous optimistic fair exchange* by Huang et al. [28], in which they proposed an AOFE with (stronger) security in the standard model. In our AOFE construction, we have an interactive version and a non-interactive one. Both of them have much shorter signatures than the previous schemes [21,28]. A partial signature in our non-interactive AOFE consists of three elements of $\mathbb{G}$ and four of $\mathbb{Z}_p$ (about 140 bytes for 80-bit security), and a full signature has only three $\mathbb{G}$ elements (about 60 bytes). In other words, a partial signature of our scheme is about 70% in size of the scheme due to Garay et al. [21], and a full signature is only about 21% in size of their scheme. The significant reduction in the full signature size is because in our scheme we do not need to include the corresponding partial signature into the full signature, as opposed to [21,28].

2   Related Work

(*Designated Confirmer Signature*). In [33], Okamoto showed that DCS is equivalent to public key encryption and proposed a concrete DCS scheme, which was later shown by Michels and Stadler [32] to be insecure that a confirmer can forge a signer's signatures. In [32], a generic DCS was also proposed using the 'commit-then-sign' paradigm: the signer computes a commitment c to the message, that can be opened by the confirmer, and generates a signature σ on c. A DCS signature ζ consists of c and σ. To confirm/disavow ζ, the confirmer proves that c is/is not a commitment to the message. In [14], Chen proposed another DCS scheme, in which a signature is a non-interactive proof showing the equality of two discrete logarithms. Camenisch and Michels [8] later showed that the confirmer in the scheme can forge. In [8], the security of DCS in a multi-user setting was formalized, where multiple signers share the same confirmer. Many DCS schemes previously proposed were found vulnerable to the signature-transformation attack which transforms one signer's signature maliciously to another signer's signature. A new DCS was also proposed in [8] and proven secure in their model under the RSA assumption.

All the works above are in the random oracle model. In [24], Goldwasser and Waisbard revised the definition and model given by Okamoto [33] to not requiring zero-knowledge proof for signature validity assertions. They also used strong witness hiding proofs of knowledge to construct DCS schemes by following the 'sign-then-encrypt' paradigm. The resulting schemes are proven in the standard model. However, the disavowal protocol still requires general zero-knowledge proofs. Gentry, Molnar and Ramzan [23] solved this problem and proposed the first DCS scheme which does not require any general zero-knowledge proof, and is provably secure in the standard model. Their scheme is based on Camenisch and Shoup's verifiable encryption scheme [9]. Later, Wang, Baek, Wong and Bao [36] showed that there are some security subtleties in the extractability and invisibility of Gentry et al.'s scheme. They proposed another DCS scheme which does not require public key encryption. Wikström [38] revisited the security definitions of DCS and proposed new ones. He also proposed a generic construction and a concrete instantiation based on strong RSA assumption, decision composite residuosity assumption and decision Diffie-Hellman assumption without random oracles. In [39], Zhang, Chen and Wei proposed a bilinear-pairing-based DCS scheme without random oracle. The security for the confirmer considered in [39] requires that no adversary can impersonate the confirmer to confirm signatures, that is strictly weaker than the invisibility defined by Camenisch et al. [8]. The scheme was recently shown to be visible [37].

(*Optimistic Fair Exchange*). Designated confirmer signature has many applications. One of them is *Optimistic Fair Exchange* (OFE) of signatures, the notion of which was introduced by Asokan, Shoup and Waidner [1] for solving the fairness problem in exchange of signatures between two parties say, Alice and Bob. Previous work either have Alice and Bob release their signatures gradually, e.g. bit-by-bit, and thus inefficient, or need a third party fully trusted by Alice and Bob. In OFE, as the initiator, Alice generates and sends her partial signature σ to Bob. Bob returns his full signature and Alice then sends her full signature ζ to Bob. If Bob does not receive ζ (e.g. due to interrupted connection or system crash), Bob will turn to a third party, called the arbitrator, show the fulfillment of his obligation, and request for resolving σ. The arbitrator first checks the validity of Bob's full signature. If it is valid, the arbitrator converts σ to ζ and sends it to Bob, and in the meanwhile, Bob's full signature will be forwarded to Alice.

Since the introduction, OFE has attracted the attention of many researchers, i.e. [2,18,19,28,29,34]. For example, Park, Chong and Siegel [34] proposed a construction using sequential two-party multisignatures. The construction was later broken and repaired by Dodis and Reyzin [19]. In [18], Dodis, Lee and Yum showed that an OFE secure in a single-user setting does not imply the security in a multi-user setting. They also proposed an efficient OFE in the multi-user setting under the random oracle model. Huang, Yang, Wong and Susilo [29] further strengthened their results by relaxing the restriction on using a public key. They demonstrated a security gap for OFE between the chosen-key model [31] (in which an adversary can use any public key) and the registered-key model [3] (in which the adversary has to prove its knowledge of the secret key before using a public key). They also proposed a generic OFE scheme secure in the multi-user setting and chosen-key model, using a standard signature and a ring signature.

(*Abuse-free/Ambiguous* OFE). In OFE, Alice's partial signature is generally self-authenticating and indicates her commitment to some message already. This may allow Bob to make use of it to convince others that Alice has already committed herself to the message; while Alice obtains nothing. This could be unfair to Alice. Garay, Jakobsson and MacKenzie [21] and Huang, Yang, Wong and Susilo [28] addressed this problem and proposed notions of *abuse-free optimistic contract signing* and *ambiguous optimistic fair exchange*, respectively. In both notions, Alice and Bob should be able to produce indistinguishable partial signatures so that given a valid partial signature from Alice, Bob cannot transfer the conviction to others. In this paper we universally call both of them as 'AOFE' in short. Garay et al. constructed an efficient AOFE from a type of signatures called '*private contract signatures*', which is similar to but different from DCS (see [21] for details). Their private contract signature scheme is built from designated-verifier signature [30], and is secure in the registered-key model with random oracles. Huang et al. [28] proposed another efficient construction of AOFE using Groth-Sahai NIWI and NIZK proofs [25]. Their scheme is secure in the chosen-key model without random oracle.

3   Definition and Security Model of Designated Confirmer Signature

3.1   Definition

In a Designated Confirmer Signature (DCS) scheme, there are three parties, a signer S, a verifier V and a designated confirmer C. A DCS scheme consists of the following (probabilistic) polynomial-time (PPT) algorithms and two protocols (which will be defined shortly). Let $k \in \mathbb{N}$ be a security parameter.

- SKg. Signer S runs it to produce a key pair, i.e. $(\mathsf{spk}, \mathsf{ssk}) \leftarrow \mathsf{SKg}(1^k)$.
- CKg. Confirmer C runs it to produce a key pair, i.e. $(\mathsf{cpk}, \mathsf{csk}) \leftarrow \mathsf{CKg}(1^k)$.
- Sig. The algorithm takes as input the signer's secret key, a message M and optionally the confirmer's public key, and outputs a standard signature ζ on M, i.e. $\zeta \leftarrow \mathsf{Sig}(\mathsf{ssk}, M, \mathsf{cpk})$.

- Ver. This is the corresponding verification algorithm. It takes as input the signer's public key, a standard signature, a message and optionally the confirmer's public key, and outputs a bit b, which is 1 for acceptance and 0 for rejection, i.e. $b \leftarrow \mathsf{Ver}(M, \zeta, \mathsf{spk}, \mathsf{cpk})$.
- DCSig. Signer S runs it to generate a DCS signature. It takes as input the signer's secret key, a message and the confirmer's public key, and outputs a DCS signature σ, i.e. $\sigma \leftarrow \mathsf{DCSig}(\mathsf{ssk}, M, \mathsf{cpk})$.
- Ext. Confirmer C runs it to extract the signer's standard signature from its DCS signature. The algorithm takes as input the confirmer's secret key, a message, a DCS signature and the signer's public key, and outputs a standard signature or $\perp$ for the failure of extraction, i.e. $\zeta / \perp \leftarrow \mathsf{Ext}(\mathsf{csk}, M, \sigma, \mathsf{spk})$.

A DCS scheme also has the following two protocols, which are for the signer or the confirmer to confirm/disavow DCS signatures. In the protocols we use P to denote a prover, which could be the signer S or the confirmer C, and use V to denote a verifier. The common input of P and V is $(M, \sigma, \mathsf{spk}, \mathsf{cpk})$, where σ is an alleged DCS signature of S on message M. P also has an auxiliary input, denoted by sk, which is either ssk if P is the signer, or csk if it is the confirmer.

- Confirm. It is for P to convince V the validity of σ. At the end of the protocol, V outputs a single bit b which is 1 for accepting σ as a valid DCS signature on M of S, and 0 otherwise. We denote an execution of the protocol by $b \leftarrow \mathsf{Confirm}_{\langle \mathsf{P(sk)}, \mathsf{V} \rangle}(M, \sigma, \mathsf{spk}, \mathsf{cpk})$.
- Disavow. It is for P to convince V the invalidity of σ. At the end of the protocol, V outputs a single bit b which is 1 for accepting σ as an invalid DCS signature on M of S, and 0 otherwise. We denote an execution of the protocol by $b \leftarrow \mathsf{Disavow}_{\langle \mathsf{P(sk)}, \mathsf{V} \rangle}(M, \sigma, \mathsf{spk}, \mathsf{cpk})$.

The correctness can be defined in a natural way, i.e. the output of the algorithms/protocols should be correct if the parties are honest, and the protocols should be sound. A DCS scheme is *extraction ambiguous* if a standard signature output by Sig is indistinguishable from that output by Ext. If the two distributions are identical, the scheme is said to be *perfectly extraction ambiguous*.

3.2 Security Model

Let $\mathcal{O} = \{O_{\mathsf{DCSig}}, O_{\mathsf{Confirm}}, O_{\mathsf{Disavow}}, O_{\mathsf{Ext}}\}$ be a collection of oracles that an adversary has access to. We consider the following security properties for DCS. Here we do not provide the oracle O_{Sig} which returns the signer's standard signatures, as it can be implemented using O_{DCSig} and O_{Ext}.

(**Security for Verifiers - Extractability**). An adversary, even after compromising the secret keys of the signer and the confirmer, should not be able to cheat the verifier, by generating a pair (M^*, σ^*) so that either σ^* is confirmable but unextractable, or disavowable but extractable. Formally, we consider the game $\mathsf{G_{S4V}}$ depicted in Fig. 1, where 'Case 1' and 'Case 2' refer to the following:

- Case 1: $\mathsf{Ver}(M^*, \mathsf{Ext}(\mathsf{csk}, M^*, \sigma^*, \mathsf{spk}), \mathsf{spk}, \mathsf{cpk}) = 0$, i.e. σ^* is unextractable.
- Case 2: $\mathsf{Ver}(M^*, \mathsf{Ext}(\mathsf{csk}, M^*, \sigma^*, \mathsf{spk}), \mathsf{spk}, \mathsf{cpk}) = 1$, i.e. σ^* is extractable.

The advantage of the adversary $\mathcal{A} = (\mathcal{A}_0, \mathcal{A}_1, \mathcal{A}_2)$ is defined as the probability that $b = 1$.

Definition 1 (Security for Verifiers). *A DCS scheme is (t, ϵ)-secure for verifiers (or* extractable*) if there is no adversary $\mathcal{A} = (\mathcal{A}_0, \mathcal{A}_1, \mathcal{A}_2)$ which runs in time t, and wins game $\mathsf{G}_{\mathsf{S4V}}$ with advantage at least ϵ.*

Game $\mathsf{G}_{\mathsf{S4V}}$:

$(\mathsf{spk}, \mathsf{ssk}) \leftarrow \mathsf{SKg}(1^k)$, $(\mathsf{cpk}, \mathsf{csk}) \leftarrow \mathsf{CKg}(1^k)$

$(M^*, \sigma^*, \tau_1, \tau_2) \leftarrow \mathcal{A}_0(\mathsf{spk}, \mathsf{ssk}, \mathsf{cpk}, \mathsf{csk})$

Case 1 : $b_1 \leftarrow \mathsf{Confirm}_{\langle \mathcal{A}_1(\tau_1), \mathsf{V} \rangle}(M^*, \sigma^*, \mathsf{spk}, \mathsf{cpk})$

Case 2 : $b_2 \leftarrow \mathsf{Disavow}_{\langle \mathcal{A}_2(\tau_2), \mathsf{V} \rangle}(M^*, \sigma^*, \mathsf{spk}, \mathsf{cpk})$

Return $b \leftarrow (b_1 \vee b_2)$.

Game $\mathsf{G}_{\mathsf{INV}}$:   **Game $\mathsf{G}_{\mathsf{S4S}}$:**

$(\mathsf{spk}, \mathsf{ssk}) \leftarrow \mathsf{SKg}(1^k)$, $(\mathsf{cpk}, \mathsf{csk}) \leftarrow \mathsf{CKg}(1^k)$ $(\mathsf{spk}, \mathsf{ssk}) \leftarrow \mathsf{SKg}(1^k)$

$(M^*, \tau) \leftarrow \mathcal{A}_1^{\mathcal{O}}(\mathsf{spk}, \mathsf{cpk})$, $b \leftarrow \{0, 1\}$ $(\mathsf{cpk}, \mathsf{csk}) \leftarrow \mathsf{CKg}(1^k)$

$\sigma^* \begin{cases} \leftarrow \mathsf{DCSig}(\mathsf{ssk}, M^*, \mathsf{cpk}), & \text{if } b = 0 \\ \leftarrow_\$ \mathcal{S} & , \text{ otherwise} \end{cases}$ $(M^*, \zeta^*) \leftarrow \mathcal{A}^{\mathcal{O}_1}(\mathsf{spk}, \mathsf{cpk}, \mathsf{csk})$

$\qquad\qquad\qquad\qquad\qquad\qquad\qquad\qquad$ Return $b \leftarrow \mathsf{Ver}(M^*, \zeta^*, \mathsf{spk}, \mathsf{cpk})$

$b' \leftarrow \mathcal{A}_2^{\mathcal{O}}(\tau, \sigma^*)$

Return $b \leftarrow [b' \overset{?}{=} b]$

Fig. 1. The Security Model of DCS

(Security for Signers - Unforgeability). Anyone, including the confirmer, should not be able to forge an (honest) signer's signatures. Formally, we consider the game $\mathsf{G}_{\mathsf{S4S}}$ depicted in Fig. 1, where $\mathcal{O}_1 = \mathcal{O} \backslash \{O_{\mathsf{Ext}}\}$. The advantage of the adversary $\mathcal{A}$ is defined as the probability that the returned bit is $b = 1$ and $\mathcal{A}$ did not query O_{DCSig} on input M^*.

Definition 2 (Security for Signers). *A DCS scheme is $(t, q_s, q_c, q_d, \epsilon)$-secure for signers (or* unforgeable*) if there is no adversary $\mathcal{A}$ which runs in time t, makes at most q_s queries to O_{DCSig}, q_c queries to O_{Confirm} and q_d queries to O_{Disavow}, and wins game $\mathsf{G}_{\mathsf{S4S}}$ with advantage at least ϵ.*

A stronger variant of the definition is *strong unforgeability*. Let σ_i be the return of O_{DCSig} on input M_i queried by $\mathcal{A}$, and ζ_i the output of the (deterministic) algorithm Ext on input $(\mathsf{csk}, M_i, \sigma_i, \mathsf{spk})$. We change the unforgeability game so that the adversary wins if $b = 1$ and $(M^*, \zeta^*) \notin \{M_i, \zeta_i\}_{i=1}^{q_s}$. A DCS scheme is $(t, q_s, q_c, q_d, \epsilon)$-*strongly unforgeable* if no adversary after running in time t can have advantage at least ϵ in this new game after making queries with the bound as above.

(Invisibility). Given a DCS signature, a verifier should not be able to tell the validity of it without the help from the signer or the confirmer. Formally, we consider the game $\mathsf{G_{INV}}$ depicted in Fig. 1, where $\mathcal{S}$ is the signature space defined by cpk and spk. The advantage of the adversary $\mathcal{A}$ is defined as the difference between one half and the probability that the returned bit is $b = 1$ and $\mathcal{A}$ did not query O_{Confirm}, O_{Disavow} and O_{Ext} on input (M^*, σ^*).

Definition 3 (Invisibility). *A DCS scheme is $(t,\ q_s, q_e, q_c, q_d, \epsilon)$-invisible if there is no adversary $\mathcal{A} = (\mathcal{A}_1, \mathcal{A}_2)$ which runs in time t, makes at most q_s queries to O_{DCSig}, q_e queries to O_{Ext}, q_c queries to O_{Confirm} and q_d queries to O_{Disavow}, and wins game $\mathsf{G_{INV}}$ with advantage at least ϵ.*

Definition 4. *A DCS scheme is said to be* secure *if it is secure for signers (Def. 2), secure for verifiers (Def. 1) and invisible (Def. 3).*

The definition of invisibility above implicitly requires the existence of an efficient algorithm which samples a random signature from the space $\mathcal{S}$ of the signer's DCS signatures. Clearly, if the sampler has the signer's secret key, it can sample a signature efficiently by signing a random message. However, if it is only given the public keys, the situation becomes different. We consider the following definition.

Definition 5 (Samplability). *A DCS scheme is* samplable *if there is a probabilistic polynomial-time algorithm, which given 1^k and public keys cpk and spk, chooses a signature σ randomly and uniformly from the signature space $\mathcal{S}$ defined by cpk and spk in time polynomial in the security parameter k.*

The schemes following the 'commit-then-sign' paradigm, e.g. [8,23,32,36], are not samplable, because a DCS signature includes a commitment and the signer's (standard) signature on the commitment, and it is infeasible to efficiently sample the commitment and the corresponding (standard) signature simultaneously. While for some other schemes, e.g. [39] and the one we propose in Sec. 5, there is an efficient sampling algorithm. We stress that for DCS schemes with samplability, although it is easy to randomly select a signature from $\mathcal{S}$, it does not imply that finding M is easy. Otherwise, the DCS scheme will be forgeable.

4   Assumptions

We assume that there is an efficient algorithm $\mathcal{IG}$ which takes as input 1^k and outputs a random instance of the bilinear groups, i.e. $(\mathbb{G}, \mathbb{G}_T, \hat{e}, p, g)$, where $\mathbb{G}, \mathbb{G}_T$ are represented as multiplicative groups of prime order p, g is a random generator of $\mathbb{G}$, and $\hat{e} : \mathbb{G} \times \mathbb{G} \to \mathbb{G}_T$ is a bilinear pairing.

Definition 6 (q-HSDH Assumption [7]). *The q-Hidden Strong Diffie-Hellman (q-HSDH) assumption (t, ϵ)-holds in $\mathbb{G}$ if there is no algorithm $\mathcal{A}$ which runs in time at most t, and satisfies the following condition:*

$$\Pr[\mathcal{A}(g, g^x, u, \{(g^{\frac{1}{x+s_i}}, g^{s_i}, u^{s_i})\}_{i=1}^q) = (g^{\frac{1}{x+s}}, g^s, u^s)] \geq \epsilon$$

where $s \in \mathbb{Z}_p$ and $s \notin \{s_1, \cdots, s_q\}$, the probability is taken over the random choices of $x, s_1, \cdots, s_q \in \mathbb{Z}_p$, $u \in \mathbb{G}$ and the random coins used by $\mathcal{A}$.

Definition 7 (q-DHSDH Assumption [27]). *The q-Decisional Hidden Strong Diffie-Hellman (q-DHSDH) assumption (t, ϵ)-holds in $\mathbb{G}$ if there is no algorithm $\mathcal{A}$ which runs in time at most t, and satisfies the following condition:*

$$|\Pr[\mathcal{A}(g, g^x, u, \mathbf{Q}, u^s, g^{\frac{1}{x+s}}) = 1] - \Pr[\mathcal{A}(g, g^x, u, \mathbf{Q}, u^s, Z) = 1]| \geq \epsilon$$

where $\mathbf{Q} = \{(g^{\frac{1}{x+s_i}}, g^{s_i}, u^{s_i})\}_{i=1}^q$, and the probability is taken over the random choices of $x, s_1, \cdots, s_q, s \in \mathbb{Z}_p$ and $u, Z \in \mathbb{G}$, and the random coins used by $\mathcal{A}$.

5   A New DCS Scheme without Random Oracle

We now propose an efficient construction of designated confirmer signature, which makes use of a programmable hash function (PHF) [26]. A PHF = (Gen, Eval, Trap) is a keyed group hash function which maps the set of arbitrarily long messages to a group $\mathbb{G}$. It behaves in two indistinguishable ways, depending on how the key κ is generated. If we use the standard key generation algorithm, the function behaves normally as prescribed. If we use the alternative trapdoor key generation algorithm which outputs a simulated key for the function (indistinguishable from a real key) and a trapdoor τ, besides the normal output, the function (on input X and τ) also outputs some secret key trapdoor information τ' dependent on two generators g, h from the group, e.g. $\tau' = (a_X, b_X)$ such that PHF.Eval$(\kappa, X) = g^{a_X} h^{b_X}$. PHF is (m, n, ϕ, φ)-programmable if the statistical distance between distributions of real keys for the function and simulated keys is bounded by ϕ, and for all choices $X_1, \cdots, X_m \in \{0, 1\}^n$ and $Z_1, \cdots, Z_n \in \{0, 1\}^n$ with $X_i \neq Z_j$, it holds that $a_{X_i} = 0$ but $a_{Z_j} \neq 0$ with probability at least φ. An instantiation of PHF is the 'multi-generator' hash function, defined as PHF.Eval$(\kappa, X) = h_0 \prod_{i=1}^n h_i^{X_i}$, where h_i's are the public generators of $\mathbb{G}$ included in κ, and $X = (X_1 \cdots X_n) \in \{0, 1\}^n$. We refer readers to [26] for details.

5.1   The Scheme

Let $(\mathbb{G}, \mathbb{G}_T, \hat{e}, p, g)$ be a random instance of bilinear groups output by $\mathcal{IG}(1^k)$, PHF be a family of programmable hash functions [26]. The new DCS scheme, denoted by $\mathcal{DCS}$, works as below, where for simplicity we write PHF.Eval(κ, M) as $H_\kappa(M)$ for a key κ.

- SKg(1^k). Choose at random $x \leftarrow_\$ \mathbb{Z}_p$, $u \leftarrow_\$ \mathbb{G}$, and set $X := g^x$. Run PHF.Gen(1^k) to produce a key κ for the programmable hash function H, Set and return $(\text{spk}, \text{ssk}) := ((X, u, \kappa), x)$.
- CKg(1^k). Choose $y \leftarrow_\$ \mathbb{Z}_p$ randomly. Return $(\text{cpk}, \text{csk}) := (Y, y) = (g^{\frac{1}{y}}, y)$.
- Sig$(\text{ssk}, M, \text{cpk})$. Pick $s \leftarrow_\$ \mathbb{Z}_p$ randomly. Return $\zeta := (H_\kappa(M)^{\frac{1}{x+s}}, g^s, u^s)$.
- Ver$(M, \zeta, \text{spk}, \text{cpk})$. Parse ζ as (δ, ν, θ), and return 1 if both $\hat{e}(\nu, u) = \hat{e}(g, \theta)$ and $\hat{e}(\delta, X\nu) = \hat{e}(H_\kappa(M), g)$ hold, and 0 otherwise.
- DCSig$(\text{ssk}, M, \text{cpk})$. Pick $s \leftarrow_\$ \mathbb{Z}_p$ randomly. Return $\sigma := (H_\kappa(M)^{\frac{1}{x+s}}, Y^s, u^s)$.
- Ext$(\text{csk}, M, \sigma, \text{spk})$. Parse σ as (δ, γ, θ). Set $\nu := \gamma^y$, and return $\zeta := (\delta, \nu, \theta)$.

(**Signature Space**). The DCS signature space with respect to cpk and spk is defined as $\mathcal{S} := \{(\delta, \gamma, \theta) : \hat{e}(\gamma, u) = \hat{e}(Y, \theta)\}$. To sample a signature at random from the space, one picks at random $s \leftarrow_{\$} \mathbb{Z}_p$, sets $\gamma = Y^s$, $\theta = u^s$, and then picks at random $\delta \leftarrow_{\$} \mathbb{G}$. The sampled signature is $\sigma = (\delta, \gamma, \theta) \in \mathcal{S}$. Note that it is hard to find message M such that $\delta = \mathrm{H}_\kappa(M)^{\frac{1}{x+s}}$.

(**Confirmation**). Let the common input be $(M, \sigma, \mathsf{spk}, \mathsf{cpk})$ where $\sigma = (\delta, \gamma, \theta)$ is a signature on M. Both the prover P, which is either the signer S or the confirmer C, and the verifier V check whether the equation $\hat{e}(\gamma, u) = \hat{e}(Y, \theta)$ holds. If not, they do nothing. Below we assume that the equation holds. Note that both S and C can verify the validity of σ. For S, it additionally checks whether $\hat{e}(\delta, u^x \theta) = \hat{e}(\mathrm{H}_\kappa(M), u)$, which is equivalent to

$$\hat{e}(\delta, u)^x = \hat{e}(\mathrm{H}_\kappa(M), u) \cdot \hat{e}(\delta, \theta)^{-1} \overset{\text{def}}{=} W_1. \tag{1}$$

For C, it additionally checks if $\hat{e}(\delta, X\gamma^y) = \hat{e}(\mathrm{H}_\kappa(M), g)$, which is equivalent to

$$\hat{e}(\delta, \gamma)^y = \hat{e}(\mathrm{H}_\kappa(M), g) \cdot \hat{e}(\delta, X)^{-1} \overset{\text{def}}{=} W_2. \tag{2}$$

Therefore, P starts an execution of the following proof of knowledge of equality of discrete logarithms, using the knowledge of either x or y.

$$\mathrm{PK}\,\{\alpha : (\hat{e}(\delta, u)^\alpha = W_1 \wedge g^\alpha = X) \vee (\hat{e}(\delta, \gamma)^\alpha = W_2 \wedge Y^\alpha = g)\} \tag{3}$$

(**Disavowal**). Let the common input be $(M, \sigma, \mathsf{spk}, \mathsf{cpk})$ where $\sigma = (\delta, \gamma, \theta)$ is an invalid signature on M of S. Both P and V check whether the equation $\hat{e}(\gamma, u) = \hat{e}(Y, \theta)$ holds. If not, they do nothing. Otherwise, they start an execution of the following proof of knowledge of inequality of discrete logarithms, using the knowledge of either x or y.

$$\mathrm{PK}\,\{\alpha : (\hat{e}(\delta, u)^\alpha \neq W_1 \wedge g^\alpha = X) \vee (\hat{e}(\delta, \gamma)^\alpha \neq W_2 \wedge Y^\alpha = g)\} \tag{4}$$

Remark 1. The confirmation/disavowal protocols given above are Σ-protocols [17] with special soundness and perfect special honest-verifier zero-knowledge. Their corresponding four-move fully fledged perfect zero-knowledge protocols can be obtained by applying the transformation in [15].

5.2   Security Analysis

Theorem 1. *DCS is $(t, 2/p)$-extractable (Def. 1).*

Theorem 2. *Let* PHF *be a family of $(m, 1, \phi, \varphi)$-programmable hash functions. Let $\mathcal{F}$ be a $(t, q_s, q_c, q_d, \epsilon)$-adversary against the unforgeability (Def. 2) of DCS. Then there exists an adversary $\mathcal{A}_1$ that (t_1, ϵ_1)-breaks the q_s-SDH assumption [4] with $t_1 \approx t$ and $\epsilon_1 \geq q_s^{-1} \varphi(\epsilon - q_s^{m+1}/p^m - \phi)$, or there exists an adversary $\mathcal{A}_2$ that (t_2, ϵ_2)-breaks the q_s-HSDH assumption and an adversary $\mathcal{A}_3$ that (t_3, ϵ_3)-breaks the Discrete Logarithm assumption in $\mathbb{G}$ with $t_2, t_3 \approx t$ and $\epsilon_2 + \epsilon_3 \geq \epsilon - \phi$.*

Theorem 3. *Let* PHF *be a family of* $(m, 1, \phi, \varphi)$-*programmable hash functions and* $\mathcal{D}$ *a* $(t, q_s, q_e, q_c, q_d, \epsilon)$-*distinguisher against invisibility (Def. 3) of* $\mathcal{DCS}$. *If* $\mathcal{DCS}$ *is* $(t_1, q_s, q_c, q_d, \epsilon_1)$-*strongly unforgeable and its confirmation and disavowal protocols are* perfect *zero-knowledge, then there exists an adversary* $\mathcal{A}$ *which* (t', ϵ')-*breaks the* (q_s+1)-*DHSDH assumption and an adversary* $\mathcal{A}'$ *which* (t'', ϵ'')-*breaks the Discrete Logarithm assumption with* t_1, t', $t'' \approx t$ *and* $\epsilon' + \epsilon'' \geq \epsilon - \varphi - \epsilon_1$.

The proofs are deferred to the full version of this paper due to page limit. We remark that a slight modification of the proof of Theorem 2 can be used to show the *strong unforgeability* of $\mathcal{DCS}$ so that even if the adversary obtained a signature on a message, it cannot forge a new signature on the same message.

5.3   Extensions

(*Signer-Convertible DCS*). The scheme $\mathcal{DCS}$ allows the signer S to confirm/ disavow DCS signatures without keeping any state information (e.g. the randomness used in signature generation), however, similar to all the existing DCS schemes, it does not let the (stateless) signer convert a valid DCS signature to a standard signature yet. Fortunately, we can extend $\mathcal{DCS}$ so to support this feature with merely a little cost, which includes an additional $\mathbb{G}$ element to S's public key and its DCS signature. The idea behind our construction is to put another 'confirmation key' into S's public key. Specifically, S selects at random $\tilde{y} \leftarrow_\$ \mathbb{Z}_p$, and puts $\tilde{Y} := g^{1/\tilde{y}}$ into spk and $\tilde{y}$ into ssk. When generating a DCS signature, besides δ, γ, θ, S also computes $\eta := (\tilde{Y})^s$ and sets $\sigma := (\delta, \gamma, \theta, \eta)$. The other algorithms and protocols of $\mathcal{DCS}$ can be modified accordingly. To convert σ to a standard signature, S computes $\nu := \eta^{\tilde{y}} = (g^{1/\tilde{y}})^{s\tilde{y}} = g^s$ and releases $\zeta = (\delta, \nu, \theta)$, which is identical to the one output by the confirmer. Clearly, the correctness of the conversion is publicly verifiable.

(*Universal Convertible DCS*). We can also borrow the notion of *universal conversion* from convertible undeniable signature [5] and apply it to the context of DCS. To universally convert DCS signatures, S only needs to publish $\tilde{y}$, which does not do any harm to C or other signers (note: C could also be the confirmer of other signers), as well as the unforgeability of S's signatures.

(*Multiple Confirmers*). The $\mathcal{DCS}$ can also be extended to support multiple confirmers so that each confirmer can convert a DCS signature to obtain the same standard signature. Suppose that there are n confirmers with public keys $Y_1, \cdots, Y_n$. To sign a message M w.r.t. these confirmers, the signer computes δ, θ as prescribed, and then for each confirmer it computes $\gamma_i = Y_i^s$. The DCS signature is $\sigma = (\delta, \gamma_1, \cdots, \gamma_n, \theta)$.

(*Threshold Conversion*). The $\mathcal{DCS}$ also supports t-out-of-n threshold conversion. The confirmers run a protocol for example, the distributed key generation protocol proposed by Canetti et al. [10], to jointly generate the confirmation public key Y so that each of them holds a share y_i of the corresponding secret key

y where $Y_i = g^{1/y_i}$ is the respective public key. To extract a signature, each confirmer uses their share y_i to compute $\nu_i = \gamma^{y_i}$. Anyone who collects at least t ν_i's can recover ν efficiently, i.e. using Lagrange interpolation.

5.4   Performance Comparison

In Camenisch-Michels DCS scheme [8], the signer's RSA signature is hidden using Cramer-Shoup encryption [16]. A DCS signature contains four elements from $\mathbb{Z}_p^*$ where p is a prime larger than the RSA modulus N, which needs about $4K$ bits for 80-bit security. In Gentry-Molnar-Ramzan scheme [23], a DCS signature consists of a commitment, a standard signature on the commitment and the Cramer-Shoup encryption [9] of the randomness used in generating the commitment, and thus has size larger than that of Camenisch-Michels scheme. The extracted signature consists of the DCS signature and the randomness encapsulated in the ciphertext. In Wang-Baek-Wong-Bao scheme [36], a DCS signature contains two elements d_1, d_2 of $\mathbb{Z}_p^*$ computed from the message (and some randomness), the signer's standard signature on d_1, d_2, and a non-interactive proof of the discrete logarithm of d_1 (w.r.t. the group generator), and an extracted signature consists of d_1, d_2, the signer's standard signature on d_1, d_2 and a non-interactive proof showing some quadruple being a Diffie-Hellman tuple. Their confirmation/disavowal protocol is very efficient, simply a proof showing that some quadruple is/is not a Diffie-Hellman tuple.

In our DCS scheme, both a DCS signature and an extracted signature contain only three elements of the bilinear group $\mathbb{G}$ (about 60 bytes for 80-bit security), shorter than any of the schemes above. Therefore, when compared with previous schemes, the signature size of this new DCS scheme is only about 12% that of the Camenisch-Michels DCS scheme [8]. When compared with the Wang-Baek-Wong-Bao DCS scheme [36] which is proven in the random oracle model, this new DCS scheme still has a significant extent of advantage. The signature size of the new DCS scheme is only about 50% that of theirs. Table 1 shows the comparison of our DCS scheme with two schemes in the literature in terms of key sizes, signature sizes and whether the security relies on the random oracle model (ROM). In the comparison we instantiate the encryption scheme and signature scheme used in [23] with [9] and [22] respectively, and instantiate the signature scheme used in [36] with Schnorr signature [35].

Inherited from the shortness of the only known '*multi-generator*' instantiation of programmable hash function [26], the public key of the signer in our DCS scheme is also long. If we choose to put the key κ for the programmable hash

Table 1. Comparison with Two Existing DCS schemes

Schemes	cpk	csk	spk	ssk	σ	ζ	ROM
[8]	$5\mathbb{Z}_p$	$5\mathbb{Z}_q$	$1\mathbb{Z}_N + k$	$1\mathbb{Z}_N$	$4\mathbb{Z}_p \approx 4K$	$1\mathbb{Z}_N \approx 1K$	$\times$
[36]	$1\mathbb{G}$	$1\mathbb{Z}_p$	$1\mathbb{G}$	$1\mathbb{Z}_p$	$2\mathbb{G} + 4\mathbb{Z}_p \approx 0.95K$	$2\mathbb{G} + 6\mathbb{Z}_p \approx 1.2K$	$\checkmark$
Ours	$1\mathbb{G}$	$1\mathbb{Z}_p$	$163\mathbb{G}$	$1\mathbb{Z}_p$	$3\mathbb{G} \approx 0.47K$	$3\mathbb{G} \approx 0.47K$	$\times$

function and u in the signer's public key to the system parameters, we obtain a DCS scheme secure in the common reference string model. The signer's public key of the resulting scheme becomes much shorter, i.e. one $\mathbb{G}$ element only (about 20 bytes), and all the signers will share the same signature space. Furthermore, it can be shown that the scheme enjoys *anonymity* [20] in the sense that two signers' DCS signatures on the same message are indistinguishable.

6   A New Construction of AOFE

As an important application, designated confirmer signature can be used for building efficient optimistic fair exchange protocols [1,14]. In this section, we show how to use our new DCS scheme to build an efficient Ambiguous Optimistic Fair Exchange (AOFE) scheme. Essentially, AOFE is a variant of the traditional OFE, in which both of the exchanging parties can produce indistinguishable signatures on the same message. An AOFE scheme consists of the following probabilistic polynomial time algorithms/protocols: PMGen for generating system parameters; Setup^{TTP} for generating a key pair for the arbitrator; Setup^{User} for generating a key pair for each user; PSig/PVer for generating and verifying a partial signature (interactively or non-interactively); Sig/Ver for generating and verifying a full signature; and Res for converting a partial signature to a full one. We refer readers to [28] for the details.

The security of AOFE was originally defined in the *chosen-key* model [28], in which the adversary is allowed to use any public key arbitrarily. In this work we consider a stronger but still practical variant of the key model, named *registered-key* model, in which the adversary has to prove its knowledge of the secret key before using a public key.

6.1   Security Model of AOFE in the Registered-Key Model

Let $\mathcal{Q}(O)$ be the set of queries that the adversary submits to oracle O, where O could be any of the oracles below. We assume that the adversary has already registered a public key to the oracle O_{KR} before using it; otherwise, no response is given to the adversary.

- O_{KR} takes as input a key pair $(\mathrm{pk}_i, \mathrm{sk}_i)$, and checks the validity of the pair. If valid, it stores the pair and returns pk_i to the adversary; otherwise, it returns $\perp$.
- O_{PSig} takes as input (M, pk_i) and returns a partial signature σ of the signer with public key pk_A, which is valid on M under pk_A, pk_i. The oracle then starts an execution of PVer with the adversary to show the validity of σ.
- O_{FakePSig} takes as input (M, pk_i) and returns a partial signature σ generated using sk_B, which is valid under pk_i, pk_B. The oracle then starts an execution of the PVer protocol with the adversary to show the validity of σ.
- O_{Res} takes as input $(M, \sigma, \mathrm{pk}_i, \mathrm{pk}_j)$ and outputs ζ if it is a valid (standard) signature on M under pk_i, and $\perp$ otherwise.

Signer Ambiguity. The signer ambiguity says that after obtaining the valid partial signature from the signer S, the verifier V cannot transfer the conviction to any third party. We require that V is able to produce signatures indistinguishable from those by S. Formally, we consider the game $\mathbf{G}_{sa}$ depicted in Fig. 2 (page 55), where $\mathcal{O}_1 = \{O_{KR}, O_{PSig}, O_{Res}\}$, and Υ is $\mathcal{D}$'s state information. Note that after sending σ^* to $\mathcal{D}$ in the game, the challenger also starts an execution of the PVer protocol with $\mathcal{D}$ to show the validity of σ^* under pk_A, pk_B. The advantage of $\mathcal{D}$, denoted by $\mathrm{Adv}_{\mathcal{D}}^{sa}(k)$, is defined to be the gap between its success probability in the game and one half, i.e. $\mathrm{Adv}_{\mathcal{D}}^{sa}(k) = |\Pr[\mathcal{D}\ \mathrm{Succ}] - 1/2|$.

Definition 8 (Signer Ambiguity). *An AOFE scheme is* signer ambiguous *if there is no probabilistic polynomial-time distinguisher $\mathcal{D}$ such that $\mathrm{Adv}_{\mathcal{D}}^{sa}(k)$ is non-negligible in k.*

Our definition of signer ambiguity is slightly different from that in [28]. There the distinguisher corrupts both pk_A and pk_B, while here it only is allowed to corrupt pk_B. We believe that this more reflects the reality, as it is unlikely for Bob to already know the secret key of Alice before/when he wants to show Alice's partial commitment to others.

Security Against Signers. This requires that (malicious) signer $\mathcal{A}$ cannot produce a partial signature, which looks good to V but cannot be resolved to a full signature by the honest arbitrator, ensuring the fairness for verifiers. V should always be able to obtain the full commitment of the signer if the signer has committed to a message. Formally, we consider the game $\mathbf{G}_{sas}$ depicted in Fig. 2, where $\mathcal{O}_2 = \{O_{KR}, O_{FakePSig}, O_{Res}\}$. The advantage of $\mathcal{A}$ in the game, denoted by $\mathrm{Adv}_{\mathcal{A}}^{sas}(k)$, is defined as its success probability.

Definition 9 (Security Against Signers). *An AOFE scheme is* secure against signers *if there is no probabilistic polynomial-time adversary $\mathcal{A}$ such that $\mathrm{Adv}_{\mathcal{A}}^{sas}(k)$ is non-negligible in k.*

Security Against Verifiers. It requires that any efficient verifier $\mathcal{B}$ should not be able to convert a partial signature into a full one with non-negligible probability if it obtains no help from the signer or the arbitrator. This ensures the fairness for the arbitrator and the signer. Formally, we consider the game $\mathbf{G}_{sav}$ depicted in Fig. 2, where $\mathcal{O}_3 = \{O_{KR}, O_{PSig}, O_{Res}\}$. The advantage of $\mathcal{B} = (\mathcal{B}_1, \mathcal{B}_2)$ in the game, denoted by $\mathrm{Adv}_{\mathcal{B}}^{sav}(k)$, is defined as its success probability.

Definition 10 (Security Against Verifiers). *An AOFE scheme is* secure against verifiers *if there is no probabilistic polynomial-time adversary $\mathcal{B}$ such that $\mathrm{Adv}_{\mathcal{B}}^{sav}(k)$ is non-negligible in k.*

Security Against the Arbitrator. This is for ensuring the unforgeability of the signer's signatures. It says that no efficient adversary $\mathcal{C}$, even the arbitrator, is able to generate with non-negligible probability a valid full signature without explicitly asking the signer for generating one. Formally, we consider the game $\mathbf{G}_{saa}$ depicted in Fig. 2, where $\mathcal{O}_4 = \{O_{KR}, O_{PSig}\}$. The advantage of $\mathcal{C}$ in this game, denoted by $\mathrm{Adv}_{\mathcal{C}}^{saa}(k)$, is defined as its success probability.

Definition 11 (Security Against the Arbitrator). *An AOFE scheme is secure against the arbitrator if there is no probabilistic polynomial-time adversary $\mathcal{C}$ such that $\mathrm{Adv}_{\mathcal{C}}^{\mathrm{saa}}(k)$ is non-negligible in k.*

Game $\mathbf{G}_{\mathrm{sa}}$:
$\quad$ $\mathrm{PM} \leftarrow \mathsf{PMGen}(1^k), \quad (\mathrm{apk}, \mathrm{ask}) \leftarrow \mathsf{Setup}^{\mathsf{TTP}}(\mathrm{PM})$

$\quad$ $(\mathrm{pk}_A, \mathrm{sk}_A) \leftarrow \mathsf{Setup}^{\mathsf{User}}(\mathrm{PM}, \mathrm{apk}), \quad (M^*, \mathrm{pk}_B, \Upsilon) \leftarrow \mathcal{D}^{\mathcal{O}_1}(\mathrm{apk}, \mathrm{pk}_A)$

$\quad$ $b \leftarrow \{0,1\}, \quad \sigma^* \leftarrow \begin{cases} \mathsf{PSig}(M^*, \mathrm{sk}_A, \mathrm{pk}_A, \mathrm{pk}_B, \mathrm{apk}) & , \text{ if } b = 0 \\ \mathsf{FakePSig}(M^*, \mathrm{sk}_B, \mathrm{pk}_A, \mathrm{pk}_B, \mathrm{apk}), & \text{ if } b = 1 \end{cases}$

$\quad$ $b' \leftarrow \mathcal{D}^{\mathcal{O}_1}(\Upsilon, \sigma^*)$

$\quad$ Succ. of $\mathcal{D} := \big[b' = b \ \wedge \ (M^*, \sigma, \{\mathrm{pk}_A, \mathrm{pk}_B\}) \notin \mathcal{Q}(O_{\mathsf{Res}})\big]$

Game $\mathbf{G}_{\mathrm{sas}}$:
$\quad$ $\mathrm{PM} \leftarrow \mathsf{PMGen}(1^k), \quad (\mathrm{apk}, \mathrm{ask}) \leftarrow \mathsf{Setup}^{\mathsf{TTP}}(\mathrm{PM})$

$\quad$ $(\mathrm{pk}_B, \mathrm{sk}_B) \leftarrow \mathsf{Setup}^{\mathsf{User}}(\mathrm{PM}, \mathrm{apk}), \quad (M^*, \mathrm{pk}_A, \sigma^*) \leftarrow \mathcal{A}^{\mathcal{O}_2}(\mathrm{apk}, \mathrm{pk}_B)$

$\quad$ $\zeta^* \leftarrow \mathsf{Res}(M^*, \sigma^*, \mathrm{ask}, \mathrm{pk}_A, \mathrm{pk}_B)$

$\quad$ Succ. of $\mathcal{A} := \big[\mathsf{PVer}(M^*, \sigma^*, \{\mathrm{pk}_A, \mathrm{pk}_B\}, \mathrm{apk}) = 1 \wedge$
$\qquad\qquad \mathsf{Ver}(M^*, \zeta^*, \mathrm{pk}_A, \mathrm{pk}_B, \mathrm{apk}) = 0 \wedge (M^*, \mathrm{pk}_A) \notin \mathcal{Q}(O_{\mathsf{FakePSig}})\big]$

Game $\mathbf{G}_{\mathrm{sav}}$:
$\quad$ $\mathrm{PM} \leftarrow \mathsf{PMGen}(1^k), \quad (\mathrm{apk}, \mathrm{ask}) \leftarrow \mathsf{Setup}^{\mathsf{TTP}}(\mathrm{PM})$

$\quad$ $(\mathrm{pk}_A, \mathrm{sk}_A) \leftarrow \mathsf{Setup}^{\mathsf{User}}(\mathrm{PM}, \mathrm{apk}), \quad (M^*, \mathrm{pk}_B, \Upsilon) \leftarrow \mathcal{B}_1^{\mathcal{O}_3}(\mathrm{apk}, \mathrm{pk}_A)$

$\quad$ $\sigma^* \leftarrow \mathsf{PSig}(M^*, \mathrm{sk}_A, \mathrm{pk}_A, \mathrm{pk}_B, \mathrm{apk}), \quad \zeta^* \leftarrow \mathcal{B}_2^{\mathcal{O}_3}(\Upsilon, \sigma^*)$

$\quad$ Succ. of $\mathcal{B} := \big[\mathsf{Ver}(M^*, \zeta^*, \mathrm{pk}_A, \mathrm{pk}_B, \mathrm{apk}) = 1 \wedge (M^*, \cdot, \{\mathrm{pk}_A, \mathrm{pk}_B\}) \notin \mathcal{Q}(O_{\mathsf{Res}})\big]$

Game $\mathbf{G}_{\mathrm{saa}}$:
$\quad$ $\mathrm{PM} \leftarrow \mathsf{PMGen}(1^k), \quad (\mathrm{apk}, \mathrm{ask}) \leftarrow \mathsf{Setup}^{\mathsf{TTP}}(\mathrm{PM})$

$\quad$ $(\mathrm{pk}_A, \mathrm{sk}_A) \leftarrow \mathsf{Setup}^{\mathsf{User}}(\mathrm{PM}, \mathrm{apk}), \quad (M^*, \mathrm{pk}_B, \zeta^*) \leftarrow \mathcal{C}^{\mathcal{O}_4}(\mathrm{ask}, \mathrm{apk}, \mathrm{pk}_A)$

$\quad$ Succ. of $\mathcal{C} := \big[\mathsf{Ver}(M^*, \zeta, \mathrm{pk}_A, \mathrm{pk}_B, \mathrm{apk}) = 1 \wedge (M^*, \mathrm{pk}_B) \notin \mathcal{Q}(O_{\mathsf{PSig}})\big]$

Fig. 2. Security Model of AOFE

Definition 12 (Secure AOFE). *An AOFE scheme is said to be* secure in the multi-user setting and registered-key model *(or simply,* secure*), if it satisfies signer ambiguity (Def. 8), security against signers (Def. 9), security against verifiers (Def. 10) and security against the arbitrator (Def. 11).*

6.2 Interactive AOFE

Our interactive AOFE scheme is conceptually simple. The arbitrator acts as the confirmer of DCS, and the signer's partial signature σ on a message M consists of its DCS signature on M and a zero-knowledge proof, i.e. FullConfirm, showing either the validity of σ or the knowledge of the verifier's secret key. This proof is also known as the *designated-verifier proof* [30]. To resolve partial signature σ, the arbitrator uses its secret key in DCS to convert σ to the signer's standard signature ζ, which is defined as the signer's full signature in the AOFE scheme. The following is the description of the AOFE scheme.

PMGen. This algorithm generates the parameters for the DCS scheme, e.g. $\mathbb{G}, \mathbb{G}_T, \hat{e} : \mathbb{G} \times \mathbb{G} \to \mathbb{G}_T, p, g$.

Setup^{TTP}. The arbitrator chooses at random $y \leftarrow_\$ \mathbb{Z}_p$, and sets its key pair as $(\mathsf{apk}, \mathsf{ask}) := (g^{1/y}, y)$.

Setup^{User}. Each user chooses at random $x \leftarrow_\$ \mathbb{Z}_p$, $u \leftarrow_\$ \mathbb{G}$, and sets $X := g^x$. It also runs PHF.Gen(1^k) to produce a key κ for the programmable hash function H. It sets its key pair as $(\mathsf{pk}, \mathsf{sk}) := ((X, u, \kappa), x)$.

PSig/PVer. Let the signer be U_i with public key $\mathsf{pk}_i = (X_i, u_i, \kappa_i)$ and the verifier be U_j with public key $\mathsf{pk}_j = (X_j, u_j, \kappa_j)$. To partially sign a message M, U_i and U_j work as follows:

1. U_i selects $s \leftarrow_\$ \mathbb{Z}_p$ at random, and sends $\sigma := (\delta, \gamma, \theta)$ to U_j, where $\delta = \mathsf{H}_{\kappa_i}(M\|\mathsf{pk}_j)^{1/(x_i+s)}$, $\gamma = \mathsf{apk}^s$, and $\theta = u_i^s$.
2. U_i starts an execution of protocol FullConfirm with U_j to show that either σ is U_i's valid signature on $M\|\mathsf{pk}_j$ or it knows the secret key of U_j.
3. U_j outputs 1 if it accepts at the end of the proof, and 0 otherwise.

Sig/Ver. To fully sign a message M, the signer U_i selects $s \leftarrow_\$ \mathbb{Z}_p$ at random and computes $\zeta := (\delta, \nu, \theta) = (\mathsf{H}_{\kappa_i}(M\|\mathsf{pk}_j)^{1/(x_i+s)}, g^s, u_i^s)$. It sends ζ to the verifier U_j, which then checks whether both $\hat{e}(\nu, u_i) = \hat{e}(g, \theta)$ and $\hat{e}(\delta, X_i\nu) = \hat{e}(\mathsf{H}_{\kappa_i}(M\|\mathsf{pk}_j), g)$ hold. If so, U_j outputs 1; otherwise it outputs 0.

Res. After receiving from U_j a signature $\sigma = (\delta, \gamma, \theta)$ and a proof transcript claimed to be U_i's partial signature, the arbitrator first checks the validity of the proof, and returns $\perp$ to U_j if it is not valid. It then computes $\nu := \gamma^y$, and returns $\zeta := (\delta, \nu, \theta)$ if ζ is U_i's valid signature on $M\|\mathsf{pk}_j$, and $\perp$ otherwise.

How to simulate a partial signature: To simulate U_i's partial signature on a message M, user U_j can randomly select a signature σ' from the space of U_i's partial signatures (see Sec. 5), and then make a FullConfirm proof using its own secret key as the witness. The invisibility of the underlying DCS scheme and the perfect zero-knowledge of FullConfirm protocol together tell that the simulated partial signature looks indistinguishable from a real one.

Theorem 4. *The interactive AOFE scheme above is secure (Def. 12) provided that HSDH and DHSDH assumptions hold in* $\mathbb{G}$.

The proof is deferred to the full version of this paper due to the lack of space.

6.3 Non-interactive AOFE

Our interactive AOFE requires seven moves in total. The signer S sends its partial signature to the verifier V in the first move, then starts an execution of the protocol FullConfirm, which needs four moves. V sends back its full signature in the sixth move, and in response, S returns its full signature in the last move.

However, it is more desirable in practice to reduce the communication cost and save the bandwidth. The optimal case is the *non-interactive* AOFE (originally considered in [28]), which requires only three moves, i.e. the partial signature of S could be directly verified by V.

By applying the Fiat-Shamir transformation to (the Σ-protocol version of) FullConfirm, we get a non-interactive zero-knowledge proof of knowledge showing either the validity of σ or the knowledge of the secret key of V. The resulting proof, denoted by π, can also be viewed as a designated verifier signature (DVS) [30], where V is the designated verifier. After receiving σ and π from S, V checks the validity of π, and accepts σ as S's valid partial signature only if π is valid.

The non-interactive proof π obtained via Fiat-Shamir transformation consists of only four elements of $\mathbb{Z}_p$, i.e. $\pi = (c_1, c_2, z_1, z_2)$. The verification is done by checking if

$$c_1 + c_2 \stackrel{?}{=} H(\overline{M}, \hat{e}(\delta, u_i^{z_1})W_1^{-c_1}, g^{z_1}X_i^{-c_1}, g^{z_2}X_j^{-c_2})$$
$$= H(\overline{M}, \hat{e}(\delta, u_i^{z_1}\theta^{c_1}) \cdot \hat{e}(\mathsf{H}_{\kappa_i}(M), u_i^{-c_1}), g^{z_1}X_i^{-c_1}, g^{z_2}X_j^{-c_2}),$$

where $\overline{M}$ is the concatenation of message M and the public keys of P and V, and $H : \{0,1\}^* \to \mathbb{Z}_p$ is a collision-resistant hash function.

Theorem 5. *The non-interactive AOFE above is secure (Def. 12) in the random oracle model provided that HSDH and DHSDH assumptions hold in* $\mathbb{G}$.

Reduce Public Key Size. Since the key for only known instantiation of programmable hash functions [26] is long, which consists of 161 elements of $\mathbb{G}$, the signer's public key in both AOFE schemes is also long. Observe that the non-interactive scheme already resorts to a random oracle, we can replace the programmable hash function with another random oracle. That is, we change the computation of δ to $\delta = \mathsf{H}'(M)^{1/(x+s)}$, where $\mathsf{H}' : \{0,1\}^* \to \mathbb{G}$ is another collision-resistant hash function and will be modeled as a random oracle in the security proofs. The size of the signer's public key in our scheme is significantly reduced from 163 $\mathbb{G}$ elements to two only, i.e. pk $= (X, u)$.

Efficiency and Comparison. In Table 2 we compare our non-interactive AOFE with Garay et al.'s scheme [21] and Huang et al.'s [28] in terms of signature sizes, the public-key model (registered-key model or chosen-key model) and whether the security relies on the random oracle heuristics. In the comparison we instantiate the one-time signature scheme used in [28] with Boneh-Boyen short signature [4]. We can learn from the table that our scheme has full signature much shorter than [21,28]. This is because schemes in [21,28] need to include the corresponding partial signature into the full signature, while ours does not.

The partial signature generation in our scheme costs seven exponentiations in $\mathbb{G}$, while the verification of a partial signature costs seven exponentiations in $\mathbb{G}$ and two pairing evaluations. Both the signature generation and verification involve the evaluation of a 'multi-generator' hash, which can be reduced to one simple hash evaluation as explained above.

Table 2. Comparison of our non-interactive AOFE with [21,28]

Schemes	PSig	Sig	Key	ROM
Ours	$3\mathbb{G} + 4\mathbb{Z}_p$	$3\mathbb{G}$	registered	$\checkmark$
[21]	$2\mathbb{G} + 8\mathbb{Z}_p$	$2\mathbb{G} + 12\mathbb{Z}_p$	registered	$\checkmark$
[28]	$45\mathbb{G} + 1\mathbb{Z}_p$	$46\mathbb{G} + 1\mathbb{Z}_p$	chosen	$\times$

6.4   On Chen's Transformation

In [14], Chen showed how to construct a (traditional) OFE from DCS. The transformation is different from ours in that after sending a DCS signature σ_i to U_j, the signer U_i executes the DCS confirmation protocol with U_j, proving only the validity of σ_i. U_i does not show its knowledge of U_j's secret key in the proof. The invisibility of the DCS tells that σ_i is indistinguishable from any random signature σ' chosen from the space of U_i's signatures. Besides, the confirmation protocol is zero-knowledge, thus U_j is also able to simulate the proof transcripts, i.e. by running the zero-knowledge simulator. Therefore, at the first sight, one may think that Chen's transformation also results in an AOFE. However, this is actually not the case as the following issue needs to be handled.

Depending on how U_j is convinced of the validity of U_i's signature, there are two kinds of signer ambiguity. One is *offline* signer ambiguity, which requires that given a message M, a signature σ_i and a proof transcript **T**, no distinguisher can tell whether σ_i was generated by U_i or U_j. The other one is *online* signer ambiguity, which in contrast, requires that even when interacting with the prover, the distinguisher still could not tell with probability greater than one-half whether σ_i was generated by U_i or U_j. The former is static and weak for practical use, because it models passive attacks only. In this work, we consider the online signer ambiguity. Our definition given in Def. 8 models this stronger type of signer ambiguity.

Although the randomly chosen signature σ' resembles a valid signature on M generated by U_i, with overwhelmingly high probability σ' would not be valid. By the soundness of the confirmation protocol, even the real signer itself cannot prove online (or, interactively) that σ' is valid. Therefore, the resulting scheme of Chen's transformation cannot satisfy the online signer ambiguity, and thus cannot be shown as a secure AOFE. On the other hand, if we compress the interactive confirmation protocol using the Fiat-Shamir heuristic, the resulting non-interactive proof reveals the identity of the prover, because no one else but the one who holds the witness, i.e. the signer, is able to generate a valid non-interactive proof. Therefore, Chen's transformation cannot be used for building a *non-interactive* AOFE scheme in this way either.

Acknowledgements

We are grateful to the anonymous reviewers of Pairing 2010 for their invaluable comments.

References

1. Asokan, N., Shoup, V., Waidner, M.: Optimistic fair exchange of digital signatures (extended abstract). In: Nyberg, K. (ed.) EUROCRYPT 1998. LNCS, vol. 1403, pp. 591–606. Springer, Heidelberg (1998)
2. Asokan, N., Shoup, V., Waidner, M.: Optimistic fair exchange of digital signatures. IEEE Journal on Selected Areas in Communication 18(4), 593–610 (2000)
3. Barak, B., Canetti, R., Nielsen, J.B., Pass, R.: Universally composable protocols with relaxed set-up assumptions. In: FOCS 2004, pp. 186–195. IEEE Computer Society, Los Alamitos (2004)
4. Boneh, D., Boyen, X.: Short signatures without random oracles. In: Cachin, C., Camenisch, J.L. (eds.) EUROCRYPT 2004. LNCS, vol. 3027, pp. 56–73. Springer, Heidelberg (2004)
5. Boyar, J., Chaum, D., Damgård, I., Pederson, T.P.: Convertible undeniable signatures. In: Menezes, A., Vanstone, S.A. (eds.) CRYPTO 1990. LNCS, vol. 537, pp. 189–205. Springer, Heidelberg (1990)
6. Boyd, C., Foo, E.: Off-line fair payment protocols using convertible signatures. In: Ohta, K., Pei, D. (eds.) ASIACRYPT 1998. LNCS, vol. 1514, pp. 271–285. Springer, Heidelberg (1998)
7. Boyen, X., Waters, B.: Full-domain subgroup hiding and constant-size group signatures. In: Okamoto, T., Wang, X. (eds.) PKC 2007. LNCS, vol. 4450, pp. 1–15. Springer, Heidelberg (2007)
8. Camenisch, J., Michels, M.: Confirmer signature schemes secure against adaptive adversaries (extended abstract). In: Preneel, B. (ed.) EUROCRYPT 2000. LNCS, vol. 1807, pp. 243–258. Springer, Heidelberg (2000)
9. Camenisch, J., Shoup, V.: Practical verifiable encryption and decryption of discrete logarithms. In: Boneh, D. (ed.) CRYPTO 2003. LNCS, vol. 2729, pp. 126–144. Springer, Heidelberg (2003)
10. Canetti, R., Gennaro, R., Jarecki, S., Krawczyk, H., Rabin, T.: Adaptive security for threshold cryptosystems. In: Wiener, M. (ed.) CRYPTO 1999. LNCS, vol. 1666, pp. 98–115. Springer, Heidelberg (1999)
11. Chaum, D.: Zero-knowledge undeniable signatures. In: Damgård, I.B. (ed.) EUROCRYPT 1990. LNCS, vol. 473, pp. 458–464. Springer, Heidelberg (1990)
12. Chaum, D.: Designated confirmer signatures. In: De Santis, A. (ed.) EUROCRYPT 1994. LNCS, vol. 950, pp. 86–91. Springer, Heidelberg (1995)
13. Chaum, D., van Antwerpen, H.: Undeniable signatures. In: Brassard, G. (ed.) CRYPTO 1989. LNCS, vol. 435, pp. 212–216. Springer, Heidelberg (1989)
14. Chen, L.: Efficient fair exchange with verifiable confirmation of signatures. In: Ohta, K., Pei, D. (eds.) ASIACRYPT 1998. LNCS, vol. 1514, pp. 286–299. Springer, Heidelberg (1998)
15. Cramer, R., Damgård, I., MacKenzie, P.: Efficient zero-knowledge proofs of knowledge without intractability assumptions. In: Imai, H., Zheng, Y. (eds.) PKC 2000. LNCS, vol. 1751, pp. 354–373. Springer, Heidelberg (2000)
16. Cramer, R., Shoup, V.: A practical public key cryptosystem provably secure against adaptive chosen ciphertext attack. In: Krawczyk, H. (ed.) CRYPTO 1998. LNCS, vol. 1462, pp. 13–25. Springer, Heidelberg (1998)
17. Damgård, I.: On Σ-protocols. Course on Cryptologic Protocol Theory, Aarhus University (2009), http://www.daimi.au.dk/~ivan/Sigma.pdf

18. Dodis, Y., Lee, P.J., Yum, D.H.: Optimistic fair exchange in a multi-user setting. In: Okamoto, T., Wang, X. (eds.) PKC 2007. LNCS, vol. 4450, pp. 118–133. Springer, Heidelberg (2007); also at Cryptology ePrint Archive, Report 2007/182
19. Dodis, Y., Reyzin, L.: Breaking and repairing optimistic fair exchange from PODC 2003. In: DRM 2003, pp. 47–54. ACM, New York (2003)
20. Galbraith, S.D., Mao, W.: Invisibility and anonymity of undeniable and confirmer signatures. In: Joye, M. (ed.) CT-RSA 2003. LNCS, vol. 2612, pp. 80–97. Springer, Heidelberg (2003)
21. Garay, J.A., Jakobsson, M., MacKenzie, P.: Abuse-free optimistic contract signing. In: Wiener, M. (ed.) CRYPTO 1999. LNCS, vol. 1666, pp. 449–466. Springer, Heidelberg (1999)
22. Gennaro, R., Halevi, S., Rabin, T.: Secure hash-and-sign signatures without the random oracle. In: Stern, J. (ed.) EUROCRYPT 1999. LNCS, vol. 1592, pp. 123–139. Springer, Heidelberg (1999)
23. Gentry, C., Molnar, D., Ramzan, Z.: Efficient designated confirmer signatures without random oracles or general zero-knowledge proofs (extended abstract). In: Roy, B. (ed.) ASIACRYPT 2005. LNCS, vol. 3788, pp. 662–681. Springer, Heidelberg (2005)
24. Goldwasser, S., Waisbard, E.: Transformation of digital signature schemes into designated confirmer signature schemes. In: Naor, M. (ed.) TCC 2004. LNCS, vol. 2951, pp. 77–100. Springer, Heidelberg (2004)
25. Groth, J., Sahai, A.: Efficient non-interactive proof systems for bilinear groups. In: Smart, N. (ed.) EUROCRYPT 2008. LNCS, vol. 4965, pp. 415–432. Springer, Heidelberg (2008)
26. Hofheinz, D., Kiltz, E.: Programmable hash functions and their applications. In: Wagner, D. (ed.) CRYPTO 2008. LNCS, vol. 5157, pp. 21–38. Springer, Heidelberg (2008)
27. Huang, Q., Wong, D.S.: New constructions of convertible undeniable signature schemes without random oracles. Cryptology ePrint Archive, Report 2009/517 (2009)
28. Huang, Q., Yang, G., Wong, D.S., Susilo, W.: Ambiguous optimistic fair exchange. In: Pieprzyk, J. (ed.) ASIACRYPT 2008. LNCS, vol. 5350, pp. 74–89. Springer, Heidelberg (2008)
29. Huang, Q., Yang, G., Wong, D.S., Susilo, W.: Efficient optimistic fair exchange secure in the multi-user setting and chosen-key model without random oracles. In: Malkin, T.G. (ed.) CT-RSA 2008. LNCS, vol. 4964, pp. 106–120. Springer, Heidelberg (2008)
30. Jakobsson, M., Sako, K., Impagliazzo, R.: Designated verifier proofs and their applications. In: Maurer, U.M. (ed.) EUROCRYPT 1996. LNCS, vol. 1070, pp. 143–154. Springer, Heidelberg (1996)
31. Lysyanskaya, A., Micali, S., Reyzin, L., Shacham, H.: Sequential aggregate signatures from trapdoor permutations. In: Cachin, C., Camenisch, J. (eds.) EUROCRYPT 2004. LNCS, vol. 3027, pp. 74–90. Springer, Heidelberg (2004)
32. Michels, M., Stadler, M.: Generic constructions for secure and efficient confirmer signature schemes. In: Nyberg, K. (ed.) EUROCRYPT 1998. LNCS, vol. 1403, pp. 406–421. Springer, Heidelberg (1998)
33. Okamoto, T.: Designated confirmer signatures and public key encryption are equivalent. In: Desmedt, Y.G. (ed.) CRYPTO 1994. LNCS, vol. 839, pp. 61–74. Springer, Heidelberg (1994)

34. Park, J.M., Chong, E.K., Siegel, H.J.: Constructing fair-exchange protocols for e-commerce via distributed computation of RSA signatures. In: PODC 2003, pp. 172–181. ACM, New York (2003)
35. Schnorr, C.: Efficient signature generation by smart cards. J. Cryptology 4(3), 161–174 (1991)
36. Wang, G., Baek, J., Wong, D.S., Bao, F.: On the generic and efficient constructions of secure designated confirmer signatures. In: Okamoto, T., Wang, X. (eds.) PKC 2007. LNCS, vol. 4450, pp. 43–60. Springer, Heidelberg (2007)
37. Wang, G., Xia, F.: A pairing based designated confirmer signature scheme with unified verification. Technical report, School of Computer Science, University of Birmingham (December 2009) ISSN: 0962-3671
38. Wikström, D.: Designated confirmer signatures revisited. In: Vadhan, S.P. (ed.) TCC 2007. LNCS, vol. 4392, pp. 342–361. Springer, Heidelberg (2007)
39. Zhang, F., Chen, X., Wei, B.: Efficient designated confirmer signature from bilinear pairings. In: ASIACCS 2008, pp. 363–368. ACM, New York (2008)

Anonymizable Signature and Its Construction from Pairings

Fumitaka Hoshino, Tetsutaro Kobayashi, and Koutarou Suzuki

NTT Information Sharing Platform Laboratories, NTT Corporation,
3-9-11 Midori-cho, Musashino-shi, Tokyo, 180-8585 Japan
{hoshino.fumitaka,kobayashi.tetsutaro,suzuki.koutarou}@lab.ntt.co.jp

Abstract. We present the notion of *anonymizable signature*, which is an extension of the ring signature [RST01, BKM06]. By using an anonymizable signature, anyone who has a signed message can convert the signature into an anonymous signature. In other words, one can leave a signed message with an appropriate agent who will later anonymize the signature.

A relinkable ring signature [SHK09] is also an extension of the ring signature by which the ring forming ability can be separated from the signing ability. In the relinkable ring signature, an agent who has a special key given by the signer can modify the membership of existing ring signatures. However, the relinkable ring signature has two problematic limitations; a signer cannot select an agent according to the worth of the signature, because there exists the unique key to modify the membership for each public key, and we cannot achieve perfect anonymity even if the agent is honest.

The proposed anonymizable signature can free one from these limitations. In the anonymizable signature scheme, each signature can be anonymized without any secret but the signature itself. Thus, the signer can delegate signature anonymization to multiple agents signature by signature. Moreover, the anonymizable signature can guarantee unconditional anonymity and be used for anonymity-sensitive purposes, e.g., voting. After providing the definition of the anonymizable signature, we also give a simple construction methodology and a concrete scheme that satisfies perfect anonymity and computational unforgeability under the gap Diffie-Hellman assumption with the random oracle model.

1 Introduction

We present the notion of *anonymizable signature*, which is an extension of the ring signature [RST01, BKM06]. By using an anonymizable signature, anyone who has a signed message can convert the signature into an anonymous signature, i.e., one can leave a signed message with an appropriate agent who will later anonymize the signature.

For example, in the case of publication of a governmental document through a "freedom of information act", the governmental staff who publicize the document need to hide the personal information of the individuals in the document. To

M. Joye, A. Miyaji, and A. Otsuka (Eds.): Pairing 2010, LNCS 6487, pp. 62–77, 2010.

hide this information, they can use a sanitizing signature [SBZ01]. However, if the document is a contract between the government and an individual, the information on the signer cannot be hidden because readers must be convinced that the contract is valid.

To hide the information on the signer, it may seem to be effective for the signer to leave an additional ring signature with the document at the signing phase, which will replace his signature at the publishing phase. However, during the long term preservation of the document, there is a large probability that the ring signature will be made invalid by leakage of a signing key of a member in the ring, even if there is little probability that a member would leak his or her signing key. In such a case, the anonymizable signature could be effective. A signer can leave a signed message with the governmental staff who will convert it into an appropriate ring signature with a valid ring at the publishing phase.

If the agent is not trustworthy, the anonymizable signature is not suitable for an application that needs highly strong anonymity, such as whistleblowing. However, if the agent can be regarded as an ideal functionality, such a signature has many applications, e.g., server-aided computation of ring signatures, sanitizing of the signer in a signed documents, dynamic group management of ring signatures, as a countermeasure against the invalidation attack by the secret key exposure, as a gradually convertible ring signature [SHK09].

1.1 Related Works

An anonymizable signature is an extension of the ring signature [RST01, BKM06]. A ring signature is a kind of anonymous signature by which one can sign anonymously without a group setup or group manager. For each time of signing, a signer of a ring signature chooses a set of appropriate members, called a ring, then signs a message on behalf of the ring by using his or her own secret key and all of the public keys of the members in the ring. A signer can form any ring that includes him or herself as long as he or she has the public keys.

A relinkable ring signature [SHK09] is also an extension of the ring signature and has a similar functionality to the anonymizable signature. By using a relinkable ring signature, one can separate the ring-forming ability from the signing ability. Besides normal ring signature algorithms, the relinkable ring signature has a special algorithm called *relink*. An agent who has a special key given by the signer can derive a ring signature with a modified ring from an existing signed message by use of the relink algorithm. The agent who has the key to modify a ring signature can just modify the ring membership of an existing signed message; he or she cannot sign a new message. Thus, the signer can transfer the ring-forming ability to an agent without leaking the signing key. An anonymizable signature can be regarded as an extension of a relinkable ring signature, s.t., for each signing phase, the signer generates a new key, namely the signature.

An anonymizable signature also can be regarded as an extension of the convertible ring signature [LWH05], by which the signer can generate a key to revoke the anonymity of the ring signature. An agent who has an anonymizable signature can gradually decrease the anonymity of signed message by decreasing the number of ring members of the signature. In an extreme case, the agent can generate a non-anonymous signature by making the ring members include only the signer. This is similar to a convertible ring signature [LWH05], although the agent never proof that the two ring signatures are generated from the same anonymizable signature.

Moreover an anonymizable signature can be regarded as an extension of the ID-based signature (IBS) [Sha84, BNN04]. The syntax of an anonymizable signature is similar to that of an IBS, although the roles of the participants are somewhat different.

1.2   Motivation

In the definition of a relinkable ring signature in [SHK09], a special secret called *relink key* is given by the signer at the key generation phase. One can modify the ring membership of an existing ring signature by use of the signer's relink key. Every public key has only one corresponding relink key. Once an agent has a relink key, it remains valid until the public key is revoked. This correspondence gives rise to many problematic limitations.

For example, a signer cannot select an agent according to the worth of the signature because any agent who has the relink key can always relink any signatures that the signer generates. Moreover, the signer cannot avoid the risk that an agent may relink unexpected ring signatures without his or her permission.

Furthermore, we cannot achieve a perfectly anonymous relinkable ring signature without strong limitations even if the agent is honest. To prohibit an agent from excluding the actual signer from the ring, the signer and the agent must share the information on the actual signer in the ring. If they share the information for each ring signature, they need large secrets whose size is proportional to the number of ring signatures. However, according to the relinkable ring signature, the only difference between the agent and the verifier is whether he or she has the relink key. It is impossible to hide the information perfectly beyond the size of the shared secret. Thus, according to their definition, it is impossible to achieve perfect anonymity without strong limitation of the number of signatures or giving some additional information to the agent.

Therefore, to construct an efficient relinkable ring signature, we must give up the idea of achieving perfect anonymity. Indeed, the concrete scheme proposed in [SHK09] just achieves a weaker notion of computational anonymity. Even though the agent is honest, an attacker with unexpected computational resources may derive the actual signer directly from a ring signature.

All of these problems are caused by the flaw in the definition of a relinkable ring signature.

1.3   Contributions

In this paper, we present an improved notion, which we call anonymizable signature. In the anonymizable signature scheme, the signing algorithm creates a signature on a message. The signature can be converted to an anonymous ring signature, while the signed message cannot be changed. By using an anonymizable signature, the signer passes a message and a signature on the message to a proxy agent. The proxy agent can convert the signature into a ring signature afterward. We provide the definition of anonymizable signature, a simple construction methodology based on the non-interactive proof of knowledge of a signature, and an anonymizable signature scheme that can be proven to be unconditionally anonymous and computationally unforgeable under the GDH assumption in the random oracle model.

2 Definition

2.1 Notations

When X is a probabilistic Turing machine, $X(Y)$ denotes that X takes Y as an input. If X has an output value, $X(Y)$ also denotes the output value of X when X takes Y as an input. To simplify the description, we will omit some inputs that are not relevant, e.g., random tape, security parameter, and common reference string.

When Y is a fixed value, $X \leftarrow Y$ denotes that a value Y is assigned to a variable X. When Y is a set, $X \xleftarrow{\$} Y$ denotes that X is uniformly selected from Y. When Y is a probabilistic Turing machine, $X \xleftarrow{\$} Y()$ denotes that X is randomly selected from the output space of Y according to the distribution of Y's output when Y's random tape is uniformly selected. $X() \xrightarrow{\$} Y$ denotes that X is a probabilistic Turing machine that outputs Y. For any binary operator $\circ$, $X \xleftarrow{\circ} Y$ denotes that a new value $X \circ Y$ is assigned to the variable X. For instance, when Y is a set, $X \xleftarrow{\cup} Y$ denotes that a value $X \cup Y$ is assigned to the variable X. $\star$ denotes an appropriate string that is not relevant. $X \overset{?}{=} Y$ denotes a Boolean value equivalent to 1 if $X = Y$ and equivalent to 0 otherwise.

2.2 Syntax

Let $k \in \mathbb{N}$ be the security parameter. Let $N = \{0, 1, \ldots\}$ be the set of signers; we denote a subset of signers by $L \subset N$. We denote by x_i the secret key and by y_i the public key of member $i \in N$. We use notation $a_L = (a_i)_{i \in L}$.

The anonymizable signature scheme Σ consists of the following four algorithms: $\Sigma = (\text{KeyGen}, \text{Sign}, \text{Anonymize}, \text{Verify})$.

Key Generation $\text{KeyGen}(1^k) \xrightarrow{\$} (x, y)$: The key generation algorithm is a probabilistic poly-time algorithm that takes a security parameter k as input and outputs a secret key x and a public key y.

Signing $\text{Sign}(x, m) \xrightarrow{\$} r$: The signing algorithm is a probabilistic poly-time algorithm that takes a secret key x and a message m as input and outputs a signature r.

Anonymization $\text{Anonymize}(r, L, y_L, m) \xrightarrow{\$} \sigma/\bot$: The anonymization algorithm is a probabilistic poly-time algorithm that takes a signature r, ring $L \subset N$, list y_L of public keys, and message m as input and outputs a ring signature σ or $\bot$ as rejection.

Verification $\mathtt{Verify}(L, y_L, m, \sigma) \xrightarrow{\$} 0/1$: The verification algorithm is a probabilistic poly-time algorithm that takes a ring $L \subset N$, list y_L of public keys, message m, and ring signature σ as input and outputs a single bit $b \in \{0, 1\}$.

Note that we can always verify i's signature r of a message m as

$$\mathtt{Verify}(\{i\}, y_i, m, \mathtt{Anonymize}(r, \{i\}, (y_i), m)) \overset{?}{=} 1.$$

To avoid violating the signer's anonymity, we must treat the signature r as a secret between the signer and the agent, in contrast to the ring signature σ. A signature r can be regarded as a secret seed of a ring signature σ.

The syntax of an anonymizable signature is similar to that of an ID-based signature (IBS) [Sha84, BNN04], although the roles of the participants are somewhat different. We can regard the signing algorithm in an anonymizable signature as the key extraction in an IBS. The key extraction in an IBS generates a signing key for each ID, while the signing algorithm in an anonymizable signature generates a signing key, i.e., an anonymizable signature, for each message. In other words, a ring signature derived from an anonymizable signature is a special case of an ID-based signature such that the ID is identical to the message if it's ring consists of only the signer. Later in this paper, we will present a concrete scheme based on the BLS signature [BLS01], which can be seen as a scheme based on the ID-based Schnorr signature [SK03].

2.3   Security

The security of an anonymizable signature scheme is defined as follows. First, we prepare definitions of some oracles. Let $L^0 = \{1, ..., n\}$ be the set of indices of initially registered public keys, $L^{sk} \subset L^0$ be the set of indices for which secret key exposure oracle $\mathcal{O}_{sk}$ is called, and $L^{kr} = \{n + 1, n + 2, ...\}$ be the set of indices of public keys registered by adversary via key registration oracle $\mathcal{O}_{kr}$.

Signing Oracle $\mathcal{O}_s(i, m) \xrightarrow{\$} \mu$: The signing oracle takes a signer $i \in L^0$ and a message m as input and outputs a document index $\mu \in \mathbb{N}$ as follows:

1. if $i \in L^0$ and m is in valid domain, then set $r \xleftarrow{\$} \mathtt{Sign}(x_i, m)$, otherwise set $r \leftarrow \bot$,
2. increment document counter $\hat{\mu} \in \mathbb{N}$ that is a state information, and set $\mu \leftarrow \hat{\mu}$
3. update list of signing oracle queries and answers as $Q_s \xleftarrow{\cup} \{(\mu, i, m, r)\}$,
4. return μ.

Anonymization Oracle $\mathcal{O}_a(\mu, L) \xrightarrow{\$} \sigma$: The anonymization oracle takes a document index $\mu \in \mathbb{N}$ and a list of ring members $L \subset L^0 \cup L^{kr}$ as input and outputs a ring signature σ as follows:

1. if μ is registered in Q_s and $L \subset L^0 \cup L^{kr}$, then find (i, m, r) s.t. $(\mu, i, m, r) \in Q_s$ and set $\sigma \xleftarrow{\$} \mathtt{Anonymize}(r, L, y_L, m)$, otherwise set $\sigma \leftarrow \bot$,

2. update list of anonymization oracle queries and answers as $Q_a \overset{\cup}{\leftarrow} \{(\mu, i, m, r, L, \sigma)\}$,
3. return σ.

Signature Exposure Oracle $\mathcal{O}_e(\mu) \overset{\$}{\rightarrow} r$: The signature exposure oracle takes a document index $\mu \in \mathbb{N}$ as input and outputs a signature r as follows:

1. if μ is registered in Q_s, then find (i, m, r) s.t. $(\mu, i, m, r) \in Q_s$, otherwise set $r \leftarrow \perp$,
2. update list of anonymization oracle queries and answers as $Q_e \overset{\cup}{\leftarrow} \{(\mu, i, m, r)\}$,
3. return r.

Secret Key Exposure Oracle $\mathcal{O}_{sk}(i) \overset{\$}{\rightarrow} x_i$: The secret key exposure oracle takes a user index $i \in \mathbb{N}$ as input and outputs secret key x_i of i-th user.

1. if $i \in L^0$, then set $sk \leftarrow x_i$, otherwise set $sk \leftarrow \perp$ and return sk,
2. update the set of indices for which secret key exposure oracle is called, $L^{sk} \overset{\cup}{\leftarrow} \{i\}$,
3. return sk.

Key Registration Oracle $\mathcal{O}_{kr}(y) \overset{\$}{\rightarrow} i$: The key registration oracle takes a public key y as input, outputs a new user index i, and register y as the public key of the i-th user.

1. if y is in valid domain, then increment counter $\hat{i} \in \mathbb{N}$ that is a state information, set $i \leftarrow \hat{i}$, and register y as the public key of the i-th user, otherwise set $i \leftarrow \perp$ and return i,
2. update the set of indices of public keys registered by adversary via key registration oracle, $L^{kr} \overset{\cup}{\leftarrow} \{i\}$,
3. return i.

We say that anonymizable signature Σ is secure if it satisfies the following three properties.

Completeness
Correctly generated signatures are accepted with overwhelming probability. We say that anonymizable signature Σ satisfies *completeness*, if

$$\Pr[(x_j, y_j) \overset{\$}{\leftarrow} \texttt{KeyGen}(1^k) \ (j \in L), r \overset{\$}{\leftarrow} \texttt{Sign}(x_i, m),$$
$$\sigma \overset{\$}{\leftarrow} \texttt{Anonymize}(r, L, y_L, m), \texttt{Verify}(L, y_L, m, \sigma) = 0]$$

is negligible in k for any message $m \in \{0, 1\}^*$, any ring $L \subset N$, and any signer $i \in L$.

Anonymity
We consider the following experiment $Exp_{k,\Sigma}^{\text{anon}}(\mathcal{A})$, where adversary $\mathcal{A}$ try to distinguish the signer of a ring signature:

1. select random bit $b \overset{\$}{\leftarrow} \{0,1\}$,
2. at the beginning of the experiment, adversary $\mathcal{A}$ generates two pairs of secret and public keys $(x_0, y_0), (x_1, y_1)$, register two public keys y_0, y_1 by key registration oracle $\mathcal{O}_{kr}$, and outputs two secret keys x_0, x_1, and the experiment is aborted if the secret keys are not correct w.r.t. the public keys,
3. adversary $\mathcal{A}$ can access key registration oracle $\mathcal{O}_{kr}(y)$ adaptively during the experiment, and we denote the set of indices of all public keys registered by adversary $\mathcal{A}$ during the experiment by $L^{kr} = \{0, 1, ...\}$,
4. adversary $\mathcal{A}$ outputs (m^*, L^*) s.t. $0, 1 \in L^* \subset L^{kr}$,
5. generate ring signature $\sigma^* \overset{\$}{\leftarrow}$ Anonymize(Sign(x_b, m^*), L^*, y_{L^*}, m^*) using the secret key of signer b, and adversary $\mathcal{A}$ is given σ^*,
6. finally, adversary $\mathcal{A}$ outputs a bit $b' \overset{\$}{\leftarrow} \mathcal{A}^{\mathcal{O}_{kr}}(x_1, ..., x_n)$,
7. return 1 if $b = b'$, 0 otherwise.

We define the advantage Adv^{anon} of the adversary $\mathcal{A}$ as

$$Adv_{k,\Sigma}^{anon}(\mathcal{A}) = \left| \Pr\left[Exp_{k,\Sigma}^{anon}(\mathcal{A}) = 1 \right] - \frac{1}{2} \right|,$$

where probability is taken over a random bit b, keys, random tapes of the oracles, random tapes of the adversary $\mathcal{A}$, and random tapes of KeyGen, Sign, Anonymize.

Definition 1 (computational anonymity w.r.t. adversarially-chosen keys). *An anonymizable signature Σ satisfies computational anonymity w.r.t. adversarially-chosen keys if $Adv_{k,\Sigma}^{anon}(\mathcal{A}_k)$ is negligible in k for any probabilistic poly-time adversary $\mathcal{A}_k$.*

In particular, we say that anonymizable signature Σ satisfies *perfect anonymity* if we have
$$\Pr[\text{Anonymize}(\text{Sign}(x_i, m), L, y_L, m) = \sigma] =$$
$$\Pr[\text{Anonymize}(\text{Sign}(x_j, m), L, y_L, m) = \sigma]$$

for any security parameter k, any message m, any ring L, any valid ring signature σ with L, any $i, j \in L$, and any keys x_i, x_j.

Unforgeability
We consider the following experiment $Exp_{k,\Sigma}^{unforge}(\mathcal{A})$, where adversary $\mathcal{A}$ try to forge a valid ring signature without knowing the corresponding signing key or signature:

1. select random bit $b \overset{\$}{\leftarrow} \{0,1\}$, and generate secret and public keys $(x_i, y_i) \overset{\$}{\leftarrow}$ KeyGen(1^k) for $i \in L^0 = \{1, ..., n\}$,
2. at the beginning of the experiment, adversary $\mathcal{A}$ is given all public keys $y_1, ..., y_n$,
3. adversary $\mathcal{A}$ can access signing oracle $\mathcal{O}_s(i, m)$, anonymization oracle $\mathcal{O}_a(\mu, L)$ for $L \subset L^0 \cup L^{kr}$, and signature exposure oracle $\mathcal{O}_e(\mu)$ adaptively during the experiment,

4. adversary $\mathcal{A}$ can access secret key exposure oracle $\mathcal{O}_{sk}(i)$ adaptively during the experiment, and we denote the set of indices of all secret keys exposed by adversary $\mathcal{A}$ during the experiment by $L^{sk} \subset L^0$,

5. adversary $\mathcal{A}$ can access key registration oracle $\mathcal{O}_{kr}(y)$ adaptively during the experiment, and we denote the set of indices of all public keys registered by adversary $\mathcal{A}$ during the experiment by $L^{kr} = \{n+1, n+2, ...\}$,

6. finally, adversary $\mathcal{A}$ outputs $(m^*, L^*, \sigma^*) \stackrel{\$}{\leftarrow} \mathcal{A}^{\mathcal{O}_s, \mathcal{O}_a, \mathcal{O}_e, \mathcal{O}_{sk}, \mathcal{O}_{kr}}(y_1, ..., y_n)$,

7. return 1 if adversary $\mathcal{A}$ wins, 0 otherwise.

Here, we say that adversary $\mathcal{A}$ wins, if adversary $\mathcal{A}$ outputs forged signature (m^*, L^*, σ^*) and the following conditions hold:

1. $\texttt{Verify}(L^*, y_{L^*}, m^*, \sigma^*) = 1$,
2. $L^* \subset L^0 - L^{sk}$,
3. $\forall i \in L^*, (\star, i, m^*, \star) \notin Q_e$, i.e., adversary $\mathcal{A}$ never ask queries $\mathcal{O}_s(i, m^*) = \mu$ and $\mathcal{O}_e(\mu)$ for all $i \in L^*$, and
4. $(\star, \star, m^*, \star, L^*, \sigma^*) \notin Q_a$, i.e., adversary $\mathcal{A}$ never ask queries $\mathcal{O}_s(i, m^*) = \mu$ and $\mathcal{O}_a(\mu, L^*) = \sigma^*$.

We define the advantage Adv^{unforge} of the adversary $\mathcal{A}$ as

$$Adv_{k,\Sigma}^{\text{unforge}}(\mathcal{A}) = \Pr\left[Exp_{k,\Sigma}^{\text{unforge}}(\mathcal{A}) = 1\right],$$

where probability is taken over a random bit b, keys, random tapes of the oracles, random tapes of the adversary $\mathcal{A}$, and random tapes of $\texttt{KeyGen}$, $\texttt{Verify}$.

Definition 2 (unforgeability). *Anonymizable signature Σ satisfies unforgeability if $Adv_{k,\Sigma}^{\text{unforge}}(\mathcal{A}_k)$ is negligible in k for any probabilistic poly-time adversary $\mathcal{A}_k$.*

3   Proposed Scheme

In this section, we provide a very simple construction methodology of the anonymizable signature from any signature scheme, then according to our methodology we give a concrete scheme based on the BLS signature [BLS01].

3.1   Construction Methodology

We can construct an anonymizable signature scheme Σ from any signature scheme as follows.

(1) At the signing phase, the signer i generates a signature r of a message m by using the normal signature scheme that the anonymizable signature is based on. The signer passes the signature r to the proxy agent. The signature r must be treated as a secret between the signer and the agent.

(2) At the anonymizing phase, the agent makes a *"non-interactive proof of knowledge"* [FFS87, GMR85] σ_i which proves that he or she knows a valid signature r of the signer i on the message m.

(3) To anonymize the proof σ_i, the agent chooses an appropriate ring and simulates the proof of knowledge with respect to other members in the ring. By use of a *"proof of partial knowledge (or-proof)"* [CDS94], a ring signature σ can be composed of the proof of knowledge and it's simulations.

3.2   Concrete Scheme

We constructed an anonymizable signature scheme based on the BLS signature [BLS01] by which we can easily construct an efficient non-interactive proof of knowledge.

Let G and G_T be cyclic groups of prime order p, G^* be the set of the generators of G, $e : G \times G \to G_T$ be a non-degenerate bilinear map, $g \in G^*$ be a generator of G, and $H : \{0,1\}^* \to G^*$ and $H' : \{0,1\}^* \to \mathbb{Z}_p$ be mutually independent random oracles. We call the common reference string $\rho = (p, G, G_T, e, g, H, H')$ a system parameter. Let k be the security parameter. We assume that the system parameter can be determined in the polynomial time of k, ρ is encoded in the polynomial size of k, and all participants in our scheme use the same ρ.

Key Generation. Algorithm $\mathtt{KeyGen}(1^k)$ takes a security parameter k, randomly chooses $x_i \in_U \mathbb{Z}_p$, and outputs secret and public keys $(sk_i = x_i, pk_i = y_i = g^{x_i})$ for signer $i \in N$.

Signing. Algorithm $\mathtt{Sign}(x_i, m)$ takes i's secret key $sk_i = x_i$ and a message m, computes $h = H(\rho, m) \in G^*$, and outputs a signature $r = h^{x_i}$.

Anonymization. Algorithm $\mathtt{Anonymize}(r, L, y_L, m)$ takes i's signature $r = h^{x_i}$, a ring $L \subset N$ s.t. $i \in L$, public keys y_L, and a message m, and outputs a ring signature σ as follows.

$h \leftarrow H(\rho, m) \in G^*$;
If $\exists i$ s.t. $e(g, r) = e(y_i, h), i \in L$, then perform the following steps, otherwise return $\perp$.
Generate a proof of the knowledge r s.t. $(\exists j \in L, r = h^{x_j})$ as the followings.

$\quad t \xleftarrow{\$} \mathbb{Z}_p$;
$\quad \tilde{a}_i \leftarrow e(g, h)^t \in G_T$;
$\quad \forall j \in L \setminus \{i\},$
$\quad\quad c_j \xleftarrow{\$} \mathbb{Z}_p, \; z_j \xleftarrow{\$} G,$
$\quad\quad \tilde{a}_j \leftarrow e(g, z_j)e(h, y_j)^{c_j} \in G_T$;
$\quad c_i \leftarrow H'(\rho, L, m, y_L, \tilde{a}_L) - \sum_{j \neq i} c_j$;
$\quad z_i \leftarrow h^t r^{-c_i} \in G$;
Return $\sigma \leftarrow (c_L, z_L)$.

Verification. $\texttt{Verify}(L, y_L, m, \sigma)$ takes a ring $L \subset N$, public keys y_L, a message m, and a ring signature σ, and outputs a bit $0/1$ as follows.

$$(c_L, z_L) \leftarrow \sigma \; ;$$
$$h \leftarrow H(\rho, m) \in G^* \; ;$$
$$\forall j \in L, \; \tilde{a}_j \leftarrow e(g, z_j)e(h, y_j)^{c_j} \in G_T \; ;$$
$$\text{return } \begin{cases} 1 \,, & \text{if } H'(\rho, L, m, y_L, \tilde{a}_L) = \sum_{j \in L} c_j \\ 0 \,, & \text{otherwise.} \end{cases}$$

4 Security of the Proposed Scheme

In this section, we show that the proposed scheme satisfies computational unforgeability and perfect anonymity. We first refer to the GDH assumption [OP01] and the BLS signature [BLS01] that is used in the security proof of computational unforgeability.

4.1 Preliminary

Gap Diffie-Hellman (GDH) Assumption [OP01]

We refer to the GDH assumption where the adversary computes a CDH answer by use of a DDH oracle.

Let G_k be a cyclic group (family) with a security parameter k. We will omit the suffix k to simplify the description. For any $g_0, g_1, g_2, g_3 \in G$, we define a DDH oracle $\mathcal{O}_{\text{ddh}}(g_0, g_1, g_2, g_3)$ as

$$\mathcal{O}_{\text{ddh}}(g_0, g_1, g_2, g_3) :=$$
$$\text{return } \begin{cases} 1 \,, & \text{if } \log_{g_0} g_1 = \log_{g_2} g_3 \\ 0 \,, & \text{otherwise.} \end{cases}$$

For any algorithm $\mathcal{C}$, we define the GDH experiment Exp^{gdh} as

$$Exp_{k,G}^{\text{gdh}}(\mathcal{C}) :=$$
$$g \xleftarrow{\$} G, \; a \xleftarrow{\$} \mathbb{Z}_p, \; b \xleftarrow{\$} \mathbb{Z}_p \; ;$$
$$h \xleftarrow{\$} \mathcal{C}^{\mathcal{O}_{\text{ddh}}}(g, g^a, g^b) \; ;$$
$$\text{return } \begin{cases} 1 \,, & \text{if } h = g^{ab} \\ 0 \,, & \text{otherwise.} \end{cases}$$

We also define the advantage of $\mathcal{C}$, Adv^{gdh}, as

$$Adv_{k,G}^{\text{gdh}}(\mathcal{C}) := \Pr\left[Exp_{k,G}^{\text{gdh}}(\mathcal{C}) = 1 \right],$$

where the sample space of the probability is the random tape of Exp^{gdh}.

Assumption 1 (the GDH assumption over G). *We say that the GDH assumption holds in G if, for any p.p.t. $\mathcal{C}_k$, $Adv_{k,G}^{gdh}(\mathcal{C}_k)$ is negligible in k.*

Hereafter, we assume that the above GDH assumption holds in pairing group G with pairing $e : G \times G \to G_T$.

BLS Signature [BLS01]

We refer to the BLS signature to which we reduce the computational unforgeability of the proposed scheme.

BLS signature $\Sigma_b = (\text{KeyGen}_{\Sigma_b}, \text{Sign}_{\Sigma_b}, \text{Verify}_{\Sigma_b})$ is defined by the following algorithms.

Key Generation. Algorithm $\text{KeyGen}_{\Sigma_b}(1^k)$ takes 1^k as input, selects random $x \xleftarrow{\$} \mathbb{Z}_p$, computes $y \leftarrow g^x \in G$, and outputs secret and public keys (x, y)

Signing. Algorithm $\text{Sign}_{\Sigma_b}(x, m)$ takes secret key x and message m as input, computes $h \leftarrow H(\rho, m) \in G^*$ and $r \leftarrow h^x \in G$, where ρ is a system parameter, outputs signature r.

Verification. Algorithm $\text{Verify}_{\Sigma_b}(y, m, r)$ takes public key y, message m, and signature r as input, computes $h \leftarrow H(\rho, m) \in G$, where ρ is a system parameter, outputs 1 if $e(h, y) = e(g, r)$, 0 otherwise.

We define BLS signature oracle $\mathcal{O}_{\Sigma_b}(m)$ as

$$\mathcal{O}_{\Sigma_b}(m) :=$$
$$r \xleftarrow{\$} \text{Sign}_{\Sigma_b}(x, m) \; ;$$
$$Q_b \xleftarrow{\cup} (m, r) \; ;$$
$$\text{return } r \; ;$$

where x is a secret key of the signer. For any algorithm $\mathcal{B}$, we define a Turing machine $Exp^{\text{bls}}_{k, \Sigma_b}(\mathcal{B})$ as

$$Exp^{\text{bls}}_{k, \Sigma_b}(\mathcal{B}) :=$$
$$\text{clear } Q_b \; ;$$
$$(x, y) \xleftarrow{\$} \text{KeyGen}_{\Sigma_b}(1^k) \; ;$$
$$(m^*, r^*) \xleftarrow{\$} \mathcal{B}^{\mathcal{O}_{\Sigma_b}, H}(y) \; ;$$
$$\text{return } \begin{cases} 1 \, , & \text{if } (m^*, r^*) \notin Q_b \wedge \text{Verify}_{\Sigma_b}(y, m^*, r^*) = 1 \\ 0 \, , & \text{otherwise.} \end{cases}$$

and the advantage of $\mathcal{B}$, Adv^{bls}, as

$$Adv^{\text{bls}}_{k, \Sigma_b}(\mathcal{B}) := \Pr\left[Exp^{\text{bls}}_{k, \Sigma_b}(\mathcal{B}) = 1 \right]$$

where the sample space of the probability is the random tape of Exp^{bls}.

Lemma 1 ([BLS01]). *Let H be a random oracle. We define ϵ' and ϵ'' as*

$$\epsilon' := Adv^{bls}_{k,\Sigma_b}(\mathcal{B}), \quad \epsilon'' := Adv^{gdh}_{k,G}(\mathcal{C}^{\mathcal{B}}).$$

Let q_e be the maximum number of queries to $\mathcal{O}_{\Sigma_b}$. For any real number $\xi \in [0,1]$, there exists a p.p.t. $\mathcal{C}$ that satisfies

$$(1 - \xi)^{q_e} \xi \epsilon' \leq \epsilon''.$$

Corollary 1. *There exists a p.p.t. $\mathcal{C}$ that satisfies*

$$\epsilon' \leq 4(q_e + 1)\epsilon''.$$

4.2 Unforgeability

The proposed scheme satisfies computational unforgeability. To show the unforgeability, we reduce the unforgeability of the proposed scheme to the unforgeability of the BLS signature. In a simulation, by the rewinding technique, we can obtain a forgery of the BLS signature.

Theorem 1. *The proposed scheme satisfies computational unforgeability under the GDH assumption in the random oracle model.*

Proof. In this proof, we construct a forger $\mathcal{B}$ against the BLS signature by use of a forger $\mathcal{A}$ against our scheme. The simulator executes the following steps.

$\mathcal{B}^{\mathcal{A},\mathcal{O}_{\Sigma_b},H}(y) :=$
    clear ν, Q_s, Q_a, Q_e ;
    $j \xleftarrow{\$} L^0$;
    $\forall i \in L^0$, if $i = j$ then $u \xleftarrow{\$} \mathbb{Z}_p$, $y_i \leftarrow yg^u$;
        else $(x_i, y_i) \xleftarrow{\$} \mathsf{KeyGen}(1^k)$;
    $(L^*, m^*, \sigma^*) \xleftarrow{\$} \mathcal{A}^{\mathcal{S}_s,\mathcal{S}_a,\mathcal{S}_e,\mathcal{S}_{sk},\mathcal{S}_{kr}}(y_{L^0})$;
    if $\mathcal{A}$ doesn't win the game then abort.
    rewind $\mathcal{A}$ to the corresponding H'.
    $\mathcal{A}$ outputs (L'^*, m'^*, σ'^*).
    if $\mathcal{A}$ doesn't win the second game then abort.
    if $(L^*, m^*) \neq (L'^*, m'^*) \vee \sigma^* = \sigma'^*$ then abort.
    $(c_{L^*}, z_{L^*}) \leftarrow \sigma^*$, $(c'_{L^*}, z'_{L^*}) \leftarrow \sigma'^*$;
    find $i \in L^*$ s.t. $c_i \neq c'_i$;
    if $i \neq j$ then abort.
    $r^* \leftarrow (z'_i/z_i)^{\frac{1}{c_i - c'_i}} (H(\rho, m^*))^{-u}$;
    return (m^*, r^*).

The simulator executes the following signing oracle simulation $\mathcal{S}_s$.

$\mathcal{S}_s(i, m) :=$
    $\mu \leftarrow \nu$++ ;

if $i \in L^0$ and m is in valid domain then
$\quad r \xleftarrow{\$} \star$;
else $r \leftarrow \perp$;
$Q_s \xleftarrow{\cup} \{(\mu, i, m, r)\}$;
return μ ;

The simulator executes the following anonymization oracle simulation $\mathcal{S}_a$.

$\mathcal{S}_a(\mu, L) :=$
$\quad$ if μ is registered in Q_s and $L \subset L^0 \cup L^{kr}$ then
$\quad\quad$ find $(i, m, \star)$ s.t. $(\mu, i, m, \star) \in Q_s$;
$\quad\quad \sigma \xleftarrow{\$} Simulate(\texttt{Anonymize}(\star, L, y_L, m))$;
$\quad$ else $\sigma \leftarrow \perp$;
$\quad Q_a \xleftarrow{\cup} \{(\mu, L, \sigma)\}$;
$\quad$ return σ ;

where $Simulate(\texttt{Anonymize}(\star, L, y_L, m))$ is the following algorithm that produces a ring signature without a signature by manipulating the value of H'.

$Simulate(\texttt{Anonymize}(\star, L, y_L, m)) :=$
$\quad h \leftarrow H(\rho, m)$;
$\quad \forall j \in L, \ c_j \xleftarrow{\$} \mathbb{Z}_p, \ z_j \xleftarrow{\$} G,$
$\quad\quad \tilde{a}_j \leftarrow e(g, z_j)e(h, y_j)^{c_j} \in G_T$;
$\quad$ if $H'(\rho, L, m, y_L, \tilde{a}_L)$ is defined then abort.
$\quad$ else $H'(\rho, L, m, y_L, \tilde{a}_L) \leftarrow \sum_j c_j$;
$\quad$ return (c_L, z_L) ;

Let q be the maximum number of queries to H' and α be the success probability that all $Simulate(\texttt{Anonymize}(\star, L, y_L, m))$ does not abort through the execution of $\mathcal{B}^{\mathcal{A}}$. α is evaluated as the following, where $q \geq 1$, $p^n \gg q$, and $n = \#L^0$.

$$\alpha > \prod_{i=0}^{q-1} \left(1 - \frac{i}{p^n}\right) > \left(1 - \frac{q}{p^n}\right)^q > 1 - \frac{q^2}{p^n}$$

The simulator executes the following signature exposure oracle simulation $\mathcal{S}_e$.

$\mathcal{S}_e(\mu) :=$
$\quad$ if μ is registered in Q_s then
$\quad\quad$ find $(i, m, \star)$ s.t. $(\mu, i, m, \star) \in Q_s$;
$\quad\quad$ if $i = j$ then $r \xleftarrow{\$} (\mathcal{O}_{\Sigma_b}(m))(H(\rho, m))^u$;
$\quad\quad$ else $r \xleftarrow{\$} \texttt{Sign}(x_i, m)$;
$\quad$ else $r \leftarrow \perp$;
$\quad Q_e \xleftarrow{\cup} \{(\mu, r)\}$;
$\quad$ return r ;

The simulator executes the following secret key exposure oracle simulation $\mathcal{S}_{sk}$.

$\mathcal{S}_{sk}(i) :=$
    if $i \in L^0$ then
        if $i = j$ then abort.
        $x \leftarrow x_i$;
        $L^{sk} \overset{\cup}{\leftarrow} \{i\}$;
    else $x \leftarrow \perp$;
    return x ;

The simulator executes the key registration oracle simulation $\mathcal{S}_{kr}$ which is identical to the key registration oracle $\mathcal{O}_{kr}$ defined in section 2.3.

We define ϵ, ϵ', and ϵ'' as

$$\epsilon := Adv_{k,\Sigma}^{unforge}(\mathcal{A}), \quad \epsilon' := Adv_{k,\Sigma_b}^{bls}(\mathcal{B}^{\mathcal{A}}), \quad \epsilon'' := Adv_{k,G}^{gdh}(\mathcal{C}^{\mathcal{B}^{\mathcal{A}}}).$$

Let q_e be the maximum number of queries to $\mathcal{S}_e$, q be the maximum number of queries to H' and assume $q \geq 1$. Let $n = \#L^0$. According to the forking lemma [BN06],

$$\alpha\epsilon(\alpha\epsilon/q - 1/p)/n \leq \epsilon'.$$

Thus,

$$\begin{aligned}
\epsilon &\leq (q/2p + \sqrt{(q/2p)^2 + qn\epsilon'})/\alpha \\
&\leq (q/2p + \sqrt{(q/2p)^2 + 4qn(q_e + 1)\epsilon''})/\alpha \\
&\leq (q/2p + \sqrt{(q/2p)^2 + 4qn(q_e + 1)\epsilon''})/(1 - q^2/p^n).
\end{aligned}$$

We assume that the maximum number of queries to the random oracles, namely q and q_e, and the size of the ring n are bounded by some polynomial of security parameter k. Let $\mathcal{A}_k$ be a polynomial time algorithm of k that attacks the unforgeability of our scheme with success probability ϵ. Immediately we have a polynomial time algorithm $\mathcal{C}^{\mathcal{B}^{\mathcal{A}_k}}$ that wins the GDH game with success probability ϵ'', which satisfies the above inequality. Thus, if the GDH assumption over G holds, for any p.p.t. $\mathcal{A}_k$, $\epsilon'' = Adv_{k,G}^{gdh}(\mathcal{C}^{\mathcal{B}^{\mathcal{A}_k}})$ is negligible. Therefore, for any p.p.t. $\mathcal{A}_k$, $\epsilon = Adv_{k,\Sigma}^{unforge}(\mathcal{A}_k)$ is negligible by the above inequality. □

4.3   Perfect Anonymity

The proposed scheme satisfies perfect anonymity. To show the perfect anonymity, we prove that for any valid ring signature σ, for any member i in the ring L, there exists a unique way to produce the ring signature σ by use of i's secret key x_i.

Theorem 2. *The proposed scheme satisfies the perfect anonymity in the random oracle model.*

Proof. For any $\sigma^* = (c_L^*, z_L^*)$ s.t. $\mathtt{Verify}(L, y_L, m, \sigma^*) = 1$, for any $i \in L$, there exists a unique assignment of the random tape t, c_j, z_j in the $\mathtt{Anonymize}$ function that satisfies $\sigma^* = \mathtt{Anonymize}(\mathtt{Sign}(x_i, m), L, y_L, m)$.

$$t = x_i c_i^* + \log_{H(\rho, m)} z_i^*, \text{ and } c_j = c_j^*, z_j = z_j^*, \forall j \neq i. \qquad \square$$

5   Conclusion

We presented a novel concept of a ring signature called anonymizable signature, by which one can convert a signature into an anonymous ring signature without any secret but the signature itself. By using an anonymizable signature, a signer can leave a signed message to a proxy agent who will convert the signature into a ring signature afterward. If the agent is not trustworthy, the anonymizable signature is not suitable for an application that needs highly strong anonymity. However, if the agent can be regarded as an ideal functionality, it has many applications. We also provided the definition of anonymizable signature, a simple construction methodology, and a concrete scheme that can be proven to be unconditionally anonymous and computationally unforgeable under the GDH assumption in the random oracle model.

References

[BKM06] Bender, A., Katz, J., Morselli, R.: Ring Signatures: Stronger Definitions, and Constructions Without Random Oracles. In: Halevi, S., Rabin, T. (eds.) TCC 2006. LNCS, vol. 3876, pp. 60–79. Springer, Heidelberg (2006)

[BLS01] Boneh, D., Lynn, B., Shacham, H.: Short signatures from the Weil pairing. In: Boyd, C. (ed.) ASIACRYPT 2001. LNCS, vol. 2248, pp. 514–532. Springer, Heidelberg (2001)

[BN06] Bellare, M., Neven, G.: Multi-Signatures in the Plain Public-Key Model and a General Forking Lemma. In: Proc. of the 13th ACM Conference on Computer and Communications Security (CCS), pp. 390–399 (2006)

[BNN04] Bellare, M., Namprempre, C., Neven, G.: Security Proofs for Identity-Based Identification and Signature Schemes. In: Cachin, C., Camenisch, J.L. (eds.) EUROCRYPT 2004. LNCS, vol. 3027, pp. 268–286. Springer, Heidelberg (2004)

[CDS94] Cramer, R., Damgård, I., Schoenmakers, B.: Proofs of Partial Knowledge and Simplified Design of Witness Hiding Protocols. In: Desmedt, Y.G. (ed.) CRYPTO 1994. LNCS, vol. 839, pp. 174–187. Springer, Heidelberg (1994)

[FFS87] Feige, U., Fiat, A., Shamir, A.: Zero Knowledge Proofs of Identity. In: Proc. of STOC 1987, pp. 210–217 (1987)

[GMR85] Goldwasser, S., Micali, S., Rackoff, C.: The Knowledge Complexity of Interactive Proof-systems. In: Proc. of STOC 1985, pp. 291–304 (1985)

[LWH05] Lee, K.-C., Wen, H.-A., Hwang, T.: Convertible ring signature. IEE Proc. of Communications 152(4), 411–414 (2005)

[OP01] Okamoto, T., Pointcheval, D.: The Gap Problems: A New Class of Problems for the Security of Cryptographic Primitives. In: Kim, K.-c. (ed.) PKC 2001. LNCS, vol. 1992, pp. 104–118. Springer, Heidelberg (2001)

[RST01] Rivest, R.L., Shamir, A., Tauman, Y.: How to Leak a Secret. In: Boyd, C. (ed.) ASIACRYPT 2001. LNCS, vol. 2248, pp. 552–565. Springer, Heidelberg (2001)

[SBZ01] Steinfeld, R., Bull, L., Zheng, Y.: Content Extraction Signatures. In: Kim, K. (ed.) ICISC 2001. LNCS, vol. 2288, pp. 285–304. Springer, Heidelberg (2001)

[Sha84]   Shamir, A.: Identity-based cryptosystems and signature schemes. In: Blakely, G.R., Chaum, D. (eds.) CRYPTO 1984. LNCS, vol. 196, pp. 47–53. Springer, Heidelberg (1984)
[SHK09]   Suzuki, K., Hoshino, F., Kobayashi, T.: Relinkable Ring Signature. In: Garay, J.A., Miyaji, A., Otsuka, A. (eds.) CANS 2009. LNCS, vol. 5888, pp. 518–536. Springer, Heidelberg (2009)
[SK03]   Sakai, R., Kasahara, M.: ID based Cryptosystems with Pairing on Elliptic Curve. Cryptology ePrint Archive: 2003/054 (2003)

Identification of Multiple Invalid Pairing-Based Signatures in Constrained Batches*

Brian J. Matt

Johns Hopkins University
Applied Physics Laboratory
Laurel, MD, 21102, USA
brian.matt@jhuapl.edu

Abstract. This paper describes a new method in pairing-based signature schemes for identifying the invalid digital signatures in a batch after batch verification has failed. The method more efficiently identifies non-trivial numbers, w, of invalid signatures in constrained sized, N, batches than previously published methods, and does not require that the verifier possess detailed knowledge of w. Our method uses "divide-and-conquer" search to identify the invalid signatures within a batch, pruning the search tree to reduce the number of pairing computations required. The method prunes the search tree more rapidly than previously published techniques and thereby provides performance gains for batch sizes of interest.

We are motivated by wireless systems where the verifier seeks to conserve computations or a related resource, such as energy, by using large batches. However, the batch size is constrained by how long the verifier can delay batch verification while accumulating signatures to verify.

We compare the expected performance of our method (for a number of different signature schemes at varying security levels) for varying batch sizes and numbers of invalid signatures against earlier methods. We find that our new method provides the best performance for constrained batches, whenever the number of invalid signatures is less than half the batch size. We include recently published methods based on techniques from the group-testing literature in our analysis. Our new method consistently outperforms these group-testing based methods, and substantially reduces the cost ($> 50\%$) when $w \leq N/4$.

1 Introduction

In many network security and E-commerce systems that use batch signature verification, the verifier does not have the freedom to accumulate arbitrarily large batches of messages and signatures to maximize the efficiency of the batch verifier. Typically the batch size is constrained by how long the verifier can delay verification of early arriving messages while waiting to accumulate additional messages for the batch. In such applications, whenever a batch fails verification,

---

* The views and conclusions contained in this paper are those of the author and should not be interpreted as representing the official policies, either expressed or implied, of the Army Research Laboratory, or the U. S. Government.

M. Joye, A. Miyaji, and A. Otsuka (Eds.): Pairing 2010, LNCS 6487, pp. 78–95, 2010.

the verifier then chooses the best method available to identify the invalid signatures. The choice is determined by the size of the batch, and perhaps based on some belief about the likely number of invalid signatures, or some estimate of the bound on the number of invalid signatures.

When the system is part of a data rate limited wireless network, the signature scheme of choice is often a communication efficient bilinear pairing-based scheme.[1] This choice is justifiable if the need for communication efficiency justifies the higher processing costs of these schemes compared to 1) conventional signature schemes such as ECDSA [11], or 2) signature schemes using implicit-certificates [1,23,4]. Such pairing-based schemes include some short signature schemes [3,5] and several bandwidth efficient identity-based signature schemes [6,5,29,5].

When batch verification fails, a number of methods have been proposed, primarily for large batches, to identify the invalid signatures in batch verifiable, pairing-based signature schemes. These proposals include "divide-and-conquer" (DC) methods such as Fast DC verifier [22] and Binary Quick Search [15], as well as methods that significantly augment DC with other techniques (i.e., hybrid methods) [18], and some specialized techniques that are practical for batches with only a very few invalid signatures [15]. Recently, methods based on group testing have been proposed [30]. However, no methods have been proposed specifically for constrained batches.

Our Contribution

In this paper, we present a new method for finding invalid signatures in pairing-based schemes based on hybrid divide-and-conquer searching. The method outperforms earlier hybrid divide-and-conquer methods when N is constrained (16 - 128) and $w < N/2$. We compare our method with earlier work for a number of pairing-based schemes and present the results using cost parameters drawn from a realization of the Cha-Cheon [6] signature scheme at the 80 bit and 192 bit security levels. Our analysis can be easily applied to other schemes and at other security levels. Our new hybrid method seeks to identify more invalid signatures in each (sub-) batch than earlier hybrid methods before resorting to sub-dividing the (sub-)batches. The new method reduces the number of computations required whenever $w < N/2$.

Recently group testing algorithms [8] have been proposed for use in identifying invalid signatures in batches [30]. However, many group algorithms assume that w (or an upper bound) is known. If the estimate d of w must be precise in order to obtain good performance, then such methods will not be useful in practice. We compare the expected performance of our method against the best methods in [30]. We find that our new method, as well as some earlier methods, always significantly outperform the proposed group testing methods, even when w is precisely known. We also examine the impact of an inaccurate estimate on

---

[1] Some examples of such systems are: some secure wireless routing protocols, secure accounting and charging schemes, authenticated localization, and safety messages (vehicular networks).

the expected performance of the method in [30] which has the best worse case performance in our setting when w is known precisely. We find that even when the estimate is good ($d = 2w$), the impact on performance is severe.

2   Notation

In this paper we assume that pairing-based schemes use bilinear pairings on an elliptic curve E, defined over $\mathbb{F}_q$, where q is a large prime. $\mathbb{G}_1$ and $\mathbb{G}_2$ are distinct subgroups of prime order r on this curve, where $\mathbb{G}_1$ is a subset of the points on E with coordinates in $\mathbb{F}_q$, and $\mathbb{G}_2$ is a subset of the points on E with coordinates in $\mathbb{F}_{q^d}$, for a small integer d (the embedding degree). The pairing e is a map from $\mathbb{G}_1 \times \mathbb{G}_2$ into $\mathbb{G}_T$ where $\mathbb{G}_T$ is a multiplicative group of order r in the field $\mathbb{F}_{q^d}$.

Once the initial batch verification is performed, the costs of the methods for finding the invalid signatures in a batch are dominated by the cost of a product of pairings computations, CstMultPair, and the cost of multiplying two elements of $\mathbb{G}_T$, CstMult$\mathbb{G}_T$.[2]

3   Background

Fiat [10] introduced batch cryptography, and the first batch verification signature scheme was that of Naccache *et al.* [21] for a variant of DSA signatures. Bellare *et al.* [2] presented three generic methods for batching modular exponentiations: *the random subset test*, the *small exponents test* (SET), and the *bucket test*, which are related to techniques in [21,27].

A number of pairing-based signature schemes have batch verifiers which use the small exponents test, many of which have the form

$$e\left(\sum_{i=1}^{N} B_i, S\right) = \prod_{h=1}^{\bar{n}-1} e\left(\sum_{i=1}^{N} D_{i,h}, T_h\right) \tag{1}$$

where S and T_h are system parameters. Examples when $\bar{n} = 2$ include Boneh, Lynn and Shacham short signatures [3], when the batch consists of messages signed by a single signer or a common message signed by different signers, the Cha-Cheon identity-based scheme [6], and the scheme of Xun Yi [28] as interpreted by Solinas [25]. Examples of schemes that have this form with $\bar{n} = 3$ include the Camenisch, Hohenberger, and Pedersen (CHP) short signature scheme (for a common time period [5]) and a recent proposal of Zhang et al. [31] for signing and batch verifying location and safety messages in vehicular networks.[3]

---

[2] A cost that can be significant in large batches for some methods is the cost of additions in $\mathbb{G}_1$, CstAdd$\mathbb{G}_1$, (or additions in $\mathbb{G}_2$). The other operations used in the methods discussed in this paper, such as exponentiation CstExpt$\mathbb{G}_T(t_1)$ in $\mathbb{F}_{q^d}$ (for small t_1), computing an inverse in $\mathbb{G}_T$ (CstInv$\mathbb{G}_T$), multiplying an element of $\mathbb{G}_1$ or $\mathbb{G}_2$ by a modest sized scalar, have minimal impact.

[3] Some of these schemes are defined for pairings where $\mathbb{G}_1 = \mathbb{G}_2$. In CHP short signatures one of the pairings has the form $e\left(T_h, \sum_{i=1}^{N} D_{i,h}\right)$. For simplicity of presentation we ignore such distinctions in the remainder of this paper.

3.1 Identifying Invalid Signatures

Methods for identifying invalid signatures fall into three categories: divide-and-conquer methods [22,15], exponent testing methods [16,17,26,15], and hybrid techniques [17,18] which combine aspects of divide-and-conquer and other methods.

Divide-and-Conquer Methods. Pastuszak *et al.* [22] first investigated methods for identifying invalid signatures within a batch. They explored divide-and-conquer methods for generic batch verifiers, i.e., methods that work with any of the three batch verifiers studied by Ballare *et al.* In these methods the set of signatures in an invalid batch is repeatedly divided into $d \geq 2$ smaller sub-batches to verify. The most efficient of their techniques, the Fast DC Verifier Method, exploits knowledge of the results of the first $d - 1$ sub-batch verifications to determine whether the verification of the dth sub-batch is necessary. Performance measurements of one of the methods of [22] for the Boneh, Lynn and Shacham (BLS) [3] signature scheme have been reported [9].

In [15] a more efficient divide-and-conquer method, called Binary Quick Search (BQS), for small exponents test based verifiers was presented. In this method a batch verifier that compares two quantities, X and Y, is replaced with an equivalent test $A = XY^{-1}$, and the batch is accepted if $A = 1$. The BQS algorithm is always as least as efficient as any $d = 2^n$ary DC Verifier [19]. The upper bound of the number of batch verifications required by BQS is half that of the Fast DC Verifier for $d = 2$ [15].

Exponent Testing Methods. The first exponent testing method, developed by Lee *et al.* [16], was capable of finding a single invalid signature within a batch of "DSA-type" signatures. Law and Matt presented two exponent testing methods for pairing-based batch signatures, the Exponentiation Method and the Exponentiation with Sectors (EwS) Method, in [15]. Both methods use exhaustive search during batch verification, resulting in exponential cost.

The Exponentiation Method replaces (1) with $\alpha_0 = \prod\limits_{h=0}^{\bar{n}-1} e\left(\sum\limits_{i=1}^{N} D_{i,h}, T_h\right)$ where $D_{i,0} = B_i$ and $T_0 = -S$. If α_0 is equal to the identity, the batch is valid. Otherwise compute α_j, for $1 \leq j \leq w$,

$$\alpha_j = \prod_{h=0}^{\bar{n}-1} e\left(\sum_{i=1}^{N} i^j D_{i,h}, T_h\right) \tag{2}$$

and perform a test on the values $\alpha_j, \alpha_{j-1}, \ldots \alpha_0$. For $j = 1$, test whether $\alpha_1 = \alpha_0^{z_1}$ has a solution for $1 \leq z_1 \leq N$ using Shanks' giant-step baby-step algorithm [24]. If successful, $w = 1$ and z_1 is the position of the invalid signature. In general the method tests whether

$$\alpha_j = \prod_{t=1}^{j} (\alpha_{j-t})^{(-1)^{t-1} p_t} \tag{3}$$

has a solution where p_t is the tth elementary symmetric polynomial in $1 \leq z_1 \leq \ldots \leq z_j \leq N$. If a test fails increment j, compute α_j, and test. When $j = w$

the test will succeed, and the values of $z_1, \ldots, z_w$ are the positions of the invalid signatures.

The Exponentiation with Sectors Method uses two stages. In the first stage, the batch is divided into approximately $\sqrt{N}$ sectors of approximately equal size and the Exponentiation Method is used, where each $D_{i,h}$ within a sector is multiplied by the same constant to identify the v invalid sectors. In the second stage, the Exponentiation Method is used to find the invalid signatures within a batch consisting of the signatures from the v invalid sectors.[4]

Hybrid DC Methods. Lee et al. [17] applied their approach for DSA-type signatures to identifying a single invalid signature in batches of RSA signatures. They addressed the problem of identifying multiple invalid RSA signatures by using their RSA method in a DC method. Each (sub-)batch is tested using their RSA method. If the (sub-)batch has multiple invalid signatures, it is divided and its child sub-batches are tested. If a (sub-)batch has a single invalid signature, that signature is identified; if a (sub-)batch has no invalid signatures, that (sub-)batch is not tested further. Otherwise the (sub-)batch is divided and its child sub-batches are tested. However, Stanek showed in [26] that their approach for RSA signatures is not secure.

In [19] Matt presented two hybrid DC methods. The first, called Single Pruning Search (SPS), uses (2) and (3) for $0 \le j \le 1$ to identify single invalid signatures in (sub-)batches until the root of every maximal sub-tree of the search tree with a single invalid signature is identified. This method is somewhat similar to the Lee et al. method for RSA signature batches with multiple invalid signatures.

The second method, Paired Single Pruning Search (PSPS), extends SPS with an additional test. When a (sub-)batch B has two or more invalid signatures, $\alpha_{0,L}$ is computed for the left child sub-batch of B, and if both child sub-batches have invalid signatures, then $\alpha_{0,R} = \alpha_0 \cdot \alpha_{0,L}^{-1}$ is calculated for the right child sub-batch. Then $\alpha_1 = \alpha_{0,L}^{z_L} \cdot \alpha_{0,R}^{z_R}$ is tested for a solution where the exponents are restricted to the set of i's used in the child sub-batches. A solution will exist whenever both child sub-batches have a single invalid signature. The additional test determines if the two child nodes are both roots of maximal sub-trees of the search tree with a single invalid signature, without computing $\alpha_{1,L}$ and $\alpha_{1,R}$.

4   A Hybrid DC Method Exploiting $w = 2$ Maximal Sub-trees

Hybrid divide-and-conquer methods operate on (for simplicity) a binary tree T with $w \ge 1$ invalid signatures whose root node is the batch, and each pair of child nodes represents the two nearly equal size sub-batches of their parent. The SPS and PSPS methods search down though T until the roots of the w maximal sub-trees ST_i, $i = 1, .., w$, of T, which represent sub-batches that have a single invalid signature are reached and tested. The Triple Pruning Search method we describe

---

[4] The EwS method is always outperformed by one or more of the other methods we discuss in this section in our setting; therefore we do not discuss the performance of this method in Section 6.

in this paper searches down through T until the roots of the maximal sub-trees $ST2_i$, $i = 1, .., v$, of T, which represent the sub-batches that have exactly two invalid signatures, and the maximal sub-trees $ST1_j$, $j = 1, .., w - 2v$, which represent sub-batches that have a single invalid signature and which are not a sub-tree of any of the $ST2_i$ sub-trees, are reached and tested.

Let B be the batch. $|X|$ is the size the (sub-)batch X of B, $lowbnd(X)$ is the index in B of the lowest position signature in X, $w(X)$ is the number of invalid signatures in X, and $invalid(X)$ is the set of invalid signatures in X. If T is a binary tree and X is sub-batch, then $\hat{X}$ is the sibling of X.

4.1   Triple Pruning Search (TPS) Method

The recursive algorithm on the next page describes the Triple Pruning Search (TPS) method on a batch B, which is a list of $N = 2^h$, $h \geq 2$, randomly ordered message / signature pairs $((m_1, s_1), \ldots, (m_N, s_N))$, where the signature components are verified elements of the appropriate groups. On the initial call to $TPS(X)$, $X = B$.

$TPS(X)$ includes the initial batch verification (lines 2 through 4). When $X = B$, $Get_0(B)$ computes $\alpha_{0,B}$ following the SET algorithm, and then computes $\alpha_{0,B}^{-1}$. The test $\alpha_{0,B} = 1$ determines whether $w(B) = 0$.

Lines 5 through 9 determine whether $w(B) = 1$, and if so they locate the invalid signature. $Get_1(B)$ computes $\alpha_{1,B}$ in about $\bar{n} \cdot N \cdot \text{CstAdd}\mathbb{G}_1 + \text{CstMultPair}$ operations using the partial results from the computation of $\alpha_{0,B}$, and then computes $\alpha_{1,B}^{-1}$. $Shanks(B)$ is used to locate a single invalid signature. $Shanks(X)$ tests whether $\alpha_{1,X} \cdot (\alpha_{0,X}^{-1})^d = (\alpha_{0,X}^s)^c$ has a solution with $l \leq d \leq s + l$ and $0 \leq c \leq t$, where $s \approx \sqrt{|X|}$, $t \approx |X|/s$ and $l = lowbnd(X)$. If $w(X) = 1$, $Shanks(X)$ returns $d + c * s$, the position of the invalid signature. If $w(X) > 1$, then it returns 0. $Shanks(X)$ uses the giant-step baby-step algorithm [24].

Lines 10 through 14 determine whether $w(B) = 2$, and if so they locate the two invalid signatures. $Get_2(X = B)$ computes $\alpha_{2,B}$ in about $\bar{n} \cdot N \cdot \text{CstAdd}\mathbb{G}_1 + \text{CstMultPair}$ operations using the partial results from the computation of $\alpha_{1,B}$ and $\alpha_{0,B}$, and then computes $\alpha_{2,B}^{-1}$. $FastFactor(B)$ is used to locate the pair of invalid signatures. $FastFactor(X)$ tests whether $\alpha_{2,X}^4 \cdot (\alpha_{1,X}^{-4})^n \cdot \alpha_{0,X}^{n^2} = \alpha_{0,X}^{m^2}$ has a solution with $2l + 1 \leq n \leq 2(l + |X|) - 1$ and $1 \leq m \leq |X| - 1$, where $l = lowbnd(X)$; if so, then $z_2 = (n + m)/2$ and $z_1 = (n - m)/2$ with $z_2 > z_1$ are the positions of the two invalid signatures. If $w(X) = 2$, $FastFactor(X)$ returns (z_1, z_2). If $w(X) > 2$, $FastFactor(X)$ returns $(0, 0)$. See [20].

In line 15, the function $TPSQuadSolver(X, Left(X), Right(X))$ determines whether X has two or fewer invalid signatures in its left sub-batch $Left(X)$ and two or fewer invalid signatures in its right sub-batch $Right(X)$. $TPSQuadSolver$ places the locations of the invalid signature it identifies in a list which $PrintList()$ outputs.

4.2 $TPSQuadSolver(Parent, Left, Right)$

The algorithm on page 85 describes the $TPSQuadSolver(Parent, Left, Right)$ function on a (sub-)batch $Parent$ with $|Parent| = 2^h$, $h \geq 2$, and $w(Parent) \geq 3$. $Left$ and $Right$ represent the two equal size sub-batches of $Parent$.

Algorithm 4.1. $TPS\ (X)$ *(Triple Pruning Search)*

Input: X – a list of message / signature pairs.
Output: A list of the invalid pairs in the batch.
1: **if** $X = B$ **then**
2:         $\alpha_{0,[B]} \leftarrow Get_0(B)$
3:         **if** $\alpha_{0,[B]} = 1$ **then**
4:                 **return**
5:         $\alpha_{1,[B]} \leftarrow Get_1(B)$
6:         $z \leftarrow Shanks(B)$
7:         **if** $z \neq 0$ **then**
8:                 **print** (m_z, s_z)
9:                 **return**
10:         $\alpha_{2,[B]} \leftarrow Get_2(B)$
11:         $(z_1, z_2) \leftarrow FastFactor(B)$
12:         **if** $z_1 \neq 0$ **then**
13:                 **print** $(m_{z_1}, s_{z_1}), (m_{z_2}, s_{z_2})$
14:                 **return**
15: $(SearchLeft, SearchRight) \leftarrow TPSQuadSolver(X, Left(X), Right(X))$.
16: **if** $SearchLeft = $ **true then**
17:         $TPS\ (Left(X))$
18: **if** $SearchRight = $ **true then**
19:         $TPS\ (Right(X))$
20: **if** $X = B$ **then**
21:         PrintList() // Prints the sorted list of invalid message / signature pairs
22: **return**

$TPSQuadSolver(Parent,\ Left,\ Right)$ uses $Get_0(Left)$ to compute $\alpha_{0,Left}$ (and $\alpha_{0,Left}^{-1}$), which requires a CstMultPair computation as well as some comparatively minor cost computations in $\mathbb{G}_1$. Lines 3 through 8 determine whether all of the invalid signatures in $Parent$ are in either $Left$, $Right$, or are divided between the two. If both $Left$ and $Right$ have at least one invalid signature then $Get_0(Right)$ is used to compute $\alpha_{0,Right}$ ($\alpha_{0,Right}^{-1}$) with negligible cost.[5]

If $TScap \geq |Parent| \geq 4$, $TriFactor(Parent)$ is used to determine whether case 1) $w(Left) = 2$ and $w(Right) = 1$, or case 2) $w(Left) = 1$ and $w(Right) = 2$; otherwise it fails. $TriFactor(Parent)$ [20], uses the function $TriSolver$ to test case 1 and if that fails, case 2.

Case 1) $TriSolver(P = Parent,\ L = Left,\ R = Right)$
If $\alpha_{2,P}^4 \cdot (\alpha_{0,R}^{-4})^{z_3^2} \cdot (\alpha_{1,P}^{-4})^{n_L} \cdot (\alpha_{0,R}^4)^{n_L \cdot z_3} \cdot \alpha_{0,L}^{n_L^2} = \alpha_{0,L}^{m_L^2}$ has a solution with $2l_L + 1 \leq n_L \leq 2(l_L + |L|) - 1$, $1 \leq m_L \leq |L| - 1$ and $l_R \leq z_3 < l_R + |R|$, where $l_L = lowbnd(L)$ and $l_R = lowbnd(R)$, then $z_2 = (n_L + m_L)/2$ and $z_1 = (n_L - m_L)/2$ where $z_2 > z_1$ are the positions of the two invalid signatures in L, and z_3 in R.

Case 2) $TriSolver(P = Parent,\ R = Right,\ L = Left)$
If $\alpha_{2,P}^4 \cdot (\alpha_{0,L}^{-4})^{z_1^2} \cdot (\alpha_{1,P}^{-4})^{n_R} \cdot (\alpha_{0,L}^4)^{n_R \cdot z_1} \cdot \alpha_{0,R}^{n_R^2} = \alpha_{0,R}^{m_R^2}$ has a solution with $2l_R + 1 \leq n_R \leq 2(l_R + |R|) - 1$, $1 \leq m_R \leq |R| - 1$ and $l_L \leq z_1 < l_L + |L|$, then

---

[5] $\alpha_{i,Right}, i = 0, 1, 2$ can be computed inexpensively if $\alpha_{i,Left}$ is known by $\alpha_{i,Right} = \alpha_{i,Parent} \cdot \alpha_{i,Left}^{-1}$.

Algorithm 4.2. $TPSQuadSolver(Parent, Left, Right)$

Input: $Parent, Left, Right$ – the lists of message / signature pairs.
Return: $(SearchLeft, SearchRight)$ – control behavior of TPS
1: $(z_1, z_2, z_3, z_4) \leftarrow (0, 0, 0, 0)$
2: $\alpha_{0,[Left]} \leftarrow Get_0(Left)$
3: **if** $\alpha_{0,[Left]} = 1$ **then**
4:       copyAlphasAndInverses($Right, Parent$);
5:       **return (false, true)**
6: **if** $\alpha_{0,[Left]} = \alpha_{0,[Parent]}$ **then**
7:       copyAlphasAndInverses($Left, Parent$)
8:       **return (true, false)**
9: $\alpha_{0,[Right]} \leftarrow Get_0(Right)$
10: **if** $TScap \geq |Parent| \geq 4$ **then** // $TScap = 8$, see Section 6.1
11:       $(z_1, z_2, z_3) \leftarrow TriFactor(Parent)$
12:       **if** $z_1 \neq 0$ **then**
13:             AddToList(z_1, z_2, z_3)
14:             **return (false, false)**
15:       **if** $|Parent| = 4$ **then**
16:             $i \leftarrow lowbnd(Parent)$; AddToList($i, i+1, i+2, i+3$)
17:             **return (false, false)**
18: $SearchLeft \leftarrow$ **false**; $SearchRight \leftarrow$ **false**
19: $\alpha_{1,[Left]} \leftarrow Get_1(Left)$; $\alpha_{1,[Right]} \leftarrow Get_1(Right)$;
20: $z_1 \leftarrow Shanks(Left)$
21: **if** $z_1 \neq 0$ **then** // $w(Left) = 1$
22:       $\alpha_{2,[Right]} \leftarrow Get_2(Right)$
23:       **if** $|Parent| > TScap$ **then**
24:             $(z_3, z_4) \leftarrow FastFactor(Right)$
25:       AddToList(z_1, z_3, z_4) // zeros are not added to the list
26:       **if** $z_3 = 0$ **then**
27:             $SearchRight \leftarrow$ **true**
28:       **return (false,** $SearchRight$**)**
29: **else** // $w(Left) \geq 2$
30:       $z_3 \leftarrow Shanks(Right)$
31:       **if** $z_3 \neq 0$ **then** // $w(Right) = 1$
32:             $\alpha_{2,[Left]} \leftarrow Get_2(Left)$
33:             **if** $|Parent| > TScap$ **then**
34:                  $(z_1, z_2) \leftarrow FastFactor(Left)$
35:             AddToList(z_1, z_2, z_3) // zeros are not added to the list
36:             **if** $z_1 = 0$ **then**
37:                  $SearchLeft \leftarrow$ **true**
38:             **return (**$SearchLeft$**, false)**
39:       **else** // $w(Left) \geq 2$ and $w(Right) \geq 2$
40:             $\alpha_{2,[Left]} \leftarrow Get_2(Left)$; $\alpha_{2,[Right]} \leftarrow Get_2(Right)$
41:             $(z_1, z_2) \leftarrow FastFactor(Left)$; $(z_3, z_4) \leftarrow FastFactor(Right)$
42:             AddToList(z_1, z_2, z_3, z_4) // zeros are not added to the list
43:             **if** $z_1 = 0$ **then**
44:                  $SearchLeft \leftarrow$ **true**
45:             **if** $z_3 = 0$ **then**
46:                  $SearchRight \leftarrow$ **true**
47:             **return (**$SearchLeft, SearchRight$**)**

$z_3 = (n_R + m_R)/2$ and $z_2 = (n_R - m_R)/2$ where $z_3 > z_2$ are the positions of the invalid signatures in R, and z_1 in L.

If $w(Parent) = 3$, $TriFactor(Parent)$ returns the positions of the three invalid signatures, which are added to the list of invalid signatures (line 13).

If $|Parent| = 4$ and $TriFactor(Parent)$ fails, then the positions of the four signatures in $Parent$ are added to the list of invalid signatures (line 16).

If $|Parent| > 4$ and $TriFactor(Parent)$ failed (or was not used), then Get_1 is used to compute $\alpha_{1,Left}$ (and $\alpha_{1,Left}^{-1}$) and $\alpha_{1,Right}$ ($\alpha_{1,Right}^{-1}$) with approximate total cost CstMultPair (line 19).

If the following $Shanks(Left)$ test succeeds, then $Get_2(Right)$ can compute $\alpha_{2,Right} = \alpha_{2,Parent} \cdot \alpha_{1,Left}^{z_1}$ and its inverse efficiently with cost $2\,\mathsf{CstMult}\mathbb{G}_T + 2\,\mathsf{CstInv}\mathbb{G}_T + \mathsf{CstExpt}\mathbb{G}_T(t_1)$, where $t_1 < \lceil log_2(N) \rceil$. This cost is much less than CstMultPair, we ignore this cost in Section 5. Next if $TriFactor(Parent)$ was not used, then $FastFactor(Right)$ is used (line 24) to test whether $w(Right) = 2$ and if so, identify the two invalid signatures in $Right$. If $TriFactor(Parent)$ was used, it must have failed, and so would $FastFactor(Right)$.

If the $Shanks(Left)$ test (line 20) fails, then $Shanks(Right)$ (line 30) is used to test the right sub-batch. If that test succeeds, then by exchanging $Left$ and $Right$, the preceding paragraph describes the function of lines 31 through 38.

If $w(Left) \geq 2$ and $w(Right) \geq 2$, then Get_2 is used to compute $\alpha_{2,Left}$ and $\alpha_{2,Right}$ and their inverses, with approximate total cost CstMultPair (line 40), followed by tests of $Left$ and $Right$ using $FastFactor$.

5   Expected Cost of the New Method

TPS requires that initial batch verification is performed using the Small Exponents Test, and for simplicity, we assume that the batch verifier is of the form $\alpha_{0,B} = \prod_{h=0}^{\bar{n}-1} e\left(\sum_{i=1}^{N} D_{i,h}, T_h\right)$. The cost of this process for Cha-Cheon signatures includes first checking that the signature components are in $\mathbb{G}_1$, then computing the terms $\sum_{i=1}^{N} D_{i,h}, h = 0, 1$ in $\mathbb{G}_1$, and finally computing $\alpha_{0,B}$ an its inverse, and testing whether $\alpha_{0,B} = 1$.

If $\alpha_{0,B} \neq 1$ (and all $D_{i,h} \in \mathbb{G}_1$), then assuming that the intermediate values $D_{i,h}$ are retained, the costs of computing $\alpha_{1,B}$ and $\alpha_{2,B}$ are each $\bar{n} \cdot |B| \cdot \mathsf{CstAdd}\mathbb{G}_1 + \mathsf{CstInv}\mathbb{G}_T + \mathsf{CstMultPair}$. Since $\mathsf{CstInv}\mathbb{G}_T \ll \mathsf{CstMultPair}$, we ignore the cost of computing the inverses.

If $w = 1$, the cost of TPS, not including initial verification, is $\bar{n} \cdot |B| \cdot \mathsf{CstAdd}\mathbb{G}_1 + \mathsf{CstMultPair}$ plus the average cost of a successful $Shanks(B)$, which is $\frac{4}{3}\sqrt{|B|}\,\mathsf{CstMult}\mathbb{G}_T$.

If $w = 2$, the cost is $2(\bar{n} \cdot |B| \cdot \mathsf{CstAdd}\mathbb{G}_1 + \mathsf{CstMultPair}) + 2\sqrt{|B|}\,\mathsf{CstMult}\mathbb{G}_T + \frac{11}{4}|B|\,\mathsf{CstMult}\mathbb{G}_T$, which is the cost of computing the two products of pairings, including their inputs, a failed $Shanks(B)$, and the average cost of a successful $FastFactor(B)$.

If $w > 2$, the cost includes $2(\bar{n}\cdot|B|\cdot\mathsf{CstAdd}\mathbb{G}_1 + \mathsf{CstMultPair}) + 2\sqrt{|B|}\,\mathsf{CstMult}\mathbb{G}_T + \frac{9}{2}|B|\,\mathsf{CstMult}\mathbb{G}_T$, which is two products of pairings, a failed $Shanks(B)$, and a failed $FastFactor(B)$. In addition the cost includes the cost generated by the recurrence relation $\mathbf{R_{(TPS)}}(w, M)$ to below, with $|B| = 2^h$, where $h \geq 2$, and on initial call $M = |B|$ and $w(B) \geq 3$. Note that $\mathsf{CstMult}\mathbb{G}_T \ll \mathsf{CstMultPair}$, so we ignore small numbers of $\mathsf{CstMult}\mathbb{G}_T$ in the table below.[6]

$\mathbf{R_{(TPS)}}(w, M) =$

$$
\begin{cases}
0, & w = 0, 1, 2, \\
& w > M; \\[2ex]
\dfrac{\left[\displaystyle\sum_{i=0}^{w} \binom{M/2}{w-i}\binom{M/2}{i}\left(\mathbf{R_{(TPS)}}(w-i, M/2) + \mathbf{R_{(TPS)}}(i, M/2)\right.\right.}{\binom{M}{w}}, & w \geq 3, \\
\end{cases}
$$

where the cost functions $\mathrm{C_{(TPS)}}$ are given in the following table.

Argument	Costs	
	CstMultPair	CstMult$\mathbb{G}_T$
$M = 4$		
$\mathrm{C_{(TPS)}}(2, 1, M/2)$	1	$\frac{14(6+\sqrt{2})}{32}M + \frac{3(7+\sqrt{2})}{32}M^2$
$\mathrm{C_{(TPS)}}(2, 2, M/2)$	1	$\frac{11}{2}M + \frac{3}{2}M^2$
$TScap \geq M > 4$		
$\mathrm{C_{(TPS)}}(w, 0, M/2)$	1	
$\mathrm{C_{(TPS)}}(2, 1, M/2)$	1	$\frac{14(6+\sqrt{2})}{32}M + \frac{3(7+\sqrt{2})}{32}M^2$
$\mathrm{C_{(TPS)}}((w-1) > 2, 1, M/2)$	2	$\frac{7}{3}\sqrt{M} + 10M + \frac{3}{2}M^2$
$\mathrm{C_{(TPS)}}(2, 2, M/2)$	3	$4\sqrt{M} + 11M + \frac{3}{2}M^2$
$\mathrm{C_{(TPS)}}((w-2) > 2, 2, M/2)$	3	$4\sqrt{M} + 22\frac{3}{4}M + \frac{3}{2}M^2$
$\mathrm{C_{(TPS)}}((w-i) > 2, i > 2, M/2)$	3	$4\sqrt{M} + 24\frac{1}{2}M + \frac{3}{2}M^2$
$M > TScap$		
$\mathrm{C_{(TPS)}}(w, 0, M/2)$	1	
$\mathrm{C_{(TPS)}}(2, 1, M/2)$	2	$\frac{7}{3}\sqrt{M} + \frac{11}{4}M$
$\mathrm{C_{(TPS)}}((w-1) > 2, 1, M/2)$	2	$\frac{7}{3}\sqrt{M} + \frac{9}{2}M$
$\mathrm{C_{(TPS)}}(2, 2, M/2)$	3	$4\sqrt{M} + \frac{11}{2}M$
$\mathrm{C_{(TPS)}}((w-2) > 2, 2, M/2)$	3	$4\sqrt{M} + 7\frac{1}{4}M$
$\mathrm{C_{(TPS)}}((w-i) > 2, i > 2, M/2)$	3	$4\sqrt{M} + 9M$

---

[6] Also note that $\mathrm{C_{(TPS)}}(w-i, i, M/2) = \mathrm{C_{(TPS)}}(i, w-i, M/2)$ in $\mathbf{R_{(TPS)}}(w, M)$ and that the value of each $\mathrm{C_{(TPS)}}(w-i, i, M/2)$ in the table is the average of $\mathrm{C_{(TPS)}}(w-i, i, M/2)$ and $\mathrm{C_{(TPS)}}(i, w-i, M/2)$.

6   Performance

All of the methods discussed in this section perform initial batch verification in a similar manner. For Cha-Cheon signatures, they all check that the signature components are in $\mathbb{G}_1$, then compute α_0 for the batch, and then test whether $\alpha_0 = 1$. There are some slight variations in how the terms $\sum_{i=1}^{N} D_{i,h}$ are summed, but the cost in each case is the same. In Sections 6.1 and 7.1, we compare the expected performance of TPS against first the methods discussed in Section 3.1 and then against the group testing based methods, once the initial batch verification has failed. See Section 5 and [20] for the derivations of the costs presented below and additional discussion of the performance of the methods.

We use Cases A and E of [12] for Cha-Cheon signatures to give an indication of how our results change with variations in the relative cost of operations.[7] In Case A, the group order r is a 160-bit value, the elliptic curve E is defined over $\mathbb{F}_q$, where q is a 160-bit value, and the embedding degree $d = 6$. In Case E, the group order r is a 384-bit value, q is a 384-bit value, and the embedding degree $d = 12$. All costs are given in terms of the number of multiplications (m) in $\mathbb{F}_q$, assuming that squaring has the same cost as multiplication, using the following estimates from Granger, Page and Smart [12], Granger and Smart [13], and Devegili et al. [7].

- For Case A, 1 double product of pairings $= 14,027m$, 1 multiplication in $F_{q^6} = 15m$, 1 addition in $\mathbb{G}_1 = 11m$.
- For Case E, 1 double product of pairings $= 104,316m$, 1 multiplication in $F_{q^{12}} = 45m$, 1 addition in $\mathbb{G}_1 = 11m$.

6.1   Comparison with Earlier Methods

Figures 1 through 4 compare methods relative to PSPS, previously the best performing method for our setting. We also include two additional divide-and-conquer methods, SPS and BQS, the Exponential method, as well as testing the signatures individually.

TPS uses successful $TriSolver$ tests to avoid computing CstMultPair for α_1's (and perhaps) α_2's for child sub-batches. We observe that the $O(N^2)$ cost of these tests requires that we restrict the use of $TriFactor$ to parent batches of size less than or equal to $TScap$ (line 10 of Algorithm 4.2); otherwise for larger batches the cost of $TriFactor$, even when successful, would become greater than the cost of the α's and their associated $Shanks$ and $FastFactor$ tests. In Figures 1 through 8, $TScap = 8$ which limits the cost of $TriFactor$ to no more than ≈ 140 CstMult$\mathbb{G}_T$. Since the ratio of CstMultPair to CstMult$\mathbb{G}_T$ for Case A is $1275 : 1$ and Case E is $2318 : 1$, the cost of $TriFactor$ does not significantly impact the overall cost of TPS in these figures. However, setting $TScap = 8$ rather than $TScap = |B|$ increases the number of product of pairing computations used by TPS.

---

[7] The most important factor in the relative performance of all the methods is the ratio of CstMultPair to CstMult$\mathbb{G}_T$. The ratio of CstMultPair to $\bar{n} \cdot |B| \cdot$ CstAdd$\mathbb{G}_1$ is much less significant in our setting.

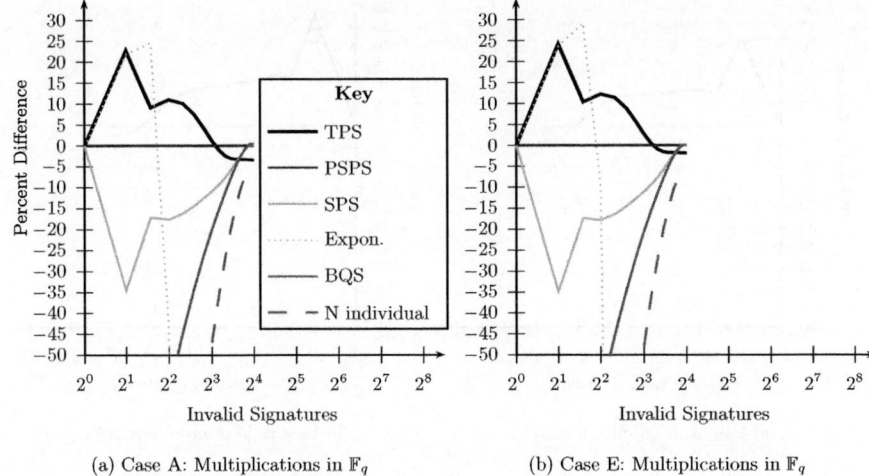

(a) Case A: Multiplications in $\mathbb{F}_q$

(b) Case E: Multiplications in $\mathbb{F}_q$

Fig. 1. Percent Difference Comparison with PSPS, $N = 16$

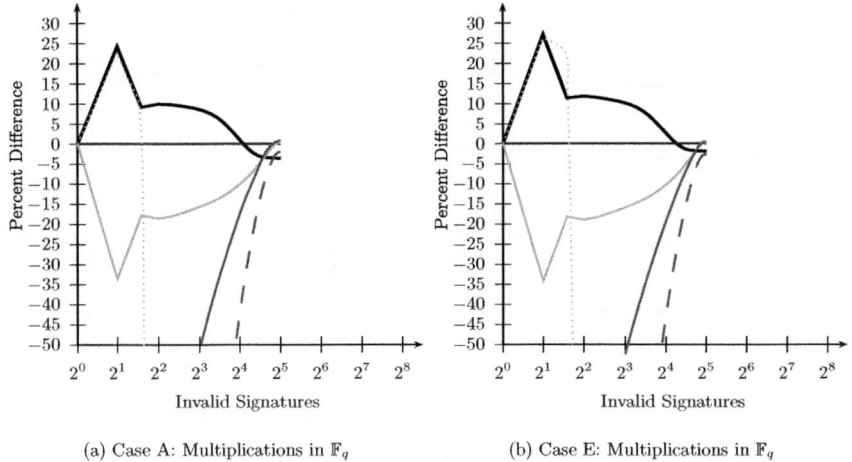

(a) Case A: Multiplications in $\mathbb{F}_q$

(b) Case E: Multiplications in $\mathbb{F}_q$

Fig. 2. Percent Difference Comparison with PSPS, $N = 32$

7   Group Testing Based Methods

Zaverucha and Stinson [30] recently examined algorithms from the group testing literature for use in identifying invalid signatures in batches. Like Pastuszak
et al., they work with generic batch verifiers. Zaverucha and Stinson state that
for single processor systems, identifying invalid signatures using Binary Splitting (same method as the Fast DC-verifier of Pastuszak *et al.* [22]) and Hwang's

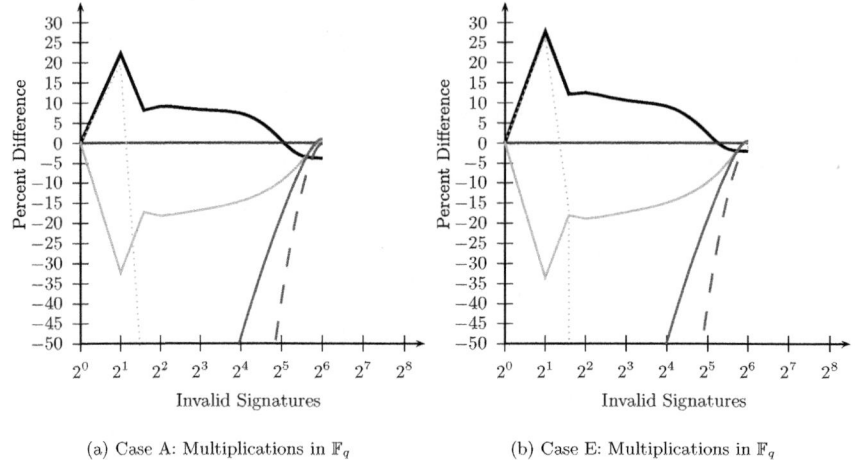

(a) Case A: Multiplications in $\mathbb{F}_q$ (b) Case E: Multiplications in $\mathbb{F}_q$

Fig. 3. Percent Difference Comparison with PSPS, $N = 64$

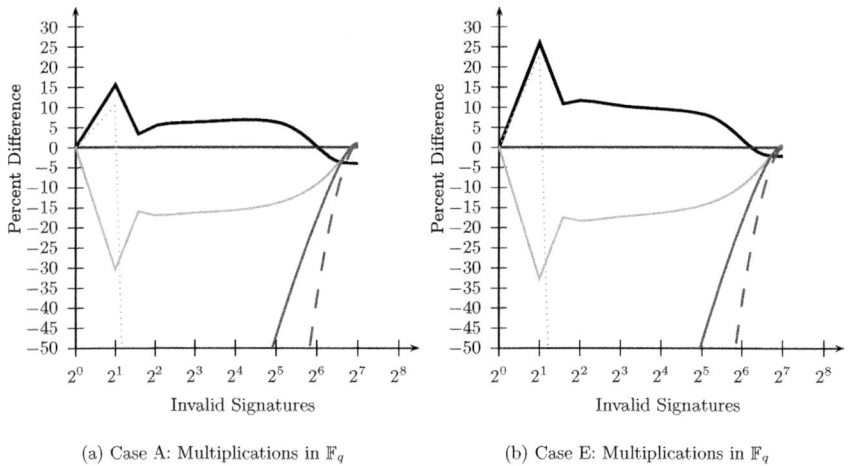

(a) Case A: Multiplications in $\mathbb{F}_q$ (b) Case E: Multiplications in $\mathbb{F}_q$

Fig. 4. Percent Difference Comparison with PSPS, $N = 128$

Generalized Binary Splitting (HGBS) methods have the lowest bounds on the worse case number of verifications. Here we examine the expected performance of these algorithms.

Binary Splitting tests an invalid (sub-)batch of size M by first testing $\lfloor \frac{M}{2} \rfloor$ signatures from the (sub-)batch. If this test indicates an invalid signature, a test of the remaining $\lceil \frac{M}{2} \rceil$ signatures is required, and the method is applied to both sub-batchs; otherwise the test is not performed on the sub-batch with the $\lceil \frac{M}{2} \rceil$ signatures and the method is applied only to that sub-batch. Binary splitting is the same method as the Fast DC-verifier of Pastuszak *et al.* [22].

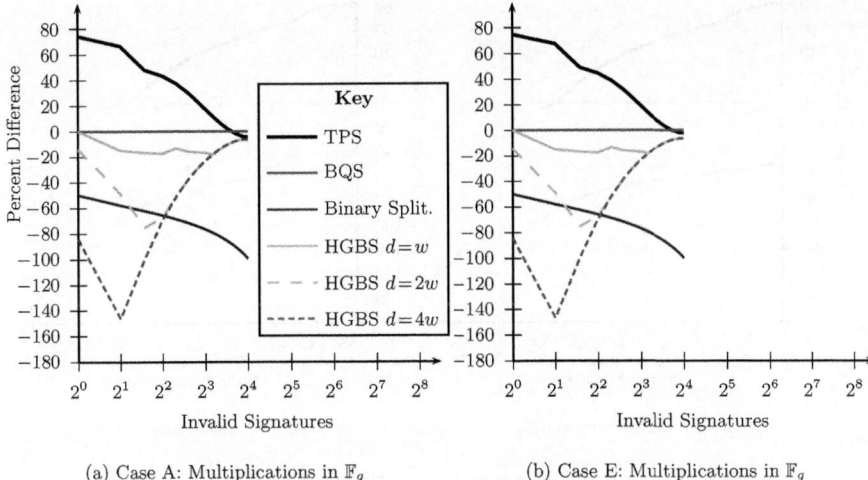

(a) Case A: Multiplications in $\mathbb{F}_q$ (b) Case E: Multiplications in $\mathbb{F}_q$

Fig. 5. Percent Difference Comparison with BQS, $N = 16$

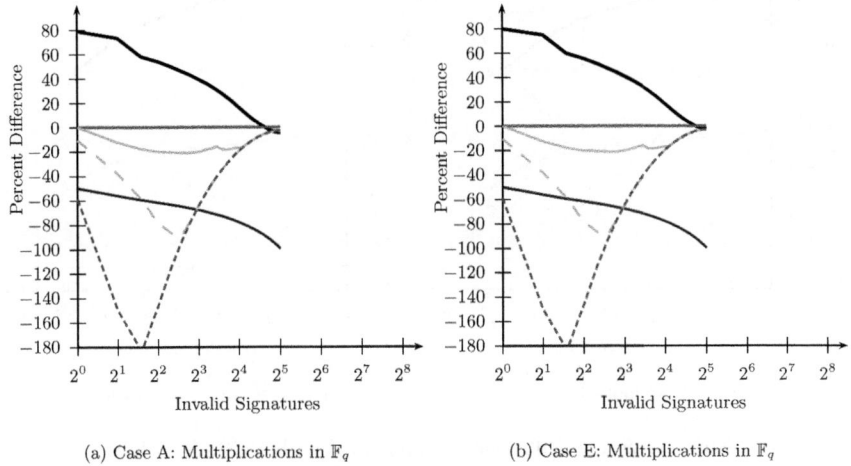

(a) Case A: Multiplications in $\mathbb{F}_q$ (b) Case E: Multiplications in $\mathbb{F}_q$

Fig. 6. Percent Difference Comparison with BQS, $N = 32$

The HGBS method [14] makes an estimate d of the number of invalid signatures in a batch of size M. The descriptions of the HGBS method which have appeared in the literature differ slightly. Here we describe the version of HGBS which appears in [14].

G1:

If $M \leq 2(d - 1)$ verify the M signatures individually, otherwise compute $a \in \mathbb{N}$ s.t. $2^{a+1} > (M - d + 1)/d \geq 2^a$ and goto **G2**.

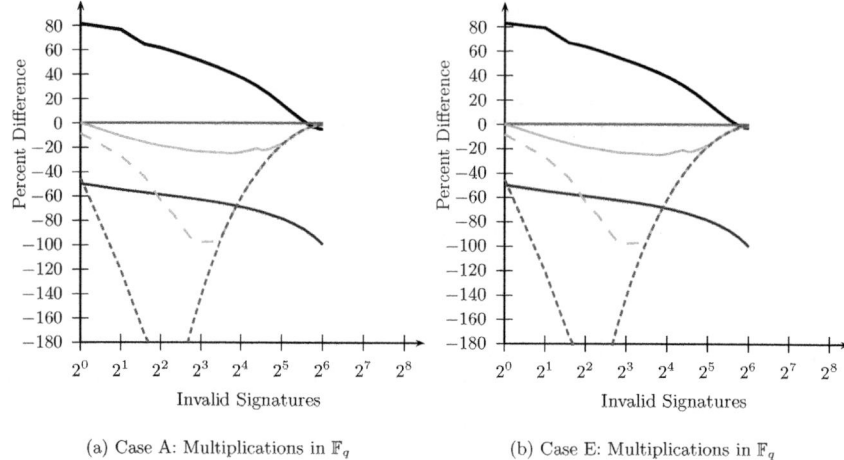

(a) Case A: Multiplications in $\mathbb{F}_q$ (b) Case E: Multiplications in $\mathbb{F}_q$

Fig. 7. Percent Difference Comparison with BQS, $N = 64$

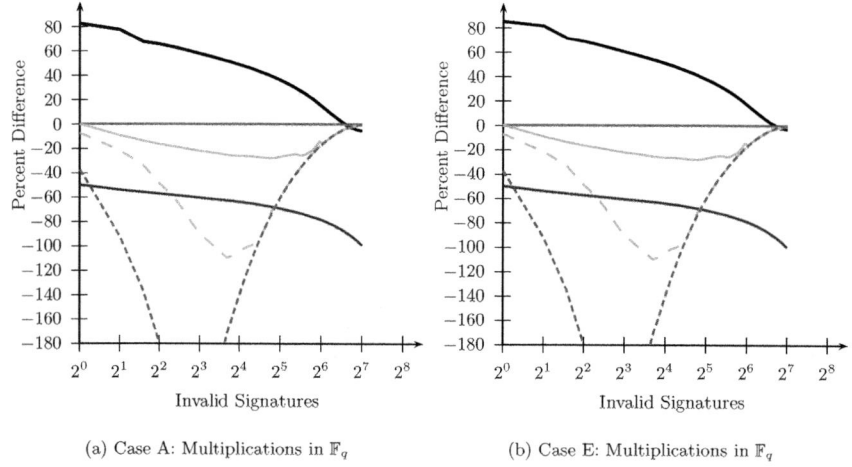

(a) Case A: Multiplications in $\mathbb{F}_q$ (b) Case E: Multiplications in $\mathbb{F}_q$

Fig. 8. Percent Difference Comparison with BQS, $N = 128$

G2:

   Test a sub-batch X of size 2^a. If all are valid $M \leftarrow M - 2^a$ and go to **G1**
   for the other signatures. If X is invalid find one invalid signature in X using
   a tests via binary search and dispose of the invalid signature and all valid
   sub-batches identified during the search. Create a new batch consisting of
   all the remaining sub-batches. Set M to the size of this batch and $d \leftarrow d - 1$
   and go to **G1**.

Note that d must be greater than or equal to w, otherwise the method is undefined.[8]

---

[8] In [14] d is an upper bound, in [8] it is the known number of invalid signatures.

7.1   Comparison of TPS with Group Testing Based Methods

In Figures 5 through 8 we compare the performance of TPS against two group testing methods, Binary Splitting and HGBS, relative to BQS. We use BQS since it only requires that a batch verifier that compares two quantities, X and Y, can be replaced with the test $A = XY^{-1}$. BQS is intermediate between the more signature scheme specific TPS method and the general group testing methods. We also show the extent to which uncertainty in the estimated (or a bound of the) number of the invalid signatures in a batch degrades performance of HGBS. Binary Splitting does not use such an estimate.

8   Conclusion

We presented the TPS method for identifying invalid signatures in pairing-based batch signature schemes using SET, and have analyzed its expected performance. The new method provides improved performance for $1 < w \leq N/2$, for the range of batch sizes of interest. The new method is the best available for our setting, constrained sized batches verified by single processor systems, when the number of invalid signatures in a batch can vary considerably but does not exceed $N/2$. The method is applicable to a number of batch verified signature schemes, those presented in [9] and that of Zhang et al. [31].

In [30] the authors investigated using generic verifier methods derived from group testing algorithms for invalid signature identification. Of the five methods they discussed, two — Binary Splitting and HGBS — were identified as the best methods for single processor verifiers. A number of group testing algorithms such as HGBS rely on an estimate, d, of the number of invalid signatures. In [30] the authors state that when d differs from w "it is unclear to what extent this will hurt the performance of the algorithm." We investigated this issue for the expected performance of HGBS and showed that the impact can be severe.

The authors also observed that more restrictive verifiers such as the Exponentiation and EwS methods of Law and Matt [15] and the hybrid methods of Matt [18] (and by extension ours) will outperform their generic verifiers for the class of signature schemes to which these methods apply. We observe that BQS assumes only a common feature of many batch verifiers, yet outperforms the generic group testing based verifiers, especially when the choice of value of d is uncertain.

Acknowledgment

Prepared in part through collaborative participation in the Communications and Networks Consortium sponsored by the U. S. Army Research Laboratory under the Collaborative Technology Alliance Program, Cooperative Agreement DAAD-19-01-2-0011. The U. S. Government is authorized to reproduce and distribute reprints for Government purposes notwithstanding any copyright notation thereon.

References

1. Arazi, B.: Certification of dl/ec keys. Submission to P1363 (August 1998) (updated May 1999), http://grouper.ieee.org/groups/1363/StudyGroup/Hybrid.html
2. Bellare, M., Garay, J., Rabin, T.: Fast batch verification for modular exponentiation and digital signatures. In: Nyberg, K. (ed.) EUROCRYPT 1998. LNCS, vol. 1403, pp. 236–250. Springer, Heidelberg (1998)
3. Boneh, D., Lynn, B., Shacham, H.: Short signatures from the weil pairing. In: Boyd, C. (ed.) ASIACRYPT 2001. LNCS, vol. 2248, pp. 514–532. Springer, Heidelberg (2001)
4. Brown, D., Gallant, R., Vanstone, S.: Provably secure implicit certificate schemes. In: Syverson, P.F. (ed.) FC 2001. LNCS, vol. 2339, pp. 105–120. Springer, Heidelberg (2001)
5. Camenisch, J., Hohenberger, S., Pedersen, M.: Batch verification of short signatures. In: Naor, M. (ed.) EUROCRYPT 2007. LNCS, vol. 4515, pp. 246–263. Springer, Heidelberg (2007)
6. Cha, J., Cheon, J.: An identity-based signature from gap diffie-hellman groups. In: Desmedt, Y. (ed.) PKC 2003. LNCS, vol. 2567, pp. 18–30. Springer, Heidelberg (2002)
7. Devegili, A.J., hÉigeartaigh, C.O., Scott, M., Dahab, R.: Multiplication and squaring on pairing-friendly fields. Technical report, Cryptology ePrint Archive, Report 2006/471 (2006), http://eprint.iacr.org/2006/471
8. Du, D., Hwang, F.K.: Combinatorial Group Testing And Its Applications, 2nd edn. World Scientific, Singapore (December 1999)
9. Ferrara, A.L., Green, M., Hohenberger, S., Pedersen, M.O.: Practical short signature batch verification. In: Fischlin, M. (ed.) CT-RSA 2009. LNCS, vol. 5473, pp. 309–324. Springer, Heidelberg (2009)
10. Fiat, A.: Batch RSA. In: Brassard, G. (ed.) CRYPTO 1989. LNCS, vol. 435, pp. 175–185. Springer, Heidelberg (1989)
11. FIPS 186-2: Digital Signature Standard (DSS). Federal Information Processing Standards Publication 186-2 (January 2000)
12. Granger, R., Page, D., Smart, N.P.: High security pairing-based cryptography revisited. In: Hess, F., Pauli, S., Pohst, M. (eds.) ANTS 2006. LNCS, vol. 4076, pp. 480–494. Springer, Heidelberg (2006)
13. Granger, R., Smart, N.P.: On computing products of pairings. Cryptology ePrint Archive, Report 2006/172 (2006), http://eprint.iacr.org/2006/172
14. Hwang, F.K.: A method for detecting all defective members in a population by group testing. Journal of the American Statistical Association 67(339) (1972)
15. Law, L., Matt, B.J.: Finding invalid signatures in pairing-based batches. In: Galbraith, S. (ed.) Cryptography and Coding 2007. LNCS, vol. 4887, pp. 35–53. Springer, Heidelberg (2007)
16. Lee, S., Cho, S., Choi, J., Cho, Y.: Batch verification with DSA-type digital signatures for ubiquitous computing. In: Hao, Y., et al. (eds.) CIS 2005. LNCS (LNAI), vol. 3802, pp. 125–130. Springer, Heidelberg (2005)
17. Lee, S., Cho, S., Choi, J., Cho, Y.: Efficient identification of bad signatures in RSA-type batch signature. IEICE Transactions on Fundamentals of Electronics, Communications and Computer Sciences E89-A(1), 74–80 (2006)
18. Matt, B.J.: Identification of multiple invalid signatures in pairing-based batched signatures. In: Jarecki, S., Tsudik, G. (eds.) PKC 2009. LNCS, vol. 5443, pp. 337–356. Springer, Heidelberg (2009)

19. Matt, B.J.: Identification of multiple invalid signatures in pairing-based batched signatures. Cryptology ePrint Archive Report 2009/097 (2009), http://eprint.iacr.org/2009/097

20. Matt, B.J.: Identification of multiple invalid pairing-based signatures in constrained batches. Cryptology ePrint Archive (2010), http://eprint.iacr.org/2010

21. Naccache, D., M'Raihi, D., Vaudenay, S., Raphaeli, D.: Can D.S.A. be improved? complexity trade-offs with the Digital Signature Standard. In: De Santis, A. (ed.) EUROCRYPT 1994. LNCS, vol. 950, pp. 77–85. Springer, Heidelberg (1995)

22. Pastuszak, J., Michalek, D., Pieprzyk, J., Seberry, J.: Identification of bad signatures in batches. In: Santis, A.D. (ed.) PKC 2000. LNCS, vol. 1751, pp. 28–45. Springer, Heidelberg (2000)

23. Pintsov, L., Vanstone, S.: Postal revenue collection in the digital age. In: Frankel, Y. (ed.) FC 2000. LNCS, vol. 1962, pp. 105–120. Springer, Heidelberg (2000)

24. Shanks, D.: Class number, a theory of factorization and genera. In: Symposium on Pure Mathematics, vol. 20, pp. 415–440. AMS, Providence (1971)

25. Solinas, J.: Identity-based digital signature algorithms. In: 7th Workshop on Elliptic Curve Cryptography, ECC 2003 (August 2003) (invited talk)

26. Stanek, M.: Attacking LCCC batch verification of RSA signatures. Cryptology ePrint Archive, Report 2006/111 (2006), http://eprint.iacr.org/2006/111

27. Yen, S., Laih, C.: Improved digital signature suitable for batch verification. IEEE Transactions on Computers 44(7), 957–959 (1995)

28. Yi, X.: An identity-based signature scheme from the weil pairing. IEEE Communications Letters 7(2) (Feburary 2003)

29. Yoon, H., Cheon, J.H., Kim, Y.: Batch verifications with ID-based signatures. In: Park, C.-s., Chee, S. (eds.) ICISC 2004. LNCS, vol. 3506, pp. 223–248. Springer, Heidelberg (2005)

30. Zaverucha, G., Stinson, D.: Group testing and batch verification. In: Kurosawa, K. (ed.) ITCS 2009. LNCS, vol. 5973, pp. 140–157. Springer, Heidelberg (2010)

31. Zhang, C., Lu, R., Lin, X., Ho, P.H., Shen, X.: An efficient identity-based batch verification scheme for vehicular sensor networks. In: The 27th IEEE International Conference on Computer Communications, INFOCOM 2008 (2008)

Oblivious Transfer with Access Control : Realizing Disjunction without Duplication

Ye Zhang[1], Man Ho Au[2], Duncan S. Wong[3], Qiong Huang[3], Nikos Mamoulis[1], David W. Cheung[1], and Siu-Ming Yiu[1]

[1] Department of Computer Science, University of Hong Kong, Hong Kong
{yzhang4,nikos,dcheung,smyiu}@cs.hku.hk
[2] School of Computer Science and Software Engineering, University of Wollongong, Australia
aau@uow.edu.au
[3] Department of Computer Science, City University of Hong Kong, Hong Kong
duncan@cityu.edu.hk, csqhuang@gmail.com

Abstract. Oblivious Transfer with Access Control (AC-OT) is a protocol which allows a user to obtain a database record with a credential satisfying the access policy of the record while the database server learns nothing about the record or the credential. The only AC-OT construction that supports policy in *disjunctive form* requires duplication of records in the database, each with a different *conjunction* of attributes (representing one possible criterion for accessing the record). In this paper, we propose a new AC-OT construction secure in the standard model. It supports policy in disjunctive form directly, without the above duplication issue. Due to the duplication issue in the previous construction, the size of an encrypted record is in $O(\prod_{i=1}^{t} n_i)$ for a CNF policy $(A_{1,1} \vee \ldots \vee A_{1,n_1}) \wedge \ldots \wedge (A_{t,1} \vee \ldots \vee A_{t,n_t})$ and in $O(\binom{n}{k})$ for a k-of-n threshold gate. In our construction, the encrypted record size can be reduced to $O(\sum_{i=1}^{t} n_i)$ for CNF form and $O(n)$ for threshold case.

1 Introduction

When a user tries to obtain a record from a database, a conventional database server knows which record is being accessed and whether the user has access right to the record. However, for some applications, user privacy is a concern. In an outsourced medical database (e.g., Google Health [7]), knowing which records a user has accessed may leak private information about the user's medical condition to the service provider (i.e., Google). Also, knowing the user's access rights may provide hints to the service provider on what records the user may want to access. If the user has access right to diabetes related records, it is very likely that the user may have related medical issues. *Oblivious Transfer with Access Control (AC-OT)* [4] is a protocol designed for providing solutions to these user privacy problems.

In an AC-OT protocol, there is a database server, an issuer and a set of users. The issuer issues credentials to users where credentials are attributes which

M. Joye, A. Miyaji, and A. Otsuka (Eds.): Pairing 2010, LNCS 6487, pp. 96–115, 2010.

specify the access rights of the users. The database server has a database of records, each record is encrypted by the database server under a record-specific access policy $\mathbb{A}$. This encrypted database is accessible to all the users, e.g., by posting it onto a public web site. A user with credentials S (i.e. an attribute set) can obtain a record R anonymously by running the AC-OT transfer protocol with the database server if S satisfies the access policy $\mathbb{A}$ of R. The database server learns nothing except the fact that the user with proper credentials has obtained a record.

The access policy supported in the previous AC-OT construction [4] is for the conjunction of attributes (e.g., Student $\wedge$ CS Dept.). To realize disjunctive policy, same records may need to be duplicated and appear for multiple times in the database. Suppose the policy of a record R is $(A_1 \wedge A_2 \wedge A_3) \vee (B_1 \wedge B_2 \wedge B_3)$. R needs to be duplicated so that one encrypted R associates with a conjunctive policy $A_1 \wedge A_2 \wedge A_3$ and the other one with a conjunctive policy $B_1 \wedge B_2 \wedge B_3$. For a CNF policy $(A_{1,1} \vee \ldots \vee A_{1,n_1}) \wedge \ldots \wedge (A_{t,1} \vee \ldots \vee A_{t,n_t})$, the scheme in [4] produces a set of encrypted duplicated records of size $O(\prod_{i=1}^{t} n_i)$. For a k-of-n threshold policy, requiring at least k attributes in the set $\{A_1, \ldots, A_n\}$, the scheme produces a set of size $O(\binom{n}{k})$.

Our Result: In this paper, we propose a new AC-OT protocol and show that it is secure in the standard model. The construction supports policy in disjunctive form directly, without duplicating records. For policies expressed in CNF or threshold gate, the construction produces a smaller size of encrypted database. Specifically, the size is in $O(\sum_{i=1}^{t} n_i)$ for CNF type and in $O(n)$ for k-of-n threshold type of policies. Our construction idea is to use a signature in an oblivious transfer protocol which is then integrated with a ciphertext-policy attribute-based encryption (CP-ABE) scheme for supporting policies in the form of CNF and k-of-n threshold.

1.1 Related Work

In [6], Coull et al. considers AC-OT in a stateful environment, in which, the access policy (e.g., Biba and Bell-LaPadual) is represented by a graph. Each node in the graph is a state and each edge models a transaction from one state to another. Each user in a stateful environment is also assigned an initial state via a stateful anonymous credential. By using zero-knowledge proof, a user proves to the database server that he/she is in possession of a state and tries to access a record where the record corresponds to an edge from the user's current state. After the user obtains the record, the database server updates the user's credential to a new state. However, the scheme will be less efficient if it is used in a stateless environment which is considered in [4] and this paper. According to [4], an access policy in a stateless environment can be represented as a graph with nodes for each subset of attributes that a user could have access to, and with a self-loop edge for each record that can be accessed using this subset. If there are C attributes and N records, the graph will have 2^C nodes and up to N self-loop edges for each node, and the encrypted database will be of size $O(2^C N)$. Also, users have to update their credentials after each access, since it is fundamentally stateful.

The AC-OT construction proposed in this paper relies on a fully-simulatable adaptive oblivious transfer (OT) protocol and a ciphertext-policy attribute-based encryption (CP-ABE). A fully-simulatable adaptive OT is an adaptive OT in which the security is defined in a simulation based model (i.e., following an ideal-world/real-world paradigm). The first fully-simulatable adaptive OT was proposed by Camenisch et al. [5]. Two other protocols are due to Green and Hohenberger [8] and Jarecki and Liu [10]. In our construction, we employ the OT of [5] due to the special use of an unforgeable signature in the OT.

In [2], Bethencourt et al. proposed the first CP-ABE. It supports monotonic access structures and the security is proven in a selective model, where an adversary submits the challenge access policy $\mathbb{A}^*$ before obtaining the public key of CP-ABE. The CP-ABE used in our construction requires full security. Lewko et al. in [11] proposed the first fully secure CP-ABE in the standard model. The scheme supports access policies with linear secret sharing scheme (LSSS) [11,1].

2   Definitions and Security Model

2.1   Syntax

Let $k \in \mathbb{N}$ be a security parameter. Let $[a, b]$ be the set $\{i \in \mathbb{N} | a \leq i \leq b\}$ where $a, b \in \mathbb{Z}$. Let $y \xleftarrow{\$} A(x)$ be the assignment of y to the output of a probabilistic algorithm A on input x and a fresh random tape. We say that a function $f(k)$ is negligible in k if for all polynomial $p(k)$, there exists k' such that $f(k) < \frac{1}{p(k)}$ when all $k > k'$. Without loss of generality, we define a universe of attributes $\mathcal{U} = \{1, \ldots, |\mathcal{U}|\}$ and denote each attribute as an element of $\mathcal{U}$. Therefore, an attribute set $S \subseteq \mathcal{U}$. We also define an access policy $\mathbb{A}$ as a collection of attribute sets, i.e. $\mathbb{A} \subseteq 2^{\mathcal{U}} \backslash \{\}$.

An Oblivious Transfer with Access Control (AC-OT) protocol is a tuple of probabilistic polynomial-time (PPT) algorithms/protocols:

$ISetup(1^k)$: This issuer setup algorithm generates a public/secret key pair (pk_I, sk_I). The issuer runs the algorithm and publishes pk_I.

$DBSetup(pk_I, DB = (R_i, AP_i)_{i=1,\ldots,N})$: In a database DB with $N \in \mathbb{N}$ records (where N is another security parameter), $R_i \in \{0, 1\}^*$ is the i-th record and AP_i is the access policy of R_i. This database setup algorithm generates a public/secret key pair (pk_{DB}, sk_{DB}) and encrypts R_i to encrypted records C_i with respect to AP_i. The database server runs this algorithm and publishes $(ER_i)_{i=1,\ldots,N} = (C_i, AP_i)_{i=1,\ldots,N}$ along with its public key pk_{DB}.

$Issue$: A user U engages in this protocol with the issuer. The inputs of U are an attribute set $S \subseteq \mathcal{U}$ and pk_I; and the input of the issuer is (pk_I, sk_I). By executing this protocol, U will obtain a credential $Creds_S \in \{0, 1\}^*$ corresponding to S. We assume that the issuer has already authenticated U with respect to S.

$Transfer$: U engages in this protocol with the database server. The inputs of U are an index $\sigma \in [1, N]$, ER_σ, $Creds_S$ (w.r.t. S), pk_I and pk_{DB} where S satisfies the access policy AP_σ of ER_σ. The input of the database server is

(pk_{DB}, sk_{DB}). By executing this protocol, U will obtain R_σ if the execution is successful, otherwise, U will get $\perp$ (an error signal).

2.2 Security Model

The security of AC-OT is defined in a simulation-based model. In the model, there is a real world and an ideal world. In the real world, all parties communicate using a real AC-OT protocol π. Some of the parties are corrupted and controlled by an adversary $\mathcal{A}$. We call these parties as "dishonest parties". Other parties follow π honestly and are called "honest parties". In the ideal world, all honest parties and an adversary $\mathcal{A}'$ communicate by sending their outputs to, and receiving inputs from, a party T, which cannot be corrupted.

We say that π *securely implements* the functionality T if for any real-world adversary $\mathcal{A}$, there exists an ideal-world adversary $\mathcal{A}'$ such that no PPT distinguisher (the environment) Z can tell whether it interacts with $\mathcal{A}$ in the real world or with $\mathcal{A}'$ in the ideal world, with non-negligible probability. The environment Z provides inputs to all parties and interacts with the adversary arbitrarily.

Real world: Now we define how honest parties in the real world follow the AC-OT protocol π. The database server and issuer do not return anything to Z; only the users return to Z.

1. The issuer I runs $ISetup(1^k)$ to generate (pk_I, sk_I) and publishes pk_I.
2. Upon receiving $(issue, S)$ from environment Z, where $S \subseteq \mathcal{U}$, an honest user U engages in the Issue protocol with I. After running the protocol, U sends a bit b to Z indicating whether the protocol run is successful ($b = 1$) or not ($b = 0$). Note that if $b = 1$, U has obtained a credential $Cred_S$ for S.
3. When receiving $(initDB, DB = (R_i, AP_i)_{i=1,\ldots,N})$ from environment Z, the database server runs $DBSetup(pk_I, DB)$ to generate (pk_{DB}, sk_{DB}) and then creates encrypted records $(C_i)_{i=1,\ldots,N}$. It publishes $\{ER_i = (C_i, AP_i)\}_{i=1,\ldots,N}$ and pk_{DB}.
4. Upon receiving $(transfer, \sigma)$ from environment Z, U checks whether it has a credential corresponding to an attribute set satisfying AP_σ. If so, U engages in the Transfer protocol with the database server. To this end, U obtains R_σ if the protocol run is successful; otherwise, U receives an error signal $\perp$ from the database server. U also sends R_σ or $\perp$ to environment Z.

Ideal world: All parties communicate with each other via T. When receiving $(issue, \ldots)$, $(initDB, \ldots)$ or $(transfer, \ldots)$ from environment Z, honest parties forward it to T, then forward the outputs of T to Z. We now define the behavior of T. T maintains an attribute set S_U which is initially empty for each user U. T also sets $DB = \perp$.

1. Upon receiving $(issue, S)$ from U, T sends $(issue, U, S)$ to issuer I who, in turn, sends back a bit b. If $b = 1$, T sets $S_U = S$. Otherwise, T does nothing.
2. Upon receiving $(initDB, (R_i, AP_i)_{i=1,\ldots,N})$ from the database server, T sets $DB = (R_i, AP_i)_{i=1,\ldots,N}$.

3. Upon receiving $(transfer, \sigma)$ from U, T checks whether $DB =\bot$. If $DB \neq\bot$, it sends $(transfer)$ to the database server. If the database server sends back $b = 1$, T checks if $\sigma \in [1, N]$ and S_U satisfies the access policy AP_σ. If so, T sends R_σ to U. Otherwise, it sends $\bot$ to U.

3   Preliminaries

3.1   Bilinear Maps and Assumptions

Let $\mathbb{G}$ and $\mathbb{G}_T$ be two cyclic multiplicative groups of order n (which can be prime or composite). A bilinear map is defined as $\hat{e} : \mathbb{G} \times \mathbb{G} \to \mathbb{G}_T$ with the following properties: (1) Bilinear: for all $u, v \in \mathbb{G}$ and $a, b \in \mathbb{Z}$, $\hat{e}(u^a, v^b) = \hat{e}(u, v)^{ab}$; (2) Non-degenerate: if g is a generator of $\mathbb{G}$, then $\hat{e}(g, g) \neq 1_{\mathbb{G}_T}$; and (3) Computable: there exists an efficient algorithm to compute $\hat{e}(u, v)$ for any $u, v \in \mathbb{G}$.

Definition 1. (ℓ-strong Diffie-Hellman (ℓ-SDH) assumption) *This assumption holds in $\mathbb{G}$ if for all PPT adversaries $\mathcal{A}$, the advantage $\boldsymbol{Adv}_{\mathbb{G}}^{\ell\text{-}SDH}(k)$:*

$$\boldsymbol{Adv}_{\mathbb{G}}^{\ell\text{-}SDH}(k) = \Pr[\mathcal{A}(g, g^x, \ldots, g^{x^\ell}) = (c, g^{1/(x+c)})]$$

is negligible in k where $g \xleftarrow{\$} \mathbb{G}^*$, $x \xleftarrow{\$} \mathbb{Z}_p$ and $c \in \mathbb{Z}_p$.

Definition 2. (ℓ-power Decisional Diffie-Hellman (ℓ-PDDH) assumption) *It holds in $(\mathbb{G}, \mathbb{G}_T)$ if for all PPT adversaries $\mathcal{A}$, the advantage $\boldsymbol{Adv}_{\mathbb{G},\mathbb{G}_T}^{\ell\text{-}PDDH}(k)$:*

$$\left| \Pr[\mathcal{A}(g, g^\alpha, \ldots, g^{\alpha^\ell}, H, H^\alpha, \ldots, H^{\alpha^\ell}) = 1] - \Pr[\mathcal{A}(g, g^\alpha, \ldots, g^{\alpha^\ell}, H, H_1, \ldots, H_\ell) = 1] \right|$$

is negligible in k where $g \xleftarrow{\$} \mathbb{G}^*$, $H, H_1, \ldots, H_\ell \xleftarrow{\$} \mathbb{G}_T^*$ and $\alpha \xleftarrow{\$} \mathbb{Z}_p$.

The ℓ-PDDH assumption is implied by $(\ell+1)$-BDHE assumption [4].

3.2   Building Blocks

In our construction, we employ a fully secure CP-ABE and a fully-simulatable adaptive OT. Below are the definitions of these two building blocks.

CP-ABE. It consists of four PPT algorithms ($Setup_{ABE}$, $GenKey_{ABE}$, Enc_{ABE}, Dec_{ABE}) [2,11]. The setup algorithm, $Setup_{ABE}(1^k)$ generates a master public/secret key pair (pk, mk). The key generation algorithm, $GenKey_{ABE}$ (mk, S), takes the master secret key mk and an attribute set $S \subseteq \mathcal{U}$ outputs a decryption key dk. The encryption algorithm, $Enc_{ABE}(pk, M, \mathbb{A})$, takes pk, a message $M \in \{0,1\}^*$ and an access policy $\mathbb{A}$, produces a ciphertext C. The decryption algorithm, $Dec_{ABE}(pk, C, dk)$, takes pk, decryption key dk and C, outputs M if S associated with dk satisfies $\mathbb{A}$, which is associated with C. The security model (full security) [2,11] is given as follows.

Definition 3. *A CP-ABE is fully secure if for all PPT adversaries $\mathcal{A}$, the advantage of $\mathcal{A}$ in the game below is negligible in k.*

Setup: The challenger runs $Setup_{ABE}(1^k)$ and gives pk to $\mathcal{A}$.

Phase 1: $\mathcal{A}$ may query for the decryption keys of attribute sets $S_1, \ldots, S_{q_1} \subseteq \mathcal{U}$.

Challenge: $\mathcal{A}$ submits two equal-length messages $M_0, M_1 \in \{0,1\}^*$ and a challenge access policy $\mathbb{A}^*$ such that none of the sets $S_1, \ldots, S_{q_1}$ satisfies $\mathbb{A}^*$. The challenger flips a random coin $b \in \{0,1\}$ and encrypts M_b with respect to $\mathbb{A}^*$. The ciphertext C^* is given to $\mathcal{A}$.

Phase 2: Same as Phase 1 with the restriction that none of the additional attribute sets $S_{q_1+1}, \ldots, S_q$ satisfies $\mathbb{A}^*$.

Guess: $\mathcal{A}$ outputs $b' \in \{0,1\}$. The advantage of $\mathcal{A}$ is defined as $|\Pr[b' = b] - \frac{1}{2}|$.

Fully-Simulatable Oblivious Transfer. We employ the fully-simulatable adaptive OT due to Camenisch et al. [5]. The sender, on input $k \in \mathbb{N}$ and messages $M_1, \ldots, M_N \in \{0,1\}^*$, randomly generates g, h from $\mathbb{G}^*$ and calculates $H = \hat{e}(g, h)$. It also randomly chooses x from $\mathbb{Z}_p$ and calculates $y = g^x$. The sender's public key $pk = (\hat{e}, \mathbb{G}, \mathbb{G}_T, p, g, y, H)$ and secret key $sk = (h, x)$. For $i = 1, \ldots, N$, the sender calculates $A_i = g^{\frac{1}{x+i}}$, $B_i = \hat{e}(h, A_i) \cdot M_i$ and sets $C_i = (A_i, B_i)$. The sender sends $C_1, \ldots, C_N$ to the receiver along with pk. The sender also shows that $PK\{(h) : H = \hat{e}(g, h)\}$.

When the receiver wants to obliviously transfer a message M_σ where index $\sigma \in [1, N]$, it randomly chooses v from $\mathbb{Z}_p$ and calculates $V = A_\sigma^v$. The receiver sends V along with $PK\{(\sigma, v) : \hat{e}(V, y) = \hat{e}(V, g)^{-\sigma}\hat{e}(g, g)^v\}$ to the sender. The sender verifies it and calculates $W = \hat{e}(h, V)$. It sends W along with $PK\{(h) : H = \hat{e}(g, h) \wedge W = \hat{e}(h, V)\}$ to the receiver. The receiver verifies it and calculates $M_\sigma = \frac{B_\sigma}{W^{1/v}}$.

The security model of Camenisch et al. scheme [5] is fully simulatable, meaning that both the sender and the receiver security are formalized by a simulation-based definition. Full simulatability is required even if the receiver can adaptively choose the message to obliviously transfer, based on those messages it has received from the sender. [5] proves that the above scheme is fully simulatable secure under the $(N+1)$-SDH and the $(N+1)$-PDDH assumptions in the standard model.

Note that $A_i = g^{\frac{1}{x+i}}$ is a modified Boneh-Boyen signature [5] on the message i with the signer's secret key x. [5] mentions that such a signature scheme is unforgeable under the weak chosen message attack [3,5] provided that the $(N+1)$-SDH assumption holds in $\mathbb{G}$, meaning that the receiver in the oblivious transfer protocol cannot forge a valid $A_i = g^{\frac{1}{x+i}}$ by herself.

4   Our Construction

In this section, we first devise a new AC-OT construction (Sec. 4.1) which employs a fully secure CP-ABE. This new AC-OT construction supports the same access policies of the underlying CP-ABE. Then, we discuss the security of the construction. At last, we instantiate the construction (Sec. 4.1) with a concrete CP-ABE.

4.1   A New AC-OT Construction

In our construction, we combine a ciphertext-policy attribute-based encryption with Camenisch et al.'s oblivious transfer to provide the access control and oblivious transfer functionality in an AC-OT protocol. Interestingly, [4] also uses Camenisch et al.'s OT to build their AC-OT construction. However, we should note that the reason why we choose Camenisch et al.'s OT is quite different from [4]. This reason also relates to the idea how we implement access control and why our AC-OT construction can support disjunction policy directly, but [4] cannot.

Recall that given the i-th record, Camenisch et al.'s OT construction generates a modified Boneh-Boyen signature $A_i = g^{\frac{1}{(i+x)}}$ (where x is the secret key). The user proves to the database server in a zero-knowledge fashion that she tries to access the i-th record. [4] extends this idea by embedding a conjunction policy $c_1 \wedge \ldots \wedge c_l$ into $A_i = g^{\frac{1}{(i+x+x_1c_1+\ldots+x_lc_l)}}$ (where now $(x, x_1, \ldots, x_l)$ is the secret key). Then, the user can prove in zero-knowledge fashion that she has the attributes $c_1, \ldots, c_l$ and tries to access the i-th record.

However, it seems hard to find an expression allowing to embed a disjunction policy into A_i. To overcome this problem, we do not embed an access policy into A_i, but instead we introduce ciphertext-policy attribute-based encryption (CP-ABE) to the methodology of devising an AC-OT construction. Specifically, we encrypt $A_i = g^{1/(i+x)}$ using CP-ABE under the access policy of the i-th record and let the database server distribute the ciphertext to the users. Now the user can conditionally release A_i provided that she has a decryption key (which acts as credentials in our construction) associated with attributes satisfy the policy. Then, the user can use the decryption result A_i in a Camenisch's oblivious transfer protocol to obtain record R_i. It is easy to see that such an AC-OT construction supports the same access policy of the underlying CP-ABE, which allows the construction to support disjunction policy directly.

The security of our AC-OT construction relies on the fact that A_i is a modified Boneh-Boyen signature, which cannot be forged by the users. Therefore, the only way to obtain A_i is done by a proper CP-ABE decryption. To our knowledge, the only fully-simulatable adaptive oblivious transfer including a signature in it is the construction proposed by Camenisch et al. [5]. That is the reason why we choose Camenisch et al.'s OT. Since the database setup algorithm is non-interactive, we postpone the proof-of-knowledge $PK\{(h) : H = \hat{e}(g, h)\}$ in [5] to the Transfer protocol (this trick is used in [4] as well.); We also encrypt $C_i = (A_i, B_i)$ in OT rather than distributing them directly to the users. The details of this construction is as follows.

$ISetup(1^k)$: Given a security parameter $k \in \mathbb{N}$, the issuer setup algorithm runs $Setup_{ABE}(1^k)$ to generate a pair of keys (pk_I, sk_I). The issuer publishes pk_I to all parties.

$Issue$: The user sends f_I, which is initially set to 0, to the issuer. If f_I is 0, the issuer gives $PK\{(sk_I) : (sk_I, pk_I)$ is a key pair$\}$ to the user. The user also updates $f_I = 1$. Then, the user sends an attribute set S to the issuer. The issuer runs $dk_S \overset{\$}{\leftarrow} GenKey_{ABE}(sk_I, S)$ to generate a decryption key. The issuer sends dk_S to the user as a credential $Creds_S$ corresponding to S.

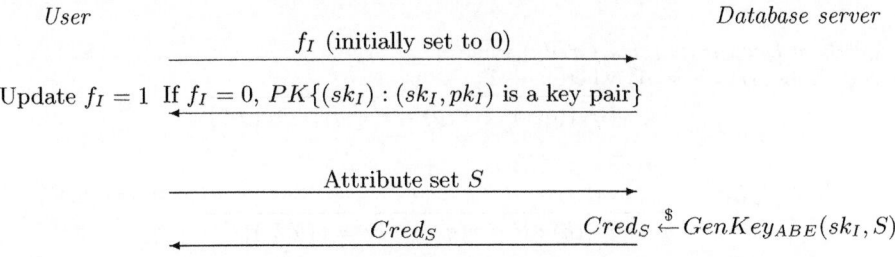

Fig. 1. Issue Protocol

To prove the issuer's knowledge on her private key is important in our security proof because we need to construct an ideal-world adversary who extracts the private key to decrypt for any CP-ABE ciphertexts.

$DBSetup(pk_I, (R_i, AP_i)_{i=1,...,N})$: The database setup algorithm first chooses $\mathbb{G}$ and $\mathbb{G}_T$ with the same prime order p. It also chooses a bilinear map $\hat{e}$: $\mathbb{G} \times \mathbb{G} \to \mathbb{G}_T$. It randomly chooses g, h from $\mathbb{G}$ and randomly chooses x from $\mathbb{Z}_p$. It also calculates $H = \hat{e}(g, h)$ and $y = g^x$. For each $i = 1, \ldots, N$, this algorithm calculates $C_i = (A_i, B_i)$ where $A_i = g^{\frac{1}{x+i}}$ and $B_i = \hat{e}(A_i, h)R_i$. It sets the public key $pk_{DB} = (\hat{e}, \mathbb{G}, \mathbb{G}_T, p, g, y, H)$ and the secret key $sk_{DB} = (h, x)$.

Then, it runs $D_i = Enc_{ABE}(pk_I, C_i, AP_i)$ which encrypts C_i under the policy AP_i. The database server publishes $(ER_i)_{i=1,...,N} = (D_i, AP_i)_{i=1,...,N}$ and pk_{DB} to all users. It stores pk_{DB} and keeps sk_{DB} as secret.

$Transfer$: It is shown by the Fig. 2. The user U first decrypts D_σ to C_σ using $Creds_S$ corresponding to the attribute set S.

Then, U randomly chooses v from $\mathbb{Z}_p$ and calculates $V = A_\sigma^v$. The user U sends V to the database server along with a zero-knowledge proof-of-knowledge $PK\{(v, \sigma) : \hat{e}(V, y) = \hat{e}(V, g)^{-\sigma}\hat{e}(g, g)^v\}$. Then, the database server verifies the proof and calculates $W = \hat{e}(h, V)$ via its $sk_{DB} = (h, x)$. The database server sends W along with $PK\{(h) : H = \hat{e}(g, h) \wedge W = \hat{e}(h, V)\}$ to U. U obtains $R_\sigma = \frac{B_\sigma}{W^{1/v}}$.

4.2 Security

Theorem 1. *The AC-OT protocol described in subsection 4.1 securely implements the AC-OT functionality, provided that the underlying CP-ABE is fully secure, the $(N + 1)$-SDH assumption holds in $\mathbb{G}$, the $(N + 1)$-PDDH assumption holds in $\mathbb{G}$ and $\mathbb{G}_T$, the knowledge error of the underlying PK is negligible and the underlying PK is perfect zero-knowledgeness.*

$$\textit{User} \hspace{8cm} \textit{Database server}$$

$$\xrightarrow{\hspace{2cm} f_{DB} \ (\text{initially set to } 0) \hspace{2cm}}$$

$$\text{Update } f_{DB} = 1 \quad \xleftarrow{\hspace{1cm} \text{If } f_{DB} = 0, \ PK\{(h) : H = \hat{e}(g,h)\} \hspace{1cm}}$$

$A_\sigma \| B_\sigma = Dec_{ABE}(pk_I, D_\sigma, Creds) \text{ where}$
$S \text{ satisfies } AP_\sigma. \ V = A_\sigma^v \text{ where } v \xleftarrow{\$} \mathbb{Z}_p. \qquad V$

$$\xrightarrow{\hspace{2cm} PK\{(\sigma, v) : \hat{e}(V, y) = \hat{e}(V, g)^{-\sigma} \hat{e}(g, g)^v\} \hspace{1cm}}$$

$R_\sigma = \dfrac{B_\sigma}{W^{1/v}} \qquad \xleftarrow{\hspace{1.5cm} W \hspace{1.5cm}} \qquad W = \hat{e}(h, V)$
$$PK\{(h) : H = \hat{e}(g,h) \wedge W = \hat{e}(h, V)\}$$

Fig. 2. Transfer Protocol

We organize the proof to Theorem 1 into different cases. In each case, some of the parties (i.e. the users, the issuer and the database server) are assumed to be corrupted and controlled by an adversary. We do not consider the cases where all parties are honest/dishonest or where the issuer is the only honest/dishonest party, as these cases do not have practical impact.

For each case, we assume that there exists a real-world adversary $\mathcal{A}$ and show how to construct an ideal-world adversary $\mathcal{A}'$ such that no PPT environment Z can tell whether it interacts with $\mathcal{A}$ in the real world or $\mathcal{A}'$ in the ideal world.

Following the strategy of game-hopping, we define a sequence of hybrid games $Game_0$ to $Game_n$. In $Game_i$ $(i = 1, \ldots, n)$, we construct a simulator $\mathcal{S}_i$ that runs $\mathcal{A}$ as a subroutine and provides the view for the environment Z. $Game_0$ models the case of the real world, while in $Game_n$, the simulator $\mathcal{S}_n$ can be used to construct an ideal-world adversary $\mathcal{A}'$. More specifically, the adversary $\mathcal{A}'$ runs $\mathcal{A}$ as a subroutine and provides the same view of $\mathcal{S}_n$ to the adversary $\mathcal{A}$. $\mathcal{A}'$ also simulates the real-world honest parties that communicate with $\mathcal{A}$. We prove that $Game_i$ and $Game_{i+1}$ are indistinguishable for $i = 0$ to $n - 1$, which means $\mathcal{A}$ in the real world ($\mathcal{S}_0$) and $\mathcal{A}'$ in the ideal world ($\mathcal{S}_n$) are indistinguishable by the environment Z. The details of the proof are given in Appendix A.

4.3   Instantiating with Concrete CP-ABE

In the above, we proposed an AC-OT protocol that employs (any fully secure) CP-ABE to encrypt $A_i \| B_i$ where $A_i \in \mathbb{G}$ and $B_i \in \mathbb{G}_T$. However, as we aware that message spaces in the most ciphertext-policy attribute-based encryption (CP-ABE) schemes are restricted to $\overline{\mathbb{G}}_T$ (which may or may not be identical to $\mathbb{G}_T$). In this subsection, we first show, given a CP-ABE scheme, how to employ the idea of "hybrid encryption" to devise a new CP-ABE which does not only support the same access policy as the original one but also supports an unbounded message space. More specifically, given a CP-ABE scheme $(Setup_{ABE}, GenKey_{ABE}, Enc_{ABE}, Dec_{ABE})$ and a data encapsulation mechanism (DEM) (Enc_{DEM}, Dec_{DEM}), we construct a new CP-ABE scheme $(Setup, GenKey, Enc, Dec)$ as follows.

1. $Setup(1^k)$: It runs $(pk, mk) \overset{\$}{\leftarrow} Setup_{ABE}(1^k)$ and outputs (pk, mk) as the public and master secret key pair.
2. $GenKey(mk, S)$: It runs $dk_S \overset{\$}{\leftarrow} GenKey_{ABE}(mk, S)$ and outputs dk_S.
3. $Enc(pk, M, \mathbb{A})$: Given $M \in \{0,1\}^*$, it first randomly chooses K from $\overline{\mathbb{G}}_T$ (i.e., the key space of DEM is assumed to be $\overline{\mathbb{G}}_T$). It computes $C_1 \overset{\$}{\leftarrow} Enc_{ABE}(pk, K, \mathbb{A})$. It also calculates $C_2 \overset{\$}{\leftarrow} Enc_{DEM}(K, M)$ and outputs the ciphertext $C = (C_1, C_2)$.
4. $Dec(pk, C, dk_S)$: Denote C as (C_1, C_2). It first runs $K = Dec_{ABE}(pk, C_1, dk_S)$. Then, it computes $M = Dec_{DEM}(K, C_2)$ and outputs M as the decryption result.

Theorem 2. *The CP-ABE construction described above is fully secure (under CPA), provided that the underlying original CP-ABE scheme is fully secure (under CPA) and DEM is one-time indistinguishability (IND-OT) [9] secure.*

More specifically, given an adversary $\mathcal{A}$ in the full security game of the new CP-ABE, we can construct an adversary $\mathcal{B}_1$ in the full security game of the original CP-ABE and an adversary $\mathcal{B}_2$ in the IND-OT game of DEM and show that $\mathbf{Adv}_{\mathcal{A}}(k) \leq 2\mathbf{Adv}_{\mathcal{B}_1}(k) + \mathbf{Adv}_{\mathcal{B}_2}(k)$. Since the original CP-ABE is fully secure (under CPA) and DEM is IND-OT secure, $\mathbf{Adv}_{\mathcal{B}_1}(k)$ and $\mathbf{Adv}_{\mathcal{B}_2}(k)$ are negligible in security parameter $k \in \mathbb{N}$, which completes the proof. The detailed proof is omitted from the paper.

Since DEM required in Theorem 2 is IND-OT secure, one-time pad is enough. $Enc_{DEM}(K, M)$ can be done by first employing a pseudo-random bit generator G to stretch the K and then outputting the ciphertext as $G(K) \oplus M$ where $\oplus$ is XOR operation. Theoretically, a pseudo-random bit generator can be built from any one-way function. Practically, we can use a block cipher with counter (CRT) mode or output feedback (OFB) mode to do this, provided that we model the block cipher as a pseudo-random permutation.

Next, we employ a concrete CP-ABE construction in [11] to instantiate the AC-OT construction in subsection 4.1. Recall that [11] supports any linear secret sharing scheme (LSSS) access policy $\mathbb{A} = (A, \rho)$ where A is an $l \times n$ matrix and ρ is a map from each row A_x of A to an attribute $\rho(x) \in \mathcal{U}$. Attribute set $S \subseteq \mathcal{U}$ satisfies $\mathbb{A} = (A, \rho)$ if and only if there exist constants $\{\omega_i\}_{\rho(i) \in S}$ such that $\sum_{\rho(i) \in S} \omega_i A_i = (1, 0, \ldots, 0)$. The AC-OT construction instantiated with Lewko et al.'s CP-ABE [11] is as follows.

$ISetup(1^k)$: Given a security parameter $k \in \mathbb{N}$, the issuer setup algorithm chooses a bilinear group $\overline{\mathbb{G}}$ with a composite order $N' = p_1 p_2 p_3$ where p_1, p_2 and p_3 are primes. It chooses $\overline{\mathbb{G}}_T$ with the same order N' and a bilinear map $\overline{e} : \overline{\mathbb{G}} \times \overline{\mathbb{G}} \to \overline{\mathbb{G}}_T$. It also randomly chooses $\alpha, a \in \mathbb{Z}_{N'}$ and randomly chooses $u \in \overline{\mathbb{G}}_{p_1}$. For each attribute $i \in \mathcal{U}$, it randomly chooses $s_i \in \mathbb{Z}_{N'}$. The public key pk_I is $(N', u, u^a, Y = \overline{e}(u, u)^\alpha, \{T_i = u^{s_i}\}_{i \in \mathcal{U}})$. The (master) secret key sk_I is (α, X_3) where X_3 is a generator of $\overline{\mathbb{G}}_{p_3}$. The issuer publishes pk_I to all parties.

Issue : The user sends f_I, which is initially set to 0, to the issuer. If f_I is 0, the issuer gives $PK\{(\alpha, p_1, p_2, p_3) : Y = \bar{e}(u,u)^\alpha \wedge N' = p_1p_2p_3\}$ to the user.[1] The user also updates $f_I = 1$. Then, the user sends an attribute set S to the issuer. The issuer randomly chooses $t \in \mathbb{Z}_{N'}$ and randomly chooses $\overline{R}_0, R'_0, \overline{R}_i$ from $\overline{\mathbb{G}}_{p_3}$ for $i \in S$. It outputs a credential $Creds_S$ as the decryption key $dk_S = (S, u^\alpha u^{at}\overline{R}_0, u^t R'_0, \{T_i^t \overline{R}_i\}_{i \in S})$. The issuer sends $Creds_S$ to the user.

$DBSetup(pk_I, (R_i, AP_i)_{i=1,...,N})$: The database setup algorithm first chooses $\mathbb{G}$ and $\mathbb{G}_T$ with the same prime order p. It also chooses a bilinear map $\hat{e} : \mathbb{G} \times \mathbb{G} \to \mathbb{G}_T$. It randomly chooses g, h from $\mathbb{G}$ and randomly chooses x' from $\mathbb{Z}_p$. It also calculates $H = \hat{e}(g, h)$ and $y = g^{x'}$. For each $i = 1, \ldots, N$, this algorithm calculates $C_i = (A_i, B_i)$ where $A_i = g^{\frac{1}{x'+i}}$ and $B_i = \hat{e}(A_i, h)R_i$. It sets the public key $pk_{DB} = (\hat{e}, \mathbb{G}, \mathbb{G}_T, p, g, y, H)$ and the secret key $sk_{DB} = (h, x')$.

For each $i = 1, \ldots, N$, it parses AP_i as (A, ρ) where A is an $l \times n$ matrix and ρ is a map from each row $\boldsymbol{A_x}$ of A to an attribute $\rho(x) \in \mathcal{U}$. Then, it randomly chooses κ from $\overline{\mathbb{G}}_T$. It also randomly chooses a vector $\boldsymbol{\nu} = (s, v_2, \ldots, v_n) \in \mathbb{Z}_{N'}^n$. For each row $\boldsymbol{A_x}$ of A, it randomly chooses $r_x \in \mathbb{Z}_{N'}$. It calculates $D_{i,1} = (A, \rho, \kappa\bar{e}(u,u)^{\alpha s}, u^s, \{u^{a\boldsymbol{A_x}\cdot\boldsymbol{\nu}}T_{\rho(x)}^{-r_x}, u^{r_x}\}_x)$. It also runs $Enc_{DEM}(\kappa, A_i\|B_i)$ to obtain $D_{i,2}$. Note that $\kappa, \boldsymbol{\nu}$ and r_x (for each x) are chosen freshly for every $i = 1, \ldots, N$.

The database server publishes $(ER_i)_{i=1,...,N} = (D_i = (D_{i,1}, D_{i,2}), AP_i)_{i=1,...,N}$ and pk_{DB} to all users. It stores pk_{DB} and keeps sk_{DB} as secret.

$Transfer$: Denote $D_{\sigma,1} = (A, \rho, C, C', \{\overline{C}_x, D_x\}_x)$ and $Creds_S = (S, K, L, \{K_i\}_{i \in S})$. The user U first computes constants $\omega_x \in \mathbb{Z}_{N'}$ such that $\sum_{\rho(x) \in S} \omega_x \boldsymbol{A_x} = (1, 0, \ldots, 0)$. It also computes $\bar{e}(C', K)/\prod_{\rho(x) \in S}(\bar{e}(\overline{C}_x, L)\bar{e}(D_x, K_{\rho(x)}))^{\omega_x} = \bar{e}(u,u)^{\alpha s}$. Then, it recovers $\kappa = C/\bar{e}(u,u)^{\alpha s}$ and runs $Dec_{DEM}(\kappa, D_{\sigma,2})$ to obtain $C_\sigma = (A_\sigma, B_\sigma)$. Note that if S satisfies (A, ρ), U will find ω_x efficiently [11].

U randomly chooses v from $\mathbb{Z}_p$ and calculates $V = A_\sigma^v$. The user U sends V to the database server along with a zero-knowledge proof-of-knowledge $PK\{(v, \sigma) : \hat{e}(V, y) = \hat{e}(V, g)^{-\sigma}\hat{e}(g, g)^v\}$. Then, the database server verifies the proof and calculates $W = \hat{e}(h, V)$ via its $sk_{DB} = (h, x')$. The database server sends W along with $PK\{(h) : H = \hat{e}(g, h) \wedge W = \hat{e}(h, V)\}$ to U. U obtains $R_\sigma = \frac{B_\sigma}{W^{1/v}}$.

It is easy to see that the security of the above AC-OT construction is a corollary of Theorem 1 and 2.

5   Performance

Given the concrete construction in subsection 4.3, we analyze the encrypted record size for the access policies of CNF formula and threshold gate. We should

---

[1] Strictly speaking, this is not identical to $PK\{(sk_I) : (sk_I, pk_I)$ is a key pair $\}$. However, to prove $N' = p_1p_2p_3$ would be efficient. More importantly, recall that in the security proof, the constructed ideal-world adversary just needs to extract sk_I from the PK. Given p_1, p_2, p_3, the adversary can generate a valid sk_I by itself.

note that our construction is not restricted to express CNF formula and threshold gate, but any policies which can be expressed by a LSSS matrix.

Given a general CNF formula $(A_{1,1} \vee \ldots \vee A_{1,n_1}) \wedge \ldots \wedge (A_{t,1} \vee \ldots \vee A_{t,n_t})$, we can first represent it by an access tree whose interior nodes are AND and OR gates and leaf nodes are attributes (e.g., $A_{1,1}$). It is easy to see that such an access tree has $n_1 + \ldots + n_t$ leaf nodes.

Lemma 1. *For an access policy which can be expressed by an access tree whose interior nodes are AND and OR gates and leaf nodes are attributes, the ciphertext size in the CP-ABE construction [11] is $O(n)$ where n is the number of leaf nodes in that access tree.*

Lemma 1 is given in [11]. The encrypted (i-th) record D_i consists of two components $D_{i,1}$ and $D_{i,2}$ where $D_{i,1}$ is a CP-ABE ciphertext under the above access tree policy for CNF formula. Therefore, $D_{i,1}$ is of size of $O(n_1 + \ldots + n_t)$. $D_{i,2}$ is a DEM ciphertext whose size is a constant (e.g, $O(1)$). To sum up, the encrypted record size in our AC-OT construction is $O(\sum_{i=1}^{t} n_i)$. We also show a comparison for our construction and [4] (and [6]) in Table 1 for completeness. Note that due to the use of duplication strategy in [4], the record of above CNF formula will appear for $n_1 \times \ldots \times n_t$ times.

Table 1. Comparison of AC-OT protocols expressing CNF formula

Protocol	Encrypted record size
[6]	$O(2^{\sum_{i=1}^{t} n_i})$
[4]	$O(\prod_{i=1}^{t} n_i)$
This paper	$O(\sum_{i=1}^{t} n_i)$

Given a threshold gate T_k^n: with at least k attributes in the set $\{A_1, \ldots, A_n\}$, we have the following lemma.

Lemma 2. *For any $1 \leq k \leq n$, a LSSS matrix A for threshold gate T_k^n can be constructed with n rows.*

Specifically, we can construct a $n \times k$ matrix $A = (a_1, \ldots, a_n)^T$ where a_i ($i = 1, \ldots, n$) are k-length vectors such that any k of them consist of a base. Therefore, if more than k attributes (without loss of generality, we assume the attributes correspond to $a_{i_1}, \ldots, a_{i_k}, \ldots, a_{i_j}$), then we can find $\omega_1, \ldots, \omega_k$ such that $(1, 0, \ldots, 0) = \sum_{l=1}^{k} \omega_k a_{i_l}$. However, if there are less than k attributes, $a_{i_1}, \ldots, a_{i_j}$ ($j < k$) is linear independent with $(1, 0, \ldots, 0)$. The proof to Lemma 2 is omitted in this paper.

In the construction (Sec. 4.3), the size of $D_{i,1}$ is $O(n)$ if the LSSS matrix A has n rows. Consequently, the encrypted record size of a T_k^n threshold gate is $O(n)$ in our AC-OT construction. The comparison for our construction and [4] (and [6]) is shown in Table 2. The record will appear for $\binom{n}{k}$ times in [4] due to its duplication strategy.

Table 2. Comparison of AC-OT protocols expressing threshold gate

Protocol	Encrypted record size
[6]	$O(2^n)$
[4]	$O(\binom{n}{k})$
This paper	$O(n)$

6   Conclusion

In this paper, we proposed a new AC-OT construction which is secure in the standard model. Our construction is based on an observation that Camenisch et al.'s OT construction contains an unforgeable signature, which allows us to conditionally release the signature with a ciphertext-policy attribute-based encryption. Without duplicating records, our construction reduces the size of the encrypted database by a substantial amount for access policies represented in CNF or k-of-n threshold gate.

Acknowledgments

This work was supported by Grant HKU 715509E from Hong Kong RGC. We thank the anonymous reviewers for their valuable comments.

References

1. Beimel, A.: Secure schemes for secret sharing and key distribution. PhD thesis, Israel Institute of Technology, Technion, Haifa, Israel (1996)
2. Bethencourt, J., Sahai, A., Waters, B.: Ciphertext-policy attribute-based encryption. In: 28th IEEE Symposium on Security and Privacy, pp. 321–334. IEEE Press, New York (2007)
3. Boneh, D., Boyen, X.: Short signatures without random oracles. In: Cachin, C., Camenisch, J. (eds.) EUROCRYPT 2004. LNCS, vol. 3027, pp. 56–73. Springer, Heidelberg (2004)
4. Camenisch, J., Dubovitskaya, M., Neven, G.: Oblivious transfer with access control. In: 16th ACM Conference on Computer and Communications Security, pp. 131–140. ACM, New York (2009)
5. Camenisch, J., Neven, G., Shelat, A.: Simulatable adaptive oblivious transfer. In: Naor, M. (ed.) EUROCRYPT 2007. LNCS, vol. 4515, pp. 573–590. Springer, Heidelberg (2007)
6. Coull, S., Green, M., Hohenberger, S.: Controlling access to an oblivious database using stateful anonymous credentials. In: Jarecki, S., Tsudik, G. (eds.) PKC 2009. LNCS, vol. 5443, pp. 501–520. Springer, Heidelberg (2009)
7. Google Inc.: Google health, https://www.google.com/health
8. Green, M., Hohenberger, S.: Practical adaptive oblivious transfer from a simple assumption. Cryptology ePrint Archive, Report 2010/109 (2010), http://eprint.iacr.org/

9. Herranz, J., Hofheinz, D., Kiltz, E.: KEM/DEM: Necessary and sufficient conditions for secure hybrid encryption. Cryptology ePrint Archive, Report 2006/265 (2006), http://eprint.iacr.org/

10. Jarecki, S., Liu, X.: Efficient oblivious pseudorandom function with applications to adaptive OT and secure computation of set intersection. In: Reingold, O. (ed.) TCC 2009. LNCS, vol. 5444, pp. 577–594. Springer, Heidelberg (2009)

11. Lewko, A., Okamoto, T., Sahai, A., Takashima, K., Waters, B.: Fully secure functional encryption: Attribute-based encryption and (hierarchical) inner product encryption. In: Gilbert, H. (ed.) EUROCRYPT 2010. LNCS, vol. 6110, pp. 62–91. Springer, Heidelberg (2010), full version http://eprint.iacr.org/2010/110

A Proof of Theorem 1

A.1 Proof of Case 1 (The Database Server and the Issuer Are Dishonest)

Lemma 3. *For all environments Z and any real-world adversary $\mathcal{A}$ controlling the database server and the issuer, there exists an ideal-world adversary $\mathcal{A}'$ such that the probability that Z can distinguish whether it interacts with $\mathcal{A}$ in the real world or it interacts with $\mathcal{A}'$ in the ideal world is negligible, provided that the knowledge error of the underlying PK is negligible and the underlying PK is perfect zero-knowledgeness.*

Game$_0$. The real world. $\mathcal{S}_0$ plays the role of the honest users.

Game$_1$. $\mathcal{S}_1$ is the same as $\mathcal{S}_0$ except that at the first time of issue query instructed by environment Z, $\mathcal{S}_1$ runs the extractor of $PK\{(sk_I) : sk_I \text{ and } pk_I \text{ is a key pair}\}$ to extract sk_I from $\mathcal{A}$. If it fails, $\mathcal{S}_1$ outputs $\bot$ to Z. Recall that sk_I and pk_I is the (master) secret key and the public key of the underlying CP-ABE.

Note that the difference between *Game$_0$* and *Game$_1$* is negligible provided that the underlying PK is sound. We believe that constructing such PK for most CP-ABE should be easy.

Game$_2$. $\mathcal{S}_2$ is the same as $\mathcal{S}_1$ except that at the first time of transfer query instructed by environment Z, $\mathcal{S}_2$ runs the extractor of $PK\{(h) : H = \hat{e}(g, h)\}$ to extract h from $\mathcal{A}$. If it fails, $\mathcal{S}_2$ outputs $\bot$ to Z.

The difference between *Game$_1$* and *Game$_2$* is negligible provided that the underlying PK is sound.

Game$_3$. $\mathcal{S}_3$ is the same as $\mathcal{S}_2$ except that at each time of transfer query instructed by environment Z, $\mathcal{S}_3$ engages in Transfer protocol with $\mathcal{A}$ to query a record randomly chosen from those which it has the necessary decryption keys, rather than querying σ instructed by Z.

Note that *Game$_2$* and *Game$_3$* is identical due to the perfect zero-knowledgeness of the underlying $PK\{(v, \sigma) : \hat{e}(V, y) = \hat{e}(V, g)^{-\sigma}\hat{e}(g, g)^{v}\}$.

Now we show how to construct the ideal-world adversary $\mathcal{A}'$ with black-box access to $\mathcal{A}$, where $\mathcal{A}'$ incorporates all steps from *Game$_3$*.

Adversary $\mathcal{A}'$ first runs $\mathcal{A}$ to obtain $\{ER_i\}$ and public parameters. At the each issue query instructed by Z, $\mathcal{A}'$ will simulate a user interacting with $\mathcal{A}$ for issuing the decryption key. If the decryption key is valid, $\mathcal{A}'$ will send $b = 1$ to T; otherwise, it will send $b = 0$. If it is the first time of Issue protocol, $\mathcal{A}'$ will also run the extractor with $\mathcal{A}$ to extract sk_I of the underlying CP-ABE (As we have shown, the probability that the extractor fails is negligible).

When receiving $(transfer)$ from T, $\mathcal{A}'$ will query a record randomly chosen from those which it has the necessary decryption keys with $\mathcal{A}$. If the Transfer protocol succeeds, $\mathcal{A}'$ will send back $b = 1$ to T; otherwise, it sends back $b = 0$. If it is the first time of transfer query instructed by Z, $\mathcal{A}'$ will also run the extractor to extract h with $\mathcal{A}$ (As we have shown, the extractor fails with only negligible probability).

If $\mathcal{A}'$ has extracted sk_I and h from $\mathcal{A}$, $\mathcal{A}'$ will use sk_I to decrypt ER_i to $A_i \| B_i$ where $B_i = \hat{e}(h, A_i)R_i$. Then, $\mathcal{A}'$ calculates $R_i = \frac{B_i}{\hat{e}(h, A_i)}$. $\mathcal{A}'$ sends $(R_i, AP_i)_{i=1,\dots,N}$ to T for initDB.

A.2   Proof of Case 2 (Only the Database Server Is Dishonest)

Lemma 4. *For all environments Z and any real-world adversary $\mathcal{A}$ controlling the database server, there exists an ideal-world adversary $\mathcal{A}'$ such that the probability that Z can distinguish whether it interacts with $\mathcal{A}$ in the real world or it interacts with $\mathcal{A}'$ in the ideal world is negligible, provided that the knowledge error of the underlying PK is negligible and the underlying PK is perfect zero-knowledgeness.*

$Game_0$. The real world. $\mathcal{S}_0$ plays the role of the honest users and the issuer to $\mathcal{A}$.

$Game_1$. $\mathcal{S}_1$ is the same as $\mathcal{S}_0$ except that at the first time of transfer query, $\mathcal{S}_1$ runs the extractor of $PK\{(h) : H = \hat{e}(g, h)\}$ with $\mathcal{A}$ to extract h. If it is failed, $\mathcal{S}_1$ outputs $\perp$ to Z.

The difference between $Game_0$ and $Game_1$ is negligible provided that the underlying PK is sound.

$Game_2$. $\mathcal{S}_2$ is the same as $\mathcal{S}_1$ except that at each time of transfer query, $\mathcal{S}_2$ engages in the Transfer protocol with $\mathcal{A}$ to query a record randomly chosen from those which it has the necessary decryption keys, rather than querying σ instructed by Z.

$Game_1$ is identical $Game_2$ due to the zero-knowledgeness of the underlying $PK\{(v, \sigma) : \hat{e}(V, y) = \hat{e}(V, g)^{-\sigma}\hat{e}(g, g)^v\}$.

Now we show how to construct the ideal-world adversary $\mathcal{A}'$. Adversary $\mathcal{A}'$ first runs $\mathcal{A}$ to obtain $\{ER_i\}$ and public parameters. At each time of transfer query, $\mathcal{A}'$ simulates a user interacting with $\mathcal{A}$ in Transfer protocol to query a record randomly chosen from those which it has the necessary decryption keys. If Transfer protocol succeeds, $\mathcal{A}'$ sends $b = 1$ to T; otherwise, it sends $b = 0$. If it is the first time of transfer query, $\mathcal{A}'$ also runs the extractor of

$PK\{(h) : H = \hat{e}(g, h)\}$ with $\mathcal{A}$ to extract h (As we have shown that the extractor fails with only negligible probability).

If $\mathcal{A}'$ has extracted h from $\mathcal{A}$, $\mathcal{A}'$ (since it also simultaneously simulates the issuer) uses its (master) secret key sk_I to decrypt $\{ER_i\}$ to $A_i\|B_i$ where $B_i = \hat{e}(h, A_i)R_i$. $\mathcal{A}'$ calculates $R_i = \frac{B_i}{\hat{e}(h, A_i)}$. $\mathcal{A}'$ also sends $(R_i, AP_i)_{i=1,\ldots,N}$ to T for initDB.

A.3   Proof of Case 3 (Only Some Users Are Dishonest)

Lemma 5. *For all environments Z and any real-world adversary $\mathcal{A}$ controlling some users, there exists an ideal-world adversary $\mathcal{A}'$ such that the probability that Z can distinguish whether it interacts with $\mathcal{A}$ in the real world or it interacts with $\mathcal{A}'$ in the ideal world is negligible, provided that the underlying CP-ABE is fully secure, the $(N + 1)$-SDH assumption and the $(N + 1)$-PDDH assumption hold, the knowledge error of the underlying PK is negligible and the underlying PK is perfect zero-knowledgeness.*

$Game_0$. The real world. $\mathcal{S}_0$ plays the role of the honest issuer and database server to $\mathcal{A}$.

$Game_1$. $\mathcal{S}_1$ follows the specification except during each transfer query, when $\mathcal{A}$ requests for a record R_σ from the database server on behalf of some user. $\mathcal{A}$ submits V along with the proof of v, σ:

$$PK\{(v, \sigma) : \hat{e}(V, y) = \hat{e}(V, g)^{-\sigma}\hat{e}(g, g)^v\}.$$

$\mathcal{S}_1$ extracts v and σ and continues the rest of the simulation only if extraction is successful. Otherwise, it outputs $\perp$ to Z.

The difference between $Game_0$ and $Game_1$ is negligible provided that the underlying PK is sound.

$Game_2$. $\mathcal{S}_2$ is the same as $\mathcal{S}_1$ except after v, σ has been extracted, $\mathcal{S}_2$ computes $\hat{A}_\sigma = V^{1/v}$. If $\mathcal{A}$ has never request the decryption key for attribute set S satisfying AP_σ from the issuer and that $\hat{A}_\sigma = A_\sigma$ (notice that $\mathcal{S}_2$ simulates the real-world honest database server to adversary $\mathcal{A}$ and $\mathcal{S}_2$ generates $A_1, \ldots, A_N$ by herself.), $\mathcal{S}_2$ outputs $\perp$ to Z.

The difference between $Game_1$ and $Game_2$ is negligible provided that the underlying CP-ABE is fully secure and the $(N + 1)$-SDH assumption holds.

If $\mathcal{S}_2$ obtains the correct A_σ from $\mathcal{A}$ such that $\mathcal{A}$ is not given the corresponding CP-ABE decryption key, we can use $\mathcal{A}$ to break the full security of the underlying CP-ABE. More specifically, we show how to construct an adversary $\mathcal{B}$ that wins the full security game of the underlying CP-ABE. $\mathcal{B}$ plays the role of $\mathcal{S}_2$ and is with black-box access to $\mathcal{A}$.

The challenger first runs $Setup_{ABE}(1^k)$ and gives the public key pk to $\mathcal{B}$. $\mathcal{B}$ submits AP_σ to the challenger where AP_i ($i = 1, \ldots, N$) is instructed by the environment Z. $\mathcal{B}$ also computes $A_i = g^{\frac{1}{x+i}}$ and $B_i = \hat{e}(h, A_i)R_i$ for $i = 1, \ldots, N$

(Note that $\mathcal{B}$ runs Z as subroutine and simulates the real-world honest database server.). Then $\mathcal{B}$ randomly chooses two random numbers r_A from $\mathbb{G}$ and r_B from $\mathbb{G}_T$. $\mathcal{B}$ sets $M_0 = A_\sigma \| B_\sigma$ and $M_1 = r_A \| r_B$. $\mathcal{B}$ submits M_0 and M_1 to the challenger. The challenger flips a coin $b \in \{0, 1\}$ and encrypts M_b to C^* under AP_σ. The ciphertext C^* is given to $\mathcal{B}$. $\mathcal{B}$ also encrypts $A_i \| B_i$ to ciphertext C_i under AP_i for all $i = 1, \ldots, \sigma - 1, \sigma + 1, \ldots, N$. $\mathcal{B}$ publishes $(C_1, AP_1), \ldots, (C_{\sigma-1}, AP_{\sigma-1}), (C^*, AP_\sigma), (C_{\sigma+1}, AP_{\sigma+1}), \ldots, (C_N, AP_N)$ to $\mathcal{A}$. When $\mathcal{A}$ asks for decryption keys, $\mathcal{B}$ will forward the requests to the challenger. Recall that $\mathcal{A}$ does not request the decryption key for attribute set S satisfying AP_σ from the issuer, the challenger will answer all requests properly. Finally, if $\mathcal{A}$ outputs A_σ, then $\mathcal{B}$ will output $b' = 0$ as its guess bit; otherwise, $\mathcal{B}$ will output $b' = 1$.

When $b = 0$ ($\Pr[b = 0] = \frac{1}{2}$), C^* is the encryption of $A_\sigma \| B_\sigma$ and this is identical to $Game_2$. In this case, $\mathcal{A}$ will output A_σ with a non-negligible probability ϵ. When $b = 1$ ($\Pr[b = 1] = \frac{1}{2}$), C^* is encryption of random $r_A \| r_B$ which is independent with A_σ. $\mathcal{A}$ would not output proper A_σ, otherwise, $\mathcal{A}$ will forge a modified Boneh-Boyen signature under weak chosen message attack. This happens with a negligible probability η under the $(N + 1)$-SDH assumption [5].

$$\Pr[b = b'] = \Pr[b = 0]\Pr[b' = 0] + \Pr[b = 1]\Pr[b' = 1]$$
$$= \frac{1}{2}\epsilon + \frac{1}{2}(1 - \eta)$$
$$= \frac{1}{2} + \frac{1}{2}(\epsilon - \eta)$$

where $\frac{1}{2}(\epsilon - \eta)$ is non-negligible.

$Game_3$. $\mathcal{S}_3$ is the same as $\mathcal{S}_2$ except for each transfer query, it computes $W = (B_\sigma / R_\sigma)^v$ and that the proof-of-knowledge protocol

$$PK\{(h) : W = \hat{e}(V, h)\}$$

becomes a simulated proof such that $\mathcal{S}_3$ does not require the knowledge of h.

The difference between $Game_2$ and $Game_3$ is negligible provided that the underlying PK is perfect zero-knowledge.

$Game_4$. $\mathcal{S}_4$ is the same as $\mathcal{S}_3$ except $\mathcal{S}_4$ now deviates from the database setup protocol by replacing the value B_i from $\hat{e}(h, A_i)R_i$ to just random elements in group $\mathbb{G}_T$.

The difference between $Game_3$ and $Game_4$ is negligible provided that the $(N + 1)$-PDDH assumption holds.

Suppose there exists an environment Z that can distinguish $Game_3$ and $Game_4$, we show how to construct an adversary $\mathcal{B}$ that solves the $(N+1)$-PDDH problem. $\mathcal{B}$ is given the problem instance $g', g'^x, g'^{x^2}, \ldots, g'^{x^{N+1}}, H, H_1, \ldots, H_{N+1}$ and its task is to tell if $H_i = H^{x^i}$ or H_i are just random elements from $\mathbb{G}_T$ for $i = 1, \ldots, N + 1$.

$\mathcal{B}$ runs the environment Z and the adversary $\mathcal{A}$ as subroutines. It plays the role of the honest issuer and database server with $\mathcal{A}$ and Z. Denote $f(x) = \prod_{i=1}^{N}(x+i)$ as the N-degree polynomial in x. Under the additive notation, $f(x) = \sum_{i=0}^{N} \beta_i x^i$. Set $g = g'^{f(x)}$ and thus $g = \prod_{i=0}^{N} (g'^{x^i})^{\beta_i}$ and is computable by $\mathcal{B}$ without the knowledge of x. Next, the value $y = g^x$ is also computable as $g'^{xf(x)}$ since $xf(x)$ is a degree $N + 1$ polynomial. $\mathcal{B}$ sets the public key of the database server as $pk_{DB} = \{g, H, y, \hat{e}, \mathbb{G}, \mathbb{G}_T, p\}$

When it is instructed by environment Z to create a database with records $(R_i, AP_i)_{i=1,\dots,N}$, it generates $A_i = g^{\frac{1}{x+i}}$ by $A_i = g'^{\frac{f(x)}{x+i}}$. This is computable by $\mathcal{B}$ without the knowledge of x since $f(x)/(x + i) = (x + 1)\cdots(x + i - 1)(x + i + 1)\cdots(x + N)$ is a polynomial of degree $N - 1$. Next, it computes B_i as $(\prod H_i^{\beta_i})R_i$.

Note that if $H_i = H^{x^i}$ for $i = 1,\dots, N$, then $\mathcal{B}$ is acting as $\mathcal{S}_3$ in $Game_3$ while if H_i are just random elements in $\mathbb{G}_T$, $\mathcal{B}$ is acting as $\mathcal{S}_4$ in $Game_4$. Thus, if the environment Z can distinguish between $Game_3$ and $Game_4$, we have solved the $(N + 1)$-PDDH problem.

Now we show how to construct the ideal-world adversary $\mathcal{A}'$ which is given black-box access to $\mathcal{A}$.

In the beginning, Z tells the ideal-world honest database server to initialize the database with inputs (R_i, AP_i) for $i = 1$ to N. The honest database server forwards this request to the trusted party T who sets database as $(R_i, AP_i)_{i=1,\dots,N}$. While AP_i is made public to all ideal world users, R_i is kept secret. T also maintains a set S_U to record what set of attributes has been issued to the user U.

$\mathcal{A}'$ setups the key pairs for the issuer as well as the database server to provide the simulation for $\mathcal{A}$. However, it has to setup the database without knowing R_i. $\mathcal{A}'$ did so by setting ER_i as (C_i, AP_i), where C_i is the encryption of $A_i\|B_i$, with B_i being random element in $\mathbb{G}_T$. (As we have shown, Z cannot distinguish this difference.)

When Z issues the message $(issue, S)$ to $\mathcal{A}'$, $\mathcal{A}'$ sends the same message to $\mathcal{A}$ (Recall that $\mathcal{A}'$ acts as the environment to $\mathcal{A}$). If $\mathcal{A}$ deviates from the protocol and does nothing (remember $\mathcal{A}'$ also acts as the issuer to $\mathcal{A}$), $\mathcal{A}'$ also did not send any message to T. Otherwise, if $\mathcal{A}$ requests a certain set of credentials from $\mathcal{A}'$, $\mathcal{A}'$ requests the same set of credentials from the trusted party T. If T sends back 1 to $\mathcal{A}'$, $\mathcal{A}'$ issues the corresponding credential to $\mathcal{A}$. On the other hand, if T returns 0, $\mathcal{A}'$ returns 0 to $\mathcal{A}$ as well.

When Z issues the message $(transfer, \sigma)$ to $\mathcal{A}'$, $\mathcal{A}'$ sends the same message to $\mathcal{A}$. If $\mathcal{A}$ deviates from the protocol and does nothing ($\mathcal{A}'$ also acts as the issuer to $\mathcal{A}$), $\mathcal{A}'$ also did not send any message to T. Otherwise, $\mathcal{A}'$ extracts σ from $\mathcal{A}$. (As we have shown, $\mathcal{A}'$ fails with a negligible probability.)

$\mathcal{A}'$ requests for credentials from T that satisfy AP_σ (if there is no any user satisfying this policy). After that, $\mathcal{A}'$ requests for the record R_σ on behalf of that user. Next, it computes $W = (B_\sigma/R_\sigma)^v$ and sends W, along with the zero-knowledge proof, back to $\mathcal{A}$ on behalf of the database server. (Again, it is different from the actual honest database server. However, we have shown that $\mathcal{A}$ (and therefore Z) cannot notice this difference.)

In fact, one can notice that the simulation provided by $\mathcal{A}'$ to $\mathcal{A}$ is the same as the simulator $\mathcal{S}_5$ provided to $\mathcal{A}$. According to the above proof, Z cannot tell any difference between $\mathcal{A}$ and $\mathcal{A}'$.

A.4   Proof of Case 4 (The Issuer and Some Users Are Dishonest)

Lemma 6. *For all environments Z and any real-world adversary $\mathcal{A}$ controlling the issuer and some users, there exists an ideal-world adversary $\mathcal{A}'$ such that the probability that Z can distinguish whether it interacts with $\mathcal{A}$ in the real world or it interacts with $\mathcal{A}'$ in the ideal world is negligible, provided that the $(N+1)$-PDDH assumption holds, the knowledge error of the underlying PK is negligible and the underlying PK is perfect zero-knowledgeness.*

$Game_0$. The real world. $\mathcal{S}_0$ plays the role of honest database server and some honest users to $\mathcal{A}$.

$Game_1$. $\mathcal{S}_1$ is the same as $\mathcal{S}_0$ except that at each time of transfer query, when $\mathcal{A}$ requests for a record R_σ from the database server on behalf of some user. $\mathcal{A}$ submits V along with the proof of v, σ:

$$PK\{(v, \sigma) : \hat{e}(V, y) = \hat{e}(V, g)^{-\sigma}\hat{e}(g, g)^v\}.$$

$\mathcal{S}_1$ extracts v and σ. If $\mathcal{S}_1$ fails, it outputs $\perp$ to the environment Z.

The difference between $Game_0$ and $Game_1$ is negligible provided that the underlying PK is sound.

$Game_2$. $\mathcal{S}_2$ is the same as $\mathcal{S}_1$ except that for each time of transfer query, it computes $W = (B_\sigma/R_\sigma)^v$ and that the proof-of-knowledge protocol

$$PK\{(h) : W = \hat{e}(V, h)\}$$

becomes a simulated proof such that $\mathcal{S}_2$ does not require any knowledge of h.

The difference between $Game_1$ and $Game_2$ is negligible provided that the underlying PK is perfect zero-knowledge.

$Game_3$. $\mathcal{S}_3$ is the same as $\mathcal{S}_2$ except that $\mathcal{S}_3$ now deviates from the database setup protocol by replacing the value B_i from $\hat{e}(h, A_i)R_i$ to just random element in $\mathbb{G}_T$.

As we have shown, the difference between $Game_2$ and $Game_3$ is negligible provided that the $(N+1)$-PDDH assumption holds.

Now we show how to construct the ideal-world adversary $\mathcal{A}'$ with black-box access to $\mathcal{A}$. At each time of transfer query, $\mathcal{A}'$ extracts (v, σ) from $\mathcal{A}$ (As we have shown it fails with only negligible probability). $\mathcal{A}'$ requests the record R_σ on behalf of the user. $\mathcal{A}'$ queries T for the decryption key of attributes satisfying AP_σ. Since $\mathcal{A}'$ simultaneously acts as the dishonest issuer, $\mathcal{A}'$ will send back $b = 1$ to T for granting the decryption key. Next, $\mathcal{A}'$ computes $W = (B_\sigma/R_\sigma)^v$ and sends W, along with the simulated zero-knowledge proof, back to $\mathcal{A}$ on behalf of the database server (It is different from the actual honest database server. However, as we have shown that $\mathcal{A}$ (and therefore Z) cannot notice the difference.).

$\mathcal{A}'$ setups the key pair of the database server to provide the simulation for $\mathcal{A}$. However, it has to setup the database without knowing R_i. $\mathcal{A}'$ did so by setting $ER_i = (C_i, AP_i)$, where C_i is the encryption of $A_i \| B_i$, with B_i being random element in $\mathbb{G}_T$. As we have shown that Z cannot distinguish this difference.

B   Lewko et al.'s Fully-Secure CP-ABE [11]

$Setup(\lambda, \mathcal{U})$: The setup algorithm chooses a bilinear group $\mathbb{G}$ of order $N = p_1 p_2 p_3$ where p_1, p_2 and p_3 are 3 distinct primes. Let $\mathbb{G}_{p_i}$ denote the subgroup of order p_i in $\mathbb{G}$. It then chooses random exponents $\alpha, a \in \mathbb{Z}_N$ and a random group element $g \in \mathbb{G}_{p_1}$. For each attribute $i \in \mathcal{U}$, it chooses a random value $s_i \in \mathbb{Z}_N$. The public parameters $PK = (N, g, g^a, \hat{e}(g, g)^\alpha, \{T_i = g^{s_i}\}_{\forall i})$. The master secret key $MSK = (\alpha, X_3)$ where X_3 is a generator of $\mathbb{G}_{p_3}$.

$KeyGen(MSK, S, PK)$: The key generation algorithm chooses a random $t \in \mathbb{Z}_N$ and random elements $R_0, R_0', R_i \in \mathbb{G}_{p_3}$. The secret key is:

$$SK = (S, K = g^\alpha g^{at} R_0, L = g^t R_0', \{K_i = T_i^t R_i\}_{i \in S}).$$

$Encrypt((A, \rho), PK, M)$: A is an $\ell \times n$ matrix and ρ is map from each row $\boldsymbol{A_x}$ of A to an attribute $\rho(x)$. The encryption algorithm chooses a random vector $v \in \mathbb{Z}_N^n$, denoted $\boldsymbol{v} = (s, v_2, \ldots, v_n)$. For each row A_x of A, it chooses a random $r_x \in \mathbb{Z}_N$. The ciphertext is:

$$CT = (C = M\hat{e}(g, g)^{\alpha s}, C' = g^s, \{C_x = g^{a\boldsymbol{A_x} \cdot \boldsymbol{v}}, D_x = g^{r_x}\}_x).$$

$Decrypt(CT, PK, SK)$: The decryption algorithm computes constants $\omega_x \in \mathbb{Z}_N$ such that $\sum_{\rho(x) \in S} \omega_x \boldsymbol{A_x} = (1, 0, \ldots, 0)$. It then computes:

$$\hat{e}(C', K) / \prod_{\rho(x) \in S} (\hat{e}(C_x, L)\hat{e}(D_x, K_{\rho(x)}))^{\omega_x} = \hat{e}(g, g)^{\alpha s}.$$

Then M can be recovered as $C/\hat{e}(g, g)^{\alpha s}$.

Increased Resilience in Threshold Cryptography: Sharing a Secret with Devices That Cannot Store Shares

Koen Simoens, Roel Peeters, and Bart Preneel

Department of Electrical Engineering – COSIC
Katholieke Universiteit Leuven and IBBT, Belgium
firstname.lastname@esat.kuleuven.be

Abstract. Threshold cryptography increases security and resilience by sharing a private cryptographic key over different devices. Many personal devices, however, are not suited for threshold schemes, because they do not offer secure storage, which is needed to store shares of the private key. We present a solution that allows to include devices without them having to store their share. Shares are stored in protected form, possibly externally, which makes our solution suitable for low-cost devices with a factory-embedded key, e.g., car keys and access cards. By using pairings we achieve public verifiability in a wide range of protocols, which removes the need for private channels. We demonstrate how to modify existing discrete-log based threshold schemes to work in this setting. Our core result is a new publicly verifiable distributed key generation protocol that is provably secure against static adversaries and does not require all devices to be present.

1 Introduction

The increased capabilities of mobile devices and connectivity with the rest of the world have made the use of these devices exceed their original purpose. Mobile phones are being used to read e-mail, authorise bank transactions or access social network sites. As a consequence, personal devices are used more and more for security-sensitive tasks. Moreover, personal data are copied to these devices and need to be protected. In both cases, by using cryptography, security reduces to the management of cryptographic keys. Although mobility is considered as a major benefit, it is a weakness in terms of security and reliability. Mobile devices are susceptible to theft, can easily be forgotten or lost, or run out of battery power. These issues can be mitigated by introducing threshold cryptography.

The aim of threshold cryptography is to protect a key by sharing it amongst a number of entities in such a way that only a subset of minimal size, namely the threshold $t + 1$, can use the key. No information about the key can be learnt from t or less shares. The setup of a threshold scheme typically involves a Distributed Key Generation (DKG) protocol. In a DKG protocol a group of entities cooperate to jointly generate a key pair and obtain shares of the private key. These shares can then be used to sign or decrypt on behalf of the group.

M. Joye, A. Miyaji, and A. Otsuka (Eds.): Pairing 2010, LNCS 6487, pp. 116–135, 2010.

The benefits of a threshold scheme are increased security, because an adversary can compromise up to t devices, and resilience, since any subset of $t + 1$ devices is sufficient. To increase resilience we want to maximise the number of devices included in the threshold scheme. However, the number of personal devices suitable for threshold schemes is limited because many of these do not incorporate secure storage, which is needed to store shares of the private key. We enlarge the group of high-end devices by also considering small devices with public-key functionality, e.g., car keys or access cards. Typically, these small devices have a factory-embedded private key, which cannot be updated and is the only object that resides in tamper-proof secure storage.

Our proposed solution allows to store shares in protected form,[1] possibly externally. These protected shares are generated through a run of our new DKG protocol. By using pairings we achieve publicly verifiability, which implies that the correctness of any device's contribution can be verified by any entity observing the DKG protocol thus eliminating private channels. As such, not every device needs to be present during the DKG protocol. We apply our setting to existing threshold schemes and we show how shares can be used implicitly without being needed in unprotected form. Furthermore, some devices can be completely ignorant of the underlying schemes and only serve as partial decryption oracles.

Organisation. Related work is surveyed in Sect. 2. In Sect. 3 we introduce some basic concepts. We give an overview of typical routines in a threshold setting and we describe our communication and adversarial model. Security definitions are given along with an overview of relevant number-theoretic assumptions and notation on bilinear pairings. In Sect. 4 we present how to protect shares and our main result, which is a new publicly verifiable DKG protocol that does not require every device to be present. In Sect. 5 we demonstrate how protected shares can easily be used in discrete-log based cryptosystems and signature schemes. More specifically, we demonstrate this for the ElGamal [14] and the Cramer-Shoup [7] cryptosystems, and the Schnorr signature scheme [25].

2   Related Work

Shamir's early idea [27] of distributing shares of a secret as evaluations of a polynomial has become a standard building block in threshold cryptography. Feldman [8] introduced verifiable secret sharing (VSS) by publishing the coefficients of this polynomial hidden in the exponent of the generator of a group in which the discrete-log assumption holds. Pedersen [22] then used this idea to construct the first distributed key generation (DKG) protocol, sometimes referred to as Joint Feldman, by having each player in a group run an instance of Feldman's protocol in parallel. Soon thereafter, Pedersen [23] produced another remarkable result. He made Feldman's VSS scheme information-theoretically

---

[1] An obvious answer would be to encrypt shares under the devices' public keys. This is undesired because at some point shares will be in the clear in unprotected memory.

secure by choosing two polynomials and broadcasting the corresponding coefficients as paired commitments, which are known as Pedersen commitments. Gennaro et al. [15] pointed out that the uniformity of the key produced by Pedersen's DKG protocol cannot be guaranteed in the context of a rushing adversary. They constructed a new DKG protocol [15] by first running Pedersen's VSS in parallel (Joint Pedersen). Since Pedersen VSS does not produce a public key, an extra round of communication, basically an instance of Joint Feldman on the first polynomial, has to be added to compute the public key. They proved their protocol secure against a static adversary by means of a simulation argument. Interestingly, Gennaro et al. showed later [16] that, despite the biased distribution of the key, certain discrete-log schemes that use Pedersen DKG can still be proved secure at the cost of an increased security parameter. Canetti et al. [6] used interactive knowledge proofs and erasures, i.e., players erase private data before commitments or public values are broadcast, in the key construction phase of the DKG of [15] to make the protocol secure against adaptive adversaries. Comparable adaptively secure threshold schemes were presented by Frankel et al. [10].

In the protocols discussed so far, it is assumed that there are private channels between each pair of players. Both [6] and [10] suggest that these channels can still be established even with an adaptive adversary using the non-committing encryption technique of Beaver and Haber [3], which assumes erasures. Jarecki and Lysyanskaya [19] criticised this erasure model and pointed out that the protocols presented in [6] and [10] are not secure in the concurrent setting, i.e., two instances of the same scheme can not be run at the same time. They solved this by introducing a "committed proof", i.e., a zero-knowledge proof where the statement that is being proved is not revealed until the end of the proof. To implement the secure channels without erasures they use an encryption scheme that is non-committing to the receiver. Abe and Fehr [1] later proposed an adaptively-secure (Feldman-based) DKG and applications with complete security proofs in the Universal Composability framework of Canetti [5]. They demonstrated that a discrete-log DKG protocol can be achieved without interactive zero-knowledge proofs. However, they still need a single inconsistent player and a secure message transmission functionality (private channels), which can be realised using a receiver non-committing transmission protocol based on [19].

As a consequence of private channels, each of the aforementioned DKG protocols has some kind of complaint procedure or dispute resolution mechanism. To get rid of these, several authors have proposed protocols that provide public verifiability. Stadler [29] was of the first to propose a publicly verifiable secret sharing (PVSS) protocol. In addition to the Feldman commitments, shares were broadcast in encrypted form and verified using a non-interactive proof of equality of (double) discrete logarithms. A more efficient protocol was presented by Fujisaki and Okamoto [11], which is secure under a modified RSA assumption. The first PVSS shown secure under the Decisional Diffie-Hellman (DDH) assumption was given by Schoenmakers [26]. The shares are broadcast in encrypted form by hiding them in the exponent of each player's individual public key, which has a different base (another generator) than the Feldman commitments. The dealer then uses non-interactive proofs of discrete-log equality. Furthermore, correct behaviour of the players is verified by extending the secret reconstruction phase with additional proofs of correctness. Based on Schoenmakers' result Heidarvand and Villar [18] presented the first PVSS protocol where verifiability is obtained from bilinear pairings over elliptic curves and no proofs are needed. Unfortunately, the scheme cannot be used to set up a DKG because the shared secret is in the co-domain of the pairing. The first full DKG that does not require private channels was given by Fouque and Stern [9]. The buildings blocks for their construction are Paillier's cryptosystem and a new non-interactive zero-knowledge proof. To deal with rushing adversar it is simply assumed that communication is completely synchronous. For participants not present during the DKG the amount of information that needs to be stored, i.e., the subshares that need to be decrypted, is linear in the number of participants that are active in the DKG.

3 Basic Concepts

Before we describe our new protocols, we give an overview of basic concepts that will be used later on.

3.1 Threshold Cryptography

Threshold cryptography typically involves routines related to setting up the group, encryption and signatures. A private key is shared amongst n devices and only a subset of at least $t+1$ devices need to employ their shares to (implicitly) use this private key in a cryptosystem or signature scheme. We define the following set of routines (threshold routines are indicated with the prefix **T**):

Pre-setup.

- **Init:** Initialise the system parameters.
- **KeyGen:** Generate key material for a device.

Setup.

- **ConstructGroup:** Given a set of n devices and their public keys, create and share a key pair for the group with a subset of the devices.

Signatures.

- **T-Sign:** At least $t+1$ devices collaborate to generate a signature on a message that is verifiable under the group's public key.
- **Verify:** Using the group's public key a signature is verified.

Encryption.

- **Encrypt:** Encrypt a message under the group's public key.
- **T-Decrypt:** At least $t+1$ devices collaborate to decrypt a given ciphertext that was encrypted under the group's public key.

3.2   Communication and Adversarial Model

We assume that n devices $\{\mathcal{D}_i\}_{i=1..n}$, of which t can be faulty, communicate over a dedicated broadcast channel.[2] By dedicated we mean that if a device $\mathcal{D}_i$ broadcasts a message, then it is received by all other devices and recognised as coming from this device. There are no private channels, all communication goes over the broadcast channel. Communication is round-synchronous, protocols run in rounds and there is a time bound on each round.

A distinction is commonly made between static and adaptive adversaries. Static means that the adversary corrupts the devices before the protocol starts, whereas adaptive means that a device can become corrupt before or at any time during execution of a protocol. We assume a malicious computationally bounded static adversary who can corrupt up to t devices. The adversary has access to all information stored by the corrupted devices and can manipulate their behaviour during the execution of a protocol in any way. The round-synchronous communication implies that the adversary could be rushing, i.e., he can wait in each round to send messages on behalf of the corrupted devices until he has received the messages from all uncorrupted devices.

3.3   Security Definitions

In a secret sharing scheme a dealer splits a secret into pieces, called shares, and distributes them amongst several parties. In a threshold setting, the secret can be reconstructed from any subset of shares of a minimum size. An early solution was given by Shamir [27], who shared a secret x by choosing a random polynomial f of degree t such that $x = f(0)$ and each share is an evaluation of this polynomial, i.e., $x_i = f(i)$. Any point on a polynomial of degree t can be reconstructed by Lagrange interpolation through at least $t + 1$ points of this polynomial. To reconstruct the secret x, the shares are combined as $x = \sum \lambda_i x_i$, with λ_i the appropriate Lagrange multipliers.

Verifiable secret sharing (VSS) allows the receivers to verify that the dealer properly shared a secret. We briefly rephrase the requirements of a secure VSS ([23] and [15, Lemma 1]).

Definition 1 (Secure VSS). *A VSS protocol is secure if it satisfies the following conditions:*

1. *Correctness. If the dealer is not disqualified then any subset of $t + 1$ honest players can recover the unique secret.*
2. *Verifiability. Incorrect shares can be detected at reconstruction time by using the output of the protocol.*
3. *Secrecy. The view of a computationally bounded static adversary $\mathcal{A}$ is independent of the secret, or, the protocol is semantically secure against $\mathcal{A}$.*

The drawback of VSS is that a single party knows the secret. This can be solved by generating and sharing the key in a distributed way. The correctness and secrecy requirements for DKG were defined by Gennaro et al. [17].

---

[2] We abstract from the actual implementation of the dedicated broadcast channel.

Definition 2 (Secure DKG). *A DKG protocol is secure if it satisfies the following conditions:*
Correctness *is guaranteed if:*
(C1) All subsets of $t+1$ shares provided by honest players define the same private key.
(C2) All honest parties know the same public key corresponding to the unique private key as defined by (C1).
(C3) The private key (and thus also the public key) is uniformly distributed.
Secrecy *is guaranteed if an adversary can learn no information about the private key beyond what can be learnt from the public key. This requirement can be further enhanced with a simulation argument: for any adversary there should be a simulator that, given a public key, simulates a run of the protocol for which the output is indistinguishable of the adversary's view of a real run of the protocol that ended with the given public key.*

3.4   Pairings and Number-Theoretic Assumptions

We review some assumptions that are relevant for this paper and we refer the reader to [21] and [28] for more details.

Pairing Notation. Let $\mathbb{G}_1$, $\mathbb{G}_2$ and $\mathbb{G}_T$ be cyclic groups of order ℓ and let $\hat{e}$ be a non-degenerate bilinear pairing

$$\hat{e} : \mathbb{G}_1 \times \mathbb{G}_2 \to \mathbb{G}_T \ .$$

A pairing is non-degenerate if for each element P in $\mathbb{G}_1$ there is a Q in $\mathbb{G}_2$ such that $\hat{e}(P,Q) \neq 1$ and vice versa for each element Q in $\mathbb{G}_2$. A pairing is bilinear if $\hat{e}(P + P', Q) = \hat{e}(P,Q)\hat{e}(P',Q)$, thus $\hat{e}(aP,Q) = \hat{e}(P,Q)^a$ with $a \in \mathbb{Z}_\ell$, and vice versa for elements in $\mathbb{G}_2$. We will use multiplicative notation for $\mathbb{G}_T$ and additive notation for $\mathbb{G}_1$ and $\mathbb{G}_2$.

Discrete Logarithm. Let P be a generator of $\mathbb{G}_1$ and let Y be a given arbitrary element of $\mathbb{G}_1$. The discrete logarithm (DL) problem in $\mathbb{G}_1$ is to find the unique integer $a \in \mathbb{Z}_\ell$ such that $Y = aP$. Similarly, the problem can be defined in $\mathbb{G}_2$ and $\mathbb{G}_T$. The DL assumption states that it is computationally hard to solve the DL problem.

Diffie-Hellman. Let P be a generator of $\mathbb{G}_1$ and let aP, bP be two given arbitrary elements of $\mathbb{G}_1$, with $a, b \in \mathbb{Z}_\ell$. The computational Diffie-Hellman (CDH) problem in $\mathbb{G}_1$ is to find abP. The tuple $\langle P, aP, bP, abP \rangle$ is called a Diffie-Hellman tuple. Given a third element $cP \in \mathbb{G}_1$, the decisional Diffie-Hellman (DDH) problem is to determine if $\langle P, aP, bP, cP \rangle$ is a valid Diffie-Hellman tuple or not. Obviously, if one can solve the DL problem then one can also solve the CDH problem. The opposite does not necessarily hold and, therefore, the CDH assumption is said to be a stronger assumption than the DL assumption. A divisional variant of the DDH problem [2], which is considered to be equivalent, is to determine if $\langle P, aP, cP, abP \rangle$ is a valid DH tuple or not, i.e., if $c = b$.

Co-Bilinear Diffie-Hellman (coBDH). For asymmetric pairings, i.e., $\mathbb{G}_1 \neq \mathbb{G}_2$, where there is no known efficiently computable isomorphism $\psi : \mathbb{G}_2 \rightarrow \mathbb{G}_1$ the following problem can be defined. The coBDH-2 problem is defined as given $P \in \mathbb{G}_1$ and $Q, aQ, bQ \in \mathbb{G}_2$, find $\hat{e}(P,Q)^{ab}$. We denote the decisional variant as coDBDH-2. A divisional variant of the coDBDH-2 problem is to determine whether $\langle P, Q, aQ, abQ, g^c \rangle$ is a valid coBDH-2 tuple.

Inversion Problems. Galbraith et al. [12] studied several inversion problems for pairings. They concluded that these problems are hard enough to rely upon. The most intuitive argument is that if one can solve a particular pairing inversion in polynomial time then one can also solve a related Diffie-Hellman problem in one of the domains or the co-domain.

3.5   Pre-setup

The pre-setup phase is straightforward and works as follows.

Init(1^k)**:** The input is a security parameter k. Let $\mathbb{G}_1$, $\mathbb{G}_2$ and $\mathbb{G}_T$ be finite cyclic groups of prime order ℓ with P, Q and $g = \hat{e}(P,Q)$ generators of the respective groups. It is assumed that there is no known efficiently computable isomorphism $\psi : \mathbb{G}_2 \rightarrow \mathbb{G}_1$. Let P' and P'' be two other generators of $\mathbb{G}_1$ for which the discrete logarithm relative to the base P is unknown and let $g_1 = g$, $g_2 = \hat{e}(P',Q)$ and $g_3 = \hat{e}(P'',Q)$. The procedure outputs the description of the groups $(\mathbb{G}_1, \mathbb{G}_2, \mathbb{G}_T)$ and the pairing $(\hat{e})$ along with the public system parameters

$$PubPar = (P, P', P'', Q) \in \mathbb{G}_1^3 \times \mathbb{G}_2 \ .$$

KeyGen(*PubPar,*$\mathcal{D}_i$**):** For the given device $\mathcal{D}_i$ a random $s_i \in_R \mathbb{Z}_\ell^*$ is chosen as private key. The corresponding public key is $S_i = s_iQ$. The procedure outputs $\mathcal{D}_i$'s key pair

$$(s_i, S_i) \in \mathbb{Z}_\ell^* \times \mathbb{G}_2 \ .$$

Note that this procedure is executed only once in the lifetime of each $\mathcal{D}_i$ and that s_i is the only secret that has to be securely stored. Typically, this routine is executed during fabrication of the device. A public key can easily be computed for a different set of system parameters if this would be required.

4   Distributed Key Generation

In this section, we present our main result, which is a new distributed key generation (DKG) protocol. Recall from the introduction that we want to set up a threshold construction without the devices having to securely store their share. Instead, the shares will be stored in protected form. This idea is put forward in Sect. 4.1. Our DKG consists of two phases. First, a private key is jointly generated and shared through the parallel execution of a new publicly verifiable secret sharing (PVSS) protocol. This PVSS protocol is described in Sect. 4.2. Second, the corresponding public key is extracted. Together, these two phases make up our new DKG protocol, which is presented in Sect. 4.3.

4.1 Protecting Shares

As mentioned in Sect. 3.5, each device $\mathcal{D}_i$ is initialised with its own key pair (s_i, S_i). If the group's private key was $x \in \mathbb{Z}_\ell$, then the device would receive a share $x_i \in \mathbb{Z}_\ell$ that has to be securely stored. Since the device does not provide secure storage it has to store its share in protected form. One option is to encrypt the share x_i. This has the disadvantages of involving a costly decryption operation and the fact that the share will at some point reside in the clear in the device's memory. Another option is to store the share as the product $x_i s_i$. The obvious disadvantage is that $t + 1$ devices can collaborate to compute another device's private key s_i.

As we do not want a device's private key s_i ever to be revealed, we combine shares with the device's public key and store these as public correction factors $C_i = x_i S_i \in \mathbb{G}_2$. A similar idea was used in [26]. However, here we use bilinear pairings to achieve public verifiability and easy integration of our scheme in existing discrete-log cryptosystems and signature schemes, without ever having to reveal the shares (see Sect. 5). We define the group's private key as $xQ \in \mathbb{G}_2$ and $y = g^x = \hat{e}(P, xQ) \in \mathbb{G}_T$ as its public key.[3] As such, the share of a device is $x_i Q = s_i^{-1} C_i \in \mathbb{G}_2$. The construct group routine is formally defined as follows.

- **ConstructGroup(***PubPar***,{$\mathcal{D}_i, S_i$},t):** A subset of the devices $\mathcal{D}_i$ generates the group's public key g^x and shares the private key xQ in the form of public correction factors $C_i = x_i S_i$ for all n devices. The procedure outputs the group's public key

$$y = g^x = \hat{e}(P, xQ) \in \mathbb{G}_T$$

and the public correction factors which are added to the public parameters

$$PubPar = (P, P', P'', Q, y, \{C_i\}_{i=1,\dots,n}) \in \mathbb{G}_1^3 \times \mathbb{G}_2 \times \mathbb{G}_T \times \mathbb{G}_2^n \ .$$

4.2 Publicly Verifiable Secret Sharing

The main building block to construct our DKG protocol is a new PVSS protocol. In this protocol, a dealer generates shares of a secret and distributes them in protected form. Any party observing the output of the protocol can verify that the dealer behaved correctly. Basically, the protocol goes as follows.

The dealer chooses uniformly at random $x \in_R \mathbb{Z}_\ell$. The actual secret that is shared at the end of the protocol is xQ. Similar to Pedersen's VSS scheme [23], the dealer chooses two random polynomials f and f' of degree t, sets the constant term of polynomial f to x and broadcasts pairwise commitments $A_k \in \mathbb{G}_1$ to the coefficients of the polynomials. Evaluations of both polynomials are combined with the public keys of the devices and broadcast in protected form. Each device verifies that all broadcast shares are correct by applying the pairing to check them against the commitments. The details of the protocol are given in Fig. 1.

---

[3] We note that we could use notation X and X_i for the private key and its shares. However, to compute the correction factor C_i, elements of $\mathbb{Z}_\ell$ will be combined with S_i, but by definition, private key material is in $\mathbb{G}_2$.

The dealer shares the secret xQ, for which he chooses $x \in_R \mathbb{Z}_\ell$:

1. The dealer constructs two polynomials $f(z)$ and $f'(z)$ of degree t by choosing random coefficients $c_k, c'_k \in_R \mathbb{Z}_\ell^*$ for $k = 0 \ldots t$, except for c_0, which is $c_0 = x$:

$$f(z) = c_0 + c_1 z + \ldots + c_t z^t \quad , \quad f'(z) = c'_0 + c'_1 z + \ldots + c'_t z^t \ .$$

The dealer broadcasts commitments

$$A_k = c_k P + c'_k P' \quad , \quad k = 0 \ldots t \ .$$

2. For each device $\mathcal{D}_i$, the dealer computes and broadcasts

$$x_i S_i \ , \ x'_i S_i \quad \text{with} \quad x_i = f(i) \ , \ x'_i = f'(i) \quad , \quad i = 1 \ldots n \ .$$

3. Each device verifies the broadcast shares for all $\mathcal{D}_i$ by checking that

$$\hat{e}(P, x_i S_i) \cdot \hat{e}(P', x'_i S_i) = \prod_{k=0}^{t} \hat{e}(A_k, S_i)^{i^k} \ . \tag{1}$$

If any of these checks fails, the dealer is disqualified.

Fig. 1. Publicly verifiable secret sharing

Private channels are avoided because the shares $x_i Q$ are broadcast in protected form $x_i S_i$. Each device could recover its share by using its private key. However, the shares are never needed in unprotected form. The protected form allows for public verifiability, since for any device $\mathcal{D}_i$ the correctness of $x_i S_i$ and $x'_i S_i$ can be verified by pairing the commitments with $\mathcal{D}_i$'s public key S_i. The dealer is disqualified, if for any $\mathcal{D}_i$ this verification fails. As a consequence, there is no need for a cumbersome complaint procedure. Moreover, not all devices need to be present during the execution of the protocol because the shares were already broadcast in the form in which they will be stored and used.

In the next theorem we will demonstrate that our new PVSS protocol satisfies the requirements of secure VSS protocol as given by Definition 1.

Theorem 1. *Our new PVSS protocol is a secure VSS protocol (Definition 1) under the divisional variant of the DDH assumption in $\mathbb{G}_2$.*

Proof (Correctness). It follows directly from Pedersen's result [23] that each subset of $t + 1$ devices can reconstruct the coefficients $c_k Q, c'_k Q$ of the polynomials $F(z) = f(z)Q$ and $F'(z) = f'(z)Q$ from their shares. If the dealer is not disqualified then (1) holds for all devices and the coefficients will successfully be verified against the commitments A_k. Hence, it can be verified that all shares are on the same (respective) polynomial and each subset of $t + 1$ devices can compute the same secret $xQ = F(0)$. $\qquad\square$

Proof (Verifiability). During reconstruction $\mathcal{D}_i$ provides $x_i Q$ and $x'_i Q$, and it can be verified that $\hat{e}(P, x_i Q) \cdot \hat{e}(P', x'_i Q) = \prod_{k=0}^{t} \hat{e}(A_k, Q)^{i^k}$. $\qquad\square$

Proof (Secrecy). Consider a worst-case static adversary $\mathcal{A}$, i.e., an adversary that corrupts t devices before the protocol starts. The protocol is semantically secure against $\mathcal{A}$, if $\mathcal{A}$ chooses two values $x_0Q, x_1Q \in \mathbb{G}_2$ and cannot determine which of these two was shared with negligible advantage over random guessing, given the output of a run of the protocol that shared either the secret x_0Q or x_1Q. We prove the semantic security by showing that no such adversary exists.

If there exists an $\mathcal{A}$ that has a non-negligible advantage in attacking the semantic security of our protocol, we can build a simulator SIM that uses $\mathcal{A}$ to solve an instance of the divisional DDH problem in $\mathbb{G}_2$ (see Sect. 3.4). Since, this is assumed to be a hard problem we conclude that no such adversary can exist.

We now describe this simulator. A tuple $\langle Q, aQ, cQ, abQ \rangle$ is given to SIM who has to decide if this is a valid DH tuple, i.e., if $cQ = bQ$.

1. The simulator SIM does the pre-setup. He chooses the system parameters *PubPar*, which contain P and $P' = \eta P$, with η known to SIM. He constructs a set of devices $\mathcal{D}_i$, of which one will be the designated device, denoted as $\mathcal{D}_d$. For each $\mathcal{D}_i \neq \mathcal{D}_d$, SIM generates a random key pair. The public key of $\mathcal{D}_d$ is set to $S_d = cQ$.
2. The adversary $\mathcal{A}$ receives *PubPar* and the set of devices along with their public keys. He announces the subset of corrupted devices, which will be denoted by $\mathcal{D}_j$ for $j = 1 \ldots t$.
3. The simulator SIM gives the private keys s_j of the corrupted devices to $\mathcal{A}$. Device $\mathcal{D}_d$ is corrupted with a worst-case probability of roughly $1/2$, in which case the simulation fails.
4. $\mathcal{A}$ outputs two values x_0Q and x_1Q, of which one has to be shared.
5. Without loss of generality, we assume SIM chooses x_0Q. The output of the VSS protocol is generated as follows.
   - SIM chooses k random $z_k \in_R \mathbb{Z}_\ell^*$ and broadcasts commitments $A_k = z_kP$.
   - SIM constructs a random polynomial $F(z)$ of degree t subject to $F(0) = x_0Q$ and $F(d) = aQ$. For (1) to hold, future shares x_iQ and $x_i'Q$ will have to satisfy

$$\alpha_iQ = x_iQ + \eta x_i'Q \quad \text{with} \quad \alpha_i = \sum_{k=0}^{t} z_k i^k \ . \tag{2}$$

   SIM evaluates the polynomial $F(z)$ and sets the shares $x_jQ = F(j)$ for each corrupted $\mathcal{D}_j$. For the non-corrupted $\mathcal{D}_i \neq \mathcal{D}_d$, SIM chooses random shares $x_iQ \in_R \mathbb{G}_2$. For $i \neq j$, the shares on the second polynomial $x_i'Q$ and $x_j'Q$ are determined by (2).
   - With the private keys s_i and s_j, SIM computes the protected shares $x_iS_i, x_i'S_i$ and $x_jS_j, x_j'S_j$.
   - For $\mathcal{D}_d$, SIM sets $x_dS_d = abQ$ and $x_d'S_d = \eta^{-1}(\alpha_dS_d - abQ)$.
   - All protected shares are broadcast by SIM.
6. The adversary outputs a guess to which of the secrets was shared. If $\mathcal{A}$ has a non-negligible advantage in determining which secret was shared then SIM concludes that $\langle Q, aQ, cQ, abQ \rangle$ must be a valid DH tuple.

The view of $\mathcal{A}$ consists of the commitments A_k, all public keys, the private keys of the corrupted devices, all protected shares and the shares of the corrupted devices. The adversary $\mathcal{A}$ can only gain an advantage in guessing which key was shared from values, other than his own shares, which were not chosen at random. This leaves him with only his shares x_jQ and the values x_dS_d and S_d. The adversary's problem of deciding which secret was shared is equivalent to deciding whether $x_dQ = x_0Q - \sum \lambda_j x_jQ$ or $x_dQ = x_1Q - \sum \lambda_j x_jQ$. Because we assume SIM chose x_0Q, $\mathcal{A}$ has to decide whether $\langle Q, x_0Q - \sum \lambda_j x_jQ, S_d, x_dS_d \rangle$ is a valid DH tuple or not.   □

We note that given the specific form in which the shares are broadcast, our PVSS protocol cannot be proved secure against an adaptive adversary by means of a simulation argument, which does not imply that it is insecure. Indeed, it was already suggested in [6] and [10] that to maintain private transmission of shares some form of non-committing encryption should be used. We insist on storing shares as x_iS_i in order to maintain the nice properties of this form, which allow integrating our construction in other threshold applications, as shown in Sect. 5.

A somewhat related PVSS scheme was presented by Heidarvand and Villar [18].[4] Our PVSS scheme differs from theirs by putting the secret in $\mathbb{G}_2$, instead of $\mathbb{G}_T$, and thus allowing it to be a building block for DKG and discrete-log constructions. Moreover, our protocol is semantically secure while the scheme in [18] is only proved to be secure under a weaker security definition, because the adversary is not allowed to choose the secrets that he has to distinguish.

4.3   Distributed Key Generation

We now establish a new DKG protocol that outputs protected shares and is publicly verifiable. Inspired by [15] and [6] the protocol consists of two phases. In the first phase, the group's private key is generated in a distributive manner and shared through a joint PVSS. In the second phase, the group's public key is computed. This phase follows to a large extent the result of Canetti et al. [6]. The protocols proceeds as follows.

Each participating device runs an instance of our new PVSS protocol. It chooses a secret $c_{i,0} \in_R \mathbb{Z}_\ell$ and broadcasts shares of that secret in protected form. These will be denoted as protected subshares. Each device, acting as a dealer, that is not disqualified is added to a set of qualified devices, denoted as QUAL. The group's private key, although never computed explicitly, is defined as $xQ = \sum_{i \in \text{QUAL}} c_{i,0}Q$. A device's protected share x_iS_i is computed as the sum of the protected subshares that were received from the devices in QUAL.

To recover the group's public key $y = g^x$, the qualified devices will expose g^{x_i} from which y can easily be computed through Lagrange interpolation.[5] Each device will prove in zero-knowledge that the exponent of g^{x_i} matches the share

---

[4] Note that we use an asymmetric pairing which is more standard (e.g., see [13]) than the symmetric form used in [18].

[5] As opposed to [6], we do not expose $g^{c_{i,0}}$, which avoids the costly reconstruction of the $g^{c_{j,0}}$ of the qualified devices that no longer participate in the second phase.

x_iQ hidden in x_iS_i, without revealing it. These interactive zero-knowledge proofs require uniformly distributed challenges, which can be the same for all devices.

A uniformly distributed challenge is generated through another run of our joint PVSS. All devices receive protected shares d_iS_i. After open reconstruction we have a uniformly distributed element $dQ \in \mathbb{G}_2$. However, the challenge needs to be some element $\tilde{d} \in \mathbb{Z}_\ell$. This implies a bijective (not necessarily homomorphic) mapping $\psi : \mathbb{G}_2 \to \mathbb{Z}_\ell$. An example of such a mapping is to take the x-coordinate of dQ modulo ℓ, as is used in ECDSA signatures. Several issues have been reported with this mapping and alternatives, e.g., taking the sum of the x and the y-coordinates modulo ℓ [20], have been proposed. We refer the reader to [4] for a more in-depth treatment of this subject.

The details of the protocols are given in Fig. 2. Note that on the one hand, at least $t + 1$ honest devices are required for the protocol to end successfully, hence we require $n > 2t$. On the other hand, since we require no explicit private channels, only a minimum of $t + 1$ honest devices must participate in the DKG.

We now prove that our new DKG protocol is a secure DKG protocol according to the requirements specified in Definition 2.

Theorem 2. *Our new DKG protocol is a secure DKG protocol (Definition 2) under the divisional variant of the coDBDH-2 assumption.*

Proof (Correctness). All honest devices construct the same set of qualified devices QUAL since this is determined by public broadcast information.

- (C1) Each $\mathcal{D}_i$ that is in QUAL at the end of phase 1 has successfully shared $c_{i,0}Q$ through a run of our PVSS protocol. Any set of $t+1$ honest devices $\mathcal{D}_i$ that combine correct shares x_jQ can reconstruct the same secret xQ since

$$xQ = \sum_{i \in \text{QUAL}} c_{i,0}Q = \sum_{i \in \text{QUAL}} \left(\sum_j \lambda_j x_{ij}Q \right)$$
$$= \sum_j \lambda_j \sum_{i \in \text{QUAL}} x_{ij}Q = \sum_j \lambda_j x_jQ .$$

  In the key extraction phase of our protocol at least $t + 1$ values g^{x_i} have been exposed and thus using interpolation g^{x_j} can be computed for any $\mathcal{D}_j$. This allows to tell apart correct shares from incorrect ones.
- (C2) This follows immediately from the key extraction phase and the relation between the $c_{i,0}Q$ and the shares x_iQ given for the previous property (C1).
- (C3) The private key is defined as $xQ = \sum_{i \in QUAL} c_{i,0}Q$ and each $c_{i,0}Q$ was shared through an instance of our PVSS. Since we proved that a static adversary cannot learn any information about the shared secret, the private key is uniformly distributed as long as one non-corrupted device successfully contributed to the sum that defines xQ. □

Uniformity. Our protocol withstands the attack of a rushing adversary that can influence the distribution of the group's key as described by Gennaro et al. [17].

1. All participating devices $\mathcal{D}_i$ run the PVSS protocol simultaneously, the protected subshares are only broadcast after receiving all commitments from all $\mathcal{D}_i$.

   (a) Each $\mathcal{D}_i$ constructs two polynomials $f_i(z)$ and $f'_i(z)$ of degree t by choosing random coefficients $c_{i,k}, c'_{i,k} \in_R \mathbb{Z}^*_\ell$ for $k = 0 \ldots t$:

   $$f_i(z) = c_{i,0} + c_{i,1}z + \ldots + c_{i,t}z^t \quad , \quad f'_i(z) = c'_{i,0} + c'_{i,1}z + \ldots + c'_{i,t}z^t ,$$

   and broadcasts commitments

   $$A_{i,k} = c_{i,k}P + c'_{i,k}P' \quad , \quad k = 0 \ldots t .$$

   (b) For each device $\mathcal{D}_j$, each $\mathcal{D}_i$ computes and broadcasts

   $$x_{ij}S_j \; , \; x'_{ij}S_j \quad \text{with} \quad x_{ij} = f_i(j) \; , \; x'_{ij} = f'_i(j) .$$

   (c) Each device verifies the broadcast shares for all $\mathcal{D}_i$ by checking that

   $$\hat{e}(P, x_{ij}S_j) \cdot \hat{e}(P', x'_{ij}S_j) = \prod_{k=0}^{t} \hat{e}(A_{i,k}, S_j)^{j^k} .$$

   Each $\mathcal{D}_i$ that is not disqualified as a dealer is added to the list of qualified devices, denoted by QUAL. The group's private key is defined as $xQ = \sum_{i \in \text{QUAL}} c_{i,0}Q$. For each $\mathcal{D}_i$ its protected share is computed as

   $$C_i = x_iS_i = \sum_{j \in \text{QUAL}} x_{ji}S_i .$$

2. The qualified devices expose g^{x_i} to compute the public key $y = g^x$.

   (a) Each $\mathcal{D}_i$ in QUAL broadcasts g^{x_i} and s_iP''. It is easily verified that $\hat{e}(s_iP'', Q) = \hat{e}(P'', S_i)$. In addition, $\mathcal{D}_i$ chooses a random $r_i \in_R \mathbb{Z}^*_\ell$ and broadcasts commitments g^{r_i} and r_iS_i.

   (b) Generation of the uniform challenge, needed in the zero-knowledge proof.
   - Devices in QUAL run a Joint PVSS and obtain protected shares d_iS_i and d'_iS_i, which are broadcast and verified. We denote the commitments of this Joint PVSS as $B_{i,k}$.
   - Open reconstruction of dQ. Devices in QUAL broadcast d_iQ and d'_iQ. These are verified by checking that

   $$\hat{e}(P, d_iQ) \cdot \hat{e}(P', d'_iQ) = \prod_{k=0}^{t} \hat{e}(B_k, Q)^{j^k} \qquad \text{for} \qquad B_k = \sum_{i \in QUAL} B_{i,k} .$$

   - Let $\tilde{d} = \psi(dQ)$, where ψ is a bijective map from $\mathbb{G}_2$ to $\mathbb{Z}_\ell$.

   (c) Each $\mathcal{D}_i$ broadcasts $Z_i = s_i^{-1}(r_iS_i + \tilde{d}C_i) = (r_i + \tilde{d}x_i)Q$. Any device can verify that

   $$\hat{e}(P, Z_i) = g^{r_i}(g^{x_i})^{\tilde{d}} \quad \text{and} \quad \hat{e}(s_iP'', Z_i) = \hat{e}(P'', r_iS_i) \cdot \hat{e}(P'', C_i)^{\tilde{d}} .$$

   (d) Public key y is computed from $t + 1$ correctly verified g^{x_i} as $y = \prod g^{x_i\lambda_i}$.

Fig. 2. Publicly verifiable DKG with protected shares

In this attack an adversary is able to compute a deterministic function of the private key from the broadcasts, before sending out his contributions. He can influence the set of qualified devices by choosing whether or not to send out proper contributions. This allows influencing the outcome of the deterministic function and thus the distribution of the private key. In our protocol, no such function can be computed before the second phase. But, because the private key and thus also the correction factors are fixed after the first phase and determined by QUAL, the adversary can no longer influence the group's key. As long as $t+1$ honest devices participate, the public key can be recovered in the second phase.

Proof (Secrecy). We describe a simulator SIM that, given a public key y, simulates a run of the protocol and produces an output that is indistinguishable from the adversary's view of a real run of the protocol that ended with the given public key. We assume that SIM knows $\eta \in \mathbb{Z}_\ell^*$ for which $P' = \eta P$.

- The first phase of the DKG is run as in the real protocol. Since SIM knows the private keys s_i of at least $t+1$ non-corrupted devices, he knows at least $t+1$ shares $x_i Q = s_i^{-1} C_i$. By interpolation of these shares, SIM learns the shares of the corrupted devices. This allows SIM to compute g^{x_i} for all devices.
- In the second phase of the DKG protocol SIM sets g^{x_i} for the non-corrupted $\mathcal{D}_i$, such that the public key will be y. The $g^{x_i^*}$ for the non-corrupted $\mathcal{D}_i$ are calculated by interpolation of the g^{x_j} of the corrupted $\mathcal{D}_j$ and $y = g^x$. For the zero knowledge proof to hold, SIM chooses a random $d^* \in_R \mathbb{Z}_\ell^*$ and forces the outcome of the open reconstruction of the challenge to $d^* Q$. For each non-corrupted $\mathcal{D}_i$, SIM computes the commitments $\beta_i^* = g^{z_i}(g^{x_i^*})^{-\tilde{d}}$ and $B_i^* = z_i S_i - \tilde{d} x_i S_i$, for random $z_i \in_R \mathbb{Z}_\ell^*$ and $\tilde{d} = \psi(d^* Q)$.

(a) SIM broadcasts $\langle g^{x_i^*}, s_i P'', \beta_i^*, B_i^* \rangle$ for each non-corrupted $\mathcal{D}_i$.

(b) All devices run the Joint PVSS and hold shares $d_i Q$ and $d_i' Q$. SIM forces the outcome of the open reconstruction of the challenge to $d^* Q$.
  - SIM computes the $d_j Q$ from the corrupted devices by interpolation of $t+1$ shares $d_i Q$ of the non-corrupted devices.
  - SIM sets the $d_i^* Q$ for the non-corrupted devices by interpolation of the $d_j Q$ of the corrupted devices and $d^* Q$.
  - By knowing η, SIM will compute $d_i'^* Q$ such that $d_i Q + \eta d_i' Q = d_i^* Q + \eta d_i'^* Q$. As such, the broadcast shares $d_i^* Q$, $d_i'^* Q$, will verify against the commitments.

(c) All $Z_i^* = z_i Q$ are broadcast and correctly verified.

(d) At the end of the protocol the public key is computed as the given y.

To prevent an adversary from being able to distinguish between a real run of the protocol and a simulation, the output distribution must be identical. The first phase, i.e., the Joint PVSS, is identical in both cases. The data that are output in the second phase and that have a potentially different distribution in a real run and simulation are given in the following table. We show that all data in this table have a uniform distribution.

REAL	SIM
1. g^{x_i}	$g^{x_i^*}$
2. $g^{r_i}, r_i S_i$	β_i^*, B_i^*
3. $d_i Q, d_i' Q$	$d_i^* Q, d_i'^* Q$
4. Z_i	Z_i^*

1. The values x_i are evaluations of a polynomial of degree t with uniformly random coefficients. The values x_i^* are evaluations of a polynomial that goes through t evaluations of the first polynomial, namely the x_j of the corrupted participants, and through the discrete logarithm of y. Since the protocol is assumed to generate a uniformly random key, the new polynomial's distribution is indistinguishable from the distribution of the first.
2. The value r_i was chosen uniformly at random. In the simulation $\beta_i^* = g^{z_i}(g^{x_i^*})^{-\tilde{d}}$ and $B_i^* = z_i S_i - \tilde{d} x_i S_i$. The value z_i is uniformly random and $\tilde{d} = \psi(d^* Q)$ is derived from the uniformly random d^*.
3. Since the following relation holds, $d_i Q + \eta d_i' Q = d_i^* Q + \eta d_i'^* Q$, it suffices to show that both $d_i Q$ and $d_i^* Q$ have identical distributions. Because d was chosen uniformly at random, the same reasoning as for the g^{x_i} holds.
4. We have that $Z_i = r_i Q + \eta x_i Q$ and $Z_i^* = z_i Q$. The values r_i, d and z_i were chosen uniformly at random.

We notice that even though the modified $g^{x_i^*}$ have the right output distribution, it is important to note that by broadcasting the modified $g^{x_i^*}$ we introduce a new assumption. Namely that an adversary cannot distinguish between $\langle P, Q, x_i s_i Q, s_i Q, g^{x_i} \rangle$ and $\langle P, Q, x_i s_i Q, s_i Q, g^{x_i^*} \rangle$. This is the divisional variant of the coDBDH-2 assumption, as defined in Sect. 3.4. An adversary cannot distinguish $\langle Q, S_i, d_i Q, d_i S_i \rangle$ from $\langle Q, S_i, d_i^* Q, d_i S_i \rangle$. This is the divisional variant of the DDH assumption, which is a weaker assumption than the coDBDH-2 assumption, meaning that if one could not solve the coDBDH-2 problem, one can also not solve the DDH problem. Knowledge of P allows to calculate g^{d_i} and $g^{d_i^*}$ and to transform this to the divisional variant of the coDBDH-2 assumption.   □

5   Threshold Applications

In this section our construction is used to turn discrete-log schemes into threshold variants with protected shares. It is not our intention to give a rigorous proof of security of these variants. We rather want to demonstrate the ease with which our construction fits into existing schemes. We do this for the ElGamal [14] and the Cramer-Shoup [7] cryptosystems, where we show how pairings allow implicit use of the shares, i.e., without having to reveal them explicitly, and the Schnorr [25] signature scheme.

5.1   ElGamal

Basic Scheme. We define the ElGamal [14] scheme in $\mathbb{G}_T$ with some minor modifications; the randomness is moved from $\mathbb{G}_T$ to $\mathbb{G}_1$ and the private key is an element of $\mathbb{G}_2$ instead of $\mathbb{Z}_\ell^*$, i.e., $xQ \in \mathbb{G}_2$ for some $x \in_R \mathbb{Z}_\ell^*$. Let $y = \hat{e}(P,Q)^x$ be the corresponding public key. Encryption and decryption are defined as follows.

- **Encrypt**(*PubPar*,*y*,*m*): To encrypt a message $m \in \mathbb{G}_T$ under the public key y, choose a random $k \in_R \mathbb{Z}_\ell^*$ and output the ciphertext

$$(R, e) = (kP, my^k) \in \mathbb{G}_1 \times \mathbb{G}_T \ .$$

- **Decrypt**(*PubPar*,*xQ*,(*R*,*e*)): To decrypt the given ciphertext (R, e) output the plaintext

$$m = \frac{e}{\hat{e}(R, xQ)} \in \mathbb{G}_T \ .$$

Threshold Variant. Encryption in the threshold variant is the same as in the basic scheme. To decrypt a given ciphertext we have to combine the randomness kP with $t + 1$ shares x_iQ, which are stored as x_iS_i. If the shares were stored as $g^{x_is_i}$, it would have been impossible to combine them with the randomiser or the ElGamal encryption and for each device $\mathcal{D}_i$ the ciphertext would contain something like $g^{x_is_ik}$. By taking advantage of the bilinearity of the pairing, the size of the ciphertext remains constant. Note that $\mathcal{D}_i$ never has to reveal his share explicitly; his private key is combined with the randomness and then paired with the correction factor. The cost of providing a partial decryption is minimal, namely one elliptic-curve point multiplication. In this way we can use small devices as partial decryption oracles. Decryption goes as follows.

- **T-Decrypt**(*PubPar*,{$\mathcal{D}_i$, S_i},(*R*,*e*)): To decrypt the ciphertext (R, e) each device $\mathcal{D}_i$ provides a partial decryption

$$D_i = s_i^{-1}R = s_i^{-1}kP \in \mathbb{G}_1 \ .$$

The combining device receives the D_i and verifies that $\hat{e}(D_i, S_i) = \hat{e}(R, Q)$. He then combines $t + 1$ contributions to output the plaintext

$$m = \frac{e}{d} \quad \text{with} \quad d = \prod \hat{e}(D_i, C_i)^{\lambda_i} \ .$$

5.2   Cramer-Shoup

Basic Scheme. Cramer and Shoup [7] presented an ElGamal based cryptosystem in the standard model that provides ciphertext indistinguishability under adaptive chosen ciphertext attacks (IND-CCA2). We define their scheme in $\mathbb{G}_T$ with the same modifications as in the ElGamal scheme; the first two (random) elements in the ciphertext are moved from $\mathbb{G}_T$ to $\mathbb{G}_1$ and the private key is a tuple from $\mathbb{G}_2^5$ instead of $(\mathbb{Z}_\ell^*)^5$. Let $H : \mathbb{G}_1 \times \mathbb{G}_1 \times \mathbb{G}_T \to \mathbb{Z}_\ell$ be an element of a family of universal one-way hash functions. The private key is

$$privK = (x_1Q, x_2Q, y_1Q, y_2Q, zQ) \in_R \mathbb{G}_2^5$$

and the public key is

$$pubK = (c, d, h) = (g_1^{x_1}g_2^{x_2}, g_1^{y_1}g_2^{y_2}, g_1^z) \in \mathbb{G}_T^3 \ .$$

Encryption and decryption are defined as follows.

- **Encrypt**(*PubPar*,*pubK*,*m*): To encrypt a message $m \in \mathbb{G}_T$ under *pubK*, choose a random $k \in_R \mathbb{Z}_\ell$ and output the ciphertext

$$(U_1, U_2, e, v) = (kP, kP', mh^k, c^k d^{k\alpha}) \in \mathbb{G}_1^2 \times \mathbb{G}_T^2 \quad \text{with} \quad \alpha = H(U_1, U_2, e) \ .$$

- **Decrypt**(*PubPar*,*privK*,(U_1, U_2, e, v)): To decrypt ciphertext (U_1, U_2, e, v), first compute $\alpha = H(U_1, U_2, e)$ and validate the ciphertext by testing if

$$\hat{e}(U_1, x_1 Q + y_1 \alpha Q) \cdot \hat{e}(U_2, x_2 Q + y_2 \alpha Q) = v \ .$$

If the test fails, the ciphertext is rejected, otherwise output the plaintext

$$m = \frac{e}{\hat{e}(U_1, zQ)} \in \mathbb{G}_T \ .$$

Threshold Variant. It is clear that the Cramer-Shoup public key is not immediately established from running five instances of our DKG protocol. The decomposition of $c = g_1^{x_1} g_2^{x_2}$ and $d = g_1^{y_1} g_2^{y_2}$ should not be known. We can solve this problem by introducing a third polynomial $f''(z)$. Each device receives three instead of two shares. The public key is extracted by revealing the third share and by proving the discrete log equality of $g_3^{x_i''}$ and $x_i'' S_i$. The DKG is thereby reduced to two runs of the variant and one run of the basic DKG protocol. This results in five protected shares $C_i^{x_1}, C_i^{x_2}, C_i^{y_1}, C_i^{y_2}$ and C_i^z for each device.

Encryption is the same as in the basic scheme. The decryption routine, which applies the same ideas as in the threshold ElGamal scheme goes as follows. Note that the cost of providing a partial decryption is minimal, namely two elliptic-curve point multiplications.

- **T-Decrypt**(*PubPar*,$\{\mathcal{D}_i, S_i\}$,(U_1, U_2, e, v)): To decrypt the given ciphertext (U_1, U_2, e, v) each device $\mathcal{D}_i$ provides $D_i = s_i^{-1} U_1$ and $D_i' = s_i^{-1} U_2$. The combining device verifies that $\hat{e}(D_i, S_i) = \hat{e}(U_1, Q)$ and $\hat{e}(D_i', S_i) = \hat{e}(U_2, Q)$. He then computes

$$v_i = \hat{e}(D_i, C_i^{x_1} + \alpha C_i^{y_1}) \cdot \hat{e}(D_i', C_i^{x_2} + \alpha C_i^{y_2}) \ .$$

and combines $t + 1$ values v_i to validate the ciphertext by testing that $v = \prod v_i^{\lambda_i}$. If validation fails, the ciphertext is rejected. The combining device combines $t + 1$ contributions to output the plaintext

$$m = \frac{e}{d} \quad \text{with} \quad d = \prod \hat{e}(D_i, C_i^z)^{\lambda_i} \ .$$

5.3   Schnorr Signatures

The Schnorr signature scheme [25] is an example of a scheme that provides existential unforgeability under an adaptive chosen-message attack in the random oracle model [24] and has been used many times to create a threshold signature scheme, e.g., in [17,1]. We will define the signature scheme in $\mathbb{G}_T$ and then extend it to a threshold variant.

Basic Scheme. Let $H' : \{0,1\}^* \times \mathbb{G}_T \to \mathbb{Z}_\ell$ be a cryptographic hash function. Let the private key be $xQ \in \mathbb{G}_2$ for $x \in_R \mathbb{Z}_\ell^*$ and $y = g^x \in \mathbb{G}_T$ the public key.

- **Sign**(*PubPar*,xQ,m): To sign a message $m \in \{0,1\}^*$ with the private key xQ choose a random $k \in_R \mathbb{Z}_\ell$, compute $r = \hat{e}(P, kQ)$ and $c = H'(m, r)$, and output the signature

$$(c, \sigma) = (H'(m, r), kQ + c\,xQ) \in \mathbb{Z}_\ell \times \mathbb{G}_2 \ .$$

- **Verify**(*PubPar*,y,(c,σ),m): To verify the signature (c, σ) on a message m compute $\tilde{r} = \hat{e}(P, \sigma)y^{-c}$ and verify equality of $c = H'(m, \tilde{r})$.

Threshold Variant. The basic scheme naturally extends to a threshold variant. As opposed to the encryption schemes, the bilinearity of the pairing is not really needed. However, the signing devices need to share some randomness and will, therefore, run the DKG protocol of Sect. 4.3. Signature verification is the same as in the basic scheme. Signing goes as follows.

- **T-Sign**(*PubPar*,$\{\mathcal{D}_i\}$,m): To sign a message $m \in \{0,1\}^*$ with the group's private key the devices $\mathcal{D}_i$ will run an instance of the DKG protocol of Sect. 4.3. Each device then holds a share $k_i S_i$ in protected form of $kQ \in \mathbb{G}_2$. Because the value $r = \hat{e}(P, kQ) \in \mathbb{G}_T$ is publicly computed at the end of the protocol, each device can compute $c = H'(m, r)$ and $\sigma_i = s_i^{-1}(k_i S_i + cC_i) = (k_i + c\,x_i)Q$, which is sent to the combining device. Note that these partial signatures can be verified since the output of the DKG protocols contained g^{k_i} and g^{x_i}. Values that were not in the output can be computed through interpolation. The combining device computes the signing equation $\sigma = \sum \sigma_i \lambda_i$ and outputs the signature

$$(c, \sigma) = (H'(m, r), kQ + c\,xQ) \in \mathbb{Z}_\ell \times \mathbb{G}_2 \ .$$

6   Conclusion

In this paper, we have shown how to increase resilience in threshold cryptography by including small devices with limited or no secure storage capabilities. Assuming these devices have some support for public-key functionality, shares can be stored in protected form. By using bilinear pairings, this particular form yields some advantages. The most important feature is public verifiability, which makes explicit private channels and cumbersome complaint procedures obsolete. Moreover, not all devices need to be present during group setup, which is performed by the DKG protocol. We have demonstrated how to adopt the protected shares in existing discrete-log based cryptosystems and signature schemes. Because shares are never needed in unprotected form, small devices can be used as decryption oracles at a minimal cost.

Acknowledgments. The authors would like to thank Frederik Vercauteren, Alfredo Rial Duran and Markulf Kohlweiss for the fruitful discussions. This work was supported in part by the Concerted Research Action (GOA) Ambiorics 2005/11 of the Flemish Government, by the IAP Programme P6/26 BCRYPT of the Belgian State (Belgian Science Policy), and in part by the European Commission through the IST programmes under contract ICT-2007-216676 ECRYPT II. Roel Peeters is funded by a research grant of the Institute for the Promotion of Innovation through Science and Technology in Flanders (IWT-Vlaanderen).

References

1. Abe, M., Fehr, S.: Adaptively secure Feldman VSS and applications to universally-composable threshold cryptography. In: Franklin, M.K. (ed.) CRYPTO 2004. LNCS, vol. 3152, pp. 317–334. Springer, Heidelberg (2004)
2. Bao, F., Deng, R.H., Zhu, H.: Variations of Diffie-Hellman problem. In: Qing, S., Gollmann, D., Zhou, J. (eds.) ICICS 2003. LNCS, vol. 2836, pp. 301–312. Springer, Heidelberg (2003)
3. Beaver, D., Haber, S.: Cryptographic protocols provably secure against dynamic adversaries. In: Rueppel, R.A. (ed.) EUROCRYPT 1992. LNCS, vol. 658, pp. 307–323. Springer, Heidelberg (1992)
4. Brown, D.R.L.: Generic groups, collision resistance, and ECDSA. Designs, Codes and Cryptography 35(1), 119–152 (2005)
5. Canetti, R.: Universally composable security: A new paradigm for cryptographic protocols. In: FOCS 2001, pp. 136–145. IEEE Computer Society, Los Alamitos (2001)
6. Canetti, R., Gennaro, R., Jarecki, S., Krawczyk, H., Rabin, T.: Adaptive security for threshold cryptosystems. In: Wiener, M.J. (ed.) CRYPTO 1999. LNCS, vol. 1666, pp. 98–115. Springer, Heidelberg (1999)
7. Cramer, R., Shoup, V.: A practical public key cryptosystem provably secure against adaptive chosen ciphertext attack. In: Krawczyk, H. (ed.) CRYPTO 1998. LNCS, vol. 1462, pp. 13–25. Springer, Heidelberg (1998)
8. Feldman, P.: A practical scheme for non-interactive verifiable secret sharing. In: FOCS 1987, pp. 427–437. IEEE Computer Society, Los Alamitos (1987)
9. Fouque, P.A., Stern, J.: One round threshold discrete-log key generation without private channels. In: Kim, K.-c. (ed.) PKC 2001. LNCS, vol. 1992, pp. 300–316. Springer, Heidelberg (2001)
10. Frankel, Y., MacKenzie, P.D., Yung, M.: Adaptively-secure distributed public-key systems. In: Nešetřil, J. (ed.) ESA 1999. LNCS, vol. 1643, pp. 4–27. Springer, Heidelberg (1999)
11. Fujisaki, E., Okamoto, T.: A practical and provably secure scheme for publicly verifiable secret sharing and its applications. In: Nyberg, K. (ed.) EUROCRYPT 1998. LNCS, vol. 1403, pp. 32–46. Springer, Heidelberg (1998)
12. Galbraith, S., Hess, F., Vercauteren, F.: Aspects of pairing inversion. IEEE Transactions on Information Theory 54(12), 5719–5728 (2008)
13. Galbraith, S., Paterson, K., Smart, N.: Pairings for cryptographers. Cryptology ePrint Archive, Report 2006/165 (2006), http://eprint.iacr.org/
14. Gamal, T.E.: A public key cryptosystem and a signature scheme based on discrete logarithms. In: Blakely, G.R., Chaum, D. (eds.) CRYPTO 1984. LNCS, vol. 196, pp. 10–18. Springer, Heidelberg (1985)

15. Gennaro, R., Jarecki, S., Krawczyk, H., Rabin, T.: Secure distributed key generation for discrete-log based cryptosystems. In: Stern, J. (ed.) EUROCRYPT 1999. LNCS, vol. 1592, pp. 295–310. Springer, Heidelberg (1999)
16. Gennaro, R., Jarecki, S., Krawczyk, H., Rabin, T.: Secure applications of Pedersen's distributed key generation protocol. In: Joye, M. (ed.) CT-RSA 2003. LNCS, vol. 2612, pp. 373–390. Springer, Heidelberg (2003)
17. Gennaro, R., Jarecki, S., Krawczyk, H., Rabin, T.: Secure distributed key generation for discrete-log based cryptosystems. J. of Cryptology 20(1), 51–83 (2007)
18. Heidarvand, S., Villar, J.L.: Public verifiability from pairings in secret sharing schemes. In: Avanzi, R., Keliher, L., Sica, F. (eds.) SAC 2008. LNCS, vol. 5381, pp. 294–308. Springer, Heidelberg (2009)
19. Jarecki, S., Lysyanskaya, A.: Adaptively secure threshold cryptography: Introducing concurrency, removing erasures. In: Preneel, B. (ed.) EUROCRYPT 2000. LNCS, vol. 1807, pp. 221–242. Springer, Heidelberg (2000)
20. Malone-Lee, J., Smart, N.P.: Modifications of ECDSA. In: Nyberg, K., Heys, H.M. (eds.) SAC 2002. LNCS, vol. 2595, pp. 1–12. Springer, Heidelberg (2002)
21. Mao, W.: Modern Cryptography: Theory and Practice. Prentice Hall, Englewood Cliffs (2003)
22. Pedersen, T.P.: A threshold cryptosystem without a trusted party (extended abstract). In: Davies, D.W. (ed.) EUROCRYPT 1991. LNCS, vol. 547, pp. 522–526. Springer, Heidelberg (1991)
23. Pedersen, T.P.: Non-interactive and information-theoretic secure verifiable secret sharing. In: Feigenbaum, J. (ed.) CRYPTO 1991. LNCS, vol. 576, pp. 129–140. Springer, Heidelberg (1992)
24. Pointcheval, D., Stern, J.: Security arguments for digital signatures and blind signatures. J. of Cryptology 13(3), 361–396 (2000)
25. Schnorr, C.P.: Efficient identification and signatures for smart cards. In: Brassard, G. (ed.) CRYPTO 1989. LNCS, vol. 435, pp. 239–252. Springer, Heidelberg (1989)
26. Schoenmakers, B.: A simple publicly verifiable secret sharing scheme and its application to electronic voting. In: Wiener, M.J. (ed.) CRYPTO 1999. LNCS, vol. 1666, pp. 148–164. Springer, Heidelberg (1999)
27. Shamir, A.: How to share a secret. Comm. of the ACM 22(11), 612–613 (1979)
28. Smart, N.P., Vercauteren, F.: On computable isomorphisms in efficient asymmetric pairing-based systems. Discrete Applied Mathematics 155(4), 538–547 (2007)
29. Stadler, M.: Publicly verifiable secret sharing. In: Maurer, U.M. (ed.) EUROCRYPT 1996. LNCS, vol. 1070, pp. 190–199. Springer, Heidelberg (1996)

Shorter Verifier-Local Revocation Group Signature with Backward Unlinkability⋆

Lingbo Wei[1,2] and Jianwei Liu[1]

[1] School of Electronics and Information Engineering, Beihang University, China
[2] State Key Laboratory of Information Security, Institute of Software,
Chinese Academy of Sciences, China
{lingbowei,liujianwei}@buaa.edu.cn

Abstract. Used as the privacy-preserving attestation by Trusted Computing effort (TCG) or the privacy-preserving authentication protocol in vehicular ad hoc networks (VANETs), group signature becomes more important than ever. Membership revocation is a delicate issue in group signatures. Verifier-local revocation (VLR) is a reasonable resolution, especially for mobile environments. Back unlinkability (BU) is a currently introduced security property providing further privacy. Based on the Decision Linear (DLIN) assumption and the q-Strong Diffie-Hellman (q-SDH) assumption, a new BU-VLR group signature scheme is proposed, which has the shortest signature size and smallest computation overhead among the previous BU-VLR group signature schemes.

Keywords: group signature, membership revocation, verifier-local revocation, backward unlinkability.

1 Introduction

As a kind of group-oriented signatures, group signature has advantages over traditional signature when used in the group environment. This concept was introduced by Chaum and Heyst [1] in 1991, which has one public key corresponding to multiple private signing keys held by each group member. It allows any group member to sign anonymously on behalf of the group. In case of dispute, the actual signer can be identified by the group manager (GM). The motivation of this kind signature is to protect the signer's anonymity.

The feature of privacy-preserving makes group signature have many applications, such as anonymous credential systems, internet voting, trust computing and bidding. Currently, an appealing application is to design secure vehicular ad hoc networks [2, 3] based on the group signatures. However, wide implementation in practice has been confined. One important reason, as pointed out in [4], is membership revocation. Revocation of a member should disable his signing

---

⋆ This work is supported by the National High Technology Research and Development Program of China under grant No. 2009AA01Z418, the China Postdoctoral Science Foundation under grant No. 20090460192, and the Fundamental Research Funds for the Central Universities under grant No. YWF1002009.

M. Joye, A. Miyaji, and A. Otsuka (Eds.): Pairing 2010, LNCS 6487, pp. 136–146, 2010.

ability in the future and preserve the anonymity of his past signatures. There are two main nontrivial resolutions. One is based on witness [5–7]. Another is based on revocation list (RL) [8–13].

In the RL-based revocation, the GM issues a revocation list of revoked identities. Any group signature can be verified that in a zero-knowledge way that the group member's identity embedded in the signature is not included in the RL. The drawback of this method is that the signature size is linearly dependent on the size of RL [8]. This method was improved in [9] that the signature size and the computation overhead were constant while complexity of verification was linearly dependent on the size of RL. It is called *Verifier-Local Revocation* and formalized by Boneh et al. in [10]. Based on the DLIN assumption and the q-SDH assumption, they also proposed a VLR group signature scheme in [10] with 1192 bits in signature size, which is the most efficient VLR scheme in performance so far due to the shortest signature length and least computation.

Nakanishi et al. [11, 12] introduced the backward unlinkability (BU) property into VLR group signature. This property means that even a member is revoked, signatures created by himself before the revocation still remain anonymous. This enhanced privacy protection is preferred under many circumstances. However, the performance (see table 1) of their scheme is not comparable to Boneh et al.'s. After that, several BU-VLR group signature schemes are proposed. The seventh scheme (ZL06) in [13] is the most efficient BU-VLR scheme with non-frameability, with the same total computation and 114.3% of the length of [10].

The motivation of our paper is to design an efficient BU-VLR group signature scheme, and a shorter BU-VLR group signature scheme is proposed using bilinear group. The signature length of our scheme is only 852 bits which is about 71.48% of that of [10], and 6 of multi-exponentiation and 1 bilinear computation are reduced in our scheme compared with [10].

Our paper is organized as follows: Section 2 is the preliminaries. The model and security definitions of BU-VLR group signature are described in section 3. Our proposed scheme is presented in section 4, and its security is proved in section 5. Section 6 is the performance comparison with previous VLR group signature schemes and conclusion is given in section 7.

2   Preliminaries

Our scheme is constructed in bilinear groups with computable bilinear map. Its security is based on the DLIN assumption and the $q-$SDH assumption. We review these preliminaries in this section.

Definition 1 (Bilinear groups [16]). (G_1, G_2) *is called a bilinear group pair, if there exists a group G_T and a bilinear map $e : G_1 \times G_2 \to G_T$ with the following properties:*

1. *$G_1 = < g_1 >, G_2 = < g_2 >$, and G_T are multiplicative cyclic groups of prime order p;*
2. *ψ is an efficiently computable isomorphism from G_2 to G_1, with $\psi(g_2) = g_1$;*

3. *e is an efficiently computable bilinear map, such that (1) Bilinear:* $\forall (u, v) \in (G_1 \times G_2), a, b \in Z, e(u^a, v^b) = e(u, v)^{ab}$; *(2) Non-degenerate:* $e(g_1, g_2) \neq 1$.

There are two frequently used types of (G_1, G_2): (1) $G_1 = G_2$, then ψ is an identity map; (2) more general case is $G_1 \neq G_2$, where the trace map can be used as homomorphism and certain families of non-supersingular elliptic curves [15] can be used to construct bilinear groups. Our scheme allows for this case.

Definition 2 (DLIN assumption). *For all PPT algorithm $\mathcal{A}$ in G_2, the probability*

$$|\Pr[\mathcal{A}(u, v, h, u^a, v^b, h^{a+b}) = 1] - \Pr[\mathcal{A}(u, v, h, u^a, v^b, h^c) = 1]|$$

is negligible, where $u, v, h \in G_2$ and $a, b, c \in Z_p^*$.

Definition 3 (q-SDH assumption). *For all PPT algorithm $\mathcal{A}$ in (G_1, G_2), the probability*

$$\Pr[\mathcal{A}(g_1, g_2, g_2^{\gamma}, ..., g_2^{(\gamma^q)}) = (g_1^{1/(\gamma+x)}, x) : x \in Z_p^*]$$

is negligible, where $\gamma \in Z_p^*$. And such a pair $(g_1^{1/(\gamma+x)}, x)$ is called a SDH pair.

3   Model and Definitions of BU-VLR Group Signature

We review the model of BU-VLR group signature in [11, 12] bellow.

Definition 4 (BU-VLR Group Signature). *A BU-VLR group signature scheme consists of the following algorithms.*

- **KeyGen**(n, T): *A probabilistic algorithm, on input the number of members n and the number of time intervals T, generates a group public key gpk, an n-element vector of members' signing keys $gsk = (gsk_1, ..., gsk_n)$ and revocation token $grt = (grt_{11}, ..., grt_{nT})$, where gsk_i is kept secret by member $i \in [1, n]$ and grt_{ij} denotes the revocation token of member $i \in [1, n]$ at time interval $j \in [1, T]$.*
- **Sign**(gpk, j, gsk_i, M): *A probabilistic algorithm generates the signature σ on a message M at the current time interval j by member i using gsk_i and gpk.*
- **Verify**$(gpk, j, RL_j, \sigma, M)$: *A deterministic algorithm includes signature check and revocation check, which can be performed by anyone to generate one bit b. If $b = 1$, it means σ is a valid signature on M at interval j by one member of the group whose revocation token is not in RL_j. If $b = 0$, then σ is invalid.*
- **Revoke**(RL_j, grt_{ij}): *This algorithm adds grt_{ij} to RL_j if member i is to be revoked at the time interval $j \in [1, T]$.*
  Sometimes, a group signature need be opened to find the actually singer. It will be shown in section 4 that an open algorithm can be constructed by using revocation check.

Definition 5 (Correctness). *For all* $(gpk, gsk, grt) = \mathbf{KeyGen}(n, T)$, *all* $j \in [1, T]$, *all* $i \in [1, n]$, *and all* $M \in \{0, 1\}^*$, *this requires that*,

$$\mathbf{Verify}(gpk, j, RL_j, \mathbf{Sign}(gpk, j, gsk_i, M), M) = 1 \Longleftrightarrow grt_{ij} \notin RL_j$$

Definition 6 (BU-anonymity). *BU-anonymity requires that for all PPT* $\mathcal{A}$, *the advantage of* $\mathcal{A}$ *on the following BU-anonymity game is negligible.*

- *Setup: The challenger runs the key generation algorithm to obtain* (gpk, gsk, grt), *and provides the adversary* $\mathcal{A}$ *with gpk.*
- *Queries: The challenger announces the beginning of every interval* $j \in [1, T]$ *to* $\mathcal{A}$, *which is incremented with time.* $\mathcal{A}$ *can request the challenger about the following queries at the current interval* j.
  - *Signing:* $\mathcal{A}$ *requests a signature of any member* i *on arbitrary message* M *at interval* j. *The corresponding signature is responded by the challenger.*
  - *Corruption:* $\mathcal{A}$ *requests the secret key of any member* i.
  - *Revocation:* $\mathcal{A}$ *requests the revocation token of any member* i *at interval* j. *The challenger responds with* grt_{ij}.
- *Challenge:* $\mathcal{A}$ *outputs some* (M, i_0, i_1, j_0) *with restriction that* i_0 *and* i_1 *have not been corrupted, and their revocation tokens have not been queried before the current interval* j_0 *(including* j_0). *The challenger randomly selects* $\phi \in \{0, 1\}$, *and responds with signature of member* i_ϕ *on* M *at interval* j_0.
- *Restricted queries:* $\mathcal{A}$ *is allowed to make queries of signing, corruption and revocation, except the corruption queries of* i_0, i_1 *and their revocation queries at interval* j_0. *Note that* $\mathcal{A}$ *can query he revocations of* i_0 *and* i_1 *at interval* $j'(j' \geq j_0$ *for the BU property.*
- *Output:* $\mathcal{A}$ *outputs a bit* ϕ' *as its guess of* ϕ.

If $\phi' = \phi$, $\mathcal{A}$ *wins the game. The advantage of* $\mathcal{A}$ *is defined as* $|\mathrm{pr}[\phi' = \phi] - 1/2|$.

Definition 7 (Traceability). *Traceability requires that for all PPT* $\mathcal{A}$, *the advantage of* $\mathcal{A}$ *on the following game is negligible.*

- *Setup: The challenger runs the key generation algorithm to obtain* (gpk, gsk, grt), *and sets* U *empty. The adversary* $\mathcal{A}$ *is provided with gpk and grt.*
- *Queries:* $\mathcal{A}$ *can request the challenger about the following queries at each interval* $j \in [1, T]$.
  - *Signing:* $\mathcal{A}$ *requests a signature of any member* i *on arbitrary message* M *at interval* j. *The corresponding signature is responded by the challenger.*
  - *Corruption:* $\mathcal{A}$ *requests the secret key of any member* i. *The challenger responds the corresponding key and adds* i *to* U.
- *Output:* $\mathcal{A}$ *outputs* $(M^*, j^*, RL_{j^*}, \sigma^*)$. $\mathcal{A}$ *wins if (1)* $\mathbf{Verify}(gpk, M^*, j^*, RL_{j^*}, \sigma^*) = 1$, *and (2)* σ^* *is traced to a member outside of* $U \setminus RL_{j^*}$ *or failure, and (3)* $\mathcal{A}$ *has not obtained* σ^* *in signing queries on message* M^*.

4   Proposed Scheme

Our scheme is constructed under the bilinear groups (G_1, G_2) with isomorphism ψ as described in Section 2.

KeyGen(n, T)

1. Select a generator $g_2 \xleftarrow{R} G_2$ and a collision resistant hash function $H : \{0,1\}^* \rightarrow Z_p^*$. Set $g_1 = \psi(g_2), G_1 = < g_1 >$.

2. Select $\gamma \xleftarrow{R} Z_p^*$ and compute $w = g_2^\gamma$.

3. Select $x_i \xleftarrow{R} Z_p^*$ and compute $A_i = g_1^{1/(\gamma + x_i)}$ for all $i \in [1, n]$.

4. Select $r_j \xleftarrow{R} Z_p^*$, then compute $h_j = g_1^{r_j}$ and $grt_{ij} = (grt_{ij}^1, grt_{ij}^2) = ((wg_2^{x_i})^{r_j}, h_j^{(-x_i)})$ for all $j \in [1, T]$.

The group public key gpk is $(g_1, g_2, h_1, ..., h_T, w)$, the private signing key of member i is (A_i, x_i), and his revocation token at interval j is grt_{ij}. Output $(gpk, gsk, grt) = (gpk, (gsk_1, ..., gsk_n), (grt_{11}, ..., grt_{nT}))$.

Remark: This algorithm can be implemented with flexibility. First, the group member can join the group in succession, i.e., the group total number n can be pre-fixed, but the private signing key $gsk[i]$ can be generated later while member i's joining. Second, there is no need to generate all the data in grt at the beginning, that is to say grt_{ij} can be generated when member i is to be revoked at interval j.

Sign$(gpk, j, gsk[i], M)$

1. Select $\alpha \xleftarrow{R} Z_p^*$ and compute $T_1 = A_i^\alpha, T_2 = h_j^{\alpha + x_i}$.

2. The signature of knowledge [5, 11, 12] is expressed by the following equation:

$$\pi = SPK\{(\alpha, x_i, A_i) : T_1 = A_i^\alpha, T_2 = h_j^{\alpha + x_i}, e(A_i, wg_2^{x_i}) = e(g_1, g_2)\}(M)$$
$$= SPK\{(\alpha, x_i, A_i) : e(T_1, w) = e(g_1, g_2)^\alpha / e(T_1, g_2)^{x_i}, T_2 = h_j^{\alpha + x_i}\}(M).$$

which is computed as follows by using Fiat-shamir heuristic method [19]:

(a) Pick blinding factors $r_\alpha, r_{x_i} \xleftarrow{R} Z_p^*$ to compute

$$R_1 = e(g_1, g_2)^{r_\alpha} / e(T_1, g_2)^{r_{x_i}} \tag{1}$$
$$R_2 = h_j^{r_\alpha + r_{x_i}}. \tag{2}$$

(b) Compute the challenge value $c = H(gpk, j, M, T_1, T_2, R_1, R_2)$, and $s_\alpha = r_\alpha + c\alpha, s_{x_i} = r_{x_i} + cx_i$.

The signature is $\sigma = (T_1, T_2, c, s_\alpha, s_{x_i})$.

Verify$(gpk, j, RL_j, \sigma, M)$

1. **Signature check.** Verify the validity of σ by checking π as follows:

(a) Retrieve

$$\bar{R}_1 = e(g_1, g_2)^{s_\alpha} (1/e(T_1, g_2))^{s_{x_i}} (1/e(T_1, w))^c \tag{3}$$
$$\bar{R}_2 = h_j^{s_\alpha + s_{x_i}} (1/T_2)^c \tag{4}$$

(b) Verify the correctness of the challenge c by checking:

$$c \stackrel{?}{=} H(gpk, j, M, T_1, T_2, \bar{R}_1, \bar{R}_2)$$

If the above equation holds, then accept; otherwise, reject.

2. **Revocation check.** For each $grt_{ij} = (grt_{ij}^1, grt_{ij}^2) \in RL_j$ at the current interval j, if $e(T_1, grt_{ij}^1) \neq e(T_2 grt_{ij}^2, g_2)$ then return 1. Otherwise return 0.

Implicit tracing algorithm. From revocation check, we can see that any trust third party who knows all $grt_{ij}(i \in [1, n], j \in [1, T])$ can play the role of the opener to find the signer of a given signature. Take the interval j for example, the opener can identify the signer is member i by checking $e(T_1, grt_{ij}^1) = e(T_2 grt_{ij}^2, g_2)$.

Revoke(RL_j, grt_{ij})

If member i is to be revoked at the time interval $j \in [1, T]$, the GM adds $grt_{ij} = (grt_{ij}^1, grt_{ij}^2) = ((wg_2^{x_i})^{r_j}, h_j^{x_i})$ into the RL_j.

Remark: Compared with the current BU-VLR group signature schemes, the shorter size is gained in our scheme due to the argument that the underlying SPK is short. Our above scheme is actually a non-interactive zero-knowledge of having a SDH pair (A_i, x_i).

5   Security Analysis

Theorem 1. *Suppose an adversary $\mathcal{A}$ breaks the BU-anonymity of the proposed scheme with advantage ϵ, after q_H hash queries and q_S signing queries, then there exists an algorithm $\mathcal{B}$ breaking the DLIN assumption in G_2 with advantage $(1/nT - q_H q_S/p)\epsilon$.*

Proof. The input of $\mathcal{B}$ is (u, v, h, u^a, v^b, Z), where $u, v, h \in G_2, a, b \in Z_p^*$ and either $Z = h^{a+b}$ or $Z = h^c(c \in Z_p)$. $\mathcal{B}$ decides which Z it is given by communicating with $\mathcal{A}$ as follows:

- **Setup.** $\mathcal{B}$ simulates **KeyGen** (n, T) as follows.
  1. $\mathcal{B}$ sets $g_2 = u$ and computes $g_1 = \psi(u)$. $\mathcal{B}$ selects $i^* \in [1, n], j^* \in [1, T]$.
  2. For all $j \in [1, T]$ except j^*, $\mathcal{B}$ selects $\gamma \stackrel{R}{\leftarrow} Z_p^*$ to set $w = g_2^\gamma$. Select $r_j \stackrel{R}{\leftarrow} Z_p^*$ to compute $h_j = g_1^{r_j} = \psi(u)^{r_j}$. For j^*, $\mathcal{B}$ sets $h_j = \psi(h)$.
  3. For all $i \in [1, n]$ except i^*, $\mathcal{B}$ selects $x_i \stackrel{R}{\leftarrow} Z_p^*$ and computes $A_i = g_1^{1/(\gamma + x_i)}$. For $i = i^*$, set $x_i = a, A_i = g_1^{1/\gamma + a}$ which is unknown to $\mathcal{B}$ for not having the value of a.
  4. $\mathcal{B}$ computes $grt_{ij} = ((wg_2^{x_i})^{r_j}, h_j^{x_i})$ for all i and j except j^*. For i^* except j^*, $\mathcal{B}$ computes

  $$grt_{i^*j} = (wg_2^{x_{i^*} r_j}, h_j^{x_{i^*}})$$
  $$= ((wg_2^a)^{r_j}, (\psi(u)^{r_j})^a)) = ((wu^a)^{r_j}, (\psi(u^a))^{r_j})$$

  For $i = i^*, j = j^*$, $\mathcal{B}$ sets $grt_{i^*j^*} = ((wg_2^a)^{r_{j^*}}, h^a)$ which is unknown to him for not knowing r_{j^*}.

- **Hash queries.** At any time, $\mathcal{A}$ can query the hash function used in π. $\mathcal{B}$ responds with random values with consistency.
- **Phase 1.** At any interval j, $\mathcal{A}$ can issue signing queries, corruption queries, and revocation queries. If $i \neq i^*$, $\mathcal{B}$ responds to queries as usual by using the secret key of member i. If $i = i^*$, $\mathcal{B}$ responds as follows.
  - **Signing queries.** $\mathcal{B}$ computes a simulated group signature of member i^* as follows.
    If $j \neq j^*$:
    1. Randomly select $\alpha \in Z_p^*, T_1 \overset{R}{\leftarrow} G_1$.
    2. Compute $T_2 = h_j^{\alpha + x_i} = h_j^\alpha (g_1^{r_j})^a = h_j^\alpha (\psi(u^a))^{r_j}$.
    3. Pick random $c, s_\alpha, s_{x_i} \in Z_p^*$ to compute R_1, R_2 by equation (3) and (4), and then set $c = H(gpk, j, M, T_1, T_2, R_1, R_2)$. Now, the simulated $\pi = (c, s_{x_i}, s_\alpha)$ is obtained. $\mathcal{B}$ outputs a random guess $\omega' \in \{0, 1\}$ and aborts if $\mathcal{A}$ has issued the hash query on $H(gpk, j, M, T_1, T_2, R_1, R_2)$. Since j is random in T, this happens with probability q_H/p.
    Finally, $\mathcal{B}$ responds signature $\sigma = (T_1, T_2, c, s_\alpha, s_{x_i})$ and the message M to $\mathcal{A}$. Each value in σ has the same distribution as in the real, for random $\alpha \in Z_p^*$ and the perfect zero-knowledge-ness of π.
    Otherwise $j = j^*$,
    1. Keep a List $\mathcal{L}$ which is initially empty.
    2. When $\mathcal{L}$ is empty then select $r \in Z_p^*, T_1 \overset{R}{\leftarrow} G$ and set $\alpha = r - a$. Compute $T_2 = h^r$, and then add (T_1, T_2, r) into $\mathcal{L}$. Otherwise, select $r' \in Z_p^*$ to compute $(T_1)^{r'}, (T_2)^{r'}$, and update $\mathcal{L}$ by date $((T_1)^{r'}, (T_2)^{r'}, rr')$.
    3. Compute a simulated π as that in the case of $j \neq j^*$. If the backpatch of the hash function causes a collision, $\mathcal{B}$ outputs a random guess $\omega' \in \{0, 1\}$ and aborts.
  - **Revocation queries.** $\mathcal{B}$ outputs a random guess $\omega' \in \{0, 1\}$ and aborts.
  - **Corruption queries.** $\mathcal{A}$ can requests the private signing key of any member i.
  - **Challenge** $\mathcal{A}$ outputs (M, j, i_0, i_1) to be challenged with restriction that the corruptions and revocations of members i_0 and i_1 must not be requested before. If $j \neq j^*$, $\mathcal{B}$ outputs a random guess $\omega' \in \{0, 1\}$ and aborts. Otherwise, $\mathcal{B}$ picks $\phi \in \{0, 1\}$. If $i_b \neq i^*$, $\mathcal{B}$ outputs a random guess $\omega' \in \{0, 1\}$ and aborts, otherwise, responds with the following simulated signature.
    1. $\mathcal{B}$ sets $\alpha = b, T_2 = Z$. Note that if $Z = h^{a+b}$, then $T_2 = h_{j^*}^{\alpha + x_i^*}$
    2. Compute the simulated π as that in phase 1. If the backpatch of the hash function causes a collision, $\mathcal{B}$ outputs a random guess $\omega' \in \{0, 1\}$ and aborts.
- **Phase 2.** This is the same as Phase 1, with restriction that the corruptions and revocations of members i_0 and i_1 at interval j must not be requested.
- **Output.** $\mathcal{A}$ outputs its guess ϕ'. If $\phi' = \phi$, $\mathcal{B}$ outputs $\omega' = 1$ (implying it guesses $Z = h^{a+b}$), and otherwise outputs $\omega' = 0$ (implying it guesses

$Z = h^c$). Let the variable $\omega \in \{0, 1\}$ represents the state of Z. If $Z = h^{a+b}$ then $\omega = 1$, otherwise $\omega = 0$. The advantage of $\mathcal{B}$ is

$$|\Pr[\mathcal{B}(u, v, h, u^a, u^b, Z = h^{a+b}) = 1] - \Pr[\mathcal{B}(u, v, h, u^a, u^b, Z = h^c) = 1]|$$
$$= |\Pr[\omega' = 1|\omega = 1] - \Pr[\omega' = 1|\omega = 0]|$$
$$= \Pr[\overline{abort}]\epsilon$$

If $\mathcal{B}$ correctly guesses the value j in setup , it only aborts when the backpath is failure, the probability of which is at most q_H/p. Therefore, the probability that $\mathcal{B}$ aborts, due to $\mathcal{A}'s$ signature queries, is at most $q_S q_H/p$. On the other hand, the probability that $\mathcal{B}$ correctly guesses the value j is $1/T$ in that $\mathcal{A}$ has no information on j^*. Thus, $\Pr[\overline{abort}] \geq 1/T - q_S q_H/p$ and this theorem is proved.

Notice that the above theorem implies the two following facts: one is that the proposed scheme satisfies BU-anonymity in the random oracle model under the DLIN assumption. The other is that the strongly $\mathcal{A}$ breaks the BU-anonymity of our scheme, the more advantage $\mathcal{B}$ breaks the DLIN assumption in G_1.

Theorem 2. *Suppose an adversary $\mathcal{A}$ breaks the traceability with advantage ϵ, after q_H hash queries and q_S signature queries, then there exists an algorithm $\mathcal{B}$ breaking $(n + 1)$-SDH assumption with advantage $(\epsilon/n - 1/p)16q_H$.*

Proof. The following is an interaction framework between $\mathcal{A}$ and $\mathcal{B}$.

Setup. $\mathcal{B}$ is given (g_1, g_2, wg_2^γ) and n SDH pairs (A_i, x_i). For each $i \in [1, n]$, either $s_i = 1$ indicating that an SDH pair (A_i, x_i) is known, or $s_i = 0$ indicating that A_i is known but x_i is unknown. Run $\mathcal{A}$ on the *gpk* and *grt* drawn from the given parameters.

Hash queries. At any time, $\mathcal{A}$ can query the hash function used in π. Random values are responded with consistency.

Signing queries. $\mathcal{A}$ requests a signature of member i on message M at interval j. If $s_i = 1$, $\mathcal{B}$ responds with the signature using the secret key (A_i, x_i). If $s_i = 0$, $\mathcal{B}$ selects $\alpha \in Z_p^*$ to compute $T_1 = A_i^\alpha, T_2 = h_j^{\alpha+x_i}$ and the simulated π. If the backpatch of the hash function causes a collision, output a random guess $\omega' \in \{0, 1\}$ and abort. Otherwise, it responds with (T_1, T_2, π).

Corruption queries. $\mathcal{A}$ requests the secret key of member i. If $s_i = 1$, $\mathcal{B}$ responds with (A_i, x_i). Otherwise, it outputs a random guess $\omega' \in \{0, 1\}$ and aborts.

Output. $\mathcal{A}$ outputs a forged signature $\sigma' = (T_1', T_2', \pi')$ with secret key $(A_{i'}, x_{i'})$. If $\mathcal{B}$ fails to identify the signer by revocation check, it outputs σ'. Otherwise, some i is identified. If $s_i = 0$ then output σ'. If $s_i = 1$, output a random guess $\omega' \in \{0, 1\}$ and abort.

There are two types of forger on the above framework. In type 1, $\mathcal{A}$ forges a signature of a member different from all i. In type 2, $\mathcal{A}$ forges a signature of the member i whose corruption is not requested.

Given q-SDH instance $(g'_1, g'_2, (g'_1)^\gamma)$, we can obtain (g_1, g_2, wg_2^γ) and $q - 1$ SDH pairs such that $e(A_i, wg_2^\gamma) = e(g_1, g_2)$, following the technique of [5]. On the other hand, any SDH pair besides these $q - 1$ pairs can be transformed a solution of the q-SDH instance, which means that the q-SDH assumption is broken.

Type 1. From $(n + 1)$-SDH instance, we obtain $(g_1, g_2, w = g_2^\gamma)$ and n SDH pairs (A_i, x_i). Then, apply the framework to $\mathcal{A}$. $\mathcal{A}$ outputs a signature with secret key $(A_{i'}, x_{i'})$ such that $A_{i'} \neq A_i$ for $i \in [1, n]$. The simulation is perfect, and $\mathcal{A}$ succeeds with advantage ϵ.

Type 2. From n-SDH instance, we obtain $(g_1, g_2, w = g_2^\gamma)$ and $n - 1$ SDH pairs (A_i, x_i), which is distributed amongst n pairs, and set $s_{i'} = 0$. For the unfilled entry at random index i', select $x_{i'} \in Z_p^*(A_{i'}$ is unknown). Apply the framework to $\mathcal{A}$. It succeeds unless $\mathcal{A}$ never requests the corruption of i' while the forged signature including $A_{i'}$. The value of i' is independent of $\mathcal{A}$'s view, thus the probability that $\mathcal{A}$ outputs the signature of member i' is at least ϵ/n.

We can obtain another SDH pair beyond the given $q - 1$ SDH pairs using the framework with Type 1 or Type 2. Rewind the framework to obtain two forged signatures on the same message M and interval j, where the commitments in the π are the same but the challenges and responses are different. As shown in [11, 12], the successful probability is at least $(\epsilon' - 1/p)^2/16q_H$ by the forking lemma, where ϵ' is the probability that the framework on each forger succeeds. Therefore, we can obtain a pair $(A_{i'}, x_{i'})$ s.t. $(A_{i'} \neq A_i, x_{i'} \neq x_i)$ for all i with the probability $(\epsilon' - 1/p)^2/16q_H$.

From above, we can solve the $(n+1)$-SDH instance with $(\epsilon - 1/p)^2/16q_H$ using Type 1. And we can solve the n-SDH instance with $(\epsilon/n - 1/p)^2/16q_H$ using Type 2. We can guess the type of forger with the probability $1/2$. Therefore, the pessimistic Type 2 proves the theorem, which implies traceability is satisfied in the random oracle model under the SDH assumption.

6   Performance Comparison

The scheme in [10] is the current most efficient VLR scheme, the schemes in [11, 12] is the first two BU-VLR ones, and the seventh scheme in [13] is the current most efficient BU-VLR one with non-frameablity. The following table shows performance comparisons between these previous VLR schemes and ours in signature size and computation overhead.

Just like the above schemes, our scheme can make use of the MNT curves [15], where p is 170 bits, elements in G_1 are 171 bits, and elements in G_3 are 1020 bits. We denote the computation of multi-exponentiation and bilinear map as ME and BM respectively. Computing isomorphism takes roughly the same time as ME [10]. Besides, the properties of back-unlinkability (BU) is also showed in table 1. Note that each ME may take different time. We ignore the differences just as the previous schemes in above schemes for convenience to statistical comparison.

Table 1. Performance comparisons

	$|\sigma|$(bits)	SIGN	VER	BU
[10]	1192	8ME+2BM	6ME+(3+$|RL|$)BM	No
[11]	2893	10ME+1BM	6ME+(2+$|RL_j|$)BM	Yes
[12]	1533	7ME+1BM	4ME+(2+2$|RL_j|$)BM	Yes
[13]-scheme 7	1364	8ME+1BM	6ME+(4+$|RL_j|$)BM	Yes
Our scheme	852	4ME+1BM	2ME+(3+$|RL_j|$)BM	Yes

From the above table, we can see that: 1) The signature length of our scheme is about 71.48% of that of [10], and 6 multi-exponentiation and 1 bilinear computation are also reduced. 2) Among the BU-VLR schemes, our scheme is more efficient.

It should be noted that the revocation mechanisms in above VLR schemes are similar. They all compute and add grt_{ij} to RL_j when member i is to be revoked at interval j. Compared with that in [10] this method has two merits: (1) There is no extra computation overhead for the unrevoked members and their group secret signing keys are still valid; (2) At any time, when the revoked members want and are approved to rejoin the group, then they can create valid group signature by using their previous group secret keys again. This can be achieved only by not publishing their revocation tokens in the RL at the corresponding intervals. The above two features imply the dynamic addition and deletion of members with flexibility, which are very important in practice.

7   Conclusion

The group signature scheme [10] proposed by D. Boneh et al. is the most efficient current VLR group signature scheme. BU-VLR group signature schemes were proposed by T. Nakanishi et al. later. But their schemes are not comparable to D. Boneh et al.'s in signature size and computation overhead. Based on the DLIN assumption and the q-SDH assumption, we propose a new BU-VLR group signature. It is more efficient than the previous VLR group signature, for its signature length is just 71.48% of D. Boneh et al.'s and the overhead in computation is also lower than previous BU-VLR schemes.

References

1. Chaum, D., Heyst, E.: Group signatures. In: Davies, D.W. (ed.) EUROCRYPT 1991. LNCS, vol. 547, pp. 257–265. Springer, Heidelberg (1991)
2. Sun, X., Lin, X., Ho, P.: Secure Vehicular Communications based on Group Signature and ID-based Signature Scheme. In: IEEE ICC 2007, Glasgow, Scotland, pp. 1539–1545 (2007)
3. Zhang, J., Ma, L., Su, W., Wang, Y.: Privacy-Preserving Authentication based on Short Group Signature in Vehicular Networks. In: IEEE Symposiun on Data, Privacy and E-Commerce 2007, Chengdu, China, pp. 138–142 (2007)

4. Ateniese, G., Tsudik, G.: Some open issues and new directions in group signature schemes. In: Franklin, M. (ed.) FC 1999. LNCS, vol. 1648, pp. 196–211. Springer, Heidelberg (1999)
5. Boneh, D., Boyen, X., Shacham, H.: Short group signatures. In: Franklin, M. (ed.) CRYPTO 2004. LNCS, vol. 3152, pp. 45–55. Springer, Heidelberg (2004)
6. Ateniese, G., Camenisch, J., Joye, M., Tsudik, G.: A practical and provably secure coalition-resistant group signature scheme. In: Bellare, M. (ed.) CRYPTO 2000. LNCS, vol. 1880, pp. 255–270. Springer, Heidelberg (2000)
7. Camenisch, J., Lysyanskaya, A.: Dynamic accumulators and application to efficient revocation of anonymous credentials. In: Yung, M. (ed.) CRYPTO 2002. LNCS, vol. 2442, pp. 61–76. Springer, Heidelberg (2002)
8. Bresson, E., Stern, J.: Efficient revocation in group signatures. In: Kim, K.-c. (ed.) PKC 2001. LNCS, vol. 1992, pp. 190–206. Springer, Heidelberg (2001)
9. Ateniese, G., Song, D., Tsudik, G.: Quasi-efficient revocation in group signatures. In: Blaze, M. (ed.) FC 2002. LNCS, vol. 2357, pp. 183–197. Springer, Heidelberg (2002)
10. Boneh, D., Shacham, H.: Group signatures with verifier-local revocation. In: ACM CCS 2004, Washington, DC, USA, pp. 168–177. ACM Press, New York (2004)
11. Nakanishi, T., Funabiki, N.: Verifier-local revocation group signature schemes with backward unlinkability from bilinear maps. In: Roy, B., et al. (eds.) ASIACRYPT 2005. LNCS, vol. 3788, pp. 533–548. Springer, Heidelberg (2005)
12. Nakanishi, T., Funabiki, N.: A short verifier-local revocation group signature schemes with backward unlinkability. IEICE Trans. Fundamentals E90-A(9), 1793–1802 (2007); Also in Yoshiura, H. et. al. (eds.) IWSEC 2006. LNCS, vol. 4266, pp. 17–32. Springer, Heidelberg (2006)
13. Zhou, S., Lin, D.: Shorter Verifier-local revocation group signatures from bilinear maps. In: Pointcheval, D., Mu, Y., Chen, K. (eds.) CANS 2006. LNCS, vol. 4301, pp. 126–143. Springer, Heidelberg (2006); Full vesion is avialable at Cryptology ePrint Archive, Report 2006/286 (2006)
14. Bellare, M., Shi, H., Zhang, C.: Foundations of group signatures: The case of dynamic groups. In: Menezes, A. (ed.) CT-RSA 2005. LNCS, vol. 3376, pp. 136–153. Springer, Heidelberg (2005)
15. Miyaji, A., Nakabayashi, M., Takano, S.: New explicit conditions of elliptic curve traces for FR-reduction. IEICE Trans. Fundamentals E84-A(5), 1234–1243 (2001)
16. Boneh, D., Lynnd, B., Shachamd, H.: Short signatures from the Weil pairing. J. of Cryptology 17(4), 297–319 (2004); Extended abstract in Boyd, C. (ed.): ASIACRYPT 2001. LNCS, vol. 2248, pp. 514–532. Springer, Heidelberg (2001)
17. Boneh, D., Boyen, X.: Short Signatures Without Random Oracles and the SDH Assumption in Bilinear Groups. J. of Cryptology 21(2), 149–177 (2008)
18. Galbraith, S., Paterson, K., Smart, N.: Pairings for cryptographers. J. of Discrete Applied Mathematics 156(16), 3113–3121 (2008)
19. Fiat, A., Shamir, A.: How to prove yourself: practical solutions to identification and signature problems. In: Odlyzko, A.M. (ed.) CRYPTO 1986. LNCS, vol. 263, pp. 186–194. Springer, Heidelberg (1987)

Strongly Secure Two-Pass Attribute-Based Authenticated Key Exchange

Kazuki Yoneyama

NTT Information Sharing Platform Laboratories
yoneyama.kazuki@lab.ntt.co.jp

Abstract. In this paper, we present a two-party attribute-based authenticated key exchange scheme secure in the stronger security model than the previous models. Our strong security model is a natural extension of the eCK model, which is for PKI-based authenticated key exchange, into the attribute-based setting. We prove the security of our scheme under the gap Bilinear Diffie-Hellman assumption. Moreover, while the previous scheme needs the three-pass interaction between parties, our scheme needs only the two-pass interaction. In a practical sense, we can use any string as an attribute in our scheme because the setup algorithm of our scheme does not depend on the number of attribute candidates (i.e., the setup algorithm outputs constant size parameters).

Keywords: authenticated key exchange, attribute-based authenticated key exchange, eCK model, large universe of attributes.

1 Introduction

The aim of Authenticated key exchange (AKE) is to share a common session key between the authenticated parties. Various variants of AKE (e.g., group setting, password-based, etc.) are used for our daily life (e.g., establishing secure channels for various web-services). As a new variant of AKE, the *attribute-based* AKE (ABAKE) is recently studied [1–5]. In ABAKE schemes, the authentication condition is different from other AKE variants. While parties (the initiator and the responder) in a session authenticate each other by their identities in most of AKE variants, parties authenticate each other by their *attributes* in ABAKE schemes. That is, the key generation center (KGC) issues the static secret key to a party by using the master secret key according to the attributes of the party in advance and parties specify their policies (i.e., the condition which the peer is expected to satisfy) respectively. If the attribute of a party satisfies the policy of the peer and vice versa, the common session key is established. Since attributes can contain an identity, the ABAKE is a generalization of ID-based AKE schemes [6] The ABAKE is useful in the situation that some sensitive information (e.g., medical history) is sent with the secure channel established by some AKE scheme. Then, parties may hope to hide their identities from the peer of the session though the peer is needed to be a qualified registered person. By using the ABAKE, parties can establish the secure channel with a qualified registered person without revealing their identities.

M. Joye, A. Miyaji, and A. Otsuka (Eds.): Pairing 2010, LNCS 6487, pp. 147–166, 2010.

1.1 Motivating Problem

Recently, some ABAKE schemes are proposed. In the context of the secret handshake, some schemes are regarded as ABAKE schemes. Ateniese, Kirsch and Blanton [7] proposed a secret handshake scheme with dynamic and fuzzy matching, where users specify the attribute of the peer. However, their scheme can deal with only the simple authentication condition of whether the attributes are matching more than a threshold. Wang, Xu and Ban [1], and Wang, Xu and Fu [2, 3] proposed simple variants of the ABAKE. In their schemes, attributes are regarded as identification strings and there is no mechanism for evaluating policy. Thus, their schemes are a kind of the ID-based AKE rather than the ABAKE. Gorantla, Boyd and Nieto [4] proposed an ABAKE scheme which provides parties with the fine-grained access control based on parties' attributes. However, the condition is common for all users. Thus, their scheme does not fit in the ABAKE scenario as each party cannot specify the condition which the peer is expected to satisfy each other in the session. Their scheme is constructed based on an attribute-based key encapsulation mechanism and the security is proved in the security model based on the BR model [8].

There is another previous study on the ABAKE with the fine-grained access control by Birkett and Stebila (BS10) [5]. The BS10 scheme is generically constructed with a predicate-based signature and parties can specify the condition which the peer is expected to satisfy each other. They prove security of their scheme without random oracles in the predicate-based key exchange security model based on the BR model. However, the BS10 scheme has several drawbacks as follows: First, the BS10 scheme does not satisfy the security against revealing some of the session variables. The major security model (called eCK model [9]) for the AKE and its variants have an additional query to allow revealing all randomness (i.e., the ephemeral secret key) used during the run in a session. Since the BS10 uses a predicate-based signature as a building block, the underlying signature scheme has to be unforgeable against revealing all randomness. But, there is no such predicate-based signature scheme as in the conclusion of [5]. To prove the security in the eCK model (eCK security), it is known that the NAXOS technique [9] is effective. The NAXOS technique means that each user applies a hash function to the static secret key and the ephemeral secret key, and computes an ephemeral public key by using the output of the function as the exponent of the ephemeral public key. The adversary cannot know the output of the function as long as the adversary cannot obtain the static secret key even if the ephemeral secret key is leaked. But, the NAXOS technique cannot be applied the predicate-based signature trivially. Secondly, the BS10 scheme needs the three-pass interaction in a session. Most of Diffie-Hellman (DH) key exchange-based AKE schemes need only the two-pass interaction in a session. This drawback comes from that the BS10 scheme adopts the signed-DH paradigm [10].

Fujioka, Suzuki and Yoneyama [11] proposed a variant of ABAKE scheme from scratch, which is secure against revealing internal states. Their scheme provides parties with the key-policy based access control, that is, the access policy of a party is fixed when the party generates his static secret key, not when the party establish a session. Thus, applications of their scheme may be different from ABAKE schemes.

1.2 Our Contribution

We introduce an ABAKE scheme which solves the drawback of the BS10 scheme. Our scheme only needs the two-pass interaction in a session and can be proved the security against revealing secret keys.

First, we extend the security model of [5] to the eCK(-like) model. By allowing the adversary to pose special queries to reveal the master secret key, static secret keys and the ephemeral secret key, the attack scenario is strengthened. Also, if the adversary obtains both the ephemeral secret key and the static secret key of an party (or the master key) together, the session key can be trivially computed by the adversary. Thus, we have to consider *freshness* of the session. Freshness means the condition of the session as the adversary cannot trivially break secrecy of the session key. Although the adversary is not allowed to reveal any secret information in the session in freshness of the security model of [5], we clarify the most of malicious behaviors with respect to revealing secret information as our freshness definition.

Secondly, we construct our ABAKE scheme based on the ciphertext-policy attribute-based encryption (ABE) by Waters [12]. Waters proposed three ABE schemes in [12]. We use the third scheme based on the decisional Bilinear Diffie-Hellman (DBDH) assumption to construct our scheme. However, we cannot trivially adopt the Waters ABE scheme to the ABAKE scheme. Since the Waters ABE[1] has a flow in the encryption procedure, ciphertexts cannot be decrypted correctly as it is. Thus, we show a repair of the encryption procedure in the Waters ABE in order to carry out the decryption procedure correctly. Also, we cannot use any string as an attribute in the Waters ABE because the universe of attributes is needed to be fixed before the setup by the KGC. By using the property of the random oracle, we modify the Waters ABE as the setup algorithm does not depend on the number of attribute candidates (i.e., the setup algorithm outputs constant size parameters) and so we can use any string as an attribute. We construct our ABAKE scheme by the combination of the modified Waters ABE and the NAXOS technique.

Finally, we prove the security of our ABAKE scheme in the proposed model under the gap Bilinear Diffie-Hellman (GBDH) assumption in the random oracle model. Informally, the GBDH problem is the problem of solving the Bilinear Diffie-Hellman (BDH) problem with the help of an oracle which solves the DBDH problem. The use of such "gap" problems was first proposed by Okamoto and Pointcheval [13] and the GBDH assumption is firstly used in [14].

1.3 Related Works

Attribute-based encryption. The ABAKE is closely related to the ciphertext-policy ABE because it is natural that access policies are not decided on key gen-

---

[1] Version: 20100330:173851.

eration but on key exchange. The first ABE scheme is proposed by Sahai and Waters [15], called the fuzzy ID-based encryption, which parties must match at least a certain threshold of attributes. Bethencourt et al. [16] proposed the first ciphertext-policy ABE scheme which allows the ciphertext policies to be very expressive, but the security proof is in the generic group model. Cheung and Newport [17] proposed a provably secure ciphertext-policy ABE scheme and their scheme deals with negative attributes explicitly and supports wildcards in the ciphertext policies. Kapadia et al. [18] and Nishide et al. [19] also proposed a ciphertext-policy ABE scheme and their scheme realizes hidden ciphertext policies in a limited way, respectively. Shi et al. [20] proposed a predicate encryption scheme that focuses on range queries over huge numbers, which can also realize a ciphertext-policy ABE scheme with range queries. Boneh and Waters [21] proposed a predicate encryption scheme based on the primitive called the hidden vector encryption. It needs bilinear groups whose order is a product of two large primes, so it needs to deal with large group elements and the number of attributes is fixed at the system setup. Katz et al. [22] proposed a novel predicate encryption scheme and their scheme is very general and can realize both key-policy and ciphertext-policy ABE schemes. Waters [12] proposed expressive and efficient ciphertext-policy ABE schemes based on noninteractive assumptions. Lewko et al. [23] proposed the first fully secure ciphertext-policy ABE scheme.

2   Preliminaries

2.1   Access Structure

We introduce the notion of the access structure to represent the access control by the policy. We show the definition given in [24].

Definition 1 (Access Structure [24]). *Let $\{P_1, P_2, \ldots, P_n\}$ be a set of parties. A collection $\mathbb{A} \subseteq 2^{\{P_1, P_2, \ldots, P_n\}}$ is monotone if $\forall Att_1, Att_2$: if $Att_1 \in \mathbb{A}$ and $Att_1 \subseteq Att_2$ then $Att_2 \in \mathbb{A}$. An access structure (resp. monotone access structure) is a collection (resp. monotone collection) $\mathbb{A}$ of non-empty subsets of $\{P_1, P_2, \ldots, P_n\}$, i.e., $\mathbb{A} \subseteq 2^{\{P_1, P_2, \ldots, P_n\}} \backslash \{\emptyset\}$. The sets in $\mathbb{A}$ are called the authorized sets, and the sets not in $\mathbb{A}$ are called the unauthorized sets.*

Though this definition restricts monotone access structures, it is also possible to (inefficiently) realize general access structures by having the not of an attribute as a separate attribute altogether. Thus, the number of attributes in the system will be doubled.

2.2   Linear Secret Sharing

We use linear secret sharing schemes (LSSSs) to obtain the fine-grained access control. The LSSS can provide arbitrary conditions for the reconstruction of the secret with monotone access structures. We show the definition given in [24].

Definition 2 (Linear Secret Sharing Schemes [24]). *A secret sharing scheme Π over a set of parties $\mathbb{P}$ is called linear (over $\mathbb{Z}_p$) if*

1. *The shares for each party form a vector over $\mathbb{Z}_p$.*
2. *There exists a matrix an M with ℓ rows and n columns called the share-generating matrix for Π. For all $i = 1, \ldots, \ell$, the ith row of M we let the function ρ defined the party labeling row i as $\rho(i)$. When we consider the column vector $v = (s, r_2, \ldots, r_n)$, where $s \in \mathbb{Z}_p$ is the secret to be shared, and $r_2, \ldots, r_n \in \mathbb{Z}_p$ are randomly chosen, then Mv is the vector of ℓ shares of the secret s according to Π. The share $(Mv)_i$ belongs to party $\rho(i)$.*

The important property of LSSSs is the linear reconstruction property, defined as follows: Suppose that Π is an LSSS for the access structure $\mathbb{A}$. Let $S \in \mathbb{A}$ be any authorized set, and let $I \subset \{1, 2, \ldots \ell\}$ be defined as $I = \{i : \rho(i) \in S\}$. Then, there exist constants $\{w_i \in \mathbb{Z}_p\}_{i \in I}$ such that, if $\{\lambda_i\}$ are valid shares of any secret s according to Π, then $\sum_{i \in I} w_i \lambda_i = s$. In [24], it is shown that these constants $\{w_i\}$ can be found in time polynomial in the size of the share generating matrix M.

Note on Convention. We note that we use the convention that vector $(1, 0, 0, \ldots, 0)$ is the "target" vector for any linear secret sharing scheme. For any satisfying set of rows I in M, we will have that the target vector is in the span of I. For any unauthorized set of rows I the target vector is not in the span of the rows of the set I. Moreover, there will exist a vector w such that $w \cdot (1, 0, 0, \ldots, 0) = -1$ and $w \cdot M_i = 0$ for all $i \in I$.

2.3 Bilinear Maps

Definition 3 (Bilinear Maps). *Let G be a cyclic group of prime order p and g is a generator of G. We say that $e : G \times G \rightarrow G_T$ is a bilinear map if the following holds:*

- *For all $X, Y \in G$ and $a, b \in \mathbb{Z}_p$, we have $e(X^a, Y^b) = e(X, Y)^{ab}$,*
- *$e(g, g) \neq 1$.*

We say that G is a bilinear group if e and the group operation in G and G_T can be computed efficiently.

2.4 Gap Bilinear Diffie-Hellman Assumption

Let κ be the security parameter and p be a κ-bit prime. Let G be a cyclic group of a prime order p with a generator g and G_T be a cyclic group of the prime order p with a generator g_T. Let $e : G \times G \rightarrow G_T$ be a bilinear map. We say that G, G_T are bilinear groups with the pairing e.

The gap Bilinear Diffie-Hellman problem is as follows. We define the BDH function $\mathsf{BDH} : G^3 \rightarrow G_T$ as $\mathsf{BDH}(g^a, g^b, g^c) = e(g, g)^{abc}$, and the DBDH predicate $\mathsf{DBDH} : G^4 \rightarrow \{0, 1\}$ as a function which takes an input $(g^a, g^b, g^c, e(g, g)^d)$ and returns 1 if $abc = d \bmod p$ and 0 otherwise. An adversary $\mathcal{A}$ is given input $\alpha = g^a, \beta = g^b, \gamma = g^c \in G$ selected uniformly random and oracle access to $\mathsf{DBDH}(\cdot, \cdot, \cdot, \cdot)$ oracle, and tries to compute $\mathsf{BDH}(\alpha, \beta, \gamma)$. For adversary $\mathcal{A}$, we define advantage

$$\mathsf{Adv}^{\mathrm{GBDH}}(\mathcal{A}) = \Pr[\alpha, \beta, \gamma \in G, \mathcal{A}^{\mathrm{DBDH}(\cdot,\cdot,\cdot,\cdot)}(g, \alpha, \beta, \gamma) = \mathsf{BDH}(\alpha, \beta, \gamma)],$$

where the probability is taken over the choices of g^a, g^b, g^c and the random tape of $\mathcal{A}$.

Definition 4 (Gap Bilinear Diffie-Hellman Assumption). *We say that the GBDH assumption in G holds if for all polynomial-time adversary $\mathcal{A}$, the advantage $\mathsf{Adv}^{\mathrm{GBDH}}(\mathcal{A})$ is negligible in security parameter κ.*

3   Security Model

In this section, we introduce an eCK security model for the ABAKE. Our attribute-based eCK (ABeCK) model is an extension of the eCK security model for conventional AKE by the LaMacchia, Lauter and Mityagin [9] to the ABAKE.

The proposed ABeCK model is different from the original eCK model in the following points: (1) the session is identified by a set of attributes $\mathbb{S}_P$ of party P, (2) freshness conditions for queries to reveal static secret keys are different, and (3) the query to reveal the master key is allowed for the adversary same as in the ID-based AKE.

Syntax. An ABAKE scheme consists of the following algorithms. We denote a party by P and his associated set of attributes by $\mathbb{S}_P$. The party P and other parties are modeled as a probabilistic polynomial-time Turing machine.

Setup. The setup algorithm **Setup** takes a security parameter κ as input, and outputs a master secret key MSK and a master public key MPK, i.e.,

$$\mathbf{Setup}(1^\kappa) \to (MSK, MPK).$$

Key Generation. The key generation algorithm **KeyGen** takes the master secret key MSK, the master public key MPK, and a set of attributes $\mathbb{S}_P$ given by a party P, and outputs a static secret key $SK_{\mathbb{S}_P}$ corresponding to $\mathbb{S}_P$, i.e.,

$$\mathbf{KeyGen}(MSK, MPK, \mathbb{S}_P) \to SK_{\mathbb{S}_P}.$$

Key Exchange. The party A and the party B share a session key by performing the following n-pass protocol. A (resp. B) selects a policy $\mathbb{A}_A$ (resp. $\mathbb{A}_B$) as an access structure, respectively.

A starts the protocol by computing the 1st message m_1 by the algorithm **Message**, that takes the master public key MPK, the set of attributes $\mathbb{S}_A$, the static secret key $SK_{\mathbb{S}_A}$ and the policy $\mathbb{A}_A$, and outputs 1st message m_1. A sends m_1 to the other party B.

For $i = 2, ..., n$, upon receiving the $(i-1)$th message m_{i-1}, the party P ($P = A$ or B) computes the ith message by algorithm **Message**, that takes the master public key MPK, the set of attributes $\mathbb{S}_P$, the static secret key $SK_{\mathbb{S}_P}$, the policy $\mathbb{A}_P$ and the sent and received messages $m_1, \ldots, m_{i-1}$, and outputs the ith message m_i, i.e.,

$$\textbf{Message}(MPK, \mathbb{S}_P, SK_{\mathbb{S}_P}, \mathbb{A}_P, m_1, \ldots, m_{i-1}) \rightarrow m_i.$$

The party P sends m_i to the other user $\bar{P}$ ($\bar{P} = B$ or A).

Upon receiving or after sending the final nth message m_n, P computes a session key by algorithm **SessionKey**, that takes the master public key MPK, the set of attributes $\mathbb{S}_P$, the static secret key $SK_{\mathbb{S}_P}$, the policy $\mathbb{A}_P$ and the sent and received messages $m_1, \ldots, m_n$, and outputs an session key K, i.e.,

$$\textbf{SessionKey}(MPK, \mathbb{S}_P, SK_{\mathbb{S}_P}, \mathbb{A}_P, m_1, \ldots, m_n) \rightarrow K.$$

Both parties A and B can compute the same session key if and only if $\mathbb{S}_A \in \mathbb{A}_B$ and $\mathbb{S}_B \in \mathbb{A}_A$.

Session. An invocation of a protocol is called a *session*. A session is activated with an incoming message of the forms $(\mathcal{I}, \mathbb{S}_A, \mathbb{S}_B)$ or $(\mathcal{R}, \mathbb{S}_B, \mathbb{S}_A, m_1)$, where $\mathcal{I}$ and $\mathcal{R}$ with role identifiers, and A and B with user identifiers. If A was activated with $(\mathcal{I}, \mathbb{S}_A, \mathbb{S}_B)$, then A is called the session *initiator*. If B was activated with $(\mathcal{R}, \mathbb{S}_B, \mathbb{S}_A, m_1)$, then B is called the session *responder*. After activated with an incoming message of the forms $(\mathcal{I}, \mathbb{S}_A, \mathbb{S}_B, m_1, \ldots, m_{k-1})$ from the responder B, the initiator A outputs m_k, then may be activated next by an incoming message of the forms $(\mathcal{I}, \mathbb{S}_A, \mathbb{S}_B, m_1, \ldots, m_{k+1})$ from the responder B. After activated by an incoming message of the forms $(\mathcal{R}, \mathbb{S}_B, \mathbb{S}_A, m_1, \ldots, m_k)$ from the initiator A, the responder B outputs m_{k+1}, then may be activated next by an incoming message of the forms $(\mathcal{R}, \mathbb{S}_B, \mathbb{S}_A, m_1, \ldots, m_{k+2})$ from the initiator A. Upon receiving or after sending the final nth message m_n, both parties A and B computes a session key K.

If A is the initiator of a session, the session is identified by $sid = (\mathcal{I}, \mathbb{S}_A, \mathbb{S}_B, m_1), (\mathcal{I}, \mathbb{S}_A, \mathbb{S}_B, m_1, m_2, m_3), \ldots, (\mathcal{I}, \mathbb{S}_A, \mathbb{S}_B, m_1, \ldots, m_n)$. If B is the responder of a session, the session is identified by $sid = (\mathcal{R}, \mathbb{S}_B, \mathbb{S}_A, m_1, m_2), (\mathcal{R}, \mathbb{S}_B, \mathbb{S}_A, m_1, m_2, m_3, m_4), \ldots, (\mathcal{R}, \mathbb{S}_B, \mathbb{S}_A, m_1, \ldots, m_n)$. We say that a session is *completed* if a session key is computed in the session. The *matching session* of a completed session $(\mathcal{I}, \mathbb{S}_A, \mathbb{S}_B, m_1, \ldots, m_n)$ is a completed session with identifier $(\mathcal{R}, \mathbb{S}_B, \mathbb{S}_A, m_1, \ldots, m_n)$ and vice versa.

Adversary. The adversary $\mathcal{A}$ that is modeled as a probabilistic polynomial-time Turing machine controls all communications between parties including the session activation by performing the following queries.

- Send(message): The message has one of the following forms: $(\mathcal{I}, \mathbb{S}_A, \mathbb{S}_B, m_1, \ldots, m_k)$, or $(\mathcal{R}, \mathbb{S}_B, \mathbb{S}_A, m_1, \ldots, m_{k+1})$. The adversary obtains the response from the party.

Revealing secret information of parties is captured via the following queries.

- SessionReveal(*sid*): The adversary obtains the session key for the session *sid* if the session is completed.

- EphemeralReveal(sid): The adversary obtains the ephemeral secret key associated with the session sid.
- StaticReveal($\mathbb{S}_P$): The adversary learns the static secret key corresponding to the set of attributes $\mathbb{S}_P$.
- MasterReveal: The adversary learns the master secret key of the system.
- Establish($P, \mathbb{S}_P$): This query allows the adversary to register a static public key corresponding to the set of attributes $\mathbb{S}_P$ on behalf of the party P; the adversary totally controls that party. If a party is established by Establish($P, \mathbb{S}_P$) query issued by the adversary, then we call the party P *dishonest*. If not, we call the party *honest*.

Freshness. For the security definition, we need the notion of freshness.

Definition 5 (Freshness). *Let* $sid^* = (\mathcal{I}, \mathbb{S}_A, \mathbb{S}_B, m_1, \ldots, m_n)$ *or* $(\mathcal{R}, \mathbb{S}_B, \mathbb{S}_A, m_1, \ldots, m_n)$ *be a completed session between an honest user* A *with the set of attributes* $\mathbb{S}_A$ *and* B *with* $\mathbb{S}_B$. *If the matching session exists, then let* $\overline{sid}^*$ *be the matching session of* sid^*. *We say* sid^* *to be* fresh *if none of the following conditions hold:*

1. *The adversary issues a* SessionReveal(sid^*) *or* SessionReveal($\overline{sid}^*$) *query if* $\overline{sid}^*$ *exists,*
2. $\overline{sid}^*$ *exists and the adversary makes either of the following queries*
   - *both* StaticReveal($\mathbb{S}$) *s.t.* $\mathbb{S} \in \mathbb{A}_B$ *and* EphemeralReveal(sid^*), *or*
   - *both* StaticReveal($\mathbb{S}$) *s.t.* $\mathbb{S} \in \mathbb{A}_A$ *and* EphemeralReveal($\overline{sid}^*$),
3. $\overline{sid}^*$ *does not exist and the adversary makes either of the following queries*
   - *both* StaticReveal($\mathbb{S}$) *s.t.* $\mathbb{S} \in \mathbb{A}_B$ *and* EphemeralReveal(sid^*),[2] *or*
   - StaticReveal($\mathbb{S}$) *s.t.* $\mathbb{S} \in \mathbb{A}_A$,

where

- *if the adversary issues* MasterReveal *query, we regard that the adversary issues* StaticReveal($\mathbb{S}$) *s.t.* $\mathbb{S} \in \mathbb{A}_A$ *and* StaticReveal($\mathbb{S}$) *s.t.* $\mathbb{S} \in \mathbb{A}_B$ *queries.*

Security Experiment. For our security definition, we consider the following security experiment. Initially, the adversary $\mathcal{A}$ is given a set of honest users, and makes any sequence of the queries described above. During the experiment, $\mathcal{A}$ makes the following query.

- Test(sid^*): Here, sid^* must be a fresh session. Select random bit $b \in \{0, 1\}$, and return the session key held by sid^* if $b = 0$, and return a random key if $b = 1$.

The experiment continues until $\mathcal{A}$ makes a guess b'. The adversary *wins* the game if the test session sid^* is still fresh and if $\mathcal{A}$'s guess is correct, i.e., $b' = b$. The advantage of $\mathcal{A}$ in the experiment with the ABAKE scheme Π is defined as

$$\mathsf{Adv}_\Pi^{\mathrm{ABAKE}}(\mathcal{A}) = \Pr[\mathcal{A} \; wins] - \frac{1}{2}.$$

---

[2] $\mathbb{A}_B$ is decided by the adversary and sent as a part of messages to A because indeed B does not exist in the session.

We define the security as follows.

Definition 6 (ABeCK Security). *We say that an ABAKE scheme Π is secure in the ABeCK model, if the following conditions hold:*

1. *If two honest parties completing matching sessions and $\mathbb{S}_A \in \mathbb{A}_B$ and $\mathbb{S}_B \in \mathbb{A}_A$ hold, then, except with negligible probability, they both compute the same session key.*
2. *For any probabilistic polynomial-time adversary $\mathcal{A}$, $\mathrm{Adv}_\Pi^{\mathrm{ABAKE}}(\mathcal{A})$ is negligible.*

Moreover, we say that the ABAKE scheme is selectively secure *in the ABeCK model, if $\mathcal{A}$ specifies $\mathbb{A}_A$ in sid^* (and $\mathbb{A}_B$ in $\overline{sid}^*$ if $\overline{sid}^*$ exists) at the beginning of the security experiment.*

4 Modification of the Waters ABE

In this section, we point out the flaw of the Waters ABE [12] and show some modification of the scheme.

4.1 Waters ABE

We review the Waters ABE based on the DBDH assumption.

The scheme is parameterized by n_{max} which specifies the maximum number of columns in share-generating matrices corresponding to access structures. "$s \in_R S$" means randomly choosing an element s of a set S. "$|V|$" means the bit length of a value V.

Setup : For input a security parameter κ and the number of attributes $\mathbb{U}$, choose p, G, G_T, g and g_T such that bilinear groups with pairing $e : G \times G \to G_T$ of order κ-bit prime p with generators g and g_T, respectively. Then, output a master public key $MPK := (g, g^r, g_T^z, (h_{1,1}, \ldots, h_{1,\mathbb{U}}), \ldots, (h_{n_{max},1}, \ldots, h_{n_{max},\mathbb{U}}))$ and a master secret key $MSK := g^z$ such that $(h_{1,1}, \ldots, h_{1,\mathbb{U}}), \ldots, (h_{n_{max},1}, \ldots, h_{n_{max},\mathbb{U}}) \in_R G^{\mathbb{U}}$ and $r, z \in_R \mathbb{Z}_p$.

Encrypt : For input the master public key MPK, a plaintext m and an LSSS access structure (M, ρ) where the injective function ρ associates rows of $\ell \times n$ share-generating matrix M to attributes, choose $u_1, \ldots, u_n \in_R \mathbb{Z}_p$. Then, output the ciphertext $CT := (X', X, \{U\})$ for $1 \leq i \leq \ell$ and $1 \leq j \leq n$ such that $X' = m \cdot (g_T^z)^{u_1}$, $X = g^{u_1}$ and $U_{i,j} = (g^r)^{M_{i,j} u_j} h_{j,\rho(i)}^{-u_1}$ (let $\{U\}$ denote the set of $U_{i,j}$ for $1 \leq i \leq \ell$ and $1 \leq j \leq n$).

KeyGen : For input the master secret key MSK and a set of attributes $\mathbb{S}$, choose $t_1, \ldots, t_{n_{max}} \in \mathbb{Z}_p$, and compute $S' = g^z g^{rt_1}$, $T_j = g^{t_j}$ for $1 \leq j \leq n_{max}$ (let $\{T\}$ denote the set of T_j for $1 \leq j \leq n_{max}$) and $S_k = \prod_{1 \leq j \leq n_{max}} h_{j,k}^{t_j}$ for $k \in \mathbb{S}$ (let $\{S\}$ denote the set of S_k for $k \in \mathbb{S}$). Then, output a secret key $SK := (S', \{T\}, \{S\})$.

Decrypt : For input a ciphertext CT for the access structure (M, ρ) and a secret key SK for a set $\mathbb{S}$, let $I \subset \{1, 2, \ldots, \ell\}$ be defined as $I = \{i : \rho(i) \in \mathbb{S}\}$. We suppose that $\mathbb{S}$ satisfies M and ρ. Then, find $\{w_i \in \mathbb{Z}_p\}_{i \in I}$ such that $\sum_{i \in I} w_i \lambda_i = s$ for valid shares $\{\lambda_i\}$ of any secret s according to M and output the plaintext m as

$$m = X' \left(\prod_{1 \le j \le n} e(T_j, \prod_{i \in I} U_{i,j}^{w_i}) \right) \prod_{i \in I} e(X, S_{\rho(i)}^{w_i}) / e(X, S').$$

4.2 Flaw of the Encryption Procedure

Since n is decided according to the access structure and $n \le n_{max}$, n will be almost smaller than n_{max}. We point out that if $n \ne n_{max}$, the decryption algorithm cannot decrypt the plaintext correctly.

The design principle of the decryption can be considered as follows: First, $\prod_{1 \le j \le n} e(T_j, \prod_{i \in I} U_{i,j}^{w_i})$ is represented as $\prod_{1 \le j \le n} e(g^{t_j}, g^{\sum_{i \in I} r M_{i,j} u_j w_i})$. $\prod_{1 \le j \le n} e(g^{t_j}, \prod_{i \in I} h_{j,\rho(i)}^{-u_1 w_i})$. If $n = n_{max}$, $\prod_{1 \le j \le n} e(g^{t_j}, \prod_{i \in I} h_{j,\rho(i)}^{-u_1 w_i})$ is canceled out by $e(X, S_{\rho(i)}^{w_i})$ because $\prod_{1 \le j \le n} e(g^{t_j}, \prod_{i \in I} h_{j,\rho(i)}^{-u_1 w_i}) = \prod_{1 \le j \le n} \prod_{i \in I} e(g, h_{j,\rho(i)})^{-u_1 t_j w_i}$ and $e(X, S_{\rho(i)}^{w_i}) = \prod_{i \in I} e(g^{u_1}, \prod_{1 \le j \le n_{max}} h_{j,\rho(i)}^{t_j w_i}) = \prod_{1 \le j \le n_{max}} \prod_{i \in I} e(g, h_{j,\rho(i)})^{u_1 t_j w_i}$. Next, $\prod_{1 \le j \le n_{max}} e(g^{t_j}, g^{\sum_{i \in I} r M_{i,1} u_1 w_i})$ is transformed into $e(g^{t_1}, g^{\sum_{i \in I} r M_{i,1} u_1 w_i}) = g_T^{u_1 r t_1}$ by the linear reconstruction property of the LSSS. Finally, since $g_T^{u_1 r t_1} / e(X, S') = g_T^{z u_1}$, m is decrypted.

However, if $n \ne n_{max}$, the first step of this procedure does not work correctly. That is, $\prod_{n+1 \le j \le n_{max}} \prod_{i \in I} e(g, h_{j,\rho(i)})^{u_1 t_j w_i}$ is not canceled out and so remains. Thus, the Waters ABE lacks completeness in almost all cases (i.e., the number of the representation of the access structure for a receiver is not equal to n_{max}).

4.3 Our Modified Waters ABE

To repair the flaw above, we modify the Waters ABE to satisfy completeness. Moreover, the original Waters ABE restricts the universe of attributes because the setup algorithm needs to fix it in advance and the master public key depends on it. We also extend the scheme to allow the large universe of attributes by using the random oracle. Thus, in our ABE scheme, we can use any string as an attribute and the master public key only needs the constant size. Note that such an extension to allow the large universe of attributes is very popular method in the researches of ABE.

We describe our ABE scheme.

Setup : For input a security parameter κ, choose p, G, G_T, g and g_T such that bilinear groups with pairing $e : G \times G \to G_T$ of order κ-bit prime p with generators g and g_T, respectively. $H : \{0, 1\}^* \to G$ is a hash function modeled as the random oracle. Then, output a master public key $MPK := (g, g^r, g_T^z)$ and a master secret key $MSK := g^z$ such that $r, z \in_R \mathbb{Z}_p$.

Encrypt : For input the master public key MPK, a plaintext m and an LSSS access structure (M, ρ) where the injective function ρ associates rows of $\ell \times n$ share-generating matrix M to attributes, choose $u_1, \ldots, u_n \in_R \mathbb{Z}_p$. Then, output the ciphertext $CT := (X', X, \{U\})$ for $1 \leq i \leq \ell$ and $1 \leq j \leq n_{max}$ such that $X' = m \cdot (g_T^z)^{u_1}$, $X = g^{u_1}$, and $U_{i,j} = (g^r)^{M_{i,j}u_j} H(j, \rho(i))^{-u_1}$ for $1 \leq j \leq n$ and $U_{i,j} = H(j, \rho(i))^{-u_1}$ for $n+1 \leq j \leq n_{max}$ (let $\{U\}$ denote the set of $U_{i,j}$ for $1 \leq i \leq \ell$ and $1 \leq j \leq n_{max}$).

KeyGen : For input the master secret key MSK and a set of attributes $\mathbb{S}$, choose $t_1, \ldots, t_{n_{max}} \in \mathbb{Z}_p$, and compute $S' = g^z g^{rt_1}$, $T_j = g^{t_j}$ for $1 \leq j \leq n_{max}$ (let $\{T\}$ denote the set of T_j for $1 \leq j \leq n_{max}$) and $S_k = \prod_{1 \leq j \leq n_{max}} H(j, k)^{t_j}$ for $k \in \mathbb{S}$ (let $\{S\}$ denote the set of S_k for $k \in \mathbb{S}$). Then, output a secret key $SK := (S', \{T\}, \{S\})$.

Decrypt : For input a ciphertext CT for the access structure (M, ρ) and a secret key SK for a set $\mathbb{S}$, let $I \subset \{1, 2, \ldots, \ell\}$ be defined as $I = \{i : \rho(i) \in \mathbb{S}\}$. We suppose that $\mathbb{S}$ satisfies M and ρ. Then, find $\{w_i \in \mathbb{Z}_p\}_{i \in I}$ such that $\sum_{i \in I} w_i \lambda_i = s$ for valid shares $\{\lambda_i\}$ of any secret s according to M and output the plaintext m as

$$m = X' \left(\prod_{1 \leq j \leq n_{max}} e(T_j, \prod_{i \in I} U_{i,j}^{w_i}) \right) \prod_{i \in I} e(X, S_{\rho(i)}^{w_i}) / e(X, S').$$

Our ABE scheme repair the flaw of the Waters ABE as follows: We add $U_{i,j} = H(j, \rho(i))^{-u_1}$ for $n+1 \leq j \leq n_{max}$ in the encryption algorithm and the product of $e(T_j, \prod_{i \in I} U_{i,j}^{w_i})$ is computed for $1 \leq j \leq n_{max}$ in the decryption algorithm. By this modification, $\prod_{1 \leq j \leq n_{max}} e(g^{t_j}, \prod_{i \in I} h_{j,\rho(i)}^{-u_1 w_i})$ is certainly canceled out by $e(X, S_{\rho(i)}^{w_i})$ regardless of n. Hence, our ABE scheme correctly works even if $n \neq n_{max}$.

Also, we remove $\{h_{j,i}\}$ in the setup algorithm and replace $\{h_{j,i}\}$ with $H(j, i)$ in the encryption and key generation algorithms. By this modification, the setup algorithm is carried out independently of the number of attributes and the master public key becomes the constant size. Since H is the random oracle, we can embed arbitrary values to outputs of H. Thus, the security proof correctly works.[3]

Theorem 1. *Suppose the DBDH assumption holds. Then, our ABE scheme is selectively secure in the random oracle model.*

Due to space limitations, we will show definitions of the selective security and the DBDH assumption, and the proof of Theorem 1 in the full version of this paper.

---

[3] Of course, we can repair the flow of the Waters ABE without extension to allowing the large universe of attributes. Then, we add $U_{i,j} = h_{j,\rho(i)}^{-u_1}$ instead of $H(j, \rho(i))^{-u_1}$ for $n+1 \leq j \leq n_{max}$ in the encryption algorithm. The security of this modification can be proved without random oracles.

5   Our Construction

In this section, we provide our ABAKE scheme that allows fine-grained access structure and large universe of attributes. Expressiveness of access structures is due to the direct application of LSSSs for the access control same as the Waters ABE. Our construction is also parameterized by n_{max} which specifies the maximum number of columns in share-generating matrices corresponding to access structures.

Setup : For input a security parameter κ, choose p, G, G_T, g and g_T such that bilinear groups with pairing $e : G \times G \to G_T$ of order κ-bit prime p with generators g and g_T, respectively. $H_1 : \{0,1\}^* \to G$, $H_2 : \{0,1\}^* \to \mathbb{Z}_p$ and $H_3 : \{0,1\}^* \to \{0,1\}^\kappa$ are hash functions modeled as random oracles. Then, output a master public key $MPK := (g, g^r, g_T^z)$ and a master secret key $MSK := g^z$ such that r, $z \in_R \mathbb{Z}_p$.

KeyGen : For input a set of attributes $\mathbb{S}_P$ from a party P, choose $t_{P_1}, \ldots, t_{P_{n_{max}}} \in \mathbb{Z}_p$, and compute $S'_P = g^z g^{rt_{P_1}}$, $T_{P_j} = g^{t_{P_j}}$ for $1 \le j \le n_{max}$ (let $\{T_P\}$ denote the set of T_{P_j} for $1 \le j \le n_{max}$) and $S_{P_k} = \prod_{1 \le j \le n_{max}} H_1(j, k)^{t_{P_j}}$ for $k \in \mathbb{S}_P$ (let $\{S_P\}$ denote the set of S_{P_k} for $k \in \mathbb{S}_P$). Then, output a static secret key $SK_P := (S'_P, \{T_P\}, \{S_P\})$.

Exchange : We suppose that the party A is the session initiator and the party B is the session responder. A has the static secret key $SK_A = (S'_A, \{T_A\}, \{S_A\})$ corresponding to the set of his attributes $\mathbb{S}_A$ and B has the static secret key $SK_B = (S'_B, \{T_B\}, \{S_B\})$ corresponding to the set of his attributes $\mathbb{S}_B$. Then, A sends to B the ephemeral public key EPK_A corresponding to the access structure $\mathbb{A}_A$, and B sends to A the ephemeral public key EPK_B corresponding to the access structure $\mathbb{A}_B$. Finally, both parties A and B compute the shared key K if and only if the set of attributes $\mathbb{S}_A$ satisfies the access structure $\mathbb{A}_B$ and the set of attributes $\mathbb{S}_B$ satisfies the access structure $\mathbb{A}_A$.

1. First, A decides an access structure $\mathbb{A}_A$ which he hopes that the set of attributes $\mathbb{S}_B$ of B satisfies $\mathbb{A}_A$. Then, A derives the $\ell_A \times n_A$ share-generating matrix M_A and the injective labeling function ρ_A in a LSSS for $\mathbb{A}_A$. A chooses at random the ephemeral secret key $\tilde{u}_1, \ldots, \tilde{u}_{n_A} \in_R \mathbb{Z}_p$. Then, A computes $u_j = H_2(S'_A, \{T_A\}, \{S_A\}, \tilde{u}_j)$ for $1 \le j \le n_A$ and $X = g^{u_1}$. Also, A computes $U_{i,j} = g^{rM_{A_{i,j}}u_j} H_1(j, \rho_A(i))^{-u_1}$ for $1 \le i \le \ell_A$ and $1 \le j \le n_A$, and $U_{i,j} = H_1(j, \rho_A(i))^{-u_1}$ for $1 \le i \le \ell_A$ and $n_{A+1} \le j \le n_{max}$ (let $\{U\}$ denote the set of $U_{i,j}$ for $1 \le i \le \ell_A$ and $1 \le j \le n_{max}$). A sends $EPK_A := (X, \{U\})$, M_A and ρ_A to B, and erases $u_1, \ldots, u_{n_A}$.

2. Upon receiving EPK_A, B checks whether the set of his attributes $\mathbb{S}_B$ satisfies the access structure M_A and ρ_A, and $X, \{U\} \in G$ holds. If not, B aborts. Otherwise, B decides an access structure $\mathbb{A}_B$ which he hopes that the set of attributes $\mathbb{S}_A$ of A satisfies $\mathbb{A}_B$. Then, B derives the $\ell_B \times n_B$ share-generating matrix M_B and the labeling function ρ_B in an LSSS for $\mathbb{A}_B$. B chooses at random the ephemeral secret key $\tilde{v}_1, \ldots, \tilde{v}_{n_B} \in_R \mathbb{Z}_p$.

Then, B computes $v_j = H_2(S'_B, \{T_B\}, \{S_B\}, \tilde{v}_j)$ for $1 \leq j \leq n_B$ and $Y = g^{v_1}$. Also, B computes $V_{i,j} = g^{r M_{B_{i,j}} v_j} H_1(j, \rho_B(i))^{-v_1}$ for $1 \leq i \leq \ell_B$ and $1 \leq j \leq n_B$, and $V_{i,j} = H_1(j, \rho_B(i))^{-v_1}$ for $1 \leq i \leq \ell_B$ and $n_{B+1} \leq j \leq n_{max}$ (let $\{V\}$ denote the set of $V_{i,j}$ for $1 \leq i \leq \ell_B$ and $1 \leq j \leq n_{max}$). B sends $EPK_B := (Y, \{V\})$, M_B and ρ_B to A.

B computes the shared secrets as follows: We suppose that $\mathbb{S}_B$ satisfies M_A and ρ_A, and let $I_B \subset \{1, 2, \ldots, \ell_A\}$ be defined as $I_B = \{i : \rho_A(i) \in \mathbb{S}_B\}$. Then, B can efficiently find $\{w_{B_i} \in \mathbb{Z}_p\}_{i \in I_B}$ such that $\sum_{i \in I_B} w_{B_i} \lambda_i = s$ for valid shares $\{\lambda_i\}$ of any secret s according to M_A.[4] Note that, if $\mathbb{S}_B$ does not satisfy M_A and ρ_A, B cannot find all w_{B_i} for $i \in I_B$ from the property of LSSSs.

Then, B sets the shared secrets

$$\sigma_1 = e(X, S'_B) / \left(\prod_{1 \leq j \leq n_{max}} e(T_{B_j}, \prod_{i \in I_B} U_{i,j}^{w_{B_i}}) \right) \prod_{i \in I_B} e(X, S_{B_{\rho_A(i)}}^{w_{B_i}}),$$
$$\sigma_2 = (g_T^z)^{v_1}, \quad \sigma_3 = X^{v_1}$$

and the session key $K = H_3(\sigma_1, \sigma_2, \sigma_3, (X, \{U\}, M_A, \rho_A), (Y, \{V\}, M_B, \rho_B))$. B completes the session with the session key K, and erases $v_1, \ldots, v_{n_B}$.

3. Upon receiving EPK_B, A checks whether the set of his attributes $\mathbb{S}_A$ satisfies the access structure M_B and ρ_B, and $Y, \{V\} \in G$ holds. If not, A aborts. Otherwise, A computes the shared secrets as follows: We suppose that $\mathbb{S}_A$ satisfies M_B and ρ_B, and let $I_A \subset \{1, 2, \ldots, \ell_B\}$ be defined as $I_A = \{i : \rho_B(i) \in \mathbb{S}_A\}$. Then, A can efficiently find $\{w_{A_i} \in \mathbb{Z}_p\}_{i \in I_A}$ such that $\sum_{i \in I_A} w_{A_i} \lambda_i = s$ for valid shares $\{\lambda_i\}$ of any secret s according to M_B.[5] Note that, if $\mathbb{S}_A$ does not satisfy M_B and ρ_B, A cannot find all w_{A_i} for $i \in I_A$ from the property of LSSSs.

Then, A sets the shared secrets

$$\sigma_2 = e(Y, S'_A) / \left(\prod_{1 \leq j \leq n_{max}} e(T_{A_j}, \prod_{i \in I_A} V_{i,j}^{w_{A_i}}) \right) \prod_{i \in I_A} e(Y, S_{A_{\rho_B(i)}}^{w_{A_i}}),$$
$$\sigma_1 = (g_T^z)^{H_2(S'_A, \{T_A\}, \{S_A\}, \tilde{u}_1)}, \quad \sigma_3 = Y^{H_2(S'_A, \{T_A\}, \{S_A\}, \tilde{u}_1)}$$

and the session key $K = H_3(\sigma_1, \sigma_2, \sigma_3, (X, \{U\}, M_A, \rho_A), (Y, \{V\}, M_B, \rho_B))$. A completes the session with the session key K.

The shared secrets that both parties compute are

$$\sigma_1 = e(X, S'_B) / \left(\prod_{1 \leq j \leq n_{max}} e(T_{B_j}, \prod_{i \in I_B} U_{i,j}^{w_{B_i}}) \right) \prod_{i \in I_B} e(X, S_{B_{\rho_A(i)}}^{w_{B_i}})$$
$$= e(X, S'_B) / \prod_{1 \leq j \leq n_A} e(g^{t_{B_j}}, g^{\sum_{i \in I_B} r M_{A_{i,j}} u_j w_{B_i}})$$
$$\cdot \prod_{1 \leq j \leq n_{max}} e(g^{t_{B_j}}, \prod_{i \in I_B} H_1(j, \rho_A(i)^{-u_1 w_{B_i}}))$$
$$\cdot \prod_{i \in I_B} e(g^{u_1}, \prod_{1 \leq j \leq n_{max}} H_1(j, \rho_A(i))^{t_{B_j} w_{B_i}})$$

---

[4] In this case, the secret corresponds to u_1 and shares correspond to $\{M_{A_{i,j}} u_j\}$.

[5] In this case, the secret corresponds to v_1 and shares correspond to $\{M_{B_{i,j}} v_j\}$.

$$= e(X, S_B') / \prod_{1 \leq j \leq n_A} e(g^{t_{B_j}}, g^{\sum_{i \in I_B} r M_{A_{i,j}} u_j w_{B_i}})$$

$$= e(g^{u_1}, g^z g^{r t_{B_1}}) / e(g^{t_{B_1}}, g^{\sum_{i \in I_B} r M_{A_{i,1}} u_1 w_{B_i}})$$

$$= g_T^{u_1(z + r t_{B_1})} / g_T^{u_1 r t_{B_1}}$$

$$= g_T^{z u_1} (= (g_T^z)^{u_1}),$$

$$\sigma_2 = e(Y, S_A') / \left(\prod_{1 \leq j \leq n_{max}} e(T_{A_j}, \prod_{i \in I_A} V_{i,j}^{w_{A_i}}) \right) \prod_{i \in I_A} e(S_{A_{\rho_B(i)}}^{w_{A_i}}, Y),$$

$$= e(Y, S_A') / \prod_{1 \leq j \leq n_B} e(g^{t_{A_j}}, g^{\sum_{i \in I_A} r M_{B_{i,j}} v_j w_{A_i}})$$

$$\cdot \prod_{1 \leq j \leq n_{max}} e(g^{t_{A_j}}, \prod_{i \in I_A} H_1(j, \rho_B(i)^{-v_1 w_{A_i}}))$$

$$\cdot \prod_{i \in I_A} e(g^{v_1}, \prod_{1 \leq j \leq n_{max}} H_1(j, \rho_B(i))^{t_{A_j} w_{A_i}})$$

$$= g_T^{z v_1} (= (g_T^z)^{v_1}),$$

$$\sigma_3 = X^{v_1} = Y^{u_1} = g^{u_1 v_1},$$

and therefore they can compute the same session key K.

Design Principle. We construct our ABAKE scheme by combining the modified Waters ABE scheme in Section 4.3 and the NAXOS technique [9]. From the structure of the ABE scheme, the ephemeral secret key needs to contain n elements. Thus, we apply the NAXOS technique to each element. Specifically, we convert $(\tilde{u}_1, \dots, \tilde{u}_{n_A})$ and $(\tilde{v}_1, \dots, \tilde{v}_{n_B})$ into $(u_1, \dots, u_{n_A})$ and $(v_1, \dots, v_{n_B})$, respectively, by using the random oracle H_2.

The decryption algorithm of the ABE scheme can be applied to the derivation of the session key. In the process of the decryption, the decryption algorithm computes $g_T^{t_1 z}$ where t_1 is the randomness in the encryption and z is the secret in the setup. In our ABAKE, we use $\sigma_1 = g_T^{u_1 z}$ and $\sigma_2 = g_T^{v_1 z}$ as a part of the seed of the session key where u_1 and v_1 are derived from the ephemeral secret keys of A and B respectively. However, only $g_T^{u_1 z}$ and $g_T^{v_1 z}$ are not enough to achieve the security in the ABeCK model in Section 3. The ABeCK model allows the adversary to reveal the master secret key. We cannot prove the security in such a case because the simulator cannot embed the BDH instance to the master secret key and cannot extract information to obtain the answer of the GBDH problem from only $g_T^{u_1 z}$ and $g_T^{v_1 z}$. Thus, we add $\sigma_3 = g^{u_1 v_1}$ to the seed of the session key in order to simulate such a case.

Note that, it would be possible to modify our ABAKE scheme to be secure under the BDH assumption by using the twin DH technique [25]. However, this modification may bring about more keys, more shared values or much computation, and, thus, it would not be suitable to construct efficient schemes.

6   Security

We prove that our ABAKE scheme is secure in the ABeCK model. Since the underlying ABE scheme just satisfies selective security, our ABAKE scheme also satisfies selective security.

Theorem 2. *Suppose the GBDH assumption holds. Then, our ABAKE scheme is selectively secure in the ABeCK model in the random oracle model.*

Proof. We will show that if a polynomially bounded adversary $\mathcal{A}$ can distinguish the session key of a fresh session from a randomly chosen session key, we can solve the GBDH problem. Let κ be the security parameter, and let $\mathcal{A}$ be a polynomially (in κ) bounded adversary. We use adversary $\mathcal{A}$ to construct a GBDH solver $\mathcal{S}$ that succeeds with non-negligible probability. Suc denotes the event that $\mathcal{A}$ wins. Let $AskH$ be the event that adversary $\mathcal{A}$ poses $(\sigma_1, \sigma_2, \sigma_3, (X, \{U\}, M_A, \rho_A), (Y, \{V\}, M_B, \rho_B))$ to H_3. Let $\overline{AskH}$ be the complement of event $AskH$. Let sid be any completed session owned by an honest party such that $sid \neq sid^*$ and sid is non-matching to sid^*. Since sid and sid^* are distinct and non-matching, the inputs to the key derivation function H_3 are different for sid and sid^*. Since H_3 is a random oracle, $\mathcal{A}$ cannot obtain any information about the test session key from the session keys of non-matching sessions. Hence, $\Pr[Suc \wedge \overline{AskH}] \leq \frac{1}{2}$ and $\Pr[Suc] = \Pr[Suc \wedge AskH] + \Pr[Suc \wedge \overline{AskH}] \leq \Pr[Suc \wedge AskH] + \frac{1}{2}$ whence $f(\kappa) \leq \Pr[Suc \wedge AskH]$. Henceforth, the event $Suc \wedge AskH$ is denoted by Suc^*.

We denote the master secret and public keys by g^z and (g, g^r, g_T^z) respectively. For party P, we denote the set of attributes by $\mathbb{S}_P$, the static secret key by $(S'_P, \{T_P\}, \{S_P\})$, the ephemeral secret key by $\tilde{u}_1, \ldots, \tilde{u}_{n_P}$, and the exponent of the ephemeral public keys by $u_j = H_2(S'_P, \{T_P\}, \{S_P\}, \tilde{u}_j)$ for $1 \leq j \leq n_P$. We also denote the session key by K. Assume that $\mathcal{A}$ succeeds in an environment with N users, activates at most L sessions within a party.

We consider the following events.

- Let $AskS$ be the event $\mathcal{A}$ poses the static secret key $(S'_P, \{T_P\}, \{S_P\})$ to H_2, *before* asking StaticReveal queries or MasterReveal query, or *without* asking StaticReveal queries or MasterReveal query.
- Let $\overline{AskS}$ be the complement of event $AskS$.

We consider the following events that cover all cases of the behavior of $\mathcal{A}$.

- Let E_1 be the event that the test session sid^* has no matching session $\overline{sid}^*$ and $\mathcal{A}$ poses StaticReveal($\mathbb{S}$) s.t. $\mathbb{S} \in \mathbb{A}_B$.
- Let E_2 be the event that the test session sid^* has no matching session $\overline{sid}^*$ and $\mathcal{A}$ poses EphemeralReveal(sid^*).
- Let E_3 be the event that the test session sid^* has matching session $\overline{sid}^*$ and $\mathcal{A}$ poses MasterReveal or poses StaticReveal($\mathbb{S}$) s.t. $\mathbb{S} \in \mathbb{A}_B$ and StaticReveal($\mathbb{S}$) s.t. $\mathbb{S} \in \mathbb{A}_A$.
- Let E_4 be the event that the test session sid^* has matching session $\overline{sid}^*$ and $\mathcal{A}$ poses EphemeralReveal(sid^*) and EphemeralReveal($\overline{sid}^*$).
- Let E_5 be the event that the test session sid^* has matching session $\overline{sid}^*$ and $\mathcal{A}$ poses StaticReveal($\mathbb{S}$) s.t. $\mathbb{S} \in \mathbb{A}_B$ and EphemeralReveal($\overline{sid}^*$).
- Let E_6 be the event that the test session sid^* has matching session $\overline{sid}^*$ and $\mathcal{A}$ poses EphemeralReveal(sid^*) and StaticReveal($\mathbb{S}$) s.t. $\mathbb{S} \in \mathbb{A}_A$.

To finish the proof, we investigate events $AskS \wedge Suc^*$ and $E_i \wedge \overline{AskS} \wedge Suc^*$ $(i = 1, \ldots, 6)$ that cover all cases of event Suc^*.

6.1   Event $AskS \wedge Suc^*$

In the event $AskS$, $\mathcal{A}$ poses the static secret key $(S'_A, \{T_A\}, \{S_A\})$ to H_2 before posing StaticReveal queries or MasterReveal query, or without posing StaticReveal queries or MasterReveal query. The solver $\mathcal{S}$ embeds instance as $g_T^z = e(\alpha, \beta)$ and extract $g^z = g^{ab}$ from S'_A and T_{A_1} because $\mathcal{S}$ knows r and can compute $S'_A/T_{A_1}^r = g^z$. Then, $\mathcal{S}$ obtains $\mathsf{BDH}(\alpha, \beta, \gamma) = e(g^z, \gamma)$.

6.2   Event $E_1 \wedge \overline{AskS} \wedge Suc^*$

In the event E_1, the test session sid^* has no matching session $\overline{sid}^*$, $\mathcal{A}$ poses StaticReveal($\mathbb{S}$) s.t. $\mathbb{S} \in \mathbb{A}_B$, and does not pose EphemeralReveal(sid^*), MasterReveal or StaticReveal($\mathbb{S}$) s.t. $\mathbb{S} \in \mathbb{A}_A$ by the condition of freshness. In the case of event $E_1 \wedge \overline{AskS} \wedge Suc^*$, $\mathcal{S}$ performs the following steps.

Init.   The GBDH solver $\mathcal{S}$ receives a BDH tuple $(g, \alpha, \beta, \gamma)$ as a challenge. Also, $\mathcal{S}$ receives (M_A^*, ρ_A^*) for $\mathbb{A}_A^*$ and (M_B^*, ρ_B^*) for $\mathbb{A}_B^*$ as a challenge access structure from $\mathcal{A}$. M_A^* is $\ell_A^* \times n_A^*$ matrix and M_B^* is $\ell_B^* \times n_B^*$ matrix.

Setup. $\mathcal{S}$ chooses $z' \in_R \mathbb{Z}_p$ and lets $g_T^z := e(\alpha, \beta)g_T^{z'}$ (i.e., $z = ab+z'$ implicitly). $\mathcal{S}$ embeds $g^r := \alpha$ and outputs the master public key $MPK = (g, g^r, g_T^z)$.

$\mathcal{S}$ randomly selects two parties A, B and integers $i_A \in_R [1, L]$ that becomes a guess of the test session with probability $1/n^2L$. $\mathcal{S}$ sets the ephemeral public key of i_Ath session of A as follows: First, $\mathcal{S}$ programs the random oracle H_1 by building a table. For each j, k pair where $1 \leq j \leq n_{max}$ and k such that $\rho_A^*(i) = k$ for $1 \leq i \leq \ell_A^*$, $\mathcal{S}$ chooses a random value $h_{j,k} \in_R \mathbb{Z}_p$. Then, let $H_1(j, k) = g^{h_{j,k}} \alpha^{M_{A_{i,j}}^*}$ for $1 \leq j \leq n_A^*$ and k such that $\rho_A^*(i) = k$ for $1 \leq i \leq \ell_A^*$. Otherwise, let $H_1(j, k) = g^{h_{j,k}}$.[6] Next, $\mathcal{S}$ lets $X := \gamma$ and chooses random values $x_2, \ldots, x_{n_A^*} \in \mathbb{Z}_p$ and sets $x_1 = 0$. Then, $\mathcal{S}$ computes $U_{i,j} = \alpha^{M_{A_{i,j}}^* x_j} \gamma^{-h_{j,\rho_A^*(i)}}$ for $1 \leq i \leq \ell_A^*$ and $1 \leq j \leq n_A^*$ (i.e., $(u_1, \ldots, u_{n_A^*}) = (c, c + x_2, \ldots, c + x_{n_A^*})$ implicitly), and $U_{i,j} = \gamma^{-h_{j,\rho^*(i)}}$ for $1 \leq i \leq \ell_A^*$ and $n_A^* + 1 \leq j \leq n_{max}$. Finally, $\mathcal{S}$ sets the ephemeral public key $EPK_A = (X, \{U\})$ of i_Ath session of A.

Simulation. $\mathcal{S}$ simulates oracle queries by $\mathcal{A}$ as follows. $\mathcal{S}$ maintains the lists $\mathcal{L}_{H_1}$, $\mathcal{L}_{H_2}$ and $\mathcal{L}_{H_3}$ that contains queries and answers of the H_1, H_2 and H_3 oracles respectively, and the list $\mathcal{L}_K$ that contains queries and answers of SessionReveal.

1. $H_1(j, k)$: If there exists a tuple $(j, k, *) \in \mathcal{L}_{H_1}$, $\mathcal{S}$ returns the registered value. Otherwise, $\mathcal{S}$ chooses $h_{j,k} \in_R \mathbb{Z}_p$, returns $g^{h_{j,k}}$ and records it to $\mathcal{L}_{H_1}$.

---

[6] All $H_1(j, k)$ are distributed randomly due to the $g^{h_{j,k}}$. Also, since ρ_A^* is injective, for any k there is at most one i such that $\rho_A^*(i) = k$.

2. $H_2(S', \{T\}, \{S\}, \tilde{u}_j)$: If there exists a tuple $(S', \{T\}, \{S\}, \tilde{u}_j, *) \in \mathcal{L}_{H_2}$, $\mathcal{S}$ returns the registered value.[7] Otherwise, $\mathcal{S}$ chooses $u_j \in_R \mathbb{Z}_p$, returns u_j and records it to $\mathcal{L}_{H_2}$.

3. $H_3(\sigma_1, \sigma_2, \sigma_3, (X, \{U\}, M_P, \rho_P), (Y, \{V\}, M_{\bar{P}}, \rho_{\bar{P}}))$:
   (a) If there exists a tuple $(\sigma_1, \sigma_2, \sigma_3, (X, \{U\}, M_P, \rho_P), (Y, \{V\}, M_{\bar{P}}, \rho_{\bar{P}}), *) \in \mathcal{L}_{H_3}$, $\mathcal{S}$ returns the registered value.
   (b) Else if there exists a tuple $(\mathcal{I}, \mathbb{S}_P, \mathbb{S}_{\bar{P}}, (X, \{U\}, M_P, \rho_P), (Y, \{V\}, M_{\bar{P}}, \rho_{\bar{P}}), *) \in \mathcal{L}_K$ or $(\mathcal{R}, \mathbb{S}_{\bar{P}}, \mathbb{S}_P, (X, \{U\}, M_P, \rho_P), (Y, \{V\}, M_{\bar{P}}, \rho_{\bar{P}}), *) \in \mathcal{L}_K$, $\mathsf{DBDH}(X, \alpha, \beta, \sigma_1) = 1$, $\mathsf{DBDH}(Y, \alpha, \beta, \sigma_2) = 1$ and $e(X, Y) = e(g, \sigma_3)$, then $\mathcal{S}$ returns the recorded value and record it in the list $\mathcal{L}_{H_3}$.
   (c) Else if $\mathsf{DBDH}(X, \alpha, \beta, \sigma_1) = 1$, $\mathsf{DBDH}(Y, \alpha, \beta, \sigma_2) = 1$, $e(X, Y) = e(g, \sigma_3)$, $P = A, \bar{P} = B$ and the session is i_A-th session of A, then $\mathcal{S}$ stops and is successful by outputting the answer of the GBDH problem $\sigma_1 = \mathsf{BDH}(\alpha, \beta, \gamma)$.
   (d) Otherwise, $\mathcal{S}$ returns a random value $K \in_R \{0, 1\}^\kappa$ and records it in the list $\mathcal{L}_{H_3}$.

4. $\mathsf{Send}(\mathcal{I}, \mathbb{S}_P, \mathbb{S}_{\bar{P}})$: If $P = A$ and the session is i_A-th session of A, $\mathcal{S}$ returns the ephemeral public key EPK_A computed in the setup. Otherwise, $\mathcal{S}$ computes the ephemeral public key EPK_P obeying the protocol, returns it and records $(\mathbb{S}_P, \mathbb{S}_{\bar{P}}, (EPK_P, M_P, \rho_P))$.

5. $\mathsf{Send}(\mathcal{R}, \mathbb{S}_{\bar{P}}, \mathbb{S}_P, (EPK_P, M_P, \rho_P))$: $\mathcal{S}$ computes the ephemeral public key $EPK_{\bar{P}}$ obeying the protocol, returns it and records $(\mathbb{S}_P, \mathbb{S}_{\bar{P}}, (EPK_P, M_P, \rho_P), (EPK_{\bar{P}}, M_{\bar{P}}, \rho_{\bar{P}}))$ as the completed session.

6. $\mathsf{Send}(\mathcal{I}, \mathbb{S}_{\bar{P}}, \mathbb{S}_P, (EPK_P, M_P, \rho_P), (EPK_{\bar{P}}, M_{\bar{P}}, \rho_{\bar{P}}))$: If $(\mathbb{S}_P, \mathbb{S}_{\bar{P}}, (EPK_P, M_P, \rho_P))$ is not recorded, $\mathcal{S}$ records the session $(\mathbb{S}_P, \mathbb{S}_{\bar{P}}, (EPK_P, M_P, \rho_P))$ is not completed. Otherwise, $\mathcal{S}$ records the session is completed.

7. $\mathsf{SessionReveal}(sid)$:
   (a) If the session sid is not completed, $\mathcal{S}$ returns an error message.
   (b) Else if sid is recorded in the list $\mathcal{L}_K$, then $\mathcal{S}$ returns the recorded value K.
   (c) Else if $(\sigma_1, \sigma_2, \sigma_3, (X, \{U\}, M_P, \rho_P), (Y, \{V\}, M_{\bar{P}}, \rho_{\bar{P}}))$ is recorded in the list $\mathcal{L}_{H_3}$, $\mathsf{DBDH}(X, \alpha, \beta, \sigma_1) = 1$, $\mathsf{DBDH}(Y, \alpha, \beta, \sigma_2) = 1$ and $e(X, Y) = e(g, \sigma_3)$, then $\mathcal{S}$ returns the recorded value K and records it in the list $\mathcal{L}_K$.
   (d) Otherwise, $\mathcal{S}$ returns a random value $K \in_R \{0, 1\}^\kappa$ and records it in the list $\mathcal{L}_K$.

8. $\mathsf{EphemeralReveal}(sid)$: $\mathcal{S}$ returns a random value $\tilde{u}_1, \ldots, \tilde{u}_n$ where n is the size of the column of M in sid and records it.

9. $\mathsf{StaticReveal}(\mathbb{S}_P)$: In the event E_1, $\mathbb{S}_P$ does not satisfy M_A^*. Without loss of generality, we can suppose that $M_{A_{i,j}}^* = 0$ for $n_A^* + 1 \leq j \leq n_{max}$. By the definition of LSSSs, $\mathcal{S}$ can efficiently find a vector $\boldsymbol{w} = (w_1, \ldots, w_{n_{max}}) \in \mathbb{Z}_p^{n_{max}}$ such that $w_1 = -1$ and for all i where $\rho_A^*(i) \in \mathbb{S}_P$ we have that

---

[7] $c, x_2, \ldots, x_{n_A^*}$ are not registered in $\mathcal{L}_{H_2}$. However, $\mathcal{A}$ does not pose $\mathsf{EphemeralReveal}(sid^*)$ and so cannot know information about $\tilde{c}, \tilde{x}_2, \ldots, \tilde{x}_{n_A^*}$ corresponding to $c, x_2, \ldots, x_{n_A^*}$. Thus, $\mathcal{A}$ cannot distinguish the real experiment from the simulation by such queries.

$\boldsymbol{w} \cdot M^*_{Ai,j} = 0$. Note that, we can simply let $w_j = 0$ and consider $M^*_{Ai,j} = 0$ for $n^*_A + 1 \le j \le n_{max}$.

$\mathcal{S}$ sets the static secret key SK_P as follows: $\mathcal{S}$ chooses random values y_1, $\ldots$, $y_{n_{max}} \in_R \mathbb{Z}_p$ and computes $S'_P := g^z \alpha^{y_1}$ and $T_{P_j} = g^{y_j} \cdot \beta^{w_j}$ (i.e., $t_j = y_j + w_j b$ implicitly). Also, $\mathcal{S}$ sets S_{P_k} for $k \in \mathbb{S}_P$ as $S_{P_k} := \prod_{i \le j \le n_{max}} T^{h_{j,k}}_{P_j}$ for $k \in \mathbb{S}_P$ where there is no i such that $\rho^*_A(i) = k$. For $k \in \mathbb{S}_P$ where there is i such that $\rho^*_A(i) = k$, $\mathcal{S}$ sets $S_{P_k} := \prod_{i \le j \le n_{max}} g^{h_{j,k}y_j} \cdot \beta^{h_{j,k}} \cdot \gamma^{M^*_{Ai,j}y_j}$.

10. MasterReveal: $\mathcal{S}$ aborts with failure.[8]
11. Establish$(P, \mathbb{S}_P)$: $\mathcal{S}$ responds to the query as the definition.
12. Test(sid): If the ephemeral public key in the session sid is not EPK_A, then $\mathcal{S}$ aborts with failure. Otherwise, responds to the query as the definition.
13. If $\mathcal{A}$ outputs a guess b', $\mathcal{S}$ aborts with failure.

Analysis. The simulation for $\mathcal{S}$ is perfect except with negligible probability. The probability that $\mathcal{A}$ selects the session, where the ephemeral public key is EPK_A, as the test session sid^* is at least $\frac{1}{N^2 L}$.

Under the event Suc^*, $\mathcal{A}$ poses correctly formed $\sigma_1, \sigma_2, \sigma_3$ to H_3. Therefore, $\mathcal{S}$ is successful and does not abort.

Hence, $\mathcal{S}$ is successful with probability $Pr[\mathcal{S}$ solves the GBDH problem$] \ge \frac{p_1}{n^2 s}$, where p_1 is probability that $E_1 \wedge \overline{AskS} \wedge Suc^*$ occurs.

6.3   Other Events

Event $E_2 \wedge \overline{AskS} \wedge Suc^*$. In the event E_2, the test session sid^* has no matching session $\overline{sid}^*$, $\mathcal{A}$ poses EphemeralReveal(sid^*), and $\mathcal{A}$ does not pose StaticReveal$(\mathbb{S})$ s.t. $\mathbb{S} \in \mathbb{A}_A$, StaticReveal$(\mathbb{S})$ s.t. $\mathbb{S} \in \mathbb{A}_B$ or MasterReveal by the condition of freshness. Thus, $\mathcal{A}$ cannot obtain no information about $u_1, \ldots, u_{n_{max}}$ except negligible guessing probability, since H_2 is the random oracle. Hence, $\mathcal{S}$ performs the reduction same as in the case of event $E_1 \wedge \overline{AskS} \wedge Suc^*$.

Event $E_3 \wedge \overline{AskS} \wedge Suc^*$. In the event E_3, the test session sid^* has the matching session $\overline{sid}^*$, $\mathcal{A}$ poses MasterReveal or poses both StaticReveal$(\mathbb{S})$ s.t. $\mathbb{S} \in \mathbb{A}_A$ and StaticReveal$(\mathbb{S})$ s.t. $\mathbb{S} \in \mathbb{A}_B$, and $\mathcal{A}$ does not pose EphemeralReveal(sid^*) and EphemeralReveal$(\overline{sid}^*)$ by the condition of freshness. $\mathcal{S}$ simulates the setup and key generations obeying the scheme. $\mathcal{S}$ embeds the BDH instance as $X = g^x = \alpha$, $Y = g^y = \beta$ in sid^*, and extracts g^{ab} from $\sigma_3 = g^{xy}$. Then, $\mathcal{S}$ obtains BDH(α, β, γ) by $e(\sigma_3, \gamma)$.

Event $E_4 \wedge \overline{AskS} \wedge Suc^*$. In the event E_4, the test session sid^* has the matching session $\overline{sid}^*$, $\mathcal{A}$ poses EphemeralReveal(sid^*) and EphemeralReveal$(\overline{sid}^*)$, and does not pose StaticReveal$(\mathbb{S})$ s.t. $\mathbb{S} \in \mathbb{A}_A$, StaticReveal$(\mathbb{S})$ s.t. $\mathbb{S} \in \mathbb{A}_B$ or MasterReveal by the condition of freshness. Then, $\mathcal{A}$ cannot obtain no information about $u_1, \ldots, u_{n_{max}}$ and $v_1, \ldots, v_{n_{max}}$ except negligible guessing probability because H_2 is the random oracle and outputs of StaticReveal are randomized. Hence, $\mathcal{S}$ performs the reduction same as in the case of event $E_3 \wedge \overline{AskS} \wedge Suc^*$.

---

[8] In the event E_1, $\mathcal{A}$ does not pose MasterReveal query.

Event $E_5 \wedge \overline{AskS} \wedge Suc^*$. In the event E_5, the test session sid^* has the matching session $\overline{sid}^*$, $\mathcal{A}$ poses StaticReveal($\mathbb{S}$) s.t. $\mathbb{S} \in \mathbb{A}_B$ and EphemeralReveal($\overline{sid}^*$), and does not pose EphemeralReveal(sid^*), StaticReveal($\mathbb{S}$) s.t. $\mathbb{S} \in \mathbb{A}_A$ or MasterReveal by the condition of freshness. Then, $\mathcal{A}$ cannot obtain no information about $v_1, \ldots, v_{n_{max}}$ except negligible guessing probability because H_2 is the random oracle and outputs of StaticReveal are randomized. Hence, $\mathcal{S}$ performs the reduction same as in the case of event $E_3 \wedge \overline{AskS} \wedge Suc^*$.

Event $E_6 \wedge \overline{AskS} \wedge Suc^*$. In the event E_6, the test session sid^* has the matching session $\overline{sid}^*$, $\mathcal{A}$ poses StaticReveal($\mathbb{S}$) s.t. $\mathbb{S} \in \mathbb{A}_A$ and EphemeralReveal(sid^*), and does not pose EphemeralReveal($\overline{sid}^*$), StaticReveal($\mathbb{S}$) s.t. $\mathbb{S} \in \mathbb{A}_B$ or MasterReveal by the condition of freshness. Then, $\mathcal{A}$ cannot obtain no information about $u_1, \ldots, u_{n_{max}}$ except negligible guessing probability because H_2 is the random oracle and outputs of StaticReveal are randomized. Hence, $\mathcal{S}$ performs the reduction same as in the case of event $E_3 \wedge \overline{AskS} \wedge Suc^*$. □

References

1. Wang, H., Xu, Q., Ban, T.: A Provably Secure Two-Party Attribute-Based Key Agreement Protocol. In: IIH-MSP 2009, pp. 1042–1045 (2009)
2. Wang, H., Xu, Q., Fu, X.: Revocable Attribute-based Key Agreement Protocol without Random Oracles. JNW 4(8), 787–794 (2009)
3. Wang, H., Xu, Q., Fu, X.: Two-Party Attribute-based Key Agreement Protocol in the Standard Model. In: ISIP 2009, pp. 325–328 (2009)
4. Gorantla, M.C., Boyd, C., Nieto, J.M.G.: Attribute-based Authenticated Key Exchange. In: Steinfeld, H. (ed.) ACISP 2010. LNCS, vol. 6168, pp. 300–317. Springer, Heidelberg (2010)
5. Birkett, J., Stebila, D.: Predicate-Based Key Exchange. In: Steinfeld, H. (ed.) ACISP 2010. LNCS, vol. 6168, pp. 282–299. Springer, Heidelberg (2010)
6. Smart, N.P.: An Identity Based Authenticated Key Agreement Protocol Based on the Weil Pairing. Electronics Letters 38(13), 630–632 (2002)
7. Ateniese, G., Kirsch, J., Blanton, M.: Secret Handshakes with Dynamic and Fuzzy Matching. In: NDSS 2007, pp. 159–177 (2007)
8. Bellare, M., Rogaway, P.: Entity Authentication and Key Distribution. In: Stinson, D.R. (ed.) CRYPTO 1993. LNCS, vol. 773, pp. 232–249. Springer, Heidelberg (1994)
9. LaMacchia, B., Lauter, K., Mityagin, A.: Stronger Security of Authenticated Key Exchange. In: Susilo, W., Liu, J.K., Mu, Y. (eds.) ProvSec 2007. LNCS, vol. 4784, pp. 1–16. Springer, Heidelberg (2007)
10. Canetti, R., Krawczyk, H.: Analysis of Key-Exchange Protocols and Their Use for Building Secure Channels. In: Pfitzmann, B. (ed.) EUROCRYPT 2001. LNCS, vol. 2045, pp. 453–474. Springer, Heidelberg (2001)
11. Fujioka, A., Suzuki, K., Yoneyama, K.: Predicate-based Authenticated Key Exchange Resilient to Ephemeral Key Leakage. In: WISA 2010 (2010)
12. Waters, B.: Ciphertext-Policy Attribute-Based Encryption: An Expressive, Efficient, and Provably Secure Realization. In: Cryptology ePrint Archive: 2008/290 (2008)

13. Okamoto, T., Pointcheval, D.: The Gap-Problems: A New Class of Problems for the Security of Cryptographic Schemes. In: Kim, K.-c. (ed.) PKC 2001. LNCS, vol. 1992, pp. 104–118. Springer, Heidelberg (2001)
14. Baek, J., Safavi-Naini, R., Susilo, W.: Efficient Multi-receiver Identity-Based Encryption and Its Application to Broadcast Encryption. In: Vaudenay, S. (ed.) PKC 2005. LNCS, vol. 3386, pp. 380–397. Springer, Heidelberg (2005)
15. Sahai, A., Waters, B.: Fuzzy Identity-Based Encryption. In: Cramer, R. (ed.) EUROCRYPT 2005. LNCS, vol. 3494, pp. 457–473. Springer, Heidelberg (2005)
16. Bethencourt, J., Sahai, A., Waters, B.: Ciphertext-Policy Attribute-Based Encryption. In: IEEE Symposium on Security and Privacy 2007, pp. 321–334 (2007)
17. Cheung, L., Newport, C.C.: Provably secure ciphertext policy ABE. In: ACM Conference on Computer and Communications Security 2007, pp. 456–465 (2007)
18. Kapadia, A., Tsang, P.P., Smith, S.W.: Attribute-Based Publishing with Hidden Credentials and Hidden Policies. In: NDSS 2007, pp. 179–192 (2007)
19. Nishide, T., Yoneyama, K., Ohta, K.: Attribute-Based Encryption with Partially Hidden Encryptor-Specified Access Structures. In: Bellovin, S.M., Gennaro, R., Keromytis, A.D., Yung, M. (eds.) ACNS 2008. LNCS, vol. 5037, pp. 111–129. Springer, Heidelberg (2008)
20. Shi, E., Bethencourt, J., Chan, H.T.H., Song, D.X., Perrig, A.: Multi-Dimensional Range Query over Encrypted Data. In: IEEE Symposium on Security and Privacy 2007, pp. 350–364 (2007)
21. Boneh, D., Waters, B.: Conjunctive, Subset, and Range Queries on Encrypted Data. In: Vadhan, S.P. (ed.) TCC 2007. LNCS, vol. 4392, pp. 535–554. Springer, Heidelberg (2007)
22. Katz, J., Sahai, A., Waters, B.: Predicate Encryption Supporting Disjunctions, Polynomial Equations, and Inner Products. In: Smart, N.P. (ed.) EUROCRYPT 2008. LNCS, vol. 4965, pp. 146–162. Springer, Heidelberg (2008)
23. Lewko, A.B., Okamoto, T., Sahai, A., Takashima, K., Waters, B.: Fully Secure Functional Encryption: Attribute-Based Encryption and (Hierarchical) Inner Product Encryption. In: Gilbert, H. (ed.) EUROCRYPT 2010. LNCS, vol. 6110, pp. 62–91. Springer, Heidelberg (2010)
24. Beimel, A.: Secure Schemes for Secret Sharing and Key Distribution. PhD thesis, Israel Institute of Technology, Technion (1996)
25. Cash, D., Kiltz, E., Shoup, V.: The Twin Diffie-Hellman Problem and Applications. J. Cryptology 22(4), 470–504 (2009)

Constructing Certificateless Encryption and ID-Based Encryption from ID-Based Key Agreement*

Dario Fiore[1,**], Rosario Gennaro[2], and Nigel P. Smart[3]

[1] École Normale Supérieure, CNRS - INRIA, Paris, France
dario.fiore@ens.fr
[2] IBM T.J. Watson Research Center, Hawthorne, New York, U.S.A.
rosario@us.ibm.com
[3] Dept. Computer Science, University of Bristol,
Woodland Road, Bristol, BS8 1UB, United Kingdom
nigel@cs.bris.ac.uk

Abstract. We discuss the relationship between ID-based key agreement protocols, certificateless encryption and ID-based key encapsulation mechanisms. In particular we show how in some sense ID-based key agreement is a primitive from which all others can be derived. In doing so we focus on distinctions between what we term *pure* ID-based schemes and *non-pure* schemes, in various security models. We present security models for ID-based key agreement which do not "look natural" when considered as analogues of normal key agreement schemes, but which look more natural when considered in terms of the models used in certificateless encryption. Our work highlights distinctions between the two approaches to certificateless encryption, and adds to the debate about what is the "correct" security model for certificateless encryption.

1 Introduction

The notion of certificateless encryption was introduced by Al-Riyami and Paterson [3] and considers the following setting, that is similar to that of identity-based encryption. Each user is represented by a string ID (his identity) and has a matching secret key produced by a Key Generation Center (KGC). Furthermore each user has also a public/secret key pair, as in the traditional public key model. The main advantages of certificateless encryption are that such public keys do not need to be certified and the KGC cannot decrypt ciphertexts of users. In general, the security of certificateless encryption schemes is formalized by two properties related to semantic security of standard encryption schemes: Type I and Type II security. Type I security considers adversaries that are able to replace the public keys of users while Type II security is stated with respect to malicious KGCs.

---

* The full version of this paper is available at http://eprint.iacr.org/2009/600
** Work partially done while student at University of Catania, Italy.

M. Joye, A. Miyaji, and A. Otsuka (Eds.): Pairing 2010, LNCS 6487, pp. 167–186, 2010.

Ever since its introduction in [3] certificateless encryption has been the subject of debate as to what is the "correct" definition. This is not only a question of the definition of the security model, but also the syntax and functionality of the schemes itself. Many papers have presented differing restrictions for the adversaries in both Type I and Type II security games, creating a lot of different security definitions, with each paper claiming theirs to be the "correct" one. Also other papers have presented new syntax (with similar claims). Most of the claims are actually related to what can be proved about the schemes the papers present, rather than some deeper philosophical discussion. We refer the reader to [14] for a balanced summary of the existing models and schemes.

1.1 Our Contribution

This paper takes a different approach to the study of certificateless schemes, by studying their relationship to identity-based encryption. We do so in order to take a step back from scheme construction and instead concentrate on what the correct security and syntactic definitions should be. To simplify our discussion we will concentrate on the simpler notion of key-encapsulation (KEM) rather than encryption.

We show two main results: (1) a natural transform of *certain* CL-KEM schemes into ID-KEM schemes. In addition there is (2) another natural transform of *all* identity-based key agreement (ID-KA) protocols into CL-KEM schemes. We note that all our security relationships under our transforms hold in the *standard model*.

The motivation for this research is twofold: (i) by analyzing these transformations we are able to get a better understanding of what are the "correct" security notions and syntaxes for CL encryption; (ii) these reductions may give us a *generic* toolbox to construct new, and potentially improved, CL and ID schemes.

PURE AND NON-PURE SCHEMES. Certificateless schemes in the literature can be syntactically classified into two large classes, which we call *pure* and *non-pure*. This distinction between pure and non-pure schemes also applies to existing ID-KA protocols. Informally, a pure ID-based key agreement (resp. certificateless scheme) is one in which the parties compute their messages *without* using their long-term secret keys (which is used only in the derivation of the shared session key). As we will show, such pure schemes allow various functionalities such as encryption into-the-future etc. Interestingly there are no-known pure schemes (either ID-KA or CL-KEM) which do not use pairing-based groups.

We show a natural *standard model* transformation from a *pure* CL-KEM to a ID-KEM and we determine the precise security properties of the CL-KEM under which the resulting ID-KEM is secure in the usual sense. The hope is that this generic transformation might in the future yield new constructions for ID-based encryption. It is worthwhile to observe that this transform does not work for non-pure CL-KEMs. This is not surprising as non-pure CL-KEMs are the only ones that can be constructed without pairings. So, in some sense this shows that certificateless encryption is a simpler primitive than ID-based encryption, although the reverse is commonly believed (as CL encryption is thought as an extension of ID-based one).

TOWARDS A CORRECT SECURITY MODEL FOR CL-KEMs AND ID-KA PRO-
TOCOLS. Next we show a natural generic transform of ID-KA protocols into
CL-KEM schemes. The goal here is to gain some understanding on the correct security models for these notions. In particular we investigate what security
models in the ID-KA setting imply, through our transform, certain specific CL-
KEM security models. For lack of space, we do not look at all CL-KEM security
models, but we do consider the main ones. Our results, all proven in the standard model, can be summarized in two distinct points. First, if one concentrates
on pure schemes [11], then the associated transforms have a tight security reduction. This supports our previous point that pure schemes have more/better
features. Second, the required security models in the ID-KA setting needed to
imply strong notions of security in the CL-KEM setting are highly non-standard
security notions for key agreement models. This last point can be interpreted in
one of two ways: either the strong security models for CL-KEM schemes are unnatural and that the weaker definitions should suffice, or the security notions for
ID-KA protocols (and by implication all other forms of key agreement protocol)
are too weak.

At the end of the paper we try to draw some conclusions as to what the "correct" models for certificateless encryption and ID-based key agreement should
be. Our conclusion is that perhaps the strong security models for certificateless
encryption are probably correct, and that it is the security models for ID-KA
protocols, and indeed standard public key or symmetric key based key agreement
protocols, which need to be strengthened.

Our main generic constructions can be summarized by reference to Figure 1,
the definitions used in the arrows will become clear as we define them in the
following pages.

$$
\text{ID-KA} \xrightarrow[\substack{\text{mk-fs } Reveal^* \\ \implies \text{ Strong Type-II} \\ \text{mk-fs } Rewind \\ \implies \text{ Weak Type-II}}]{\substack{\text{ka } Reveal^* \implies \text{ Strong Type-I}^* \\ \text{ka } Rewind \implies \text{ Weak Type-Ib}^*}} \text{CL-KEM} \xrightarrow[\substack{\text{Strong Type-I}^* \\ \implies \text{ ID-IND-CCA}}]{\text{Pure Only}} \text{ID-KEM}
$$

Fig. 1. Relationships Between Schemes

As a final side-result of independent interest, as part of our analysis we consider a weakened notion of Type-I security for certificateless schemes (which we
denote by Type-I* etc). This is because we have discovered an overlap in the
standard security definitions for Strong Type-I and Strong Type-II security for
CL-KEMs. By weakening the definition of Type-I security slightly, we remove
this overlap and at the same time simplify a number of our security proofs, whilst
not reducing the overall security result for the resulting CL-KEMs.

OTHER RELATED WORK. Our results are similar to the work of Paterson and Srinivasan [17] on the link between ID-based non-interactice key distribution (NIKD) and ID-based encryption. In [17] the authors present a security model for ID-based NIKD and provide a transform from an ID-based NIKD to an ID-based encryption scheme. We note that the extension of this result to constructing ID-KEMs is immediate. However, this transform is not generic in that it requires special syntactic properties of the base ID-based NIKD scheme. Our transforms from ID-KA protocols (i.e. interactive protocols) to CL-KEMs and ID-KEMs are generic and do not require any special syntactic properties of the underlying ID-KA protocol. In addition the transform of [17] results in ID-IND-CPA ID-KEMs/ID-based encryption schemes. Indeed to obtain full CCA secure KEMs/encryption schemes it is easy to see that one needs to extend the security model in [17] for ID-based NIKD schemes in such a way as to provide the adversary with an analogue of our *Reveal** oracles. Thus whilst our results are syntactically stronger than those of [17], the security results are roughly equivalent. That we can achieve more syntactically is due to us considering interactive, as opposed to non-interactive, protocols as our starting point.

2 Identity-Based Key Agreement

In this section we present the notion of ID-based key agreement. We will only consider two pass ID-based key agreement protocols in this paper as this simplifies the algorithm descriptions somewhat.

ID-Based Key Agreement Definition. A two-pass ID-based key agreement protocol is specified by six algorithms which run in polynomial time in the security parameter. The two passes are illustrated in Figure 2. We let $\mathcal{ID}$ denote the set of possible user identities and $\mathbb{K}_{\mathsf{KA}}(mpk^{\mathsf{KA}})$ be the set of valid session keys for the public parameter mpk^{KA}.

- KASetup(1^t) is a PPT algorithm that takes as input the security parameter 1^t and returns the master public key mpk^{KA} and the master secret key msk^{KA}.
- KeyDer(msk^{KA}, ID) is the private key extraction algorithm. It takes as input msk^{KA} and $ID \in \mathcal{ID}$ and it returns the associated private key d_{ID}. This algorithm may be deterministic or probabilistic.
- Initiate(mpk^{KA}, d_I). This is a PPT algorithm run by the initiator, with identity I, of the key agreement protocol which produces the ephemeral public key epk_I for transmission to another party. The algorithm stores esk_I, the corresponding ephemeral private key, for use later[1].
- Respond(mpk^{KA}, d_R). This is a PPT algorithm run by the responder, with identity R, of the key agreement protocol which produces the ephemeral public/private key (epk_R, esk_R).

---

[1] Notice that we refer to the messages exchanged by the parties as *public keys*, and their secret states after the computation of the message as *secret keys*. Jumping ahead, this is because that's the role these values play in our transformation from KA to CL scheme.

- $\mathsf{Derive}_I(mpk^{\mathsf{KA}}, d_I, esk_I, epk_R, R)$. This is a (possibly probabilistic) algorithm run by the initiator to derive the session key $K_I \in \mathbb{K}_{\mathsf{KA}}(mpk^{\mathsf{KA}})$ with R.
- $\mathsf{Derive}_R(mpk^{\mathsf{KA}}, d_R, esk_R, epk_I, I)$. This is a (possibly probabilistic) algorithm run by the responder to derive the session key $K_R \in \mathbb{K}_{\mathsf{KA}}(mpk^{\mathsf{KA}})$ with I.

Initiator		Responder
d_I , mpk^{KA}		d_R , mpk^{KA}
$(epk_I, esk_I) \leftarrow \mathsf{Initiate}(mpk^{\mathsf{KA}}, d_I)$	$\xrightarrow{epk_I}$	
	$\xleftarrow{epk_R}$	$(epk_R, esk_R) \leftarrow \mathsf{Respond}(mpk^{\mathsf{KA}}, d_R)$
$K_I \leftarrow \mathsf{Derive}_I(mpk^{\mathsf{KA}}, d_I, esk_I, epk_R, R)$		$K_R \leftarrow \mathsf{Derive}_R(mpk^{\mathsf{KA}}, d_R, esk_R, epk_I, I)$

Fig. 2. Diagrammatic view of two-pass ID-KA protocols

For correctness we require that in a valid run of the protocol we have that $K_I = K_R$. Notice, that the creation of the ephemeral public/private key pairs does not depend on the intended recipient. Most ID-KA protocols are of this form. For example in [11] ID-based key agreement protocols based on pairings are divided into four Categories. Only in Categories 2 and 4 does the emphemeral key pair depend on the intended recipient, these being protocols in the Scott [18] and McCullagh–Barreto [16] families. The majority of pairing-based ID-based key agreement protocols lie in the Smart [20] family (denoted Category 1 in [11]), with Category 3 (the Chen–Kudla family [12]) also sharing this property. The non-pairing based protocol of Fiore and Gennaro [15] also has this property.

If the algorithms Initiate and Respond do not require access to d_I and d_R respectively, then we call the protocol a *pure* identity based key agreement protocol. This is because the ephemeral public keys can be created *before* the sender knows his long term secret key. This therefore allows forms of sending-into-the-future which are common in many IBE style schemes. We shall return to this distinction below when discussing the conversion of ID-KA protocols into certificateless schemes. Indeed identifying differences between these two forms of ID-KA protocols and certificateless schemes, forms a significant portion of the current paper. In the categorization of [11] Categories 1, 3 and 4 are all pure ID-based key agreement protocols, whilst Category 2 and the non-pairing based FG protocol are non-pure.

A key agreement protocol is said to be role symmetric if algorithm Initiate is identical to algorithm Respond and algorithm Derive_I is identical to algorithm Derive_R. The FG protocol is role symmetric, but role symmetry is a more complex property to determine for pairing-based protocols. For example whether a scheme is role symmetric can depend on whether one instantiates the protocol with symmetric or asymmetric pairings. For the schemes in [11] (and focusing solely on the more practical scenario of asymmetric pairings) all those in Categories

2 and 4 are role symmetric, those in Category 3 are not, whereas half of those in Category 1 are. Of particular importance in Category 1 is the SCK protocols (which are a combined version of the Smart and Chen–Kudla protocol), these are highly efficient and role symmetric.

Defining Security for ID-Based Key Agreement. We will be using a modified version of the Bellare–Rogaway key exchange model, as extended to an identity-based setting. Our model is an extension of the model contained in Chen *et al.* [11], but we extend it in various ways which we will describe later. So as to be precise we describe the model in more formal details than that used in [11], however we shall (as stated above) be focusing solely on two-pass protocols, which explains some of our specifications in what follows.

Security of a protocol is defined by a game between an adversary A and a challenger E. At the start of the game the adversary A is passed the master public key mpk^{KA} of the key generation centre. During the game the adversary is given access to various oracles $\mathcal{O}$ which maintain various meta-variables, including

- $role_{\mathcal{O}} \in \{initiator, responder, \perp\}$. This records the type of session to which the oracle responds.
- $pid_{\mathcal{O}} \in \mathcal{U}$. This keeps track of the intended partner of the session maintained by $\mathcal{O}$.
- $\delta_{\mathcal{O}} \in \{\perp, accepted, error\}$. This determines whether the session is in a finished state or not.
- $\gamma_{\mathcal{O}} \in \{\perp, corrupted, revealed\}$. This signals whether the oracle has been corrupted or not.
- $s_{\mathcal{O}}$. This denotes the session key of the protocol if the protocol has completed.

The adversary can execute a number of oracle queries which we now describe.

- *NewSession*(U, V) This creates a new oracle, to represent the new session, which we shall denote by $\mathcal{O} = \Pi_{U,V}^{i}$, where i denotes this is the ith session for the user with identity U, and that the indented partner is V. After calling this oracle we have

$$pid_{\mathcal{O}} = V \text{ and } s_{\mathcal{O}} = role_{\mathcal{O}} = \delta_{\mathcal{O}} = \gamma_{\mathcal{O}} = \perp .$$

However, if any other oracle with identity U has been corrupted then we set $\gamma_{\mathcal{O}} = corrupted$.
- *Send*$(\mathcal{O}, role, msg)$. Recall we are only modelling two-pass protocols, hence the functionality of this oracle can be described as follows:
  - If $\delta_{\mathcal{O}} \neq \perp$ then do nothing.
  - If $role = initiator$ then
    * If $msg = \perp$, $\delta_{\mathcal{O}} = \perp$ and $role_{\mathcal{O}} = \perp$ then set $role_{\mathcal{O}} = initiator$ and output a message (i.e. send the first message flow in the protocol);
    * If $msg \neq \perp$ and $role_{\mathcal{O}} = initiator$ (i.e. msg is the second message flow in the protocol) then compute $s_{\mathcal{O}}$ and set $\delta_{\mathcal{O}} = accepted$;
    * Else set $\delta_{\mathcal{O}} = error$ and return $\perp$
  - If $role = responder$ then

* If $msg \neq \perp$ and $role_{\mathcal{O}} = \perp$ then compute $s_{\mathcal{O}}$, set $\delta_{\mathcal{O}} = accepted$, $role_{\mathcal{O}} = responder$ and respond with a message (i.e. send the second message flow in the protocol).
  * Else set $\delta_{\mathcal{O}} = error$ and return $\perp$.
- $Reveal(\mathcal{O})$. If $\delta_{\mathcal{O}} \neq accepted$ or $\gamma_{\mathcal{O}} = corrupted$ then this returns $\perp$, otherwise it returns $s_{\mathcal{O}}$ and we set $\gamma_{\mathcal{O}} = revealed$.
- $Corrupt(U)$. This returns d_U and sets all oracles $\mathcal{O}$ in the game (both now and in the future) belonging to party U to have $\gamma_{\mathcal{O}} = corrupted$. Notice, that this is equivalent to the extract secret key query in security games for other types of identity based primitives. Note, that we do not assume that the rest of the internal state of the oracles belonging to U are turned over to the adversary.
- $Test(\mathcal{O}^*)$. This oracle may only be called once by the adversary during the game. It takes as input a *fresh oracle* (see below for the definition of freshness). The challenger E then selects a bit $b \in \{0,1\}$. If $b = 0$ then the challenger responds with the value of $s_{\mathcal{O}^*}$, otherwise it responds with a random key chosen from the space of session keys. We call the oracle on which $Test$ is called the Test-oracle.

At the end of the game the adversary outputs its guess b' as to the bit b used by the challenger in the $Test$ query. We define the advantage of the adversary by

$$\text{Adv}_{ID-\text{KA}}(A) = |2\Pr[b' = b] - 1|.$$

We now explain the $Test(\mathcal{O}^*)$ query in more detail. An oracle $\mathcal{O}^* = \Pi^i_{U^*, V^*}$ is said to be *fresh* if: (1) $\delta_{\mathcal{O}^*} = accepted$, (2) $\gamma^*_{\mathcal{O}} \neq revealed$, (3) Party V^* is not corrupted and (4) there is no oracle $\mathcal{O}'$ with $\gamma_{\mathcal{O}'} = revealed$ with which $\mathcal{O}^*$ has had a matching conversation. After the $Test(\mathcal{O}^*)$ query has been made the adversary can continue making queries as before, except that it cannot: corrupt party V^*, call a reveal query on $\mathcal{O}^*$'s partner oracle if it exists, call reveal on $\mathcal{O}^*$.

Definition 1. *A protocol Π is said to be a secure ID-KA protocol (or more simply ka secure) if*

1. *In the presence of a benign adversary, which faithfully conveys messages, on $\Pi^s_{i,j}$ and $\Pi^t_{j,i}$, both oracles always accept holding the same session key, and this key is distributed uniformly on $\{0,1\}^k$;*
2. *For any polynomial time adversary A, $\text{Adv}_{ID-\text{KA}}(A)$ is negligible.*

FORWARD SECRECY. We also define a notion of *master-key forward secrecy*, (or mk-fs secure) following [11]. In this model the adversary is also given the master secret key msk^{KA}. Thus the adversary can compute the private key d_{ID} of any party. The security game is the same as above, except that instead of a fresh oracle for the test session it chooses an oracle $\mathcal{O}^*$ which satisfies:

1. $\delta_{\mathcal{O}^*} = accepted$
2. $\gamma_{\mathcal{O}^*} \neq revealed$

3. There is an oracle $\mathcal{O}'$ with which $\mathcal{O}^*$ has had a matching conversation and $\delta_{\mathcal{O}'} = accepted$ and $\gamma_{\mathcal{O}'} \neq revealed$.

Weaker notions of forward-secrecy are implied by the above, for example *perfect forward secrecy* gives the adversary access to a *Corrupt* oracle for any $ID \in \mathcal{ID}$ but does not give the adversary access to msk^{KA}. A weaker form of simply *forward secrecy* is then implied where the adversary can only call the *Corrupt* oracle on one party in the test session, i.e. we must have either $\gamma_{\mathcal{O}^*} = \bot$ or $\gamma_{\mathcal{O}'} = \bot$.

The advantage for forward secrecy of an adversary is defined in the same way as above and is denoted by one of

$$\mathrm{Adv}_{ID-\mathsf{KA}}^{mk-fs}(A), \quad \mathrm{Adv}_{ID-\mathsf{KA}}^{p-fs}(A) \text{ or } \mathrm{Adv}_{ID-\mathsf{KA}}^{fs}(A),$$

as appropriate.

For non-pure ID-based key agreement protocols we can consider an additional notion of forward secrecy, which we call *active* perfect forward secrecy (resp. *active* forward secrecy). In this model we drop the third condition above that there exists another oracle $\mathcal{O}'$ with which $\mathcal{O}^*$ has had a matching conversation. This means that the adversary could have been active before corrupting the parties, i.e. he sent one of the two message flows.

It is interesting to observe that such notion cannot be achieved by any pure ID-based KA protocol because of the following attack. Assume the adversary acts as initiator and computes $epk_I \leftarrow \mathsf{Initiate}(mpk^{\mathsf{KA}})$ (he can do that without d_I as the protocol is pure). He can initiate a new session oracle setting epk_I as first message, then ask for the second message and later make a test query on this oracle. When the adversary corrupts I then he will have all the informations needed to compute the correct session key and so he will be able to distinguish wether he received the real session key or a random one. It is easy to see that such attack does not apply to the case of non-pure protocols as the private key is needed to produce protocol's messages.

OUR AUGMENTED SECURITY MODEL. In our analysis of converting ID-based key agreement protocols into certificateless schemes we will require stronger security notions in which the adversary will have access to additional oracles. We define three such oracles, the first one is relatively standard, whilst the second two are new. The second can be motivated by similar arguments one uses to motivate resettable zero-knowledge [9], whilst the third oracle is a natural analogue in the key agreement setting of the strong adversarial powers one gives an adversary for certificateless schemes. One may therefore consider the extreme nature of the third oracle as an additional argument as to why the certificateless strongest security model looks excessive.

- *StateReveal($\mathcal{O}$)*. If $role_{\mathcal{O}} = \bot$ then do nothing. Otherwise return the value of the ephemeral secret key held within the oracle.
- *Rewind($\mathcal{O}$)*. If $role_{\mathcal{O}} = initiator$ and $\delta_{\mathcal{O}} = accepted$ then this returns $\mathcal{O}$ to the state it was in before it received its last message, i.e. it sets $\delta_{\mathcal{O}} = s_{\mathcal{O}} = \bot$. If we have $\gamma_{\mathcal{O}} = revealed$ then we also reset $\gamma_{\mathcal{O}}$ to $\bot$.

– $Reveal^*(I, R, epk_I, epk_R)$. This is a stronger version of the $Reveal$ query in that it is not associated to an oracle, but simply takes the two message flows and returns the associated agreed shared secret assuming these messages had been transmitted between party I and party R. There is an obvious restriction in that the adversary is not allowed to call this oracle on the message flows used in the $Test$ query, nor (for role-symmetric protocols) with the message flows used in the $Test$ query but with the roles of initiator and responder swapped.

The $StateReveal(\mathcal{O})$ query corresponds to an adversarial power which can partially corrupt a party, but which does not allow the adversary to obtain the long term secret. This power has been used in numerous works starting with [10], and is often considered to be the main distinction between the CK model and the BR model for key exchange [13].

The presence of the $Rewind(\mathcal{O})$ oracle enables the adversary to extract more information for a particular set of ephemeral and static public key pairs. To intuitively see what the $Rewind(\mathcal{O})$ oracle provides us, imagine a standard key agreement protocol based on standard Diffie–Hellman, for example the Station-to-Station protocol. Usually one reduces the security of this protocol to the decisional Diffie–Hellman problem (DDH). But with the presence of a $Rewind(\mathcal{O})$ oracle the adversary can take a test oracle (which has output the ephemeral public key g^x) and obtain, using a combination of the $Rewind(\mathcal{O})$ and $Reveal(\mathcal{O})$ oracles, values of the form h^x for values of h of the adversaries choosing. This means the simulator is essentially solving the DDH problem with access to a static-Diffie–Hellman oracle.

The $Reveal^*(I, R, epk_I, epk_R)$ is a very strong oracle. As we will show later, if a protocol is secure even when an adversary is given such an oracle we are able to transform the protocol into a certificateless encryption scheme which also satisfies a strong security notion.

We say a protocol is a secure ID-KA protocol in the $Rewind$-model (resp. $Reveal^*$-model) if it is secure as ID-based key agreement protocol where we give the adversary access to a $Rewind$ (resp. $Reveal^*$) oracle. If we require access to two of these oracles we will call the model, for instance, the ($StateReveal$, $Rewind$)-model, We call these extra models, augmented models, since they augment the standard security model with extra functionality. Similarly we can define augmented notions for master-key forward secrecy.

3   From Mutual to One-Way Authentication

In many key agreement protocols one is only interested in one-way authentication. SSL/TLS is a classic example of this, where the server is always authenticated but the user seldom is. We overview in this section the modifications to the previous syntax of ID-KA protocols which are needed to ensure only one-way authentication and show how to convert a mutually authenticated identity-based key agreement protocol into one which is only one-way authenticated. The reason for introducing only one-way authentication is that this enables us to make the jump to certificateless encryption conceptually easier, and can also result in simpler schemes. We assume the responder in a protocol is the one who is *not* authenticated, this is to simplify notation in what follows. The scheme definitions are then rather simple to extend.

We note that any protocol proved to be secure for mutual authentication, can be simplified and remain secure in the context of one-way authentication. The transformation from mutual to one-way authentication is performed as follows. An identity is selected, let us call it R_0, which acts as a "dummy" responder identity. A "dummy" secret key is then created for this user and this is published along with the master public key. Notice, that by carefully selecting the dummy secret key one can often obtain efficiency improvements. The protocol is then defined as before except that R_0 is always used as the responding party, and we drop any reference to d_{R_0}. Thus we call $\mathsf{Respond}(mpk^{\mathsf{KA}})$ rather than $\mathsf{Respond}(mpk^{\mathsf{KA}}, d_{R_0})$. Similarly we call

$$\mathsf{Derive}_R(mpk^{\mathsf{KA}}, esk_{R_0}, epk_{ID}, ID) \text{ and } \mathsf{Derive}_I(mpk^{\mathsf{KA}}, d_{ID}, esk_{ID}, epk_{R_0})$$

rather than

$$\mathsf{Derive}_R(mpk^{\mathsf{KA}}, d_{R_0}, esk_{R_0}, epk_{ID}, ID) \text{ and } \mathsf{Derive}_I(mpk^{\mathsf{KA}}, d_{ID}, esk_{ID}, epk_{R_0}, R_0).$$

In the security model all oracles either have R_0 as an intended partner, or the oracle belongs to R_0. If the oracle belongs to R_0 then it is corrupted, since R_0's secret key is public. This means that only oracles belonging to R_0 may be used in the *Test* queries.

We argue that if the original protocol is secure then its one-way version (obtained as described above) is also one-way secure. To see this observe that an adversary $\mathcal{A}$ that breaks the security of the one-way protocol can be turned into an adversary $\mathcal{B}$ against the original protocol. Assume $\mathcal{A}$ breaks the security choosing a test session that involves a user ID (and the dummy identity R_0). Then $\mathcal{B}$ can trivially choose a test oracle $\Pi^s_{R_0, ID}$ and forward the obtained key to $\mathcal{A}$.

4   Certificateless Key Encapsulation Mechanisms

In this section we discuss various aspects of certificateless KEMs. The reader is referred to [8] and [14] for further details.

CL-KEM Definition: A CL-KEM scheme is specified by seven polynomial time algorithms:

- $\mathsf{CLSetup}(1^t)$ is a PPT algorithm that takes as input 1^t and returns the master public keys mpk^{CL} and the master secret key msk^{CL}.
- $\mathsf{Extract\text{-}Partial\text{-}Private\text{-}Key}(msk^{\mathsf{CL}}, ID)$. If $ID \in \mathcal{ID}$ is an identifier string for party ID this (possibly probabilistic) algorithm returns a partial private key d_{ID}.
- $\mathsf{Set\text{-}Secret\text{-}Value}$ is a PPT algorithm that takes no input (bar the system parameters) and outputs a secret value s_{ID}.

- Set-Public-Key is a deterministic algorithm that takes as input s_{ID} and outputs a public key pk_{ID}.
- Set-Private-Key(d_{ID}, s_{ID}) is a deterministic algorithm that returns sk_{ID} the (full) private key.
- Enc(mpk^{CL}, pk_{ID}, ID) is the PPT encapsulation algorithm. On input of pk_{ID}, ID and mpk^{CL} this outputs a pair (C, K) where $K \in \mathbb{K}_{CL-KEM}(mpk^{CL})$ is a key for the associated DEM and $C \in \mathbb{C}_{CL-KEM}(mpk^{CL})$ is the encapsulation of that key.
- Dec(mpk^{CL}, sk_{ID}, C) is the deterministic decapsulation algorithm. On input of C and sk_{ID} this outputs the corresponding K or a failure symbol $\perp$.

Baek *et al.* gave in [5] a different formulation where the Set-Public-Key algorithm takes the partial private key d_{ID} as an additional input. In this case it is possible to combine the Set-Secret-Value, Set-Public-Key and Set-Private-Key algorithms into a single Set-User-Keys algorithm that given as input the partial private key d_{ID} of ID outputs pk_{ID} and sk_{ID}. While the Baek *et al.* formulation may seem at first glance to be a simplification, it stops various possible applications of certificateless encryption, such as encrypting into the future. Extending our definition of *pure* and *non-pure* ID-based key agreement protocols to this situation, we shall call certificateless schemes which follow the original formulation as *pure*, and those which follow the formulation of Baek *et al.* as *non-pure*.

4.1   CL-KEM Security Model

To define the security model for CL-KEMs we simply adapt the security model of Al-Riyami and Paterson [3] into the KEM framework, as explained in [8]. The main issue with certificateless encryption is that, since public keys lack authenticating information, an adversary may be able to replace users' public keys with public keys of its choice. This appears to give adversaries enormous power. However, the crucial part of the certificateless framework is that to compute the full private key of a user, knowledge of the partial private key is necessary.

To capture the scenario above, Al-Riyami and Paterson [2,3,4] consider a security model in which an adversary is able to adaptively replace users' public keys with (valid) public keys of its choice. Such an adversary is called a Type-I adversary below.

Since the KGC is able to produce partial private keys, we must of course assume that the KGC does not replace users public keys itself. By assuming that a KGC does not replace users public keys itself, a user is placing a similar level of trust in a KGC that it would in a PKI certificate authority: it is always assumed that a CA does not issue certificates for individuals on public keys which it has maliciously generated itself! We do however treat other adversarial behaviour of a KGC: eavesdropping on ciphertexts and making decryption queries for example. Such an adversarial KGC is referred to as a Type-II adversary below.

Below we present a game to formally define what an adversary must do to break a certificateless KEM [8]. This is a game run between a challenger and a two stage adversary $\mathcal{A} = (\mathcal{A}_1, \mathcal{A}_2)$. Note that X can be instantiated with I or II in the description below and that the master secret msk^{CL} is only passed to the adversary in the case of Type-II adversaries.

Type-X Adversarial Game
1. $(mpk^{\text{CL}}, msk^{\text{CL}}) \leftarrow \text{CLSetup}(1^t)$.
2. $(ID^*, s) \leftarrow \mathcal{A}_1^{\mathcal{O}}(mpk^{\text{CL}}, msk^{\text{CL}})$.
3. $(K_0, C^*) \leftarrow \text{Enc}(mpk^{\text{CL}}, pk^*, ID^*)$.
4. $K_1 \leftarrow \mathbb{K}_{\text{CL-KEM}}(mpk^{\text{CL}})$.
5. $b \leftarrow \{0, 1\}$.
6. $b' \leftarrow \mathcal{A}_2^{\mathcal{O}}(C^*, s, ID^*, K_b)$.

When performing the encapsulation, in line three of both games, the challenger uses the *current* public key pk^* of the entity with identifier ID^*. The adversary's advantage in such a game is defined to be

$$\text{Adv}_{\text{CL-KEM}}^{\text{Type-X}}(\mathcal{A}) = |2\Pr[b' = b] - 1|$$

where X is either I or II. A CL-KEM is considered to be secure, in the sense of IND-CCA2, if for all PPT adversaries $\mathcal{A}$, the advantage in both the games is a negligible function of t.

The crucial point of the definition above is to specify which oracles the adversary is given access and which are the restrictions of the game. According to such specifications one can obtain different levels of security. A detailed discussion about all possible security definitions is given by Dent in [14]. In the following we describe the various oracles $\mathcal{O}$ available to the adversaries, we then describe which oracles are available in which game and any restrictions on these oracles.

- **Request Public Key:** Given an ID this returns to the adversary a value for pk_{ID}.
- **Replace Public Key:** This allows the adversary to replace user ID's public key with any (valid) public key of the adversaries choosing.
- **Extract Partial Private Key:** Given an ID this returns the partial private key d_{ID}.
- **Extract Full Private Key:** Given an ID this returns the full private key sk_{ID}.
- **Strong Decap:** Given an encapsulation C and an identity ID, this returns the encapsulated key. If the adversary has replaced the public key of ID, then this is performed using the secret key corresponding to the new public key. Note, this secret key may not be known to either the challenger or the adversary, hence this is a very strong oracle.
- **Weak SV Decap:** This takes as input an encapsulation C, an identity ID and a secret value s_{ID}. The challenger uses s_{ID} to produce the corresponding full secret key of ID that is used to decapsulate C. Note, that s_{ID} may not correspond to the actual current public key of entity ID. Also note that one can obtain this functionality using the Strong Decap oracle when the certificateless scheme is pure.

- **Decap**: On input of an encapsulation C and an identity ID it outputs the session key obtained decapsulating C with the original secret key created by ID. One can obtain this functionality using a Strong Decap oracle if the scheme is pure.

Using these oracles we can now define the following security models for certificateless KEMs, see [14] for a full discussion.

Strong Type-I Security: This adversary has the following restrictions to its access to the various oracles.

- $\mathcal{A}$ cannot extract the full private key for ID^*.
- $\mathcal{A}$ cannot extract the full private key of any identity for which it has replaced the public key.
- $\mathcal{A}$ cannot extract the partial private key of ID^* if $\mathcal{A}_1$ replaced the public key (i.e. the public key was replaced before the challenge was issued).
- $\mathcal{A}_2$ cannot query the Strong Decap oracle on the pair (C^*, ID^*) unless ID^*'s public key was replaced after the creation of C^*.
- $\mathcal{A}$ may not query the Weak SV Decap or the Decap oracles (although for pure schemes, one can always simulate these using the Strong Decap oracle).

We note that this security notion is often considered to be incredibly strong, hence often one finds it is weakened in the following manner.

Weak Type-Ia Security: Dent describes in [14] a weaker security definition called *Weak Type-Ia* that was also used in [8]. Weak Type-Ia security does not allow the adversary to make decapsulation queries against identities whose public keys have been replaced. In this case the restrictions on the adversaries oracle access is as follows:

- $\mathcal{A}$ cannot extract the full private key for ID^*.
- $\mathcal{A}$ cannot extract the full private key of any identity for which it has replaced the public key.
- $\mathcal{A}$ cannot extract the partial private key of ID^* if $\mathcal{A}_1$ replaced the public key (i.e. the public key was replaced before the challenge was issued).
- $\mathcal{A}$ may not query the Strong Decap oracle at any time.
- $\mathcal{A}_2$ cannot query the Weak SV Decap oracle on the pair (C^*, ID^*) if the attacker replaced the public key of ID^* before the challenge was issued.
- $\mathcal{A}_2$ cannot query the Decap oracle on the pair (C^*, ID^*) unless the attacker replaced the public key before the challenge was issued.

Though this notion is clearly weaker than Strong Type-I, it still looks reasonable for practical purposes. In fact Strong Type-I gives to the adversary as much power as possible, but it is unclear whether a real adversary can obtain decapsulations in practice from users whose public keys have been replaced by the adversary itself.

We pause to note that there are weaker forms of Type-I security called Weak Type-Ib and Weak Type-Ic security. In Weak Type-Ib security access to the

Weak SV Decap oracle is denied to the adversary, whereas in Weak Type-Ic security not only denies access to the Weak SV Decap oracle, but it also denies the ability to the replace public keys entirely. We also can define a CPA like-notion, which we call Weak Type-I-CPA which denies access to all forms of decapsulation oracle (this is a notion which is not used in other papers, but which will be useful when we present our conclusions).

In addition, for each definition of Type-I security we can define a slightly weaker variant denoted by $*$ (e.g. Strong Type-I$*$) in which the adversary cannot query the partial private key of the target identity ID^* at any point. This weaker variant will simplify somewhat our security theorems. But, it still allows us to obtain a final non-weakened result due to the combination with security theorems for Type-II security, which we define below.

Strong Type-II Security: In the Type-II game the adversary has access to the master secret key msk^{CL} and so can create partial private keys itself. The strong version of this security model enables the adversary to query the various oracles with the following restrictions:

- $\mathcal{A}$ cannot extract the full private key for ID^*.
- $\mathcal{A}$ cannot extract the full private key of any identity for which it has replaced the public key.
- $\mathcal{A}_1$ cannot output an identity ID^* for which it has replaced the public key.
- $\mathcal{A}$ cannot query the partial private key oracle at all.
- $\mathcal{A}_2$ cannot query the Strong Decap oracle on the pair (C^*, ID^*) unless the public key used to create C^* has been replaced.
- $\mathcal{A}$ may not query the Weak SV Decap or the Decap oracles (although for pure schemes, one can always simulate these using the Strong Decap oracle).

Note, because we assume in this case that the adversary *is* the KGC, the adversary does not have access to the partial private key oracle since all partial private keys are ones which he can compute given msk^{CL}. This applies even in the case where generation of the partial private key from msk^{CL} and ID is randomised.

Weak Type-II Security: As for the case of Type-I security one can consider a weaker variant of Type-II security In this notion the adversary is not allowed to replace public keys at any point and thus it cannot make decapsulation queries on identities whose public keys have been replaced. This is the traditional form of Type-II security, and is aimed at protecting the user against honest-but-curious key generation centres. Again a weak form, which we call Weak Type-II-CPA, can be defined which gives no access to any decapsulation oracle, this form of security will only be needed in the discussion leading up to our conclusions. There are other strengthenings of the Type-II model which try to model completely malicious key generation centres, see [14] for a discussion of these models. But we will not consider these in this paper.

Full Type-I security from Type-I$*$ security and Strong Type-II security: In this section we justify our consideration of Type-I$*$ security by showing that proving a scheme Type-I$*$ secure is sufficient to get "full" Type-I security if such a scheme also satisfies the strongest notion of Type-II security. In some sense this says that the definitions Type-I and Strong Type-II overlap in a specific case.

For ease of presentation we prove the theorem for the case of Strong Type-I security, but it is easy to see that it holds even if the scheme is Weak-Type-Ia*, Weak Type-Ib*, Weak Type-Ic* or Weak Type-I-CPA*. In this case one obtains the corresponding level of security (e.g. Weak Type-Ia). To complete the picture we recall that Dent noted in [14] that Weak Type-II security implies Weak Type-Ic security. We can state the following theorem whose proof can be found in the full version of the paper.

Theorem 1. *If a CL-KEM is Strong-Type-I* and Strong Type-II secure then it is Strong Type-I secure*

5   Generic Construction of CL-KEM from ID-KA

In this section we show our main result, namely a generic transform of any ID-KA protocol into a CL-KEM scheme.

Suppose we are given algorithms for a one-way authenticated ID-KA protocol (KASetup, KeyDer, Initiate, Respond, $Derive_I$, $Derive_R$). Given a one-way identity-based key agreement protocol KA, we let CL(KA) denote the derived certificateless KEM obtained from the following algorithms.

- CLSetup(1^t). We run $(mpk^{KA}, msk^{KA}) \leftarrow$ KASetup(1^t) and then set: $mpk^{CL} \leftarrow mpk^{KA}$ and $msk^{CL} \leftarrow msk^{KA}$.
- Extract-Partial-Private-Key(msk^{CL}, ID). We set $d_{ID} \leftarrow$ KeyDer(msk^{KA}, ID).
- The pair Set-Secret-Value and Set-Public-Key are defined by running

$$(epk_{ID}, esk_{ID}) \leftarrow \text{Initiate}(mpk^{KA}, [d_{ID}]).$$

  The output of Set-Secret-Value is defined to be $s_{ID} = esk_{ID}$ and the output of Set-Public-Key is defined to be $pk_{ID} = epk_{ID}$.
- Set-Private-Key(d_{ID}, s_{ID}) creates sk_{ID} by setting $sk_{ID} = (d_{ID}, s_{ID})$.
- Enc(mpk^{CL}, pk_{ID}, ID). This runs as follows:
  - $(epk_0, esk_0) \leftarrow$ Respond(mpk^{KA}).
  - $K \leftarrow Derive_R(mpk^{KA}, esk_0, pk_{ID}, ID)$.
  - $C \leftarrow epk_0$.
- Dec(mpk^{CL}, sk_{ID}, C). Decapsulation is obtained by executing

$$K \leftarrow Derive_I(mpk^{KA}, d_{ID}, sk_{ID}, C).$$

In the above construction if the underlying ID-based key agreement protocol is *pure* (resp. *non-pure*), then we will obtain a *pure* (resp. *non-pure*) certificateless KEM, i.e. it will follow the original formulation of Al-Riyami and Paterson (resp. Baek *et al.*). To see this, notice that the Set-Public-Key function calls the Initiate($mpk^{KA}, [d_I]$) operation, which itself may require d_I.

5.1   Security Results on the ID-KA to CL-KEM Transforms

Once we have defined our black-box construction of CL-KEM from ID-KA protocols we prove its security in the theorems below. As one can see, the theorems show that the resulting CL-KEM can achieve different types of security according to the security of the underlying ID-KA protocol. As already discussed in the introduction, this relationship between the security models of ID-KA and CL-KEM sheds light on understanding which are the correct notion of security for the two primitives.

Theorem 2 (Type-I Security). *Consider the certificateless KEM CL(KA) derived from the one-way ID-based key agreement protocol KA as above:*

- *If KA is secure in the Reveal*-model then CL(KA) is Strong Type-I* secure as a certificateless KEM.*
- *If KA is secure in the Rewind model then CL(KA) is Weak Type-Ib* secure as a certificateless KEM.*
- *If KA is secure in the normal model then CL(KA) is Weak Type-I-CPA* secure as a certificateless KEM.*

In particular if $\mathcal{A}$ is an adversary against the CL(KA) scheme (in the above sense) then there is an adversary $\mathcal{B}$ against the KA scheme (also in the above sense) such that for pure *schemes we have*

$$\mathrm{Adv}_{\mathrm{CL-KEM}}^{\mathrm{Type-I}}(A) = \mathrm{Adv}_{ID-\mathrm{KA}}(B)$$

and for non-pure *schemes we have*

$$\mathrm{Adv}_{\mathrm{CL-KEM}}^{\mathrm{Type-I}}(A) \leq e \cdot (q_{pk} + 1) \cdot \mathrm{Adv}_{ID-\mathrm{KA}}(B)$$

where q_{pk} is the maximum number of extract public key queries issued by algorithm $\mathcal{B}$.

The proof of this theorem can be found in the full version of the paper.

We notice that the proof technique does not allow the simulator to provide the partial private key of the challenge identity ID^*. Which is why our theorem is stated for the case of Strong Type-I* (resp. Weak Type-Ib* or Weak Type-I-CPA*). If we then apply the result of Theorem 1, along with the following theorems, we obtain full Strong Type-I security (resp. Weak Type-Ib or Weak Type-I-CPA) for the scheme CL(KA).

In looking at Type-II security we present two security theorems. The first one (Theorem 3) is conceptually simpler but requires our underlying identity based key agreement scheme to have a strong security property (i.e. it must support state reveal queries). The second theorem (Theorem 4) is more involved and does not provide such a tight reduction. On the other hand the second theorem requires less of a security guarantee on the underlying key agreement scheme. The proofs of both theorems can be found in the full version of the paper.

Theorem 3 (Type-II Security – Mk I). *Consider the certificateless KEM CL(KA) derived from the one-way ID-based key agreement protocol KA as above:*

- *If KA satisfies master-key forward secrecy in the (StateReveal, Reveal*)-model then CL(KA) is Strong Type-II secure as a certificateless KEM.*

- If KA satisfies master-key forward secrecy in the (StateReveal, Rewind)-model then CL(KA) is Weak Type-II secure as a certificateless KEM.
- If KA satisfies master-key forward secrecy in the StateReveal-model then CL(KA) is Weak Type-II-CPA secure as a certificateless KEM.

In particular if $\mathcal{A}$ is an adversary against the CL(KA) scheme (in the sense described above) then there is an adversary $\mathcal{B}$ against the master-key forward secrecy of the KA scheme (also in the above sense) such that

$$\text{Adv}_{CL-KEM}^{Type-II}(A) = \text{Adv}_{ID-KA}^{mk-fs}(B).$$

We now turn to showing that one does not necessarily need the *StateReveal* query to prove security, although the complication in the proof results in a less tight reduction.

Theorem 4 (Type-II Security – Mk II). *Consider the certificateless KEM CL(KA) derived from the one-way ID-based key agreement protocol KA as above:*

- If KA satisfies master-key forward secrecy in the Reveal*-model then CL(KA) is Strong Type-II secure as a certificateless KEM.
- If KA satisfies master-key forward secrecy in the Rewind model then CL(KA) is Weak Type-II secure as a certificateless KEM.
- If KA satisfies master-key forward secrecy in the normal model then CL(KA) is Weak Type-II-CPA secure as a certificateless KEM.

In particular if $\mathcal{A}$ is an adversary against the CL(KA) scheme (in the above sense) then there is an adversary $\mathcal{B}$ against the KA scheme (also in the above sense) then we have

$$\text{Adv}_{CL-KEM}^{Type-II}(A) \le e \cdot (q_{pk} + 1) \cdot \text{Adv}_{ID-KA}^{mk-fs}(B)$$

where q_{pk} is the maximum number of extract public key queries issued by algorithm $\mathcal{B}$.

6   Identity-Based Key Encapsulation Mechanisms

In this section we are going to show the relationship between CL-KEM and identity-based KEMs. In particular we will give a generic transformation from any pure CL-KEM into an ID-KEM. As in the case of ID-KA and CL-KEM, here it is also interesting to observe how the different security models of CL-KEM transform into analogous models for ID-KEM. We defer the reader to [8] for further details on the definitions and security models of ID-KEMs.

GENERIC CONSTRUCTION OF ID-KEM FROM PURE CL-KEM. To construct an ID-KEM from a CL-KEM the obvious solution is to set the user public/private keys to be trivial and known to all parties. This however can only be done for pure CL-KEMs since in non-pure schemes one does not have complete control over the public/private keys, since they depend on the partial private key d_{ID}. We call the resulting scheme the ID(CL) scheme, as it is an ID-KEM built from a CL-KEM. Now we can state the following theorem whose proof, for lack of space, appears in the full version.

Theorem 5. *Consider the pure ID-KEM ID(CL) derived from the pure CL-KEM scheme CL as above. Then if CL is Strong Type-I* secure then ID(CL) is ID-IND-CCA secure. In particular if $\mathcal{A}$ is an adversary against the ID(CL) scheme then there is an adversary $\mathcal{B}$ against the CL-KEM scheme such that*

$$\mathrm{Adv}_{ID-\mathrm{KEM}}^{ID-\mathrm{IND-CCA}}(A) = \mathrm{Adv}_{\mathrm{CL-KEM}}^{\mathrm{Strong-Type-I*}}(B).$$

7   Conclusion: Which Certificateless Model Is Correct?

In this section we summarize the conclusions we have drawn from our analysis. It is worth pointing out that these are personal conclusions, and we leave the reader to draw their own analysis.

Firstly, all our conclusions are predicated on the assumption that our transforms are all "natural", in that they are the obvious way to convert an ID-KA protocol into a CL-KEM and a CL-KEM into an ID-KEM. If these are the natural transformations then the underlying security and syntactic models should also transform naturally.

Pure vs Non-Pure. First we discuss the issue of pure vs non-pure certificateless schemes. Our transform from CL-KEMs to ID-KEMs requires the underlying CL-KEM to be pure. This is not surprising as an essential feature of ID-based cryptography is that of the identity (and hence the associated secret key) being independent of all parameters bar the actual identity. It is not surprising even because non-pure CL-KEMs are the only ones that can be constructed without pairings.

We draw two conclusions from this. First, the pure syntax is more powerful as it enables functionalities such as encryption-into-the-future (a.k.a. workflow). Second, we can say that certificateless encryption is a primitive simpler than ID-based encryption, although people have usually thought at the former as an extension of the latter. When ID-based encryption was proposed [19], one of its main motivations was to avoid the certificates management issues of standard public key encryption. Then it took almost twenty years to have IBE schemes, basically thanks to the idea of exploiting pairings. From our considerations we can say that the "hard part" of constructing ID-based encryption is not avoiding certificates, but achieving those additional properties (e.g. workflows); i.e. technically speaking, having a user's public key independent of the scheme parameters.

CPA Security. Before turning to CCA security of certificateless encryption we first consider the simpler case of CPA security. We remarked in the introduction that the [17] construction of ID-based encryption from ID-based NIKD schemes only produces CPA secure schemes, unless one assumes an oracle equivalent to our *Reveal** oracle.

Similar considerations apply in our case. The construction of ID-KEMs from CL-KEMs will produce a CPA secure ID-KEM if the underlying CL-KEM is Weak Type-I-CPA* secure. Note, that we only require Weak Type-I-CPA* and not Weak Type-I-CPA security. In constructing CL-KEMs from ID-KA protocols we need to consider what security is required of the underlying ID-KA protocol to ensure Weak Type-I-CPA and Weak Type-II-CPA security of the CL-KEM. Our theorems show that a sufficient condition is that the underlying ID-KA protocol is secure in the standard sense, i.e. with no *Reveal**, *Rewind* or *StateReveal* oracles. Although the security reduction is tighter if we assume the adversary has access to *StateReveal* oracles, i.e. we use a CK-like security model for ID-based key agreement. We note that the security reductions go through more naturally when one considers the CL-KEM to have Weak Type-I-CPA* security and Weak Type-II-CPA security. We then obtain the full Weak Type-I-CPA by appealing to the analogue of Theorem 1.

CCA Security. Our theorems show that to obtain full Strong Type-I and Strong Type-II security of the derived CL-KEM we require the ID-based key agreement security model to give the adversary access to our *Reveal** oracle. This is a very non-standard oracle for key agreement protocols, but this should not be surprising. Essentially CCA security for an encryption scheme means the adversary has to be able to open anything, even something created in an illegitimate way (even if the opening results in the $\perp$ symbol). All our *Reveal** oracle does is to provide the adversary against the ID-based key agreement scheme with an oracle to open anything. A similar remark as to Strong Type-I* as opposed to Strong Type-I security as mentioned in the above comments on CPA security also applies in this case.

Summary. So in summary we believe the correct syntactic security definitions for CL-KEMs should be schemes with Strong Type-I* and Strong Type-II security where the pure syntax allows for more properties. By using Strong Type-I* as the security definition instead of Strong Type-I we obtain a natural seperation between the two security notions, rather than dealing with the cases in the intersection twice. However, our construction from ID-based key agreement schemes would seem to imply that the correct security definition should be one which uses *StateReveal* queries (i.e. one which follows the analogue of CK-security). However, it also implies that the model also includes *Reveal** queries, which seems to provide an extreme form of security definition for key agreement schemes. Since it would seem silly to define security for normal key agreement schemes and ID-based key agreement schemes in a different manner, this would imply that standard key agreement schemes should also be defined to be secure in the presence of a *Reveal** oracle. This final conclusion is somewhat unsatifactory, and we hope our work will inspire other researchers to investigate this connection.

Acknowledgements. The third author was supported by a Royal Society Wolfson Research Merit Award.

References

1. Abdalla, M., Bellare, M., Rogaway, P.: The oracle Diffie–Hellman assumptions and an analysis of DHIES. In: Naccache, D. (ed.) CT-RSA 2001. LNCS, vol. 2020, pp. 143–158. Springer, Heidelberg (2001)
2. Al-Riyami, S.S.: Cryptographic schemes based on elliptic curve pairings. Ph.D. Thesis, University of London (2004)

3. Al-Riyami, S.S., Paterson, K.G.: Certificateless public key cryptography. In: Laih, C.-S. (ed.) ASIACRYPT 2003. LNCS, vol. 2894, pp. 452–473. Springer, Heidelberg (2003)
4. Al-Riyami, S.S., Paterson, K.G.: CBE from CL-PKE: A generic construction and efficient schemes. In: Vaudenay, S. (ed.) PKC 2005. LNCS, vol. 3386, pp. 398–415. Springer, Heidelberg (2005)
5. Baek, J., Safavi-Naini, R., Susilo, W.: Certificateless public key encryption without pairing. In: Zhou, J., López, J., Deng, R.H., Bao, F. (eds.) ISC 2005. LNCS, vol. 3650, pp. 134–148. Springer, Heidelberg (2005)
6. Bellare, M., Rogaway, P.: Entity authentication and key distribution. In: Stinson, D.R. (ed.) CRYPTO 1993. LNCS, vol. 773, pp. 232–249. Springer, Heidelberg (1993)
7. Blake-Wilson, S., Johnson, D., Menezes, A.: Key agreement protocols and their security analysis. In: Darnell, M.J. (ed.) Cryptography and Coding 1997. LNCS, vol. 1355, pp. 30–45. Springer, Heidelberg (1997)
8. Bentahar, K., Farshim, P., Malone-Lee, J., Smart, N.P.: Generic constructions of identity-based and certificateless KEMs. J. Cryptology 21, 178–199 (2008); Full version at IACR e-print 2005/058
9. Canetti, R., Goldreich, O., Goldwasser, S., Micali, S.: Resettable Zero-Knowledge. Weizmann Science Press, Israel (1999)
10. Canetti, R., Krawczyk, H.: Analysis of key-exchange protocols and their use for building secure channels. In: Pfitzmann, B. (ed.) EUROCRYPT 2001. LNCS, vol. 2045, pp. 453–474. Springer, Heidelberg (2001)
11. Chen, L., Cheng, Z., Smart, N.P.: Identity-based key agreement protocols from pairings. Int. J. Inf. Security 6, 213–241 (2007)
12. Chen, L., Kudla, C.: Identity based authenticated key agreement from pairings. In: IEEE Computer Security Foundations Workshop, pp. 219–233 (2003); The modified version of this paper is available at Cryptology ePrint Archive, Report 2002/184
13. Choo, K.-K.R., Boyd, C., Hitchcock, Y.: Examining indistinguishabilit-based proof models for key establishment protocols. In: Roy, B. (ed.) ASIACRYPT 2005. LNCS, vol. 3788, pp. 585–604. Springer, Heidelberg (2005)
14. Dent, A.: A Survey of Certificateless Encryption Schemes and Security Models. International Journal of Information Security 7, 347–377 (2008)
15. Fiore, D., Gennaro, R.: Making the Diffie–Hellman protocol identity-based. In: Pieprzyk, J. (ed.) CT-RSA 2010. LNCS, vol. 5985, pp. 165–178. Springer, Heidelberg (2010)
16. McCullagh, N., Barreto, P.S.L.M.: A new two-party identity-based authenticated key agreement. In: Menezes, A. (ed.) CT-RSA 2005. LNCS, vol. 3376, pp. 262–274. Springer, Heidelberg (2005)
17. Paterson, K., Srinivasan, S.: On the relations between non-interactive key distribution, identity based-based encryption and trapdoor discrete log groups. Designs, Codes and Cryptography 52, 219–241 (2009)
18. Scott, M.: Authenticated ID-based key exchange and remote log-in with insecure token and PIN number. Cryptology ePrint Archive, Report 2002/164
19. Shamir, A.: Identity-Based Cryptosystems and Signature Schemes. In: Blakely, G.R., Chaum, D. (eds.) CRYPTO 1984. LNCS, vol. 196, pp. 47–53. Springer, Heidelberg (1985)
20. Smart, N.P.: An identity based authenticated key agreement protocol based on the Weil pairing. Electronics Letters 38, 630–632 (2002)

Ephemeral Key Leakage Resilient and Efficient ID-AKEs That Can Share Identities, Private and Master Keys

Atsushi Fujioka, Koutarou Suzuki, and Berkant Ustaoğlu

NTT Information Sharing Platform Laboratories
3-9-11 Midori-cho Musashino-shi Tokyo 180-8585, Japan
{fujioka.atsushi,suzuki.koutarou,ustaoglu.berkant}@lab.ntt.co.jp

Abstract. One advantage of identity-based (ID-based) primitives is the reduced overhead of maintaining multiple static key pairs and the corresponding certificates. However, should a party wish to participate in more than one protocol with the same identity (ID), say email address, the party has to share a state between distinct primitives which is contrary to the conventional *key separation* principle. Thus it is desirable to consider security of protocols when a public identity and a corresponding private key are utilized in different protocols.

We focus on authenticated key exchange (AKE) and propose a pair of two-party ID-based authenticate key exchange protocols (ID-AKE) that are secure even if parties use the same IDs, private keys and master keys to engage in either protocol. To our knowledge the only ID-AKE protocol formally resilient to ephemeral key leakage is due to Huang and Cao (the HC protocol), where a party's static key consists of two group elements. Our proposed protocols provide similar assurances and require a single group element both for static and ephemeral keys, and in that sense are optimal. From an efficiency perspective, they have the same number of pairing computations as the HC protocol. The security of all these protocols is established in the random oracle.

Keywords: ID-based AKE, shared keys, combined keys, pairings.

1 Introduction

In 1984 Shamir [23] proposed the idea of ID-based primitives, whereby a static public key consists of a party's undeniable identifier. Consequently, in ID-based schemes parties are not required to maintain public key certificates: the public keys are available as soon as the identities become known. This is advantageous since parties do not have to manage certificates.

In a typical ID-based protocol a key generation center creates a static private key corresponding to an identity. However, most ID-based protocols and their analyses do not account for the fact that a party would often use the same identifier in many different settings. For example, a party, that identifies itself via an email (or a web) address, is unlikely to maintain different addresses for different

M. Joye, A. Miyaji, and A. Otsuka (Eds.): Pairing 2010, LNCS 6487, pp. 187–205, 2010.

primitives such as signatures encryption. It may be the case that obtaining private keys for different identities is more prohibitive than obtaining certificates with different public keys bound to the same identity. Furthermore, users may be attached to their identity string and be reluctant to use other identifiers. Consequently, a party might use the same private key in multiple different protocols. Such reuse of private material goes against conventional cryptographic wisdom of separating keys for different primitives.

Key Separation. Key separation can be achieved by appending each identity string with information describing the primitive or assigning a single identity string with different private keys for different protocols; for each key there would be a different key generation center dealing with a particular protocol. Compared with certificates such solutions do not reduce the number of static private keys that a party has to manage and so reduce the attractiveness of ID-based solutions. Furthermore, for a given primitive such as signatures or key establishment, parties are usually given the choice of more than one protocol. For example, the NIST's SP800-56A standard [20] defines two key agreement protocols and allows parties to use the same certificate and static public key to engage in either of those protocol variants. There is little reason to use different static keys if two protocols achieve similar goals. But as shown by Chatterjee, Menezes and Ustaoğlu [9], sharing static information between authenticated key exchange protocols can have a negative effect on overall security even if the protocols are individually secure. It is therefore not clear a priori that two ID-AKE protocols can share identities without affecting each others security.

Compositions. Protocol composition can be achieved by describing conditions that if *violated* would break security; however users find surprising ways around such lists. Kelsey, Schneier and Wagner [15] outline *chosen protocol attacks* in which, given a target secure protocol, an attacker can create a different and stand-alone secure protocol which, when sharing state with the target protocol, results in security breaches. We focus on static information reuse that defines what is *allowed*, effectively adopting a conservative approach to shared states.

Pairings. Menezes, Okamoto and Vanstone [19] used pairing to solve the discrete logarithm problem on some elliptic curves. Sakai, Ohgishi and Kasahara [22] used them to devise an ID-based variant of the well-known Diffie-Hellman key agreement protocol [11]. With the work of Boneh and Franklin [3], ID-based primitives utilizing pairings gained widespread attention. They are used in ID-based encryption, signatures, signcryption and ring-signcryption protocols. However, to our knowledge the only result that is concerned with reusing the same public ID and the corresponding private key for two distinct primitives is due to González Vasco, Hess and Steinwandt [12]. We are not aware of any previous work dealing with shared IDs in ID-AKE.

Key Establishment. Key establishment is a fundamental cryptographic primitive and it is important to devise efficient protocols that satisfy strong security requirements. Security definitions for key agreement protocols were initially developed for two-party protocols by Bellare and Rogaway [1] and Blake-Wilson, Johnson and Menezes [2] in the shared-secret and public-key setting, respectively. Recent developments in two-party authenticated key exchange have improved the security models and definitions. These models better represent environments where protocols are deployed. Security definitions such as [16,17] allow leakage of information related to the test session; in addition, [18] accounts for relative timing of information leakage; lastly, [9] models the fact that government standards allow users to share static keys between different protocols. Security considerations for AKE protocols are similar to ID-AKE protocols. Therefore it is natural to adapt the strongest model to the ID-based setting and design protocols secure within these definitions.

While such an adoption is natural, much work remains to be done. There are alternatives to Sakai, Ohgishi and Kasahara [22], see for example [10,24]. Boyd and Choo [4] observe that many existing ID-based protocols are not as secure as we expect them to be. Also, to the best of our knowledge, the only ID-AKE work that formally considers ephemeral key leakage is due to Huang and Cao [14]; their protocols is henceforth referred to as the HC protocol. Unfortunately, the public keys of the HC protocol consists of two group elements so it is worthwhile to develop efficient and secure ID-AKE protocols resilient to ephemeral key leakage using shorter public keys. The protocols proposed by Boyd, Cliff, González-Nieto and Paterson [5] do not necessarily require pairing-based ID primitives, but in any case are not ephemeral key leakage resilient.

Universally composable security notion of key exchange is studied in [7], that considers security of key exchange composing with any protocol, while our security model considers only security of combination of two authenticated key exchange protocols. On the other hand, our security model captures leakage of static and ephemeral keys of test session, that is not captured in [7].

Our Contribution. In this work we extend the shared model of Chatterjee, Menezes and Ustaoğlu [9] to the ID-based setting. We propose two novel protocols that satisfy the new security definition and show that parties can safely reuse private keys in these two protocols. As the HC protocol we require rather strong assumption for formal security. Finally, we provide the efficiency comparison to the HC protocol.

Organization. In Section 2, we recall the bilinear group and the gap BDH assumption. In Section 3, we briefly outline our combined model. In Section 4, we propose our new protocols, give comments related to their security arguments and describe their design principles. In Section 5, we compare our protocols with existing relevant protocols. We conclude the paper in Section 6.

2 Preliminaries

Let κ be the security parameter and q be a 2κ-bit prime. Let $G = \langle g \rangle$ and G_T be cyclic groups of prime order q with generators g and g_T, respectively. Let

$e : G \times G \to G_T$ be a polynomial-time computable bilinear non-degenerate map called a pairing. We say that (G, G_T) is are bilinear groups with pairing e.

The gap BDH (Bilinear Diffie-Hellman) problem is as follows. The computational BDH function BDH : $G^3 \to G_T$ is $\mathrm{BDH}(U, V, W) = e(g, g)^{\log U \log V \log W}$ and the decisional BDH (Bilinear Diffie-Hellman) predicate BDDH : $G^4 \to \{0, 1\}$ is a function which takes an input $(g^u, g^v, g^w, e(g, g)^x)$ and returns the bit 1 if $uvw = x \bmod q$ and the bit 0 otherwise. An adversary $\mathcal{A}$ is given input $U, V, W \in_U G$ selected uniformly random and oracle access to $\mathrm{BDDH}(\cdot, \cdot, \cdot, \cdot)$ oracle, and tries to compute $\mathrm{BDH}(U, V, W)$. For adversary $\mathcal{A}$, we define advantage

$$Adv^{\mathrm{gapBDH}}(\mathcal{A}) = \Pr[U, V, W \in_R G, \mathcal{A}^{\mathrm{BDDH}(\cdot, \cdot, \cdot, \cdot)}(U, V, W) = \mathrm{BDH}(U, V, W)],$$

where the probability is taken over the choices of U, V, W and $\mathcal{A}$'s random tape.

Definition 1 (gap BDH assumption). *We say that G satisfy the gap BDH assumption if, for all polynomial-time adversaries $\mathcal{A}$, advantage $Adv^{\mathrm{gapBDH}}(\mathcal{A})$ is negligible in security parameter κ.*

3   Shared Security Model for ID-Based AKE

Our model extends the shared static key model of Chatterjee, Menezes and Ustaoğlu [9] in a similar manny in which Huang and Cao [14] extend the LaMacchia, Lauter and Mityagin [17] model. Unlike Huang and Cao [14] we also allow our adversary to obtain private or public ephemeral session information before a session is initiated. Thus our model encompasses relative timing of ephemeral leakage.

We denote a party by U_i and the identifier of U_i by ID_i. We outline our model for two different two-pass Diffie-Hellman protocols, where parties U_A and U_B exchange ephemeral public keys X_A and X_B, i.e., U_A sends X_A to U_B and U_B sends X_B to U_A, and thereafter compute a session key. The session key depends on the exchanged ephemeral keys, identities of the parties, the static keys corresponding to these identities and the protocol instance that is used. We note that the order in which messages are exchanged is not important in practice as long as the session peers have consistent views about the information exchanged. However to simplify the exposition we assume a fixed order of message delivery. The model can be adapted to different protocols and number of rounds.

In the model, each party is a probabilistic polynomial-time Turing machine in security parameter κ and obtains a static private key corresponding to its identity string from a key generation center (KGC) via a secure and authentic channel. The center KGC uses a master secret key to generate individual private keys. We assume that the KGC never reveals the static private key for an identity string ID_i to two different parties. In other words, a malicious entity cannot obtain a static key corresponding to ID_i, unless the malicious entity is bound to ID_i.

Session. An invocation of a protocol is called a *session*. A session is activated via an incoming message of the forms $(\Pi_c, \mathcal{I}, ID_A, ID_B)$ or $(\Pi_c, \mathcal{R}, ID_A, ID_B, X_B)$, where Π_c is a protocol identifier. If U_A was activated with $(\Pi_c, \mathcal{I}, ID_A, ID_B)$, then U_A is the session *initiator*, otherwise the session *responder*. In the activation, Π_c identifies which protocol the party should execute. After activation, U_A appends an ephemeral public key X_A to the incoming message and sends it as an outgoing response. If U_A is the responder, U_A computes a session key. A party U_A that has been successfully activated via $(\Pi_c, \mathcal{I}, ID_A, ID_B)$, can be further activated via $(\Pi_c, \mathcal{R}, ID_A, ID_B, X_A, X_B)$ to compute a session key. We say that U_A is *owner* of session sid if the third coordinate of session sid is ID_A. We say that U_A is *peer* of session sid if the fourth coordinate of session sid is ID_A. We say that a session is *completed* if its owner computes a session key.

A session initiator U_A identifies the session via $(\Pi_c, \mathcal{I}, ID_A, ID_B, X_A, \times)$ or $(\Pi_c, \mathcal{I}, ID_A, ID_B, X_A, X_B)$. If U_A is the responder, the session is identified via $(\Pi_c, \mathcal{R}, ID_A, ID_B, X_B, X_A)$. For session $(\Pi_c, \mathcal{I}, ID_A, ID_B, X_A, X_B)$ the *matching session* has identifier $(\Pi_c, \mathcal{R}, ID_B, ID_A, X_A, X_B)$ and vice versa. From now on we omit $\mathcal{I}$ and $\mathcal{R}$ since these "role markers" are implicitly defined by the order of X_A and X_B. For further details on session activation, abortion and matching sessions with incomplete identifiers we refer to [9].

Adversary. The adversary $\mathcal{A}$ is modeled as a probabilistic Turing machine that controls all communications between parties including session activation, performed via a Send(message) query. The message has one of the following forms: (Π_c, ID_A, ID_B), (Π_c, ID_A, ID_B, X_A), or $(\Pi_c, ID_A, ID_B, X_A, X_B)$; Π_c is a protocol identifier. Each party submits its responses to the adversary, who decides the global delivery order. Note that the adversary does not control the communication between parties and the key generation center. It is possible to incorporate into the model the ability of the adversary to time when a party obtains a static private key. For simplicity, we assume that identities and corresponding static keys are part of $\mathcal{A}$'s input.

A party's private information is not accessible to the adversary; however, leakage of private information is captured via the following adversary queries. For details relating to incomplete session identifiers and discussion related to EphemeralPublicKeyReveal we refer the reader to [18]:

- SessionKeyReveal(sid). The adversary obtains the session key for the session sid, provided that the session holds a session key.
- EphemeralPublicKeyReveal(ID_i). The adversary obtains the ephemeral public key that ID_i will use when a session is next activated at ID_i.
- EphemeralKeyReveal(sid). The adversary obtains the ephemeral secret key associated with the session sid.
- StaticKeyReveal(ID_i). The adversary learns the static secret key of party U_i.
- MasterKeyReveal(). The adversary learns the master secret key of the system.
- EstablishParty(ID_i). This query allows the adversary to register a static public key on behalf of a party U_i; the adversary totally controls that party. If

a party pid is established by an EstablishParty(ID_i) query issued by the adversary, then we call the party *dishonest*. If not, we call the party *honest*. This query models malicious insiders.

Our security definition requires the notion of "freshness". The protocols we consider have the same security attributes and a single definition suffices.

Definition 2 (Freshness). *Let* sid* *be the session identifier of a completed session, owned by an honest party* U_A *with peer* U_B, *who is also honest. If the matching session exists, then let* $\overline{\text{sid}^*}$ *be the session identifier of the matching session of* sid*. *Define* sid* *to be fresh if none of the following conditions hold:*

1. *A issues* SessionKeyReveal(sid*) *or* SessionKeyReveal($\overline{\text{sid}^*}$) *(if* $\overline{\text{sid}^*}$ *exists).*
2. *$\overline{\text{sid}^*}$ exists and A makes either of the following queries*
   - *both* StaticKeyReveal(ID_A) *and* EphemeralKeyReveal(sid*), *or*
   - *both* StaticKeyReveal(ID_B) *and* EphemeralKeyReveal($\overline{\text{sid}^*}$).
3. *$\overline{\text{sid}^*}$ does not exist and A makes either of the following queries*
   - *both* StaticKeyReveal(ID_A) *and* EphemeralKeyReveal(sid*), *or*
   - StaticKeyReveal(ID_B).

Note that if A issues a MasterKeyReveal() *query, we regard A as having issued both a* StaticKeyReveal(ID_A) *query and a* StaticKeyReveal(ID_B) *query.*

Security Experiment. The adversary A starts with a set of honest parties, for whom A adaptively selects identifiers. The adversary makes an arbitrary sequence of the queries described above. During the experiment, A makes a special query Test(sid*) and is given with equal probability either the session key held by sid* or a random key; the query does not terminate the experiment. The experiment continues until A makes a guess whether the key is random or not. The adversary *wins* the game if the test session sid* is fresh at the end of A's execution and if A's guess was correct. Formally,

Definition 3 (security). *The advantage of the adversary A in the experiment with AKE protocols Π_1 and Π_2 is defined as*

$$\text{Adv}^{\text{AKE}}_{\Pi_1\Pi_2}(A) = \Pr[A \text{ wins}] - \frac{1}{2}.$$

We say that Π_1 and Π_2 are secure AKE protocols in the shared identity-based model if the following conditions hold:

1. *If two honest parties complete matching Π_c-sessions, then, except with negligible probability in security parameter κ, they both compute the same session key.*
2. *For any probabilistic polynomial-time bounded adversary A, $\text{Adv}^{\text{AKE}}_{\Pi_1\Pi_2}(A)$ is negligible in security parameter κ.*

Remark. If the adversary initiates sessions of only one protocol, issues neither EphemeralPublicKeyReveal(ID_i) queries and nor EphemeralKeyReveal(sid) queries except during the time between session initiation and completion, then the model is equivalent to the Huang-Cao [14] model.

4   Proposed ID-Based AKE Protocols

This section describes our ID-based protocols. We let $H : \{0,1\}^* \rightarrow \{0,1\}^k$, $H_1 : \{0,1\}^* \rightarrow G$, and $H_2 : \{0,1\}^* \rightarrow \mathbb{Z}_q$ be cryptographic hash function modeled as random oracles.

Key Generation Center. The KGC randomly selects master secret key $z \in_R \mathbb{Z}_q$ and publishes master public key $Z = g^z \in G$.

Private Key Generation. Given ID string $ID_i \in \{0,1\}^*$ of user U_i, the KGC computes $Q_i = H_1(ID_i)$ and returns static secret key $D_i = Q_i^z$.

4.1   Proposed ID-Based AKE Protocol 1 – Π_1

In this section, we describe the actions required to execute a Π_1 session.

Key Exchange. User U_A is the session initiator and user U_B is the session responder.

1. U_A chooses an ephemeral private key $x_A \in_R \mathbb{Z}_q$, computes the ephemeral public key $X_A = g^{x_A}$ and sends (Π_1, ID_A, ID_B, X_A) to U_B.
2. Upon receiving (Π_1, ID_A, ID_B, X_A), U_B chooses an ephemeral private key $x_B \in_R \mathbb{Z}_q$, computes the ephemeral public key $X_B = g^{x_B}$ and responds to U_A with $(\Pi_1, ID_A, ID_B, X_A, X_B)$.
   U_B also computes $e_A = H_2(X_A), e_B = H_2(X_B)$, the shared secrets

$$\sigma_1 = e(Q_A^{e_A} X_A, D_B Z^{x_B}), \quad \sigma_2 = e(Q_A X_A, D_B^{e_B} Z^{x_B}), \quad \sigma_3 = X_A^{x_B},$$

   the session key $K = H(\sigma_1, \sigma_2, \sigma_3, \Pi_1, ID_A, ID_B, X_A, X_B)$. U_B completes the session with session key K.
3. Upon receiving $(\Pi_1, ID_A, ID_B, X_A, X_B)$, U_A computes $e_A = H_2(X_A), e_B = H_2(X_B)$, the shared secrets

$$\sigma_1 = e(D_A^{e_A} Z^{x_A}, Q_B X_B), \quad \sigma_2 = e(D_A Z^{x_A}, Q_B^{e_B} X_B), \quad \sigma_3 = X_B^{x_A},$$

   the session key $K = H(\sigma_1, \sigma_2, \sigma_3, \Pi_1, ID_A, ID_B, X_A, X_B)$. U_A completes the session with session key K.

Both parties compute the shared secrets

$$\sigma_1 = g_T^{z(e_A \log(Q_A) + x_A)(\log(Q_B) + x_B)},$$
$$\sigma_2 = g_T^{z(\log(Q_A) + x_A)(e_B \log(Q_B) + x_B)},$$
$$\sigma_3 = g^{x_A x_B}$$

and therefore compute the same session key K.

It is worth noting that we could also construct secure ID-based AKE by modifying σ_1 and σ_2 to the following values

$$\sigma_1 = g_T^{z(e_A \log(Q_A) + x_A)(e_B \log(Q_B) + x_B)}, \ \sigma_2 = g_T^{z(\log(Q_A) + x_A)(\log(Q_B) + x_B)};$$

however, we opt to define our second protocol via slightly more efficient algorithms.

4.2   Proposed ID-Based AKE Protocol 2 – Π_2

In this section, we describe the actions required to execute a Π_2 session.

Key Exchange. User U_A is the session initiator and user U_B is the session responder.

1. U_A chooses an ephemeral private key $x_A \in_R \mathbb{Z}_q$, computes the ephemeral public key $X_A = g^{x_A}$ and sends (Π_2, ID_A, ID_B, X_A) to U_B.
2. Upon receiving (Π_2, ID_A, ID_B, X_A), U_B chooses an ephemeral private key $x_B \in_R \mathbb{Z}_q$, computes the ephemeral public key $X_B = g^{x_B}$ and responds to U_A with $(\Pi_2, ID_A, ID_B, X_A, X_B)$.
   U_B also computes the shared secrets

   $$\sigma_1 = e(Q_A X_A, D_B Z^{x_B}), \ \sigma_2 = e(Q_A, D_B), \ \sigma_3 = X_A^{x_B}$$

   the session key $K = H(\sigma_1, \sigma_2, \sigma_3, \Pi_2, ID_A, ID_B, X_A, X_B)$. U_B completes the session with session key K.
3. Upon receiving $(\Pi_2, ID_A, ID_B, X_A, X_B)$, U_A computes the shared secrets

   $$\sigma_1 = e(D_A Z^{x_A}, Q_B X_B), \ \sigma_2 = e(D_A, Q_B), \ \sigma_3 = X_B^{x_A}$$

   the session key $K = H(\sigma_1, \sigma_2, \sigma_3, \Pi_2, ID_A, ID_B, X_A, X_B)$. U_A completes the session with session key K.

Both parties compute the shared secrets

$$\sigma_1 = g_T^{z(\log(Q_A) + x_A)(\log(Q_B) + x_B)}, \ \sigma_2 = g_T^{z \log(Q_A) \log(Q_B)}, \ \sigma_3 = g^{x_A x_B},$$

and therefore compute the same session key K.

4.3   Security

The security of the proposed ID-AKE protocols is established by the following theorem.

Theorem 1. *If (G, G_T) are groups where gap Bilinear Diffie-Hellman assumption holds and H, H_1 and H_2 are random oracles, the ID-AKE Protocols Π_1 and Π_2 are secure in the shared static key model described in Section 3.*

The proof of Theorem 1 is provided in Appendix A.

4.4 Shared Secrets Design

Protocol Π_1. Protocol Π_1 is an adaptation of the Unified Protocol in [25] to the ID-based scenario. We refer the reader to [25] for the rationale behind the definitions of σ_1 and σ_2. Unlike the certificated-based scenario in the ID-based scenario without σ_3 the protocol has no forward secrecy with respect to the master secret key. To provide assurance that the KGC cannot compute the session key we include σ_3. There are other viable alternatives; however, we opt for the basic Diffie-Hellman value due to its relative simplicity.

Protocol Π_2. Protocol Π_2 is an adaptation of the Unified Model protocol [2,20]. More precisely, σ_2 and σ_3 are the static and ephemeral shared secrets, respectively, that Alice and Bob can compute in the ID-based scenario. The certificate-based variant is vulnerable to some attacks such as key compromise impersonation, which are carried over to the ID-based setting. We include the value σ_1 to remove these Unified Model protocol drawbacks.

Computational cost. The naive count of operations for Π_1 and Π_2 show that the former requires $2P + 4E$ and the latter $2P + 2E$, where P stands for pairing computation and E stands for group exponentiation. We do not take into account the exponentiation required to prepare the outgoing ephemeral public key since in our analysis these can be pre-computed; it is also not included in the protocol comparison in Section 5. Since pairing computations are more costly than exponentiations [8,13] the computational cost of the two protocols is effectively the same. We include both measures because novel techniques may reduce the cost of pairing computations significantly.

Efficiency. The certificate-based motivating protocol for Π_1 is less efficient but provides stronger security assurances than the certificate-based motivating protocol for Π_2. Surprisingly, in the ID-based scenario Π_2 is marginally more efficient than Π_1 and has the same security attributes. This suggest that while certificate-based protocols can help in the design of ID-based protocols, it is not necessary that more efficient certificate based protocols result in more efficient ID-based protocols. A single pairing to produce a session key is a viable option. However, as in the certificate-based setting better efficiency may require larger group sizes for comparable security levels. In particular, an ID-based adaptation of the HMQV protocol [16] would suffer from a less tight security argument because of forking arguments. We therefore opt for protocols with security arguments that do not require forking lemma type arguments.

Protocol choice. One may naturally ask why use Π_1 when Π_2 already provides the same security attributes and has some efficiency advantages. SP800-56A [20] provides two techniques for key establishment and the "unified model" protocol is both less efficient and has fewer security attributes. But, it is advantageous to offer fall-back alternatives to accommodate any unexpected environmental reasons that prevent the use of a particular protocol. Our arguments show that sharing identities for Π_1 and Π_2 is cryptographically sound.

4.5   Further Observations and Comments

Theorem 1 implies that the protocols are secure even if they are used separately. The security argument for protocol Π_1 say is a specialization of the execution model where the adversary does not activate a session of protocol Π_2. Thus, the theorem provides stronger assurance than separate security arguments.

The essential set of message that Alice (U_A) sends to Bob (U_B) consists of four messages ($\Pi_c, role, ID_A, X_A$) – the protocol identifier, the role Alice views for herself, the identifier she uses and her ephemeral public key. The response from Bob consists of a similar tuple ($\Pi'_c, role', ID_B, X_B$). The first two entries in Bob's case simply affirm that he agrees to Alice's choice of protocol and role. Our analysis is carried out only in the case where Alice sends her tuple at once and *then* receives all of Bob's response. It is possible that all these messages are interleaved: for example in the first flow Alice can send only ($\Pi_c, role, U_A$) to Bob and reveal her identifier ID_A only after receiving Bob's response. For simplicity we have omitted those technical details required in the model description. However it appears that the only requirement is that parties associate correctly incoming messages with sessions.

As in the HC protocol, Alice does not need to know her static private key to complete the message exchange in our protocols. This is unlike the Okamoto-Tanaka protocol [21], where a party needs its static private key before being able to compute outgoing messages. The feature is useful in scenarios where users select identities "on the fly" depending on some ephemeral system setting such as the day or time.

5   Comparison

Next we compare our protocols with related ID-AKE protocols in terms of underlying assumption, computational efficiency, and security model. In Table 1, number of pairing computations, the number of exponentiations in G, number of static public keys in terms of group elements, and number of ephemeral public key in terms of group elements are denoted by P, E, #sPK, and #ePK, respectively. Furthermore, id-CK and id-eCK denotes ID-based versions of the well-know Canetti-Krawzcyk [6] (CK) and LaMacchia, Lauter and Mityagin [17] (eCK) security models, respectively. Our model is denoted by id-eCK*; KCI denotes key compromise impersonation resistance and mSk-fs denotes master secret key forward security.

For further comparison we refer the reader to [14, Table 1]. It is plausible that the HC protocol is also secure if ephemeral public keys are revealed to the adversary before used in a session. Our protocols are as efficient as the HC protocol (the cost of a single exponentiation can be safely ignored here) and therefore the trade-off lies in the size of the private keys and the underlying assumptions. Such a comparison is very subjective but we believe that for practical purposes the use of shorter keys justifies the invocation of the gap assumption.

Table 1. Comparison with the existing protocols

Protocol	Computation	#sPk	#ePk	Security Model	Assumption
SCK [10]	2P+3E	1	1	id-CK,KCI,mSk-fs	BDH
SYL [10]	1P+3E	1	1	id-CK,KCI,mSk-fs	BDH
HC [14]	2P+3E	2	1	id-eCK	BDH
Π_1	2P+4E	1	1	id-eCK*	gap BDH
Π_2	2P+2E	1	1	id-eCK*	gap BDH

6 Conclusion

In this paper, we proposed the first secure ID-based AKE protocols that used a single group element as a static secret key, resist ephemeral key leakage, and can safely share static private information. Moreover, our protocols are efficient in terms of communication and computations and thus are suitable for practical applications. It is an interesting problem to consider developing a protocol with the same setup and communication message, but which requires a single pairing operation, such as an adaptation of the HMQV [16] protocol to the ID-based setting. Additionally, it is worth developing a new protocol that can be used in conjunction with protocol Π_2, and uses a single random oracle call during the session key computation stage.

Acknowledgments

We thank Alfred Menezes for helpful comments.

References

1. Bellare, M., Rogaway, P.: Entity authentication and key distribution. In: Stinson, D.R. (ed.) CRYPTO 1993. LNCS, vol. 773, pp. 232–249. Springer, Heidelberg (1994); Full version available at http://www.cs.ucdavis.edu/~rogaway/papers/eakd-abstract.html
2. Blake-Wilson, S., Johnson, D., Menezes, A.: Key agreement protocols and their security analysis. In: Darnell, M. (ed.) Cryptography and Coding 1997. LNCS, vol. 1355, pp. 30–45. Springer, Heidelberg (1997)
3. Boneh, D., Franklin, M.: Identity-based encryption from the weil pairing. In: Kilian, J. (ed.) CRYPTO 2001. LNCS, vol. 2139, pp. 213–229. Springer, Heidelberg (2001)
4. Boyd, C., Choo, K.-K.R.: Security of two-party identity-based key agreement. In: Dawson, E., Vaudenay, S. (eds.) MYCRYPT 2005. LNCS, vol. 3715, pp. 229–243. Springer, Heidelberg (2005)
5. Boyd, C., Cliff, Y., González Nieto, J.M., Paterson, K.: Efficient one-round key exchange in the standard model. In: Mu, Y., Susilo, W., Seberry, J. (eds.) ACISP 2008. LNCS, vol. 5107, pp. 69–83. Springer, Heidelberg (2008); Full version available at http://eprint.iacr.org/2008/007/

6. Canetti, R., Krawczyk, H.: Analysis of key-exchange protocols and their use for building secure channels. In: Pfitzmann, B. (ed.) EUROCRYPT 2001. LNCS, vol. 2045, pp. 453–474. Springer, Heidelberg (2001); Full version available at http://eprint.iacr.org/2001/040/

7. Canetti, R., Krawczyk, H.: Universally composable notions of key exchange and secure channels. In: Knudsen, L. (ed.) EUROCRYPT 2002. LNCS, vol. 2332, pp. 337–351. Springer, Heidelberg (2002)

8. Chatterjee, S., Hankerson, D., Knapp, E., Menezes, A.: Comparing two pairing-based aggregate signature schemes. Designs, Codes and Cryptography 55(2), 141–167 (2010)

9. Chatterjee, S., Menezes, A., Ustaoğlu, B.: Reusing static keys in key agreement protocols. In: Roy, B., Sendrier, N. (eds.) INDOCRYPT 2009. LNCS, vol. 5922, pp. 39–56. Springer, Heidelberg (2009)

10. Chen, L., Cheng, Z., Smart, N.P.: Identity-based key agreement protocols from pairings. International Journal of Information Security 6(4), 213–241 (2007)

11. Diffie, W., Hellman, M.E.: New directions in cryptography. IEEE Transactions on Information Theory IT-22(6), 644–654 (1976)

12. González Vasco, M.I., Hess, F., Steinwandt, R.: Combined (identity-based) public key schemes. Cryptology ePrint Archive, Report 2008/466 (2008), http://eprint.iacr.org/2008/466

13. Hankerson, D., Menezes, A., Scott, M.: Software implementation of pairings. In: Joye, M., Neven, G. (eds.) Identity-Based Cryptography. Cryptology and Information Security, vol. 2, ch. XII, pp. 188–206. IOS Press, Amsterdam (2008)

14. Huang, H., Cao, Z.: An id-based authenticated key exchange protocol based on bilinear diffie-hellman problem. In: Safavi-Naini, R., Varadharajan, V. (eds.) ASIACCS 2009: Proceedings of the 2009 ACM Symposium on Information, Computer and Communications Security, New York, NY, USA, pp. 333–342 (2009)

15. Kelsey, J., Schneier, B., Wagner, D.: Protocol interactions and the chosen protocol attack. In: Christianson, B., Crispo, B., Lomas, M., Roe, M. (eds.) SP 1997. LNCS, vol. 1361, pp. 91–104. Springer, Heidelberg (1998)

16. Krawczyk, H.: HMQV: A high-performance secure Diffie-Hellman protocol. In: Cramer, R. (ed.) CRYPTO 2005. LNCS, vol. 3621, pp. 546–566. Springer, Heidelberg (2005)

17. LaMacchia, B., Lauter, K., Mityagin, A.: Stronger security of authenticated key exchange. In: Susilo, W., Liu, J.K., Mu, Y. (eds.) ProvSec 2007. LNCS, vol. 4784, pp. 1–16. Springer, Heidelberg (2007)

18. Menezes, A., Ustaoğlu, B.: Comparing the pre- and post-specified peer models for key agreement. International Journal of Applied Cryptography (IJACT) 1(3), 236–250 (2009)

19. Menezes, A.J., Okamoto, T., Vanstone, S.A.: Reducing elliptic curve logarithms to logarithms in a finite field. IEEE Transactions on Information Theory 39(5), 1639–1646 (1993)

20. NIST National Institute of Standards and Technology. Special Publication 800-56A, Recommendation for Pair-Wise Key Establishment Schemes Using Discrete Logarithm Cryptography (March 2007), http://csrc.nist.gov/publications/PubsSPs.html

21. Okamoto, E., Tanaka, K.: Key distribution system based on identification information. IEEE Journal on Selected Arean in Communications 7(4), 481–485 (1989)

22. Sakai, R., Ohgishi, K., Kasahara, M.: Cryptosystems based on pairings. In: The 2000 Symposium on Cryptography and Information Security (2000)

23. Shamir, A.: Identity-based cryptosystems and signature schemes. In: Blakley, G.R., Chaum, D. (eds.) CRYPTO 1984. LNCS, vol. 196, pp. 47–53. Springer, Heidelberg (1984)
24. Smart, N.P.: Identity-based authenticated key agreement protocol based on weil pairing. Electronic Letters 38(13), 630–632 (2002)
25. Ustaoğlu, B.: Comparing *SessionStateReveal* and *EphemeralKeyReveal* for Diffie-Hellman protocols. In: Pieprzyk, J., Zhang, F. (eds.) ProvSec 2009. LNCS, vol. 5848, pp. 183–197. Springer, Heidelberg (2009)

A   Proof of Theorem 1

We call the variant of the gap BDH assumption where one tries to compute $BDH(U, U, W)$ instead of $BDH(U, V, W)$ as the square gap BDH assumption. The variant is equivalent to the gap BDH assumption as follows. Given a challenge U, W of the square gap BDH assumption, one sets $V = U^s$ for random integers $s \in_R [1, p-1]$, and then can compute $BDH(U, V, W)^{1/s} = BDH(U, U, W)$. Given a challenge (U, V, W) of the gap BDH assumption, one can set $U_1 = UVW$, $U_2 = UVW^{-1}$, $U_3 = UV^{-1}W$, $U_4 = UV^{-1}W^{-1}$, and then compute $BDH(U, V, W)$ from $BDH(U_i, U_i, U_i^s)^{1/s}$, $i = 1, \ldots, 4$.

We will show that no polynomially bounded adversary can distinguish the session key of a fresh session from a randomly chosen session key.

Let κ denote the security parameter, and let $\mathcal{A}$ be a polynomially (in κ) bounded adversary. We use $\mathcal{A}$ to construct a gap BDH solver $\mathcal{S}$ that succeeds with non-negligible probability. The adversary $\mathcal{A}$ is said to be successful with non-negligible probability if $\mathcal{A}$ wins the distinguishing game with probability $\frac{1}{2} + f(\kappa)$, where $f(\kappa)$ is non-negligible, and the event M denotes a successful $\mathcal{A}$.

Assume that $\mathcal{A}$ succeeds in an environment with n users, activates at most s sessions within a user, makes at most q_H, q_{H_1}, q_{H_2} queries to oracles H, H_1, H_2, respectively.

We denote the master secret and public keys by $z, Z = g^z$. For user U_i, we denote the identity by ID_i, the static secret and public keys by $D_i = Q_i^z, Q_i = H_1(ID_i)$, and the ephemeral secret and public keys by $x_i, X_i = g^{x_i}$. We also denote the session key by K.

Let the test session be $\mathtt{sid}^* = (\Pi_c, ID_A, ID_B, X_A, X_B)$, where users U_A, U_B are initiator and responder of the test session $\mathtt{sid}^*$. Let H^* be the event that $\mathcal{A}$ queries $(\sigma_1, \sigma_2, \sigma_3, \mathtt{sid}^*)$ to H. Let $\overline{H^*}$ be the complement of event H^*. Let $\mathtt{sid}$ be any completed session owned by an honest user such that $\mathtt{sid} \neq \mathtt{sid}^*$ and $\mathtt{sid}$ is non-matching to $\mathtt{sid}^*$. Since $\mathtt{sid}$ and $\mathtt{sid}^*$ are distinct and non-matching, the inputs to the key derivation function H are different for $\mathtt{sid}$ and $\mathtt{sid}^*$. Since H is a random oracle, $\mathcal{A}$ cannot obtain any information about the test session key from the session keys of non-matching sessions. Hence $\Pr(M \wedge \overline{H^*}) \leq \frac{1}{2}$ and $\Pr(M) = \Pr(M \wedge H^*) + \Pr(M \wedge \overline{H^*}) \leq \Pr(M \wedge H^*) + \frac{1}{2}$, whence $\Pr(M \wedge H^*) \geq p(\kappa)$. Henceforth the event $M \wedge H^*$ is denoted by M^*.

We will consider the not exclusive classification of all possible events in the following Table 2 and 3. Table 2 classifies events when Q_A, Q_B are distinct, and Table 3 classifies events when $Q_A = Q_B$. We denote by z master secret key, and by $(D_i, x_i)_{i=A,B}$ static and ephemeral secret keys of users U_A, U_B who are initiator and responder of the test session sid*. In these tables, "ok" means the static key is not revealed, or the matching session exists and the ephemeral key is not revealed. "r" means the static or ephemeral key may be revealed. "r/n" means the ephemeral key may be revealed if the matching session exists or no matching session exists. "instance embedding" row shows how simulator embeds a instance of gap BDH problem. "succ. prob." row shows the probability of success of solver $\mathcal{S}$, where $p_{xy} = Pr(E_{xy} \wedge M^*)$ and n and s is the number of parties and sessions.

Since the classification covers all possible events, at least one event $E_{xy} \wedge M^*$ in the tables occurs with non-negligible probability, if event M^* occurs with non-negligible probability. Thus, the gap BDH problem can be solved with non-negligible probability, and that means we shows that the proposed protocol is secure under the gap BDH assumption. We will investigate each of these events in the following subsections.

Table 2. Classification of attacks, when Q_A, Q_B are distinct. "ok" means the static key is not revealed, or the matching session exists and the ephemeral key is not revealed. "r" means the static or ephemeral key may be revealed. "r/n" means the ephemeral key may be revealed if the matching session exists or no matching session exists. "instance embedding" row shows how simulator embeds a instance of gap BDH problem. "succ. prob." row shows the probability of success of solver $\mathcal{S}$, where $p_{xy} = Pr(E_{xy} \wedge M^*)$ and n and s is the number of parties and sessions.

	z	D_A	x_A	D_B	x_B	instance embedding	succ. prob.
E_{1a}	ok	r	ok	ok	r/n	$Z = U, X_A = V, Q_B = W$	$p_{1a}/n^2 s$
E_{1b}	ok	ok	r	ok	r/n	$Z = U, Q_A = V, Q_B = W$	p_{1b}/n^2
E_{2a}	r	r	ok	r	ok	$X_A = V, X_B = W$	$p_{2a}/n^2 s^2$
E_{2b}	ok	ok	r	r	ok	$Z = U, Q_A = V, X_B = W$	$p_{2b}/n^2 s$

Table 3. Classification of attacks, when $Q_A = Q_B$

	z	D_A	x_A	D_A	x_B	instance embedding	succ.prob.
E'_{1b}	ok	ok	r	ok	r/n	$Z = U, Q_A = V$	p_{1b}/n
E'_{2a}	r	r	ok	r	ok	$X_A = V, X_B = W$	$p_{2a}/n^2 s^2$

A.1 E_{1a}

Setup. The gap BDH solver $\mathcal{S}$ begins by establishing n honest users that are assigned random static key pairs. For each honest user U_i, $\mathcal{S}$ maintains list L_{EK} of at most s ephemeral key pairs, and two markers – a user marker and an adversary marker. The list L_{EK} is initially empty, and the markers initially point to the first entry of the list L_{EK}. Whenever U_i is activated to create a new session, $\mathcal{S}$ checks if the user marker points to an empty entry. If so, $\mathcal{S}$ selects a new ephemeral key pair on behalf of U_i as described in the Π_c ($c = 1, 2$) protocol.

If the list entry is not empty, then S uses the ephemeral key pair in that list entry for the newly created session. In either case the user marker is updated to point to the next list entry, and the adversary marker is also advanced if it points to an earlier entry. If A issues an EphemeralPublicKeyReveal query, then S selects a new ephemeral key pair on behalf of U_i as described in the Π_c protocol. S stores the key pair in the entry pointed to by the adversary marker, returns the public key as the query response, and advances the adversary marker.

In addition to the above steps, S embeds instance (U, V, W) of gap BDH problem as follows. S randomly selects two users U_A, U_B and integer $t \in_R [1, s]$. Public master key Z is chosen to be U, and D_A is chosen to be Z^{q_A} for randomly selected $q_A = \log(Q_A)$. S simulates oracle H_1 by selecting a random integer c in the interval $[1, q]$ and setting $H_1(ID_C) = g^c$; the static private key corresponding to ID_C is U^c. S selects static and ephemeral key pairs on behalf of honest users as described above with the following exceptions. The i-th ephemeral public key X selected on behalf of U_A is chosen to be V, and the static public key Q_B selected on behalf of U_B is chosen to be W. S does not possess the corresponding static and ephemeral private keys.

The algorithm S activates A on this set of users and awaits the actions of A. We next describe the actions of S in response to user activations and oracle queries.

Simulation. The algorithm S maintains list L_H that contains queries and answers of H oracle and list L_S that contains SessionKeyReveal queries and answers, and simulates oracle queries as follows.

1. Send($\Pi_c, \mathcal{I}, ID_i, ID_j$): S picks ephemeral public key X_i from the list L_{EK}, and records (Π_c, ID_i, ID_j, X_i) and returns it.
2. Send($\Pi_c, \mathcal{R}, ID_j, ID_i, X_i$): S picks ephemeral public key X_j from the list L_{EK}, and records $(\Pi_c, ID_i, ID_j, X_i, X_j)$ and returns it.
3. Send($\Pi_c, \mathcal{I}, ID_i, ID_j, X_i, X_j$): If (Π_c, ID_i, ID_j, X_i) is not recorded, then S records the session $(\Pi_1, \mathcal{I}, ID_i, ID_j, X_i, X_j)$ as not completed. Otherwise, S records the session as completed.
4. $H(\sigma_1, \sigma_2, \sigma_3, \Pi_1, ID_i, ID_j, X_i, X_j)$:
   (a) If $(\sigma_1, \sigma_2, \sigma_3, \Pi_1, ID_i, ID_j, X_i, X_j)$ is recorded in list L_H, then S returns recorded value K.
   (b) Else if the session $(\Pi_1, \mathcal{I}, ID_i, ID_j, X_i, X_j)$ or $(\Pi_1, \mathcal{R}, ID_j, ID_i, X_i, X_j)$ is recorded in list L_H, then S checks that $\sigma_1, \sigma_2, \sigma_3$ are correctly formed, i.e.,

$$\mathrm{BDDH}(Z, Q_i^{e_i} X_i, Q_j X_j, \sigma_1) = 1,$$
$$\mathrm{BDDH}(Z, Q_i X_i, Q_j^{e_j} X_j, \sigma_2) = 1$$

and $e(X_i, X_j) = \sigma_3$. If $\sigma_1, \sigma_2, \sigma_3$ are correctly formed, then S returns recorded value K and records it in list L_H.

(c) Else if $i = A, j = B, X_A = V, Q_B = W$, then $\mathcal{S}$ checks that $\sigma_1, \sigma_2, \sigma_3$ are correctly formed, i.e.,

$$\mathrm{BDDH}(Z, Q_i^{e_i} X_i, Q_j X_j, \sigma_1) = 1,$$
$$\mathrm{BDDH}(Z, Q_i X_i, Q_j^{e_j} X_j, \sigma_2) = 1,$$

and $e(X_i, X_j) = \sigma_3$. If $\sigma_1, \sigma_2, \sigma_3$ are correctly formed, then since $\mathcal{S}$ knows $\log(Q_A)$, $\mathcal{S}$ can compute the answer to the gap BDH instance

$$((\sigma_1')(\sigma_2')^{-1})^{1/(1-e_B)} = g_T^{zx_A \log(Q_B)} = \mathrm{BDH}(Z, X_A, Q_B),$$

where

$$\sigma_1' = \sigma_1 e(Z, Q_B X_B)^{-e_A \log(Q_A)} = g_T^{zx_A(\log(Q_B)+x_B)},$$
$$\sigma_2' = \sigma_2 e(Z, Q_B^{e_B} X_B)^{-\log(Q_A)} = g_T^{zx_A(e_B \log(Q_B)+x_B)},$$

and stops successfully by outputting the answer.

(d) Otherwise, $\mathcal{S}$ returns random value K and records it in list L_H.

5. $H(\sigma_1, \sigma_2, \sigma_3, \Pi_2, ID_i, ID_j, X_i, X_j)$:

(a) If $(\sigma_1, \sigma_2, \sigma_3, \Pi_2, ID_i, ID_j, X_i, X_j)$ is recorded in list L_H, then $\mathcal{S}$ returns recorded value K.

(b) Else if the session $(\Pi_2, \mathcal{I}, ID_i, ID_j, X_i, X_j)$ or $(\Pi_2, \mathcal{R}, ID_j, ID_i, X_i, X_j)$ is recorded in list L_H, then $\mathcal{S}$ checks that $\sigma_1, \sigma_2, \sigma_3$ are correctly formed, i.e.,

$$\mathrm{BDDH}(Z, Q_i X_i, Q_j X_j, \sigma_1) = 1,$$
$$\mathrm{BDDH}(Z, Q_i, Q_j, \sigma_2) = 1,$$

and $e(X_i, X_j) = \sigma_3$. If $\sigma_1, \sigma_2, \sigma_3$ are correctly formed, then $\mathcal{S}$ returns recorded value K and records it in list L_H.

(c) Else if $i = A, j = B, X_A = V, Q_B = W$, then $\mathcal{S}$ checks that $\sigma_1, \sigma_2, \sigma_3$ are correctly formed, i.e.,

$$\mathrm{BDDH}(Z, Q_i X_i, Q_j X_j, \sigma_1) = 1,$$
$$\mathrm{BDDH}(Z, Q_i, Q_j, \sigma_2) = 1,$$

and $e(X_i, X_j) = \sigma_3$. If $\sigma_1, \sigma_2, \sigma_3$ are correctly formed, then since $\mathcal{S}$ knows $\log(Q_A)$, $\mathcal{S}$ can compute the answer of the gap BDH instance

$$\sigma_1 \tau_1^{-1} \tau_2^{-1} \tau_3^{-1} = g_T^{zx_A \log(Q_B)} = \mathrm{BDH}(Z, X_A, Q_B),$$

where

$$\tau_1 = \sigma_2 = g_T^{z \log(Q_A) \log(Q_B)},$$
$$\tau_2 = e(Z, \sigma_3) = g_T^{zx_A x_B},$$
$$\tau_3 = e(Z, X_B)^{\log(Q_A)} = g_T^{z \log(Q_A)x_B},$$

and stops successfully by outputting the answer.

(d) Otherwise, $\mathcal{S}$ returns random value K and records it in list L_H.

6. SessionKeyReveal($(\Pi_1, \mathcal{I}, ID_i, ID_j, X_i, X_j)$ or $(\Pi_1, \mathcal{R}, ID_j, ID_i, X_i, X_j)$):

   (a) If the session sid is not completed, $\mathcal{S}$ returns error.
   (b) Else if the session sid is recorded in list L_S, then $\mathcal{S}$ returns recorded value K.
   (c) Else if $(\sigma_1, \sigma_2, \sigma_3, \Pi_1, ID_i, ID_j, X_i, X_j)$ is recorded in list L_H, then $\mathcal{S}$ checks that $\sigma_1, \sigma_2, \sigma_3$ are correctly formed, i.e.,

   $$\mathrm{BDDH}(Z, Q_i^{e_i} X_i, Q_j X_j, \sigma_1) = 1,$$
   $$\mathrm{BDDH}(Z, Q_i X_i, Q_j^{e_j} X_j, \sigma_2) = 1,$$

   and $e(X_i, X_j) = \sigma_3$. If $\sigma_1, \sigma_2, \sigma_3$ are correctly formed, then $\mathcal{S}$ returns recorded value K and records it in list L_S.
   (d) Otherwise, $\mathcal{S}$ returns random value K and records it in list L_S.

7. SessionKeyReveal($(\Pi_2, \mathcal{I}, ID_i, ID_j, X_i, X_j)$ or $(\Pi_2, \mathcal{R}, ID_j, ID_i, X_i, X_j)$):

   (a) If the session sid is not completed, $\mathcal{S}$ returns error.
   (b) Else if the session sid is recorded in list L_S, then $\mathcal{S}$ returns recorded value K.
   (c) Else if $(\sigma_1, \sigma_2, \sigma_3, \Pi_2, ID_i, ID_j, X_i, X_j)$ is recorded in list L_H, then $\mathcal{S}$ checks that $\sigma_1, \sigma_2, \sigma_3$ are correctly formed, i.e.,

   $$\mathrm{BDDH}(Z, Q_i X_i, Q_j X_j, \sigma_1) = 1,$$
   $$\mathrm{BDDH}(Z, Q_i, Q_j, \sigma_2) = 1,$$

   and $e(X_i, X_j) = \sigma_3$. If $\sigma_1, \sigma_2, \sigma_3$ are correctly formed, then $\mathcal{S}$ returns recorded value K and records it in list L_S.
   (d) Otherwise, $\mathcal{S}$ returns random value K and records it in list L_S.

8. $H_1(ID_C)$: If $C = B$, $\mathcal{S}$ returns $Q_B = W$, otherwise selects a random integer c in the interval $[1, q]$ and return g^c.

9. $H_2(X_i)$: $\mathcal{S}$ simulates random oracle in the usual way.

10. EphemeralPublicKeyReveal(sid): $\mathcal{S}$ picks ephemeral public key X from the list L_{EK}, and returns X.

11. EphemeralKeyReveal(sid): $\mathcal{S}$ picks ephemeral secret key x from the list L_{EK}. If the corresponding ephemeral public key is V, then $\mathcal{S}$ aborts with failure, otherwise returns x.

12. StaticKeyReveal(ID_i): If static public key Q_i of user U_i is W, then $\mathcal{S}$ aborts with failure, otherwise responds to the query faithfully.

13. MasterKeyReveal(): $\mathcal{S}$ aborts with failure.

14. EstablishParty(ID_i): $\mathcal{S}$ responds to the query faithfully.

15. Test(sid): If ephemeral public key of a user is not V and static public key of the other user not W in session sid, then $\mathcal{S}$ aborts with failure, otherwise responds to the query faithfully.

16. If $\mathcal{A}$ outputs a guess γ, $\mathcal{S}$ aborts with failure.

Analysis. The simulation of $\mathcal{A}$ environment is perfect except with negligible probability. The probability that $\mathcal{A}$ selects the session, where ephemeral public key of a user is V and static public keys of the other users are W, as the test session sid^* is at least $\frac{1}{n^2 s}$. Suppose this is indeed the case, $\mathcal{S}$ does not abort in Step 15, and suppose event $E_{1a} \wedge M^*$ occurs, $\mathcal{S}$ does not abort in Step 11, Step 12, and Step 13.

Under event M^* except with negligible probability, $\mathcal{A}$ queries H with correctly formed $\sigma_1, \sigma_2, \sigma_3$. Therefore $\mathcal{S}$ is successful as described in Step 4c or Step 5c and does not abort as in Step 16.

Hence, $\mathcal{S}$ is successful with probability $Pr(S) \geq \frac{p_{1a}}{n^2 s}$, where p_{1a} is probability that $E_{1a} \wedge M^*$ occurs.

A.2 E_{1b}

Same as the event $E_{1a} \wedge M^*$ in Subsection A.1, except the following points.

In Setup, $\mathcal{S}$ embeds gap BDH instance (U, V, W) as $Z = U, Q_A = V, Q_B = W$, and selects randomly x_A and simulates as $X_A = g^{x_A}$.

In Simulation of H, $\mathcal{S}$ extracts $BDH(U, V, W)$ as follows. In Step 4c, since $\mathcal{S}$ knows x_A, $\mathcal{S}$ extracts

$$((\sigma_1')^{1/e_A}(\sigma_2')^{-1})^{1/(1-e_B)} = g_T^{z \log(Q_A) \log(Q_B)} = BDH(Z, Q_A, Q_B),$$

where

$$\sigma_1' = \sigma_1 e(Z, Q_B X_B)^{-x_A} = g_T^{z e_A \log(Q_A)(\log(Q_B) + x_B)},$$
$$\sigma_2' = \sigma_2 e(Z, Q_B^{e_B} X_B)^{-x_A} = g_T^{z \log(Q_A)(e_B \log(Q_B) + x_B)}.$$

In Step 5c, $\mathcal{S}$ extracts

$$\sigma_2 = g_T^{z \log(Q_A) \log(Q_B)} = BDH(Z, Q_A, Q_B).$$

A.3 E_{2a}

Same as the event $E_{1a} \wedge M^*$ in Subsection A.1, except the following points.

In Setup, $\mathcal{S}$ embeds gap CDH instance (V, W) as $X_A = V, X_B = W$, selects uniformly at random master private key z and computes the corresponding master public key $Z = g^z$. The oracle H_1 is simulated honestly private keys of all parties are computed using z.

In Simulation of H, $\mathcal{S}$ extracts $BDH(U, V, W)$ as follows. In Step 4c, since $\mathcal{S}$ knows $\log(Q_A)$, $\mathcal{S}$ extracts

$$((\sigma_1')^{e_B}(\sigma_2')^{-1})^{1/(e_B - 1)} = g_T^{z x_A x_B} = BDH(Z, X_A, X_B),$$

where

$$\sigma_1' = \sigma_1 e(Z, Q_B X_B)^{-e_A \log(Q_A)} = g_T^{z x_A(\log(Q_B) + x_B)},$$
$$\sigma_2' = \sigma_2 e(Z, Q_B^{e_B} X_B)^{- \log(Q_A)} = g_T^{z x_A(e_B \log(Q_B) + x_B)}.$$

In Step 5c, $\mathcal{S}$ extracts

$$e(Z, \sigma_3) = g_T^{z x_A x_B} = BDH(Z, X_A, X_B).$$

A.4 E_{2b}

Same as the event $E_{1a} \wedge M^*$ in Subsection A.1, except the following points.

In Setup, $\mathcal{S}$ embeds gap BDH instance (U, V, W) as $Z = U, Q_A = V, X_B = W$, and selects randomly $x_A, q_B = \log(Q_B)$ and simulates as $X_A = g^{x_A}, D_B = Z^{q_B}$.

In Simulation of H, $\mathcal{S}$ extracts BDH(U, V, W) as follows. In Step 4c, since $\mathcal{S}$ knows x_A, $\mathcal{S}$ extracts

$$((\sigma_1')^{e_B/e_A}(\sigma_2')^{-1})^{1/(e_B-1)} = g_T^{z\log(Q_A)x_B} = \text{BDH}(Z, Q_A, X_B),$$

where

$$\sigma_1' = \sigma_1 e(Z, Q_B X_B)^{-x_A} = g_T^{ze_A \log(Q_A)(\log(Q_B)+x_B)},$$
$$\sigma_2' = \sigma_2 e(Z, Q_B^{e_B} X_B)^{-x_A} = g_T^{z\log(Q_A)(e_B \log(Q_B)+x_B)}.$$

In Step 5c, since $\mathcal{S}$ knows x_A, $\mathcal{S}$ extracts

$$\sigma_1 \tau_1^{-1} \tau_2^{-1} \tau_3^{-1} = g_T^{z\log(Q_A)x_B} = \text{BDH}(Z, Q_A, X_B),$$

where

$$\tau_1 = \sigma_2 = g_T^{z\log(Q_A)\log(Q_B)},$$
$$\tau_2 = e(Z, \sigma_3) = g_T^{zx_A x_B},$$
$$\tau_3 = e(Z, Q_B)^{x_A} = g_T^{zx_A \log(Q_B)}.$$

A.5   Other Cases

Event E_{1b}' in Table 3 can be handled same as event E_{1b} in Table 2, with condition $Q_A = Q_B$ under the square gap BDH assumption that is equivalent to the gap BDH assumption.

Events E_{2a}', E_3' in Table 3 can be handled same as events E_{2a}, E_3 in Table 2, with condition $Q_A = Q_B$ under the gap BDH assumption.

Pairing-Based Non-interactive
Zero-Knowledge Proofs

Jens Groth

University College London
j.groth@ucl.ac.uk

Abstract. A non-interactive zero-knowledge proof permits the construction of a proof of the truth of a statement that reveals nothing else but the fact that the statement is true. Non-interactive zero-knowledge proofs are used in the construction of numerous cryptographic schemes such as public-key cryptosystems and advanced digital signatures. The only practically efficient constructions of non-interactive zero-knowledge proofs that are based on standard intractability assumptions come from pairing based-cryptography. These pairing-based non-interactive zero-knowledge proofs integrate smoothly with other pairing-based cryptographic schemes making the combined schemes quite efficient.

We will sketch how to construct non-interactive zero-knowledge and non-interactive witness-indistinguishable proofs from modules with a bilinear map. The general approach based on modules with a bilinear map implies that different types of groups with pairings can be used in the construction of non-interactive cryptographic proofs and that security can be based on a number of different decisional assumptions.

M. Joye, A. Miyaji, and A. Otsuka (Eds.): Pairing 2010, LNCS 6487, p. 206, 2010.

Designing a Code Generator for Pairing Based Cryptographic Functions

Luis J. Dominguez Perez* and Michael Scott**

School of Computing
Dublin City University
Ireland
{ldominguez,mike}@computing.dcu.ie

Abstract. Pairing-Based Cryptography has become relevant in industry mainly because of the increasing interest in Identity-Based protocols. A major deterrent to the general use of pairing-based protocols is the complex nature of such protocols; efficient implementation of pairing functions is often difficult as it requires more knowledge than previous cryptographic primitives. In this paper we present a tool for automatically generating optimized code for pairing functions.

Our cryptographic compiler chooses the most appropriate pairing function for the target family of curves, either the Tate, ate, R-ate or Optimal pairing function, and generates its code. It also generates optimized code for the final exponentiation using the parameterisation of the chosen pairing-friendly elliptic curve.

Keywords: pairings, implementation, code generator.

1 Introduction

Interest in pairing-based cryptography has been growing since the arrival of the new millennium, and since the development of many constructive protocols, for example those given in [29], [13]. The usefulness of these protocols has caught the attention of industry.

Traditional cryptographic protocols, such as RSA, are well established and seen as "good enough" for the immediate future, but have limited functionality. Pairing-based cryptography is slowly being seen as a viable option.

The main disadvantage of implementing pairing-based protocols instead of these commercial solutions is the deeper mathematical background required to produce an efficient implementation. Every year new improvements on how to compute pairings appear. A pairing-based protocol designer may prefer to focus on the proof and formalization of the protocol itself rather than on the physical

* This author acknowledge support from the Consejo Nacional de Ciencia y Tecnología.
** This author acknowledge support from the Science Foundation Ireland under Grant No. 06/MI/006.

M. Joye, A. Miyaji, and A. Otsuka (Eds.): Pairing 2010, LNCS 6487, pp. 207–224, 2010.

construction of the primitives upon which it relies. Given the many improvements, it is easy to lose track of the most "up-to-date" optimizations and use a less than optimal implementation.

The aim of a code generator for cryptographic pairing functions is the following: to decide (or suggest) which family of pairing friendly elliptic curves to choose (for examples we refer to Freeman et al. [22]), to find a low Hamming-weight x-parameter for the definition of the system parameters; to generate the elliptic curve with a subgroup size corresponding to the desired security level; to choose the pairing function that best suits the family of curves to which the chosen curve belongs, and it's representation; to add supportive functions; and, optionally, to generate a sample "playground" for testing the code.

In order for a code generator to be flexible, support for several multiprecision libraries should be included. Some characteristics of the code rely on the programming language itself, others rely on the library. Some operations use an in-fixed operator, depending on the library. For some, this may only be possible with the use of a map and with explicit intermediate storage, where for others, the compiler can handle it.

Naehrig, Niederhagen and Schwabe in [39], described the abstraction level of optimization of their BN curves implementation in high, mid and low levels. This research currently focuses on the high-level. We are not dealing with fine tuning-up the finite field arithmetic, modular reduction method, NAF representation, elliptic curve arithmetic costs, or making use of processor specific registers. Also, we focus only on non-supersingular curves.

The paper continues as follows: section 2 contains a brief introduction to the Tate, ate, R-ate and Optimal pairings. In section 3 an introduction to addition chains [20] and addition sequences is presented with an artificial immune system algorithm to find them. This is relevant to the compiler as it is a key component in the computation of the final exponentiation of the pairing function, introduced in section 4.

Galbraith and Scott [23] presented a novel way to perform both scalar multiplication of elements of G_2 and exponentiation in G_T. In section 5 we present a shorter vector than that presented by the authors. This vector is used during the precomputation stage.

Section 7 describes the target libraries used for this work. Section 9 contains timings of the generated code for the BN $k = 12$ curves and the KSS $k = 18$ curves. The appendixes contain some sample output code.

2   Pairing Functions

Traditionally there are two cryptographic pairings: the Weil and the Tate-Lichtenbaum pairings. The Weil pairing requires two *Miller loops* [44, III.§8]. The Tate pairing requires only one application of the *Miller loop*, but with more complex arithmetic. Recent research efforts have concentrated on finding viable pairings with shorter loops and on simplifying the underlying arithmetic. This research focuses on the Tate pairings (and variations thereof) as it is known to be faster [26].

The best known method for computing pairings (both Weil and Tate variants) is based on Miller's algorithm. This is a standard method, and is a double-and-add and line-and-tangent algorithm. [35] [36]

The computation of pairings basically involves as input elements from two groups, G_1 and G_2. These groups use an additive notation and at least one of them is of prime order r. The pairing will map elements from both groups to a multiplicatively written group, also of prime order r, and denoted as G_T.

Let k be the *embedding degree*, which sometimes is referred to as the *security multiplier*, of an elliptic curve E defined over a finite field $\mathbb{F}_p$, and let r be the large prime number that divides $\#E$ such that r divides $(p^k - 1)$. We prefer k to be even as this will lead to improvements that will become obvious in the following sections.

Many constructions of pairing-friendly elliptic curves have been proposed by different authors; examples are MNT curves [37], Freeman curves [21] and BN [8] curves. These families of curves produce ideal pairing-friendly elliptic curves for a given security level. These curves have group size approximately equal to the size of the underlying field, making them very efficient for implementation. This ratio is known as the ρ-value, and it is defined as: $\rho = \frac{\deg p(x)}{\deg r(x)}$ (assuming that p and r are represented as polynomials $p(x)$ and $r(x)$ respectively).

2.1   The Tate Pairing

The Tate pairing is defined as follows [12]:

Let $P \in E(\mathbb{F}_p)[r]$ and let $Q \in E(\mathbb{F}_{p^k})$. Let $f_{a,P}$ be a function with a divisor $(f_{a,P}) = a(P) - (aP) - (a-1)(\mathcal{O})$ for $a \in \mathbb{Z}$. The non-degenerate, bilinear Tate pairing is defined as a map:

$$e_r : E(\mathbb{F}_p)[r] \times E(\mathbb{F}_{p^k})/rE(\mathbb{F}_{p^k}) \to \mathbb{F}_{p^k}^*/(\mathbb{F}_{p^k}^*)^r$$
$$(P, Q) \mapsto \langle P, Q \rangle_r = f_{r,P}(Q)$$

For practical and security purposes it is preferred to raise the value of the pairing to the power of $(p^k - 1)/r$ to obtain a unique representative of the class, i.e

$$e_r : (P, Q) \mapsto f_{r,P}(Q)^{(p^k-1)/r}.$$

We refer the interested reader to [12] for further details. It is also advantageous for Q to be defined as a point over a twist of E.

2.2   The Ate Pairing

The ate pairing [28] is a variant of the Tate pairing and a generalization of the Eta pairing [6] for ordinary pairing-friendly elliptic curves. The ate pairing is particularly suitable for pairing-friendly elliptic curves with small values of the trace of the Frobenius, given by $t = t(x)$ for a chosen value of the x parameter.

In practice, the reduced ate pairing is preferred

$$e_T : (Q, P) \mapsto f_{T,Q}(P)^{(p^k-1)/r}$$

where $T = t - 1$ and $P \in E(\mathbb{F}_p)$ and $Q \in E'(\mathbb{F}_{p^{k/d}})[r]$ with d the degree of the twist. In contrast to the Tate pairing, the ate pairing requires curve arithmetic in the full finite field extension (over the twist, in this case), but enjoys the advantage of a shorter loop in the Miller operation. This shortening, however, depends on the family of pairing friendly curves itself. This can be measured as $\omega = \frac{\log r}{\log |t|}$. The bigger the ω, the shorter the Miller loop of the ate pairing compared with the Tate pairing. Note the change of arguments of the f function between the pairings.

2.3   R-Ate Pairing

The R-ate pairing is a generalization of the ate [28] and ate$_i$ [46] pairings, improving their computation efficiency [32]. The computation uses up to three short Miller loops instead of a single Miller loop. The aim is to chose the parameters so that the three loops together are shorter that a typical ate pairing loop. Corollary 3.3 from [32] defines four cases of the R-ate pairing. Case 1 matches the Miller loop length of the ate$_i$ pairing. Case 2 requires a Miller loop length of the field size, which is not optimal, and case 4 matches the Miller loop length of the Tate pairing. We prefer the R-ate pairing case 3. If there is no bilinear and non-degenerate construction for our curve on this type, we can safely fall back to case 1, which presents the same length as the ate$_i$ pairing.

The R-ate pairing is defined as follows [32]:

$$e_{A,B} : (Q, P) \mapsto f_{a,BQ}(P) \times f_{b,Q}(P) \times G_{aBQ,bQ}(P),$$

where $A, B, a, b, \in \mathbb{N}$, non-trivial, given by $T_i = a \cdot T_j + b$, $A = T_i$, $B = T_j$, $T_i \equiv p^i \bmod r$, $T_j \equiv p^j \bmod r$, for some i, j, $1 \leq i, j \leq k$.

The three Miller loop calls in the R-ate pairing are:

$$M(Q, P, m_2), \ M(m_2 Q, P, c), \ \text{and} \ M(Q, P, d)$$

where Q and P are defined as for the ate pairing and M is the Miller algorithm. Here, $m_1 = \text{Max}(A, B)$, $m_2 = \text{Min}(A, B)$, $c = [m_1/m_2]$ and $d = m_1 - c \cdot m_2$. A combination of the a, b, A, B parameters close enough to, or far enough from, each other may produce a compatible combination.

The parameters of the first and the third Miller loop are the same except for the length. If the bit representation of d is the same as the higher bits of m_2, then we can reuse an intermediate value of the Miller function. This condition, however, is not always satisfied and must be verified on a case-by-case basis for each curve.

Some $T_i - T_j$ combinations require one Miller loop to be executed, others require three. The fastest R-ate pairing, in terms of the computation cost of its Miller loops, may not be the one with the lowest footprint; if memory consumption is an issue for a particular implementation, a slightly slower (that is, with more iterations) R-ate pairing may be preferred over three shorter Miller loops.

2.4 Pairing Lattices

The family of ate pairings [28,25,46,34,32] are optimized versions of the Tate pairing restricted to the eigenspaces of the Frobenius.

Let $s \in \mathbb{Z}$, $h = \sum_{i=0}^{d} h_i x_i \in \mathbb{Z}[x]$ with $h(s) \equiv 0 \bmod r$ and $d = \varphi(k)$, with k the embedding degree, and $Q \in E(\mathbb{F}_{p^k})[r]$, then:

$$(f_{s,h,Q}) = \sum_{i=0}^{d} h_i((s^i Q) - (\mathcal{O})).$$

Defining $\|h\|_1 = \sum_{i=0}^{d} |h_i|$ we have that if s is a primitive k^{th} root of unity modulo r^2, and if $h(s) \equiv 0 \bmod r$ but $h(s) \not\equiv 0 \bmod r^2$, then

$$e_{s,h} : (Q, P) \mapsto f_{s,h,Q}(P)^{(p^k - 1)/r}$$

defines a bilinear and non-degenerate pairing [27].

Choosing s. For the choice of s, following the ate pairing definition, we can take $s = r$, the subgroup size. We prefer to take $s = T = t - 1$, which is already an improvement with respect to the Tate pairing.

Constructing h. For the case of h, we construct a $m \times m$ matrix, with $m = \varphi(k)$:

$$M = \begin{pmatrix} r & 0 & \cdots & 0 \\ -T & 1 & 0 & \cdots & 0 \\ -T^2 & 0 & 1 & \cdots & 0 \\ \vdots & & & \ddots & \vdots \\ -T^{m-1} & 0 & \cdots & 1 \end{pmatrix} \tag{1}$$

Let $w = (w_0, w_1, \ldots, w_{m-1})$ be the shortest $\mathbb{Z}$-linear combination of the rows of M, then we can construct $h = \sum_{i=0}^{m-1} w_i x^i$. We have to LLL-reduce the matrix M to get the shortest vector. The explicit construction will be covered in section 2.6.

2.5 Weak Popov Form

Barreto [5] suggested the use of the *Weak Popov Form* of a matrix to get a reduced vector. This matrix construction, presented in [38] by Mulders and Stojohann, and recalled in [7] by Barreto, Lindner, and Misoczki in the context of Coding Theory, has more relaxed conditions than a so-called quasi-echelon form. In this construction, the pivot is any non-zero element at the end of the vector, and a pivot is unique in the column.

Mulders and Stojohann presented a few algorithms to transform a matrix into Weak Popov form [38, Lemmas 2.1-2.5]. An excellent description of the Weak Popov Algorithm is presented in Appendix A and Algorithm 2 of [7]. A modified version of this algorithm, may be required to force the coefficients to be in $\mathbb{Z}$. This can be done after the transformation of the first kind step in the algorithm. We can transform the M matrix from Equation 1 to be in Weak-Popov form to get the shortest vector and construct the pairing function.

2.6   Optimal Pairing

Let $\lambda^i \equiv p^i \bmod r$ and $r | \Phi_{k/d}(\lambda_i)$ with $d =\gcd(i, k)$, an ate pairing can be defined as

$$e_{\lambda_i} : (Q, P) \mapsto f_{\lambda_i, Q}(P)^{(p^k-1)/r}$$

this implies that the minimum value of λ_i is $r^{1/\varphi(k/d)}$.

Definition 1. *A pairing function $e(\cdot, \cdot)$ is called* Optimal Pairing *if it can be computed in $\log_2 r/\varphi(k) + \varepsilon(k)$ basic Miller iterations, with $\varepsilon(k) \leq \log_2 k$ [45].*

The optimal pairing construction reduces the number of iterations of the Miller loop by decomposing a multiple of r as a sum of the Frobenius endomorphism. As we suggested for the Pairing Lattice, we prefer to use $T = t - 1$, where t is the trace of the curve.

The pairing is defined as:

$$(Q, P) \mapsto \left(\prod_{i=0}^{l} f_{c_i, Q}^{p^i}(P) \cdot \prod_{i=0}^{l} G_{[s_i+1]Q, [c_i q^i]Q}(P) \right)^{(p^k-1)/r}$$

with $s_i = \sum_{j=i}^{l} c_j p^j$, G the line function, and k even.

To find the expansion with short coefficients, Vercauteren [45] uses the ShortestVectors() Magma function. We prefer to transform the matrix into a Weak Popov Form, which yields similar results.

BN $k = 12$ Curves. The matrix M from section 2.4 for these curves is:

$$M_{BN} = \begin{pmatrix} 3x & 4x+2 & 1 & x \\ x & 3x+1 & x+1 & 0 \\ -1 & 6x+2 & 2 & -1 \\ 6x+2 & 1 & -1 & 1 \end{pmatrix} \tag{2}$$

The optimal pairing can be constructed $s = t - 1$, and with $h(s) = (6x + 2, 1, -1, 1)$ as follows:

$$e_{s,h}(\cdot, \cdot) : (Q, P) \mapsto f_{6x-2, Q}(P) \cdot G_{[6x+2]Q, Q_1}(P) \cdot G_{[6x+2]Q+Q_1, -Q_2}(P).$$

where $Q_i = Q^{p^i}$, computed using the p-power Frobenius.

KSS $k = 18$ Curves. The matrix M for this family of curves is:

$$M_{k=18} = \begin{pmatrix} 1 & 0 & 5x/7 & 1 & 0 & -x/7 \\ -5x/7 & -2 & 0 & x/7 & 1 & 0 \\ 0 & 2x/7 & 1 & 0 & x/7 & 0 \\ 1 & 0 & x & 2 & 0 & 0 \\ -x & -3 & 0 & 0 & 1 & 0 \\ 0 & -x & -3 & 0 & 0 & 1 \end{pmatrix} \tag{3}$$

The optimal pairing can be constructed with $s = t - 1$, and $h(s) = (-x, -3, 0, 0, 1, 0)$ as follows:

$$e_{s,h}(\cdot, \cdot) :(Q, P) \mapsto 1/f_{x,Q}(P) \cdot 1/f_{3,Q}(P) \cdot G_{Q_4, -[3]Q_1}(P) \cdot G_{Q_4-[3]Q_1, -[x]Q}(P)$$

3   Addition Chains

One of the difficult parts of the Tate (and similar) pairing calculation that requires specific optimization is the so-called "final exponentiation" performed after the Miller loop. This is an exponentiation by $(p^k - 1)/r$ of an element in $\mathbb{F}_{p^k}$. This exponentiation can be simplified by using multiplication and squaring operations following an addition chain pattern, making the calculation faster by reusing intermediate values of the computation.

Definition 2. *Addition chain. An* addition chain *for a given integer e is a sequence $U = (u_0, u_1, u_2, \ldots, u_l)$ such that $u_0 = 1$, $u_l = e$ and $u_k = u_i + u_j$ for $k \leq l$ and some i, j with $0 \leq i \leq j$.*

Finding the shortest addition chain for a given positive integer is an NP-complete problem [20]. It is clear that a short addition chain for an integer e gives a faster method for computing $f^e \in \mathbb{F}_{p^k}$. There are special cases where the shortest addition chain does not give the best speed up. For example, this might be the case if one can exchange slower operations for a few extra faster ones (i.e. exchange an addition for a few doublings, if they are faster). This idea will be explored in §4.

Definition 3. *Addition sequence. Given a list of integers $\Gamma = \{v_1, .., v_l\}$ where $v_l \geq v_i \forall i = 1, .., l - 1$, an* addition sequence *for Γ is an addition chain for v_l containing all elements of Γ.*

Addition sequences, otherwise known as multi-addition-chains, are used to speed up the final exponentiation [43] and for hashing to a point in G_2 [42]. To use these implementation improvements it is necessary to have code to generate the addition sequence from a given list of integers.

Another generalization of the addition chains is the addition-subtraction chain, where the elements can also be constructed by subtraction of the previous elements. We preferred to use the simpler addition chain.

To automate the addition sequence code generation we use the Dominguez Perez and Scott [19] suggestion on generalizing the Cruz-Cortés et al. [16] and Bos and Coster [14] method for generating addition-chains. We construct the code using vectorial addition chains as in [40]. Selecting which elements must remain in the sequence and which to discard will continually improve the sequence.

As constructing optimal addition-chains is an NP-complete problem, we therefore propose to limit the search to a reduced number of improvements. Then the user can decide if they want to continue the search.

We compared Bernstein's method [10] to find a short addition sequence for the KSS $k = 36$ curves. His method is very fast, however, it generated 363 elements, whereas our method shows that only 174 are needed. This was thanks to the artificial intelligence nature of our method.

4   Final Exponentiation

One of the most expensive operations in the pairing computation is the final exponentiation by $(p^k - 1)/r \in \mathbb{F}_{p^k}$, which is required at the end of the computation of the Tate family of pairings. This computation eliminates the r-th powers and returns a unique r-th roots of unity.

Devegili et al. [17] observed that if the exponent $(p^k - 1)$ is appropriately factored, then one can perform the "easy part" of the exponentiation exploiting the Frobenius and then perform the "hard part" separately.

The idea is to separate the exponent into 3 pieces:

$$(p^k - 1)/r \Rightarrow (p^{\frac{k}{2}} + 1) \cdot (p^{\frac{k}{2}} - 1)/\Phi_k(p) \cdot (\Phi_k(p))/r.$$

The first 2 parts can be easily executed as described. The third part, the "hard part", can be executed by the Scott et al. method. [43]

The Scott et al. method [43] requires an addition chain, and the Olivos method [40] (Also see [15, §9.2]). Table 1 present the typical output of these methods. The t_i's represent temporary variables, and the x_i's is the corresponding element in the base $p(x)$ representation of the hard part of the final exponentiation and the i-th element in the initial addition-chain.

Table 1. Final exponentiation code from the Olivos and Scott et al method. BN $k = 12$ curve

$$t_0 \leftarrow x_6 * x_6$$
$$t_1 \leftarrow t_0 * x_4$$
$$t_2 \leftarrow t_1 * x_5$$
$$t_3 \leftarrow x_3 * x_5$$
$$t_4 \leftarrow t_3 * t_2$$
$$\ldots$$

Table 2. Final exponentiation code for a BN $k = 12$ curve, with a reduced number of t_i elements

$$t_0 \leftarrow x_6 * x_6$$
$$t_0 \leftarrow t_0 * x_4$$
$$t_1 \leftarrow t_0 * x_5$$
$$t_0 \leftarrow x_3 * x_5$$
$$t_0 \leftarrow t_0 * t_1$$
$$\ldots$$

While constructing the code sequence with the Olivos method, it is not possible to know when any t_i (temporary variable) is no longer used in the computation of the addition sequence. For simplicity, we assign a new temporary variable for each operation required to compute the addition sequence. We then optimize the memory usage by scanning from bottom-to-top and defining a group of related operations, recycling the variables whenever possible.

Another improvement is to exchange the memory usage of the x_i elements for a small extra computation. We move all of the x_i elements, denoted as R-value[1] (Table 2), to a just-in-time L-value assignment.

---

[1] In computer science, an L-value is referred to represent the address of the identified memory allocation, where areas an R-value is referred to represent its contents. In other words: let $a = b + c$, a is an L-value that represents the memory address where the addition of the b and c, R-values, is going to be stored.

5   Galbraith and Scott GLV-Like Method

Gallant, Lambert and Vanstone [24] introduced a method to speed up general scalar point multiplication of a point P by a scalar n, when there is an efficiently computable endomorphism ψ on E defined over $\mathbb{F}_p$ such that $\psi(P) = \lambda P$. Galbraith and Scott [23] extended this idea for the G_2 and G_T groups using higher dimensions of the endomorphism. They presented a reduced matrix for the BN $k = 12$ curves case. We can also use the Weak Popov transformation from section 2.5 with the Galbraith-Scott method.

For the BN $k = 12$ curves, the vector $\mathbf{v}$ generated from Equation 2, which can be used for precomputing is as follows:

$$\mathbf{v} = (6x^3 + 6x^2 + 2x, -(6x^3 + 6x^2 + x), -(2x + 1), -(5x^2)) \cdot \tfrac{n}{r}.$$

which is smaller than that presented by Galbraith and Scott [23].

To compute a fast exponentiation of a random scalar n by P in G_2: we multiply the vector $\mathbf{v}$ by the matrix M from §2.5 and form a vector $\mathbf{u}$, we then multiply each $\mathbf{u}_i$ coefficient by $\psi^{i-1}(P)$. For an exponentiation of a random element n by $f \in G_T$: we exponentiate $\pi^{i-1}(f)$ by $\mathbf{u}_i$.

6   Tower Construction Dependent Code

Sections 4 present constructions that rely on addition sequence generation, which depends on the parameters of the curve. Benger and Scott [9] defined some criteria for choosing a towering construction for the extension field $\mathbb{F}_{p^k}$. They present a new method for the towering construction.

For the code construction of the tower of extension fields, we can use [9, Table 1] but also accept a user supplied towering construction. Some multiprecision libraries have finite field arithmetic specific for a finite field extension. They will use their own towering construction rather than that recommended by Benger and Scott. For example, for implementing pairings over the BN curves [8], RELIC uses a towering method of $\mathbb{F}_p \rightarrow \mathbb{F}_{p^2} \rightarrow \mathbb{F}_{p^6} \rightarrow \mathbb{F}_{p^{12}}$, whereas MIRACL recommends $\mathbb{F}_p \rightarrow \mathbb{F}_{p^2} \rightarrow \mathbb{F}_{p^4} \rightarrow \mathbb{F}_{p^{12}}$. An element of $\mathbb{F}_{p^{12}}$ may be composed by two elements in $\mathbb{F}_{p^6}$ or three elements in $\mathbb{F}_{p^4}$. In practice it is easy to switch between the two representations.

7   Multiprecision Libraries

When coding cryptographic code from scratch, one question is to decide whether to create our own set of functions, or use an already standard multiprecision library.

According to this survey: [1], the main libraries were NTL, Lidia and MIRACL. Official support of Lidia has recently ended as shown on its web-page [4]. Abusharekh, in his MS Thesis [2], presents a more detailed survey of the libraries available at that time. In his conclusions, he shows that MIRACL is on average the fastest, with OpenSSL a close contender. Both surveys are somewhat outdated, though.

Some recent speed records on the pairing computation are not based on a particular multiprecision library, but on a hand-crafted set of finite field arithmetic functions, such as: [39,11,31]. In any of the previous cases, the set of instructions is only valid for the specific family of curves, and we want a more general approach. We also wanted the elliptic curve arithmetic already implemented.

For this code generator, we decided to use RELIC and MIRACL as the target libraries. Some Magma code is also generated and used at runtime.

MIRACL [41] is a multiprecision library which has been on the market for many years. It is free for academic purposes. It also supports a PBC implementation including the BN curves [8]. It contains prime field arithmetic and arithmetic for several extensions fields. It is possible to implement PBC with several families of pairing-friendly curves. It has recently added support for the KSS curves with $k = 18$ embedding degree [30].

The RELIC library [3] is currently at an early stage of development. To date it includes multiprecision integer arithmetic and prime and extension field arithmetic, among other features. In particular, it has support for the BN $k = 12$ curves [8].

Other libraries can be easily added to the generator, but not limited to the ones based on C/C++ code. Currently, we have not added support to the code generator for libraries written in languages that use the indentation level of the code to group their statements.

7.1   Multilibrary Management

To output code for essentially different libraries, we preferred to use associative arrays in Magma, which are arrays with string identifiers as indexes. These arrays contain an identifier associated with the different operations, datatype names, and some basic syntax rules for the corresponding arithmetic. Our code generator constructs the code for the operations depending on the description of the operators and operands. Some special circumstances were considered, such as: operator overloading, temporary assignments, initialization of variables, to name a few.

For the final exponentiation code, we used an array defining the datatype names, inversion operator, inversion position (prefix, suffix, infix, circumfix, and map, etc), accumulator operator (and position), initialization requirements (allocate memory, zeroing, try-and-catch support, freeing the variables), how can we store the output data, etc.

We also created similar arrays for the pairing and the line function. Support for other multiprecision libraries can be added if we supply this information to the compiler.

8   The Code Generator Sequence

The code generator has the code sequence shown in Algorithm 1.

Algorithm 1. Automatic code generator pseudo-code

Input: Security level. **Optional**: curve to use (see [22]), a and b from the short Weier-strass representation of the curve, word size of the target environment, precomputed addition sequences, pairing-friendly family x-parameter generator algorithm to use (or user supplied x), method to construct the tower of extension fields, the settings for the artificial immune system part of the code for addition sequence generation, and the compiling directory.

Defaults to BN curves [8] with a security level of 128 bits with a random x-value.

Output: Compressed directory with the final exponentiation and hashing to G2 code, supporting code functions depending on the x-value and on the towering construction, the pairing function code, a test bed, and a script to compile a test bed.

1: Verify the default values and the user supplied parameters.
2: Get the parameters of the curve, the polynomials describing the family, degree of the twist admitted, the towering construction, the x-value, among others.
3: Construct the final exponentiation and the code for hashing to G_2 for the code generator (in Magma) and for the requested multiprecision library.
4: Verify that the irreducible polynomial generates the twisted curve of the right order
5: Compute the cost of the optimal pairing, construct the R-ate pairing in Magma to check bilinearity of the parameters, and compute its cost.
6: Compare and decide which pairing function to construct for the desired target library: Optimal, R-ate, ate or Tate pairing. The user may have requested an specific pairing function.
7: Generate the code, update the template files, pack the code and compress it.
8: Optionally: compile an example program.

---

9   Timing the Output Code

We now compare timings from our automatically generated code (for the MIR-ACL library), with the general purpose PBC library [33], and with the highly optimized hand-written implementation bundled with the MIRACL library. For comparison purposes we choose a BN $k = 12$ curve at the 128-bit security level. We note that the overall operation count for the MIRACL implementation is very close to that reported recently by Beuchat et al., [11] in their recent record setting implementation (although the latter uses a significantly faster specially tailored implementation of $\mathbb{F}_{p^2}$ arithmetic). The timings were collected on an Intel Core 2 Duo E6850 3GHz and are presented in table 3.

This demonstrates the validity of our approach, and shows that automatically generated code is nearly as fast as carefully hand-crafted code, and is significantly faster than a generic pairing implementation.

Table 3. Comparison of the generated code, BN curves. CPU cycles in millions.

Library	CPU cycles	Time
PBC	160.22	53.37 ms
This paper	10.85	3.63 ms
MIRACL	9.24	3.17 ms

Table 4. Timings for the KSS $k = 18$ curve. CPU cycles in millions.

Pairing function	CPU cycles			Time in ms
	M	FE	Tot	
This paper	23.8	51.3	72.5	25.1
MIRACL	23.5	49.8	73.3	24.5

In table 4, the timings for the generated code for a R-ate pairing and the MIRACL implementation are presented.

The CPU cycles are described in table 4 as follows: column M shows the millions of CPU cycles for the Miller function and related operations; the column FE, for the final exponentiation; and TOT, is the total. The last column is the time in milliseconds.

We have managed to generate the code for a KSS curve with k = 36. However, we currently do not have access to the required finite extension field arithmetic to run the code. The construction time took several hours with a prescribed addition sequence for the final exponentiation (section 4). This however, was an extreme case. The typical construction time ranges from seconds to a couple of minutes.

10   Conclusions and Future Work

In this work we have presented some constructions for the implementation of cryptographic pairings necessary to build an automatic code generator.

The generated code uses a slightly larger number of variables in the overall pairing computation compared to hand-written implementations.

As it was shown in section 9, the generated code is competitive in terms of speed and number of CPU cycles. The generation time for the addition sequence in some cases can be very long, but as it was shown for a particular case in section 3, it can be worth the wait.

This work can be extended to other multiprecision libraries or for its use over a set of specialized hand-crafted functions. An interesting target library would be one based on JavaME. We also recommend as a future work, the inclusion of automatic finite field arithmetic generation.

The source code of this project is available at: [18]

Acknowledgements

We would like to thank Paulo Barreto for suggesting the use of the Weak Popov form, and for providing an example implementation of the algorithm.

References

1. Multi-precision libraries for submissions for the Nessie project, https://www.cosic.esat.kuleuven.be/nessie/call/mplibs.html
2. Abusharekh, A.: Comparative analysis of multi-precision arithmetic libraries for Public Key Cryptography. Master's thesis, George Mason University (April 2004)
3. Aranha, D.F., Porto Lopes Gouvêa, C.: RELIC: an efficient library for cryptography, http://code.google.com/p/relic-toolkit/
4. LG at the Darmstadt University of Technology. Lidia library, http://www.cdc.informatik.tu-darmstadt.de/TI/LiDIA/
5. Barreto, P.S.L.M.: Personal communication
6. Barreto, P.S.L.M., Galbraith, S.D., Héigeartaigh, C.Ó., Scott, M.: Efficient pairing computation on supersingular abelian varieties. Des. Codes Cryptography 42(3), 239–271 (2007)
7. Barreto, P.S.L.M., Lindner, R., Misoczki, R.: Decoding square-free Goppa codes over $\mathbb{F}_p$. Cryptology ePrint Archive, Report 2010/372 (2010), http://eprint.iacr.org/
8. Barreto, P.S.L.M., Naehrig, M.: Pairing-friendly elliptic curves of prime order. In: Preneel, B., Tavares, S. (eds.) SAC 2005. LNCS, vol. 3897, pp. 319–331. Springer, Heidelberg (2006)
9. Benger, N., Scott, M.: Constructing tower extensions for the implementation of Pairing-Based Cryptography. In: Hasan, M.A., Helleseth, T. (eds.) WAIFI 2010. LNCS, vol. 6087, pp. 180–195. Springer, Heidelberg (2009)
10. Bernstein, D.J.: Optimizing linear maps modulo 2. In: Workshop: Software Performance Enhancement for Encryption and Decryption and Cryptographic Compilers, SPEED-CC 2009, pp. 3–18 (2009), http://www.hyperelliptic.org/SPEED/record09.pdf
11. Beuchat, J.-L., González Díaz, J.E., Mitsunari, S., Okamoto, E., Rodríguez-Henríquez, F., Teruya, T.: High-speed software implementation of the optimal ate pairing over Barreto-Naehrig curves. Cryptology ePrint Archive, Report 2010/354 (2010), http://eprint.iacr.org/ (to appear in Pairing 2010)
12. Blake, I., Seroussi, G., Smart, N.P. (eds.): Advances in Elliptic Curve Cryptography. London Mathematical Society. Lecture Note Series. Cambridge University Press, Cambridge (2005)
13. Boneh, D., Lynn, B., Shacham, H.: Short signatures from the Weil pairing. In: Boyd, C. (ed.) ASIACRYPT 2001. LNCS, vol. 2248, pp. 514–532. Springer, Heidelberg (2001)
14. Bos, J., Coster, M.: Addition chain heuristics. In: Goos, G., Hartmanis, J. (eds.) CRYPTO 1989. LNCS, vol. 435, pp. 400–407. Springer, Heidelberg (1989)
15. Cohen, H., Frey, G.: Hanbook of Elliptic and Hyperelliptic Curve Cryptography. Chapman & Hall/CRC (2006)
16. Cruz-Cortés, N., Rodríguez-Henriquez, F., Coello Coello, C.: An Artificial Immune System heuristic for generating short addition chains. IEEE Transactions on Evolutionary Computation 12(1), 1–24 (2008)
17. Devegili, A.J., Scott, M., Dahab, R.: Implementing cryptographic pairings over Barreto-Naehrig curves. In: Takagi, T., Okamoto, T., Okamoto, E., Okamoto, T. (eds.) Pairing 2007. LNCS, vol. 4575, pp. 197–207. Springer, Heidelberg (2007)
18. Dominguez Perez, L.J.: Automatic code generator website, http://www.computing.dcu.ie/~ldominguez/phdproject.html

19. Dominguez Perez, L.J., Scott, M.: Automatic generation of optimised cryptographic pairing functions. In: SPEED-CC 2009 Workshop Memories (2009), http://www.hyperelliptic.org/SPEED/record09.pdf
20. Downey, P., Leong, B., Sethi, R.: Computing sequences with addition chains. SIAM Journal on Computing 10(3), 638–646 (1981)
21. Freeman, D.: Constructing pairing-friendly elliptic curves with embedding degree 10. In: Sha, E., Han, S.-K., Xu, C.-Z., Kim, M.H., Yang, L.T., Xiao, B. (eds.) EUC 2006. LNCS, vol. 4096, pp. 452–465. Springer, Heidelberg (2006); In: Algorithmic Number Theory Symposium ANTS-VII
22. Freeman, D., Scott, M., Teske, E.: A taxonomy of pairing-friendly elliptic curves. Journal of Cryptology 23(2), 224–280 (2010)
23. Galbraith, S.D., Scott, M.: Exponentiation in pairing-friendly groups using homomorphisms. In: Galbraith, S.D., Paterson, K.G. (eds.) Pairing 2008. LNCS, vol. 5209, pp. 211–224. Springer, Heidelberg (2008)
24. Gallant, R.P., Lambert, R.J., Vanstone, S.A.: Faster point multiplication on elliptic curves with efficient endomorphisms. In: Kilian, J. (ed.) CRYPTO 2001. LNCS, vol. 2139, pp. 190–200. Springer, Heidelberg (2001)
25. Granger, R., Hess, F., Oyono, R., Thériault, N., Vercauteren, F.: Ate pairing on hyperelliptic curves. In: Naor, M. (ed.) EUROCRYPT 2007. LNCS, vol. 4515, pp. 430–447. Springer, Heidelberg (2007)
26. Granger, R., Page, D., Smart, N.P.: High security Pairing-Based Cryptography revisited. In: Hess, F., Pauli, S., Pohst, M. (eds.) ANTS 2006. LNCS, vol. 4076, pp. 480–494. Springer, Heidelberg (2006)
27. Hess, F.: Pairing lattices. In: Galbraith, S.D., Paterson, K.G. (eds.) Pairing 2008. LNCS, vol. 5209, pp. 18–38. Springer, Heidelberg (2008), http://www.math.tu-berlin.de/~hess/personal/pairing-lattice.pdf
28. Hess, F., Smart, N.P., Vercauteren, F.: The Eta pairing revisited. IEEE Trans. Information Theory 52(10), 4595–4602 (2006)
29. Joux, A.: A one round protocol for tripartite Diffie–Hellman. In: Bosma, W. (ed.) ANTS 2000. LNCS, vol. 1838, pp. 385–394. Springer, Heidelberg (2000)
30. Kachisa, E., Schaeffer, E.F., Scott, M.: Constructing Brezing-Weng pairing friendly elliptic curves using elements in the cyclotomic field. In: Galbraith, S.D., Paterson, K.G. (eds.) Pairing 2008. LNCS, vol. 5209, pp. 126–135. Springer, Heidelberg (2008)
31. Lauter, K., Montgomery, P.L., Naehrig, M.: An analysis of affine coordinates for pairing computation. Cryptology ePrint Archive, Report 2010/363 (2010), http://eprint.iacr.org/ (to appear in Pairing 2010)
32. Lee, H.-S., Lee, E., Park, C.-M.: Efficient and generalized pairing computation on Abelian varieties. Cryptology ePrint Archive, Report 2008/040 (2008), http://eprint.iacr.org/
33. Lynn, B.: PBC library - the Pairing-Based Cryptography library, http://crypto.stanford.edu/pbc/
34. Matsuda, S., Kanayama, N., Hess, F., Okamoto, E.: Optimised versions of the ate and twisted ate pairings. Cryptology ePrint Archive, Report 2007/013 (2007), http://eprint.iacr.org/
35. Miller, V.S.: Short programs for functions on curves (1986), http://crypto.stanford.edu/Miller/Miller.ps
36. Miller, V.S.: The Weil pairing, and its efficient calculation. J. Cryptol. 17(4), 235–261 (2004)
37. Miyaji, A., Nakabayashi, M., Takano, S.: New explicit conditions of elliptic curve traces for FR-reduction. IEICE Trans. Fundamentals E84, 1234–1243 (2001)

38. Mulders, T., Storjohann, A.: On lattice reduction for polynomial matrices. J. Symb. Comput. 35(4), 377–401 (2003)
39. Naehrig, M., Niederhagen, R., Schwabe, P.: New software speed records for cryptographic pairings. In: Abdalla, M. (ed.) LATINCRYPT 2010. LNCS, vol. 6212, pp. 109–123. Springer, Heidelberg (2010)
40. Olivos, J.: On vectorial addition chains. Journal of Algorithms 2, 13–21 (1981)
41. Scott, M.: MIRACL – Multiprecision Integer and Rational Arithmetic C/C++ Library, http://ftp.computing.dcu.ie/pub/crypto/miracl.zip
42. Scott, M., Benger, N., Charlemagne, M., Dominguez Perez, L.J., Kachisa, E.J.: Fast hashing to G_2 on pairing-friendly curves. In: Shacham, H., Waters, B. (eds.) Pairing 2009. LNCS, vol. 5671, pp. 102–113. Springer, Heidelberg (2009)
43. Scott, M., Benger, N., Charlemagne, M., Dominguez Perez, L.J., Kachisa, E.J.: On the final exponentiation for calculating pairings on ordinary elliptic curves. In: Shacham, H., Waters, B. (eds.) Pairing 2009. LNCS, vol. 5671, pp. 78–88. Springer, Heidelberg (2009)
44. Silverman, J.H.: The Arithmetic of Elliptic Curves. Springer, Heidelberg (1986)
45. Vercauteren, F.: Optimal pairings. IEEE Transactions on Information Theory 56(1), 455–461 (2010), http://www.cosic.esat.kuleuven.be/publications/article-1039.pdf
46. Zhao, C.-A., Zhang, F., Huang, J.: A note on the ate pairing. Cryptology ePrint Archive, Report 2007/247 (2007), http://eprint.iacr.org/

A   Sample Final Exponentiation in MIRACL

The following code was manually re-arranged for formatting purposes.

Listing 1.1. Sample Final Exponentiation for MIRACL, KSS k=18 curves

```
// INPUT <- f3x0, the value to exponentiate, X is the Frobenius
      constant. The x-parameter of the curve.
// OUTPUT -> r, the value of the pairing after the hard part of the
      final exponentiation.
#include "FEc6.121MIRACL.h"
void HardExpo(ZZn18 &r, ZZn18 &f3x0, ZZn3 &X, Big &x){
  ZZn18 xA;   ZZn18 xB;   ZZn18 t0;   ZZn18 t1;
  ZZn18 t2;   ZZn18 t3;   ZZn18 t4;   ZZn18 t5;
  ZZn18 t6;   ZZn18 t7;   ZZn18 f3x1;   ZZn18 f3x2;
  ZZn18 f3x3;   ZZn18 f3x4;   ZZn18 f3x5;   ZZn18 f3x6;
  ZZn18 f3x7;

   f3x1=pow(f3x0,x);   f3x2=pow(f3x1,x);   f3x3=pow(f3x2,x);
   f3x4=pow(f3x3,x);   f3x5=pow(f3x4,x);   f3x6=pow(f3x5,x);
   f3x7=pow(f3x6,x);

   xA=Frobenius(inverse(f3x1),X,2); xB=Frobenius(inverse(f3x0),X,2);
   t0=xA*xB;
   xB=Frobenius(inverse(f3x2),X,2);   t1=t0*xB;
```

```
t0=t0*t0;
xB=Frobenius(inverse(f3x0),X,2);   t0=t0*xB;
xB=Frobenius(f3x1,X,1);   t0=t0*xB;
xA=Frobenius(inverse(f3x5),X,2)*Frobenius(f3x4,X,4)*Frobenius(
   f3x2,X,5);
xB=Frobenius(f3x1,X,1);
t5=xA*xB;   t0=t0*t0;   t3=t0*t1;
xA=Frobenius(inverse(f3x4),X,2)*Frobenius(f3x1,X,5);
xB=Frobenius(f3x2,X,1);
t1=xA*xB;
xA=Frobenius(f3x2,X,1);   xB=Frobenius(f3x2,X,1);   t0=xA*xB;
xB=Frobenius(f3x2,X,4);   t0=t0*xB;
xB=Frobenius(f3x1,X,4);   t2=t3*xB;
xB=Frobenius(inverse(f3x1),X,2);   t4=t3*xB;
t2=t2*t2;
xB=Frobenius(inverse(f3x2),X,3);   t3=t0*xB;
xB=inverse(f3x2);   t0=t3*xB;
t4=t3*t4;
xB=Frobenius(inverse(f3x3),X,3);   t0=t0*xB;
t3=t0*t2;
xB=Frobenius(inverse(f3x3),X,2)*Frobenius(f3x0,X,5);
t2=t3*xB;   t3=t3*t5;   t5=t3*t2;
xB=inverse(f3x3);   t2=t2*xB;
xA=Frobenius(inverse(f3x6),X,3);   xB=inverse(f3x3);   t3=xA*xB;
t2=t2*t2;   t4=t2*t4;
xB=Frobenius(f3x3,X,1);   t2=t1*xB;
xA=Frobenius(f3x3,X,1);   xB=Frobenius(inverse(f3x2),X,3);
t1=xA*xB;   t6=t2*t4;
xB=Frobenius(f3x4,X,1);   t4=t2*xB;
xB=Frobenius(f3x3,X,4);   t2=t6*xB;
xB=Frobenius(inverse(f3x5),X,3)*Frobenius(f3x5,X,4);
t7=t6*xB;   t4=t2*t4;
xB=Frobenius(f3x6,X,1);   t2=t2*xB;
t4=t4*t4;   t4=t4*t5;
xA=inverse(f3x4);   xB=Frobenius(inverse(f3x4),X,3);
t5=xA*xB;
xB=Frobenius(inverse(f3x4),X,3);   t3=t3*xB;
xA=Frobenius(f3x5,X,1);   xB=Frobenius(f3x5,X,1);   t6=xA*xB;
t7=t6*t7;
xB=Frobenius(f3x0,X,3);   t6=t5*xB;
t4=t6*t4;
xB=Frobenius(inverse(f3x7),X,3);   t6=t6*xB;
t0=t4*t0;
xB=Frobenius(f3x6,X,4);   t4=t4*xB;
t0=t0*t0;
xB=inverse(f3x5);   t0=t0*xB;
t1=t7*t1;   t4=t4*t7;   t1=t1*t1;   t2=t1*t2;   t1=t0*t3;
xB=Frobenius(inverse(f3x3),X,3);   t0=t1*xB;
t1=t1*t6;   t0=t0*t0;   t0=t0*t5;
xB=inverse(f3x6);   t2=t2*xB;
t2=t2*t2;   t2=t2*t4;   t0=t0*t0;   t0=t0*t3;
```

```
t1=t2*t1;   t0=t1*t0;
xB=inverse(f3x6);   t1=t1*xB;
t0=t0*t0;   t0=t0*t2;
xB=f3x0*inverse(f3x7);   t0=t0*xB;
xB=f3x0*inverse(f3x7);   t1=t1*xB;
t0=t0*t0;   t0=t0*t1;
r=t0;}
```

B   Sample Pairing Construction for MIRACL, KSS k=18

Listing 1.2 shows the code for the R-ate pairing. Listing 1.3 present the code construction for a shared Miller loop. Listing 1.4 show the line function code. The code was re-arranged for formatting purposes.

Listing 1.2. R-ate pairing for the MIRACL library. KSS k=18 curves

```
// INPUT <- Point P in G_2. Point Q in G_1 with the x and y
//     coordinates in Qx and Qy respectively. X, d2, d3, the Frobenius
//     constant, and its quadratic and cubic powers. The x-parameter
//     of the curve. m2 and d, the Miller loop length parameters of
//     the R-ate pairing.
// OUTPUT -> r, the value of the pairing after the hard part of the
//     final exponentiation.
#include "RATEc6.121MIRACL.h"
BOOL Pairing(ECn3 &P, ZZn &Qx, ZZn &Qy, ZZn18 &r, ZZn3 &X, ZZn3 &d2
    , ZZn3 &d3, Big &x,Big &m2,Big &d){
  ZZn18 fm2;   ZZn18 fm1;
  ECn3 m1P;   ECn3 m2P;
  int nb;   int i;
  ZZn18 fd;
  ECn3 dP;
  Miller(P,Qx,Qy,fm2,m2P,m2,fd,dP,d);
  fd*=fm2;
  m1P=m2P;
  fd*=g(m1P,dP,Qx,Qy);
  fm2*=Frobenius(fd,X,6);
  m1P=psi(m1P,d2,d3,6);
  fm2*=g(m1P,m2P,Qx,Qy);
  SoftExpo(fm2,X);
  HardExpo(r,fm2,X,x);
  if (fm2.iszero()) {
    return FALSE;
  }
  return TRUE;}
```

Listing 1.3. Miller function implementation

```
// INPUT <- Point P in G_2. Point Q in G_1 with the x and y
      coordinates in Qx and Qy respectively. m2 and d, the Miller
      loop length parameters of the R-ate pairing. The loop length d
      needs to be embedable into m2.
// OUTPUT -> f3 and fd, the value of the Miller function with loop
      lengths m2 and d respectively. m2P, dP the scalar-point
      multiplication of m2.P and d.P respectively.
void Miller(ECn3 &P, ZZn &Qx, ZZn &Qy, ZZn18 &f3, ECn3 &m2P, Big &
      m2, ZZn18 &fd, ECn3 &dP, Big &d){
  int i;
  int nb=bits(m2);    int nb2=bits(d);
  f3=1;   m2P=P;
  for (i=nb-2;i>=0;i--) {
    f3*=f3;
    f3*=g(m2P,m2P,Qx,Qy);
    if (bit(m2,i)) f3*=g(m2P,P,Qx,Qy);
    if (i==nb-nb2) {dP=m2P; fd=f3;}
  }
}
```

Listing 1.4. Line function

```
// INPUT <- A=T+P, where T and P are two points in G_2. The slope
      of the line crossing through the point T and P. A fixed point Q
      in G_1, with Qx and Qy representing its x,y coordinates
      respectively.
// OUTPUT -> The distance between the point Q and the line formed
      by the points T,P.
ZZn18 line(ECn3 &A, ZZn3 &slope, ZZn &Qx, ZZn &Qy){
  ZZn18 w;   ZZn6 a;
  ZZn6 b;   ZZn6 c;
  ZZn3 X,Y;     ZZn3 t0;   ZZn3 t1;
  A.get(X,Y);
  t0=-Qy;   t1=slope*X;   t1=Y-t1;
  a.set(t0,t1);
  t0=slope*Qx;
  b.set(t0);
  w.set(a,b);
  return w;}
```

Efficient Generic Constructions of Timed-Release Encryption with Pre-open Capability

Takahiro Matsuda*, Yasumasa Nakai, and Kanta Matsuura

The University of Tokyo, Japan
{tmatsuda,kanta}@iis.u-tokyo.ac.jp

Abstract. Timed-release encryption with pre-open capability (TRE-PC), introduced by Hwang et al. in 2005, is a cryptosystem with which a sender can make a ciphertext so that a receiver can decrypt it by using a timed-release key provided from a trusted time-server, or by using a special information called pre-open key provided from the sender before the release-time, and thus adds flexibility to ordinary TRE schemes in many practical situations. Recently, Nakai et al. proposed a generic construction of a TRE-PC scheme from a public-key encryption scheme, an identity-based encryption scheme (with some special property), and a signature scheme. Concrete TRE-PC schemes derived via their generic construction are, however, not so practical because of the used building block primitives. Motivated by this situation, in this paper we propose two new generic constructions of TRE-PC schemes. Both of our constructions follow the basic idea behind the generic construction by Nakai et al. but overcome its inefficiency without losing "generality" for the used building block primitives. Concrete TRE-PC schemes derived from our generic constructions are comparable to or more efficient than the currently known TRE-PC schemes in terms of ciphertext overhead size and computation costs.

Keywords: timed-release encryption, pre-open capability, generic construction, tag-KEM.

1 Introduction

Background and Motivation. Timed-release encryption (TRE) is a kind of encryption system introduced by May [23] in 1993. Roughly speaking, in TRE, a message can be encrypted in such a way that it cannot be decrypted even by a legitimate receiver who owns a decryption key for the ciphertext until the time (called release-time) that is specified by an encryptor. Many of practical applications/situations where TRE schemes can be used have been considered so far, such as sealed-bid auctions, electronic voting, contents predelivery systems, and on-line examinations. Although there are several known approaches and models

---

* Takahiro Matsuda is supported by JSPS Research Fellowships for Young Scientists.

M. Joye, A. Miyaji, and A. Otsuka (Eds.): Pairing 2010, LNCS 6487, pp. 225–245, 2010.

for realizing TRE systems, as with the most recent works regarding TRE, in this paper we only focus on the *public key* TRE systems which consider the existence of a trusted agent called *time-server* that publishes a global system parameter and periodically issues a timed-release key that is used by each receiver for decrypting a ciphertext together with the receiver's private key, and in which a sender and a receiver need not interact with the time-server [13,15,16].

In 2005, Hwang et al. [22] introduced an additional functionality called *pre-open capability* in TRE schemes. Roughly, in TRE with pre-open capability (TRE-PC), a sender can make a ciphertext so that a receiver can decrypt it by using a timed-release key provided from the time-server at a predetermined release-time specified by the sender, or by using a special information called *pre-open key* provided from the sender before the release-time. This pre-open key is generated as a by-product when generating a ciphertext. It is naturally required that even if a honest-but-curious time-server gets a ciphertext together with a corresponding pre-open key, it should not be able to learn any information on a plaintext, and even a legitimate receiver should not be able to learn any information from a ciphertext without a timed-release key or a pre-open key.

This pre-open capability adds more flexibility to TRE in many practical situations and also increases applications of TRE. For example, Hwang et al. [22] showed that TRE-PC can be used to realize certified email systems. Dent and Tang [20] exemplified how TRE-PC is useful for timed-disclosure of governmental documents which must be kept secret until a certain period of time. See these papers for more details about the applications of TRE-PC.

Recently, Nakai et al. [24] proposed a generic construction (we call it the *NMKM construction*) of a TRE-PC scheme from a chosen ciphertext secure (CCA-secure) public-key encryption (PKE) scheme, a chosen plaintext secure (CPA-secure) identity-based encryption (IBE) scheme with a special property called target collision resistance for randomness, and a one-time signature scheme. Their construction is essentially the generic construction of a TRE scheme (without pre-open capability) by Cheon et al. [15,16]. Although their construction enables us to instantiate a number of TRE-PC schemes based on existing basic primitives, concrete TRE-PC schemes derived via their generic construction are not so practical because of the used building block primitives (especially the use of a one-time signature), and are not more efficient than the existing TRE-PC scheme by Dent and Tang [20] and the one by Chow and Yiu [18].

The main motivation of this paper is to show more efficient generic constructions of TRE-PC schemes from existing basic primitives that lead to practical TRE-PC schemes.

Our Contribution. In this paper we propose two new generic constructions of TRE-PC schemes from existing basic primitives that are secure in the model of [20] and are more efficient than the NMKM construction. Both of our generic constructions follow the basic idea behind the NMKM construction, but overcome its inefficiency without losing "generality" of the building block primitives in the sense that we can construct TRE-PC schemes from a combination of ordinary KEMs, a wide class (explained below) of identity-based KEMs (IBKEMs), and symmetric-key primitives, via our constructions. Concretely, our generic constructions employ the hybrid encryption structure (i.e., in a KEM/DEM approach [19,4], where KEM stands for key encapsulation mechanism and DEM stands for data encapsulation mechanism), and use different approaches from the NMKM construction for avoiding a one-time signature scheme, which is one of the main obstacles that makes the NMKM construction less practical compared to the existing concrete TRE-PC schemes [20,18].

Our first construction is based on a CCA-secure *tag-KEM* [4], a CPA-secure IBKEM with target collision resistance for randomness (originally introduced in [24] for IBE schemes and can be naturally considered for IBKEMs), and a passively secure DEM. A tag-KEM is a primitive for a secure hybrid encryption introduced by Abe et al. [4] and has been widely studied. Although one might think that a tag-KEM is not a basic primitive, Abe et al. show that CCA-secure tag-KEMs can be generically built from any CCA-secure ordinary (i.e. non-tag-)KEMs and one-time secure message authentication codes (MACs), and thus can be achieved generically from existing basic primitives. Moreover, many practical (direct) constructions of tag-KEMs are also known (with and without random oracles), e.g. [4,3,5,2]. Our first construction might also be interesting as a concrete application of tag-KEMs for a different purpose than constructing hybrid PKE schemes.

Our second construction is simpler than the above construction and is based on a CCA-secure (ordinary) KEM, a CPA-secure IBKEM with the same security as above, and a CCA-secure DEM, but it also requires a random oracle. Therefore, this construction is suitable for constructing TRE-PC schemes based on existing building block primitives that already use random oracles.

A number of practical TRE-PC schemes can be obtained via our generic constructions. Concretely, we show that concrete TRE-PC schemes derived from our generic constructions are comparable to or more efficient than the currently known TRE-PC schemes [20,18] in terms of ciphertext overhead size and computation costs. See Section 5 for details. We also discuss the extensions of the proposed constructions for TRE-PC that supports *release-time confidentiality* [12] and for public key *time-specific encryption* [25]. See *Related Work* paragraph below and Section 4.4 for these. We believe that the constructions obtained via our generic constructions can also be used as "benchmarks" to compare with (and evaluate) concrete TRE-PC schemes that will be proposed in the future.

As in the NMKM construction, the IBKEM used in our constructions is required to satisfy the non-standard security called *target collision resistance for randomness*. However, this security is satisfied unconditionally by many existing pairing-based IBKEM schemes. See a more detailed explanation in Section 2.2.

Related Work.   There are two major approaches for realizing TRE. One approach is to use *time-lock puzzles* [26]. In this approach, a sender makes a ciphertext which cannot be finished decrypting before the release-time in a receiver's environment, even if the receiver keeps computing to decrypt the ciphertext after he receives it. This imposes heavy computation cost on the receiver, and it is difficult to precisely estimate the required time for decryption.

The other approach is to use a trusted agent (i.e. time-server) which periodically generates a time specific information (timed-release key) needed to encrypt a message and/or decrypt a ciphertext. Earlier TRE schemes (e.g. [23,26]) adopted a model in which the time-server and system users (senders and/or receivers) need to interact. Chan and Blake [13] and Cheon et al. [15,16] independently introduced a model in which no interaction between the time-server and users is required. Most of the works about TRE schemes (e.g. [12,22,20,17,18,24]) after [13,15,16] adopt the model of [13,15,16]. Cheon et al. [15,16] and Cathalo et al. [12] independently proposed formal security definitions for TRE.

Several additional functionality and security properties have also been introduced for TRE. Cheon et al. [15,16] introduced a TRE scheme with authentication, which is a TRE-version of public key authenticated encryption [7]. Cathalo et al. [12] formalized a notion of *release time confidentiality* for TRE, in which a ciphertext does not leak any information about its release-time for other entities than a legitimate receiver.

Very recently, Paterson and Quaglia [25] introduced a new primitive, *time-specific encryption* (TSE), which in some sense is a generalization of TRE. Roughly, in TSE, an encryptor chooses not a release-time, but a release-time "interval" $[T_{from}, T_{to}]$, and the ciphertext can be decrypted by using a time instant key (TIK) (which corresponds to a timed-release key in TRE) only when the time T associated with the TIK satisfies $T \in [T_{from}, T_{to}]$. They consider plain setting in which a ciphertext is not specific to any receiver, and public-key and identity-based settings in which each ciphertext is specified to some receiver (who has his own secret-key). We will explain in Section 4.4 how our constructions can be used to construct public-key TSE with pre-open capability.

2   Preliminaries

In this section, we review the primitives that are used as building blocks in our TRE-PC constructions. The primitives not reviewed here appear in Appendix A, which are public-key KEM and DEM. (Due to space limitations, some standard security definitions are omitted, which will be given in the full version.)

Notation. Throughout this paper, "$x \leftarrow y$" denotes that x is chosen uniformly at random from y if y is a finite set, x is output from y if y is an algorithm, or y is assigned to x otherwise. "$x||y$" denotes a concatenation of x and y. "PPTA" denotes *probabilistic polynomial time algorithm*. If $\mathcal{A}$ is a PPTA then "$\mathcal{A}^{\mathcal{O}}$" denotes that $\mathcal{A}$ has oracle access to $\mathcal{O}$, and "$y \leftarrow \mathcal{A}(x; r)$" denotes that $\mathcal{A}$ computes y as output, taking x as input and using r as randomness. "κ" always denotes the security parameter. We say that a function $f(\kappa)$ is negligible (in κ) if $f(\kappa) < 1/p(\kappa)$ for any positive polynomial $p(\kappa)$ and all sufficiently large κ.

2.1   Tag-KEM

A tag-KEM Π_T consists of the following four PPTAs (TKG, TSKey, TEncap, TDecap): TKG is a key generation algorithm that takes 1^κ as input, and outputs a public/secret key pair (pk, sk); TSKey is a session-key generation algorithm that takes pk as input, and outputs a session-key $K \in \mathcal{K}$ ($\mathcal{K}$ is a session key space of Π_T) and a corresponding state information ω that is used in the encapsulation algorithm; TEncap is an encapsulation algorithm that takes a tag tag and ω as input, and outputs a ciphertext c, which is an encapsulation of a session-key K under pk (where information about K and pk is implicitly transmitted by ω); TDecap is a deterministic decapsulation algorithm that takes sk, tag, and c as input, and outputs $K \in \mathcal{K} \cup \{\perp\}$. We require, for all $(pk, sk) \leftarrow \text{TKG}(1^\kappa)$, all $(K, \omega) \leftarrow \text{TSKey}(pk)$, all tag, and all $c \leftarrow \text{TEncap}(\text{tag}, \omega)$, that $\text{TDecap}(sk, \text{tag}, c) = K$.

Definition 1. *We say that a tag-KEM Π_T is indistinguishable against chosen ciphertext attacks (IND-CCA secure) if for any PPTA $\mathcal{A} = (\mathcal{A}_1, \mathcal{A}_2)$ the following IND-CCA advantage $\text{Adv}_{\Pi_T, \mathcal{A}}^{\text{IND-CCA}}(\kappa)$ is negligible:*

$$\text{Adv}_{\Pi_T, \mathcal{A}}^{\text{IND-CCA}}(\kappa) = |\Pr[(pk, sk) \leftarrow \text{TKG}(1^\kappa); K_0^* \leftarrow \mathcal{K}; (K_1^*, \omega^*) \leftarrow \text{TSKey}(pk);$$

$$b \leftarrow \{0, 1\}; (\text{tag}^*, \text{st}_\mathcal{A}) \leftarrow \mathcal{A}_1^{\text{TDecap}(sk, \cdot, \cdot)}(pk, K_b^*); c^* \leftarrow \text{TEncap}(\text{tag}^*, \omega^*);$$

$$b' \leftarrow \mathcal{A}_2^{\text{TDecap}(sk, \cdot, \cdot)}(c^*, \text{st}_\mathcal{A}) : b' = b] - \frac{1}{2}|,$$

where $\mathcal{A}_2$ is not allowed to submit the challenge tag/ciphertext pair (tag^*, c^*) to its given decapsulation oracle $\text{TDecap}(sk, \cdot, \cdot)$.

2.2 Identity-Based Key Encapsulation Mechanism

An identity-based key encapsulation mechanism (IBKEM) Π_I consists of the following four PPTAs (ISetup, IExt, IEncap, IDecap): ISetup is a setup algorithm that takes 1^κ as input, and outputs a pair of global parameters prm and a master secret key msk; IExt is an extraction algorithm that takes prm, msk, and an identity ID as input, and outputs a decapsulation key dk_{ID} corresponding to ID; IEncap is an encapsulation algorithm that takes prm and ID as input, and outputs a ciphertext/session-key pair (c, K); IDecap is a deterministic decapsulation algorithm that takes prm, dk_{ID}, and c as input, and outputs $K \in \mathcal{K} \cup \{\perp\}$ ($\mathcal{K}$ is a session-key space of Π_I). We require, for all $(\text{prm}, \text{msk}) \leftarrow \text{ISetup}(1^\kappa)$, all ID, all $dk_{\text{ID}} \leftarrow \text{IExt}(\text{prm}, \text{msk}, \text{ID})$, and all $(c, K) \leftarrow \text{IEncap}(\text{prm}, \text{ID})$, that $\text{IDecap}(\text{prm}, dk_{\text{ID}}, c) = K$.

We will need an IBKEM that satisfies indistinguishability against adaptive-identity, chosen plaintext attacks (IND-ID-CPA security) for our first construction and an IBKEM that satisfies one-wayness under the same attacks (OW-ID-CPA security) for our second construction. Since these security definitions are standard, we omit the definitions here.

Target Collision Resistance for Randomness. We define target collision resistance for randomness of an IBKEM here. This security is previously defined by Nakai et al. in [24] for IBE schemes, but we naturally adopt it for IBKEMs.

Definition 2. *We say that an IBKEM Π_I (whose randomness space of IEncap is $\mathcal{R}_{\text{IEncap}}$) satisfies target collision resistance for randomness if for any PPTA $\mathcal{A}$ the following advantage function $\text{Adv}_{\Pi_I,\mathcal{A}}^{\text{Rand}}(\kappa)$ is negligible:*

$$\text{Adv}_{\Pi_I,\mathcal{A}}^{\text{Rand}}(\kappa) =$$
$$\Pr[(\text{prm}, \text{msk}) \leftarrow \text{ISetup}(1^\kappa); R^* \leftarrow \mathcal{R}_{\text{IEncap}}; (\text{ID}', R') \leftarrow \mathcal{A}(\text{prm}, \text{msk}, R^*) :$$
$$\text{IEncap}(\text{prm}, \text{ID}'; R') = \text{IEncap}(\text{prm}, \text{ID}'; R^*) \wedge R' \neq R^*].$$

Whether this security is satisfied or not (possibly under some hardness assumption) depends on concrete instantiations. However, we note that this security is satisfied *unconditionally* by most existing pairing-based IBKEMs, such as [9,27,21,14,28] and their variants. These IBKEMs have the property that under fixed global parameters and a fixed identity, if a different randomness is used in the encapsulation algorithm, then a ciphertext/session-key pair will always be different. As discussed in [24, Section 2.2], many practical pairing-based IBKEMs (and IBE schemes) have this property, and therefore satisfy target collision resistance for randomness unconditionally.

3   TRE-PC Scheme and TRE-PC KEM

In this section, we review the definitions of algorithms and security of TRE-PC schemes and TRE-PC KEMs. We adopt the models by Dent and Tang [20].

A timed-release encryption with pre-open capability (TRE-PC) scheme Γ consists of the following six PPTAs.

TRE.Setup: A setup algorithm that takes 1^κ as input, and outputs a pair of global parameters prm and a master secret key msk.

TRE.Ext: A timed-release key extraction algorithm that takes prm, msk, and a time T as input, and outputs a timed-release key trk corresponding to T.

TRE.UKG: A user key generation algorithm that takes prm as input, and outputs a user's public/secret key pair (upk, usk).

TRE.Enc: An encryption algorithm that takes prm, T, upk, and a plaintext $m \in \mathcal{M}$ as input, and outputs a ciphertext c and a corresponding pre-open key pok.

TRE.Dec$_{\text{TR}}$: A deterministic release-time decryption algorithm that takes prm, usk, trk, and c as input, and outputs $m \in \mathcal{M} \cup \{\bot\}$.

TRE.Dec$_{\text{PO}}$: A deterministic pre-open decryption algorithm that takes prm, usk, pok, and c as input, and outputs $m \in \mathcal{M} \cup \{\bot\}$.

In the above, $\mathcal{M}$ is a plaintext space of Γ.

We require, for all $(\text{prm}, \text{msk}) \leftarrow \text{TRE.Setup}(1^\kappa)$, all $(upk, usk) \leftarrow \text{TRE.UKG}(\text{prm})$, all T, all $trk \leftarrow \text{TRE.Ext}(\text{prm}, \text{msk}, T)$, all m, and all $(c, pok) \leftarrow \text{TRE.Enc}(\text{prm}, T, upk, m)$, that $\text{TRE.Dec}_{\text{TR}}(\text{prm}, usk, trk, c) = \text{TRE.Dec}_{\text{PO}}(\text{prm}, usk, pok, c) = m$.

A TRE-PC KEM is a natural KEM analogue of a TRE-PC scheme and consists of six algorithms (we will use the prefix "TRKEM" for denoting algorithms

$\mathsf{Expt}_{\Gamma,\mathcal{A}}^{\mathrm{IND\text{-}TR\text{-}CCA}_{TS}}(\kappa)$:
  $(prm, msk) \leftarrow$ TRE.Setup(1^κ)
  $(upk, usk) \leftarrow$ TRE.UKG(prm)
  $(m_0, m_1, T^*, st_\mathcal{A})$
    $\leftarrow \mathcal{A}_1^{\mathcal{O}_{\mathrm{Dec_{TR}}}, \mathcal{O}_{\mathrm{Dec_{PO}}}}(prm, msk, upk)$
  $b \leftarrow \{0, 1\}$
  $(c^*, pok^*) \leftarrow$ TRE.Enc(prm, T^*, upk, m_b)
  $b' \leftarrow \mathcal{A}_2^{\mathcal{O}_{\mathrm{Dec_{TR}}}, \mathcal{O}_{\mathrm{Dec_{PO}}}}(c^*, pok^*, st_\mathcal{A})$
  If $b' = b$ then return 1 else return 0

$\mathsf{Expt}_{\Gamma,\mathcal{A}}^{\mathrm{IND\text{-}TR\text{-}CPA}_{IS}}(\kappa)$:
  $(prm, msk) \leftarrow$ TRE.Setup(1^κ)
  $(upk, usk) \leftarrow$ TRE.UKG(prm)
  $(m_0, m_1, T^*, st_\mathcal{A})$
    $\leftarrow \mathcal{A}_1^{\mathcal{O}_{\mathrm{Ext}}}(prm, upk, usk)$
  $b \leftarrow \{0, 1\}$
  $(c^*, pok^*) \leftarrow$ TRE.Enc(prm, T^*, upk, m_b)
  $b' \leftarrow \mathcal{A}_2^{\mathcal{O}_{\mathrm{Ext}}}(c^*, st_\mathcal{A})$
  If $b' = b$ then return 1 else return 0

$\mathsf{Expt}_{\Gamma,\mathcal{A}}^{\mathrm{Binding}}(\kappa)$:
  $(prm, msk) \leftarrow$ TRE.Setup(1^κ)
  $(upk, usk) \leftarrow$ TRE.UKG(prm)
  (c^*, T^*, pok^*)
    $\leftarrow \mathcal{A}^{\mathcal{O}_{\mathrm{Ext}}, \mathcal{O}_{\mathrm{Dec_{TR}}}, \mathcal{O}_{\mathrm{Dec_{PO}}}}(prm, upk)$
  $trk^* \leftarrow$ TRE.Ext(prm, msk, T^*)
  $m_{tr}^* \leftarrow$ TRE.Dec$_{\mathrm{TR}}(prm, usk, trk^*, c^*)$
  $m_{po}^* \leftarrow$ TRE.Dec$_{\mathrm{PO}}(prm, usk, pok^*, c^*)$
  If $\perp \neq m_{tr}^* \neq m_{po}^* \neq \perp$
    then return 1 else return 0

Definitions of Oracles

Oracle	Input	Output
$\mathcal{O}_{\mathrm{Ext}}$	T	TRE.Ext(prm, msk, T)
$\mathcal{O}_{\mathrm{Dec_{TR}}}$	(T, c)	TRE.Dec$_{\mathrm{TR}}(prm, usk, trk, c)$ where $trk \leftarrow$ TRE.Ext(prm, msk, T)
$\mathcal{O}_{\mathrm{Dec_{PO}}}$	(pok, c)	TRE.Dec$_{\mathrm{PO}}(prm, usk, pok, c)$

Fig. 1. Security experiments for a TRE-PC scheme Γ. The table (right-bottom) is the definitions of oracles in the experiments.

of a TRE-PC KEM), i.e., TRKEM.Setup, TRKEM.Ext, TRKEM.UKG, TRKEM.Encap, TRKEM.Decap$_{\mathrm{TR}}$, and TRKEM.Decap$_{\mathrm{PO}}$. Since the interface of these algorithms are easily inferred from that of a TRE-PC scheme, we omit it.

3.1   Security Requirements

Here, we review the three kinds of security requirements defined in [20] for a TRE-PC scheme, which are *time-server security*, *insider security*, and *binding*. Though here we only mention the security definitions for a TRE-PC scheme, those for a TRE-PC KEM is a natural KEM analogue of the definitions for a TRE-PC scheme, and are easily inferred.

Time-Server Security. This security protects message confidentiality against a curious time-server who owns a master secret key of a TRE-PC scheme. In [20], it was shown that the security against an outsider who does not have access to a master secret key nor a user's secret key is also captured by this security. Formally, we define the security experiment $\mathsf{Expt}_{\Gamma,\mathcal{A}}^{\mathrm{IND\text{-}TR\text{-}CCA}_{TS}}(\kappa)$ for time-server security (IND-TR-CCA$_{TS}$ security) of a TRE-PC scheme Γ in which an adversary $\mathcal{A} = (\mathcal{A}_1, \mathcal{A}_2)$ is run as in Fig. 1 (left-top). It should be noted that the second stage adversary $\mathcal{A}_2$ is given not only a challenge ciphertext but also its corresponding pre-open key. We make several restrictions in the experiment: $\mathcal{A}$'s challenge plaintexts must satisfy $|m_0| = |m_1|$; $\mathcal{A}_2$ is not allowed to submit the challenge time/ciphertext pair (T^*, c^*) to $\mathcal{O}_{\mathrm{Dec_{TR}}}$, and also is not allowed to submit the challenge pre-open key/ciphertext pair (pok^*, c^*) to $\mathcal{O}_{\mathrm{Dec_{PO}}}$.

Definition 3. *We say that a TRE-PC scheme Γ is IND-TR-CCA$_{TS}$ secure if $\mathsf{Adv}_{\Gamma,\mathcal{A}}^{\mathrm{IND\text{-}TR\text{-}CCA_{TS}}}(\kappa) = |\Pr[\mathsf{Expt}_{\Gamma,\mathcal{A}}^{\mathrm{IND\text{-}TR\text{-}CCA_{TS}}}(\kappa) = 1] - \frac{1}{2}|$ is negligible for any PPTA $\mathcal{A}$.*

Insider Security. This security protects the message confidentiality against a malicious receiver who has a user's secret key and tries to obtain some information about a plaintext from a ciphertext without a timed-release key or a pre-open key. Formally, we define the security experiment $\mathsf{Expt}_{\Gamma,\mathcal{A}}^{\mathrm{IND\text{-}TR\text{-}CPA_{IS}}}(\kappa)$ for insider security (IND-TR-CPA$_{IS}$ security) of a TRE-PC scheme Γ in which an adversary $\mathcal{A} = (\mathcal{A}_1, \mathcal{A}_2)$ is run as in Fig. 1 (right-top). We make several restrictions in the experiment: $\mathcal{A}_1$'s challenge time T^* must satisfy $T^* > T$ for any time T that is submitted to $\mathcal{O}_{\mathsf{Ext}}$ by $\mathcal{A}_1$; $\mathcal{A}$'s challenge plaintexts must satisfy $|m_0| = |m_1|$; $\mathcal{A}_2$ must not issue a time T satisfying $T^* \leq T$ to $\mathcal{O}_{\mathsf{Ext}}$.

Definition 4. *We say that a TRE-PC scheme Γ is IND-TR-CPA$_{IS}$ secure if $\mathsf{Adv}_{\Gamma,\mathcal{A}}^{\mathrm{IND\text{-}TR\text{-}CPA_{IS}}}(\kappa) = |\Pr[\mathsf{Expt}_{\Gamma,\mathcal{A}}^{\mathrm{IND\text{-}TR\text{-}CPA_{IS}}}(\kappa) = 1] - \frac{1}{2}|$ is negligible for any PPTA $\mathcal{A}$.*

As noted in [20], the term "CPA" is used because an adversary in the experiment does not have access to decryption oracles $\mathcal{O}_{\mathsf{Dec_{TR}}}$ and $\mathcal{O}_{\mathsf{Dec_{PO}}}$. This is simply because the decryption oracles can be simulated by an adversary who owns user's secret key and has access to $\mathcal{O}_{\mathsf{Ext}}$. See [20] for details.

Binding. Binding protects a receiver from a malicious sender who tries to make a ciphertext that decrypts to some plaintext with TRE.Dec$_{\mathsf{TR}}$ but can be pre-opened to another plaintext with TRE.Dec$_{\mathsf{PO}}$. Formally, we define the security experiment $\mathsf{Expt}_{\Gamma,\mathcal{A}}^{\mathrm{Binding}}(\kappa)$ for binding of a TRE-PC scheme Γ in which an adversary $\mathcal{A}$ is run as in Fig. 1 (left-bottom).

Definition 5. *We say that a TRE-PC scheme Γ satisfies binding if $\mathsf{Adv}_{\Gamma,\mathcal{A}}^{\mathrm{Binding}}(\kappa) = \Pr[\mathsf{Expt}_{\Gamma,\mathcal{A}}^{\mathrm{Binding}}(\kappa) = 1]$ is negligible for any PPTA $\mathcal{A}$.*

4   Proposed Generic Constructions

In this section, we show two new generic constructions of TRE-PC schemes. We first explain the basic construction idea in Section 4.1 for better understanding of our constructions. Then in Sections 4.2 and 4.3, we show the details of our generic constructions. Finally we discuss the extensions of our constructions in Section 4.4.

4.1   Basic Construction Idea

As mentioned earlier, the basic idea for both of our generic constructions follows the idea behind the NMKM construction: We employ "PKE"-like and "IBE"-like primitive as main building blocks; A message m is encrypted for the release-time T in such a way that (a) it can be decrypted (with release-time decryption algorithm) only when the user's secret key (which is a decryption key of

"PKE"-part) and the timed-release key (which is a decryption key of "IBE-part" under the "identity" T) are simultaneously available, (b) with some mechanism such that the ciphertexts from each building block component are strongly bound together into a single TRE-PC ciphertext. In the NMKM construction, "(a)" is achieved by adopting multiple encryption by PKE and IBE schemes together with a 2-out-of-2 secret sharing of m, and "(b)" is achieved using a one-time signature.

However, the naive use of the 2-out-of-2 secret sharing of a message m and the use of "encryption" schemes as the underlying building blocks cause a ciphertext overhead (the difference between the ciphertext size and the plaintext size) by exactly the size of the message m itself, which can lead to a long ciphertext if m is long, and thus we avoid this by adopting "KEM/DEM" approaches appropriately. Moreover, a one-time signature also causes a large ciphertext overhead as well as relatively large computational costs in encryption and decryption algorithms because of signing and verification of signatures. Therefore, in order to avoid using one-time signatures but still achieve a similar "binding" mechanism of ciphertext components, each of our proposed constructions employs different ideas, which will be explained in detail in the following subsections.

As in the NMKM construction, in order to realize pre-open decryption and achieve binding security, we will use the randomness used to generate the IBE-part ciphertext as a pre-open key. In the pre-open decryption procedure, this randomness is used to check the validity of the IBE-part ciphertext and to recover the "information" hidden in the IBE-part ciphertext.

4.2   Proposed TRE-PC Scheme from Tag-KEM, IBKEM, and DEM

Our first construction utilizes a tag-KEM as the "PKE"-part and an IBKEM as the "IBE"-part. An actual message is encrypted with a DEM where the session-key for the DEM is the XOR of the session-keys from the tag-KEM and the IBKEM. Specifically, let $\Pi_T = (\text{TKG, TSKey, TEncap, TDecap})$ be a tag-KEM, $\Pi_I = (\text{ISetup, IExt, IEncap, IDecap})$ be an IBKEM whose randomness space of IEncap is $\mathcal{R}_{\text{IEncap}}$, and $D = (\text{DEnc, DDec})$ be a DEM. Then the proposed TRE-PC scheme Γ_1 is constructed as in Fig. 2. For simplicity, we assume that the session key space of Π_T, that of Π_I, and the key space of D are all $\{0,1\}^\kappa$ where κ is a security parameter (we can always achieve this using an appropriate key derivation function and/or a pseudorandom generator).

Intuition. Recall that when one constructs a CCA-secure hybrid PKE scheme via the tag-KEM/DEM composition paradigm established in [4], the DEM ciphertext is regarded as a tag for a CCA-secure tag-KEM, and then it is input into the encapsulation algorithm of the tag-KEM. Roughly, because of this "feedback" structure, the DEM is only required to be passively secure (i.e. IND-OT security) while a DEM ciphertext is strongly tied to a tag-KEM ciphertext, and the entire hybrid PKE satisfies CCA-security. We use this feedback structure of a tag-KEM to bind all the ciphertext components together into one single TRE-PC ciphertext. More specifically, the IBKEM ciphertext c_2, the DEM ciphertext

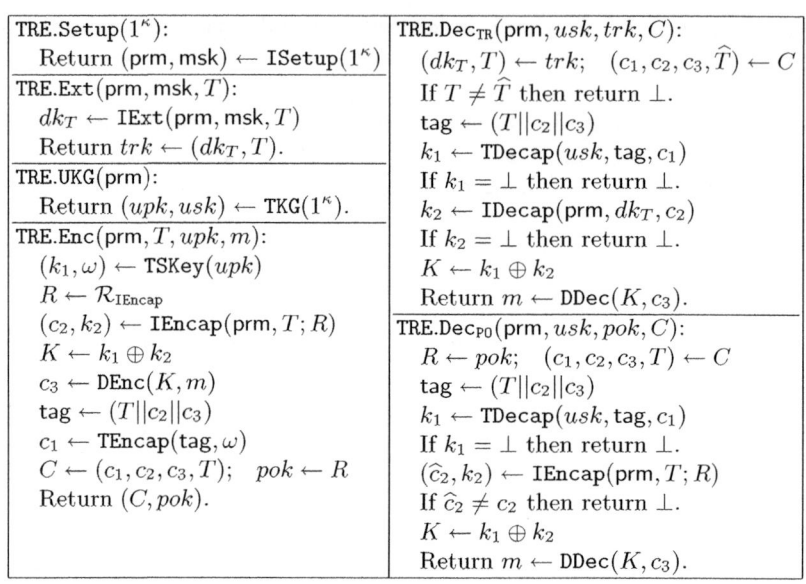

Fig. 2. Proposed TRE-PC Scheme Γ_1

c_3, and a time T are regarded as a tag for the tag-KEM, and are input into the encapsulation algorithm of the tag-KEM to generate the tag-KEM ciphertext c_1. Since an entire TRE-PC ciphertext is of the form (c_1, c_2, c_3, T) where (c_2, c_3, T) are treated as a tag for the tag-KEM, an adversary of the time-server security experiment cannot gain any useful information of the session-key k_1 of the tag-KEM, and thus that of the session-key $K = k_1 \oplus k_2$ used in the DEM, without breaking the IND-CCA security of the tag-KEM. If not the information on K, the adversary cannot gain any useful information about the message without breaking the IND-OT security of the DEM. This intuitively ensures the time-server security (i.e. IND-TR-CCA$_{TS}$ security) (we actually need target collision resistance for randomness of the IBKEM to deal with a certain kind of pre-open decryption queries, as is same with the NMKM construction). The insider security (i.e. IND-TR-CPA$_{IS}$ security) is straightforwardly achieved by a combination of the IND-ID-CPA security of the IBKEM and the IND-OT security of the DEM. In our construction, the pre-open key is the randomness used to generate the IBKEM ciphertext c_2. The appropriate re-computation and the equality check of the IBKEM ciphertext c_2 in the pre-open decryption algorithm ensure binding security.

The security of the proposed TRE-PC scheme Γ_1 is guaranteed by the following theorems.

Theorem 1. *If the tag-KEM Π_T is IND-CCA secure, the DEM D is IND-OT secure, and the IBKEM Π_I satisfies target collision resistance for randomness, then the proposed TRE-PC scheme Γ_1 is IND-TR-CCA$_{TS}$ secure.*

Proof. Let $\mathcal{A} = (\mathcal{A}_1, \mathcal{A}_2)$ be any PPTA IND-TR-CCA$_{TS}$ adversary against the proposed TRE-PC scheme Γ_1. We consider the following sequence of games. (The values with asterisk (*) are the ones used for generating the challenge ciphertext $C^* = (c_1^*, c_2^*, c_3^*, T^*)$ and its corresponding pre-open key pok^*.)

Game 1: This is the ordinary IND-TR-CCA$_{TS}$ experiment regarding our proposed TRE-PC scheme Γ_1.

Game 2: Same as Game 1, except that if $\mathcal{A}_2$ issues a pre-open decryption query of the form (pok, C^*) with $pok \neq pok^*$, it is responded with $\perp$.

Game 3: Same as Game 2, except that when the challenge ciphertext C^* is generated, a random session-key $K^* \leftarrow \{0,1\}^*$ is used to generate c_3^* instead of using $K^* = k_1^* \oplus k_2^*$.

Let Succ$_i$ be the event that $\mathcal{A}$ succeeds in guessing the challenge bit (i.e. $b' = b$ occurs) in Game i, and let POQuery$_i$ be the event that $\mathcal{A}_2$ submits at least one pre-open decryption query (pok, C^*) that satisfies $pok \neq pok^*$ and TRE.Dec(prm, $usk, pok, C^*) \neq \perp$ in Game i.

Then, the IND-TR-CCA$_{TS}$ advantage of the adversary $\mathcal{A}$ is estimated as:

$$\mathrm{Adv}_{\Gamma_1, \mathcal{A}}^{\mathrm{IND\text{-}TR\text{-}CCA_{TS}}}(\kappa) = |\Pr[\mathsf{Succ}_1] - \frac{1}{2}|$$

$$\leq \sum_{i \in \{1,2\}} |\Pr[\mathsf{Succ}_i] - \Pr[\mathsf{Succ}_{i+1}]| + |\Pr[\mathsf{Succ}_3] - \frac{1}{2}| \quad (1)$$

In the following, we will upperbound the above terms.

Lemma 1. $|\Pr[\mathsf{Succ}_1] - \Pr[\mathsf{Succ}_2]|$ *is negligible.*

Proof of Lemma 1. Notice that Game 1 and Game 2 proceed identically until the event POQuery$_1$ = POQuery$_2$ occurs. Therefore, we have

$$|\Pr[\mathsf{Succ}_1] - \Pr[\mathsf{Succ}_2]| \leq \Pr[\mathsf{POQuery}_1] = \Pr[\mathsf{POQuery}_2].$$

Hence, it is sufficient to show that $\Pr[\mathsf{POQuery}_2]$ is negligible.

Towards a contradiction assume that $\Pr[\mathsf{POQuery}_2]$ is non-negligible. Then we can construct another adversary $\mathcal{B}$ that can break target collision resistance for randomness of the IBKEM Π_I with advantage exactly $\Pr[\mathsf{POQuery}_2]$, using $\mathcal{A}$. Given (prm, msk, R^*), $\mathcal{B}$ generates a user key pair $(upk, usk) \leftarrow \mathsf{TKG}(1^\kappa)$, and then starts simulating Game 2 for $\mathcal{A}$ by running $\mathcal{A}_1$ on input (prm, msk, upk). Since $\mathcal{B}$ possesses msk and usk, it can do a perfect simulation of Game 2 (in particular can respond to all decryption queries from $\mathcal{A}$ as in Game 2). When $\mathcal{B}$ generate the challenge ciphertext $C^* = (c_1^*, c_2^*, c_3^*, T^*)$, it uses R^* as a randomness to generate c_2^* and moreover sets $pok^* = R^*$ as a pre-open key for C^* and gives them to $\mathcal{A}$. Now, if $\mathcal{A}_2$ issues a pre-open decryption query of the form $(pok = R', C^* = (c_1^*, c_2^*, c_3^*, T^*))$ that causes POQuery$_2$, it holds that $(c_2^*, k_2^*) = \mathsf{IEncap}(\mathsf{prm}, T^*; R') = \mathsf{IEncap}(\mathsf{prm}, T^*; R^*)$ and $R^* \neq R'$, where the latter is due to the restriction on $\mathcal{A}$ in the IND-TR-CCA$_{TS}$ experiment. Therefore, whenever POQuery$_2$ occurs, $\mathcal{B}$ can break target collision resistance for randomness by outputting (T^*, R'). This means $\mathcal{B}$'s advantage is exactly $\Pr[\mathsf{POQuery}_2]$ that is non-negligible, which contradicts target collision resistance for randomness of the IBKEM Π_I. This completes the proof of Lemma 1. $\qquad\square$

Lemma 2. $|\Pr[\mathsf{Succ}_2] - \Pr[\mathsf{Succ}_3]|$ *is negligible.*

Proof of Lemma 2. Assume towards a contradiction that $|\Pr[\mathsf{Succ}_2] - \Pr[\mathsf{Succ}_3]|$ is not negligible. Then we show that we can construct another PPTA adversary $\mathcal{B} = (\mathcal{B}_1, \mathcal{B}_2)$ that uses $\mathcal{A}$ as a subroutine and has non-negligible IND-CCA advantage against the tag-KEM Π_T. The description of $\mathcal{B}$ is as follows.

$\mathcal{B}_1^{\mathsf{TDecap}(usk, \cdot, \cdot)}(upk, k_{1,b}^*)$: (Here, $k_{1,1}^*$ is a real session-key corresponding to some state information ω^*, $k_{1,0}^*$ is a random session-key, and b is a bit that $\mathcal{B}$ has to guess.) $\mathcal{B}_1$ first generates $(\mathsf{prm}, \mathsf{msk}) \leftarrow \mathsf{ISetup}(1^\kappa)$, and runs $\mathcal{A}_1$ with input $(\mathsf{prm}, \mathsf{msk}, upk)$. When $\mathcal{A}_1$ issues a release-time decryption query $(T', C = (c_1, c_2, c_3, T))$, $\mathcal{B}_1$ first generates $trk \leftarrow \mathsf{TRE.Ext}(\mathsf{prm}, \mathsf{msk}, T')$ and then decrypts C by faithfully following the procedure of $\mathsf{TRE.Dec_{TR}}(\mathsf{prm}, usk, trk, C)$, except that if $\mathcal{B}_1$ has to run $\mathsf{TDecap}(usk, \mathsf{tag}, c_1)$, then $\mathcal{B}_1$ submits (tag, c_1) to $\mathcal{B}$'s decapsulation oracle and obtains k_1. Pre-open decryption queries $(pok, C = (c_1, c_2, c_3, T))$ from $\mathcal{A}$ are answered similarly by using the decapsulation oracle. When $\mathcal{A}_1$ terminates with output $(m_0, m_1, T^*, \mathsf{st}_A)$, $\mathcal{B}_1$ picks a coin $\beta \in \{0, 1\}$ and a randomness $R^* \in \mathcal{R}_{\mathsf{IEncap}}$ uniformly at random, and runs $(c_2^*, k_2^*) \leftarrow \mathsf{IEncap}(\mathsf{prm}, T^*; R^*)$. Next, $\mathcal{B}_1$ sets $K^* \leftarrow k_{1,b}^* \oplus k_2^*$ and computes $c_3^* \leftarrow \mathsf{DEnc}(K^*, m_\beta)$. Then $\mathcal{B}_1$ sets the challenge tag $\mathsf{tag}^* \leftarrow (T^* \| c_2^* \| c_3^*)$ and the state information st_B that consists of all the information known to $\mathcal{B}_1$. Finally $\mathcal{B}_1$ terminates with output $(\mathsf{tag}^*, \mathsf{st}_B)$.

$\mathcal{B}_2^{\mathsf{TDecap}(usk, \cdot, \cdot)}(c_1^*, \mathsf{st}_B)$: $\mathcal{B}_2$ first sets $C^* \leftarrow (c_1^*, c_2^*, c_3^*, T^*)$ and $pok^* \leftarrow R^*$, and then runs $\mathcal{A}_2$ with input $(C^*, pok^*, \mathsf{st}_A)$. All the decryption queries from $\mathcal{A}_2$ are answered in the same way as $\mathcal{B}_1$ does, except that if $\mathcal{A}_2$ submits a release-time decryption query of the form (T, C^*) with $T \neq T^*$ or a pre-open decryption query of the form (pok, C^*) with $pok \neq pok^*$, then $\mathcal{B}_2$ immediately returns $\perp$ to $\mathcal{A}_2$. When $\mathcal{A}_2$ terminates with output a guess bit β', $\mathcal{B}_2$ sets $b' \leftarrow 1$ if $\beta' = \beta$ or $b' \leftarrow 0$ otherwise. Finally, $\mathcal{B}_2$ terminates with output b' as its guess for b.

Note that $\mathcal{B}_2$'s decapsulation queries do not contain the prohibited query (tag^*, c_1^*) where $\mathsf{tag}^* = (T^* \| c_2^* \| c_3^*)$. This is because $\mathcal{A}_2$'s release-time decryption query (T, C) always satisfies $(T, C) \neq (T^*, C^*)$, and a problematic query of the form (T, C^*) with $T \neq T^*$ is answered with $\perp$ without using $\mathcal{B}$'s decapsulation oracle. This is an appropriate answer for $\mathcal{A}$ in Game 2 and Game 3, since if $T \neq T^*$, the release-time decryption result of our scheme Γ_1 returns $\perp$ as well. Similarly, $\mathcal{A}_2$'s pre-open decryption query (pok, C) always satisfies $(pok, C) \neq (pok^*, C^*)$, and a problematic query of the form (pok, C^*) with $pok \neq pok^*$ is answered with $\perp$ without using $\mathcal{B}$'s decapsulation oracle. Note that this answer is appropriate in Game 2 and Game 3.

Then, we estimate the IND-CCA advantage of $\mathcal{B}$.

$$\mathsf{Adv}_{\Pi_T,\mathcal{B}}^{\mathrm{IND\text{-}CCA}}(\kappa) = |\Pr[b' = b] - \frac{1}{2}| = \frac{1}{2}|\Pr[b' = 1|b = 1] - \Pr[b' = 1|b = 0]|$$

$$= \frac{1}{2}|\Pr[\beta' = \beta|b = 1] - \Pr[\beta' = \beta|b = 0]|$$

Now, consider the case when $b = 1$, i.e., $k_{1,b}^* = k_{1,1}^*$ is a real session-key corresponding to c_1^* under the tag $\mathsf{tag}^* = (T^*||c_2^*||c_3^*)$. Then, it is easy to see that $\mathcal{B}$ perfectly simulates Game 2 for $\mathcal{A}$ in which the challenge bit for $\mathcal{A}$ is β. Specifically, $\mathcal{B}$'s responses to decryption queries are perfectly answered as in Game 2, and the challenge ciphertext C^* for $\mathcal{A}$ is generated as in Game 2 (c_3^* is computed with a correct session-key $K^* = k_{1,1}^* \oplus k_2^*$ as in the proposed TRE-PC scheme Γ_1). Under this situation, the event $\beta' = \beta$ corresponds to the event Succ_2, i.e., $\Pr[\beta' = \beta|b = 1] = \Pr[\mathsf{Succ}_2]$.

When $b = 0$, i.e., $k_{1,b}^* = k_{1,0}^*$ is a uniformly random value in $\{0,1\}^\kappa$, on the other hand, $\mathcal{B}$ perfectly simulates Game 3 for $\mathcal{A}$ in which the challenge bit for $\mathcal{A}$ is β. Specifically, the difference between Game 2 and Game 3 is only in the generation of c_3^* in the challenge ciphertext C^* for $\mathcal{A}$, and c_3^* is computed with a session-key $K^* = k_{1,0}^* \oplus k_2^*$ that is a uniformly random value in $\{0,1\}^\kappa$ because $k_{1,0}^*$ is chosen uniformly at random. Therefore the challenge ciphertext C^* for $\mathcal{A}$ is computed as in Game 3. Under this situation, the event $\beta' = \beta$ corresponds to the event Succ_3, i.e., $\Pr[\beta' = \beta|b = 0] = \Pr[\mathsf{Succ}_3]$.

In summary, we have $\mathsf{Adv}_{\Pi_T,\mathcal{B}}^{\mathrm{IND\text{-}CCA}}(\kappa) = \frac{1}{2}|\Pr[\mathsf{Succ}_2] - \Pr[\mathsf{Succ}_3]|$, which is not negligible according to the assumption we made at the beginning of this proof. Since this contradicts the IND-CCA security of the tag-KEM Π_T, $|\Pr[\mathsf{Succ}_2] - \Pr[\mathsf{Succ}_3]|$ must be negligible. This completes the proof of Lemma 2.   □

Lemma 3. $|\Pr[\mathsf{Succ}_3] - \frac{1}{2}|$ *is negligible.*

We omit the proof of this lemma, because it is almost obvious from the IND-OT security of the DEM D. Note that in Game 3, the session-key for the DEM ciphertext in $\mathcal{A}$'s challenge ciphertext is a random value. If $|\Pr[\mathsf{Succ}_3] - \frac{1}{2}|$ is not negligible, we can use $\mathcal{A}$ to break the IND-OT security of the DEM D. (For an IND-OT attacker using $\mathcal{A}$, to simulate Game 3 for $\mathcal{A}$ is very easy because it can generate msk and usk for Game 3.)

The inequality (1) and Lemmas 1 to 3 imply that $\mathsf{Adv}_{\Gamma_1,\mathcal{A}}^{\mathrm{IND\text{-}TR\text{-}CCA_{TS}}}(\kappa)$ is negligible for any PPTA $\mathcal{A}$. This completes the proof of Theorem 1.   □

Theorem 2. *If the IBKEM Π_I is IND-ID-CPA secure and the DEM D is IND-OT secure, then the proposed TRE-PC scheme Γ_1 is IND-TR-CPA$_{IS}$ secure.*

Proof. Let $\mathcal{A} = (\mathcal{A}_1, \mathcal{A}_2)$ be any PPTA IND-TR-CPA$_{IS}$ adversary against the proposed TRE-PC scheme Γ_1. We consider the following sequence of games. (The values with asterisk (*) are the ones generated during the computation the challenge ciphertext $C^* = (c_1^*, c_2^*, c_3^*, T^*)$.)

Game 1: This is the ordinary IND-TR-CPA$_{IS}$ experiment regarding our proposed TRE-PC scheme Γ_1.

Game 2: Same as Game 1, except that when the challenge ciphertext C^* is generated, a random session-key $K^* \leftarrow \{0,1\}^\kappa$ is used to generate c_3^* instead of using $K^* = k_1^* \oplus k_2^*$.

Let Succ_i be the event that $\mathcal{A}$ succeeds in guessing the challenge bit (i.e. $b' = b$ occurs) in Game i.

The IND-TR-CPA$_{IS}$ advantage of $\mathcal{A} = (\mathcal{A}_1, \mathcal{A}_2)$ is estimated as:

$$\mathsf{Adv}_{\Gamma_1,\mathcal{A}}^{\mathrm{IND\text{-}TR\text{-}CPA_{IS}}}(\kappa) = |\Pr[\mathsf{Succ}_1] - \frac{1}{2}|$$

$$\leq |\Pr[\mathsf{Succ}_1] - \Pr[\mathsf{Succ}_2]| + |\Pr[\mathsf{Succ}_2] - \frac{1}{2}| \qquad (2)$$

In the following, we will upperbound the above terms.

Lemma 4. $|\Pr[\mathsf{Succ}_1] - \Pr[\mathsf{Succ}_2]|$ *is negligible.*

It is almost obvious from the IND-ID-CPA security of the IBKEM Π_I, and thus we omit the proof (which is given in the full version). Note that the difference between Game 1 and Game 2 is only in the generation of the challenge ciphertext for $\mathcal{A}$. If the adversary $\mathcal{A}$'s success probability (in guessing the challenge bit) is non-negligibly different between Game 1 and Game 2, we can use $\mathcal{A}$ to construct another PPTA adversary $\mathcal{B}$ that can distinguish a real-session key corresponding to the challenge IBKEM ciphertext from a random in the IND-ID-CPA security of Π_I with non-negligible advantage. Recall that an IND-ID-CPA adversary is given access to the extraction oracle, and thus $\mathcal{B}$ can perfectly simulate the responses to timed-release key extraction queries from $\mathcal{A}$.

Lemma 5. $|\Pr[\mathsf{Succ}_2] - \frac{1}{2}|$ *is negligible.*

We again omit the proof of this lemma, because it is almost obvious from the IND-OT security of the DEM D. Note that in Game 2, the session-key for the DEM ciphertext in $\mathcal{A}$'s challenge ciphertext is a random value, and thus almost the same explanation mentioned for Lemma 3 is applicable here.

The inequality (2) and Lemmas 4 and 5 imply that $\mathsf{Adv}_{\Gamma_1,\mathcal{A}}^{\mathrm{IND\text{-}TR\text{-}CPA_{IS}}}(\kappa)$ is negligible for any PPTA $\mathcal{A}$. This completes the proof of Theorem 2.   □

Theorem 3. *The proposed TRE-PC scheme Γ_1 satisfies binding against any (even computationally unbounded) adversary.*

Proof. Suppose that an adversary $\mathcal{A}$ in the binding experiment regarding our proposed TRE-PC scheme Γ_1 outputs a ciphertext $C^* = (c_1^*, c_2^*, c_3^*, \widehat{T}^*)$, a time T^*, and a pre-open key $pok^* = R^*$. Let $trk^* = (dk_{T^*}, T^*) \leftarrow$ TRE.Ext(prm, msk, T^*), $m_{tr}^* =$ TRE.Dec$_{\mathrm{TR}}$(prm, usk, trk^*, C^*), and $m_{po}^* =$ TRE.Dec$_{\mathrm{PO}}$(prm, usk, pok^*, C^*), all of which are computed in the binding experiment, and let $\mathsf{tag}^* = (\widehat{T}^*||c_2^*||c_3^*)$.

TRKEM.Setup(1^κ):	TRKEM.Decap$_{\text{TR}}$(prm, usk, trk, C):
Return (prm, msk) $\leftarrow$ ISetup(1^κ).	$(dk_T, T) \leftarrow trk;$ $(c_1, c_2, \widehat{T}) \leftarrow C$
TRKEM.Ext(prm, msk, T):	If $T \neq \widehat{T}$ then return $\perp$.
$dk_T \leftarrow$ IExt(prm, msk, T)	$k_1 \leftarrow$ PDecap(usk, c_1)
Return $trk \leftarrow (dk_T, T)$.	If $k_1 = \perp$ then return $\perp$.
TRKEM.UKG(prm):	$k_2 \leftarrow$ IDecap(prm, dk_T, c_2)
Return $(upk, usk) \leftarrow$ PKG(1^κ).	If $k_2 = \perp$ then return $\perp$.
TRKEM.Encap(prm, T, upk):	Return $K \leftarrow H(c_1, c_2, k_1, k_2, T)$.
$(c_1, k_1) \leftarrow$ PEncap(upk)	TRKEM.Decap$_{\text{PO}}$(prm, usk, pok, C):
$R \leftarrow \mathcal{R}_{\text{IEncap}}$	$R \leftarrow pok;$ $(c_1, c_2, T) \leftarrow C$
$(c_2, k_2) \leftarrow$ IEncap(prm$_I, T; R$)	$k_1 \leftarrow$ PDecap(usk, c_1)
$K \leftarrow H(c_1, c_2, k_1, k_2, T)$	If $k_1 = \perp$ then return $\perp$.
$C \leftarrow (c_1, c_2, T);$ $pok \leftarrow R$	$(\widehat{c_2}, k_2) \leftarrow$ IEncap(prm, $T; R$)
Return (C, K, pok).	If $\widehat{c_2} \neq c_2$ then return $\perp$.
	Return $K \leftarrow H(c_1, c_2, k_1, k_2, T)$

Fig. 3. Proposed TRE-PC KEM Γ_2

On the one hand, the event $[m_{tr}^* \neq \perp]$ implies $\widehat{T}^* = T^*$, $\texttt{TDecap}(usk, \texttt{tag}^*, c_1^*) = k_1^* \neq \perp$, $\texttt{IDecap}(\text{prm}, dk_{T^*}, c_2^*) = k_{2,tr}^* \neq \perp$, $K_{tr}^* = k_1^* \oplus k_{2,tr}^*$, and $\texttt{DDec}(K_{tr}^*, c_3^*) = m_{tr}^* \neq \perp$. On the other hand, the event $[m_{po}^* \neq \perp]$ implies $\texttt{TDecap}(usk, \texttt{tag}^*, c_1^*) = k_1^* \neq \perp$, $\texttt{IEncap}(\text{prm}, \widehat{T}^*; R^*) = (c_2^*, k_{2,po}^*)$, $K_{po}^* = k_1^* \oplus k_{2,po}^*$, and $\texttt{DDec}(K_{po}^*, c_3^*) = m_{po}^* \neq \perp$. Then, consider the event $[m_{tr}^* \neq \perp \wedge m_{po}^* \neq \perp]$. Since $T^* = \widehat{T}^*$, the correctness of the IBKEM Π_I implies $k_{2,tr}^* = k_{2,po}^*$. Therefore, $K_{tr}^* = (k_1^* \oplus k_{2,tr}^*) = (k_1^* \oplus k_{2,po}^*) = K_{po}^*$ holds, which in turn implies $m_{tr}^* = \texttt{DDec}(K_{tr}^*, c_3^*) = \texttt{DDec}(K_{po}^*, c_3^*) = m_{po}^*$. Hence, we have

$$\mathsf{Adv}_{\Gamma_1, \mathcal{A}}^{\text{Binding}}(\kappa) = \Pr[\perp \neq m_{tr}^* \neq m_{po}^* \neq \perp] \leq \Pr[m_{tr}^* = m_{po}^* \wedge m_{tr}^* \neq m_{po}^*],$$

where the right hand side of the above inequality is clearly zero, which proves the theorem. Note that the above holds regardless of $\mathcal{A}$'s running time. □

4.3 Proposed TRE-PC KEM from PKKEM, IBKEM, and Random Oracle

Here, we show a generic construction of a TRE-PC KEM that uses a PKKEM, an IBKEM and a random oracle. Let $\Pi_P = (\text{PKG}, \text{PEncap}, \text{PDecap})$ be a PKKEM, $\Pi_I = (\text{ISetup}, \text{IExt}, \text{IEncap}, \text{IDecap})$ be an IBKEM whose randomness space of IEncap is $\mathcal{R}_{\text{IEncap}}$, and $H : \{0,1\}^* \to \{0,1\}^\kappa$ be a cryptographic hash function that is modeled as a random oracle. Then the proposed TRE-PC KEM Γ_2 is constructed as in Fig. 3.

Intuition. The key point of the second construction is to hash all the ciphertext components with a random oracle to derive the session-key for the DEM. Intuitively, this "hashing" strongly ties all the ciphertext components via the derived TRE-PC KEM session-key. Constructed like this, any modification to a TRE-PC KEM ciphertext leads either to an invalid ciphertext or to a ciphertext that decrypts to a completely unpredictable session-key. An adversary against time-server security cannot distinguish a real TRE-PC KEM session-key that corresponds to the given challenge ciphertext from a randomness unless he by himself generates a valid ciphertext/session-key pair (c_1, k_1) of the PKKEM which is used in the challenge ciphertext. However, the adversary cannot produce such a valid ciphertext/session-key pair unless it breaks the OW-CCA security of the PKKEM, which ensures the time-server security. (As in the first construction, we actually need target collision resistance for randomness for dealing with a certain type of pre-open decapsulation queries.) With a similar reason, insider security is guaranteed from the OW-ID-CPA security of the IBKEM. As in our first construction Γ_1, the pre-open key is a randomness used to generate the IBKEM ciphertext c_2, and binding is achieved essentially in the same way.

The security of the proposed TRE-PC KEM Γ_2 is guaranteed by the following theorems. (Due to space limitations, we will provide the proofs in the full version.) By the TRE-PC KEM/DEM composition result by Dent and Tang [20], by combining our TRE-PC KEM Γ_2 with an IND-CCA secure DEM, we obtain a full TRE-PC scheme.

Theorem 4. *If the PKKEM Π_P is OW-CCA secure and the IBKEM Π_I satisfies target collision resistance for randomness, then the proposed TRE-PC KEM Γ_2 is IND-TR-CCA$_{TS}$ secure in the random oracle model where H is modeled as a random oracle.*

Theorem 5. *If the IBKEM Π_I is OW-ID-CPA secure, then the proposed TRE-PC KEM Γ_2 is IND-TR-CPA$_{IS}$ secure in the random oracle model where H is modeled as a random oracle.*

Theorem 6. *The proposed TRE-PC KEM Γ_2 satisfies binding against any (even computationally unbounded) adversary.*

4.4   Extensions

TRE-PC with Release Time Confidentiality.   The ciphertexts in both of our constructions contain the release-time T directly and thus do not provide *release-time confidentiality* [12]. However, we can extend our constructions to achieve it as well, by modifying our constructions slightly so that T and the IBKEM ciphertext c_2 are encrypted with the DEM by using a part of the "PKE"-part session-key k_1 as a DEM-key. The modification to each construction requires no additional building block primitives (such as anonymous IBKEM [21]), and introduces no ciphertext size overhead and no computation overhead other than the additional DEM computation. We will discuss this in detail in the full version.

Public Key Time-Specific Encryption with Pre-open Capability.   Very recently, Paterson and Quaglia [25] introduced a new primitive called *time-specific encryption* (TSE) (see the explanation in the last paragraph in Section 1). We remark that we can construct a public-key TSE scheme with pre-open capability by replacing the IBKEM in our constructions with a KEM analogue of a CPA-secure plain TSE scheme which satisfies target collision resistance for randomness (which can be defined similarly to that for IBKEMs). The public-key

TSE schemes based on our generic constructions satisfy CCA-security against the time-server in the model extended from that of [25] so that the model also deals with pre-open keys and pre-open decryption oracles as in [20], and CPA security against a curious receiver (where CPA security against a curious receiver is sufficient with a similar reason to that for insider security for TRE-PC).

5 Comparison

In this section, we compare concrete TRE-PC schemes derived from our generic constructions in Section 4 with the existing TRE-PC schemes: the original scheme by Dent and Tang [20] (DT), the same scheme which is instantiated with bilinear groups with an asymmetric pairing (DT'), and the scheme by Chow and Yiu [18] (CY)[1]. They are summarized in Table 1.

In Table 1, "Γ_1(XXX, YYY)" denotes a concrete TRE-PC scheme derived from our first generic construction Γ_1 where the tag-KEM is instantiated by "XXX" scheme and the IBKEM is instantiated by "YYY" scheme, and similar notation is used for Γ_2. There, AKO denotes the tag-KEM by Abe et al. [5], tWaters denotes the tag-KEM obtained from the (IND-ID-CPA secure version of) IBKEM by Waters [27] via the method described in [2, Section 7.3][2], and tBMW denotes the tag-KEM obtained from the combination of the CCA-secure PKKEM by Boyen et al. [10] and a one-time MAC via the method described in [4]. DHIES denotes the PKKEM part of the DHIES scheme [1]. BF denotes the IBKEM part of the basic IBE scheme by Boneh and Franklin [9], CCMS denotes the OW-ID-CPA secure version[3] of the IBKEM by Chen et al. [14], and Waters denotes the Waters IBKEM [27] as above.

We stress that the concrete schemes listed in Table 1 are just small examples derived via our generic constructions, and various combinations of a tag-KEM and an IBKEM via the first construction Γ_1 and those of a PKKEM and an IBKEM via the second construction Γ_2 are possible. For example, all the standard model schemes in Table 1 have large parameter sizes ($\mathcal{O}(\kappa)$ group elements) for prm, upk, and/or usk because of the Waters hash functions [27]. If one wants to achieve a standard model TRE-PC scheme whose parameter sizes are simultaneously constant, one can combine the tag-KEM tBMW and the IBKEM by Gentry [21] at the cost of basing the security on a somewhat stronger assumption, or combine tBMW and the IBKEM recently proposed by Waters [28] at the cost of relatively large computation costs. As noted earlier, the IBKEMs from [21] and [28] also satisfy target collision resistance for randomness unconditionally.

---

[1] It seems not possible, at least with a minor modification, to implement CY with bilinear groups with an asymmetric paring.

[2] Their method turns an IBKEM with a special structural property and slightly stronger security than IND-ID-CPA into an IND-CCA secure tag-KEM without any overhead, and the Waters IBKEM [27] satisfies the requirements.

[3] For IBKEMs, OW-ID-CPA security can be strengthened into IND-ID-CPA security by hashing the session-key with a random oracle. We assume this modification in Γ_1(AKO, CCMS) because we need an IND-ID-CPA secure IBKEM in Γ_1.

Random Oracle Schemes. Compared to DT and DT', all of our schemes have smaller ciphertext size, and provide smaller computation costs for encryption and similar amount costs for release-time decryption. $\Gamma_2(\mathsf{DHIES}, \mathsf{BF})$ has a slightly worse computation cost than DT and DT' for pre-open decryption. The computation costs for pre-open decryption in $\Gamma_2(\mathsf{DHIES}, \mathsf{CCMS})$ and $\Gamma_1(\mathsf{AKO}, \mathsf{CCMS})$ and those in DT and in DT' would be difficult to compare without further specifying how the bilinear groups are implemented. The schemes $\Gamma_1(\mathsf{AKO}, \mathsf{CCMS})$ and $\Gamma_2(\mathsf{DHIES}, \mathsf{CCMS})$ require no pairing computation in encryption and pre-open decryption, and thus may be suitable in applications in which pre-open decryption is frequently performed compared to release-time decryption. (Such situation would occur in the certified email system based on TRE-PC [22] if the arbitrating procedure occurs infrequently.) Note that when the SDH assumption [1] is considered in bilinear groups with a symmetric pairing (resp. an asymmetric pairing), it is equivalent to the CDH assumption (resp. the co-CDH assumption [1]). Hence, the assumptions for time-server security of the random oracle model schemes in Table 1 can be considered to be essentially equivalent.

Standard Model Schemes. Compared to CY, our concrete TRE-PC schemes derived from Γ_1 provide shorter ciphertext overhead, taking into account the currently known instantiations of symmetric/asymmetric pairings. Moreover, our concrete TRE-PC schemes offer smaller computation costs, and enable us to prove their security on weaker assumptions (the 3DDH and the 3MDDH assumptions [18] required for CY are, if considered in bilinear groups, strictly stronger than the decisional bilinear DH (DBDH) assumption). However, for fair comparison we note that CY is designed to be secure in a stronger security model (which considers a similar attack model adopted for *certificateless encryption*

Table 1. Efficiency Comparison among Concrete TRE-PC Schemes

Scheme	Random Oracle?	Ciphertext Size $\|C\| - (\|m\| + \|T\|)$	(Bits)	Assumptions TS / Insider	Computation Costs Enc	$\mathsf{Dec}_{\mathsf{TR}}$ / $\mathsf{Dec}_{\mathsf{PO}}$
DT [20]	yes	$2\|g_s\| + \kappa$	(554)	$\mathsf{CDH_b}$ / BDH	1P, 5E, 1H	1P, 1E / 1P, 1E
DT'	yes	$2\|g_a\| + \kappa$	(400)	co-CDH / BDH	1P, 5E, 1H	1P, 1E / 1P, 1E
$\Gamma_1(\mathsf{AKO}, \mathsf{CCMS})$	yes	$\|g_{np}\| + \|g_a\|$	(320)	SDH / q-BDHI	5E	1P, 1E / 1E
$\Gamma_2(\mathsf{DHIES}, \mathsf{BF})$	yes	$\|g_{np}\| + \|g_a\|$	(320)	SDH / BDH	1P, 4E, 1H	1P, 1E / 1P, 3E, 1H
$\Gamma_2(\mathsf{DHIES}, \mathsf{CCMS})$	yes	$\|g_{np}\| + \|g_a\|$	(320)	SDH / q-BDHI	4.5E	1P, 1E / 3.5E
CY [18]	no	$3\|g_s\|$	(711)	$\mathsf{3DDH_b}$ / $\mathsf{3MDDH_b}$	6E, 2W	5P, 1E, 2W / 6P, 2W
$\Gamma_1(\mathsf{tWaters}, \mathsf{Waters})$	no	$4\|g_a\|$	(640)	DBDH / DBDH	6E, 2W	3P, 1E / 1P, 4E, 1W
$\Gamma_1(\mathsf{tBMW}, \mathsf{Waters})$	no	$4\|g_a\| + \|\mathsf{Mac}\|$	(720)	DBDH / DBDH	6.5E, 1W	3P, 1E / 1P, 4E, 1W

In "Ciphertext Size" column, $|\mathsf{Mac}|$ denotes the size of a one-time MAC tag. $|g_a|$ (resp. $|g_s|$) denotes the bit length of elements in bilinear groups with asymmetric pairing (resp. symmetric pairing), and $|g_{np}|$ denotes the bit length of elements in ordinary groups without pairings. We assume that an IND-CCA secure DEM for Γ_2 is redundancy-free, i.e., $|\mathsf{DEnc}(K, m)| = |m|$ for any K and m (e.g. a strong pseudorandom permutation). For concrete bit size examples for 80-bit security, we set $\kappa = |\mathsf{Mac}| = 80$, $|g_s| = 237$ [8], and $|g_a| = |g_{np}| = 160$. In "Assumptions" column, the subscript "b" means that the assumption is considered in bilinear groups. In "Computation Costs" columns, P, E, H, and W denote the numbers of pairings, exponentiations, "map-to-point" hashings (hashing into group elements), and Waters' hash computations [27], respectively, and other computations are ignored. We count one multi-exponentiation as 1.5E. Moreover, we assume that each scheme is implemented so that computation costs become as small as possible. For example, for CY, we do not count the pairing computations for the validity check of usk because in practice it needs to be done only once.

[6] and which we think is too strong for TRE-PC schemes used for practical scenarios) than the model in [20] and provide release-time confidentiality [12], both of which are not achieved by our concrete constructions. Thus, it can be considered as a tradeoff between efficiency and stronger security.

Acknowledgement

The authors would like to thank anonymous reviewers of Pairing 2010 for their invaluable comments.

References

1. Abdalla, M., Bellare, M., Rogaway, P.: The oracle Diffie-Hellman assumptions and an analysis of DHIES. In: Naccache, D. (ed.) CT-RSA 2001. LNCS, vol. 2020, pp. 143–158. Springer, Heidelberg (2001)
2. Abe, M., Cui, Y., Imai, H., Kiltz, E.: Efficient hybrid encryption from ID-based encryption. Designs, Codes and Cryptography 54(3), 205–240 (2010)
3. Abe, M., Cui, Y., Imai, H., Kurosawa, K.: Tag-KEM from set partial domain one-way permutations. In: Batten, L.M., Safavi-Naini, R. (eds.) ACISP 2006. LNCS, vol. 4058, pp. 360–370. Springer, Heidelberg (2006)
4. Abe, M., Gennaro, R., Kurosawa, K.: Tag-KEM/DEM: A new framework for hybrid encryption. J. of Cryptology 21(1), 97–130 (2008)
5. Abe, M., Kiltz, E., Okamoto, T.: Compact CCA-secure encryption for messages of arbitrary length. In: Jarecki, S., Tsudik, G. (eds.) PKC 2009. LNCS, vol. 5443, pp. 377–392. Springer, Heidelberg (2009)
6. Al-Riyami, S., Paterson, K.: Certificateless public key cryptography. In: Laih, C.-S. (ed.) ASIACRYPT 2003. LNCS, vol. 2894, pp. 452–473. Springer, Heidelberg (2003)
7. An, J.H., Dodis, Y., Rabin, T.: On the security of joint signature and encryption. In: Knudsen, L.R. (ed.) EUROCRYPT 2002. LNCS, vol. 2332, pp. 83–107. Springer, Heidelberg (2002)
8. Boldyreva, A., Gentry, C., O'Neill, A., Yum, D.H.: Ordered multisignatures and identity-based sequential aggregate signatures, with applications to secure routing. In: CCS 2007, pp. 276–285 (2007)
9. Boneh, D., Franklin, M.: Identity-based encryption from the Weil pairing. SIAM J. Computing 32(3), 585–615 (2003)
10. Boyen, X., Mei, Q., Waters, B.: Direct chosen ciphertext security from identity-based techniques. Updated version of [11]. Cryptology ePrint Archive: Report 2005/288 (2005), http://eprint.iacr.org/2005/288/
11. Boyen, X., Mei, Q., Waters, B.: Direct chosen ciphertext security from identity-based techniques. In: CCS 2005, pp. 320–329 (2005)
12. Cathalo, J., Libert, B., Quisquater, J.-J.: Efficient and non-interactive timed-release encryption. In: Qing, S., Mao, W., López, J., Wang, G. (eds.) ICICS 2005. LNCS, vol. 3783, pp. 291–303. Springer, Heidelberg (2005)
13. Chan, A.C.-F., Blake, I.F.: Scalable, server-passive, user-anonymous timed release cryptography. In: ICDCS 2005, pp. 504–513 (2005)

14. Chen, L., Cheng, Z., Malone-Lee, J., Smart, N.P.: Efficient ID-KEM based on the Sakai-Kasahara key construction. IEE Proceedings - Information Security 153(1), 19–26 (2006)

15. Cheon, J.H., Hopper, N., Kim, Y., Osipkov, I.: Timed-release and key-insulated public key encryption. Cryptology ePrint Archive: Report 2004/231 (2004), http://eprint.iacr.org/2004/231/

16. Cheon, J.H., Hopper, N., Kim, Y., Osipkov, I.: Provably secure timed-release public key encryption. ACM Trans. Inf. Syst. Secur. 11(2) (2008)

17. Chow, S.S.M., Roth, V., Rieffel, E.G.: General certificateless encryption and timed-release encryption. In: Ostrovsky, R., De Prisco, R., Visconti, I. (eds.) SCN 2008. LNCS, vol. 5229, pp. 126–143. Springer, Heidelberg (2008)

18. Chow, S.S.M., Yiu, S.M.: Timed-release encryption revisited. In: Baek, J., Bao, F., Chen, K., Lai, X. (eds.) ProvSec 2008. LNCS, vol. 5324, pp. 38–51. Springer, Heidelberg (2008)

19. Cramer, R., Shoup, V.: Design and analysis of practical public-key encryption schemes secure against adaptive chosen ciphertext attack. SIAM J. Computing 33(1), 167–226 (2003)

20. Dent, A.W., Tang, Q.: Revisiting the security model for timed-release encryption with pre-open capability. In: Garay, J.A., Lenstra, A.K., Mambo, M., Peralta, R. (eds.) ISC 2007. LNCS, vol. 4779, pp. 158–174. Springer, Heidelberg (2007)

21. Gentry, C.: Practical identity-based encryption without random oracles. In: Vaudenay, S. (ed.) EUROCRYPT 2006. LNCS, vol. 4004, pp. 445–464. Springer, Heidelberg (2006)

22. Hwang, Y.H., Yum, D.H., Lee, P.J.: Timed-release encryption with pre-open capability and its application to certified e-mail system. In: Zhou, J., López, J., Deng, R.H., Bao, F. (eds.) ISC 2005. LNCS, vol. 3650, pp. 344–358. Springer, Heidelberg (2005)

23. May, T.: Timed-release crypto (1993) (manuscript)

24. Nakai, Y., Matsuda, T., Kitada, W., Matsuura, K.: A generic construction of timed-release encryption with pre-open capability. In: Takagi, T., Echizen, I. (eds.) IWSEC 2009. LNCS, vol. 5824, pp. 53–70. Springer, Heidelberg (2009)

25. Paterson, K.G., Quaglia, E.A.: Time-specific encryption. In: Garay, J.A., De Prisco, R. (eds.) SCN 2010. LNCS, vol. 6280, pp. 1–16. Springer, Heidelberg (2010)

26. Rivest, R.L., Shamir, A., Wagner, D.A.: Time-lock puzzles and timed-release crypto. Technical Report MIT/LCS/TR-684, Massachusetts Institute of Technology (1996)

27. Waters, B.: Efficient identity-based encryption without random oracles. In: Cramer, R. (ed.) EUROCRYPT 2005. LNCS, vol. 3494, pp. 114–127. Springer, Heidelberg (2005)

28. Waters, B.: Dual system encryption: Realizing fully secure IBE and HIBE under simple assumptions. In: Halevi, S. (ed.) EUROCRYPT 2009. LNCS, vol. 5677, pp. 619–636. Springer, Heidelberg (2009)

A   Other Building Block Primitives

(Public Key) Key Encapsulation Mechanism. A public key (i.e. neither tag-nor identity-based) key encapsulation mechanism (PKKEM) Π_P consists of the following three PPTAs (PKG, PEncap, PDecap): PKG is a key generation algorithm that takes 1^κ as input, and outputs a public/secret key pair (pk, sk); PEncap is an

encapsulation algorithm that takes pk as input, and outputs a ciphertext/session-key pair (c, K); PDecap is a deterministic decapsulation algorithm that takes sk and c as input, and outputs $K \in \mathcal{K} \cup \{\perp\}$ ($\mathcal{K}$ is a session-key space of Π_P). We require, for all $(pk, sk) \leftarrow \mathsf{PKG}(1^\kappa)$ and all $(c, K) \leftarrow \mathsf{PEncap}(pk)$, that PDecap$(sk, c) = K$.

We will need a PKKEM that is one-way against chosen ciphertext attacks (OW-CCA) for our second construction. Since the definition of OW-CCA security is standard, we omit the definition here.

Data Encapsulation Mechanism. A data encapsulation mechanism (DEM) D consists of the following two PPTAs (DEnc, DDec): DEnc is an encryption algorithm that takes a key $K \in \{0, 1\}^\kappa$ and a plaintext m as input, and outputs a ciphertext c; DDec is a deterministic decryption algorithm that takes K and c as input, and outputs $m \in \mathcal{M} \cup \{\perp\}$ ($\mathcal{M}$ is a plaintext space of D). We require, for all K and all m, that DDec$(K, \mathsf{DEnc}(K, m)) = m$.

We will need a DEM with very weak security, indistinguishability against one-time attacks (IND-OT security) [4]. This is even a weaker security notion than IND-CPA security for a symmetric-key encryption and can be achieved by a one-time pad of a message and a session-key.

Optimal Authenticated Data Structures with Multilinear Forms

Charalampos Papamanthou[1], Roberto Tamassia[1], and Nikos Triandopoulos[2]

[1] Department of Computer Science, Brown University, Providence RI, USA
[2] RSA Laboratories, Cambridge MA, USA

Abstract. Cloud computing and cloud storage are becoming increasingly prevalent. In this paradigm, clients outsource their data and computations to third-party service providers. Data integrity in the cloud therefore becomes an important factor for the functionality of these web services. Authenticated data structures, implemented with various cryptographic primitives, have been widely studied as a means of providing efficient solutions to data integrity problems (e.g., Merkle trees). In this paper, we introduce a new authenticated dictionary data structure that employs *multilinear forms*, a cryptographic primitive proposed by Silverberg and Boneh in 2003 [10], the construction of which, however, remains an open problem to date. Our authenticated dictionary is *optimal*, that is, it does not add any extra asymptotic cost to the plain dictionary data structure, yielding proofs of *constant* size, i.e., asymptotically equal to the size of the answer, while maintaining other relevant complexities *logarithmic*. Instead, solutions based on cryptographic hashing (e.g., Merkle trees) require proofs of logarithmic size [40]. Because multilinear forms are not known to exist yet, our result can be viewed from a different angle: if one could prove that optimal authenticated dictionaries cannot exist in the computational model, irrespectively of cryptographic primitives, then our solution would imply that cryptographically interesting multilinear form generators cannot exist as well (i.e., it can be viewed as a reduction). Thus, we provide an alternative avenue towards proving the nonexistence of multilinear form generators in the context of general lower bounds for authenticated data structures [40] and for memory checking [18], a model similar to the authenticated data structures model.

Keywords: authenticated dictionary, multilinear forms.

1 Introduction

Recently, there has been an increasing interest in *remote storage* of information and data. People outsource their personal files at service providers that offer huge storage space and fast network connections (e.g., Amazon S3). In this way, clients create *virtual* hard drives consisting of online storage units that are operated by remote and geographically dispersed servers. In addition to a convenient solution to space-shortage, data-archiving or back-up issues, remote storage allows for load-balanced distributed data management (e.g., database outsourcing). In such settings, the ability to check the integrity of remotely stored data is an important security property, or otherwise a malicious server can easily tamper with the client's data. When this data is structured, we

M. Joye, A. Miyaji, and A. Otsuka (Eds.): Pairing 2010, LNCS 6487, pp. 246–264, 2010.

need to provide solutions for *authenticated data structures* [39], which offer a computational model where untrusted entities answer queries on a data structure on behalf of
a trusted source and provide proof of validity of the answer to the user.

In this paper we study two-party authenticated data structures, a model closely related to memory checking [6], where a client decides to outsource his data, which is
stored in a data structure, to an untrusted server. However, the client needs to make sure
that whenever he retrieves his data back, he is able to verify its validity, i.e., that nobody
has tampered with it. In existing literature, a variety of authenticated data structures offer solutions that use different cryptographic primitives, such as cryptographic hashing
(e.g., [6]), accumulators (e.g., [35]) and lattices (e.g., [34]). The choice of the specific
cryptographic primitive has a drastic impact on the efficiency of the authenticated data
structure. For example it is known that, when using generic collision-resistant hash
functions[1] the best one can hope for are logarithmic complexities (in the size of the
data structure) [40]. On the other hand, using accumulators, which favor constant size
proofs, has so far resulted only in sublinear solutions [35], i.e., of $O(n^\epsilon)$ complexities
(see Table 1). Deriving for example logarithmic time query algorithms (time to construct the proof) that come with constant size proofs and constant time verification has
been an open problem.

This work begins by defining the notion of "optimal" authenticated data structures
(see Definition 7), i.e., authenticated data structures that do not add any asymptotic
overhead to the respective "plain" non-authenticated data structures.[2] Then we present
an authenticated dictionary data structure that is based on a new cryptographic primitive
that was recently proposed by Silverberg and Boneh, namely *multilinear forms* [10],
the construction of which remains however an open problem to date. The use of such a
primitive gives an authenticated dictionary with constant communication and constant
verification complexity, while maintaining all other complexities logarithmic. To the
best of our knowledge, this is the *first* optimal authenticated dictionary to appear in the
literature, as it exactly matches the respective complexities[3] (update time, query time,
answer size) of the optimal dictionary data structure (e.g., implemented as a red-black
tree).

The multilinear form cryptographic primitive that is used in our construction can
be described as the "multi" version of the well-known *bilinear map*. Although initially
used to attack elliptic curve systems [28], bilinear maps, being literally an efficient a tool
for solving the decisional Diffie-Hellman problem, eventually proved to be a very useful
tool in cryptography (e.g., [7,8,9]) after their first appearance in the literature for a
"good purpose" [24]. However, the main limitation of bilinear maps is the fact that they
cannot be applied twice, i.e., the output element cannot be fed back into the map $e(.,.)$
in an efficient way. Finding such maps, i.e., self-bilinear maps, which could be used
in a recursive way to construct multilinear forms, was recently proved to be infeasible

---

[1] Generic collision-resistant (denoted with GCR in Table 1) hash functions are functions that are
believed to be collision-resistant in practice, e.g., the SHA family of functions.

[2] We note here that for the dictionary problem, no such authenticated data structure is known to
exist to date.

[3] In this line of work, the asymptotic complexities always refer to the *size* of the data structure
n and not to the security parameter t, i.e., $t = O(1)$ (see Table 1).

for groups that are of interest in cryptography, i.e., groups where the computational Diffie-Hellman problem is hard [13].

However, since *cryptographically interesting* multilinear form generators[4] are not known to exist to date, one can view our work from a different (and more theoretical) angle: A proof through a complexity lower bound of the nonexistence of *optimal* authenticated dictionaries would imply the nonexistence of cryptographically interesting multilinear form generators (see Theorem 2). This reveals yet another important relation between two fields—combinatorics and cryptography—and becomes more promising (towards proving nonexistence of cryptographically interesting multilinear form generators) given recent advances in the derivation of general complexity lower bounds for memory checking [18] and authenticated data structures [40].

1.1 Related Work

Multilinear forms were proposed as a possible useful tool in cryptography in 2003 [10] by Silverberg and Boneh. Since then, no efficient construction of interest in cryptography has appeared. A work similar in nature with ours—where an efficient construction for a cryptographic application based on multilinear forms is proposed— appears in [25]. The impossibility of deriving multilinear forms through *self-bilinear* maps is investigated in [13].

In the context of *authenticated data structures*, several authenticated data structures based on cryptographic hashing have been developed, first being the well-known Merkle trees [29]. Blum et al. develop a similar solution [6], which is dynamized in sake of efficient certificate revocation by Naor and Nissim in [30]. *Authenticated skip lists* in the two-party model are presented in [33]. Authenticated multi-dimensional range-searching and external memory data structures (I/O efficient) are presented in [27]. Queries over distributed hash tables are efficiently authenticated in [41]. In the context of databases, authenticated inclusive and exclusive (join, projection) queries are discussed [17], where the size of the proof is linear in the size of the answer. Atallah et al. [2] present authenticated data structures for efficient 1D and 2D authenticated range search queries (query time and proof size are constant) and also for authenticating tree hierarchies [43]. Also, logarithmic lower bounds for hash-based methods are shown in [40]. Finally, in the context of memory checking, a linear lower bound on the product of reliable space and query time for online memory checkers in the information theoretic model is given [31], whereas in [18], an $\Omega(\log n / \log \log n)$ lower bound on query time for online memory checkers in the semantic security model has been shown.

Solutions for authenticated data structures in various settings using other cryptographic primitives, namely *one-way accumulators*, were introduced by Benaloh and de Mare [5]. Subsequently, refinements of the RSA accumulator [4,20,38] were shown to achieve collision resistance as well. Dynamic accumulators were introduced in [12], where, assuming an honest prover, the time to update the accumulated value or a witness is independent on the number of the accumulated elements. A first step towards a different direction, where we assume that we cannot trust the prover and therefore the trapdoor information is kept secret, was made in [21] (however this work is not

---

[4] I.e., multilinear form generators for groups where the discrete log problem is hard, e.g., elliptic curve groups. We call these generators *admissible* later in the paper.

applicable to the two-party model), achieving $O(n^\epsilon)$ bounds. An authenticated data structure that combines hierarchical hashing with the accumulation-based scheme of [21] is presented in [22] and accumulators using other cryptographic primitives (e.g., general groups with bilinear pairings) the security of which is based on other assumptions (e.g., hardness of strong Diffie-Hellman problem) are introduced in [11,32,42]. Non-membership proofs for accumulators are presented in [3,16,26]. Finally, *authenticated hash tables* that use the RSA accumulator are introduced in [35]. In particular, for *authenticated membership queries*, there has been a lot of work using different algorithmic and cryptographic approaches. A summary and qualitative comparison can be found in Table 1.

Table 1. Comparison of existing schemes for authenticating dictionary queries in the two-party model for a set of size n w.r.t. used techniques and various asymptotic complexity measures. Here, $0 < \epsilon < 1$ is a fixed constant, d can be any function of n ($d \leq n$), "GCR" stands for "generic collision-resistant", OWF for "one-way function", "q-SDH" for "q-strong Diffie-Hellman", "SRSA" for "strong RSA" and "q-SMDH" for "q-strong multilinear Diffie-Hellman". All complexity measures refer to n (not to the security parameter t, which is taken to be a constant) and are asymptotic values. In all schemes, the server space is $O(n)$ and the trusted space needed at the client is $O(1)$.

reference	assumption	verification proof size	consistency proof size	query time	update time	verification time
[6,27,30,33]	GCR	$\log n$	$\log n$	$\log n$	$\log n$	$\log n$
[18]	OWF	$\log_d n$	$d \log_d n$	$\log_d n$	$d \log_d n$	$\log_d n$
[12,38]	SRSA	1	1	1	$n \log n$	1
[32]	q-SDH	1	1	1	n	1
[35]	SRSA	1	1	1	$n^\epsilon \log n$	1
[35]	SRSA	1	1	n^ϵ	1	1
[36]	q-SDH	1	1	1	n^ϵ	1
this work	q-SMDH, GCR	1	$\log n$	$\log n$	$\log n$	1

1.2   Contributions

The contributions of this paper are as follows:

1. We formally (and in a rather intuitive way) define the notion of *optimal authenticated data structures*;
2. We present the first two-party authenticated dictionary that is based on *multilinear forms*. Our construction has constant verification time and constant communication complexity, while keeping all the other complexities logarithmic in the size of the structure, n (see Table 1). To the best of our knowledge, this is the first *optimal* authenticated dictionary to appear in the literature (Theorem 1);
3. We identify an important connection between lower bounds for authenticated data structures and the existence of cryptographically interesting multilinear form generators, giving new directions and intuition for deciding this problem (Theorem 2).

2   Preliminaries

In this section we introduce the main cryptographic primitive that we use in our solution, the *k-multilinear form*. We also introduce the notion of *cryptographic accumulators*, an instantiation of which will lie at the heart of our solution. Finally we formally define two-party authenticated data structures. Before we proceed, we give the definition of *negligible* functions. If t denotes the security parameter, then we have the following:

Definition 1. *We say that a real-valued function $\nu(t)$ over natural numbers is $\mathsf{neg}(t)$ if for any nonzero polynomial p, there exists m such that $\forall n > m$, $|\nu(n)| < \frac{1}{p(n)}$.*

2.1   Multilinear Forms

Let t be the security parameter. Let now $\mathbb{G}$, $\mathbb{G_T}$ be two cyclic groups of prime order p and g be a generator of $\mathbb{G}$. We let the bit-size of p (the order of both $\mathbb{G}$ and $\mathbb{G_T}$) to be a polynomial of the security parameter t, i.e., $\log p = O(\mathrm{poly}(t))$. In our context, any polynomial in the security parameter is regarded as a constant, since the main dimension of our problem is the size of the authenticated data structure, n. We are now ready to define a k-multilinear form. The definition is similar to the one presented in [10]:

Definition 2. *We say that a map $e : \mathbb{G}^k \to \mathbb{G_T}$ is a k-multilinear form if it satisfies the following properties:*

1. *$\mathbb{G}$ and $\mathbb{G_T}$ are cyclic groups of the same prime order p;*
2. *For all $a_1, a_2, \ldots, a_k \in \mathbb{Z}_p^*$ and $x_1, x_2, \ldots, x_k \in \mathbb{G}$ it is*

$$e(x_1^{a_1}, x_2^{a_2}, \ldots, x_k^{a_k}) = e(x_1, x_2, \ldots, x_k)^{a_1 a_2 \cdots a_k} \in \mathbb{G_T};$$

3. *The map is non-degenerate: If $g \in \mathbb{G}$ generates $\mathbb{G}$ then $e(g, g, \ldots, g) \in \mathbb{G_T}$ generates $\mathbb{G_T}$.*

Note now that since both $\mathbb{G}$ and $\mathbb{G_T}$ are cyclic groups of prime order p the discrete logarithm problem is hard for both $\mathbb{G}$ and $\mathbb{G_T}$. Following we continue with the definition of an *admissible k-mulitlinear form*, that is going to be useful in our context:

Definition 3. *We say that a k-multilinear form $e : \mathbb{G}^k \to \mathbb{G_T}$ is admissible if it satisfies the following properties:*

1. *The bit-size of the elements in $\mathbb{G}$ and $\mathbb{G_T}$ is independent of k, i.e., it is $O(\mathrm{poly}(t)) = O(1)$;*
2. *Group operations (exponentiation, multiplication, inversion) in $\mathbb{G}$ and $\mathbb{G_T}$ are independent of k, i.e., they take time $O(\mathrm{poly}(t)) = O(1)$;*
3. *The multilinear form computation $e(x_1, x_2, \ldots, x_k)$ for $x_1, x_2, \ldots, x_k \in \mathbb{G}$ takes time $O(\mathrm{poly}(t)k) = O(k)$;*
4. *The discrete logarithm problem is hard both in $\mathbb{G}$ and $\mathbb{G_T}$.*

We call the groups $\mathbb{G}$ and $\mathbb{G_T}$ for which there exists an admissible multilinear form *admissible multilinear groups*. Finally we define the admissible multilinear form generator:

Definition 4. *An admissible multilinear form generator* $\mathcal{G}(1^t, k)$ *is an algorithm that runs in polynomial time and for any k, outputs a description of admissible multilinear groups $\mathbb{G}$ and $\mathbb{G}_T$ (along with algorithms for group operations) and the admissible k-multilinear form $e : \mathbb{G}^k \to \mathbb{G}_T$.*

2.2 The Bilinear-Map Accumulator

Let $\mathbb{G}$ be a cyclic group of prime order p that has generator g. The bilinear map accumulator [32] is an efficient way to provide short proofs of membership for elements that belong to a set. The bilinear-map accumulator works as follows. It accumulates a set of elements $\mathcal{X}$ in $\mathbb{Z}_p^*$ and the accumulation value $\mathrm{acc}(\mathcal{X})$ is an element in $\mathbb{G}$. Given a set of n elements $\mathcal{X} = \{x_1, x_2, \ldots, x_n\}$ the accumulation value $\mathrm{acc}(\mathcal{X})$ is defined as

$$\mathrm{acc}(\mathcal{X}) = g^{(x_1+s)(x_2+s)\cdots(x_n+s)} \in \mathbb{G},$$

where $s \in \mathbb{Z}_p^*$ is a randomly chosen value that constitutes the trapdoor in the scheme, i.e., the security of the scheme is based on the fact that s is kept secret from the adversary. The proof of membership for an element x_i that belongs to set $\mathcal{X}$ will be the witness

$$\mathcal{W}_{x_i} = g^{\prod_{x_j \in \mathcal{X} : x_j \neq x_i}(x_j+s)} \in \mathbb{G}.$$

Accordingly, a verifier can test set membership for x_i by testing the relationship

$$\mathcal{W}_{x_i}^{(x_i+s)} \overset{?}{=} \mathrm{acc}(\mathcal{X}). \tag{1}$$

If (1) holds, then a verifier outputs "accept", else he outputs "reject". For a proof of non-membership, instead of explicitly storing the elements $x_{j_1} < x_{j_2} < \ldots < x_{j_n}$, the respective *hashed intervals* $h(-\infty, x_{j_1}), h(x_{j_1}, x_{j_2}), \ldots, h(x_{j_n}, +\infty)$ can be stored: Then a proof of non-membership for element x is the proof of membership of the hashed interval $h(x_{j_k}, x_{j_{k+1}})$ such that $x_{j_k} < x < x_{j_{k+1}}$. Alternatively (and less efficiently), one can explicitly compute non-membership witnesses, as described in [3,16]. In the following we present the computational assumption on which the security of the bilinear-map accumulator is based, the q-strong Diffie-Hellman assumption. This assumption was introduced in [7] and holds over general cyclic prime-order groups:

Definition 5 (q-**strong Diffie-Hellman assumption**). *Let $\mathbb{G}$ be a cyclic group of prime order p, generated by an element $g \in \mathbb{G}$. Given the elements of $\mathbb{G}$ $g, g^s, g^{s^2}, \ldots, g^{s^q}$ for some s chosen at random from $\mathbb{Z}_p^*$, the probability that a computationally bounded adversary Adv finds $c \in \mathbb{Z}_p^*$ and outputs*

$$g^{\frac{1}{s+c}} \in \mathbb{G}$$

is $\mathrm{neg}(t)$, *where the probability is taken over the random choices of $s \in \mathbb{Z}_p^*$.*

The security proof argument [32] for the accumulator that has just been presented is as follows: The adversary is given the elements $g, g^s, g^{s^2}, \ldots, g^{s^q} \in \mathbb{G}$. Then, given a set of elements $\mathcal{X} = \{x_1, x_2, \ldots, x_n\}$ and the respective accumulation value $\mathrm{acc}(\mathcal{X})$, we can prove that if a computationally bounded adversary has an algorithm to find a witness $\mathcal{W}_{x_i}$ that passes the verification test of Equation 1 for an element $x_i \notin \mathcal{X}$, then the adversary can break the q-strong Diffie-Hellman assumption (see [32] for the proof).

2.3   Two-Party Authenticated Data Structures

In the two-party authenticated data structures model, a *client* fully *outsources* the data structure to an untrusted *server*, keeping locally only the data structure digest (e.g., digest of the Merkle tree). The digest (or succinct state) d of an authenticated data structure is computed by augmenting the data structure with an *authentication structure* that uses a certain cryptographic primitive. Every solution for an authenticated data structure needs to define d. Computed on the correct data, d will serve as a secure data structure description subject to which the answer to a data structure query will be verified at the client by means of a corresponding proof. The digest d, for a given authenticated data structure (e.g., for the well-known Merkle trees [29], the digest is the cryptographic hash of the root of the tree), should have the following, fundamental property: If t is the security parameter, a computationally bounded adversary should not be able to find two different instances of the authenticated data structure (say for example two different graphs) that have the same digest with probability more than 2^{-t}. In order for the client to verify a query on a data structure, he needs to make sure that the digest d he possesses is the correct one (i.e., it corresponds to the most fresh version of the data structure). Therefore the client has to make sure it is consistent with the history of updates.

We now continue with the security definition of a two-party authenticated data structure: Suppose D_h is the data structure (at time h) we wish to authenticate and let t be the security parameter. A two-party authenticated data structure is a collection of the following algorithms (we assume that the algorithms run by the client have always access to the secret key sk):

1. $\{\text{sk}, \text{pk}\} \leftarrow \text{genkey}(1^t)$. It outputs the secret (known only to the client) and public key (information known to the adversary) on input the security parameter. This procedure is executed by the client;
2. $\{\text{auth}(D_0), d_0\} \leftarrow \text{setup}(D_0, \text{pk})$: This algorithm outputs the authenticated data structure $\text{auth}(D_0)$, and the respective digest d_0. This procedure is executed by either the client or the server;
3. $\{\Pi(o), \alpha(o), D_{h+1}, \text{auth}(D_{h+1}), d_{h+1}\} \leftarrow \text{operate}(o, D_h, \text{auth}(D_h), d_h)$, where $\Pi(o)$ is the proof returned (to the client) concerning an operation o (issued by the client) and $\alpha(o)$ is the answer to the operation o. We distinguish two cases:
   - If o is a query, $\Pi(o)$ is called **verification proof** and $\alpha(o)$ is the answer to the query o;
   - If o is an update, $\Pi(o)$ is called **consistency proof** and $\alpha(o)$ is the "answer" to the update, i.e., the portion of the data structure that has changed due to the update.
   The quantities $D_{h+1}, d_{h+1}, \text{auth}(D_{h+1})$ take values only in the case of updates (i.e., when the data structure changes). This procedure is executed by the server;
4. $\{\text{accept}, \text{reject}, d_{h+1}\} \leftarrow \text{verify}(o, \alpha(o), \Pi(o), d_h, \text{sk})$, where d_h is the current digest of D_h. If $\Pi(o)$ is a verification proof, then it outputs either accept or reject. If $\Pi(o)$ is a consistency proof it outputs either accept or reject and also outputs the new digest d_{h+1}. We say that the client *accepts* the update (see security definition) if this algorithm outputs "accept". Note that this method does not have access to the whole data structure but only to the proof $\Pi(o)$ and is run by the client;

5. $\{accept, reject\} = check(q, \alpha(q), D_h)$. This method decides whether $\alpha(q)$ is a right answer for query q on data structure D_h. Note that here q is an actual query, i.e., an operation that does not change the state of the data structure.

We can now state the formal security definition:

Definition 6 (Security). *Let t be the security parameter and* Adv *be a computationally-bounded adversary that is given the public key* pk, *output by* keygen(). *The adversary chooses the initial state of our data structure D_0 and the client computes the respective digest d_0. The adversary* Adv *is given access to D_0 and d_0. For $i = 0, \ldots, h - 1 = $ poly(t) the adversary* Adv *issues an update u_i in the data structure D_i and computes D_{i+1} and d_{i+1}. The client accepts updates u_i, for all $i = 0, \ldots, h - 1$, by running algorithm* verify(). *At the end of this game of polynomially-many rounds, the adversary* Adv *enters the attack stage where he chooses a query q and computes an answer $\alpha(q)$ and a verification proof $\Pi(q)$. We say that the authenticated data structure is secure if for any computationally-bounded adversary* Adv, *for any query q and for any series of updates it is*

$$
\Pr \left[\begin{array}{l} \{q, \Pi(q), \alpha(q)\} \leftarrow \mathsf{Adv}(1^t, \mathsf{pk}); \\ accept \leftarrow \mathsf{verify}(q, \alpha(q), \Pi(q), d_h, \mathsf{sk}); \\ reject = \mathsf{check}(q, \alpha(q), D_h); \\ \mathsf{digest}(D_h) = d_h. \end{array} \right] \leq \nu(t),
$$

where $\nu(t)$ is negligible in the security parameter t.

We note here that the authenticated data structures model is different (achieving stronger guarantees) than the *verifiable computation model* [1,14,19,23]. Important properties such as efficient updates of queried data, and unlimited queries—as opposed to one-time queries or many queries admitting well-formed (verifying) answers—are all supported in the authenticated data structures model.

2.4 Optimality in Authenticated Data Structures

In this paper we are concerned with the notion of optimality of authenticated data structures, and we indeed present one such construction (i.e., an optimal *authenticated dictionary*) that is based on *multilinear forms*. But what does it mean for an authenticated data structure to be "optimal"?

Let D_h be a plain (non-authenticated) data structure, designed for efficiently answering some type of query. Denote with $|D_h|$ the space needed by D_h and with $\{\alpha(o), D_{h+1}\} \leftarrow \mathsf{OPERATE}(o, D_h)$ the following procedure:

- If o is a query, then OPERATE(o, D_h) is the algorithm that produces the answer $\alpha(o)$ to the query o. In this case $D_{h+1} = D_h$;
- If o is an update, then OPERATE(o, D_h) is the algorithm that executes the update. The answer $\alpha(o)$ is defined in a similar way as before as the "answer" to the update, i.e., the portion of the data structure that has changed due to the update. For example, in a red-black tree data structure, the size of the "update" answer can be $O(\log n)$, since information along a logarithmic-sized path can change.

We are now ready to define an *optimal* authenticated data structure:

Definition 7. *Let D_h be a data structure and let $\mathsf{auth}(D_h)$ be a respective authenticated data structure, along with algorithms $\{\mathsf{genkey}(), \mathsf{setup}(), \mathsf{operate}(), \mathsf{verify}()\}$, as defined in Section 2.3. We say that $\mathsf{auth}(D_h)$ is optimal if and only if*

1. *The authenticated data structure $\mathsf{auth}(D_h)$ is secure according to Definition 6;*
2. *It is $|\mathsf{auth}(D_h)| = O(|D_h|)$;*
3. *For both queries and updates o, the asymptotic time complexity of the algorithm $\{\Pi(o), \alpha(o), D_{h+1}, \mathsf{auth}(D_{h+1}), d_{h+1}\} \leftarrow \mathsf{operate}(o, D_h, \mathsf{auth}(D_h), d_h)$ is no more than the asymptotic time complexity of the respective plain data structure algorithm $\{\alpha(o), D_{h+1}\} \leftarrow \mathsf{OPERATE}(o, D_h);$*
4. *For both queries and updates o it holds:*
   (a) *The size of the proof is asymptotically no more than the size of the answer, i.e., $|\Pi(o)| = O(|\alpha(o)|);$*
   (b) *The asymptotic time complexity of the algorithm $\{\mathsf{accept}, \mathsf{reject}, d_{h+1}\} \leftarrow \mathsf{verify}(o, \alpha(o), \Pi(o), d_h, \mathsf{sk})$ is $O(|\alpha(o)|).$*

Note that Property 4 requires that both the size of the proof for a certain operation returned by $\mathsf{auth}(D_h)$ and the time to verify it are no more (asymptotically) than the size of the answer (computed by D_h) to the respective operation. This property greatly relates to recent work that has appeared on *super-efficient* authenticated data structures [22], where range search queries with proofs asymptotically *less* than the size of the answer are authenticated, and *operation-sensitive* authenticated data structures [37], where fundamental set operations with proofs asymptotically equal to the size of the answer are authenticated. However, none of the authenticated data structures in [22,37] is optimal, due to increased update costs.

The above definition is rather intuitive. It implies that, in order for an authenticated data structure to be *optimal*, it should not be adding any *extra* asymptotic overhead to the *plain* (non-authenticated) data structure. So far in the literature (see Table 1), and specifically for the *authenticated dictionary* problem, no *optimal* authenticated data structure has been constructed. For example, traditional hash-based methods built with Merkle trees (e.g., [6,27,30,33]) fail to achieve Property 4 from Definition 7. Indeed, although the answer is of constant size (i.e., either "yes, the element is contained" or "no, the element is not contained"), the proof for that answer is asymptotically larger than $O(1)$, i.e., it is $O(\log n)$. To achieve this property, other solutions (e.g., [32,35]) have used accumulators. However, accumulator-based solutions, although succeed in satisfying Property 4, violate Property 3 from Definition 7, since update or query time is not logarithmic (e.g., it is $O(\sqrt{n})$). In this paper, we show how to construct the first optimal authenticated dictionary, using a cryptographic primitive the construction of which is, however, still an open problem.

3   Multilinear Form Authenticated Structures

In this section we present a two-party authenticated dictionary based on multilinear forms. We begin with the main building block, the *multilinear form* accumulator.

3.1   A Multilinear Form Accumulator

Let t be the security parameter, $\mathcal{X} = \{x_1, x_2, \ldots, x_n\}$ be a set of elements and $\mathbb{G}$, $\mathbb{G}_T$ be two cyclic groups of prime order p for which there exists an admissible multilinear form $e : \mathbb{G}^t \rightarrow \mathbb{G}_T$ as defined in Definition 3. Note that we require the number of the inputs of the multilinear form to be equal to the security parameter t. Group $\mathbb{G}$ is generated by g and group $\mathbb{G}_T$ is generated by $e(g, g, \ldots, g)$. As in Section 2.2, we can define a new accumulator that is similar to the bilinear-map accumulator with the difference that the base of exponentiation is the generator $e(g, g, \ldots, g)$ of the cyclic group $\mathbb{G}_T$, i.e.,

$$\mathsf{acc}(\mathcal{X}) = e(g, g, \ldots, g)^{(x_1+s)(x_2+s)\ldots(x_n+s)} \in \mathbb{G}_T, \tag{2}$$

where s is a randomly chosen element of $\mathbb{Z}_p^*$. The proof of membership for an element x_i that belongs to set $\mathcal{X}$ will be the witness

$$\mathcal{W}_{x_i} = e(g, g, \ldots, g)^{\prod_{x_j \in \mathcal{X} : x_j \neq x_i}(x_j+s)} \in \mathbb{G}_T. \tag{3}$$

Accordingly, a verifier can test set membership for x_i by computing $\mathcal{W}_{x_i}^{(x_i+s)}$ and checking that this equals $\mathsf{acc}(\mathcal{X})$. The q-strong Diffie-Hellmann assumption can be adjusted to the multilinear form setting, leading to the q-strong multilinear Diffie-Hellmann assumption (see q-SMDH in Table 1), as follows:

Definition 8 (q-strong multilinear Diffie-Hellman assumption). *Let* $\mathbb{G}$, $\mathbb{G}_T$ *be cyclic groups of prime order p such that there exists an admissible multilinear form $e : \mathbb{G}^t \rightarrow \mathbb{G}_T$. Let g be the generator of $\mathbb{G}$. Given the elements of $\mathbb{G}$ $g, g^s, g^{s^2}, \ldots, g^{s^q}$ for some s chosen at random from $\mathbb{Z}_p^*$, the probability that a computationally bounded adversary Adv finds $c \in \mathbb{Z}_p^*$ and outputs*

$$e(g, g, \ldots, g)^{\frac{1}{c+s}} \in \mathbb{G}_T$$

is $\mathsf{neg}(t)$*, where the probability is taken over the random choices of $s \in \mathbb{Z}_p^*$.*

We now prove security (similar to [32]) of the multilinear form accumulator based on the q-strong multilinear Diffie-Hellman assumption:

Lemma 1. *Let* $\mathcal{X} = \{x_1, x_2, \ldots, x_n\}$*, t be the security parameter, $\mathbb{G}$, $\mathbb{G}_T$ be cyclic groups of prime order p such that there exists an admissible multilinear form $e : \mathbb{G}^t \rightarrow \mathbb{G}_T$ and g be the generator of $\mathbb{G}$. Under the q-strong multilinear Diffie-Hellman assumption, the probability that a computationally bounded adversary Adv that is given the elements $g, g^s, g^{s^2}, \ldots, g^{s^q} \in \mathbb{G}$ can find a valid witness $\mathcal{W}_x$ for an element $x \notin \mathcal{X}$ is* $\mathsf{neg}(t)$*.*

Proof. By Equation 3 a computationally bounded adversary Adv finds a witness $\mathcal{W}_x$ such that $\mathcal{W}_x^{(x+s)} = e(g, g, \ldots, g)^{(x_1+s)(x_2+s)\ldots(x_n+s)}$. Since $x \notin \{x_1, x_2, \ldots, x_n\}$ we can write $(x_1+s)(x_2+s)\ldots(x_n+s) = \mathcal{P}(x+s) + \lambda$, where the coefficients of polynomial $\mathcal{P}$ and quantity λ are computable in polynomial time in n (polynomial division). Therefore Adv can compute $e(g, g, \ldots, g)^{(x+s)^{-1}} = [\mathcal{W}_x[e(g, g, \ldots, g)^{\mathcal{P}}]^{-1}]^{\lambda^{-1}}$, since $e(g, g, \ldots, g)^{s^i} \in \mathbb{G}_T$ can efficiently be computed from $g^{s^i} \in \mathbb{G}$ by using the admissible multilinear form $e : \mathbb{G}^k \rightarrow \mathbb{G}_T$ for all $i = 0, \ldots, q$. This breaks the q-strong multilinear strong Diffie-Hellman assumption. $\qquad\square$

3.2   A Two-Party Multilinear Form Authenticated Dictionary Construction

In this section we describe a two-party authenticated dictionary based on admissible multilinear forms that achieves constant communication complexity and constant verification complexity. Let $\mathcal{X} = \{x_1, x_2, \ldots, x_n\}$ be the elements contained in the dictionary, where $x_1 < x_2 < \ldots < x_n$. The actual set we are going to store, in order to also support efficient range search and non-membership queries, is the set of intervals, i.e., the set $\mathcal{A} = \{a_1, a_2, \ldots, a_{n-1}\}$, where $a_i = x_i \| x_{i+1}$ is simply the concatenation of the binary strings x_i, x_{i+1} of bit length $2t$ (we use t bits for each x_i).

In our construction we use a red-black tree, with data at the leaves (i.e., internal nodes data navigates the searches and does not correspond to actual data) [15]. We store key-value pairs at the leaves and the ordering is according to the keys. For an interval $a_i = x_i \| x_{i+1}$, the key is x_i and the value is x_{i+1}, where x_i and x_{i+1} are successive elements of our set. We recall that a red-black tree implementation of a dictionary supports operations in $O(\log n)$ time in the *worst* case.

Define now k, the number of the inputs to the admissible multilinear form that we are going to use to be t, i.e., equal to the security parameter. Note that since we are in the computational model, it is always the case that $t > \log n$, where n is the total number of elements that we store in the dictionary. In the construction that follows we refer to operations on intervals $a_i = x_i \| x_{i+1}$ (i.e., insert an interval $x_i \| x_{i+1}$), instead of explicit elements x_i and then show in the proof of Theorem 1 how, by supporting operations for intervals, we can support operations for distinct elements. We define all four algorithms as described in Section 2.3:

1. $\{\mathsf{sk}, \mathsf{pk}\} \leftarrow \mathsf{genkey}(1^t)$. The secret sk is the trapdoor $s \in \mathbb{Z}_p^*$, which is picked randomly. Procedure $\mathsf{genkey}(1^t)$ also calls $\mathcal{G}(1^t, t)$ from Definition 4 and outputs the public key pk which is the description of the groups $\mathbb{G}$ and $\mathbb{G}_T$, the admissible multilinear form $e : \mathbb{G}^t \rightarrow \mathbb{G}_T$ and the elements $g, g^s, \ldots, g^{s^q} \in \mathbb{G}$. It also outputs a description of a collision-resistant hash function h that takes three inputs and outputs a hash of t bits. We recall that the number of the inputs of the admissible multilinear form that we are using is t, equal to the security parameter.

2. $\{\mathsf{auth}(D_0), d_0\} \leftarrow \mathsf{setup}(D_0, \mathsf{pk})$. The digest d_0 of the authenticated data structure is defined as the tuple $\{\mathsf{acc}(\mathcal{A}), \mathsf{hash}(\mathcal{A})\}$. It is

$$\mathsf{acc}(\mathcal{A}) = e(g, g, \ldots, g)^{(a_1+s)(a_2+s)\ldots(a_{n-1}+s)},$$

while $\mathsf{hash}(\mathcal{A})$ is computed as the digest of the well-known Merkle tree (description follows). Both are stored locally by the client and used for verification and updates. Let now T be the red-black tree built on top of the intervals $a_i = x_i \| x_{i+1}$ for $i = 1, \ldots, n-1$ and where $x_1 < x_2 < \ldots < x_n$. Let $v_1, v_2, \ldots, v_{n-1}$ be the leaves of the tree, storing the intervals $a_1, a_2, \ldots, a_{n-1}$ respectively. We define the *label* and the *hash* of v_i as

$$\mathsf{label}(v_i) = g^{a_i+s} \in \mathbb{G}, \tag{4}$$

$$\mathsf{hash}(v_i) = h(\mathsf{null}, \mathsf{label}(v_i), \mathsf{null}), \tag{5}$$

where h is the collision-resistant hash function. Let now v_A be the internal node of T that is the root of the subtree of T that contains the elements of some $A \subseteq \mathcal{A}$

(i.e., from v_A you can reach elements in A by following the two downward paths). Then

$$\text{label}(v_A) = g^{\prod_{a \in A}(a+s)} \in \mathbb{G}, \tag{6}$$

$$\text{hash}(v_A) = h(\text{hash}(\text{lchild}(v_A)), \text{label}(v_A), \text{hash}(\text{rchild}(v_A))), \tag{7}$$

where $\text{lchild}(v)$ and $\text{rchild}(v)$ are the left and the right child of node v in the tree. Note that all the labels of the internal nodes of T can be computed in polynomial time in n without the use of the trapdoor s, and only by using the public key. Also, as we will see later, in sake of maintaining constant verification and communication complexity, the hashes $\text{hash}(.)$ are used only for updates. Finally, $\text{hash}(\mathcal{A})$ is defined to be the $\text{hash}(.)$ value of the root of the tree T, recursively computed by means of the above equations.

3. $\{\Pi(o), \alpha(o), D_{h+1}, \text{auth}(D_{h+1}), d_{h+1}\} \leftarrow \text{operate}(o, D_h, \text{auth}(D_h), d_h)$.
   (a) **Verification proof case.** Suppose o is a query for the interval a_j, stored at node v_j. Let $\pi(a_j) = v_j, v_{j1}, v_{j2}, \ldots, v_{jl}$ be the path of T from node v_j that refers to interval a_j to the child of the root of T, where $l = O(\log n)$. The verification proof $\Pi(a_j)$ is defined as

$$e(\text{label}(\text{sib}(v_j)), \text{label}(\text{sib}(v_{j1})), \ldots, \text{label}(\text{sib}(v_{jl})), g, \ldots, g) \in \mathbb{G}_T, \tag{8}$$

where $\text{sib}(v)$ defines the sibling of node v in the red-black tree T, and function $\text{label}()$ is defined in Equations 4 and 6. The answer $\alpha(o)$ will be the element a_j. Note now that

$$\Pi(a_j) = \mathcal{W}_{a_j} = e(g, g, \ldots, g)^{\prod_{a \in A: a \neq a_j}(a+s)} \in \mathbb{G}_T,$$

as required by Equation 3. Moreover, it is computable in time $O(\log n)$ since we have to collect $O(\log n)$ labels and feed them as input in the admissible multilinear form $e()$. Note that the remaining inputs of the admissible multilinear form (i.e., $t - \log n$) are set equal to g, the generator of group $\mathbb{G}$. Moreover the size of the witness is $O(1)$ (only one group element of $\mathbb{G}_T$). The presented method is the first one to construct a witness for an accumulator in logarithmic time. The straightforward method takes linear time;
   (b) **Consistency proof case.** Suppose o is an update (either an insertion or a deletion) of an interval a_j. In accordance with path $\pi(a_j)$ in the verification proof, whenever there is an update, let $\pi(a_j)$ be the portion of the red-black tree (that also contains structure) that is accessed (and eventually changes) due to the update of the interval a_j. Denote with $\Pi_1(a_j)$ the set of those labels $\text{label}(v)$ such that $v \in \pi(a_j)$ and with $\Pi_2(a_j)$ the set of those hashes $\text{hash}(v)$ such that $v \in \pi(a_j)$. The consistency proof is denoted with $\Pi(a_j) = \{\Pi_1(a_j), \Pi_2(a_j)\}$. We finally note that the time for the construction of the consistency proof is $O(\log n)$ and its size is also $O(\log n)$, since a red-black tree insertion or deletion takes $O(\log n)$ time in the worst case.
   We note here that, although this algorithm outputs D_{h+1}, it does not output $\text{auth}(D_{h+1})$. The *updated* authenticated data structure (i.e., the new $O(\log n)$-sized portion of it) will be sent by the client to the server during the execution

of the verify() algorithm. This is done primarily for efficiency reasons, since auth(D_{h+1}) could have been computed by the server anyways, by using the public key, but in a less efficient way[5].

4. {accept, reject, d_{h+1}} $\leftarrow$ verify($o, \alpha(o), \Pi(o), d_h, $sk).

(a) **Verification proof case.** If $\Pi(o)$ is a verification proof, i.e., $\Pi(o) = \Pi(a_j)$ (Equation 8), then clearly the procedure outputs "accept" (see Lemma 1) if and only if

$$\Pi(a_j)^{a_j+s} = \text{acc}(\mathcal{A}),$$

else it outputs "reject". The digest of the structure remains the same. Note that the client does not use the multilinear form for verification since he has access to the trapdoor s. The verification involves only one exponentiation in the group $\mathbb{G}_T$ and therefore takes time $O(1)$.

(b) **Consistency proof case.** If $\Pi(o)$ is a consistency proof, then the client has to update both acc($\mathcal{A}$) and hash($\mathcal{A}$) to acc($\mathcal{A}'$) and hash($\mathcal{A}'$) respectively. The digest acc($\mathcal{A}$) is updated in constant time by setting acc($\mathcal{A}'$) = acc($\mathcal{A}$)$^{a_j+s}$ if a_j is inserted and acc($\mathcal{A}'$) = acc($\mathcal{A}$)$^{(a_j+s)^{-1}}$ if a_j is deleted. For the update of hash($\mathcal{A}$) to hash($\mathcal{A}'$), the client performs the following step:

i. Initially he verifies the correctness of the labels in $\Pi_1(a_j)$ by recomputing the digest hash($\mathcal{A}$) by means of elements in $\Pi_1(a_j)$ and $\Pi_2(a_j)$. If this computation succeeds (Merkle tree verification) then the client is assured with probability $1 - \text{neg}(t)$, due to collision resistance, that the labels in $\Pi_1(a_j)$ belong to the correct portion of red-black tree T before the certain update, i.e., they belong to the portion that is accessed due to this update;

If the above test succeeds then the procedure outputs "accept", else it outputs "reject". If it accepts, with probability $1 - \text{neg}(t)$, $\Pi_1(a_j)$ is the set of labels that is accessed during the update and needs to be updated. Each label in this set can be updated in constant time, since the trapdoor s is known by the client. While the labels are updated, the new hashes are also computed and finally hash($\mathcal{A}'$) is updated. Basically the client performs a red-black tree insertion/deletion locally, doing the necessary rotations, updating at the same time the information label() and hash(). We conclude that hash($\mathcal{A}'$) is the updated digest with probability $1 - \text{neg}(t)$, since it is a function of $\Pi_1(a_j)$. After the procedure finishes, the client sends the updated labels, hashes and the updated digests (acc($\mathcal{A}'$) and hash($\mathcal{A}'$)) back to the server (i.e., the new portion of the authenticated data structure auth(D_{h+1})): The server therefore only has to perform a logarithmic time writing of the new information to the data structure. Note that processing the consistency proof takes logarithmic time. Similar solutions for two-party authenticated dictionaries using only collision-resistant hashing have been explored in the literature (e.g., [33]).

Observations. Before presenting the main result of this section we make an important observation. The hashing structure on top of the red-black tree (i.e., the additional hash() label in the construction) is used only for efficiency reasons and not for security

---

[5] It is an open problem—even with the use of multilinear forms—to construct an optimal authenticated dictionary that avoids this interaction.

reasons. Namely, if we had not used the hashing structure, the untrusted server, having access to the public information $g^s, g^{s^2}, \ldots, g^{s^q}$, could update all the labels $\mathsf{label}(v)$ for all affected nodes v. However, this would take $O(n \log n)$ time ($O(n)$ time for each one of the $O(\log n)$ nodes of the path) by using well-known methods by means of Vieta's formulas (see for example [36]). In this case however, the update time at the client would be constant, since all needed would be one exponentiation for the update of $\mathsf{acc}(\mathcal{A})$.

However the labels of the affected nodes v, can easily be updated (in $O(1)$ time per node) by knowing the trapdoor s, something that only the client has access to. Therefore, with the hashing structure, we authenticate the "affected paths" so that the client can verify which labels are affected. Then the client efficiently performs the updates locally, and sends back the new values to be used in the future. This does not violate security since the new information provided to the server by the client is computable by the server in polynomial time anyways.

3.3 Main Results

We now present the main results of this section.

Theorem 1. *Assume the existence of an admissible multilinear form generator and collision-resistant hash functions. Then there exists an optimal two-party authenticated dictionary storing n elements with the following properties:*

1. *It is secure under the q-strong multilinear Diffie-Hellman assumption and according to Definition 6;*
2. *The size of the verification proof is $O(1)$ both for a (non-)membership query and for a range search query of ℓ elements;[6]*
3. *The query time at the server is $O(\log n)$ for a (non-)membership query and $O(\log n + \ell)$ for a range search query of ℓ elements;*
4. *The verification time at the client is $O(1)$ for a (non-)membership query and $O(\ell)$ for a range search query of ℓ elements;*
5. *The size of the consistency proof is $O(\log n)$;*
6. *The update time at the server and at the client is $O(\log n)$;*
7. *The server uses $O(n)$ space;*
8. *The client uses $O(1)$ space.*

Proof. **(Security)** (1) We prove security according to Definition 6. Given the security parameter t, the client runs algorithm $\{\mathsf{sk}, \mathsf{pk}\} \leftarrow \mathsf{genkey}(1^t)$. Then the adversary picks a data structure D_0, runs $\{\mathsf{auth}(D_0), d_0\} \leftarrow \mathsf{setup}(D_0, \mathsf{pk})$ and produces an empty authenticated data structure. The adversary chooses a polynomial (in t) number of updates (say h) to the data structure and turns it into a data structure D_h, with d_h being the tuple of digests $\mathsf{acc}(\mathcal{A})$ and $\mathsf{hash}(\mathcal{A})$, as defined in the description of the algorithms and where $\mathcal{A}$ is the current set of elements. Let now $\nu(t)$ be a function that is $\mathsf{neg}(t)$. Since the client has accepted all the updates (see security definition), this means, by construction, that $d_h = \{\mathsf{hash}(\mathcal{A}), \mathsf{acc}(\mathcal{A})\}$ is the correct digest of the data structure. Specifically, $\mathsf{hash}(\mathcal{A})$ is correct with probability $1 - \nu(t)$ (by collision resistance) and $\mathsf{acc}(\mathcal{A})$

---

[6] Note that *super-efficiency* [22] is achieved for range search queries besides optimality.

is correct with probability 1 (since $\text{acc}(\mathcal{A})$ is updated with only one exponentiation at every update). Therefore, if D_h is the data structure after the update phase and d_h the tuple of the updated digests, we have $\Pr[\text{digest}(D_h) = d_h] = (1 - \nu(t)) \times 1 = 1 - \nu(t)$. Therefore, the probability of Definition 6 is written

$$\Pr \begin{bmatrix} \{q, \Pi(q), \alpha(q)\} \leftarrow \text{Adv}(1^t, \text{pk}); \\ \text{accept} \leftarrow \text{verify}(q, \alpha(q), \Pi(q), d_h); \\ \text{reject} = \text{check}(q, \alpha(q), D_h); \\ \text{digest}(D_h) = d_h. \end{bmatrix} =$$

$$\Pr \begin{bmatrix} \{q, \Pi(q), \alpha(q)\} \leftarrow \text{Adv}(1^t, \text{pk}); \\ \text{accept} \leftarrow \text{verify}(q, \alpha(q), \Pi(q), \text{digest}(D_h)); \\ \text{reject} = \text{check}(q, \alpha(q), D_h). \end{bmatrix} \times \Pr[\text{digest}(D_h) = d_h] =$$

$$\Pr \begin{bmatrix} \{a, \Pi(a), x\} \leftarrow \text{Adv}(1^t, \text{pk}); \\ \Pi(a)^{a+s} = e(g, g, \ldots, g)^{(a_1+s)(a_2+s)\ldots(a_n+s)}; \\ a \notin \{a_1, a_2, \ldots, a_n\}. \end{bmatrix} \times (1 - \nu(t)).$$

The first term, by Lemma 1 is $\text{neg}(t)$. Therefore the whole probability is $\text{neg}(t) \times (1 - \nu(t))$, which is $\text{neg}(t)$.

(**Complexity**) (2-4) First of all we show equivalence of (non-)membership proofs of elements with membership proofs of intervals, that are actually stored in the authenticated data structure: A (non-)membership proof for element x is equivalent with a membership proof of the interval $a_i = x_i || x_{i+1}$ such that $x_i \leq x \leq x_{i+1}$ (note that for non-membership proofs it is $x_i < x < x_{i+1}$). Additionally, by Equation 8, we have that a verification proof for an interval is only one element of $\mathbb{G}_T$ and is computed by applying the admissible multilinear form $e()$. Therefore the size of the verification proof for an element is $O(1)$ and the time to compute it is $O(\log n)$, since $O(\log n)$ elements of $\mathbb{G}$ along the red-black tree path have to be collected and then be fed into the admissible multilinear form $e(.)$. The time to verify involves one exponentiation (see verify() algorithm) and therefore is $O(1)$. A range proof of ℓ elements consists of one membership proof of an interval (instead of one element in the exponent, we omit all the elements of the respective interval). Therefore its size is $O(1)$, it can be computed in $O(\log n + \ell)$ time and it can be verified in $O(\ell)$ time.

(5-6) We now show equivalence of elements updates with intervals updates. Suppose the client wants to insert x. Firstly the client verifies the non-membership of x by verifying the membership of the interval $a_i = x_i || x_{i+1}$ such that $x_i < x < x_{i+1}$. After interval a_i has been verified the client issues the following updates with this order: delete(a_i), insert($x_i || x$), insert($x || x_{i+1}$). For deletion of element x, the client first verifies the membership of intervals $x_i || x$ and $x || x_{i+1}$ and then issues the following updates in this order: delete($x_i || x$), delete($x || x_{i+1}$), insert($x_i || x_{i+1}$). Since the cost of these individual updates is $O(\log n)$, we conclude that any update will cost $O(\log n)$ in the worst case.

(7-8) Finally, the extra space needed to store the labels and the hashes on the red-black tree is $O(n)$, while the space needed by the client is $O(1)$, since the client only needs to store $\text{acc}(\mathcal{A})$, $\text{hash}(\mathcal{A})$ and the secret trapdoor, s.

(**Optimality**) The optimal plain dictionary data structure is the red-black tree, achieving worst-case complexities $O(\log n)$ [15] and answers of constant size.

Therefore our authenticated data structure satisfies Definition 7 since it has logarithmic complexities in the worst case, constant verification proof size and constant verification time.   □

We now present the final result of our paper that relates optimality of an authenticated dictionary with the existence of an admissible multilinear form generator.

Theorem 2. *If optimal authenticated dictionaries do not exist, then admissible multilinear form generators do not exist either.*

Proof. Let's assume this is not the case and admissible multilinear form generators do exist in the absence (through a generic lower bound proof—see for example a similar result for memory checking in [18]) of optimal authenticated dictionaries. This is a contradiction since we can use the construction of Theorem 1—which will give us a secure construction since we can use an admissible multilinear form generator for $k = t$—to derive an optimal authenticated dictionary.   □

Finally we need to make the following important observation. Theorem 2 does not exclude the existence of some instance of a multilinear form, even in the absence of optimal authenticated dictionaries (say for example an instance of a multilinear form for $k = 5$). The result holds for all admissible k-multilinear forms, i.e., for the existence of an admissible multilinear form generator, as defined in Definition 4.

4   Conclusions

In this paper, we have presented the first *optimal* authenticated dictionary with constant-size verification proof, constant-time verification, and logarithmic query/update costs. Its design is based on multilinear forms, a recently-proposed cryptographic primitive [10] whose construction remains an open problem to date.

However, since multilinear forms are not known to exist yet, this work can be viewed from a different angle: if one could prove that optimal authenticated dictionaries cannot exist in the computational model, irrespectively of cryptographic primitives, then our result would imply that cryptographically interesting multilinear form generators cannot exist as well (i.e., it can be viewed as a reduction). Thus, we provide an alternative avenue towards proving the nonexistence of multilinear form generators in the context of general lower bounds for authenticated data structures [40] and for memory checking [18].

Acknowledgments

This research was supported in part by the U.S. National Science Foundation under grants CNS–1012060 and CNS–1012798 and by the Center for Geometric Computing and the Kanellakis Fellowship at Brown University. We thank Alice Silverberg for many useful comments and discussions as well as the conference reviewers for providing us with very insightful feedback. This work was performed while the third author was with Boston University. The views in this paper do not necessarily reflect the views of the sponsors.

References

1. Applebaum, B., Ishai, Y., Kushilevitz, E.: From secrecy to soundness: Efficient verification via secure computation. In: Proc. Int. Colloquium on Automata, Languages and Programming (ICALP), pp. 152–163 (2010)
2. Atallah, M.J., Cho, Y., Kundu, A.: Efficient data authentication in an environment of untrusted third-party distributors. In: Proc. Int. Conference on Data Engineering (ICDE), pp. 696–704 (2008)
3. Au, M.H., Tsang, P.P., Susilo, W., Mu, Y.: Dynamic universal accumulators for DDH groups and their application to attribute-based anonymous credential systems. In: Fischlin, M. (ed.) CT-RSA 2009. LNCS, vol. 5473, pp. 295–308. Springer, Heidelberg (2009)
4. Baric, N., Pfitzmann, B.: Collision-free accumulators and fail-stop signature schemes without trees. In: Fumy, W. (ed.) EUROCRYPT 1997. LNCS, vol. 1233, pp. 480–494. Springer, Heidelberg (1997)
5. Benaloh, J., de Mare, M.: One-way accumulators: A decentralized alternative to digital signatures. In: Helleseth, T. (ed.) EUROCRYPT 1993. LNCS, vol. 765, pp. 274–285. Springer, Heidelberg (1993)
6. Blum, M., Evans, W.S., Gemmell, P., Kannan, S., Naor, M.: Checking the correctness of memories. Algorithmica 12(2/3), 225–244 (1994)
7. Boneh, D., Boyen, X.: Short signatures without random oracles and the SDH assumption in bilinear groups. J. Cryptology 21(2), 149–177 (2008)
8. Boneh, D., Franklin, M.K.: Identity-based encryption from the Weil pairing. In: Kilian, J. (ed.) CRYPTO 2001. LNCS, vol. 2139, pp. 213–229. Springer, Heidelberg (2001)
9. Boneh, D., Mironov, I., Shoup, V.: A secure signature scheme from bilinear maps. In: Joye, M. (ed.) CT-RSA 2003. LNCS, vol. 2612, pp. 98–110. Springer, Heidelberg (2003)
10. Boneh, D., Silverberg, A.: Applications of multilinear forms to cryptography. Contemporary Mathematics 324(1), 71–90 (2003)
11. Camenisch, J., Kohlweiss, M., Soriente, C.: An accumulator based on bilinear maps and efficient revocation for anonymous credentials. In: Proc. Public Key Cryptography (PKC), pp. 481–500 (2009)
12. Camenisch, J., Lysyanskaya, A.: Dynamic accumulators and application to efficient revocation of anonymous credentials. In: Yung, M. (ed.) CRYPTO 2002. LNCS, vol. 2442, pp. 61–76. Springer, Heidelberg (2002)
13. Cheon, J.H., Lee, D.H.: A note on self-bilinear maps. Korean Mathematical Society 46(2), 303–309 (2009)
14. Chung, K.-M., Kalai, Y., Vadhan, S.: Improved delegation of computation using fully homomorphic encryption. In: Rabin, T. (ed.) CRYPTO 2010. LNCS, vol. 6223, pp. 483–501. Springer, Heidelberg (2010)
15. Cormen, T.H., Leiserson, C.E., Rivest, R.L., Stein, C.: Introduction to Algorithms, 2nd edn. MIT Press, Cambridge (2001)
16. Damgård, I., Triandopoulos, N.: Supporting non-membership proofs with bilinear-map accumulators. Cryptology ePrint Archive, Report 2008/538 (2008)
17. Devanbu, P., Gertz, M., Kwong, A., Martel, C., Nuckolls, G., Stubblebine, S.: Flexible authentication of XML documents. Journal of Computer Security 6, 841–864 (2004)
18. Dwork, C., Naor, M., Rothblum, G.N., Vaikuntanathan, V.: How efficient can memory checking be? In: Reingold, O. (ed.) TCC 2009. LNCS, vol. 5444, pp. 503–520. Springer, Heidelberg (2009)

19. Gennaro, R., Gentry, C., Parno, B.: Non-interactive verifiable computing: Outsourcing computation to untrusted workers. In: Rabin, T. (ed.) CRYPTO 2010. LNCS, vol. 6223, pp. 465–482. Springer, Heidelberg (2010)
20. Gennaro, R., Halevi, S., Rabin, T.: Secure hash-and-sign signatures without the random oracle. In: Stern, J. (ed.) EUROCRYPT 1999. LNCS, vol. 1592, pp. 123–139. Springer, Heidelberg (1999)
21. Goodrich, M.T., Tamassia, R., Hasic, J.: An efficient dynamic and distributed cryptographic accumulator. In: Chan, A.H., Gligor, V.D. (eds.) ISC 2002. LNCS, vol. 2433, pp. 372–388. Springer, Heidelberg (2002)
22. Goodrich, M.T., Tamassia, R., Triandopoulos, N.: Super-efficient verification of dynamic outsourced databases. In: Malkin, T.G. (ed.) CT-RSA 2008. LNCS, vol. 4964, pp. 407–424. Springer, Heidelberg (2008)
23. Hohenberger, S., Lysyanskaya, A.: How to securely outsource cryptographic computations. In: Kilian, J. (ed.) TCC 2005. LNCS, vol. 3378, pp. 264–282. Springer, Heidelberg (2005)
24. Joux, A.: A one-round protocol for tripartite Diffie-Hellman. J. Cryptology 17(4), 263–276 (2004)
25. Lee, H.-M., Ha, K.J., Ku, K.-M.: ID-based multi-party authenticated key agreement protocols from multilinear forms. In: Zhou, J., López, J., Deng, R.H., Bao, F. (eds.) ISC 2005. LNCS, vol. 3650, pp. 104–117. Springer, Heidelberg (2005)
26. Li, J., Li, N., Xue, R.: Universal accumulators with efficient nonmembership proofs. In: Katz, J., Yung, M. (eds.) ACNS 2007. LNCS, vol. 4521, pp. 253–269. Springer, Heidelberg (2007)
27. Martel, C., Nuckolls, G., Devanbu, P., Gertz, M., Kwong, A., Stubblebine, S.G.: A general model for authenticated data structures. Algorithmica 39(1), 21–41 (2004)
28. Menezes, A., Vanstone, S., Okamoto, T.: Reducing elliptic curve logarithms to logarithms in a finite field. In: Proc. Symposium on Theory of Computing (STOC), pp. 80–89 (1991)
29. Merkle, R.C.: A certified digital signature. In: Brassard, G. (ed.) CRYPTO 1989. LNCS, vol. 435, pp. 218–238. Springer, Heidelberg (1989)
30. Naor, M., Nissim, K.: Certificate revocation and certificate update. In: Proc. USENIX Security Symposium (USENIX), pp. 217–228 (1998)
31. Naor, M., Rothblum, G.N.: The complexity of online memory checking. J. ACM 56(1) (2009)
32. Nguyen, L.: Accumulators from bilinear pairings and applications. In: Menezes, A. (ed.) CT-RSA 2005. LNCS, vol. 3376, pp. 275–292. Springer, Heidelberg (2005)
33. Papamanthou, C., Tamassia, R.: Time and space efficient algorithms for two-party authenticated data structures. In: Qing, S., Imai, H., Wang, G. (eds.) ICICS 2007. LNCS, vol. 4861, pp. 1–15. Springer, Heidelberg (2007)
34. Papamanthou, C., Tamassia, R.: Update-optimal authenticated structures based on lattices. Cryptology ePrint Archive, Report 2010/128 (2010)
35. Papamanthou, C., Tamassia, R., Triandopoulos, N.: Authenticated hash tables. In: Proc. ACM Conference on Computer and Communications Security (CCS), pp. 437–448 (2008)
36. Papamanthou, C., Tamassia, R., Triandopoulos, N.: Cryptographic accumulators for authenticated hash tables. Cryptology ePrint Archive, Report 2009/625 (2009)
37. Papamanthou, C., Tamassia, R., Triandopoulos, N.: Optimal authentication of set operations on dynamic sets. Cryptology ePrint Archive, Report 2010/455 (2010)
38. Sander, T., Ta-Shma, A., Yung, M.: Blind, auditable membership proofs. In: Proc. Financial Cryptography (FC), pp. 53–71 (2001)

39. Tamassia, R.: Authenticated data structures. In: Di Battista, G., Zwick, U. (eds.) ESA 2003. LNCS, vol. 2832, pp. 2–5. Springer, Heidelberg (2003)
40. Tamassia, R., Triandopoulos, N.: Computational bounds on hierarchical data processing with applications to information security. In: Caires, L., Italiano, G.F., Monteiro, L., Palamidessi, C., Yung, M. (eds.) ICALP 2005. LNCS, vol. 3580, pp. 153–165. Springer, Heidelberg (2005)
41. Tamassia, R., Triandopoulos, N.: Efficient content authentication in peer-to-peer networks. In: Katz, J., Yung, M. (eds.) ACNS 2007. LNCS, vol. 4521, pp. 354–372. Springer, Heidelberg (2007)
42. Wang, P., Wang, H., Pieprzyk, J.: A new dynamic accumulator for batch updates. In: Qing, S., Imai, H., Wang, G. (eds.) ICICS 2007. LNCS, vol. 4861, pp. 98–112. Springer, Heidelberg (2007)
43. Yuan, H., Atallah, M.J.: Efficient distributed third-party data authentication for tree hierarchies. In: Proc. Int. Conference on Distributed Computing Systems (ICDCS), pp. 184–193 (2008)

Deterministic Encoding and Hashing
to Odd Hyperelliptic Curves

Pierre-Alain Fouque and Mehdi Tibouchi

École normale supérieure
Département d'informatique, Équipe de cryptographie
45 rue d'Ulm, F-75230 Paris CEDEX 05, France
{pierre-alain.fouque,mehdi.tibouchi}@ens.fr

Abstract. In this paper we propose a very simple and efficient encoding function from $\mathbb{F}_q$ to points of a hyperelliptic curve over $\mathbb{F}_q$ of the form $H\colon y^2 = f(x)$ where f is an odd polynomial. Hyperelliptic curves of this type have been frequently considered in the literature to obtain Jacobians of good order and pairing-friendly curves.

Our new encoding is nearly a bijection to the set of $\mathbb{F}_q$-rational points on H. This makes it easy to construct well-behaved hash functions to the Jacobian J of H, as well as injective maps to $J(\mathbb{F}_q)$ which can be used to encode scalars for such applications as ElGamal encryption.

The new encoding is already interesting in the genus 1 case, where it provides a well-behaved encoding to Joux's supersingular elliptic curves.

Keywords: Hyperelliptic Curve Cryptography, Deterministic Encoding, Hashing.

1 Introduction

Hashing into elliptic and hyperelliptic curves. Many cryptosystems based on discrete log-related hardness assumptions, especially in pairing-based cryptography, involve hashing into a group, usually instantiated as the group of points of an elliptic curve or the Jacobian of a hyperelliptic curve. For example in the Boneh-Franklin IBE scheme [4], the public-key for identity $id \in \{0,1\}^*$ is an element $Q_{id} = H_1(id)$ of the group. This is also the case in many other pairing-based cryptosystems including IBE and HIBE schemes [1,15,17], signature and identity-based signature schemes [3,5,6,10,28] and identity-based signcryption schemes [8,23].

Those cryptosystems are proved to be secure when the hash function is modeled as a random oracle into the group, and it is not obvious how to instantiate such a function in practice (when the group is an elliptic curve or a Jacobian) so that the security proof can go through. As discussed in by Brier *et al.* [9], it is sometimes sufficient to use relatively simple constructions that do not behave like random oracles at all, owing to random self-reducibility properties of the underlying problems, but it is generally desirable to have proper hash functions

M. Joye, A. Miyaji, and A. Otsuka (Eds.): Pairing 2010, LNCS 6487, pp. 265–277, 2010.

that can be plugged into any cryptosystem that requires hashing into elliptic and hyperelliptic curves while not compromising proofs of security in the random oracle model.

Deterministic encodings. The basic building block for constructing such hash functions is an encoding from a set that is easy to enumerate, such as $\{0,1\}^n$ or $\mathbb{F}_q$, into the elliptic or hyperelliptic curve group. If the encoding has suitable properties, combining it with a standard hash function may provide a robust construction for hashing into the group.

Generic encodings, such as $t \mapsto t \cdot G$ where G is a group generator, will not work, since they leak the discrete logarithm (as the hash value in the group is usually obtained as from public data, such as the identity in IBE schemes). Thus, the particular form of the group elements intervenes in the encoding.

In the case of elliptic curves, the classical approach is inherently probabilistic: one will first compute an integer hash value $h(m)$ and add a short counter to get $x = 0^{\log k} \| h(m)$. If x is the abscissa of a point on the elliptic curve $y^2 = x^3 + ax + b$, this gives the desired point; otherwise, one increments the counter and tries again. Each step succeeds with probability about $1/2$, so if k is the security parameter, k steps are heuristically enough to construct a point except with negligible probability.

However, the length of the hash computation depends on the message m, which can lead to side-channel attacks [7], unless all k steps are run for all messages, and Legendre symbols and square roots are computed in constant time, in which case computational cost becomes prohibitive. More importantly for pairing-based cryptography, it is difficult to assess the security of a scheme in which such a "probabilistic" hash function is used, even when the underlying integer hash function h is considered ideal.

Therefore, it has been desirable to devise point construction algorithms on elliptic and hyperelliptic curves that are more robust, easier to analyze, and *deterministic*. Algorithms proposed so far fit in two families:

- SWU-like encodings, similar to those proposed by Shallue and van de Woestijne in [26]. They are based on the construction of explicit rational curves on a surface associated to the target curve.
- Icart-like encodings, similar to Icart's function [18]. They are obtained by writing down a root of the curve equation using radicals of degrees prime to the order of the multiplicative group. This is only possible if the curve equation is solvable.

Hyperelliptic curve encodings. While there are now rather general and efficient constructions for elliptic curves (although some important curves remain intractable with current techniques), encodings to hyperelliptic curves are scarce. The first such encoding was proposed by Ulas in [27], for curves of the form $y^2 = x^n + ax + b$ or $y^2 = x^n + ax^2 + bx$. Kammerer, Lercier and Renault, in their recent paper [20], have presented several additional families of hyperelliptic curves for which an Icart-like encoding can be constructed, but the target curves are still of a special form and may not be convenient to use for cryptographic applications. Efficiency is also a problem for both of these constructions.

Moreover, all of these algorithms construct points on the curve itself, whereas the relevant object in cryptography is the group attached to it, namely its Jacobian variety. Very recently, Farashahi et al. [12] have demonstrated how to build a well-behaved hash function to the Jacobian based on a point-construction algorithm to the curve. Their framework apply to the functions proposed by Ulas and Kammerer et al., but with some difficulties and somewhat coarse bounds due to their relatively complex geometric descriptions.

Admissible encodings and indifferentiability. To obtain their well-behaved hash function construction to the Jacobian, Farashahi et al. rely on the results by Brier et al. [9], which give sufficient conditions for a hash function construction of the form $H(m) = F(h(m))$ to be plugged into any cryptosystem using H as a random oracle provided that h behaves as a random oracle. Basically, Brier et al.'s result states that this construction is indistinguishable from a random oracle as soon as F is an *admissible* encoding in the following sense.

A function $F : S \to R$ between finite sets is an admissible encoding if it satisfies the following properties:
1. Computable: F is computable in deterministic polynomial time.
2. Regular: for s uniformly distributed in S, the distribution of $F(s)$ is statistically indistinguishable from the uniform distribution in R.
3. Samplable: there is an efficient randomized algorithm $\mathcal{I}$ such that for any $r \in R$, $\mathcal{I}(r)$ induces a distribution that is statistically indistinguishable from the uniform distribution in $F^{-1}(r)$.

Our contribution. This paper presents a new encoding for hyperelliptic curves of the form $H : y^2 = f(x)$ where f is an odd polynomial over $\mathbb{F}_q$, with $q = 3 \bmod 4$. From this encoding to the curve H, we also deduce efficient injective encodings and well-behaved hash functions to its Jacobian.

The new encoding has the following desirable properties:

- it can be very efficiently computed using one exponentiation and no division, in constant time and without branching;
- the encoding is an efficiently invertible bijection: thus, it is possible to encode messages as points on the curve and recover them. This has numerous applications, e.g. to encryption;
- in genus 1, it provides an encoding to supersingular elliptic curves, similar to Boneh and Franklin's construction [4], but for different base fields;
- in higher genus, many cryptographically interesting curves are of the form H, including the curves considered in [14,16,25];
- many constructions of pairing-friendly hyperelliptic curves yield curves of the form H [21,13];
- since the encoding has a simple geometric description, it is easy to obtain well-behaved hash functions from it, and the corresponding regularity bounds are optimally tight.

2   Odd Hyperelliptic Curves

Let f be an odd monic polynomial over a finite field $\mathbb{F}_q$ with $q \equiv 3 \pmod 4$, which has simple roots in $\overline{\mathbb{F}_q}$. We denote its degree by $2g + 1$, and consider the hyperelliptic curve over $\mathbb{F}_q$ defined by:

$$H : y^2 = f(x) = x^{2g+1} + a_1 x^{2g-1} + \cdots + a_g x$$

Let us call such curves *odd hyperelliptic curves*. Many hyperelliptic curves relevant to cryptography, and particularly pairing-based cryptography, are of this form. For example:

- the supersingular elliptic curves of Joux [19]: $y^2 = x^3 + ax$;
- the genus 2 curves studied by Furukawa *et al.* [14] and their extension to genus g by Haneda *et al.* [16]: $y^2 = x^{2g+1} + ax$ (for which one can compute the zeta function);
- in particular, the Type II pairing-friendly curves of genus 2 constructed by Kawazoe and Takahashi [21];
- the genus 2 hyperelliptic curves for which Satoh [25] gave an efficient class group counting algorithm: $y^2 = x^5 + ax^3 + bx$;
- in particular, some of the pairing-friendly genus 2 curves constructed by Freeman and Satoh [13] (although the case $q \equiv 1 \pmod 4$ is more common).

Additionally, odd hyperelliptic curves and their Jacobians admit an automorphism of order 4 over $\mathbb{F}_{q^2}$ (namely $(x, y) \mapsto (-x, \sqrt{-1} \cdot y)$) which can be used to map points over $\mathbb{F}_q$ to linearly independent points over $\mathbb{F}_{q^2}$, another useful property for pairings.

Remark 1. A hyperelliptic curve over $\mathbb{F}_q$ is birational to an odd hyperelliptic curve when the set of points in $\mathbb{P}^1$ over which it is ramified is invariant under an automorphism of $\mathbb{P}^1$ of order 2 fixing two of them, both $\mathbb{F}_q$-rational. For example, hyperelliptic curves of the form:

$$H' : y^2 = x^6 + ax^5 + bx^4 - bx^2 - ax - 1$$

are birational to odd hyperelliptic curves, since they are ramified over a set of points invariant under $x \mapsto 1/x$ and containing ± 1. One possible change of variables is $x \mapsto (x - 1)/(x + 1)$.

This remark shows that the coarse moduli space of odd hyperelliptic curves of genus g over $\overline{\mathbb{F}_q}$ is a subvariety of dimension $g - 1$ of the dimension $2g - 1$ moduli space of genus g hyperelliptic curves.

3   Our New Encoding

3.1   Definition

Let $H : y^2 = f(x)$ be an odd hyperelliptic curve over $\mathbb{F}_q$. Denote by $\sqrt{\cdot}$ the usual square root function on the set of quadratic residues in $\mathbb{F}_q$ (exponentiation by $(q + 1)/4$), and by $\left(\frac{\cdot}{q} \right)$ the Legendre symbol over $\mathbb{F}_q$.

Over $\mathbb{F}_q$, -1 is a quadratic nonresidue, and for any $t \in \mathbb{F}_q$, we have $f(-t) = -f(t)$, so unless $f(t) = 0$, exactly one of $f(t)$ or $f(-t)$ is a square. In other words, exactly one of t or $-t$ is the abscissa of an $\mathbb{F}_q$-rational point on H.

This observation allows us to define a point encoding function F to $H(\mathbb{F}_q)$ as follows:

$$F \colon \mathbb{F}_q \longrightarrow H(\mathbb{F}_q)$$
$$t \longmapsto \left(\varepsilon(t) \cdot t \; ; \; \varepsilon(t)\sqrt{\varepsilon(t) \cdot f(t)} \right) \tag{1}$$

where $\varepsilon(t) = \left(\frac{f(t)}{q} \right)$. We claim that this function is well-defined and "almost" a bijection.

More precisely, recall that a *Weierstrass point* of H is a point where the rational function y is ramified: these are the points $(x, 0)$ for x a root of f together with the point at infinity ∞. Then, let $W \subset H(\mathbb{F}_q)$ be the set of $\mathbb{F}_q$-rational Weierstrass points on H, and $T \subset \mathbb{F}_q$ the set of roots of f.

Theorem 1. *The function F given by (1) is well-defined, maps all points in T to $(0,0) \in W$, and induces a bijection $\mathbb{F}_q \setminus T \to H(\mathbb{F}_q) \setminus W$.*

Proof. For $t \in T$, we have $\varepsilon(t) = 0$, hence $F(t) = (0,0) \in W$. Now let $t \in \mathbb{F}_q \setminus T$, and $x = \varepsilon(t) \cdot t$. Since f is odd and $\varepsilon(t) = \pm 1$, $f(x) = \varepsilon(t) \cdot f(t)$. In particular, recalling that $\left(\frac{-1}{q} \right) = -1$, we can write:

$$\left(\frac{f(x)}{q} \right) = \left(\frac{\varepsilon(t) \cdot f(t)}{q} \right) = \varepsilon(t) \cdot \left(\frac{f(t)}{q} \right) = \varepsilon(t)^2 = 1$$

Thus, the second component $y = \varepsilon(t)\sqrt{\varepsilon(t) \cdot f(t)}$ of $F(t)$ is well-defined, and we have $y^2 = \varepsilon(t) \cdot f(t) = f(x)$, so $F(t)$ is an affine point on $H(\mathbb{F}_q)$ as required. The condition $t \notin T$ further implies that $f(t) \neq 0$, so $y \neq 0$. Therefore, $F(t) \in \mathbb{F}_q \setminus W$.

Let us show that the restriction of F to $\mathbb{F}_q \setminus T$ is injective. Indeed, suppose $F(t) = F(u)$ with $t, u \notin T$. Equating x-coordinates, we get $\varepsilon(t) \cdot t = \varepsilon(u) \cdot u$, hence $u = \pm t$. If $u = -t$, then comparing the y-coordinates, we obtain

$$\varepsilon(t)\sqrt{\varepsilon(t) \cdot f(t)} = \varepsilon(u)\sqrt{\varepsilon(u) \cdot f(u)}$$
$$= \varepsilon(-t)\sqrt{\varepsilon(-t) \cdot f(-t)} = -\varepsilon(t)\sqrt{\varepsilon(t) \cdot f(t)}$$

which is a condraction. Therefore, $t = u$ and F is injective on $\mathbb{F}_q \setminus T$.

Finally, $F(\mathbb{F}_q \setminus T) = H(\mathbb{F}_q) \setminus W$. To see this, take $(x, y) \in H(\mathbb{F}_q) \setminus W$ and let $t = \delta \cdot x$, where $\delta = \pm 1$ is defined by $y = \delta\sqrt{f(x)}$. We have

$$\varepsilon(t) = \left(\frac{f(\delta x)}{q} \right) = \left(\frac{\delta \cdot f(x)}{q} \right) = \delta \cdot \left(\frac{f(x)}{q} \right) = \delta$$

since $f(x) = y^2$ is a square. Thus:

$$F(t) = \left(\delta^2 \cdot x \; ; \; \delta\sqrt{\delta \cdot f(\delta x)} \right) = \left(x \; ; \; \delta\sqrt{f(x)} \right) = (x; y)$$

as required. □

Corollary 1. *The cardinal of $H(\mathbb{F}_q)$ is $q + 1$.*

Proof. From the above, we get $\#H(\mathbb{F}_q) = \#(\mathbb{F}_q \setminus T) + \#W = q - \#T + \#W$. But W consists of the point at infinity on H, and all points of the form $(x, 0)$, $x \in T$. Thus, $\#W = \#T + 1$, and $\#H(\mathbb{F}_q) = q + 1$.   □

Remark 2. – Since F is an efficiently computable bijection between all of $\mathbb{F}_q$ and $H(\mathbb{F}_q)$ except at most $2g+2$ points on both sides, with an efficiently computable inverse (namely $(x, y) \mapsto \left(\frac{y}{q}\right) x$), it is a very well-behaved encoding function.

In particular, it is clear that if t is uniformly distributed in $\mathbb{F}_q$, the distribution of $F(t)$ in $H(\mathbb{F}_q)$ is statistically indistinguishable from the uniform distribution. According to the results of Brier *et al.* [9], it follows that if $m \mapsto h(m)$ is a hash function to $\mathbb{F}_q$ modeled as a random oracle, then $F(h(m))$ is a function into $H(\mathbb{F}_q)$ that is indifferentiable from a random oracle. When the genus of H is at least 2, however, one is usually interested in hashing to the Jacobian of H rather than H itself. This will be discussed in §4.

The fact that F is injective, unlike most other constructions, makes it possible to also use it for other purposes than hashing, such as encoding a message to be encrypted, for example with ElGamal.
- Since $\#T = \#(W \setminus \{\infty\})$, it is in fact easy to modify the definition of F to obtain a bijection $F' : \mathbb{F}_q \to H(\mathbb{F}_q) \setminus \{\infty\}$ which misses only one rational point on H. It is slightly less efficient to compute, however, and using one or the other makes no difference in practice (as one is not concerned with a few exceptional points), so we shall stick to F as defined by (1).
- When H is in fact an elliptic curve E (i.e. $\deg f = 3$), Corollary 1 says that E is supersingular. These are in fact the supersingular elliptic curves $y^2 = x^3 + ax$ discussed by Joux in [19]. Thus, the function F provides a convenient way to encode points into supersingular elliptic curves over $\mathbb{F}_q$ with $q \equiv 3 \pmod 4$. This is an interesting addition to the original encoding of Boneh and Franklin [4], which applies to supersingular curves of the form $y^2 = x^3 + b$ over fields $\mathbb{F}_q$ with $q \equiv 2 \pmod 3$. In particular, our encoding can be used in characteristic 3.
- In the general case, we see that $\#H(\mathbb{F}_{q^n}) = q^n + 1$ for any odd extension degree n. This gives some constraints on the zeta function of H, but in genus $g \geq 2$, many isogeny classes are possible for the Jacobian J of H nonetheless, so the proposed encoding applies to a wide range of curves. It is not always easy to determine the order of $J(\mathbb{F}_q)$: an approach is given by Satoh in [25] for $g = 2$.

3.2   Efficient Computation

The definition of F involves a generalized Legendre symbol and one square root, which suggests that its computation might be costly, especially if it is to be done in constant time, an important property in settings where side-channel attacks are a concern. However, it is actually possible to compute F with a single exponentiation, a few multiplications and no division, making it one of the most efficient deterministic encoding function proposed to date. One such implementation is described as Algorithm 1. Note that this implementation is also branch-free, contrary to what happens for encodings such as the one by Shallue and van de Woestijne [26]; this also prevents certain active side-channel attacks.

Algorithm 1. Constant-time, single-exponentiation implementation of the encoding F. The constant r is $(q-3)/4$ if $q \equiv 3 \pmod 8$, $(q-3)/4 + (q-1)/2$ otherwise.

1: **function** $F(t)$
2:     $\alpha \leftarrow f(t)$
3:     $\beta \leftarrow \alpha^r$
4:     **return** $(\alpha\beta^2 t, \alpha\beta)$
5: **end function**

To see that this implementation is correct, consider α and β as defined in Algorithm 1. For $t \in T$, we have $\alpha = 0$, hence the procedure returns $F(t) = (0,0)$ as required. Now let $t \notin T$. We have

$$\beta^2 = \alpha^{\frac{q-3}{2}} = \frac{1}{\alpha}\left(\frac{\alpha}{q}\right) = \frac{\varepsilon(t)}{\alpha}$$

In particular, $\alpha\beta^2 t = \varepsilon(t) \cdot t$ is indeed the abscissa of $F(t)$.

Moreover, suppose $q \equiv 3 \pmod 8$. Then $(q+1)/4$ is odd and $\varepsilon(t) = \pm 1$, so we have

$$\alpha\beta = \alpha^{\frac{q-3}{4}+1} = \varepsilon(t) \cdot \varepsilon(t) \cdot f(t)^{\frac{q+1}{4}}$$
$$= \varepsilon(t) \cdot \left(\varepsilon(t) \cdot f(t)\right)^{\frac{q+1}{4}} = \varepsilon(t)\sqrt{\varepsilon(t) \cdot f(t)}$$

so the algorithm is correct.

Similarly, when $q \equiv 7 \pmod 8$, $(q+1)/4$ is odd and we obtain

$$\alpha\beta = \alpha^{\frac{q-1}{2}+\frac{q-3}{4}+1} = \varepsilon(t) \cdot f(t)^{\frac{q+1}{4}}$$
$$= \varepsilon(t) \cdot \left(\varepsilon(t) \cdot f(t)\right)^{\frac{q+1}{4}} = \varepsilon(t)\sqrt{\varepsilon(t) \cdot f(t)}$$

which concludes.

4   Mapping to the Jacobian

In the previous section, we have constructed a function $F \colon \mathbb{F}_q \to H(\mathbb{F}_q)$ which is efficiently computable and has a number of desirable properties. For cryptographic purposes, however, we are usually interested in obtaining elements of a group attached to the curve, namely the Jacobian, rather than points on the curve itself. In the case of elliptic curves, the curve and its Jacobian are isomorphic so no further work is needed, but for curves of genus $g \geq 2$, they are quite different objects.

In the following, we always denote the Jacobian of H by J, and we regard H as embedded in J via the map $H \to J$ sending a point P to the class of the degree 0 divisor $(P) - (\infty)$. In particular, if P, Q are points in $H(\mathbb{F}_q)$, $P + Q$ denotes the class of $(P) + (Q) - 2(\infty)$.

We propose two constructions of maps to $J(\mathbb{F}_q)$ to accommodate for different use cases: an injective map with large image, which can be used to encode scalars as group elements (e.g. for encryption), and a map defining an essentially uniform distribution on $J(\mathbb{F}_q)$, to obtain well-behaved hash functions.

4.1 Injective Encoding to the Jacobian

Let us first recall a few facts about hyperelliptic curves, for which we refer for example to [24]. Elements of $J(\mathbb{F}_q)$ are classes of $\mathbb{F}_q$-divisors on H and admit a canonical representation as so-called *reduced divisors* defined over $\mathbb{F}_q$. Let $\tilde{\ }$ denote the hyperelliptic involution on H, $(x, y) \mapsto (x, -y)$. A divisor $D = P_1 + \cdots + P_r$ (where the P_i are not necessarily distinct points in $H(\overline{\mathbb{F}_q})$) is said to be reduced when r is less than or equal to the genus g of H, and $P_i \neq \widetilde{P_j}$ for all $i \neq j$. The reduced divisors D and D' defined by $P_1, \ldots, P_r$ and $P'_1, \ldots, P'_r$ are distinct and non-equivalent as soon as the multisets $\{P_1, \ldots, P_r\}$ and $\{P'_1, \ldots, P'_r\}$ are different. Each divisor class in $J(\mathbb{F}_q)$ contains a unique reduced divisor defined over $\mathbb{F}_q$.

Now, with the notations of §3, the encoding $F \colon \mathbb{F}_q \to H(\mathbb{F}_q)$ defined by (1) satisfies that for all $t \in \mathbb{F}_q \setminus T$, the only u such that $F(u) = \widetilde{F(t)}$ is $u = -t$. Therefore, if $(t_1, \ldots, t_g)$ is any tuple of g elements of $\mathbb{F}_q \setminus T$ (g being the genus of H) such that $t_i + t_j \neq 0$ for all i, j, then $F(t_1) + \cdots + F(t_g)$ is a reduced divisor. In particular, consider the set X of g-element subsets of $\mathbb{F}_q \setminus T$ not containing any two opposite elements. Then it is immediate from the facts above that the map:

$$F_{\mathrm{inj}} \colon X \longrightarrow J(\mathbb{F}_q)$$
$$\{t_1, \ldots, t_g\} \longmapsto F(t_1) + \cdots + F(t_g)$$

is injective. We have

$$\#X = 2^g \binom{(q - \#T)/2}{g} = \frac{1 - o(1)}{g!} \cdot q^g \geq c_g \cdot \#J(\mathbb{F}_q)$$

for some constant $c_g > 0$ depending only on g. Thus, F_{inj} is an injective mapping to $J(\mathbb{F}_q)$ covering a large portion of all points. It is also very easy to compute since points in the image are directly given as reduced divisors, so no actual arithmetic on the Jacobian is needed.

In the case that is most relevant for cryptographic applications, namely $g = 2$, we can define an even simpler injective encoding, from the set Y of 2-element subsets of $\mathbb{F}_q \setminus T$, which may be easier to manipulate than X:

$$F'_{\text{inj}} : Y \longrightarrow J(\mathbb{F}_q)$$
$$\{t_1, t_2\} \longmapsto F(t_1) + F(-t_2)$$

This function injective, easy to compute, and reaches roughly one half of all points in $J(\mathbb{F}_q)$.

4.2   Indifferentiable Hashing to the Jacobian

One can also use F to construct well-behaved hash functions to $J(\mathbb{F}_q)$. For this purpose, Brier *et al.* [9] have shown how one could use functions to $J(\mathbb{F}_q)$ with good regularity properties, and Farashahi *et al.* [12] have proposed a framework based on character sums to prove such regularity properties for functions of the form:

$$F^{\otimes s} : (\mathbb{F}_q)^s \longrightarrow J(\mathbb{F}_q)$$
$$(t_1, \ldots, t_s) \longmapsto F(t_1) + \cdots + F(t_s)$$

Since F is so simple, we do not really need to rely on the entire framework of [12]. Indeed, the following bound, which in the terminology of Farashahi *et al.* says that F is a $(2g - 2 + \varepsilon)$-well-distributed encoding, can be proved using classical results on characters on algebraic curves. Note that this bound is very tight: it gives a better well-distributedness bound for F in genus up to 6 than can be established for Icart's function in genus 1.

Lemma 1. *For any character χ of the abelian group $J(\mathbb{F}_q)$, let*

$$S(\chi) = \sum_{t \in \mathbb{F}_q} \chi(F(t))$$

Then, whenever χ is nontrivial, we have

$$|S(\chi)| \leq (2g - 2)\sqrt{q} + 4g + 3$$

Proof. A nontrivial character χ of $J(\mathbb{F}_q)$ is also a nontrivial, unramified Artin character of H (see [22, §2] or [12, §4]). In particular, the Riemann hypothesis for the L-function on H associated with χ gives:

$$\left| \sum_{P \in H(\mathbb{F}_q)} \chi(P) \right| \leq (2g - 2)\sqrt{q}$$

The result then follows from the observation that

$$\sum_{t \in \mathbb{F}_q} \chi(F(t)) = \#T \cdot \chi((0,0)) + \sum_{P \in H(\mathbb{F}_q) \setminus W} \chi(P)$$

$$= \#T \cdot \chi((0,0)) - \sum_{P \in W} \chi(P) + \sum_{P \in H(\mathbb{F}_q)} \chi(P)$$

since $\#T + \#W \leq 4g + 3$. $\square$

We can then proceed like in [12] and deduce from this lemma a bound on the statistical distance between the distribution defined on $J(\mathbb{F}_q)$ by $F^{\otimes s}$ and the uniform distribution.

For any $D \in J(\mathbb{F}_q)$, let $N_s(D)$ denote the number of preimages of D under $F^{\otimes s}$:

$$N_s(D) = \#\{(t_1, \ldots, t_s) \in (\mathbb{F}_q)^s \mid D = F(t_1) + \cdots + F(t_s)\}$$

Then we have the following result:

Theorem 2. *The statistical distance between the distribution defined by $F^{\otimes s}$ and the uniform distribution on $J(\mathbb{F}_q)$ is bounded as:*

$$\sum_{D \in J(\mathbb{F}_q)} \left| \frac{N_s(D)}{q^s} - \frac{1}{\#J(\mathbb{F}_q)} \right| \leq \frac{\left(2g + 2 + (4g + 3)q^{-1/2}\right)^s \sqrt{\#J(\mathbb{F}_q)}}{q^{s/2}}$$

Proof. This results from [12, Theorem 2]. We can give a quick recap of the proof for the reader's convenience.

Note first that one can write $N_s(D)$ in terms of the character sums $S(\chi)$ as follows:

$$N_s(D) = \sum_{t_1, \ldots, t_s \in \mathbb{F}_q} \frac{1}{\#J(\mathbb{F}_q)} \sum_{\chi} \chi\left(F(t_1) + \cdots + F(t_s) - D\right)$$

$$= \sum_{\chi} \frac{\chi(-D)}{\#J(\mathbb{F}_q)} \sum_{t_1, \ldots, t_s \in \mathbb{F}_q} \chi\left(F(t_1) + \cdots + F(t_s)\right)$$

$$= \sum_{\chi} \frac{\chi(-D)}{\#J(\mathbb{F}_q)} S(\chi)^s$$

Putting the trivial character aside, this yields:

$$\frac{N_s(D)}{q^s} - \frac{1}{\#J(\mathbb{F}_q)} = \frac{\chi(-D)}{q^s \#J(\mathbb{F}_q)} \sum_{\chi \neq 1} S(\chi)^s$$

Then, we consider the sum of squares of this expression as D varies along $J(\mathbb{F}_q)$. Let

$$V_s = \sum_{D \in J(\mathbb{F}_q)} \left| \frac{N_s(D)}{q^s} - \frac{1}{\#J(\mathbb{F}_q)} \right|^2$$

We have

$$V_s = \sum_D \frac{1}{q^{2s} \#J(\mathbb{F}_q)^2} \sum_{\chi, \chi' \neq 1} \chi(-D)\overline{\chi'}(-D) \cdot S(\chi)^s \cdot \overline{S(\chi')^s}$$

$$= \frac{1}{q^{2s} \#J(\mathbb{F}_q)^2} \sum_{\chi, \chi' \neq 1} \left(\sum_D \chi(D)\overline{\chi'}(D) \right) S(\chi)^s \cdot \overline{S(\chi')^s}$$

$$= \frac{1}{q^{2s} \#J(\mathbb{F}_q)} \sum_{\chi \neq 1} |S(\chi)|^{2s} \leq \frac{\left((2g + 2)\sqrt{q} + 4g + 3\right)^{2s}}{q^{2s}}$$

since the sum over D of $\chi(D)\overline{\chi'}(D)$ vanishes if $\chi \neq \chi'$. Finally, the Cauchy-Schwarz inequality gives:

$$\sum_{D \in J(\mathbb{F}_q)} \left| \frac{N_s(D)}{q^s} - \frac{1}{\#J(\mathbb{F}_q)} \right| \leq \sqrt{V_s} \cdot \sqrt{\#J(\mathbb{F}_q)}$$

$$\leq \frac{\left(2g + 2 + (4g+3)q^{-1/2}\right)^s \sqrt{\#J(\mathbb{F}_q)}}{q^{s/2}}$$

as required.   □

Note that $\#J(\mathbb{F}_q) \sim q^g$, so that the bound we get on the statistical distance is in $O(q^{(g-s)/2})$. Therefore, as soon as $s > g$, the distribution defined by $F^{\otimes s}$ on $J(\mathbb{F}_q)$ is statistically indistinguishable from the uniform distribution. In particular, in the terminology of Brier *et al.* [9] which we recalled in the introduction, the encoding $F^{\otimes(g+1)}$ to $J(\mathbb{F}_q)$ is regular. It is also obviously computable and samplable, so $F^{\otimes(g+1)}$ is an admissible encoding to $J(\mathbb{F}_q)$.

This provides a simple, well-behaved hash function construction to the Jacobian of H. Indeed, it follows that the function

$$m \mapsto F(h_1(m)) + \cdots + F(h_{g+1}(m))$$

is indifferentiable from a random oracle if $h_1, \ldots, h_{g+1}$ are seen as independent random oracles into $\mathbb{F}_q$.

5   Conclusion

In this paper, we provide a very efficient construction of a deterministic encoding into odd hyperelliptic curves. Odd hyperelliptic curves are a simple and relatively large class of hyperelliptic curves, compared to the families of curves covered by previous deterministic encodings. They also include many curves of cryptographic interest (because of efficient point-counting on the Jacobian, or pairing-friendliness), even in the elliptic curve case.

This encoding is almost a bijection, which can be useful for a number of applications, such as encryption, and allows us to construct the first efficient injections with large image to the Jacobians of odd hyperelliptic curves, as well as indifferentiable hash functions to these Jacobians with particularly tight regularity bounds.

Acknowledgments. We are grateful to Reza Farashahi and anonymous referees for useful comments, and to Masayuki Abe, Jean-Sébastien Coron and Thomas Icart for earlier discussions that inspired this paper. This work was partly supported by the French ANR-07-TCOM-013-04 PACE Project and by the European Commission through the IST Program under Contract ICT-2007-216646 ECRYPT II.

References

1. Baek, J., Zheng, Y.: Identity-based threshold decryption. In: Bao, et al. (eds.) [2], pp. 262–276
2. Bao, F., Deng. R., Zhou, J. (eds.): PKC 2004. LNCS, vol. 2947. Springer, Heidelberg (2004)
3. Boldyreva, A.: Threshold signatures, multisignatures and blind signatures based on the gap-diffie-hellman-group signature scheme. In: Desmedt (ed.) [11], pp. 31–46
4. Boneh, D., Franklin, M.K.: Identity-based encryption from the weil pairing. In: Kilian, J. (ed.) CRYPTO 2001. LNCS, vol. 2139, pp. 213–229. Springer, Heidelberg (2001)
5. Boneh, D., Gentry, C., Lynn, B., Shacham, H.: Aggregate and verifiably encrypted signatures from bilinear maps. In: EUROCRYPT, pp. 416–432 (2003)
6. Boneh, D., Lynn, B., Shacham, H.: Short signatures from the weil pairing. In: Boyd, C. (ed.) ASIACRYPT 2001. LNCS, vol. 2248, pp. 514–532. Springer, Heidelberg (2001)
7. Boyd, C., Montague, P., Nguyen, K.Q.: Elliptic curve based password authenticated key exchange protocols. In: Varadharajan, V., Mu, Y. (eds.) ACISP 2001. LNCS, vol. 2119, pp. 487–501. Springer, Heidelberg (2001)
8. Boyen, X.: Multipurpose identity-based signcryption (a swiss army knife for identity-based cryptography). In: Boneh, D. (ed.) CRYPTO 2003. LNCS, vol. 2729, pp. 383–399. Springer, Heidelberg (2003)
9. Brier, E., Coron, J.-S., Icart, T., Madore, D., Randriam, H., Tibouchi, M.: Efficient indifferentiable hashing into ordinary elliptic curves. In: Rabin, T. (ed.) CRYPTO 2010. LNCS, vol. 6223, pp. 237–254. Springer, Heidelberg (2010)
10. Cha, J.C., Cheon, J.H.: An identity-based signature from gap diffie-hellman groups. In: Desmedt (ed.) [11], pp. 18–30
11. Desmedt, Y. (ed.): PKC 2003. LNCS, vol. 2567. Springer, Heidelberg (2002)
12. Farashahi, R.R., Fouque, P.-A., Shparlinski, I., Tibouchi, M., Voloch, F.: Indifferentiable deterministic hashing to elliptic and hyperelliptic curves. Preprint (2010), http://www.di.ens.fr/~tibouchi/research.html
13. Freeman, D.M., Satoh, T.: Constructing pairing-friendly hyperelliptic curves using weil restriction. Cryptology ePrint Archive, Report 2009/103 (2009), http://eprint.iacr.org/
14. Furukawa, E., Kawazoe, M., Takahashi, T.: Counting points for hyperelliptic curves of type $y^2 = x^5 + ax$ over finite prime fields. In: Matsui, M., Zuccherato, R.J. (eds.) SAC 2003. LNCS, vol. 3006, pp. 26–41. Springer, Heidelberg (2003)
15. Gentry, C., Silverberg, A.: Hierarchical id-based cryptography. In: Zheng (ed.) [29], pp. 548–566
16. Haneda, M., Kawazoe, M., Takahashi, T.: Suitable curves for genus-4 HCC over prime fields. In: Caires, L., Italiano, G.F., Monteiro, L., Palamidessi, C., Yung, M. (eds.) ICALP 2005. LNCS, vol. 3580, pp. 539–550. Springer, Heidelberg (2005)
17. Horwitz, J., Lynn, B.: Toward hierarchical identity-based encryption. In: Knudsen, L.R. (ed.) EUROCRYPT 2002. LNCS, vol. 2332, pp. 466–481. Springer, Heidelberg (2002)
18. Icart, T.: How to hash into elliptic curves. In: Halevi, S. (ed.) CRYPTO 2009. LNCS, vol. 5677, pp. 303–316. Springer, Heidelberg (2009)
19. Joux, A.: The weil and tate pairings as building blocks for public key cryptosystems. In: Fieker, C., Kohel, D.R. (eds.) ANTS 2002. LNCS, vol. 2369, pp. 20–32. Springer, Heidelberg (2002)

20. Kammerer, J.-G., Lercier, R., Renault, G.: Encoding points on hyperelliptic curves over finite fields in deterministic polynomial time. CoRR, abs/1005.1454 (2010)
21. Kawazoe, M., Takahashi, T.: Pairing-friendly hyperelliptic curves with ordinary jacobians of type $y^2 = x^5 + ax$. In: Galbraith, S.D., Paterson, K.G. (eds.) Pairing 2008. LNCS, vol. 5209, pp. 164–177. Springer, Heidelberg (2008)
22. Kohel, D.R., Shparlinski, I.: On exponential sums and group generators for elliptic curves over finite fields. In: Bosma, W. (ed.) ANTS 2000. LNCS, vol. 1838, pp. 395–404. Springer, Heidelberg (2000)
23. Libert, B., Quisquater, J.-J.: Efficient signcryption with key privacy from gap diffie-hellman groups. In: Bao, et al. (eds.) [2], pp. 187–200
24. Menezes, A.J., Wu, Y.-H., Zuccherato, R.J.: An elementary introduction to hyperelliptic curves. In: Koblitz, N. (ed.) Algebraic Aspects of Cryptography. Algorithms and Computation in Mathematics, vol. 3, pp. 155–178. Springer, Heidelberg (1998)
25. Satoh, T.: Generating genus two hyperelliptic curves over large characteristic finite fields. In: Joux, A. (ed.) EUROCRYPT 2009. LNCS, vol. 5479, pp. 536–553. Springer, Heidelberg (2010)
26. Shallue, A., van de Woestijne, C.: Construction of rational points on elliptic curves over finite fields. In: Hess, F., Pauli, S., Pohst, M.E. (eds.) ANTS 2006. LNCS, vol. 4076, pp. 510–524. Springer, Heidelberg (2006)
27. Ulas, M.: Rational points on certain hyperelliptic curves over finite fields. Bull. Polish Acad. Sci. Math. 55(2), 97–104 (2007)
28. Zhang, F., Kim, K.: Id-based blind signature and ring signature from pairings. In: Zheng (ed.) [29], pp. 533–547
29. Zheng, Y. (ed.): ASIACRYPT 2002. LNCS, vol. 2501. Springer, Heidelberg (2002)

Encoding Points on Hyperelliptic Curves over Finite Fields in Deterministic Polynomial Time

Jean-Gabriel Kammerer[1,2], Reynald Lercier[1,2], and Guénaël Renault[3]

[1] DGA MI, La Roche Marguerite, F-35174 Bruz Cedex, France
[2] Institut de recherche mathématique de Rennes, Université de Rennes 1
Campus de Beaulieu, F-35042 Rennes Cedex, France
jean-gabriel.kammerer@m4x.org,
reynald.lercier@m4x.org
[3] LIP6, Université Pierre et Marie Curie, INRIA/LIP6 SALSA Project-Team
Boite courrier 169, 4 place Jussieu, F-75252 Paris Cedex 05, France
guenael.renault@lip6.fr

Abstract. We provide new hash functions into (hyper)elliptic curves over finite fields. These functions aim at instantiating in a secure manner cryptographic protocols where we need to map strings into points on algebraic curves, typically user identities into public keys in pairing-based IBE schemes.

Contrasting with recent Icart's encoding, we start from "easy to solve by radicals" polynomials in order to obtain models of curves which in turn can be deterministically "algebraically parameterized". As a result of this strategy, we obtain a low degree encoding map for Hessian elliptic curves, and for the first time, hashing functions for genus 2 curves. More generally, we present for any genus (more narrowed) families of hyperelliptic curves with this property.

The image of these encodings is large enough to be "weak" encodings in the sense of Brier et al. As such they can be easily turned into admissible cryptographic hash functions.

Keywords: deterministic encoding, elliptic curves, Galois theory, hyperelliptic curves.

1 Introduction

Many asymmetric cryptographic mechanisms are based on the difficulty of the discrete logarithm problem in finite groups. Among these groups, algebraic curves on finite fields are of high interest because of the small size of keys needed to achieve good security. Nonetheless it is less easy to encode a message into an element of the group.

Let $\mathbb{F}_q$ be a finite field of odd characteristic p, and $H/\mathbb{F}_q : y^2 = f(x)$ where $\deg f = d$ be an elliptic (if $d = 3$ or 4) or hyperelliptic (if $d \geqslant 5$) curve, we consider the problem of computing points on H in deterministic polynomial time. In cryptographic applications, computing a point on a (hyper)elliptic curve is a prerequisite for encoding a message into its Jacobian group. In this regard,

M. Joye, A. Miyaji, and A. Otsuka (Eds.): Pairing 2010, LNCS 6487, pp. 278–297, 2010.

pairing-based cryptosystems do not make exception. Boneh-Franklin Identity-Based Encryption scheme [3] requires for instance to associate to any user identity a point on an elliptic curve.

In the case of elliptic curves, we may remark that it is enough to compute one rational point G, since we can have other points $t\,G$ from integers t (at least if G is of large enough order). To compute such a G, one might test random elements $x \in \mathbb{F}_q$ until $f(x)$ is a square. But without assuming GRH, we have no guarantee of finding a suitable x after a small enough number of attempts, and no deterministic algorithm is known for computing square roots when $p \equiv 1 \bmod 4$. Moreover, encoding t into $t\,G$ voids the security of many cryptographic protocols [13].

Maybe a more serious attempt in this direction for odd degrees d is due to Atkin and Morain [1]. They remark that if x_0 is any element of $\mathbb{F}_q$ and $\lambda = f(x_0)$, then the point $(\lambda x_0, \lambda^{(d+1)/2})$ is on the curve $Y^2 = \lambda^d f(X/\lambda)$. But again, the latter can be either isomorphic to the curve or its quadratic twist, following that λ is a quadratic residue or not, and we have no way to control this in deterministic time.

In 2006, Shallue and Woestjine [16] proposed the first practical deterministic algorithm to encode points into an elliptic curve, quickly generalized by Ulas [17] to the family of hyperelliptic curves defined by $y^2 = x^n + ax + b$ or $y^2 = x^n + ax^2 + bx$. Icart [13] proposed in 2009 another deterministic encoding for elliptic curves, of complexity $\mathcal{O}(\log^{2+o(1)} q)$, provided that the cubic root function, inverse of $x \mapsto x^3$ on $\mathbb{F}_q^*$, is a group automorphism. This is equivalent to $q \equiv 2 \bmod 3$. This encoding uses Cardano-Tartaglia's formulae to parameterize the points $(x : y : 1)$ on any elliptic curve $E : x^3 + ax + b = y^2$.

1.1   Contribution

In this paper, we propose a strategy for finding other families with such properties (Section 2). As an example, we first show how the strategy works for genus 1 curves and come to a new encoding map for Hessian elliptic curves in finite fields of odd characteristic with $q \equiv 2 \bmod 3$ (Section 3.1). Hessian curves are defined by $E_d : x^3 + y^3 + 1 = 3dxy$ with $d \neq 1$, but admit an equivalent model $C_{0,a} : y^3 + xy + ay = x^3$ with $a \neq 0, 1/27$. Our encoding function is $2 : 1$: each point of its image admits exactly 2 preimages compared to the 1 to potentially 4 in Icart's construction.

We then study more carefully genus 2 curves, still over finite fields $\mathbb{F}_q$ of odd characteristic with $q \equiv 2 \bmod 3$, and exhibit several large families (Section 3.2), namely the curves $H_{1,a,b} : y^2 = (x^3 + 3ax + 2)^2 + 8bx^3$ (which we call type A) and $H_{2,\lambda,\mu,a,v,w} : y^2/\lambda = (x^3 + 3\mu x + 2a)^2 + 4b$ (type B). Both of these families have dimension 2 over the moduli space.

Finally for all genus $g \geqslant 2$, we propose families of hyperelliptic curves which admit an efficient deterministic encoding function (Section 4), again provided some conditions on q (typically q odd, $q \equiv 2 \bmod 3$ and q coprime to $2g + 1$).

Remark 1. In the paper, we use indifferently the words "parameterization" or "encoding", even if, strictly speaking, we do not have fully parameterized curves.

We are aware that these maps are at least improperly parameterizations since there might correspond more than one parameter to one point. There are numerous points which lie outside the image of our maps too.

Remark 2. Our main goal is to find the biggest possible families for which there exists an efficient deterministic encoding. It might be possible to reduce the number of finite field operations in the description of our algorithms. However, they all use an exponentiation step like every other known encoding which dominates by far the computation time, both from a practical and a complexity viewpoint.

Remark 3. Each of our encodings is a *weak encoding* in the sense of [6]. Combined with a cryptographic hash function, we can thus construct hash functions into the set of rational points of these curves that are indifferentiable from a random oracle.

1.2   Related Work

Other papers about encoding into curves have been independently prepublished soon after the publication of a preversion of this article on the arXiv.org e-Print archive (ref 1005.1454).

Especially, Fouque and Tibouchi proposed an injective encoding over odd hyperelliptic curves, that is curves of the form $y^2 = f(x)$ with $f(-x) = -f(x)$ [12]. For any genus g, this encoding targets a dimension $g-1$ subspace of hyperelliptic curves, so of dimension 1 for genus 2 curves. Farashahi proposed an other $2:1$ encoding into Hessian curves too [11].

Because of square or cubic rooting steps, all these encodings have the same asymptotic complexity as ours, namely $\mathcal{O}(\log^{2+o(1)} q)$.

2   A Strategy

Given a genus g, we describe a strategy for finding curves of genus g which admit a deterministic encoding for a large subset of their points.

It's worth noting first that only genus 0 curves are *rationally* parameterizable. That is, any curve which admits a rational parameterization shall be a conic, see [15, Theorem 4.11]. Encoding maps into higher genus curves shall thus be *algebraic*. We are then reduced to the parameterization of roots of polynomials. Hence, the main idea of our general strategy is to start from polynomials with roots which are easily parameterizable and then deduce curves with deterministic encoding.

2.1   Solvable Polynomials

Classical Galois theory offers a large family of polynomials with easily parameterized roots: polynomials with roots that can be written as radicals, which are polynomials with solvable Galois group. Our strategy is based on these polynomials.

More precisely, let $f_{\underline{a}}(X)$ be a family of parameterized polynomials (where $\underline{a}$ denotes a k-tuple $(a_1, a_2, \ldots, a_k)$ of parameters) with solvable Galois group. We are interested in such parametric polynomials but also in the parametric radical expression of their roots $\chi_{\underline{a}}$. For instance $f_A(X) = X^2 + A$ in degree 2, or more interestingly $f_{A,B}(X) = X^3 + AX + B$ in degree 3, are such polynomials with simple radical formulae for their roots. The former verifies $\chi_A = \sqrt{-A}$ and a root of the second one is given by the well-known Cardano-Tartaglia's formulae (see [8]). The application of our general strategy to this family of degree 3 polynomials with the parameterization of its roots is described in Section 3.

Let us note that we might use the classical field machinery to construct new solvable polynomials from smaller ones. Look for instance at De Moivre's polynomials of degree d: we start from the degree 2 field extension $\theta^2 + B\theta - A^d$, followed by the degree d Kümmer extension $\gamma^d - \theta = 0$. Then the element $X = \gamma - A/\gamma$ is defined in a degree d subfield of the degree $2d$ extension. The defining polynomial of this extension is given by the minimal polynomial of X, which is equal to the De Moivre's polynomial,

$$X^d + dAX^{d-2} + 2dA^2X^{d-4} + 3dA^3X^{d-6} + \cdots + 2dA^{(d-1)/2-1}X^3 + dA^{(d-1)/2}X + B.$$

A more straightforward similar construction is to consider Kümmer extensions over quadratic (or small degree) extensions, which yields $X^{2d} + AX^d + B$. From these two specific families of solvable polynomials, we provide, in Section 4.1 and 4.2, hyperelliptic curves for all genus $g \geqslant 2$ which admit an efficient deterministic encoding function.

2.2 Rational and Deterministic Parameterizations

Given a parameterized family of solvable polynomials $f_{\underline{t}}(X)$, and a genus g, we now substitute a rational function $F_i(Y)$ in some variable Y for each parameter a_i in $\underline{a}$.

Let $\underline{F}(Y)$ denote the k-tuples of rational functions $(F_1(Y), F_2(Y), \ldots, F_k(Y))$. The equation $f_{\underline{F}(Y)}(X)$ now defines a plane algebraic curve C, with variables (X, Y). The genus of C increases when the degrees of $\underline{F}(Y)$ in Y increase. So if we target some fixed genus g for C, only few degrees for the numerators and denominators of $\underline{F}(Y)$ can occur. Since we can consider coefficients of these rational functions as parameters $\underline{a} = (a_1, \ldots, a_{k'})$, this yields a family of curves $C_{\underline{a}}$.

Less easily, it remains then to determine among these $\underline{F}(Y)$ the ones which yield roots $\chi_{\underline{F}(Y)}$ which can be computed in deterministic time. The easiest case is probably when no square root occurs in the computation of $\chi_{\underline{t}}$, since then any choice for $\underline{F}(Y)$ will work, at the expense of some constraint on the finite field. But this is usually not the case, and we might try instead to link these square roots to some algebraic parameterization of an auxiliary algebraic curve

2.3 Minimal Models

In some case (typically hyperelliptic curves), it is worth to derive from the equation for $C_{\underline{a}}$ a minimal model (typically of the form $y^2 = g_{\underline{a}}(x)$). In order to still have a deterministic encoding with the minimal model, we need explicit birational maps $x = \Lambda_{\underline{a}}(X, Y)$, $y = \Omega_{\underline{a}}(X, Y)$ too. For hyperelliptic curves, the usual way for this is to work with homomorphic differentials defined by $C_{\underline{a}}$. This method is implemented in several computer algebra systems, for instance MAPLE [14] or MAGMA [5]. All in all, we obtain the following encoding for a minimal model $g_{\underline{a}}$:

- Fix some Y as a (non-rational) function of some parameter t so that all the square roots appearing in the expression of $\chi_{\underline{F(Y)}}$ are well defined;
- Compute $X = \chi_{\underline{F}(Y)}$;
- Compute $x = \Lambda_{\underline{a}}(X, Y)$ and $y = \Omega_{\underline{a}}(X, Y)$.

2.4   Cryptographic Applications

Once we will have found an encoding, it is important for cryptographic applications to study the cardinality of the subset of the curve that we parameterize. This ensures that we obtain convenient weak encodings for hashing into curves primitives (see [6]).

In the degree 3 examples given below, as in the higher genus family given in Section 4.2, we always will be able to deduce from the encoding formulae (sometimes after some resultant computations), a polynomial relation $P_{\underline{a}}(Y, t)$ between any Y of a point of the image and its preimages. Then the number of possible preimages is at most the t-degree of $P_{\underline{a}}(Y, t)$. Factorizing $P_{\underline{a}}(Y, t)$ over $\mathbb{F}_q$ gives then precisely the number of preimages. We detail this process for the genus 1 application of our method in Section 3.1 and sketch how to obtain such a polynomial in other sections.

Hyperelliptic curve cryptography relies on hashing into the Jacobian of a curve. Section 4.3 presents two different ways for accomplishing this task, which rely on encoding into the set of rational points of the curve.

We also need to know *in advance* which values of $\mathbb{F}_q$ cannot be encoded using such functions, in order to deterministically handle such cases. In the genus 1 as in other sections of our paper, this subset is always quite small (never more than several hundred elements) compared to cryptographic sizes, and only depends on the once and for all fixed curve parameters, therefore it can be taken into account and handled appropriately when setting up the cryptosystem. Furthermore, cryptographic encodings of [6] make a heavy use of hash functions onto the finite field before encoding on the curve; the output of the hash function can then be encoded with overwhelming probability.

3   Degree 3 Polynomials

In this section, we consider degree 3 polynomials. After easy changes of variables, any cubic can be written in its "depressed form" $X^3 + 3AX + 2B$, one root of which is

$$\chi_{A,B} = \sqrt[3]{-B + \sqrt{A^3 + B^2}} - \frac{A}{\sqrt[3]{-B + \sqrt{A^3 + B^2}}}.$$

In order to make use of this root while avoiding square roots, aiming at (non-rationally) parameterizing curves of positive genus, we first restrict to finite fields $\mathbb{F}_q$ with q odd and $q \equiv 2 \bmod 3$, so that computing cubic roots can be done thanks to a deterministic exponentiation to the e-th power, $e = 1/3 \bmod q - 1$. We then need to consider rational functions A and B in Y such that the curve $A(Y)^3 + B(Y)^2 - Z^2$ can be parameterized too.

For non-zero A, let $A(Y) = T(Y)$ for some T and $B(Y) = T(Y)S(Y)$ for some S, this problem is then the same as parameterizing the curve

$$T(Y) + S^2(Y) = Z^2. \tag{3.1}$$

This can be done with rational formulae when this curve is of genus 0, or with non-rational Icart's formulae when this curve is of genus 1. In the case of irreducible plane curves, this means that T and S are of low degree. Instead of parameterizing an auxiliary curve, we could have directly chosen T and S such that $T(Y) + S(Y)^2 = Z(Y)^2$ for some rational function Z. With comparable degrees for T and S as in the rest of the section, we obtain only genus 0 curves. Thus we have to greatly increase the degree of S and T in order to get higher genus curves. Those curves then have high degree but small genus: they have many singularities.

So, we finally consider in the following degree 3 equations of the form

$$X^3 + 3T(Y)X + 2S(Y)T(Y) = 0. \tag{3.2}$$

We could have considered the case $A = 0$ too, that is polynomials of the form $f_B = X^3 + 2B$. Our experiments in genus 1 and genus 2 yield curves that are isomorphic to hyperelliptic curves of any genus constructed from De Moivre's polynomials given in Section 4.2. We thus do not study this case further.

3.1   Genus 1 Curves

Parameterization. We made a systematic study of Curves (3.2) of (generic) genus 1 as a function of the degree of the numerators and the denominators of the rational function $S(Y)$ and $T(Y)$. Results are in Tab. 1, where we put altogether columns of compatible degrees. Typically, the first column (S a polynomial of degree at most 2 and T a constant) is a subcase of the second column (S of degree at most 3 and T a constant).

The only case of interest is when $S(Y)$ is a polynomial of degree at most 1 and $T(Y)$ is a polynomial of degree at most 2. When $q \equiv 2 \bmod 3$, these elliptic curves all have a $\mathbb{F}_q$-rational 3-torsion point, coming from $X = 0$.

Elliptic curves with a $\mathbb{F}_q$-rational 3-torsion point are known to have very fast addition formulae when given in "generalized" or "twisted" Hessian forms [9, 2]. Since $q \equiv 2 \bmod 3$, we even restrict in the following to classical Hessian elliptic curves.

Let us start from $S(Y) = 3(Y + a)/2$, $T(Y) = -Y/3$, that is curves of the type

$$C_{0,a} : Y^2 + XY + aY = X^3, \ a \neq 0, 1/27. \tag{3.3}$$

Table 1. Degrees of $S(Y)$ and $T(Y)$ for genus 1 plane curves given by Eq. (3.2)

		Degrees										
$S(Y)$	Num.	2	3	2	0	1	0	1	0	0	0	0
	Den.	0	0	0	1	0	0	0	1	0	1	0
$T(Y)$	Num.	0	0	1	1	1	2	2	0	0	0	0
	Den.	0	0	0	0	0	0	0	1	2	2	3
Genus of Eq. (3.1)		1	2	1	1	0	0	0	1	1	1	2

Then, the conic $S^2(Y) + T(Y) = 9/4\,Y^2 + (9/2\,a - 1/3)\,Y + 9/4\,a^2 = Z^2$ can be classically parameterized "by line" as

$$Y = \frac{12\,t^2 - 27\,a^2}{36\,t - 4 + 54\,a}, \quad Z = \frac{36\,t^2 + (-8 + 108\,a)\,t + 81\,a^2}{72\,t - 8 + 108\,a},$$

so that $X = \Delta/6 + 2Y/\Delta$ where $\Delta = \sqrt[3]{36Y\,(3\,Y + 3\,a + 2\,Z)}$.

Besides, Curve (3.3) is birationally equivalent to the Hessian model

$$E_d : x^3 + y^3 + 1 = 3\,dxy, \ d \neq 1, \tag{3.4}$$

with $a = (d^2 + d + 1)/3\,(d + 2)^3$ and

$$x = \frac{3\,(d + 2)^2\,(Y\,(d + 2) + X)}{3\,(d + 2)^2\,X + d^2 + d + 1}, \quad y = -\frac{d^2 + d + 1 + 3\,(d + 1)\,(d + 2)^2\,X + 3\,(d + 2)^3\,Y}{3\,(d + 2)^2\,X + d^2 + d + 1}. \tag{3.5}$$

The only remaining case is $d = -2$, that is the Hessian curve E_{-2} (the quadratic twist of the curve E_0, both have their j-invariant equal to 0). This curve is for instance isomorphic to a curve of the type (3.2) with $S = (1 - 7Y)/4$ and $T = -26\,(3\,Y^2 + 1)/27$. We might use this to parameterize E_{-2}, but it is much simpler to start from the curve $Y^2 + Y = X^3$, which can be much more easily parameterized with $Y = t$, $X = \sqrt[3]{t^2 + t}$. This curve is isomorphic to E_{-2} with $x = (X + 1)/(X + Y)$, $y = (-Y + X - 1)/(X + Y)$.

We summarize these calculations in Algorithm 1.

In addition, we have proved what follows.

Theorem 1. Let $\mathbb{F}_q$ be the finite field with q elements. Suppose q odd and $q \equiv 2 \bmod 3$. Let $E_d/\mathbb{F}_q$ be the elliptic curve defined by Eq. (3.4).

Then Algorithm 1 computes a deterministic encoding e_d to E_d, from $\mathbb{F}_q^*$ if $d = -2$ and from $\mathbb{F}_q \setminus \left\{ \frac{(2\,d+1)(d^2+d+7)}{18\,(d+2)^3} \right\}$ otherwise, in time $\mathcal{O}(\log^{2+o(1)} q)$.

Remark 4. This encoding is not defined on all of $\mathbb{F}_q$, however we can map the missing value to the point at infinity on the curve.

Algorithm 1: HessianEncode

---

input : A Hessian elliptic curve
$$E_d/\mathbb{F}_q : x^3 + y^3 + 1 = 3\,dxy,\ d \neq 1,\text{ and } t \in \mathbb{F}_q.$$
output: A point $(x_t : y_t : 1)$ on E_d.

if $d = -2$ **then**   /* $t \neq 0$ */

   | $Y := t;\ X := (t + t^2)^{1/3 \bmod q-1}$;

   | $x_t := (X + 1)/(X + Y);\ y_t := (-Y + X - 1)/(X + Y)$;

   | **return** $(x_t : y_t : 1)$

$a := \dfrac{d^2 + d + 1}{3\,(d+2)^3}$;   /* $t \neq \dfrac{(2\,d+1)(d^2+d+7)}{18\,(d+2)^3}$ */

if $t = \pm 3a/2$ **then**

   | $Y := 0;\ X := 0$;

else   /* $Y \neq 0$ */

   | $Y := \dfrac{12\,t^2 - 27\,a^2}{36\,t + 54\,a - 4};\ \Delta := (36\,Y\,(2\,t + 3\,a))^{1/3 \bmod q-1}$;

   | $X := \Delta/6 + 2\,Y/\Delta$;

$x_t := \dfrac{3\,(d+2)^2\,(Y\,(d+2) + X)}{3\,(d+2)^2\,X + d^2 + d + 1}$;

$y_t := -\dfrac{3\,(d+1)\,(d+2)^2\,X + 3\,(d+2)^3\,Y + d^2 + d + 1}{3\,(d+2)^2\,X + d^2 + d + 1}$;

return $(x_t : y_t : 1)$

Fig. 1. Encoding on Hessian elliptic curves

Number of curves. A way quantifying the number of curves defined by Eq. (3.4) is to compute their j-invariant. Here, we obtain

$$j_{E_d} = 27\,d^3\,\frac{(d+2)^3\left(d^2 - 2\,d + 4\right)^3}{(d-1)^3\left(d^2 + d + 1\right)^3}. \tag{3.6}$$

When $q \equiv 2 \bmod 3$, there are exactly $\lfloor q/2 \rfloor$ distinct such invariants. Additionally, one can show that there exists $q-1$ distinct $\mathbb{F}_q$-isomorphic classes of Hessian elliptic curves (see [9]).

Cardinality of the image. It is obvious to see that $|\operatorname{Im} e_{-2}| = q - 1$, simply because $Y = t \neq 0$. Now, determining $|\operatorname{Im} e_d|$ for $d \neq 1, -2$ needs some more work, but can still be evaluated exactly.

Theorem 2. *Let $d \neq 1, -2$, then $|\operatorname{Im} e_d| = (q + 1)/2$ if $(d - 1)/(d + 2)$ is a quadratic residue in $\mathbb{F}_q$ and $|\operatorname{Im} e_d| = (q - 1)/2$ otherwise.*

Proof. Let $(x : y : 1)$ be a point on E_d, then there exists a unique point $(X : Y : 1)$ on $C_{0,a}$ sent by Isomorphism (3.5) to $(x : y : 1)$.

Viewed as a polynomial in t, the equation $12\,t^2 - 36\,Yt - 54\,Ya - 27\,a^2 + 4\,Y$ has 0 or 2 solutions except when $27\,Y^2 + (-4 + 54\,a)\,Y + 27\,a^2 = 0$. The latter has no root in Y if $1 - 27\,a = (d - 1)^3/(d + 2)^3$ is a quadratic non-residue, and two distinct roots denoted Y_0 and Y_1 otherwise (if $a = 1/27$, the curve $C_{0,a}$ degenerates into a genus 0 curve).

Let us summarize when $(d-1)/(d+2)$ is a quadratic residue in $\mathbb{F}_q$.

- (1 element) If $t \in \left\{ \frac{(2d+1)(d^2+d+7)}{18(d+2)^3} \right\}$, then t is not encodable by e_d;
- (2 elements) If $t \in \{\pm \frac{d^2+d+1}{2(d+2)^3}\}$, then $e_d(t) = (0 : -1 : 1)$;
- (2 elements) If t_i is a (double) root of $12\,t^2 - (36\,t - 4 + 54\,a)\,Y_i - 27\,a^2$ with $i = 0, 1$, we obtain two distinct points $e_d(t_i) = (x_{t_i} : y_{t_i} : 1)$;
- ($q - 5$ elements) Else, for each remaining t, there exists exactly one other t' such that $e_d(t) = e_d(t') = (x_t : y_t : 1)$.

We thus obtain $(q - 5)/2 + 2 + 1 = (q + 1)/2$ distinct rational points on the curve. Similarly if $(d-1)/(d+2)$ is a quadratic non-residue in $\mathbb{F}_q$, we obtain $(q - 1)/2$ distinct rational points on E_d. □

Related work. Compared to Icart's formulae [13], this encoding has two drawbacks of limited practical impact:

- it does not work for any elliptic curves, but only for Hessian curves;
- the subset of the curve which can be parameterized is slightly smaller than in Icart's case: we get $\simeq q/2$ points against approximately $5/8 \#E \pm \lambda\sqrt{q}$.

Nonetheless, it has three major practical advantages:

- recovering the parameter t from a given point $(x : y : 1)$ is much easier: we only have to find the roots of a degree 2 equation instead of a degree 4 one;
- the parameter t only depends on y: we can save half of the bandwidth of a protocol by sending only y and not the whole point $(x : y : 1)$;
- Y is computable using only simple (rational) finite field operations: no exponentiation is required, but it carries the whole information on the encoded point. It is thus preferable for encoding purposes to work on the $C_{0,a}$ model rather than on the Hessian model[1].

3.2   Genus 2 Curves

Parameterizations. In the same spirit as in Section 3.1, we made a systematic study of Curves (3.2) of (generic) genus 2 as a function of the degree of the numerators and the denominators of the rational function $S(Y)$ and $T(Y)$. Results are in Tab. 2.

We can see that there are three cases of interest:

- $S(Y)$ and $T(Y)$ be both a rational function of degree 1 ;
- $S(Y)$ be a rational function of degree 2 and $T(Y)$ be a constant ;
- $S(Y)$ be a constant and $T(Y)$ be a rational function of degree 2.

---

[1] For example, we could imagine that a limited power device computes the encoded y and sends it to an other device specialized in curve operations, which in turn computes the associated x and realizes the group operations.

Table 2. Degrees of $S(Y)$ and $T(Y)$ for genus 2 plane curves given by Eq. (3.2)

		Degrees																	
$S(Y)$	Num.	2	0	1	2	2	2	1	1	0	1	1	1	1	2	0	0	0
	Den.	1	2	2	2	0	1	1	0	1	1	1	0	0	0	0	0	0
$T(Y)$	Num.	0	0	0	0	0	0	0	1	1	1	1	0	1	0	1	2	2
	Den.	0	0	0	0	1	1	1	1	1	1	0	1	2	2	2	1	2
Genus of Eq. (3.1)		1	1	1	1	2	2	1	1	1	1	1	2	2	3	1	1	1

We now study the two first cases. We omit the third one because it turns out that it yields curves already obtained in the second case.

$S(Y)$ and $T(Y)$ rational functions of degree 1. Let $S(Y) = (\alpha Y + \beta)/(\gamma Y + \delta)$ and $T(Y) = (\varepsilon Y + \varphi)/(\mu Y + \nu)$, then Curve (3.2) is birationally equivalent to curves of the form $y^2/d^2 = (x^3 + 3\,ax + 2\,c)^2 + 8\,bx^3$ where

$$a = \frac{\delta\varepsilon - \gamma\varphi}{\delta\mu - \gamma\nu}, \quad b = \frac{(\alpha\delta - \gamma\beta)(\mu\varphi - \varepsilon\nu)}{(\delta\mu - \gamma\nu)^2}, \quad c = \frac{\beta\varepsilon - \alpha\varphi}{\delta\mu - \gamma\nu} \quad \text{and} \quad d = (\delta\mu - \gamma\nu).$$

Many of theses curves are isomorphic to each other and, without any loss of generality, we can set $c = 1$ and $d = 1$. We thus finally restrict to $S(Y) = -Y$, $T(Y) = (a^2Y + a)/(aY + b + 1)$, so that, when $4\,a^6b^3 - b^3\,(b^2 + 20\,b - 8)\,a^3 + 4\,b^3\,(b+1)^3 \neq 0$, Curve (3.2) is birationally equivalent to the Weierstrass model of a genus 2 curve,

$$H_{1,a,b} : y^2 = (x^3 + 3\,ax + 2)^2 + 8\,bx^3 \,, \tag{3.7}$$

with $x = X$ and $y = -4\,aY + X^3 + 3\,aX - 2$.

Besides, Curve

$$S^2(Y) + T(Y) = Y^2 + (a^2\,Y + a)/(aY + 1 + b) = Z^2 \tag{3.8}$$

is birationally equivalent to the Weierstrass elliptic curve

$$V^2 = U^3 + (-a^6 + 2\,(b+1)(2\,b - 1)a^3 - (b+1)^4)\frac{U}{3}$$
$$+ \frac{1}{27}\,(2\,a^9 + 3\,(2 - 2\,b + 5\,b^2)a^6 - 6\,(2\,b - 1)(b+1)^3a^3 + 2\,(b+1)^6). \tag{3.9}$$

The latter can now be parameterized with Icart's method. This yields

$$U = \frac{1}{6}\sqrt[3]{\frac{2\delta}{t^2}} + \frac{t^2}{3}, \quad V = \frac{1}{6}\sqrt[3]{2\delta t} + \frac{t^3}{6} + \frac{1}{6t}\,(-a^6 + 2\,(b+1)(2\,b - 1)a^3 - (b+1)^4)$$

with

$$\delta = -t^8 + (-12\,(b+1)(2\,b - 1)a^3 + 6\,a^6 + 6\,(b+1)^4)t^4 + (12\,(2\,b - 5\,b^2 - 2)a^6 - 8\,(b+1)^6$$
$$- 8\,a^9 + 24\,(2\,b - 1)(b+1)^3a^3)t^2 + 3\,(a^6 - 2\,(b+1)(2\,b - 1)a^3 + (b+1)^4)^2$$

Algorithm 2: Genus2TypeAEncode

---

input : A curve $H_{1,a,b}$ defined by Eq. (3.7) on $\mathbb{F}_q$, an element $t \in \mathbb{F}_q \setminus S_1$
output: A point $(x_t : y_t : 1)$ on $H_{1,a,b}$

$\delta := -t^8 + (-12(b+1)(2b-1)a^3 + 6a^6 + 6(b+1)^4)t^4 + (12(2b-5b^2-2)a^6 - 8(b+1)^6$
$\qquad -8a^9 + 24(2b-1)(b+1)^3a^3)t^2 + 3(a^6 - 2(b+1)(2b-1)a^3 + (b+1)^4)^2;$
$U := ((2\delta/t^2)^{1/3 \bmod q-1} + 2t^2)/6;$
$V := (2\delta t)^{1/3 \bmod q-1}/6 + t^3/6 + (-a^6 + 2(b+1)(2b-1)a^3 - (b+1)^4)/6t;$
$W := -3Ua + a((b+1)^2 + a^3);\ Y := (3(b+1)U + (2b-1)a^3 - (b+1)^3)/W;\ Z := 3V/W;$
$T := (a^2Y + a)/(aY + b + 1);\ \Delta := (T(Z+Y))^{1/3 \bmod q-1};$
$x_t := \Delta - T/\Delta;\ y_t := -4aY + X^3 + 3aX - 2;$
return $(x_t : y_t : 1)$

Fig. 2. Encoding on genus 2 curves (of the type A)

Now, back by the birational change of variables between Curve (3.9) and Curve (3.8), we get Y and Z from U and V (*cf.* Algorithm 2 for precise formulae). Let now $\Delta = \sqrt[3]{T(Y)(Z - S(Y))}$, then $X = \Delta - T(Y)/\Delta$.

So, we obtain the following theorem.

Theorem 3. *Let $\mathbb{F}_q$ be the finite field with q elements. Suppose q odd and $q \equiv 2$ mod 3. Let $H_{1,a,b}/\mathbb{F}_q$ be the hyperelliptic curve of genus 2 defined by Eq. (3.7).*

Then, Algorithm 2 computes a deterministic encoding $e_{1,a,b} : \mathbb{F}_q^* \setminus S_1 \to H_{1,a,b}$, where S_1 is a subset of $\mathbb{F}_q$ of size at most 35, in time $\mathcal{O}(\log^{2+o(1)} q)$.

Proof. The previous formulae define a deterministic encoding provided that t, W, $aY + b + 1$ and Δ are not 0.

The condition $W = 0$ yields a polynomial in t of degree 8, we thus have at most 8 values for which $W = 0$. Similarly, the condition $aY + b + 1 = 0$ yields at most 8 additional values for which $W = 0$.

Now $\Delta = 0$ if and only if $T = 0$ or $Z = -Y$. The condition $T = 0$ yields 8 additional values. Similarly, the condition $Z + Y = 0$ yields a polynomial in t of degree 10, we thus have in this case at most 18 values for which $\Delta = 0$.

The total number of field elements which cannot be encoded finally amounts to at most 35. $\qquad\square$

Cardinality of the image. Let (X, Y) be a rational point on a $C_{1,a,b,c}$ curve, let t be a possible preimage of (X, Y) by our encoding $e_{1,a,b}$. Then there exists a polynomial relation in Y and t of degree at most 8 in t (*cf.* Algorithm 2). Hence (X, Y) has at most 8 preimages by $e_{1,a,b}$. Therefore, $|\operatorname{Im} e_{1,a,b}| \geqslant (q - 35)/8$.

Number of curves. Igusa invariants of these curves are equal to

$$J_2 = 2^6 \, 3 \, (-9a^3 + 4b^2 + 4b - 9),$$
$$J_4 = 2^{10} \, 3 \, (-9b(4b - 15)a^3 + 4b(b+1)(2b^2 + 2b - 27)),$$
$$J_6 = 2^{14} \, (729 \, a^6 b^2 - 216 \, b^2 \, (2b^2 + 3b + 21) \, a^3 + 16 \, b^2 \, (4b^2 + 4b + 81) \, (b+1)^2),$$
$$J_8 = 2^{18} \, 3 \, (-6561 \, a^9 b^2 + 2916 \, b^2 \, (-7 + b^2 + 13b) \, a^6$$
$$\qquad -144 \, b^2 \, (4b^4 + 63 \, b^3 + 450 \, b^2 - 149 \, b - 810) \, a^3$$
$$\qquad +64 \, b^2 \, (b^4 + 2b^3 + 154 \, b^2 + 153 \, b - 729) \, (b+1)^2),$$
$$J_{10} = 2^{28} \, 3^6 \, (4 \, a^6 b^3 - b^3 \, (b^2 + 20b - 8) \, a^3 + 4 \, b^3 \, (b+1)^3).$$

The geometric locus of these invariants is a surface of dimension 2 given by a homogeneous equation of degree 90 (which is far too large to be written here). Consequently, Eq. (3.7) defines $\mathcal{O}(q^2)$ distinct curves over $\mathbb{F}_q$.

$S(Y)$ be a rational function of degree 2. Let now $S(Y) = (\alpha Y^2 + \beta Y + \gamma) / (\delta Y^2 + \varepsilon Y + \varphi)$ and $T(Y) = \kappa$, then Curve (3.2) is birationally equivalent to curves of the form $y^2/\lambda = (x^3 + 3\mu x + 2a)^2 + 4b$ where

$$\lambda = \varepsilon^2 - 4\varphi\delta\,, \ \mu = \kappa\,, \ a = \frac{\kappa}{\lambda}(\varepsilon\beta - 2\delta\gamma - 2\varphi\alpha) \ \text{and} \ b = \frac{\kappa^2}{\lambda}(\beta^2 - 4\alpha\gamma) - a^2\,.$$

Many of theses curves are isomorphic to each other and, without any loss of generality, we can set λ and μ to be either any quadratic residues (for instance $\lambda, \mu = 1$) or any non-quadratic residues (for instance $\lambda, \mu = -3$ because $q \equiv 2 \bmod 3$).

We finally arrive to

$$S(Y) = \frac{\lambda(a-u)Y^2 - 4vY - 4(a+u)}{\mu(\lambda Y^2 - 4)} \ \text{and} \ T(Y) = \mu\,,$$

where $u = \mu^3/2\,w - w/2 - a$ for some $w \in \mathbb{F}_q^*$. Then, when $b^3\lambda^{10}(\mu^6 + 2\mu^3 a^2 - 2b\mu^3 + a^4 + 2ba^2 + b^2) \neq 0$, Curve (3.2) is birationally equivalent to the Weierstrass model of a genus 2 curve,

$$H_{2,\lambda,\mu,a,v,w} : y^2/\lambda = (x^3 + 3\mu x + 2a)^2 + 4b\,, \tag{3.10}$$

where $b = v^2/\lambda - u^2$ for some v in $\mathbb{F}_q$, $x = X$ and $y = \lambda(X^3/2 + 3\mu X/2 + a - u)Y - 2v$.

We may remark that computing v and w from b is the same as computing a point $(v : w : 1)$ on the elliptic curve $v^2/\lambda - (\mu^3/2\,w - w/2 - a)^2 - b = 0$. This can be done in deterministic time from Icart's formulae when one can exhibit a $\mathbb{F}_q$-rational bilinear change of variable between this curve and a cubic Weierstrass model, typically when $\lambda = 1$ (but no more when $\lambda = -3$).

Besides, let $z = w/2 + r^3/2w$ and thus $(u + a)^2 + r^3 = z^2$, then

$$\mu^2(\lambda Y^2 - 4)^2(S(Y)^2 + T(Y)) = -\lambda^2(4\,ua - z^2)Y^4 - 8\lambda v(a-u)Y^3$$
$$- 8\lambda(4\mu^3 - 3z^2 - 2b + 6\,ua + 4a^2)Y^2 + 32\,v(u+a)Y + 16\,z^2 = Z^2 \tag{3.11}$$

is birationally equivalent to the Weierstrass elliptic curve

$$V^2 = U^3 + 2^8\lambda^2(-\mu^6 + (b - 2a^2)\mu^3 - (a^2 + b)^2)U/3 +$$
$$2^{12}\lambda^3(2\mu^9 + (6a^2 - 3b)\mu^6 - 3(a^2 + b)(b - 2a^2)\mu^3 + 2(a^2 + b)^3)/3^3\,. \tag{3.12}$$

The latter can now be parameterized with Icart's method. This yields

$$U = \frac{1}{6}\sqrt[3]{\frac{2\delta}{t^2}} + \frac{t^2}{3}, \ V = \frac{1}{6}\sqrt[3]{2\delta t} + \frac{t^3}{6} + 128\,(-\mu^6 + (b - 2a^2)\mu^3 - (b + a^2)^2)\frac{\lambda^2}{3t}$$

with

$$\delta = -t^8 + 2^9\, 3\, (\,\mu^6 + (-b + 2\,a^2)\mu^3 + (a^2 + b)^2)\lambda^2 t^4 +$$
$$2^{14}(-2\,\mu^9 - (6\,a^2 - 3\,b)\mu^6 + 3\,(a^2 + b)(b - 2\,a^2)\mu^3 - 2\,(a^2 + b)^3)\lambda^3 t^2 +$$
$$2^{16}\, 3(\,\mu^{12} + (-2\,b + 4\,a^2)\mu^9 + (3\,b^2 + 6\,a^4)\mu^6 + 2\,(a^2 + b)^2(-b + 2\,a^2)\mu^3 + (a^2 + b)^4)\lambda^4\,.$$
$$(3.13)$$

Again, back by a birational change of variables between Curves (3.12) and (3.11), we get Y and Z from U and V (*cf.* Algorithm 3 for precise formulae). Let now $\Delta = \sqrt[3]{T(Y)\,(Z/\mu(\lambda Y^2 - 4) - S(Y))}$, then $X = \Delta - T(Y)/\Delta$.

Algorithm 3: Genus2TypeBEncode

input : A curve $H_{2,\lambda,\mu,a,v,w}$ defined by Eq. (3.10) on $\mathbb{F}_q$, an element $t \in \mathbb{F}_q \setminus S_2$.
output: A point $(x_t : y_t : 1)$ on $H_{2,\lambda,\mu,a,v,w}$

$u := -(2\,aw + w^2 - r^3)/2w;\ b := v^2/t - u^2;\ z := (w^2 + r^3)/2w;$
$\delta := -t^8 + 2^9\,3\,(\mu^6 + (-b + 2\,a^2)\mu^3 + (a^2 + b)^2)\lambda^2 t^4 +$
$\quad 2^{14}(-2\,\mu^9 - (6\,a^2 - 3\,b)\mu^6 + 3\,(a^2 + b)(b - 2\,a^2)\mu^3 - 2\,(a^2 + b)^3)\lambda^3 t^2 +$
$\quad 2^{16}\,3\,(\mu^{12} + (-2\,b + 4\,a^2)\mu^9 + (3\,b^2 + 6\,a^4)\mu^6 + 2\,(a^2 + b)^2(-b + 2\,a^2)\mu^3 + (a^2 + b)^4)\lambda^4;$
$U := ((2\,\delta/t^2)^{1/3\bmod q-1} + 2t^2)/6;$
$V := (2\delta t)^{1/3\bmod q-1}/6 + t^3/6 + 128\,(-\mu^6 + (b - 2\,a^2)\mu^3 - (b + a^2)^2)\lambda^2/3t;$
$W := -9\,U^2 - 48\,\lambda(-3\,z^2 - 2\,b + 6\,ua + 4\,a^2 + 4\,\mu^3)U + 256\,(-4\,\mu^6 + (6\,z^2 + a^2 - 12\,ua + 4\,b)\mu^3 +$
$\quad (b + a^2)(5\,a^2 + 6\,ua - b - 3\,z^2))\lambda^2;$
$Y := (-288\,v(u + a)U - 72\,zV + 1536\,\lambda v(bu + a^3 - 2\,\mu^3 u + ab + a\mu^3 + ua^2))/W;$
$Z := -(-324\,z\,U^4 + (6912\,\lambda\mu^3 z + 1728\,\lambda z(-3\,z^2 - 2\,b + 6\,ua + 4\,a^2))U^3 - 2592\,v(u + a)U^2 V$
$\quad + (-27648\,\lambda^2 z(b + a^2)(2\,a^2 + 6\,ua - 4\,b - 3\,z^2) + 193536\,\lambda^2 z\mu^6 - 27648\,\lambda^2 z(-5\,a^2 - 12\,ua + 6\,z^2 + 7\,b)\mu^3)U^2$
$\quad + (27648\,\lambda v(-2\,u + a)\mu^3 + 27648\,\lambda v(b + a^2)(u + a))UV + (49152\,\lambda^3 z(36\,a^3 u - 18\,a^2 z^2 + 12\,a^4 + 9\,z^2 b + 30\,b^2$
$\quad - 12\,a^2 b - 18\,aub)\mu^3 + 49152\,\lambda^3 z(-6\,b + 18\,ua + 12\,a^2 - 9\,z^2)\mu^6 + 49152\,\lambda^3 z(b + a^2)^2(4\,a^2 + 18\,ua$
$\quad - 14\,b - 9\,z^2) + 196608\,\lambda^3\mu^9 z)U + (-73728\,v\lambda^2(b + a^2)^2(u + a) - 73728\,v\lambda^2(4\,u - 8\,a)\mu^6 - 73728\,v\lambda^2$
$\quad (-4\,bu + 9\,z^2 a - 7\,a^3 - 13\,ua^2 + 2\,ab)\mu^3)V - 7340032\,\lambda^4\mu^{12} z - 262144\,\lambda^4 z(60\,ua - 56\,b + 85\,a^2 - 30\,z^2)\mu^9$
$\quad - 262144\,\lambda^4 z(b + a^2)(31\,a^4 + 72\,a^3 u - 10\,a^2 b - 36\,a^2 z^2 + 18\,aub + 13\,b^2 - 9\,z^2 b)\mu^3 - 262144\,\lambda^4 z(b + a^2)^3$
$\quad (a^2 + 6\,ua - 5\,b - 3\,z^2) - 262144\,\lambda^4 z(15\,b^2 + 87\,a^4 - 63\,a^2 z^2 + 45\,z^2 b - 90\,aub - 33\,a^2 b + 126\,a^3 u)\mu^6)/W^2;$
$S := (-u + a)Y^2\lambda - 4\,vY - 4\,a - 4\,u;\ \Delta := \sqrt[3]{(Z - S)/(\lambda Y^2 - 4)};$
$x_t := \Delta - \mu/\Delta;\ y_t := \lambda\,(X^3/2 + 3\,\mu X/2 + a - u)\,Y - 2\,v;$
return $(x_t : y_t : 1)$

Fig. 3. Encoding on genus 2 curves (of the type B)

So, we obtain the following theorem.

Theorem 4. *Let $\mathbb{F}_q$ be the finite field with q elements. Suppose q odd and $q \equiv 2 \mod 3$. Let $H_{2,\lambda,\mu,a,v,w}/\mathbb{F}_q$ be the hyperelliptic curve of genus 2 defined by Eq. (3.10).*

Then, Algorithm 3 computes a deterministic encoding $e_{2,\lambda,\mu,a,v,w} : \mathbb{F}_q^* \setminus S_2 \to H_{2,\lambda,\mu,a,v,w}$, where S_2 is a subset of $\mathbb{F}_q$ of size at most 233, in time $\mathcal{O}(\log^{2+o(1)} q)$.

Proof. The previous formulae defines a deterministic encoding provided that t, W, $\lambda Y^2 - 4$ and $Z - S$ are not 0.

The condition $W = 0$ yields a polynomial in U of degree 2, we thus have at most 2 values for U for which $W = 0$. Each value of U then yields a polynomial in t, derived from δ, of degree 8. We thus have at most 16 values for t to avoid in this case.

The condition $\lambda Y^2 - 4 = 0$ similarly yields 2 values for Y. Each such value yields in return a polynomial of degree 2 in U, and degree 1 in V, which can be seen as a curve in t and $\tau = \sqrt[3]{2 t \delta}$ of degree at most 6. Besides $\tau^3 = 2 t \delta$ is a curve of degree at most 9. Bezout's theorem yields thus a maximal number of $2 \times 6 \times 9 = 108$ intersection points, or equivalently values for t, to avoid in this case.

Finally, the condition $Z = S$ can be seen as a curve in t and τ of degree 12. Thus, this yields a maximal number of $12 \times 9 = 108$ values too.

So, the total number of field elements which cannot be encoded finally amounts to at most $1 + 16 + 2 \times 108 = 233$. $\square$

Cardinality of the image. Let (X, Y) be a rational point on $H_{2,\lambda,\mu,a,v,w}$ and t a preimage by $e_{2,\lambda,\mu,a,v,w}$. Then we have seen in the proof of Theorem 4 that t and $\tau = \sqrt[3]{2 t \delta}$ are defined as intersection points of two curves, one of degree 7 parameterized by Y and the other one of degree 9 from the definition of δ. In full generality, this might yield for some curves and some of their points a total number of at most 54 t's. Therefore, $|\operatorname{Im} e_{1,a,b}| \geqslant (q - 233)/63$.

Number of curves. Igusa invariants of these curves are equal to

$$J_2 = -2^6 \, 3 \, \lambda^2 (9 \, \mu^3 + 9 \, a^2 + 10 \, b),$$
$$J_4 = 2^9 \, 3 \, b \lambda^4 (297 \, \mu^3 + 54 \, a^2 + 55 \, b),$$
$$J_6 = 2^{14} \, b^2 \lambda^6 (-6480 \, \mu^3 + 81 \, a^2 + 80 \, b),$$
$$J_8 = -2^{16} \, 3 \, b^2 \lambda^8 (31347 \, \mu^6 - 134136 \, \mu^3 a^2 - 158310 \, b \mu^3 + 11664 \, a^4 + 23940 \, b a^2 + 12275 \, b^2),$$
$$J_{10} = -2^{24} \, 3^6 \, b^3 \lambda^{10} (\mu^6 + 2 \, \mu^3 a^2 - 2 \, b \mu^3 + a^4 + 2 \, b a^2 + b^2).$$

Here, the geometric locus of these invariants is a surface of dimension 2 given by a homogeneous equation of degree 30,

$$11852352 \, J_2{}^5 J_{10}{}^2 + 196992 \, J_2{}^5 J_4 \, J_6 \, J_{10} - 362998800 \, J_2{}^3 J_4 \, J_{10}{}^2 + 64 \, J_2{}^6 J_6{}^3 - 636672 \, J_2{}^4 J_6{}^2 J_{10}$$
$$- \, 895349625 \, J_2{}^2 J_6 \, J_{10}{}^2 - 64097340625 \, J_{10}{}^3 - 373248 \, J_2{}^4 J_4{}^3 J_{10} - 4466016 \, J_2{}^3 J_4{}^2 J_6 \, J_{10}$$
$$+ \, 2903657625 \, J_2 \, J_4{}^2 J_{10}{}^2 - 3984 \, J_2{}^4 J_4 \, J_6{}^3 + 606810 \, J_2{}^2 J_4 \, J_6{}^2 J_{10} + 3383973750 \, J_4 \, J_6 \, J_{10}{}^2 + 1647 \, J_2{}^3 J_6{}^4$$
$$+ \, 49583475 \, J_2 \, J_6{}^3 J_{10} + 11290752 \, J_2{}^2 J_4{}^4 J_{10} + 38072430 \, J_2 \, J_4{}^3 J_6 \, J_{10} + 76593 \, J_2{}^2 J_4{}^2 J_6{}^3$$
$$- \, 115457700 \, J_4{}^2 J_6{}^2 J_{10} + 20196 \, J_2 \, J_4 \, J_6{}^4 - 530604 \, J_6{}^5 - 85386312 \, J_4{}^5 J_{10} - 468512 \, J_4{}^3 J_6{}^3.$$

This shows that Eq. (3.10) defines $\mathcal{O}(q^2)$ distinct curves over $\mathbb{F}_q$.

4 Hyperelliptic Curves of Any Genus

In this section, we present two families of parametric polynomials which provide deterministic parameterizable hyperelliptic curves of genus $g \geqslant 2$.

4.1 Quasiquadratic Polynomials

Curves of the form $y^2 = f(x^d)$ where f is a family of solvable polynomials whatever is its constant coefficient may yield parameterizable hyperelliptic curves. Typically, we may consider polynomials f of degree 2, 3 or 4 or some solvable families of higher degree polynomials. Here, we restrict ourselves to quadratic polynomials since it yields non trivial hyperelliptic curves for any genus.

We define quasiquadratic polynomials as follows.

Definition 1 (Quasiquadratic polynomials). *Let $\mathbb{K}$ be a field and d be an integer coprime with* char $\mathbb{K}$. *The family of quasiquadratic polynomials $q_{a,b}(x) \in \mathbb{K}[x]$ of degree $2d$ is defined for $a, b \in \mathbb{K}$ by $q_{a,b}(x) = x^{2d} + ax^d + b$.*

Quasiquadratic polynomials define an easily parameterized family of hyperelliptic curves $y^2 = q_{a,b}(x)$ (see Algorithm 4). When d does not divide $q - 1$ and when $a \neq 0$, these curves are isomorphic to curves $y^2 = q_{1,a}(x)$ by the variable substitution $x \to a^{1/d}x$. When $a = 0$, we are reduced to the unique well-known curve $y^2 = x^{2d} + b$ which can be parameterized by $t \mapsto \left(\sqrt[d]{(-b + t^2)/(2t)}, (b + t^2)/(2t) \right)$.

Algorithm 4: QuasiQuadraticEncode

input : A curve $H_a : x^{2d} + x^d + a = y^2$, and $t \in \mathbb{F}_q \setminus \{1/2\}$.
output: A point $(x_t : y_t : 1)$ on H_a

$\alpha := (t^2 - a)/(1 - 2t)$;
$x_t := \alpha^{1/d}$; $y_t := (-a + t - t^2)/(1 - 2t)$;
return $(x_t : y_t : 1)$

Fig. 4. Encoding on quasiquadratic curves

Theorem 5. *Let $\mathbb{F}_q$ be the finite field with q elements. Suppose q odd and $q \neq 2, 3$ and d coprime with $q - 1$. Let $H_a/\mathbb{F}_q : y^2 = x^{2d} + x^d + a$ be an hyperelliptic curve where a is such that the quasiquadratic polynomial $q_{1,a}$ has a non-zero discriminant over $\mathbb{F}_q$.*

Algorithm 4 computes a deterministic encoding $e_a : \mathbb{F}_q^* \setminus \{1/2\} \to H_a$ in time $\mathcal{O}(\log^{2+o(1)} q)$.

Genus of H_a. Let $q_{1,a} \in \mathbb{F}_q[X]$ and $H_a : q_{1,a}(x) = y^2$, where $q_{1,a}$ has degree $2d$. We have requested that the discriminant of $q_{1,a}$ is not 0. This implies that $q_{1,a}$ has exactly $2d$ distinct roots. Thus H_a has genus $d - 1$ provided H_a has no singularity except at the point at infinity.

It remains to study the points of the curve where both derivatives in x and y are simultaneously 0. This implies $y = 0$. Thus the only singular points are the common roots of $q_{1,a}(x)$ and its derivative. Since we request that the discriminant of $q_{1,a}$ is not 0, there are no singular points.

For $d = 3$, H_a is the well known family of genus 2 curves with automorphism group D_{12} [7]. The geometric locus of these curves is a one-dimensional variety in the moduli space. Moreover, when $x \to x^d$ is invertible over $\mathbb{F}_q$, these curves all have exactly $q + 1$ $\mathbb{F}_q$-points (but they have a much better distributed number of $\mathbb{F}_{q^2}$-points).

The encoding. The parameterization is quite simple. Let $H_a : x^{2d} + x^d + a = y^2$ be a quasiquadratic hyperelliptic curve. Setting $x = \alpha^{1/d}$ reduces the parameterization of H_a to the parameterization of the conic $\alpha^2 + \alpha + a - y^2 = 0$, which easily gives $\alpha = (-a + t^2)/(1 - 2t)$ and $y = (-a + t - t^2)/(1 - 2t)$ for some parameter t. We finally obtain Algorithm 4.

Cardinality of the image.

Theorem 6. *Given a rational point $(x : y : 1)$ on $H_a : q_{1,a}(x) = y^2$, the equation $e_a(t) = (x : y : 1)$ has exactly 1 solution. Thus, $|\operatorname{Im} e_a| = q - 1$*

Proof. Let $\alpha = x^d$, then t is a solution of the degree 1 equation $y + \alpha = ta/(a - 2t)$. □

4.2 De Moivre's Polynomials

This well-known family of degree 5 polynomials was first introduced by De Moivre for the study of trigonometric equalities and its study from a Galoisian point of view was done by Borger in [4]. This definition can be easily generalized for any odd degree.

Definition 2 (De Moivre's polynomials). *Let $\mathbb{K}$ be a field and d be an odd integer coprime with* char $\mathbb{K}$*. The family of De Moivre's polynomials $p_{a,b}(x) \in \mathbb{K}[x]$ of degree d is defined for $a, b \in \mathbb{K}$ by*

$$p_{a,b}(x) = x^d + dax^{d-2} + 2da^2x^{d-4} + 3da^3x^{d-6} + \cdots + 2da^{(d-1)/2-1}x^3 + da^{(d-1)/2}x + b\,.$$

Examples. De Moivre's polynomials of degree 5 are $x^5 + 5ax^3 + 5a^2x + b$. De Moivre's polynomials of degree 13 are $x^{13} + 13ax^{11} + 26a^2x^9 + 39a^3x^7 + 39a^4x^5 + 26a^5x^3 + 13a^6x + b$.

Borger proved in [4] that De Moivre's polynomials of degree 5 are solvable by radical, the same is true for De Moivre's polynomials of any degree.

Lemma 1 (Resolution of De Moivre's polynomials). *Let $p_{a,b}$ be a De Moivre's polynomial of degree d, let θ_0 and θ_1 be the roots of $q_{a,b}(\theta) = \theta^2 + b\theta - a^d$, then the roots of $p_{a,b}$ are*

$$(\omega_k \theta_0^{1/d} + \omega_k^{d-1} \theta_1^{1/d})_{0 \leqslant k < d}$$

where $(\omega_k)_{0 \leqslant k < d}$ are the d-th roots of unity.

Proof. As in the case of degree 5 (see [4]), we do the variable substitution $x = \gamma - a/\gamma$, then γ^d is a root of the polynomial $q_{a,b}(\theta)$. □

De Moivre's polynomials also define a family of deterministically parameterized hyperelliptic curves for any genus.

Algorithm 5: DeMoivreEncode

---

input : A curve $H : p_{a,b}(x) - y^2 = 0$ and $t \in \mathbb{F}_q^* \setminus \mathcal{S}$.
output: A point $(x_t : y_t : 1)$ on H

if $a = 0$ **then**
    | **return** $\left((t^2 - b)^{1/d \bmod q-1} : t : 1\right)$

$\delta := -(3a^d + b^2 + t^4)/6t - 2b^3/27 - a^d b/3 - t^6/27; \quad A := \delta^{1/3 \bmod q-1} + t^2/3;$
$Y := tA - (3a^d + b^2 + t^4)/(6t);$
$\alpha := 3a^d/(-3A + b);$
$y_t := -3Y/(-3A + b); \quad x_t := \alpha^{1/d \bmod q-1} + (-a^d/\alpha)^{1/d \bmod q-1};$
return $(x_t : y_t : 1)$

Fig. 5. Encoding on De Moivre's curves

Theorem 7. *Let* $\mathbb{F}_q$ *be the finite field with* q *elements. Suppose* q *odd and* $q \equiv 2$ mod 3 *and* d *coprime with* $q - 1$. *Let* $H_{a,b}/\mathbb{F}_q : y^2 = p_{a,b}(x)$ *be the hyperelliptic curve where* $p_{a,b}$ *is a De Moivre polynomial defined over* $\mathbb{F}_q$ *with non-zero discriminant.*

Algorithm 5 computes a deterministic encoding $e_{a,b} : \mathbb{F}_q^* \setminus \mathcal{S} \to H_{a,b}$, *where* $\mathcal{S}$ *is a subset of* $\mathbb{F}_q$ *of size at most 7, in time* $\mathcal{O}(\log^{2+o(1)} q)$.

Conversely, given a point on H we study how many elements in $\mathbb{F}_q$ yield this point.

Theorem 8. *Given a point* $(x : y : 1) \in H_{a,b}(\mathbb{F}_q)$, *we can compute the solutions* s *of the equation* $e_{a,b}(s) = (x : y : 1)$ *in time* $\mathcal{O}(\log^{2+o(1)} q)$. *There are at most* 8 *solutions to this equation.*

We give below proofs of these two theorems.

Genus and dimension of $H_{a,b}$. As in Section 4.1, since we request the discriminant of $q_{a,b}$ to be nonzero, there is no singularity except the point at infinity. Thus the genus of $H_{a,b}$ is $(d-1)/2$.

The encoding. Thanks to Lemma 1, parameterizing rational points on $H_{a,b}$: $p_{a,b}(x) = y^2$ amounts to finding roots of $\theta^2 + (b - y^2)\theta - a^d$. Let them be α, α', then we have $x = \alpha^{1/d} + \alpha'^{1/d}$, $\alpha\alpha' = -a^d$ and $\alpha + \alpha' = y^2 - b$. Thus $\alpha^2 - a^d = \alpha y^2 - b\alpha$. This is a genus 1 curve with variable α, y which is birationally equivalent to $Y^2 = A^3 + (-a^d - \frac{1}{3}b^2)A + \frac{2}{27}b^3 + \frac{1}{3}a^d b$, with $\alpha = 3a^d/(-3A + b)$ and $y = -3Y/(-3A + b)$.

This curve can be parameterized with Icart's method. This yields $A = \sqrt[3]{\delta} + t^2/3$, $Y = tA - (3a^d + b^2 + t^4)/6t$ where $\delta = (-53a^d + b^2 + t^4)/6t - 2b^3/27 - a^d b/3 - t^6/27$. We finally obtain Algorithm 5.

Restrictions. Previous necessary conditions on an encoding are also sufficient to give an encoding for $t \in \mathbb{F}_q$ provided that every variable substitution is computable.

In order to compute A and Y using the encoding from [13], we need $t \neq 0$. Then computing y and α from A and Y we also request $-3A + b \neq 0$, that is $\delta \neq (b/3 - t^2/3)^3$. This amounts to a degree 7 equation, thus at most 7 elements of $\mathbb{F}_q$ are not encodable.

Complexity. Our encoding function uses one Icart's encoding, of complexity $\mathcal{O}(\log^{2+o(1)} q)$ operations in $\mathbb{F}_q$, two exponentiations for computing d-th roots and a constant number of field operations. The total amounts to $\mathcal{O}(\log^{2+o(1)} q)$ running time.

Computation of $e_{a,b}^{-1}$. Let $(x : y : 1)$ be a point on $H_{a,b}$. The polynomial $\beta^2 + x\beta - \sqrt[d]{(-a^d)}$ has at most two roots. Let β be one, and $\alpha = \beta^5$. Let then $A = 1 - 3(b\alpha - 3a^d)/\alpha$ and $Y = -ya^d/\alpha$, we are reduced to finding the solutions of an Icart's encoding. It admits at most 4 solutions per α, thus there are at most 8 solutions to the equation $e_{a,b}(t) = (x : y : 1)$.

Genus 2 case. In this case we are interested in the dimension of the family of curves defined by De Moivre's polynomials, $H : y^2 = x^5 + 5ax^3 + 5a^2x + b$. We have computed their Igusa invariants,

$$J_2 = 700\,a^2\,, \quad J_4 = 13750\,a^4\,, \quad J_6 = -2500\,a(3\,a^5 + 32\,b^2)\,,$$
$$J_8 = -15625\,a^3(3109\,a^5 + 896\,b^2)\,, \quad J_{10} = 800000\,(4\,a^5 + b^2)^2\,,$$

from which it is easy to derive numerous algebraic relations. This reduces the set of curves from an expected q^2 because of the two parameters a and b to a set of cardinality $\mathcal{O}(q)$.

4.3   Encoding into the Jacobian of an Hyperelliptic Curve

Let H be a genus g hyperelliptic curve defined over a finite field $\mathbb{F}_q$ arising from the families defined in the previous sections 3.2, 4.1 and 4.2. We provide deterministic functions e_H which construct rational points on H from elements in $\mathbb{F}_q \setminus \mathcal{S}$, where $\mathcal{S}$ is a small subset of $\mathbb{F}_q$ which depends on the definition of H. In this section, we present two straightforward strategies for encoding divisors in $\mathcal{J}_H(\mathbb{F}_q)$ the Jacobian of H. This problem was also studied by Farashahi et al. [10].

Recall that each class in $\mathcal{J}_H(\mathbb{F}_q)$ can be uniquely represented by a reduced divisor. A divisor D is said to be reduced when it is a formal sum of points $\sum_{i=1}^{r} P_i - rP_\infty$ with $r \leqslant g$, $P_i \neq -P_j$ for $i \neq j$ and this sum is invariant under the action of the Galois group $\mathrm{Gal}(\overline{\mathbb{F}}_q/\mathbb{F}_q)$.

Encoding 1-smooth reduced divisors. There is a particular subset, denoted by $\mathcal{D}_1$, of reduced divisors which are called 1-smooth. These divisors are the ones with only rational points in their support. From our encoding function e_H, one easily deduces a function providing elements in $\mathcal{D}_1$: in a first step, a set of $r \leqslant g$ points (none of these points in this set is the opposite of another one) is produced then a divisor is constructed from this set. This first step can be done deterministically by computing g points with e_H and eliminating possible collisions after negation. When q is large enough, the proportion of $\mathcal{D}_1$ in $\mathcal{J}_H(\mathbb{F}_q)$ is $\approx 1/g!$ moreover, since e_H is not surjective, this function may be not surjective too. If one wants to construct more general reduced divisors, another strategy has to be used.

Extension of the base field and encoding. In the definition of the encoding e_H, we assume specific conditions on the base field $\mathbb{F}_q$ so that some power functions are deterministically bijective. If one wants to directly encode in the Jacobian of an hyperelliptic curve H defined over $\mathbb{F}_q$, one can change the conditions in the following way. These specific conditions are now assumed for the extension field $\mathbb{F}_{q^g}$ (and thus no more on $\mathbb{F}_q$). The function e_H becomes an encoding e'_H from $\mathbb{F}_{q^g} \setminus \mathcal{S}'$ (where the set $\mathcal{S}'$ can be computed in the same manner as $\mathcal{S}$) to the set of $\mathbb{F}_{q^g}$-rational points of H. From this new function e'_H one can compute a set of k points in $H(\mathbb{F}_{q^g})$ such that the sum of their degree over $\mathbb{F}_q$ is less than g. By constructing the $\mathbb{F}_q$-conjugates of these points and eliminating the possible collision after negation, we deduce a reduced divisor of $\mathcal{J}_H(\mathbb{F}_q)$. This second strategy is more general than the former but it does not assume the same conditions on the field $\mathbb{F}_q$.

Like the previous encodings, these two presented here are clearly "weak encodings" in the sense of [6].

5   Conclusion and Future Work

We have almost extensively studied families of genus 1 and 2 curves which admit a deterministic algebraic encoding using the resolution of a degree 3 polynomial. We come to a new encoding map for Hessian elliptic curves and we give, for the first time to our knowledge, encoding maps for large families of genus 2 curves. We have also sketched families of higher genus hyperelliptic curves whose deterministic algebraic parameterization is based on solvable polynomials of higher degree arising from Kümmer theory.

On-going work is being done to extend these families to finite fields of small characteristic. A natural question is to generalize the method to solvable degree 5 polynomials too, in the hope to first find a deterministic algebraic parameterization of every genus 2 curve, then of families of higher genus curves.

Acknowledgments

This work was partially supported by the French ANR under the ANR-09-BLAN-0020-01 CHIC, ANR-09-BLAN-0371-01 EXACTA and ANR-09-JCJCJ-0064-01 CAC projects.

References

[1] Atkin, A.O.L., Morain, F.: Elliptic curves and primality proving. Mathematics of Computation 61(203), 29–68 (1993)

[2] Bersntein, D.J., Kohel, D., Lange, T.: Twisted Hessian curves, http://www.hyperelliptic.org/EFD/g1p/auto-twistedhessian.html

[3] Boneh, D., Franklin, M.: Identity-Based Encryption from the Weil Pairing. In: Kilian, J. (ed.) CRYPTO 2001. LNCS, vol. 2139, pp. 213–229. Springer, Heidelberg (2001)

[4] Borger, R.L.: On De Moivre's quintic. The American Mathematical Monthly 15(10), 171–174 (1908)

[5] Bosma, W., Cannon, J., Playoust, C.: The Magma Algebra System I: The user language. J. Symb. Comput. 24(3/4), 235–265 (1997)

[6] Brier, E., Coron, J.-S., Icart, T., Madore, D., Randriam, H., Tibouchi, M.: Efficient indifferentiable hashing into ordinary elliptic curves. In: Rabin, T. (ed.) CRYPTO 2010. LNCS, vol. 6223, pp. 237–254. Springer, Heidelberg (2010), http://eprint.iacr.org/2009/340/

[7] Cardona, G., Quer, J.: Curves of genus 2 with group of automorphisms isomorphic to D_8 or D_{12}. Trans. Amer. Math. Soc. 359, 2831–2849 (2007)

[8] Cox, D.A.: Galois theory. In: Pure and Applied Mathematics (New York). Wiley-Interscience [John Wiley & Sons], Hoboken (2004)

[9] Farashahi, R.R., Joye, M.: Efficient Arithmetic on Hessian Curves. In: Nguyen, P.Q., Pointcheval, D. (eds.) PKC 2010. LNCS, vol. 6056, pp. 243–260. Springer, Heidelberg (2010)

[10] Farashahi, R.R., Fouque, P.-A., Shparlinski, I.E., Tibouchi, M., Felipe Voloch, J.: Indifferentiable deterministic hashing to elliptic and hyperelliptic curves. Preprint (2010), http://www.ma.utexas.edu/users/voloch/Preprints/welldistributed.pdf

[11] Farashahi, R.R.: Hashing into hessian curves. Cryptology ePrint Archive, Report 2010/373 (2010), http://eprint.iacr.org/

[12] Fouque, P.-A., Tibouchi, M.: Deterministic encoding and hashing to odd hyperelliptic curves. Cryptology ePrint Archive, Report 2010/382 (2010), http://eprint.iacr.org/

[13] Icart, T.: How to Hash into Elliptic Curves. In: Halevi, S. (ed.) CRYPTO 2009. LNCS, vol. 5677, pp. 303–316. Springer, Heidelberg (2009)

[14] Waterloo Maple Incorporated. Maple. Waterloo, Ontario, Canada, http://www.maplesoft.com/

[15] Sendra, J.R., Winkler, F., Prez-Diaz, S.: Rational Algebraic Curves: A Computer Algebra Approach. Springer Publishing Company, Incorporated, Heidelberg (2007)

[16] Shallue, A., van de Woestijne, C.: Construction of Rational Points on Elliptic Curves over Finite Fields. In: Hess, F., Pauli, S., Pohst, M.E. (eds.) ANTS 2006. LNCS, vol. 4076, pp. 510–524. Springer, Heidelberg (2006)

[17] Ulas, M.: Rational points on certain hyperelliptic curves over finite fields. Bull. Polish Acad. Sci. Math. (55), 97–104 (2007)

A New Method for Constructing Pairing-Friendly Abelian Surfaces

Robert Dryło[*]

Institute of Mathematics, Polish Academy of Sciences,
ul. Śniadeckich 8, 00-956 Warszawa, Poland
Instytut Matematyki, Uniwersytet Humanistyczno-Przyrodniczy w Kielcach,
ul. Świetokrzyska 15, 25-406 Kielce, Poland
r.drylo@impan.gov.pl

Abstract. We present a new method for constructing simple ordinary abelian surfaces with a small embedding degree. To a quartic CM field K, we associate a quadric surface $H \subset \mathbb{P}^3(\mathbb{Q})$ and use its parametrization to determine Weil numbers in K corresponding in the sense of Honda-Tate theory to such surfaces. In general, the resulting surfaces have parameter $\rho \approx 8$. However, if there exist rational lines on H, they can be used to achieve $\rho \approx 4$. We give examples of non-primitive quartic CM fields such that H has rulings by rational lines. Furthermore, we show how our method can be used to construct parametric families of pairing-friendly surfaces.

1 Introduction

A fundamental problem in pairing-based cryptography is to construct abelian varieties that are suitable for applications. Such varieties, commonly called *pairing-friendly*, should contain a subgroup of a large prime order r with a reasonably small embedding degree k, which allows one to efficiently compute a pairing, and provides desired security level. Furthermore, to speed up arithmetic on an abelian variety A over a finite field $\mathbb{F}_q$, it is desirable that the bit size of r is close to that of the group order $|A(\mathbb{F}_q)|$. The ratio of these quantities is closely approximated by the parameter $\rho = g\log q/\log r$, where g is the dimension of A. Thus, the main challenge is constructing abelian varieties with a prescribed embedding degree k and ρ-value as close to 1 as possible.

In the case of pairing-friendly elliptic curves (i.e., one-dimensional abelian varieties) many successful constructions have been found (for an excellent survey see [5]). Natural examples of higher-dimensional pairing-friendly abelian varieties come from supersingular varieties, which for every dimension have bounded embedding degrees (e.g., $k \leq 12$ for supersingular abelian surfaces), and can provide groups with $\rho \approx 1$ (see [7,11]).

Freeman [3] and Freeman, Stevenhagen and Streng [6] set a mathematical framework for constructing ordinary pairing-friendly abelian varieties. In brief,

[*] Research supported by the Polish Minister of Science as a project nr 0 R00 004307 in years 2009-2011.

M. Joye, A. Miyaji, and A. Otsuka (Eds.): Pairing 2010, LNCS 6487, pp. 298–311, 2010.

given a CM field K, we have to find a Weil q-number in K corresponding in the sense of Honda-Tate theory to a simple ordinary abelian variety over $\mathbb{F}_q$ with embedding degree k. Then the complex multiplication (CM) method is used to construct an algebraic curve (if it exists) whose Jacobian is isogenous to that variety. However, due to the specificity of the CM method only suitably small CM fields K are allowed.

Thus, the primary goal is to find methods for determining suitable Weil numbers. The first two such methods for constructing pairing-friendly surfaces and arbitrary abelian varieties were given in [3] and [6], respectively. However, the resulting varieties are not "optimal" in general (e.g., $\rho \approx 8$ for abelian surfaces). A standard approach to improve ρ-value is to use parametric families of abelian varieties. A well-known method for constructing parametric families of elliptic curves is due to Brezing-Weng [1], and its higher-dimensional analogue, based on the method in [6], was given by Freeman [2]. Using parametric families one can obtain abelian surfaces with $\rho < 8$, but generically ρ is still close to 8. Recently, Freeman and Satoh [4] used the Weil restriction to adopt the elliptic curve methods for constructing simple pairing-friendly surfaces, that are not absolutely simple. In general, the resulting surfaces have $\rho \approx 4$, but it is possible to obtain substantially smaller ρ-values. For example, for $k = 27$ they found a family with $\rho \approx 2.2$, which improves the previous record due to Kawazoe and Takahashi [9].

Let us note that methods for constructing p-rank 1 pairing-friendly abelian surfaces, and a suitable variant of the CM method were developed by Hitt O'Connor et al. [8].

In this note we present a new method for constructing simple ordinary abelian surfaces with small embedding degrees, which makes use of some well-known properties of quadric surfaces. Its outline is as follows. To any quartic CM field K, we associate a quadric surface $H \subset \mathbb{P}^3(\mathbb{Q})$ such that the algebraic integers $\pi \in O_K$ with $\pi\bar{\pi} \in \mathbb{Z}$ correspond to integral points on the cone $\tilde{H} \subset \mathbb{Q}^4$ over H. To determine suitable Weil numbers in K, we use a parametrization of H and proceed similarly as in the Cocks-Pinch method for elliptic curves (see [5, Theorem 4.1]). Furthermore, this method can be extended to construct families of abelian surfaces (Section 4).

Similarly as for the methods [3,6], the resulting surfaces usually have $\rho \approx 8$, but in special situations we can achieve $\rho \approx 4$. This is the case if there exist rational lines on the quadric H. Such lines seem to be very rare in general, but for some CM fields there is abundance of them. For example, H has rulings by rational lines for non-primitive CM fields $K = \mathbb{Q}(\sqrt{-a + 2\sqrt{d}})$, where a, d are positive integers such that $a^2 - 4d$ is a square in $\mathbb{Z}$ (e.g., $a = d + 1$).

In this paper, however, we restrict yourself only to giving parameters of abelian surfaces, because the CM method is developed mainly for primitive CM fields. Under certain assumptions, the recent method of Freeman and Satoh [4] can be alternatively used to find curves whose Jacobian has complex multiplications in the non-primitive CM field $K = \mathbb{Q}\left(\sqrt{-(d+1) + 2\sqrt{d}}\right) = \mathbb{Q}(i, \sqrt{d})$, and we give an example of a curve whose Jacobian realizes parameters found by our method, which was suggested by the referee.

2   Background

This section summarizes mathematical foundations for constructing pairing-friendly abelian varieties. For more details we refer to the papers [3,5,6]. (For an overview of the theory of abelian varieties see [10] and [16].)

Let A be a g-dimensional abelian variety over the finite field $\mathbb{F}_q$, and π_A be its qth power Frobenius endomorphism. To A is associated the characteristic polynomial $P_A \in \mathbb{Z}[x]$, which is of the form

$$P_A = x^{2g} + a_1 x^{2g-1} + \cdots + a_{g-1} x^{g+1} + a_g x^g + a_{g-1} q x^{g-1} + \cdots + a_1 q^{g-1} x + q^g,$$

and satisfies $P_A(\pi_A) = 0$ and $\#A(\mathbb{F}_q) = P_A(1)$. Furthermore, A is ordinary if and only if $\gcd(a_g, q) = 1$. (Let us recall that $\#A[p] = p^\nu$ for some $0 \le \nu \le g$, where $p = \operatorname{char} \mathbb{F}_q$ and $A[p]$ is the group of p-torsion points on A over an algebraic closure $\overline{\mathbb{F}}_q$. If $\nu = g$, then A is called *ordinary*.) By a theorem of Weil, all roots of P_A are Weil q-numbers (i.e., an algebraic integer π is called a *Weil q-number*, if for every embedding $\varphi : \mathbb{Q}(\pi) \to \mathbb{C}$ we have $|\varphi(\pi)| = \sqrt{q}$). The main theorem of Honda-Tate theory [14] precisely describes the correspondence between Weil q-numbers and simple abelian varieties (i.e., A is *simple* if it has no proper nonzero abelian subvarieties over $\mathbb{F}_q$).

Theorem 1. *The map that associates to a simple abelian variety over $\mathbb{F}_q$ its Frobenius endomorphism gives a one-to-one correspondence between the $\mathbb{F}_q$-isogeny classes of simple abelian varieties over $\mathbb{F}_q$ and the $\operatorname{Gal}(\overline{\mathbb{Q}}/\mathbb{Q})$-conjugacy classes of Weil q-numbers.*

Let $\operatorname{End}(A)$ be the ring of $\mathbb{F}_q$-endomorphisms on A, and $\operatorname{End}^0(A)$ denote the endomorphism algebra $\operatorname{End}(A) \otimes \mathbb{Q}$. The following theorem summarizes basic properties of this algebra for simple abelian varieties (see Waterhouse and Milne [16] and Waterhouse [15]).

Theorem 2. *Let A be a simple abelian variety over $\mathbb{F}_q$ with the endomorphism algebra $D = \operatorname{End}^0(A)$, and let $K = \mathbb{Q}(\pi_A) \subset D$. Then*

(1) *D is a division algebra, whose center is the subfield K.*
(2) *$P_A = m_A^e$ for some $e \in \mathbb{Z}$, where m_A is the minimal polynomial of π_A. Furthermore, $e[K : \mathbb{Q}] = 2 \dim A$ and $[D : K] = e^2$.*
(3) *D contains a CM field of degree $2 \dim A$.*
(4) *A is ordinary of dimension g if and only if K is a CM field of degree $2g$, and $\pi_A + \overline{\pi}_A$ and q are relatively prime in O_K.*
(5) *If A is ordinary, then $D = K$.*

(Let us recall that a number field K is a *CM field*, if K is a totally imaginary quadratic extension of a totally real field. A CM field has an automorphism (denoted by bar in the sequel), that commutes with every embedding $K \to \mathbb{C}$ and the complex conjugation.)

The above theorem implies that if A is simple and ordinary, then $\text{End}^0(A)$ is a field, so $P_A = m_A$. It follows that

$$\#A(\mathbb{F}_q) = P_A(1) = N_{K/\mathbb{Q}}(1 - \pi_A).$$

2.1 Pairing-Friendly Abelian Varieties

Let us recall that *the embedding degree k* of an abelian variety A over $\mathbb{F}_q$ with respect to its subgroup of prime order r, $r \nmid q$, is defined as the degree of the field extension $\mathbb{F}_q \subset \mathbb{F}_q(\zeta_r)$, where ζ_r is an rth primitive root of unity. Equivalently, k is the smallest integer $l \geq 1$ such that $r|(q^l - 1)$, or in other words, k is the multiplicative order of q (mod r). Furthermore, if $r \nmid kq$, then k is the unique integer satisfying $\Phi_k(q) \equiv 0$ (mod r), where Φ_k is the kth cyclotomic polynomial (see [5, Proposition 2.4]).

From the above theory we easily obtain the following useful fact due to Freeman [3] and Freeman, Stevenhagen and Streng [6].

Proposition 3. *Let $k \geq 1$ be an integer, and r be a prime such that $k|(r-1)$. Let K be a CM field of degree $2g$, and A be a simple abelian variety over $\mathbb{F}_q$ corresponding to a Weil q-number $\pi \in K$. Suppose that $r \nmid kq$. Then $r|\#A(\mathbb{F}_q)$, A has embedding degree k with respect to r, and A is ordinary of dimension g if and only if*

(1) $N_{K/\mathbb{Q}}(1 - \pi) \equiv 0$ (mod r),
(2) $\Phi_k(\pi\overline{\pi}) \equiv 0$ (mod r),
(3) $K = \mathbb{Q}(\pi)$, *and $\pi + \overline{\pi}$ and q are relatively prime in O_K.*

3 Our Solution for Abelian Surfaces

In this section we give a method for finding Weil numbers in a quartic CM field, that satisfy the conditions of Proposition 3.

Let $K = \mathbb{Q}(\sqrt{-a + b\sqrt{d}})$ be a quartic CM field, where a, b, d are positive integers such that d is not a square and $-a + b\sqrt{d} < 0$. Given an integral basis b_1, b_2, b_3, b_4 of O_K (for explicit formulas on such bases see [13]), we will determine the coordinates in this basis of Weil q-numbers $\pi \in O_K$ corresponding to ordinary abelian surfaces with embedding degree k.

Let us consider the quadratic form on $\mathbb{Z}^4$ given by $\mathbf{x} \mapsto \pi(\mathbf{x})\overline{\pi(\mathbf{x})}$, where $\pi(\mathbf{x}) = \sum_{i=1}^4 x_i b_i$ for $\mathbf{x} = (x_1, x_2, x_3, x_4) \in \mathbb{Z}^4$. Clearly, we can write

$$\pi(\mathbf{x})\overline{\pi(\mathbf{x})} = F_1(\mathbf{x}) + F_2(\mathbf{x})\sqrt{d}, \qquad (1)$$

where $F_1, F_2 \in \mathbb{Q}[x, y, z, t]$ are homogeneous quadratic forms. Thus, $\pi(\mathbf{x})\overline{\pi(\mathbf{x})} \in \mathbb{Z}$ if and only if $\mathbf{x}$ lies on the quadric $F_2 = 0$. In order to determine points corresponding to Weil numbers in question, we will use a parametrization of this quadric, which can be obtained by the following well-known fact:

Lemma 4. *Let $H \subset \mathbb{P}^n(\mathbb{Q})$ be a nonsingular quadric hypersurface with a rational point P. Then the projection from P to any hyperplane in $\mathbb{P}^n(\mathbb{Q})$ not containing P induces a birational isomorphism $G = (G_0, \ldots, G_n) : \mathbb{P}^{n-1}(\mathbb{Q}) \dashrightarrow H$. Furthermore, the components $G_0, \ldots, G_n$ can be taken as relatively prime quadratic forms in $\mathbb{Z}[X_0, \ldots, X_{n-1}]$.*

Proof. For completeness we include the proof. Let $F = 0$ be an equation of H, where $F \in \mathbb{Q}[X_0, \ldots, X_n]$ is a homogeneous quadratic form, and $\Pi \subset \mathbb{P}^n(\mathbb{Q})$ be a hyperplane such that $P \notin \Pi$. After a linear change of variables, we may assume that Π does not coincide with the hyperplane $X_0 = 0$ and $P = (a_0 : \cdots : a_n)$ with $a_0 \neq 0$. Let H' and Π' be the affine subsets of H and Π in the affine space $X_0 \neq 0$ (identified with $\mathbb{Q}^n$), and let $P' = (b_1, \ldots, b_n)$ with $b_i = a_i/a_0$. Let us take any affine parametrization $\varphi = (\varphi_1, \ldots, \varphi_n) : \mathbb{Q}^{n-1} \longrightarrow \Pi'$, and consider the lines through P' and $\varphi(a)$, $a \in \mathbb{Q}^{n-1}$, with the parametrization $x_i(a,t) = b_i + t(\varphi_i(a) - b_i)$, $1 \leq i \leq n$, $t \in \mathbb{Q}$. Substituting their parametrization into the equation of H' yields $F(1, x_1(a,t), \ldots, x_n(a,t)) = t^2 g(a) + t h(a) = 0$, where $g, h \in \mathbb{Q}[X_1, \ldots, X_{n-1}]$ are of degree 2 and 1, respectively. Since H is nonsingular, it follows that for a generic $a \in \mathbb{Q}^{n-1}$ the line through P' and $\varphi(a)$ meets H' at a unique point other than P', which corresponds to $t = -h(a)/g(a)$. Therefore the map $f = (f_1, \ldots, f_n) : \mathbb{Q}^{n-1} \dashrightarrow H'$ with $f_i = b_i - (\varphi_i - b_i)h/g$ is birational. Let us write $f_i = g_i/g_0$, where $g_i \in \mathbb{Z}[X_1, \ldots, X_n]$, $i \geq 0$, are of degree ≤ 2 and the coefficients of all g_i are relatively prime. Then $G = (G_0, \ldots, G_n) : \mathbb{P}^{n-1}(\mathbb{Q}) \dashrightarrow H$ with $G_i = X_0^2 g_i(X_1/X_0, \ldots, X_n/X_0)$ is a desired map.

We apply the above fact to the quadric surface $H := \{F_2 = 0\} \subset \mathbb{P}^3(\mathbb{Q})$. Since $1 \in O_K$, it can be written $1 = \sum a_i b_i$ with respect to the basis $\{b_i\}$, so $P = (a_1 : a_2 : a_3 : a_4) \in H$. To show that H is nonsingular, let us write the form $\mathbb{Q}^4 \ni \mathbf{x} \mapsto \pi(\mathbf{x}) \overline{\pi(\mathbf{x})} \in \mathbb{Q}(\sqrt{d})$ with respect to the basis $1, \sqrt{d}, \sqrt{u}, \sqrt{du}$ of K, where $u = -a + b\sqrt{d}$. We have

$$
\begin{aligned}
\pi(\mathbf{x})\overline{\pi(\mathbf{x})} &= \left(x + y\sqrt{d} + z\sqrt{u} + t\sqrt{du}\right)\left(x + y\sqrt{d} - z\sqrt{u} - t\sqrt{du}\right) \\
&= \left(x + y\sqrt{d}\right)^2 - u\left(z + t\sqrt{d}\right)^2 \\
&= \left(x^2 + dy^2 + az^2 + adt^2 - 2bdzt\right) + \left(2xy + 2azt - bz^2 - bdt^2\right)\sqrt{d}.
\end{aligned}
$$

Hence,

$$
F_2 = 2xy + 2azt - bz^2 - bdt^2 = 2xy - b\left((z - at/b)^2 + (d - (a/b)^2)t^2\right). \qquad (2)
$$

Since $d - (a/b)^2 < 0$, it follows that we can bring F_2 by a linear change of variables over $\mathbb{R}$ to the form $x^2 + y^2 - z^2 - t^2$, which is nonsingular.

Let $G : \mathbb{P}^2(\mathbb{Q}) \dashrightarrow H$ be a parametrization as in the lemma. Obviously, G induces the map (which we also denote by G)

$$
G : \mathbb{Z}^3 \longrightarrow \widetilde{H} \cap \mathbb{Z}^4,
$$

where $\widetilde{H} \subset \mathbb{Q}^4$ is the cone over H. Since the components of G are relatively prime forms in $\mathbb{Z}[x, y, z]$, we expect that in the image $G(\mathbb{Z}^3)$ there exist sufficiently many points corresponding to Weil numbers in O_K.

Let us turn now to our main goal. Let $N \in \mathbb{Q}[x, y, z, t]$ be defined by

$$N(\mathbf{x}) = N_{K/\mathbb{Q}}(1 - \pi(\mathbf{x})) \tag{3}$$

for $\mathbf{x} = (x_1, x_2, x_3, x_4) \in \mathbb{Z}^4$ and $\pi(\mathbf{x}) = \sum x_i b_i$. If $\mathbf{x} \in G(\mathbb{Z}^3)$ corresponds to the Weil number of an abelian surface with embedding degree k with respect to a prime r, then according to Proposition 3, $\mathbf{x}$ satisfies the system

$$N \circ G \equiv 0 \pmod{r}, \quad \Phi_k \circ F_1 \circ G \equiv 0 \pmod{r}. \tag{4}$$

Therefore, to determine points in $G(\mathbb{Z}^3)$ corresponding to abelian surfaces with embedding degree k, we choose a prime r of a given bit size such that $k | (r - 1)$, and find some number of solutions of the above system over $\mathbb{F}_r$. Then we check whether for lifts $\mathbf{x} \in \mathbb{Z}^3$ of these solutions, the value $F_1(G(\mathbf{x}))$ is prime. If this is the case, then $\sum G_i(\mathbf{x}) b_i$ is a desired Weil number. To find some solutions over $\mathbb{F}_r$, we evaluate one of the variables at various points in $\mathbb{F}_r$, and solve the system in the remaining two variables. In details the method is as follows.

Algorithm 5.

Input: A quartic CM field $K = \mathbb{Q}(\sqrt{-a + b\sqrt{d}})$, an integral basis $\{b_i\}$ of O_K, an embedding degree k, a prime r such that $k | (r - 1)$, and two integers $n, m \geq 0$.

Output: Either the empty set, or a Weil q-number $\pi \in K$ of a simple ordinary abelian surface over $\mathbb{F}_q$ with embedding degree k with respect to r.

1. Compute the polynomials $F_1, F_2, N \in \mathbb{Q}[x, y, z, t]$ satisfying (1) and (3).
2. Apply the method from the proof of Lemma 4, to compute a map $G = (G_1, \ldots, G_4) : \mathbb{Z}^3 \to \mathbb{Z}^4$ such that $F_2 \circ G = 0$, where $G_1, \ldots, G_4 \in \mathbb{Z}[x, y, z]$ are relatively prime quadratic forms.
3. Repeat n-times the following procedure: choose a random $c \in \mathbb{F}_r$, evaluate one of the variables at c, say z, and compute $f(x, y) = \Phi_k(F_1(G(x, y, c)))$ and $g(x, y) = N(G(x, y, c)) \pmod{r}$.
4. If the system $f = g = 0$ has finitely many solutions over an algebraic closure $\overline{\mathbb{F}}_r$, then determine the set Z of all its solutions over $\mathbb{F}_r$ (see, e.g., Lemma 9 and Algorithm 10).
5. For each $(a_1, a_2) \in Z$, let $x_1, x_2, x_3 \in [0, r)$ be the lifts of a_1, a_2, c, respectively. For $i_1, i_2, i_3 \in [0, m]$, let $y_j = x_j + i_j r$, $j = 1, 2, 3$, and $\mathbf{y} = (y_1, y_2, y_3)$. Put $\pi = \sum G_i(\mathbf{y}) b_i$ and $q = \pi \overline{\pi} = F_1(G(\mathbf{y}))$. If q is prime, $K = \mathbb{Q}(\pi)$, and $\pi + \overline{\pi}$ and q are relatively prime, then return π and terminate the algorithm.

Remark. The correctness of this algorithm follows from the above discussion and Proposition 3. We expect that a generic solution of system (4) is approximately of the same bit size as r, which implies that $\rho \approx 2 \deg(F_1 \circ G) = 8$ in general.

3.1   Improving ρ-Value

In order to achieve $\rho \approx 4$ using the above method we need an injective linear map $L : \mathbb{Z}^2 \to \mathbb{Z}^4 \cap \tilde{H}$. We require that L is defined on $\mathbb{Z}^2$ in order to find solutions of the system like (4) with L in place of G. Clearly, such maps correspond to rational lines on the quadric $H \subset \mathbb{P}^3(\mathbb{Q})$. Therefore, it would be of interest to know when there exist rational lines on a quadric in $\mathbb{P}^3(\mathbb{Q})$. Here we restrict yourself to giving some examples of quartic CM fields for which the quadric H contains many such lines.

Remark. Let us note that to look for rational lines on an algebraic surface $S \subset \mathbb{P}^3(\mathbb{Q})$ it may be convenient to use the Plücker coordinates of the lines in $\mathbb{P}^3(\mathbb{Q})$. Then the lines lying on S form an algebraic set in $\mathbb{P}^5(\mathbb{Q})$ (see, e.g., [12, p. 62]), and so one can search them by enumerating a part of coordinates and solving a system with respect to the remaining variables.

Let us recall that a quadric $S \subset \mathbb{P}^3(\mathbb{Q})$ has *a ruling*, if there exists a family of pairwise disjoint rational lines on S, which cover S. For example, the quadric $x^2 + y^2 - z^2 - t^2 = 0$ (which in the affine set $t \neq 0$ looks like a hyperboloid of one sheet) has two rulings given by the equations $x - z = (t - y)s$, $(x + z)s = t + y$, and $x - t = (z - y)s$, $(x + t)s = z + y$, where s parametrizes lines in the ruling. Thus, any quadric in $\mathbb{P}^3(\mathbb{Q})$, which is projectively equivalent to the quadric $x^2 + y^2 - z^2 - t^2 = 0$ also has rulings by rational lines.

Let us apply this fact to our quadric H. It follows from (2) that H is projectively equivalent in $\mathbb{P}^3(\mathbb{Q})$ to the quadric $x^2 + y^2 - z^2 - t^2 = 0$, if it comes from the CM field $K = \mathbb{Q}(\sqrt{-a + b\sqrt{d}})$ such that $b = 2$ and $(a/b)^2 - d$ is a square in $\mathbb{Q}$ (e.g., $K = \mathbb{Q}(\sqrt{-3 + 2\sqrt{2}}), \mathbb{Q}(\sqrt{-4 + 2\sqrt{3}}), \mathbb{Q}(\sqrt{-6 + 2\sqrt{5}}), \mathbb{Q}(\sqrt{-5 + 2\sqrt{6}})$, etc.). Note, however, that CM fields of this form are not primitive, and so abelian surfaces with complex multiplications in K may be not absolutely simple.

Remark. Let us note that pairing-friendly abelian surfaces with $\rho \leq 4$ having complex multiplications in the CM fields $K = \mathbb{Q}\left(\sqrt{-(d+1) + 2\sqrt{d}}\right) = \mathbb{Q}(i, \sqrt{d})$ were considered from another point of view by Freeman and Satoh [4], and in the particular case of $d = 2$ and $K = \mathbb{Q}(\zeta_8)$ by Kawazoe and Takahashi [9]. Under certain assumptions, their methods allow one to find, without using the genus 2 CM method, a hyperelliptic curve whose Jacobian has given parameters.

Note that given a rational line $E \subset H$ with a linear parametrization $L : \mathbb{P}^1(\mathbb{Q}) \to E$ whose components are relatively prime linear forms in $\mathbb{Z}[x, y]$, we should check whether the form $F_1 \circ L$ may take "many" prime values. Conjecturally, a polynomial $f \in \mathbb{Q}[x]$ takes infinitely many prime values on integers if and only if it is irreducible, has positive leading coefficient, $f(x) \in \mathbb{Z}$ for some $x \in \mathbb{Z}$, and $\gcd(\{f(x) : x, f(x) \in \mathbb{Z}\}) = 1$ (see [5, Section 2.1]). If f satisfies these conditions, we say that it *represents primes*. In the obvious way we can extend this definition and conjecture on quadratic forms in $\mathbb{Q}[x, y]$. Let us note that our form $F_1 \circ L$ is positive-definite and integer-valued, since $F_1(L(\mathbf{x})) = \pi(L(\mathbf{x}))\overline{\pi(L(\mathbf{x}))} \in \mathbb{Z}$ for $\mathbf{x} \in \mathbb{Z}^2$. Thus $F_1 \circ L$ represents primes, if it is irreducible in $\mathbb{Q}[x, y]$ and $\gcd(\{F_1(L(\mathbf{x})) : \mathbf{x} \in \mathbb{Z}^2\}) = 1$. The following algorithm summarizes the above discussion.

Algorithm 6.

Input: A quartic CM field $K = \mathbb{Q}\left(\sqrt{-a + b\sqrt{d}}\right)$, an integral basis $\{b_i\}$ of O_K, an embedding degree k, a prime r such that $k|(r-1)$, and an integer $m \geq 0$. Furthermore, we are given a rational line E on the quadric $\{F_2 = 0\} \subset \mathbb{P}^3(\mathbb{Q})$, where F_2 is given by (1).

Output: Either the empty set, or a Weil q-number $\pi \in K$ of a simple ordinary abelian surface with embedding degree k with respect to r.

1. Compute the polynomials $F_1, N \in \mathbb{Q}[x, y, z, t]$ satisfying (1) and (3).
2. Find a parametrization $L = (L_1, \ldots, L_4) : \mathbb{P}^1(\mathbb{Q}) \to E$, where $L_1, \ldots, L_4 \in \mathbb{Z}[x, y]$ are relatively prime linear forms, and consider L as a map $\mathbb{Z}^2 \to \mathbb{Z}^4$.
3. If the form $F_1 \circ L$ represents primes, and the system $N \circ L = \Phi_k \circ F_1 \circ L = 0$ has finitely many solutions over $\overline{\mathbb{F}}_r$, then determine the set Z of all its solutions in $\mathbb{F}_r^2$.
4. For each $(a_1, a_2) \in Z$, let $x_1, x_2 \in [0, r)$ be the lifts of a_1, a_2, respectively. For $i_1, i_2 \in [0, m]$, let $y_j = x_j + i_j r$, $j = 1, 2$, and let $\mathbf{y} = (y_1, y_2)$. Put $\pi = \sum L_i(\mathbf{y}) b_i$ and $q = \pi\overline{\pi} = F_1(L(\mathbf{y}))$. If q is prime, $K = \mathbb{Q}(\pi)$, and $\pi + \overline{\pi}$ and q are relatively prime, then return π and terminate the algorithm.

Remark. Arguing analogously as for the previous algorithm, we expect that the ρ-value is generically around $2 \deg(F_1 \circ L) = 4$.

Example 7. As an application of the above method, we find parameters of abelian surfaces with $\rho \approx 4$ for embedding degrees $k = 13, 31, 43$, which have complex multiplications in the CM field $K = \mathbb{Q}\left(\sqrt{-3 + 2\sqrt{2}}\right)$. Here, we restrict yourself only to giving parameters, because the CM method is developed mainly for primitive fields. In some cases the recent method of Freeman and Satoh [4] can be alternatively used to find curves whose Jacobian has complex multiplications in the CM field $K = \mathbb{Q}\left(\sqrt{-(d+1) + 2\sqrt{d}}\right)$.

For simplicity, we will look for Weil numbers in the suborder of O_K with the basis $1, u, v, uv$, where $u = \sqrt{2}$ and $v = \sqrt{-3 + 2\sqrt{2}}$. We have $F_1 = x^2 + 2y^2 + 3z^2 - 8zt + 6t^2$, and $F_2 = 2xy - 2z^2 + 6zt - 4t^2$. On the quadric $H = \{F_2 = 0\} \subset \mathbb{P}^3(\mathbb{Q})$ we find the line with the parametrization $L(x, y) = (0, y, -x, -x)$. Then the form $q(x, y) = F_1(L(x, y)) = x^2 + 2y^2$ represents primes. For an embedding degree k, we choose a random prime r of a given bit size such that $k|(r-1)$. Then using Algorithm 6, we look for coordinates $x_1, \ldots, x_4 \in \mathbb{Z}$ of Weil q-numbers $\pi = x_1 + x_2 u + x_3 v + x_4 uv$.

(1) $k = 13$, $\rho \approx 3.99$,

$r = 23405321049561401799446475832547771376750288802937$ (165 bits)

$x_1 = 0$,
$x_2 = 5179309762621573363923518967361040684119787041950$,
$x_3 = x_4 = -1607763891249945974894902131024174445878961361 0329$,

$\pi = (-1607763891249945974894902131024174445878961361 0329u - 1607763891249945974894902131\backslash$
$31024174445878961361 0329)v + 5179309762621573363923518967361040684119787041950u$,

$q = 31214097223509108798865628081419919734462954621988332885421909689882196377198886 09\backslash$
13019059975093241.

These parameters are realized by the Jacobian of the hyperelliptic curve

$$y^2 = x^5 + 3x$$

over $\mathbb{F}_q$, which was observed by the referee.

(2) $k = 31$, $\rho \approx 4.01$,

$r = 9140865813990849114396596864612922226505149321073424084520116$ (206 bits)

$x_1 = 0$,
$x_2 = 7250261696294234401171066330356959098457677269892057214340228$,
$x_3 = x_4 = -4453454327317792327353435007136492902634802507844996620560431$,

$\pi = (-4453454327317792327353435007136492902634802507844996620560431u - 445345432731779\backslash$
$2327353435007136492902634802507844996620560431)v + 7250261696294234401171066330356959909\backslash$
$8457677269892057214340228u$,

$q = 10533092188395775416759526783096451516527482756829401394370152257246074831840331 80\backslash$
$3581120165870579841701126653040599782957353$.

(3) $k = 43$, $\rho \approx 3.99$,

$r = 1289030776758495577285183860231666901856028991763686681169311027544 84928281$ (247 bits)

$x_1 = 0$,
$x_2 = 3721556596747093460362478336211022630007920972210907100465244 84502335172$,
$x_3 = x_4 = -87592687840749493706069759576404629264659124533660844895753302243637137369$,

$\pi = (-87592687840749493706069759576404629264659124533660844895753302243637137369u - 87\backslash$
$59268784074949370606975957640462926465912453366084489575330224363713 7369)v + 3721556596\backslash$
$747093460362478336211022630007920972210907100465244 84502335172u$,

$q = 767275596283704003823632884930005248169977481154544212850806286268172243395\,2440746\backslash$
$6211839919816760016789152195753440014603617321089336676030327 81329$.

4   Parametric Families of Abelian Surfaces

In this section we outline how to extend our method to construct parametric families of abelian surfaces. We start with the definition, which was introduced by Freeman [2].

Definition 8. Let K be a CM field of degree $2g$, let $\pi(x) \in K[x]$, and let $r(x) \in \mathbb{Q}[x]$. We say that $(\pi(x), r(x))$ *represents a family of g-dimensional abelian varieties with embedding degree k* if:

(1) $q(x) = \pi(x)\bar{\pi}(x) \in \mathbb{Q}[x]$.
(2) $q(x)$ represents primes.
(3) $r(x)$ is non-constant, irreducible, integer-valued, and has positive leading coefficient.
(4) $r(x) \mid N_{K/\mathbb{Q}}(1 - \pi(x))$.
(5) $r(x) \mid \Phi_k(q(x))$.

Remark. Let us explain the above notation. If K' is a normal closure of K, and $\sigma : K \to K'$ is an embedding, let us denote also by σ the extension of σ,

$K[x] \to K'[x]$, which is constant on x. If σ is the complex conjugation on K, we simply write $\overline{f}$ for $\sigma(f)$, $f \in K[x]$. The norm $N_{K/\mathbb{Q}}(f) \in \mathbb{Q}[x]$ is defined as the product of $\sigma(f)$ over all embeddings $\sigma : K \to K'$.

Note that it follows from (3) that $r(x) = d\widetilde{r}(x)$ for some $d \in \mathbb{N}$ and $\widetilde{r}(x) \in \mathbb{Q}[x]$, which is integer-valued an represents primes. In order to find an abelian variety in the family, we search for $x_0 \in \mathbb{Z}$ such that $\widetilde{r}(x_0)$ and $q(x_0)$ both are prime. If additionally $\pi(x_0)$ is an algebraic integer, $K = \mathbb{Q}(\pi(x_0))$, and $\pi(x_0) + \overline{\pi}(x_0)$ and $q(x_0)$ are relatively prime in O_K, then $\pi(x_0)$ is a Weil $q(x_0)$-number of a g-dimensional simple ordinary abelian variety with embedding degree k with respect to $\widetilde{r}(x_0)$. For large x_0, the ρ-value of the resulting varieties is close to the parameter ρ of the family

$$\rho = \frac{g \deg q(x)}{\deg r(x)}.$$

Let us now assume that K is a quartic CM field. We will keep the notation from the previous section. Let $\{b_i\}$ be an integral basis of O_K. Then $\{b_i\}$ is also a basis of $K[x]$ over $\mathbb{Q}[x]$. We will look for families of abelian surfaces $(\pi(x), r(x))$ with $\pi(x) = \sum G_i(l_1((x), l_2(x), l_3(x))b_i$ for some $l_i(x) \in \mathbb{Q}[x]$, $i = 1, 2, 3$. For this purpose, we have to find an irreducible polynomial $r(x) \in \mathbb{Q}[x]$ such that the system

$$N \circ G = 0, \quad \Phi_k \circ F_1 \circ G = 0 \qquad (5)$$

has a solution (c_1, c_2, c_3) over the number field $L = \mathbb{Q}[x]/(r(x))$. Then to define $\pi(x)$ we take the lifts $l_i(x)$ of c_i with $\deg l_i < \deg r$, $i = 1, 2, 3$. In general, if a number field L is given in advance, it may be very difficult to find an L-rational solution, but here we can in fact construct a suitable field together with a solution. This can be easily done using resultants. We start with any number field L_0, choose an element $c \in L_0$, and evaluate at c one of the variables in the system (5). If the system in the remaining two variables has finitely many solutions over $\overline{\mathbb{Q}}$, then the following well-know fact allows us to solve the system, and implies that all solutions are defined over number fields.

Lemma 9. *Let $\mathbb{F}$ be an arbitrary field with an algebraic closure $\overline{\mathbb{F}}$, and let $f, g \in \mathbb{F}[x, y]$ both depend on y. Then the system $f = g = 0$ has finitely many solutions over $\overline{\mathbb{F}}$ if and only if $\mathrm{Res}_y(f, g) \neq 0$ and the coefficients of f and g with respect to y are relatively prime in $\mathbb{F}[x]$.*

Proof. The resultant $\mathrm{Res}_y(f, g)$ vanishes if and only if f and g have a common factor in $\mathbb{F}(x)[y]$, and hence in $\mathbb{F}[x, y]$. The system $f = g = 0$ has infinitely many solutions if and only if f and g have a common factor $h \in \overline{\mathbb{F}}[x, y]$. Such a h exists and depends on y if and only if $\mathrm{Res}_y(f, g) = 0$. If $h \in \overline{\mathbb{F}}[x]$, then it must divide all the coefficients of f and g with respect to y.

It follows a simple algorithm for determining zeros of two polynomials in $\mathbb{F}[x, y]$, which requires factorization of polynomials over finite extensions of $\mathbb{F}$.

Algorithm 10.

Input: A field $\mathbb{F}$, and two polynomials $f, g \in \mathbb{F}[x, y]$ such that the system $f = g = 0$ has finitely many solutions over $\overline{\mathbb{F}}$.

Output: All solutions $(x_1, y_1), \ldots, (x_n, y_n)$ of the system $f = g = 0$, and the fields $\mathbb{F}(x_i, y_i)$ (which are finite over $\mathbb{F}$).

1. If f and g both depend on y, compute $\mathrm{Res}_y(f, g)$.
2. For each irreducible factor $h(x) \in \mathbb{F}[x]$ of $\mathrm{Res}_y(f, g)$, let $\mathbb{F}' := \mathbb{F}[x]/(h(x))$, and let x' denote the residue class of x.
3. For each irreducible factor $p(y) \in \mathbb{F}'[y]$ of $\gcd(f(x', y), g(x', y))$, let $\mathbb{F}'' := \mathbb{F}'[y]/(p(y))$, and let y' denote the residue class of y. Return (x', y') and $\mathbb{F}''$.
4. If $f \in \mathbb{F}[x]$ and $g \in \mathbb{F}[y]$, then for each irreducible factor $h(x) \in \mathbb{F}[x]$ of $f(x)$, let $\mathbb{F}' := \mathbb{F}[x]/(h(x))$, and let x' denote the residue class of x.
5. For each irreducible factor $p(y) \in \mathbb{F}'[y]$ of $g(y)$, let $\mathbb{F}'' := \mathbb{F}'[y]/(p(y))$, and let y' denote the residue class of y. Return (x', y') and $\mathbb{F}''$.
6. If either f and g both depend on x, but not on y, or $g \in \mathbb{F}[x]$ and $f \in \mathbb{F}[y]$, then we proceed similarly with x in place of y.

We are now in a position to give an analogue of Algorithm 5 for constructing families.

Algorithm 11.

Input: A quartic CM field K with an integral basis $\{b_i\}$, and an embedding degree k.

Output: Either the empty set, or a family of abelian surfaces $(\pi(x), r(x))$ with embedding degree k and $\pi(x) \in K[x]$.

1. Compute the polynomials $F_1, F_2, N \in \mathbb{Q}[x, y, z, t]$ satisfying (1) and (3).
2. Apply the method from the proof of Lemma 4, to find a map $G = (G_1, \ldots, G_4)$ satisfying $F_2 \circ G = 0$, where $G_1, \ldots, G_4 \in \mathbb{Z}[x, y, z]$ are relatively prime quadratic forms.
3. Choose a number field L_0, and any $c \in L_0$, and evaluate at c one of the variables, say z.
4. If the system $N(G(x, y, c)) = \Phi_k(F_1(G(x, y, c))) = 0$ has finitely many solutions over $\overline{\mathbb{Q}}$, then apply Algorithm 10 to $\mathbb{F} := L_0$ to determine all its solutions (x_j, y_j) and the number fields $L_j = L_0(x_j, y_j)$, $j = 1, \ldots, n$.
5. For $j = 1, \ldots, n$, find a polynomial $r(x) \in \mathbb{Q}[x]$ such that $L_j = \mathbb{Q}[x]/(r(x))$, and take the lifts $l_1(x), l_2(x), l_3(x) \in \mathbb{Q}[x]$ of x_j, y_j, c, respectively, with $\deg l_i(x) < \deg r(x)$. If $q(x) = F_1(G(l_1(x), l_2(x), l_3(x)))$ represents primes, then return $(\pi(x), r(x))$, where $\pi(x) = \sum G_i(l_1(x), l_2(x), l_3(x))b_i$, and terminate the algorithm.

Remark. Let us note that since the polynomials $l_i(x)$ in step 5 satisfy $\deg l_i(x) < \deg r(x)$, it follows that $\deg q(x) < 4 \deg r(x)$, so we obtain families with $\rho < 8$.

Similarly as in the previous section, we can use rational lines on the quadric $H \subset \mathbb{P}^3(\mathbb{Q})$ to construct families with $\rho < 4$. We left to the reader extending Algorithm 6. The following two examples are based on this idea. For simplicity, we consider quartic CM fields $K = \mathbb{Q}\left(\sqrt{-a + b\sqrt{d}}\right)$ with the basis $1, u, v, uv$ of the suborder $\mathbb{Z}[u, v]$, where $u = \sqrt{d}$ and $v = \sqrt{-a + b\sqrt{d}}$.

Example 12. For $k = 12$ and the field $K = \mathbb{Q}\left(\sqrt{-4 + 2\sqrt{3}}\right)$ we obtain the family with $\rho = 3$.

We have $F_1 = x^2 + 3y^2 + 4z^2 - 12zt + 12t^2$, $F_2 = 2xy - 2z^2 + 8zt - 6t^2$, and $N = x^4 - 4x^3 - 6x^2y^2 + 8x^2z^2 - 24x^2zt + 24x^2t^2 + 6x^2 + 12xy^2 + 24xyz^2 - 96xyzt + 72xyt^2 - 16xz^2 + 48xzt - 48xt^2 - 4x + 9y^4 + 24y^2z^2 - 72y^2zt + 72y^2t^2 - 6y^2 - 24yz^2 + 96yzt - 72yt^2 + 4z^4 - 24z^2t^2 + 8z^2 - 24zt + 36t^4 + 24t^2 + 1$.

Let us take the line with the parametrization $E(x, y) = (0, y, -x, -x)$ on the quadric $H = \{F_2 = 0\} \subset \mathbb{P}^3(\mathbb{Q})$. Applying Algorithm 10, we find that the system $N \circ E = \Phi_k \circ F_1 \circ E = 0$ has solutions over a number field L of degree 4. If $(x_1, y_1) \in L^2$ satisfies this system, then $F_1(E(x_1, y_1))$ is a 12th primitive root of unity. Since $[\mathbb{Q}(\zeta_{12}) : \mathbb{Q}] = 4$, it follows that $L = \mathbb{Q}(\zeta_{12})$. Thus, we can take $r(x) = \Phi_{12}(x) = x^4 - x^2 + 1$. We find that $\pi(x) \in K[x]$ has the coordinates $\left(0, \frac{1}{6}(x^3 - x^2 - 2x + 2), -\frac{1}{4}(x^3 + x^2), -\frac{1}{4}(x^3 + x^2)\right)$ with respect to the basis $1, u, v, uv$. This yields the following family:

$$r(x) = x^4 - x^2 + 1,$$

$$\pi(x) = \tfrac{1}{12}\left((3(-u - 1)v + 2u)x^3 - (3(u + 1)v + 2u)x^2 - 4ux + 4u\right),$$

$$q(x) = \tfrac{1}{3}\left(x^6 + x^5 + 2x^3 - 2x + 1\right).$$

Let us give some $x_0 \in \mathbb{Z}$ such that $r(x_0)$ and $q(x_0)$ both are prime, and $\pi(x_0)$ is a Weil $q(x_0)$-number:

$x_0 = 4$,
$r(x_0) = 241$,
$\pi(x_0) = (-20u - 20)v + 7u$,
$q(x_0) = 1747$,
$\rho \approx 2.86$.

$x_0 = 115348$,
$r(x_0) = 177027311990089337713$,
$\pi(x_0) = (-383684257046324u - 383684257046324)v + 255785069605399u$,
$q(x_0) = 785132441919874946203437499507$,
$\rho \approx 2.95$.

Example 13. For $k = 24$ and the field $K = \mathbb{Q}\left(\sqrt{-3 + 2\sqrt{2}}\right)$ we obtain the family with $\rho = 3$.

We have $F_1 = x^2 + 2y^2 + 3z^2 - 8zt + 6t^2$, $F_2 = 2xy - 2z^2 + 6zt - 4t^2$, and $N = x^4 - 4x^3 - 4x^2y^2 + 6x^2z^2 - 16x^2zt + 12x^2t^2 + 6x^2 + 8xy^2 + 16xyz^2 - 48xyzt + 32xyt^2 - 12xz^2 + 32xzt - 24xt^2 - 4x + 4y^4 + 12y^2z^2 - 32y^2zt + 24y^2t^2 - 4y^2 - 16yz^2 + 48yzt - 32yt^2 + z^4 - 4z^2t^2 + 6z^2 - 16zt + 4t^4 + 12t^2 + 1$.

On the quadric $H = \{F_2 = 0\} \subset \mathbb{P}^3(\mathbb{Q})$, we find the same line as above $E(x, y) = (0, y, -x, -x)$. Arguing similarly as above, we conclude that the system $N \circ E = \Phi_k \circ F_1 \circ E = 0$ has solutions over the cyclotomic field $L = \mathbb{Q}(\zeta_{24})$. Thus we can take $r(x) = \Phi_{24}(x) = x^8 - x^4 + 1$. We find that $\pi(x)$ has the coordinates $\left(0, \frac{1}{4}(-x^5 - x^4 + x^3 + x^2 + x + 1), \frac{1}{2}(x^6 - x^5), \frac{1}{2}(x^6 - x^5)\right)$ with respect to the basis $1, u, v, uv$. This yields the following family:

$$r(x) = x^8 - x^4 + 1,$$

$$\pi(x) = \frac{1}{4}\left(2(u+1)vx^6 - (2(u+1)v + u)x^5 - ux^4 + ux^3 + ux^2 + ux + u\right),$$

$$q(x) = \frac{1}{8}\left(2x^{12} - 4x^{11} + 3x^{10} + 2x^9 - x^8 - 4x^7 - 3x^6 - 2x^5 + x^4 + 4x^3 + 3x^2 + 2x + 1\right).$$

Let us give some $x_0 \in \mathbb{Z}$ such that the values $r(x_0)$ and $q(x_0)$ both are primes, and $\pi(x_0)$ is a Weil $q(x_0)$-number:

$x_0 = 187,$
$r(x_0) = 1495315557957352561,$
$\pi(x_0) = (21266253242751u + 21266253242751)v - 57471411577u,$
$q(x_0) = 4522601329111114728793301859,$
$\rho \approx 2.93.$

$x_0 = 10215,$
$r(x_0) = 11855147663017807952826751614000 1,$
$\pi(x_0) = (5680122639965599963753125u + 5680122639965599963753125)v - 27808295685150835046u,$
$q(x_0) = 3226379335971003482719098711847396222981509 09857,$
$\rho \approx 2.96.$

Acknowledgements. The author would like to thank the anonymous referees for the above mentioned example, and their helpful comments on the earlier version of this paper.

References

1. Brezing, F., Weng, A.: Elliptic curves suitable for pairing based cryptography. Des. Codes Cryptogr. 37, 133–141 (2005)
2. Freeman, D.: A generalized Brezing-Weng algorithm for constructing pairing-friendly ordinary abelian varieties. In: Galbraith, S.D., Paterson, K.G. (eds.) Pairing 2008. LNCS, vol. 5209, pp. 146–163. Springer, Heidelberg (2008)
3. Freeman, D.: Constructing pairing-friendly genus 2 curves with ordinary Jacobians. In: Takagi, T., Okamoto, T., Okamoto, E., Okamoto, T. (eds.) Pairing 2007. LNCS, vol. 4575, pp. 152–176. Springer, Heidelberg (2007)
4. Freeman, D., Satoh, T.: Constructing pairing-friendly hyperelliptic curves using Weil restriction (to appear in Journal of Number Theory)
5. Freeman, D., Scott, M., Teske, E.: A taxonomy of pairing-friendly elliptic curves. J. Cryptol. 23, 224–280 (2010)
6. Freeman, D., Stevenhagen, P., Streng, M.: Abelian varieties with prescribed embedding degree. In: van der Poorten, A.J., Stein, A. (eds.) ANTS-VIII 2008. LNCS, vol. 5011, pp. 60–73. Springer, Heidelberg (2008)
7. Galbraith, S.: Supersingular curves in Cryptography. In: Boyd, C. (ed.) ASIACRYPT 2001. LNCS, vol. 2248, pp. 495–513. Springer, Heidelberg (2001)
8. Hitt O'Connor, L., McGuire, G., Naehrig, M., Streng, M.: ACM construction for curves of genus 2 with p-rank 1. Preprint, http://eprint.iarc.org/2008/491/
9. Kawazoe, M., Takahashi, T.: Pairing-friendly ordinary hyperelliptic curves with ordinary Jacobians of type $y^2 = x^5 + ax$. In: Galbraith, S.D., Paterson, K.G. (eds.) Pairing 2008. LNCS, vol. 5209, pp. 164–177. Springer, Heidelberg (2008)

10. Milne, J.S.: Abelian varieties. In: Cornell, G., Silverman, J. (eds.) Arithmetic Geometry, pp. 103–150. Springer, New York (1986)
11. Rubin, K., Silverberg, A.: Using abelian varieties to improve pairing-based cryptography. J. Cryptol. 22, 330–364 (2009)
12. Shavarevich, I.: Basic Algebraic Geometry. Springer, New York (1977)
13. Spearman, B.K., Williams, K.S.: Relative integral bases for quartic fields over quadratic subfields. Acta Math. Hungar. 70, 185–192 (1996)
14. Tate, J.: Classes d'isogénie des variétés abéliennes sur un corps fini (d'aprés T. Honda.) Séminarie Bourbaki 1968/69, exposé 352. Lect. Notes in Math., vol. 179, pp. 95–110. Springer, Heidelberg (1971)
15. Waterhouse, W.C.: Abelian varieties over finite fields. Ann. Sci. École Norm. Sup. 2, 521–560 (1969)
16. Waterhouse, W.C., Milne, J.S.: Abelian varieties over finite fields. In: Proc. Symp. Pure Math., vol. 20, pp. 53–64 (1971)

Generating More Kawazoe-Takahashi Genus 2 Pairing-Friendly Hyperelliptic Curves

Ezekiel J Kachisa*

School of Computing
Dublin City University
Ireland
ekachisa@computing.dcu.ie

Abstract. Constructing pairing-friendly hyperelliptic curves with small ρ-values is one of challenges for practicability of pairing-friendly hyperelliptic curves. In this paper, we describe a method that extends the Kawazoe-Takahashi method of generating families of genus 2 ordinary pairing-friendly hyperelliptic curves by parameterizing the parameters as polynomials. With this approach we construct genus 2 ordinary pairing-friendly hyperelliptic curves with $2 < \rho \leq 3$.

Keywords: pairing-friendly curves, hyperelliptic curves.

1 Introduction

Efficient implementation of pairing-based protocols such as one round three way key exchange [16], identity based encryption [3] and digital signatures [4], depends on what are called *pairing-friendly curves*. These are special curves with a large prime order subgroup, so that protocols can resist the known attacks, and small embedding degree for efficient finite field computations.

Even though there are many methods for constructing pairing-friendly elliptic curves [14], there are very few methods that address the problem of constructing ordinary pairing-friendly hyperelliptic curves of higher genus. The first explicit construction of ordinary hyperelliptic curve was shown by David Freeman [11]. Freeman modeled the Cocks-Pinch method [8] to construct ordinary hyperelliptic curves of genus 2. His algorithm produce curves over prime fields with prescribed embedding degree k with ρ-value ≈ 8. Kawazoe and Takahashi [18] constructed pairing-friendly hyperelliptic curves of the form $y^2 = x^5 + ax$ which produced Jacobian varieties with ρ-values between 3 and 4. Recently, Freeman and Satoh [15] proposed algorithms for generating pairing-friendly hyperelliptic curves. In their construction it was shown that if an elliptic curve, E, is defined over a finite field, $\mathbb{F}_p$, and $\mathcal{A}$ is abelian v! ariety isogenous over $\mathbb{F}_{p^d}$ to a product of two isomorphic elliptic curves then the abelian variety, $\mathcal{A}$, is isogenous over $\mathbb{F}_p$ to a primitive subvariety of the Weil restriction of E from $\mathbb{F}_{p^d}$ to $\mathbb{F}_p$. Notably,

* This author acknowledge support from the Science Foundation Ireland under Grant No. 06/MI/006 through Claude Shannon Institute.

M. Joye, A. Miyaji, and A. Otsuka (Eds.): Pairing 2010, LNCS 6487, pp. 312–326, 2010.

the Freeman-Satoh algorithm produces hyperelliptic curves with better ρ value than previously reported. The best, for example, achieves a ρ-value of $20/9$ for embedding degree $k = 27$. However, the ρ-values of most embedding degrees for ordinary hyperelliptic curves remain too high for an efficient implementation.

For a curve to be suitable for implementation it should possess desirable properties which include efficient implementation of finite field arithmetic and the order of the Jacobian having a large prime factor.

In this paper we generate more Kawazoe-Takahashi genus 2 ordinary pairing-friendly hyperelliptic curves. In particular, we construct curves of embedding degrees $2, 7, 8, 10, 11, 13, 22, 26, 28, 44$ and 52 with ρ-value between 2 and 3.

We proceed as follows: In Section 2 we present mathematical background and facts on constructing pairing-friendly hyperelliptic curves while in Section 3 we discuss the construction of pairing-friendly hyperelliptic curves based on the Kawazoe-Takahashi algorithms and in Section 4 we present the generalization of Kawazoe-Takahashi algorithms for constructing pairing-friendly hyperelliptic curves and we give explicit examples. The paper is concluded in Section 5.

2   Pairing-Friendly Hyperelliptic Curves

2.1   Mathematical Background

Let $p > 2$ be a prime, let r be prime distinct from p. We denote a hyperelliptic curve of genus g defined over a finite field $\mathbb{F}_p$ by C. This is a non-singular projective model of the affine curve of the form:

$$y^2 = f(x) \tag{1}$$

where $f(x)$ is a monic polynomial of degree $2g + 1$, has its coefficients in $\mathbb{F}_p[x]$ and has no multiple roots in $\bar{\mathbb{F}}_p$. We denote the Jacobian of C by J_C and a group of the $\mathbb{F}_p$-rational points of the Jacobian of C by $J_C(\mathbb{F}_p)$. This group is isomorphic to degree zero divisor class group of C over $\mathbb{F}_p$.

As in the elliptic curve case the *embedding degree* of Jacobian variety is defined as follows:

Definition 1 ([11]). *Let C be an hyperelliptic curve defined over a prime finite field $\mathbb{F}_p$. Let r be a prime dividing $\#J_C(\mathbb{F}_p)$. The embedding degree of J_C with respect to r is the smallest positive integer k such that $r \mid p^k - 1$ but $r \nmid p^i - 1$ for $0 < i < k$.*

The definition, as in the elliptic curve case, explains that k is the smallest positive integer such that the extension field $\mathbb{F}_{p^k}$, contains a set of rth roots of unity. Hence we refer to a curve C as having embedding degree k with respect to r if and only if a subgroup of order r of its Jacobian J_C does. As such, for an efficient arithmetic implementation curves must have small embedding degree so that arithmetic in $\mathbb{F}_{p^k}$ is feasible. Furthermore, we require that the size of the finite field, $\mathbb{F}_p$, be as small as possible in relation to the the size of the prime order subgroup r. This is measured by a parameter known as the ρ-value. For a g-dimensional abelian variety defined over $\mathbb{F}_p$ this parameter is defined as:

$$\rho = \frac{g \log (\mathrm{p})}{\log (\mathrm{r})}.$$

In the ideal case the abelian varieties of dimension g have a prime number of points in which case $\rho \approx 1$. For pairing-friendly one-dimensional abelian varieties one can reach the ideal case by using the constructions in [19], [6] and [10]. However, this proves not be the case with higher dimensional abelian varieties. Hence the interest has been to construct higher dimensional abelian varieties with low embedding degrees and small ρ-values. And the same time, for security reasons we require r large enough so that discrete logarithm problem (DLP) in the subgroup of prime order r is suitably hard and k sufficiently large enough so that the (DLP) in $\mathbb{F}_{p^k}^*$, withstand the known attacks.

There are two main cryptographic pairings, the Weil and the Tate. In both cases the basic idea is to embed the cryptographic group of order r into a multiplicative group of rth roots of unity, μ_r. A non-degenerate, bilinear map for the Tate pairing, for example, is defined by the following map:

$$t_r : J_C(\mathbb{F}_{p^k})[r] \times J_C(\mathbb{F}_{p^k})/J_C(\mathbb{F}_{p^k}) \longrightarrow (\mathbb{F}_{p^k}^*)/(\mathbb{F}_{p^k}^*)^r.$$

3   Kawazoe-Takahashi Hyperelliptic Curves

Kawazoe and Takahashi [18] presented an algorithm which constructed hyperelliptic curves of the form $y^2 = x^5 + ax$ with ordinary Jacobians. Their construction used two approaches, one was based on the Cocks-Pinch method [8] of constructing ordinary pairing-friendly elliptic curves and the other was based on cyclotomic polynomials. This idea was first proposed by Brezing and Weng in [7]. However, both approaches are based on the predefined sizes of the Jacobians as presented in [9]. The order of the Jacobian, $\#J_C$, is closely related to the characteristic polynomial, $\chi(t)$, of the Frobenius endomorphism, π.

Consequently, for genus 2 curves the $\chi(t)$ of the Frobenius is a polynomial known to have the following form:

$$\chi(t) = t^4 - a_1 t^3 + a_2 t^2 - a_1 p t + p^2 \tag{2}$$

within $a_1, a_2 \in \mathbb{F}_p$ and furthermore $\mid a_1 \mid \leq 4p$ and $\mid a_2 \mid \leq 6p$. Hence, $\#J_C$ is determined from Equation 2 by the following relation:

$$\#J_C = \chi(1) = 1 - a_1 + a_2 - a_1 p + p^2. \tag{3}$$

The Hasse-Weil bound describes the interval in which the order of the Jacobian is found as follows:

$$\lceil (\sqrt{p} - 1)^{2g} \rceil \leq \#J_C \leq \lfloor (\sqrt{p} + 1)^{2g} \rfloor \tag{4}$$

Algorithm 1. Kawazoe-Takahashi Type I pairing-friendly Hyperelliptic curves with $\#J_C = 1 - 4d + 8d^2 - 4dp + p^2$

Input: $k \in \mathbb{Z}$.
Output: a hyperelliptic curve defined by $y^2 = x^5 + ax$ with Jacobian group having a prime subgroup of order r.

1. Choose r a prime such that $lcm(8, k)$ divides $r - 1$.
2. Choose ζ a primitive kth root of unity in $(\mathbb{Z}/r\mathbb{Z})^\times, \omega$ a positive integer such that $\omega^2 \equiv -1 \bmod r$ and σ a positive integer such that $\sigma^2 \equiv 2 \bmod r$.
3. Compute integers, c, d such that:

   - $c \equiv (\zeta + \omega)(\sigma(\omega + 1))^{-1} \bmod r$ and $c \equiv 1 \bmod 4$
   - $d \equiv (\zeta\omega + 1)(2(\omega + 1))^{-1} \bmod r$.

4. Compute a prime $p = (c^2 + 2d^2)$ such that $p \equiv 1 \bmod 8$.
5. Find $a \in \mathbb{F}_p$ such that:

   - $a^{(p-1)/2} \equiv -1 \bmod p$ and $2(-1)^{(p-1)/8}d \equiv (a^{(p-1)/8} + a^{3(p-1)/8})c \bmod p$.

6. Define a hyperelliptic curve C by $y^2 = x^5 + ax$.

Theorem 1 below outlines the characteristic polynomials which defines hyperelliptic curves, C, of the form $y^2 = x^5 + ax$ defined over $\mathbb{F}_p$. The J_C of C for these cases is a simple ordinary Jacobian over $\mathbb{F}_p$.

Theorem 1 ([9],[18]). *Let p be an odd prime, C a hyperelliptic curve defined over $\mathbb{F}_p$ by equation $y^2 = x^5 + ax$, Jc the Jacobian variety of C and $\chi(t)$ the characteristic polynomial of the pth power Frobenius map of C. Then the following holds: (In the following c, d are integers such that $p = c^2 + 2d^2$ and $c \equiv 1 \pmod 4$, $d \in \mathbb{Z}$ (such c and d exists if and only if $p \equiv 1, 3 \pmod 8$).*

1) *If $p \equiv 1 \bmod 8$ and $a^{(p-1)/2} \equiv -1 \bmod p$, then $\chi(t) = t^4 - 4dt^3 + 8d^2t^2 - 4dpt + p^2$ and $2(-1)^{(p-1)/8}d \equiv (a^{(p-1)/8} + a^{3(p-1)/8})c \bmod p$*
2) *If $p \equiv 1 \bmod 8$ and $a^{(p-1)/4} \equiv -1 \bmod p$, or if $p \equiv 3 \bmod 8$ and $a^{(p-1)/2} \equiv -1 \bmod p$, then $\chi(t) = t^4 + (4c^2 - 2p)t^2 + p^2$*

Using the formulae in Theorem 1 Kawazoe and Takahashi developed a Cocks-Pinch-like method to construct genus 2 ordinary pairing-friendly hyperelliptic curves of the form $y^2 = x^5 + ax$. As expected the curves generated by the Cocks-Pinch-like method had their ρ-values close to 4. Furthermore, they also presented cyclotomic families. With this method they managed to construct a $k = 24$ curve with $\rho = 3$. In both cases the ultimate goal is to find integers c and d such that there is a prime $p = c^2 + 2d^2$ with $c \equiv 1 \pmod 4$ and $\chi(1)$ having a large prime factor. Algorithms 1 and 2 developed from Theorem 1 construct individual genus 2 pairing-friendly hyperelliptic curves with $\rho \approx 4$.

Remark 1. The key feature in both algorithms is that r is choosen such that $r - 1$ is divisible by 8 so that $\mathbb{Z}/r\mathbb{Z}$ contains both $\sqrt{-1}$ and $\sqrt{2}$ for both c and d to satisfy the conditions in the algorithm.

Algorithm 2. Kawazoe-Takahashi Type II pairing-friendly Hyperelliptic curves with $\#J_C = 1 + (4c^2 - 2p) + p^2$

Input: $k \in \mathbb{Z}$.
Output: a hyperelliptic curve defined by $y^2 = x^5 + ax$ with Jacobian group having a prime subgroup of order r.

1. Choose r a prime such that $lcm(8, k)$ divides $r - 1$.
2. Choose ζ a primitive kth root of unity in $(\mathbb{Z}/r\mathbb{Z})^\times$, ω positive integer such that $\omega^2 \equiv -1 \bmod r$ and σ a positive integer such that $\sigma^2 \equiv 2 \bmod r$.
3. Compute integers, c, d such that:

   - $c \equiv 2^{-1}(\zeta - 1)\omega) \bmod r$ and $c \equiv 1 \bmod 4$
   - $d \equiv (\zeta + 1)(2\sigma)^{-1} \bmod r$.

4. Compute a prime $p = (c^2 + 2d^2)$ such that $p \equiv 1, 3 \bmod 8$ and for some integer δ satisfying $\delta^{(p-1)/2} \equiv -1 \bmod p$ and
5. Find $a \in \mathbb{F}_p$ such that:

   - $a = \delta^2$ when $p \equiv 1 \bmod 8$ or $a = \delta$ when $p \equiv 3 \bmod 8$.

6. Define a hyperelliptic curve C by $y^2 = x^5 + ax$.

---

4 Our Generalization

We observe that one can do better if the algorithms are parametrized by polynomials in order to construct curves with specified bit size. We represent *families* of pairing-friendly curves for which parameters c, d, r, p are parametrized as polynomials $c(z), d(z), r(z), p(z)$ in a variable z. In fact this idea of using polynomials was used in other constructions for pairing-friendly curves such as in [19],[2] [21] and [7].

When working with the polynomials we consider polynomials with rational coefficients. The definitions below describes a family of Kawazoe-Takahashi-type of pairing-friendly hyperelliptic curves.

Definition 2 ([14]). *Let* $g(z) \in \mathbb{Q}[z]$. *We say that* $g(z)$ *represents primes if the following are satisfied:*

- $g(z)$ *is non constant irreducible polynomial.*
- $g(z)$ *has a positive leading coefficient.*
- $g(z)$ *represents integers i.e for* $z_0 \in \mathbb{Z}$, $g(z_0) \in \mathbb{Z}$.
- $gcd(\{g(z) : z, g(z) \in \mathbb{Z}\}) = 1$

Definition 3. *Let* $c(z), d(z), r(z)$ *and* $p(z)$ *be non-zero polynomials with rational coefficients. For a given positive integer* k *the couple* $(r(z), p(z))$ *parameterizes a family of Kawazoe-Takahashi type of hyperelliptic curves with Jacobian* J_C *whose embedding degree is* k *if the following conditions are satisfied:*

(i) $c(z)$ *represents integers such that* $c(z) \equiv 1 \bmod 4$;
(ii) $d(z)$ *represents integers;*

(iii) $p(z) = c(z)^2 + 2d(z)^2$ *represents primes;*
(iv) $r(z)$ *represents primes;*
 (v) $r(z)|1 - 4d(z) + 8d(z)^2 - 4d(z)p(z) + p(z)^2$ *or* $r(z)|1 + (4c(z)^2 - 2p(z)) + p(z)^2$
(vi) $\Phi_k(p(z)) \equiv 0$ *mod* $r(z)$, *where* Φ_k *is the kth cyclotomic polynomial.*

And we define the ρ-value of this family as $\rho = \frac{2 \deg(p(z))}{\deg(r(z))}$.

In [9] they showed that there exists a simple ordinary abelian variety surface with characteristic polynomials of Frobenius $t^4 - 4d + 8d^2 - 4dp + p^2 \in \mathbb{Z}[t]$ or $t^4 + (4c^2 - 2p) + p^2 \in \mathbb{Z}[t]$ with certain conditons on c and d. Hence Definition 3 part (i) and (ii) ensures that the polynomial representation of c and d conforms with the conditions. While condition (v) of Definition 3 ensures that for a given z for which $p(z)$ and $r(z)$ represents prime $r(z)$ divides $\#J_C(z)$. In otherwords, the order of the Jacobian of the constructed curve has a prime order subgroup of size $r(z)$. Finaly, condition (vi) of Definition 3 ensures that the Jacobian of the constructed curve has embedding degree k.

With these definitions we now adapt Algorithms 1 and 2 to the polynomial context. This can be seen in Algorithms 3 and 4 below generalizing Algorithms 1 and 2 respectively. In particular we construct our curves by taking a similar approach as described in [17] for constructing pairing-friendly elliptic curves.

In general this method uses minimal polynomials rather than a cyclotomic polynomial in defining the size of the prime order subgroup. The difficult part is the choosing the right polynomial for representing the size of the cryptographic group.

With this approach, apart from reconstructing the Kawazoe-Takahashi genus 2 curves, we discover new families of pairing-friendly hyperelliptic curve of embedding degree $k = 2, 7, 8, 10, 11, 13, 22, 26, 28, 44$ and 52 with $2 < \rho \le 3$.

The success depends on the the choice of the number field, K. Thus, in the initial step we set K to be isomorphic to a cyclotomic field $\mathbb{Q}(\zeta_\ell)$ for some $\ell = lcm(8, k)$. The condition on ℓ ensures $\mathbb{Q}[z]/r(z)$ contains square roots of -1 and 2. We take the approach as described in [17] for constructing pairing-friendly elliptic curves for defining the irreducible polynomial $r(z)$. Even though this method is time consuming as it involves searching for a right element, it mostly gives a favorable irreducible polynomial $r(z)$, which defines the size of the prime order subgroup . Here we find a minimal polynomial of an element $\gamma \in \mathbb{Q}(\zeta_\ell)$ and call it $r(z)$, where γ is not in any proper subfield of $\mathbb{Q}(\zeta_\ell)$. Since γ is in no proper subfield, then we have $\mathbb{Q}(\zeta_\ell) = \mathbb{Q}(\gamma)$, where the degree of $\mathbb{Q}(\gamma)$ over $\mathbb{Q}$ is $\varphi(\ell)$, where $\varphi(.)$ is *Euler totient function.*

However, with most values of $k > 10$ which are not multiples of 8, the degree of $r(z)$ tends to be large. As observed in [14], for such curves this limits the number of usable primes. The current usable size of r is in the range $[2^{160}, 2^{512}]$.

4.1   The Algorithm Explained

Step 1: Set up. This involves initializing the algorithm by setting $\mathbb{Q}(\zeta_\ell)$ defined as $\mathbb{Q}[z]/\Phi_\ell(z)$. The Choice of this field ensures that it contains ζ_k and $\sqrt{-1}$ and $\sqrt{2}$. The ideal choice, in such a case, is $\mathbb{Q}(\zeta_8, \zeta_k) = \mathbb{Q}(\zeta_{lcm(k,8)})$.

Algorithm 3. Our generalization for finding pairing-friendly Hyperelliptic curves with $\#J_C(z) = 1 - 4d(z) + 8d(z)^2 - 4d(z)p(z) + p(z)^2$

Input: $k \in \mathbb{Z}, \ell = lcm(8, k), K \cong \mathbb{Q}[z]/\Phi_\ell(z)$
Output: Hyperelliptic curve of genus 2 defined by $y^2 = x^5 + ax$.

1. Choose an irreducible polynomial $r(z) \in \mathbb{Z}[z]$.
2. Choose polynomials $s(z), \omega(z)$ and $\sigma(z)$ in $\mathbb{Q}[z]$ such that $s(z)$ is a primitive kth root of unity, $\omega(z) = \sqrt{-1}$ and $\sigma(z) = \sqrt{2}$ in K.
3. Compute polynomials, $c(z), d(z)$ such that:

   - $c(z) \equiv (s(z) + \omega(z))(\sigma(z)(\omega(z) + 1))^{-1}$ in $\mathbb{Q}[z]/r(z)$.
   - $d(z) \equiv (s(z)\omega(z) + 1)(2(\omega(z) + 1))^{-1}$ in $\mathbb{Q}[z]/r(z)$.

4. Compute a polynomial, $p(z) = c(z)^2 + 2d(z)^2$.
5. For $z_0 \in \mathbb{Z}$ such that:
   - $p(z_0)$ and $r(z_0)$ represents primes and $p(z_0) \equiv 1 \mod 8$ and
   - $c(z_0), d(z_0)$ represents integers and $c(z_0) \equiv 1 \mod 4$.
   find $a \in \mathbb{F}_{p(z_0)}$ satisfying:

   - $a^{(p(z_0)-1)/2} \equiv -1 \mod p(z_0)$ and
   - $2(-1)^{(p(z_0)-1)/8}d(z_0) \equiv (a^{(p(z_0)-1)/8} + a^{3(p(z_0)-1)/8})c(z_0) \mod p(z_0)$.

6. Output $(r(z_0), p(z_0), a)$
7. Define a hyperelliptic curve C by $y^2 = x^5 + ax$.

Step 2: Representing ζ_k, $\sqrt{-1}$ and $\sqrt{2}$. We search for a favorable element, $\gamma \in \mathbb{Q}(\zeta_\ell)$ such that the minimal polynomial of γ has degree $\varphi(\ell)$ and we call this $r(z)$. We redefine our field to $\mathbb{Q}[z]/r(z)$. In this field we find a polynomial that represents ζ_k, $\sqrt{-1}$ and $\sqrt{2}$.

For ζ_k there are $\varphi(k)$ numbers of primitive kth roots of unity. In fact if $\gcd(\alpha, k) = 1$ then ζ_k^α is also primitive kth root of unity. To find the polynomial representation of $\sqrt{-1}$ and $\sqrt{2}$ in $\mathbb{Q}[z]/r(z)$ we find the solutions of the polynomials $z^2 + 1$ and $z^2 - 2$ in the number field isomorphic to $\mathbb{Q}[z]/r(z)$ respectively.

Steps 3,4,5: Finding the family. All computations in the algorithm are done modulo $r(z)$ except when computing $p(z)$. It is likely that polynomials $p(z), c(z)$ and $d(z)$ have rational coefficient. At this point polynomials are tested to determine whether they represent intergers or primes as per Definition 3.

4.2 New Curves

We now present a series of new curves constructed using the approach described above. Proving the theorems is simple considering γ has minimal polynomial $r(z)$. We give a proof of Theorem 2. For the other curves the proofs are similar.

We start by constructing a curve of embedding degree, $k = 7$. It is interesting to note that here we get a family with $\rho = 8/3$.

Algorithm 4. Our generalization for finding pairing-friendly Hyperelliptic curves with $\#J_C(z) = 1 + (4c(z)^2 - 2p(z)) + p(z)^2$

Input: $k \in \mathbb{Z}, \ell = lcm(8, k), K \cong \mathbb{Q}[z]/\Phi_\ell(z)$
Output: Hyperelliptic curve of genus 2 defined by $y^2 = x^5 + ax$.

1. Choose an irreducible polynomial $r(z) \in \mathbb{Z}[z]$.
2. Choose polynomials $s(z), \omega(z)$ and $\sigma(z)$ in $\mathbb{Q}[z]$ such that $s(z)$ is a primitive kth root of unity, $\omega(z) = \sqrt{-1}$ and $\sigma(z) = \sqrt{2}$ in K.
3. Compute polynomials, $c(z), d(z)$ such that

   - $c(z) \equiv 2^{-1}(s(z) - 1)\omega(z)) \bmod r(z)$
   - $d(z) \equiv (z(z) + 1)(2\sigma(z))^{-1} \bmod r(z)$

4. Compute an irreducible polynomial $p(z) = (c(z)^2 + 2d(z)^2)$
5. For $z_0 \in \mathbb{Z}$ such that:
   − $p(z_0)$ and $r(z_0)$ represents primes and $p(z_0) \equiv 1, 3 \bmod 8$ and
   − $c(z_0), d(z_0)$ represents integers and $c(z_0) \equiv 1 \bmod 4$.
6. Find $a \in \mathbb{F}_p(z_0)$ such that:

   - $a = \delta^2$ when $p(z_0) \equiv 1 \bmod 8$ or
   - $a = \delta$ when $p(z_0) \equiv 3 \bmod 8$.

7. Output $(r(z_0), p(z_0), a)$.
8. Define a hyperelliptic curve C by $y^2 = x^5 + ax$.

Theorem 2. *Let* $k = 7, \ell = 56$. *Let* $\gamma = \zeta_\ell + 1 \in \mathbb{Q}(\zeta_\ell)$ *and define polynomials* $r(z), p(z), c(z), d(z)$ *by the following:*

$$
\begin{aligned}
r(z) = \ & z^{24} - 24z^{23} + 276z^{22} - 2024z^{21} + 10625z^{20} - 42484z^{19} \\
& + 134406z^{18} - 344964z^{17} + 730627z^{16} - 1292016z^{15} + 1922616z^{14} \\
& - 2419184z^{13} + 2580005z^{12} - 2332540z^{11} + 1784442z^{10} - 1150764z^9 \\
& + 621877z^8 - 279240z^7 + 102948z^6 - 30632z^5 + 7175z^4 - 1276z^3 + 162z^2 - 12z + 1
\end{aligned}
$$

$$
\begin{aligned}
p(z) = \ & (z^{32} - 32z^{31} + 494z^{30} - 4900z^{29} + 35091z^{28} - 193284z^{27} + \\
& 851760z^{26} - 3084120z^{25} + 9351225z^{24} - 24075480z^{23} + 53183130z^{22} - \\
& 101594220z^{21} + 168810915z^{20} - 245025900z^{19} + 311572260z^{18} - \\
& 347677200z^{17} + 340656803z^{16} - 292929968z^{15} + 220707810z^{14} - 145300540z^{13} + \\
& 83242705z^{12} - 41279004z^{11} + 17609384z^{10} - 6432920z^9 + 2023515z^8 \\
& - 569816z^7 + 159446z^6 - 49588z^5 + 16186z^4 - 4600z^3 + 968z^2 - 128z + 8)/8
\end{aligned}
$$

$$
c(z) = (-z^9 + 9z^8 - 37z^7 + 91z^6 - 147z^5 + 161z^4 - 119z^3 + 57z^2 - 16z + 2)/2
$$

$$
\begin{aligned}
d(z) = \ & (z^{16} - 16z^{15} + 119z^{14} - 546z^{13} + 1729z^{12} - 4004z^{11} + 7007z^{10} \\
& - 9438z^9 + 9867z^8 - 8008z^7 + 5005z^6 - 2366z^5 + 819z^4 - 196z^3 + 28z^2)/4
\end{aligned}
$$

Then $(r(2z), p(2z))$ *constructs a genus 2 hyperelliptic curves. The* ρ-*value of this family is* 8/3.

Proof. Since $\zeta_\ell + 1 \in \mathbb{Q}(\zeta_\ell)$ has minimal polynomial $r(z)$, we apply Algorithm 3 by working in $\mathbb{Q}(\zeta_{56})$ defined as $\mathbb{Q}[z]/r(z)$. We choose $\zeta_7 \mapsto (z-1)^{16}, \sqrt{-1} \mapsto$

$(z-1)^{14}$ and $\sqrt{2} \mapsto z(z-1)^7(z-2)(z^6 - 7z^5 + 21z^4 - 35z^3 + 35z^2 - 21z + 7)(z^6 - 5z^5 + 11z^4 - 13z^3 + 9z^2 - 3z + 1)$. Applying Algorithm 3 we find $p(z)$ as stated. Computations with PariGP [23], show that both $r(2z)$ and $p(2z)$ represents primes and $c(2z)$ represents integers such that it is equivalent to 1 modulo 4. Furthermore, by Algorithm 3 the Jacobian of our hypothetical curve has a large prime order subgroup of order $r(z)$ and embedding degree, $k = 7$.

Considering $z_0 = 758$ we give an example of a 254- bit prime subgroup that is constructed using the parameters in Theorem 2.

Example 1.

$r = 213748555325666652890713665865251428761742681841141544849244\backslash$
  05425230130090001

$p = 741504661189142770769829861344257948821797401549707353154351\backslash$
  $08095481642765042445975666095781797666897$

$c = -21022477149693687350103984375$

$d = 192549300334893812717931530445605096860437011144944$

$a = 3$

$\rho = 2.646.$

$C : y^2 = x^5 + 3x$

The next curve is of embedding degree $k = 8$. According to [25] this family of curves admits higher order twists. This means that it is possible to have both inputs to a pairing defined over a base field. The previous record on this curve was $\rho = 4$. In Theorem 3 below we outline the parameters that defines a family of hyperelliptic curves with $\rho = 3$.

Theorem 3. *Let* $k = \ell = 8$. *Let* $\gamma = \zeta_\ell^3 + \zeta_\ell^2 + \zeta_\ell + 3 \in \mathbb{Q}(\zeta_8)$ *and define polynomials* $r(z), p(z), c(z), d(z)$ *by the following:*

$r(z) = z^4 - 12z^3 + 60z^2 - 144z + 136$
$p(z) = (11z^6 - 188z^5 + 1460z^4 - 6464z^3 + 17080z^2 - 25408z + 16448)/64$
$c(z) = (3z^3 - 26z^2 + 92z - 120)/8$
$d(z) = (-z^3 + 8z^2 - 26z + 32)/8$

Then $(r(32z)/8, p(32z))$ constructs a genus 2 hyperelliptic curves with embedding degree 8. The ρ-value of this family is 3.

This type of a curve is recommended at the 128 bit security level, see Table 3.1 in [1]. Below we give an example obtained using the above parameters.

Example 2.

$r = 131072000000009898508288000280324362739203528331792090742\backslash$
  $477643363528725893137 (257 bits)$

$p = 1845493760000020905654747136986742251766767879474504560418\backslash$
  $2525326695069336429048851161837661576412771127129831728 84737$

$$c = 1228800000000069598899200001314020933668808269532200344 0625$$
$$d = -4096000000000231996416000004380073001064027565137751569916$$
$$a = 3$$
$$\rho = 3.012$$
$$C : y^2 = x^5 + 3x$$

Theorem 4. *Let $k = 10, \ell = 40$. Let $\gamma = \zeta_\ell + 1 \in \mathbb{Q}(\zeta_\ell)$ and define polynomials $r(z), p(z), c(z), d(z)$ by the following:*

$$r(z) = z^{16} - 16z^{15} + 120z^{14} - 560z^{13} + 1819z^{12} - 4356z^{11} + 7942z^{10} -$$
$$11220z^9 + 12376z^8 - 10656z^7 + 7112z^6 - 3632z^5 + 1394z^4 - 392z^3 + 76z^2 - 8z + 1$$
$$p(z) = (z^{24} - 24z^{23} + 274z^{22} - 1980z^{21} + 10165z^{20} - 39444z^{19}$$
$$+120156z^{18} - 294576z^{17} + 591090z^{16} - 981920z^{15} + 1360476z^{14} -$$
$$1578824z^{13} + 1536842z^{12} - 1253336z^{11} + 853248z^{10} - 482384z^9 +$$
$$225861z^8 - 88872z^7 + 31522z^6 - 11676z^5 + 4802z^4 - 1848z^3 + 536z^2 - 96z + 8)/8$$
$$c(z) = (-z^7 + 7z^6 - 22z^5 + 40z^4 - 45z^3 + 31z^2 - 12z + 2)/2$$
$$d(z) = (z^{12} - 12z^{11} + 65z^{10} - 210z^9 + 450z^8 - 672z^7 + 714z^6 - 540z^5 + 285z^4 -$$
$$100z^3 + 20z^2)/4$$

Then $(r(4z), p(4z))$ constructs a genus 2 hyperelliptic curve. The ρ-value of this family is 3.

Below is a curve of embedding degree 10 with a prime subgroup of size 249 bits. The ρ-value of its J_C is 3.036.

Example 3.

$$r = 4745749105410301406815931235596753944430110861981481 0948\backslash$$
$$2797931132143318041$$
$$p = 3392680476835482274427348989075071521908024843148191 25499\backslash$$
$$3934108021750448229282701596660539123994672109536233 56417$$
$$c = -11897241590353385507970614067112 95$$
$$d = 41186651216355781032109778827651005272746978660218968 4736$$
$$a = 3$$
$$\rho = 3.036$$
$$C : y^2 = x^5 + 3x$$

Theorem 5. *Let $k = 28, \ell = 56$. Let $\gamma = \zeta_\ell + 1 \in \mathbb{Q}(\zeta_\ell)$ and define polynomials $r(z), p(z), c(z), d(z)$ by the following:*

$$r(z) = z^{24} - 24z^{23} + 276z^{22} - 2024z^{21} + 10625z^{20} - 42484z^{19} +$$
$$134406z^{18} - 344964z^{17} + 730627z^{16} - 1292016z^{15} + 1922616z^{14} -$$
$$2419184z^{13} + 2580005z^{12} - 2332540z^{11} + 1784442z^{10} - 1150764z^9 + 621877z^8 -$$
$$279240z^7 + 102948z^6 - 30632z^5 + 7175z^4 - 1276z^3 + 162z^2 - 12z + 1$$

$$p(z) = (z^{36} - 36z^{35} + 630z^{34} - 7140z^{33} + 58903z^{32} - 376928z^{31} +$$
$$1946800z^{30} - 8337760z^{29} + 30188421z^{28} - 93740556z^{27} + 252374850z^{26} -$$
$$594076860z^{25} + 1230661575z^{24} - 2254790280z^{23} + 3667649460z^{22} -$$
$$5311037640z^{21} + 6859394535z^{20} - 7909656300z^{19} + 8145387218z^{18} -$$
$$7487525484z^{17} + 613613430z^{16} - 4473905808z^{15} + 2893567080z^{14} - 1653553104z^{13} +$$
$$830662287z^{12} - 364485108z^{11} + 138635550z^{10} - 45341540z^{9} + 12681910z^{8} -$$
$$3054608z^{7} + 660688z^{6} - 141120z^{5} + 32008z^{4} - 7072z^{3} + 1256z^{2} - 144z + 8)/8$$
$$c(z) = (-z^{11} + 11z^{10} - 55z^{9} + 165z^{8} - 331z^{7} + 469z^{6} - 483z^{5} + 365z^{4} - 200z^{3} +$$
$$76z^{2} - 18z + 2)/2$$
$$d(z) = (z^{18} - 18z^{17} + 153z^{16} - 816z^{15} + 3059z^{14} - 8554z^{13} + 18473z^{12} - 31460z^{11}$$
$$+42757z^{10} - 46618z^{9} + 40755z^{8} - 28392z^{7} + 15561z^{6} - 6566z^{5} + 2058z^{4} -$$
$$448z^{3} + 56z^{2})/4$$

Then $(r(2z), p(2z))$ constructs a genus 2 hyperelliptic curve. The ρ-value of this family is $\rho \approx 3$.

Here is a curve with a 255 bit prime subgroup constructed from the above parameters:

Example 4.

$$r = 42491960053938594435112219237666767431311006357122111696\backslash$$
$$690362883228500208481$$
$$p = 109488916950130503728824712394480136647965331684153539280\backslash$$
$$56833619302663216719518472851456451963664706050519126 3121$$
$$c = -66111539648877169993055611952337239$$
$$d = 73989498224454294419334385377521846525339047033183899 8400$$
$$a = 23$$
$$\rho = 2.972$$
$$C : y^{2} = x^{5} + 23x$$

The following family for $k = 24$ has a similar ρ-value as to a family of $k = 24$ reported in [18]. One can use the following parameters to construct a *Kawazoe-Takahashi Type II* pairing-friendly hyperelliptic curve of embedding degree $k = 24$ with $\rho = 3$.

Theorem 6. *Let $k = \ell = 24$. Let $\gamma = \zeta_{24} + 1 \in \mathbb{Q}(\zeta_{24})$ and define polynomials $r(z), p(z), c(z), d(z)$ by the following:*

$$r(z) = z^{8} - 8z^{7} + 28z^{6} - 56z^{5} + 69z^{4} - 52z^{3} + 22z^{2} - 4z + 1$$
$$p(z) = (2z^{12} - 28z^{11} + 179z^{10} - 688z^{9} + 1766z^{8} - 3188z^{7} +$$
$$4155z^{6} - 3948z^{5} + 2724z^{4} - 1336z^{3} + 443z^{2} - 88z + 8)/8$$
$$c(z) = (-z^{6} + 7z^{5} - 20z^{4} + 30z^{3} - 25z^{2} + 11z - 2)/2$$
$$d(z) = (z^{5} - 4z^{4} + 5z^{3} - 2z^{2} - z)/4$$

Then $(r(8z+4)/8, p(8z+4))$ constructs a complete ordinary pairing-friendly genus 2 hyperelliptic curves with embedding degree 24. *The ρ-value of this family is* 3.

The following family is of embedding degree $k = 2$ with $\rho = 3$. In this case the parameters corresponds to a quadratic twist C' of the curve C whose order of J_C has a large prime of size r.

Theorem 7. *Let $k = 2$, $\ell = 8$. Let $\gamma = \zeta_8^2 + \zeta_8 + 1 \in \mathbb{Q}(\zeta_8)$ and define polynomials $r(z), p(z), c(z), d(z)$ by the following:*

$$r(z) = z^4 - 4z^3 + 8z^2 - 4z + 1$$
$$p(z) = (17z^6 - 128z^5 + 480z^4 - 964z^3 + 1089z^2 - 476z + 68)/36$$
$$c(z) = (z^3 - 4z^2 + 7z - 2)/2$$
$$d(z) = (-2z^3 + 7z^2 - 14z + 4)/6$$

Then $(r(36z + 8)/9, p(36z + 8))$ constructs a genus 2 hyperelliptic curve. The ρ-value of this family is 3.

Here is a curve with a 164 bit prime subgroup constructed from the above parameters:

Example 5.

$r = 18662407671139230451673881592011637799903138004697$

$p = 10279256257891516489822674213746873409099825032 5265\backslash$
$\qquad 616516412999094595596 79217$

$c = 23328007191686179030939068128424560723$

$d = -15552004794459612687736644908426134338$

$a = 10$

$\rho = 3.049$

Here our genus 2 hyperelliptic equation is $C' : y^2 = x^5 + 10x$ and hence $C : y^2 = 20(x^5 + 10x)$ is the curve whose $\#J_C$ has a large prime r and its embedding degree is 2 with repect to r.

We now present pairing-friendly hyperelliptic curves of embedding k whose polynomial that defines the prime order subgroup $r(z)$, has its degree greater or equal

Table 1. Families of curves, whose $deg(r(z)) \geq 40$

k	γ	Degree($r(z)$)	Degree($p(z)$)	ρ-value	Modular class
11	ζ_ℓ	40	48	2.400	3 mod 4
13	$\zeta_\ell + 1$	48	64	2.667	4 mod 8
22	$\zeta_\ell + 1$	40	56	2.800	0 mod 4
26	ζ_ℓ	48	56	2.333	3 mod 4
44	$\zeta_\ell + 1$	48	64	2.600	0 mod 4
52	$\zeta_\ell + 1$	48	60	2.500	0 mod 4

to 40. The polynomials that defines some of these curves can be found in Appendix A. Currently these curves, as already pointed out, are only of theoretical interest. In this table $\ell = lcm(k, 8)$.

5 Conclusion

We have presented an algorithm that produces more Kawazoe-Takahashi type of genus 2 pairing-friendly hyperelliptic curves. In addition we have presented new curves with better ρ-values. A problem with some of the reported curves is that the degree of the polynomial $r(z)$, which defines the prime order subgroup, is too large and hence a very small number, if any, of usable curves could be found. Table 2 summarises the the curves reported in this paper. Curves with $1 \le \rho \le 2$ remain elusive.

Table 2. Families of curves, $k < 60$, with $2.000 < \rho \le 3.000$

k	Degree($r(z)$)	Degree($p(z)$)	ρ-value
2	4	6	3.000
7	24	32	2.667
8	4	6	3.000
10	16	24	3.000
11	40	48	2.400
13	48	64	2.667
22	40	56	2.800
24	8	12	3.000
26	48	56	2.333
28	24	36	3.000
44	48	64	2.600
52	48	60	2.500

References

1. Balakrishnan, J., Belding, J., Chisholm, S., Eisenträger, K., Stange, K., Teske, E.: Pairings on hyperelliptic curves (2009), http://www.math.uwaterloo.ca/~eteske/teske/pairings.pdf
2. Barreto, P.S.L.M., Lynn, B., Scott, M.: Constructing elliptic curves with prescribed embedding degree. In: Cimato, S., Galdi, C., Persiano, G. (eds.) SCN 2002. LNCS, vol. 2576, pp. 263–273. Springer, Heidelberg (2002)
3. Boneh, D., Franklin, M.: Identity-based encryption from the Weil pairing. SIAM Journal of Computing 32(3), 586–615 (2003)
4. Boneh, D., Lynn, B., Shacham, H.: Short signatures from the Weil Pairing. In: Boyd, C. (ed.) ASIACRYPT 2001. LNCS, vol. 2248, pp. 514–532. Springer, Heidelberg (2001)
5. Bosma, W., Cannon, J., Playoust, C.: The Magma algebra system. I. The user language. J. Symbolic Comput. 24(3-4), 235–265 (1997)
6. Barreto, P.S.L.M., Naehrig, M.: Pairing-friendly elliptic curves of prime order. In: Preneel, B., Tavares, S. (eds.) SAC 2005. LNCS, vol. 3897, pp. 319–331. Springer, Heidelberg (2006)

7. Brezing, F., Weng, A.: Elliptic curves suitable for pairing based cryptography. Designs Codes and Cryptography 37(1), 133–141 (2005)
8. Cocks, C., Pinch, R.G.E.: Identity-based cryptosystems based on the Weil pairing (2001) (unpublished manuscript)
9. Furukawa, E., Kawazoe, M., Takahashi, T.: Counting points for hyperelliptic curves of type $y^2 = x^5 + ax$ over finite prime fields. In: Matsui, M., Zuccherato, R.J. (eds.) SAC 2003. LNCS, vol. 3006, pp. 26–41. Springer, Heidelberg (2004)
10. Freeman, D.: Constructing pairing-friendly elliptic curves with embedding Degree 10. In: Hess, F., Pauli, S., Pohst, M. (eds.) ANTS-VII 2006. LNCS, vol. 4076, pp. 452–465. Springer, Heidelberg (2006)
11. Freeman, D.: A generalized Brezing-Weng method constructing ordinary pairing-friendly abelian varieties. In: Galbraith, S.D., Paterson, K.G. (eds.) Pairing 2008. LNCS, vol. 5209, pp. 148–163. Springer, Heidelberg (2008)
12. Freeman, D.: Constructing pairing-friendly genus 2 curves with ordinary Jacobians. In: Takagi, T., Okamoto, T., Okamoto, E., Okamoto, T. (eds.) Pairing 2007. LNCS, vol. 4575, pp. 152–176. Springer, Heidelberg (2007)
13. Freeman, D., Stevenhagen, P., Streng, M.: Abelian varieties with prescribed embedding degree. In: van der Poorten, A.J., Stein, A. (eds.) ANTS-VIII 2008. LNCS, vol. 5011, pp. 60–73. Springer, Heidelberg (2008)
14. Freeman, D., Scott, M., Teske, E.: A Taxonomy of pairing-friendly elliptic curves. Journal of Cryptology 23(2) (2009)
15. Freeman, D., Satoh, T.: Constructing pairing-friendly hyperelliptic curves using Weil restrictions. Cryptography ePrint Archive, Report 2009/103 (2009), http://eprint.iacr.org
16. Joux, A.: A One Round Protocol for Tripartite Diffie-Hellman. In: Bosma, W. (ed.) ANTS 2000. LNCS, vol. 1838, pp. 385–394. Springer, Heidelberg (2000)
17. Kachisa, E., Schaeffer, E.F., Scott, M.: Constructing Brezing-Weng pairing friendly elliptic curves using elements in the cyclotomic field. In: Galbraith, S.D., Paterson, K.G. (eds.) Pairing 2008. LNCS, vol. 5209, pp. 126–135. Springer, Heidelberg (2008)
18. Kawazoe, M., Takahashi, T.: Pairing-friendly hyperelliptic curves with ordinary Jacobians of type $y^2 = x^5 + ax$. In: Galbraith, S.D., Paterson, K.G. (eds.) Pairing 2008. LNCS, vol. 5209, pp. 164–177. Springer, Heidelberg (2008)
19. Miyaji, A., Nakabayashi, M., Takano, S.: New explicit conditions of elliptic curve traces for FR-reduction. IEICE Trans. Fundamentals E84, 1234–1243 (2001)
20. Sakai, R., Ohgishi, K., Kasahara, M.: Cryptosystems based on pairing. In: Symposium on Cryptography and Information Security (SCIS 2000), Okinawa, Japan, January 26-28 (2000)
21. Scott, M., Barreto, P.S.: Generating more MNT elliptic curves. Designs, Codes and Cryptography 38, 209–217 (2006)
22. Shamir, A.: Identity-Based cryptosystems and signature schemes. In: Blakely, G.R., Chaum, D. (eds.) CRYPTO 1984. LNCS, vol. 196, pp. 47–53. Springer, Heidelberg (1985)
23. PARI-GP, version 2.3.2, Bordeaux (2006), http://pari.math.u-bordeaux.fr/
24. Silverman, J.H.: The arithmetic of elliptic curves. Springer, Heidelberg (1986)
25. Zhang, F.: Twisted Ate pairing on hyperelliptic curves and application. Cryptology ePrint Archive Report 2008/274 (2008), http://eprint.iacr.org/2008/274/

Appendix A: More Examples

Here we include the polynomials that define curves of some of the embedding degrees in Table 1.

Theorem 8. *Let $k = 11, \ell = 88$. Let $\gamma = \zeta_\ell \in \mathbb{Q}(\zeta_\ell)$ and define polynomials $r(z), p(z), c(z), d(z)$ by the following:*

$$r(z) = z^{40} - z^{36} + z^{32} - z^{28} + z^{24} - z^{20} + z^{16} - z^{12} + z^8 - z^4 + 1$$
$$p(z) = 1/8(z^{48} - 2z^{46} + z^{44} + 8z^{24} + z^4 - 2z^2 + 1)$$
$$c(z) = -1/2(z^{13} + z^{11})$$
$$d(z) = 1/4(z^{24} - z^{22} - z^2 + 1)$$
$$\rho = 12/5$$

Family $(r(4z + 3)/89, p(4z + 3))$

Theorem 9. *Let $k = 13, \ell = 104$. Let $\gamma = \zeta_\ell + 1 \in \mathbb{Q}(\zeta_\ell)$ and define polynomials $r(z), p(z), c(z), d(z)$ by the following:*

$$r(z) = z^{48} - 48z^{47} + 1128z^{46} + \ldots + 2z^2 - 24z + 1$$
$$p(z) = (z^{64} - 64z^{63} + 2016z^{62} - \ldots + 4040z^2 - 256z + 8)/8$$
$$c(z) = -(z^{19} - 19z^{18} + 171z^{17} + \ldots + 249z^2 - 32z + 2)/2$$
$$d(z) = (z^{32} - 32z^{31} + 496z^{30} - \ldots + 20995z^4 - 2340z^3 + 156z^2)/4$$
$$\rho = 8/3$$

Family $(r(8z + 4), p(8z + 4)$

Theorem 10. *Let $k = 22, \ell = 88$. Let $\gamma = \zeta_\ell \in \mathbb{Q}(\zeta_\ell)$ and define polynomials $r(z), p(z), c(z), d(z)$ by the following:*

$$r(z) = z^{40} - z^{36} + z^{32} - z^{28} + z^{24} - z^{20} + z^{16} - z^{12} + z^8 - z^4 + 1$$
$$p(z) = (z^{56} - 2z^{50} + z^{44} + z^{28} + z^{12} - 2z^6 + 1)/8$$
$$c(z) = -(z^{17} + z^{11})/2$$
$$d(z) = (z^{34} - z^{22} + z^{12} + 1)/4$$
$$\rho = 14/5$$

Family $(r(4z + 3)/89, p(4z + 3))$

Theorem 11. *Let $k = 26, \ell = 104$. Let $\gamma = \zeta_\ell \in \mathbb{Q}(\zeta_\ell)$ and define polynomials $r(z), p(z), c(z), d(z)$ by the following:*

$$r(z) = z^{48} - z^{44} + z^{40} - z^{36} + z^{32} - z^{28} + z^{24} - z^{20} + z^{16} - z^{12} + z^8 - z^4 + 1$$
$$p(z) = (z^{56} - 2z^{54} + z^{52} + 8z^{28} + z^4 - 2z^2 + 1)/8$$
$$c(z) = -(z^{15} + z^{13})/2$$
$$d(z) = (z^{28} - z^{26} - z^2 + 1)/4$$
$$\rho = 7/3$$

Family $(r(4z + 3), p(4z + 3))$

New Identity-Based Proxy Re-encryption Schemes to Prevent Collusion Attacks

Lihua Wang[1], Licheng Wang[1,2], Masahiro Mambo[3], and Eiji Okamoto[3]

[1] Information Security Research Center
National Institute of Information and Communications Technology
Tokyo 184-8795, Japan
wlh@nict.go.jp

[2] Information Security Center, Beijing University of Posts and Telecommunications
Beijing 100876, P.R. China
wanglc.cn@gmail.com

[3] Graduate School of Systems and Information Engineering
University of Tsukuba, Tsukuba 305-8573, Japan
mambo@cs.tsukuba.ac.jp, okamoto@risk.tsukuba.ac.jp

Abstract. In this paper, we propose two new constructions of identity-based proxy re-encryption (IB-PRE). The most important feature of our schemes is that we no longer need the semi-trust assumption on the proxy. Moreover, we describe the IND-PrID-CCA/CPA security models for an IB-PRE in a single-hop scenario, and then give a general analysis on the relationship between the IND-PrID-CPA security model and the desirable PRE properties: unidirectionality, collusion "safeness" and non-transitivity. Our first scheme has no ciphertext expansion through the re-encryption and is proven IND-PrID-CPA secure in the random oracle model. The second one achieves the IND-PrID-CCA security.

1 Introduction

Background and Motivations. In 1998, Blaze et al. [3] first proposed the primitive of proxy re-encryption (PRE), in which a proxy with specific information (re-encryption key) can translate a ciphertext for the original decryptor, Alice (delegator), to another ciphertext with the same plaintext for the delegated decryptor, Bob (delegatee). The proxy cannot, however, access the plaintext. PREs have many intriguing applications, such as email forwarding, law enforcement, and secure network file storage [2,3]. This concept quickly gained popularity, and many PRE schemes with identity-based settings (IB-PRE) have also appeared [10,14,8,18].

The idea of identity-based cryptography can be traced back to the 1984 study by Shamir [16]. In identity-based settings, the public key of a user is only his/her identity, such as a social service number, e-mail address, or IP address, while the corresponding secret key is generated by a *trusted* private key generator (PKG) and transmitted to the user via a *secure* channel. Identity-based framework alleviates the burden for certification management and thus identity-based proxy re-encryption (IB-PRE) schemes are more desirable than non-identity-based ones.

M. Joye, A. Miyaji, and A. Otsuka (Eds.): Pairing 2010, LNCS 6487, pp. 327–346, 2010.

Therefore in this study we mainly focus on IB-PREs. Note that there is another approach to constructing PRE schemes that do not require the ID-based setting. Please refer to recent literatures [1,7,9,12,13,17], etc. in this approach.

In 2007, Green and Ateniese [10] first presented two IB-PRE schemes: One is IND-PrID-CPA secure and another is IND-PrID-CCA secure. Shortly afterwards, Matsuo [14] also presented two PRE schemes: a hybrid PRE scheme and an IB-PRE scheme that was allegedly be CPA-secure in the standard model. However, in Matsuo's IB-PRE scheme the re-encryption key contains no information about the delegator. As a result, once a single re-encryption key and the secret key of the corresponding delegatee were leaked, original ciphertexts for any users, including the target user, in the system could be decrypted by the adversary. In the same year, Chu and Tzeng [8] also proposed two IB-PRE schemes and claimed that both were secure in the standard models. They also claimed that the second scheme was CCA-secure. However, Shao and Cao [17] showed that the second scheme failed to achieve CCA security since the transformed ciphertext is, in essence, malleable. With further checking, we also find that the reduction on the CPA security of their first IB-PRE scheme involved the coin-tossing technique similar to that used by Boneh and Franklin in [5]. We know that this private coin-tossing technique is essentially a random oracle – more precisely, a one-bit random oracle. This suggests that at the present it is still a challenge to design an IB-PRE scheme in the standard model, even just achieving CPA security. In [18], Tang et al. proposed a new inter-domain IB-PRE scheme. Though it is only CPA secure in the random oracle model, a remarkable advantage of the scheme is that it allows the delegator and delegatee to be in different domains, thus it is extremely useful in certain scenarios.

However, all these IB-PRE schemes need the assumption of the proxy to be semi-trusted, i.e., the proxy does not collude with the delegatees and other proxies to attack on the delegator, in order to protect the delegator's secret key from collusion attacks. This assumption is unnatural in many cases when the proxy cannot be assumed as trusted, such as in an open cloud system [11,19]. As far as we know, the first PRE scheme that tries to remove the semi-trust assumption is that in [2]. The PRE scheme is not in identity-based framework and we are not aware of any IB-PRE scheme without a semi-trust assumption.

Many interesting properties have also been surmised for the primitive of PREs. In particular, the following properties become greatly sensitive when we remove the semi-trust assumption. In the following context, let $X + Y \rightarrow Z$ (resp. $X + Y \nrightarrow Z$) denote Z can be derived from X and Y (resp. Z cannot be derived from X and Y).

1. *Collusion "safeness"*. This property is first proposed in [2], indicating that even if Bob and the proxy collude, they cannot extract Alice's secret key. More specifically, this property can be denoted by $rk_{A \rightarrow B} + sk_B \nrightarrow sk_A$, where $rk_{A \rightarrow B}$ means the re-encryption key for transferring user A's ciphertext into user B's ciphertext, and sk_A, sk_B are user A's secret key and user B's secret key, respectively.

2. *Non-transferability.* This property is also first proposed in [2] which requires that the proxy and a set of colluding delegatees cannot re-delegate decryption rights [2]. More specifically, this property can be denoted by $rk_{A\rightarrow B} + sk_B \nrightarrow rk_{A\rightarrow C}$ for any user C other than A and B.

Here we need to point out *in which scenarios the above properties are extremely useful.*

- Regarding the property of collusion "safeness", we know that in general scenarios, knowing re-encryption key $rk_{A\rightarrow B}$ and secret key sk_B allows one to decrypt user A's ciphertexts; so in some sense the functionality of the combination of $rk_{A\rightarrow B}$ and sk_B, denoted by $F_{rk_{A\rightarrow B},sk_B}$, *seems* to imply the functionality of sk_A, denoted by F_{sk_A}. However, the following cases suggest that $F_{rk_{A\rightarrow B},sk_B}$ does not imply F_{sk_A}:
  - If the sender has the choice to generate a ciphertext for which re-encryption is not permitted, then one can decrypt user A's ciphertexts by using sk_A, but cannot do the same work by using the combination of $rk_{A\rightarrow B}$ and sk_B.
  - If the scenario is also single-hop[1] and $rk_{C\rightarrow A}$ is given, then by using sk_A one can decrypt user C's ciphertexts, but cannot do the same work by using the combination of $rk_{A\rightarrow B}$ and sk_B.

- Regarding the property of non-transferability, we know that in general scenarios, the proxy, colluding with the delegatee, can decrypt the delegator's ciphertexts and then can encrypt the messages under others' public keys; so in some sense $F_{rk_{A\rightarrow B},sk_B}$ *seems* to imply the functionality of re-encryption key $rk_{A\rightarrow C}$, denoted by $F_{rk_{A\rightarrow C}}$. However, the latter pattern requires that the delegatee remains an *active, online* participant. According to [2], the property of non-transferability is meaningful in the sense that the only way for the delegatee to transfer offline decryption capabilities to others is to expose his/her own secret key.

Up to now, we have neither found a secure IB-PRE scheme with collusion "safeness", nor known any secure PRE/IB-PRE scheme with non-transferability. Therefore *our second motivation is to design IB-PRE schemes that are collusion-"safe"*. Also, it is desirable that *the schemes are non-transferable*.

In addition, we would like to address another important property of PRE/IB-PRE:

3. *Non-transitivity.* This property means that two consecutive proxies cannot re-delegate decryption rights [2]. More specifically, this property can be denoted by $rk_{A\rightarrow B} + rk_{B\rightarrow C} \nrightarrow rk_{A\rightarrow C}$.

---

[1] In the single-hop scenario, re-encrypting a re-encrypted ciphertext leads to an invalid ciphertext. In the opposite (i.e., multi-hop) scenario, Bob's ciphertext that is re-encrypted from Alice's ciphertext by using the re-encryption key $rk_{A\rightarrow B}$ could be further converted into Charlie's ciphertext by using the re-encryption key $rk_{B\rightarrow C}$. Interested readers can refer to [2,7] for more details related to the definitions of "single-hop" and "multi-hop".

Suppose that user A's true intention is merely to enable a proxy possessing $rk_{A\to B}$ to convert his ciphertexts only to user B, and not to any other users. Suppose further, by coincidence user B allows a proxy who knows $rk_{B\to C}$ to convert his ciphertexts to user C. Now, let us consider the single-hop scenario. If a PRE/IB-PRE scheme is not non-transitive, then user A's ciphertexts could be converted into user C's ciphertexts by collusion of these two proxies. This is unexpected from user A's perspective. Therefore *the property of non-transitivity is also desirable in certain scenarios.*

Remark 1. Regarding the above mentioned properties of PRE/IB-PRE, we would like to clarify the following points:

(1) The property of non-transitivity is much more desirable for a PRE/IB-PRE scheme in the single-hop scenario than in the multi-hop scenario, because for a multi-hop scenario if two consecutive re-encryption keys, e.g., $rk_{A\to B}$ and $rk_{B\to C}$, are given, then re-encrypting transformation from user A's ciphertexts into user C's ciphertexts is an authorized functionality, instead of an attack, though this approach requires that the proxies remain an active, online participant.
(2) In the single-hop scenario, the property of collusion "safeness" becomes a necessary condition of non-transitivity. The property of non-transitivity implies that by using the combination of two consecutive re-encryption keys $rk_{A\to B}, rk_{B\to C}$ and secret key sk_C, one cannot decrypt a ciphertext for user A, while if a PRE/IB-PRE scheme is not collusion-safe, by using $rk_{A\to B}$, $rk_{B\to C}$ and sk_C, one could derive sk_B (or even sk_A), which enables him/her to decrypt a ciphertext for user A. This is contradictory.

Contributions and Organizations. To safely remove the semi-trust assumption on the proxy, we let the PKG, which has already been assumed as trustable, take part in generating the re-encryption keys. From the standpoint of the delegator, this is a *less desirable but tolerable* action since he/she cannot prevent the PKG from letting others decrypt his/her ciphertexts. By doing so, we propose two new constructions of IB-PRE. Our first scheme has no ciphertext expansion after re-encryption operation and the receiver need not know whether the ciphertext is re-encrypted. Our enhanced version achieves IND-PrID-CCA security by using the Canetti-Halevi-Katz (CHK) technique [6]. In addition, both constructions are *unidirectional,* meaning "delegation from $A \to B$ does not allow re-encryption from $B \to A$", which is another desirable property of PRE/IB-PRE (refer to [2] for details).

The rest of the paper is organized as follows. In Section 2, we recall the related preliminary and cryptographic assumptions. In Section 3, we first reformulate the definition of IB-PRE scheme, then propose the security models that reflect the removal of the semi-trust assumption, and finally give an analysis on the relationship between the IND-PrID-CPA security model and the desirable PRE properties: unidirectionality, collusion "safenes" and non-transitivity. Concrete constructions of IB-PRE schemes and the corresponding provable security reductions are given in Section 4. Concluding remarks are presented in Section 5.

2   Bilinear Pairings and Complexity Assumptions

Our scheme is based on an admissible pairing that was first used to construct cryptosystems independently by Sakai et al. [15] and Boneh et al. [5]. The modified Weil pairing and Tate pairing associated with supersingular elliptic curves are examples of such admissible pairings. However, we describe pairings and the related mathematics in a more general format here.

Let $\mathbb{G}_1$, $\mathbb{G}_2$ be two multiplicative groups with prime order p and g as a generator of $\mathbb{G}_1$. $\mathbb{G}_1$ has an admissible bilinear map into $\mathbb{G}_2$, $\hat{e} : \mathbb{G}_1 \times \mathbb{G}_1 \longrightarrow \mathbb{G}_2$, if the following three conditions hold:

(1) *Bilinear.* $\hat{e}(g^a, g^b) = \hat{e}(g, g)^{ab}$ for all $a, b \in \mathbb{Z}_p^*$.
(2) *Non-degenerate.* $\hat{e}(g, g) \neq 1_{\mathbb{G}_2}$.
(3) *Computable.* There is an efficient algorithm to compute $\hat{e}(g_1, g_2)$ for $\forall g_1, g_2 \in \mathbb{G}_1$.

Bilinear Diffie-Hellman (BDH) Parameter Generator [5,4]: A randomized algorithm IG is a BDH parameter generator if IG takes a security parameter $k > 0$, runs in time polynomial in k, and outputs the description of two groups $\mathbb{G}_1$ and $\mathbb{G}_2$ of the same prime order p and the description of an admissible pairing $\hat{e} : \mathbb{G}_1 \times \mathbb{G}_1 \to \mathbb{G}_2$.

Discrete Logarithm Problem: Given $g, g^a \in \mathbb{G}_1$, or $\mu, \mu^a \in \mathbb{G}_2$, find $a \in \mathbb{Z}_p^*$.
Computational Diffie-Hellman (CDH) Problem [5]: Given $g, g^a, g^b \in \mathbb{G}_1$, find $g^{ab} \in \mathbb{G}_1$.
Bilinear Computational Diffie-Hellman (BCDH) Problem [4]: Given $g, g^a, g^b, g^c \in \mathbb{G}_1$, find $\hat{e}(g, g)^{abc} \in \mathbb{G}_2$.
Bilinear Decisional Diffie-Hellman (BDDH) Problem [4]: Given g, g^a, g^b, $g^c \in G_1$ and $T \in \mathbb{G}_2$, determine whether $T = \hat{e}(g, g)^{abc}$.

The BDDH (resp. CDH or BCDH) assumption says that the BDDH (resp. CDH or BCDH) problem is intractable; i.e., there is no polynomial time algorithm that can solve the BDDH (resp. CDH or BCDH) problem with non-negligible probability. Clearly the BDDH assumption implies the BCDH assumption, which in turn implies the CDH assumption.

3   Definitions, Models and Properties of IB-PRE

In this section, we first give a review of the definition of the IB-PRE scheme and then propose the security models of IB-PRE without the semi-trust assumption. Our model has some differences from the models defined in [10] since those are based on semi-trust assumption on the proxy. Finally, we further explain the proposed CPA security model and some aforementioned properties.

3.1   Definition

Definition 1 (IB-PRE). *An identity-based proxy re-encryption (IB-PRE) system consists of seven algorithms (**Setup**, **Ext**, **RKG**, **Enc**, **REnc**, **Dec**, **RDec**):*

Setup (setup): *This algorithm takes a security parameter 1^k as input and returns the system parameters params and the master secret key msk to the private key generator (PKG); params are shared by all users in the system.*

Extract (Ext): *This algorithm takes params, msk, and the user's identity id as inputs. It outputs the user's secret key sk_{id}.*

ReKeyGen (RKG): *This algorithm takes params, msk, and a pair of identities (id, id') as inputs. It outputs a proxy re-encryption key $rk_{id \to id'}$, which enables the proxy to convert a ciphertext for user id into a ciphertext for user id'.*

Encryption (Enc): *This algorithm takes a plaintext m and user id's identity id as inputs. It outputs a ciphertext C of m.*

Re-encryption (REnc): *This algorithm takes a ciphertext C of the message m for user id and the re-encryption key $rk_{id \to id'}$ as inputs. It outputs a re-encrypted ciphertext C' of the same message m for user id'.*

Decryption (Dec): *This algorithm takes an original ciphertext C and user id's secret key sk_{id} as inputs, and it outputs plaintext m corresponding to C.*

Re-Decryption (RDec): *This algorithm takes a re-encrypted ciphertext C' for user id' and the secret key $sk_{id'}$ as inputs. It outputs the plaintext m corresponding to C'.*

The consistency of an IB-PRE scheme is defined as follows: Given msk, $params$, and two arbitrary users' identities id and id', if

(1) $sk_{id} \leftarrow \text{Ext}(params, msk, id)$,
(2) $sk_{id'} \leftarrow \text{Ext}(params, msk, id')$,
(3) $rk_{id \to id'} \leftarrow \text{RKG}(params, msk, id, id')$,

then the following conditions must hold:

(1) Consistency between encryption and decryption; i.e.,

$$\text{Dec}_{id}(sk_{id}, \text{Enc}(m, id)) = m, \quad \forall m \in \mathcal{M}.$$

(2) Consistency among encryption, proxy re-encryption and re-decryption; i.e.,

$$\text{RDec}_{id'}(sk_{id'}, \text{REnc}(rk_{id \to id'}, \text{Enc}(m, id))) = m, \quad \forall m \in \mathcal{M}.$$

3.2 Security Models

In a sequel, we focus on identity-based unidirectional proxy re-encryption (IB-PRE) schemes and the so-called single-hop scenario is taken into consideration. We start from an intuitive analysis. The security of an IB-PRE scheme requires the following condition to be satisfied: given a challenge ciphertext C^* for identity id_i, the adversary without knowing

(1) secret key sk_{id_i}, and
(2) secret key sk_{id_j} and proxy re-encryption key $rk_{id_i \to id_j}$

is not allowed to obtain any information about the corresponding plaintext. In order to describe this condition more formally, we introduce relation matrix R and secret key vector s as follows: Suppose there are n users in the system, R and s are defined as follows:

$$R = (R_{id_1}, ..., R_{id_n})^T = (r_{ij}) \triangleq \begin{cases} 1, \text{ if } i = j, \text{ or the re-encryption key} \\ \quad \text{from } id_i \text{ to } id_j \text{ is issued;} \\ 0, \text{ otherwise,} \end{cases}$$

where $R_{id_i} = (r_{i1}, ..., r_{in})^T$ denotes the delegation relations from id_i to others and r_{ii} always equals 1 for the trivial relation from id_i to itself.

$$s = (s_i) \triangleq \begin{cases} 1, \text{ if the secret key of the user } id_i \text{ is issued;} \\ 0, \text{ otherwise.} \end{cases}$$

Clearly, $R_i^T s = 0$ if and only if the above condition holds, i.e., the adversary knows NEITHER "secret key sk_{id_i}" NOR "secret key sk_{id_j} and proxy re-encryption key $rk_{id_i \to id_j}$". (Note that in the definition of R, if the phrase "*from id_i to id_j*" is replaced with the phrase "*between id_i and id_j*", then the bidirectional model is formulated.)

Now, the security models of an IB-PRE scheme can be formally defined via **Game CCA** and **Game CPA**, each of which involves an adversary $\mathcal{A}$ and challenger $\mathcal{C}$.

Game CCA

- **Setup.** $\mathcal{C}$ generates system parameters *params* and master key *msk*, and then transmits *params* to $\mathcal{A}$ while hiding *msk*. $\mathcal{C}$ also initializes R as an identity matrix and s as a zero vector.
- **Phase 1.** $\mathcal{A}$ is permitted to ask the following queries and $\mathcal{C}$ responds to them.
  - *Secret-key query* on input id_i: $\mathcal{C}$ returns sk_{id_i} and sets $s_i = 1$.
  - *Proxy re-encryption key query* on input (id_i, id_j): $\mathcal{C}$ returns proxy re-encryption key $rk_{id_i \to id_j}$ and sets $r_{ij} = 1$.
  - *Re-encryption query* on input (id, id', C): $\mathcal{C}$ returns corresponding re-encrypted ciphertext C'.
  - *Decryption query* on input (id, C): $\mathcal{C}$ returns the corresponding plaintext.
  - *Re-decryption query* on input (id', C'): $\mathcal{C}$ returns the corresponding plaintext.
- **Challenge.** When the adversary $\mathcal{A}$ decides to end phase 1, it submits a target identity id^* and messages (m_0, m_1) of equal lengths. Then $\mathcal{C}$ responds as follows:
  - $\mathcal{C}$ rejects $\mathcal{A}$'s challenge targets and the game ends[2] if $R_{id^*}^T s \neq 0$; otherwise, the simulation continues.
  - Flips a fair binary coin κ, constructs a challenge ciphertext C^* for the message m_κ under the identity id^*, and finally sends C^* to $\mathcal{A}$.

---

[2] This occurs when $\mathcal{A}$ submits an invalid challenge identity.

- **Phase 2.** Phase 1 is repeated with the following restrictions.
  - Whenever $\mathcal{A}$ makes a new secret key query on some id_i or a new proxy re-encryption key query on $(id_i, \ id_j)$, $\mathcal{C}$ performs the following steps:
    * First, prepares makes a *virtual response* (i.e., without sending the response to $\mathcal{A}$) as in Phase 1.
    * Checks whether $R^T_{id^*} s \neq 0$ holds. If it does, $\mathcal{C}$ removes all modifications that have just been made and rejects $\mathcal{A}$'s query. Otherwise, $\mathcal{C}$ converts this virtual response into a real response (i.e., sends the response to $\mathcal{A}$).
  - $\mathcal{A}$ cannot ask a re-encryption query on (id^*, id, C^*) if $\mathcal{A}$ obtained the secret key sk_{id}.
  - $\mathcal{A}$ cannot ask a decryption query on (id^*, C^*) or a re-decryption query on (id, C) which C for id has been re-encrypted from C^* for id^*.
- **Guess.** Finally, $\mathcal{A}$ outputs a guess $\kappa' \in \{0, 1\}$.

Definition 2 (IND-PrID-CCA). *In* **Game CCA**, $\mathcal{A}$ *wins if* $\kappa' = \kappa$, *and an IB-PRE scheme is said to be indistinguishable against an adaptively chosen identity and ciphertext attacks (IND-PrID-CCA) if for an arbitrary polynomial time adversary* $\mathcal{A}$, *the probability* $|\Pr[\kappa' = \kappa] - 1/2|$ *is negligible.*

Game CPA

- **Setup.** Identical to **Setup** in **Game CCA**.
- **Phase 1.** Identical to **Phase 1** in **Game CCA** but with the following restriction:
  - $\mathcal{A}$ cannot make decryption and re-decryption queries on *any* inputs.
  - $\mathcal{A}$ cannot make re-encryption queries on *any* inputs.
- **Challenge.** Identical to **Challenge** in **Game CCA**.
- **Phase 2.** Phase 1 is repeated with the restriction that whenever $\mathcal{A}$ makes a new secret key query on some id_i or a new proxy re-encryption key query on $(id_i, \ id_j)$, $\mathcal{C}$ performs the following steps:
  - First, prepares a *virtual response* (i.e., without sending the response to $\mathcal{A}$) as in Phase 1.
  - Checks whether $R^T_{id^*} s \neq 0$ holds. If it does, $\mathcal{C}$ removes all modifications that it has just made and rejects $\mathcal{A}$'s query. Otherwise, $\mathcal{C}$ converts this virtual response into a real response (i.e., sends the response to $\mathcal{A}$).
- **Guess.** Identical to **Guess** in **Game CCA**.

Definition 3 (IND-PrID-CPA). *In* **Game CPA**, *adversary* $\mathcal{A}$ *wins if* $\kappa' = \kappa$, *and an IB-PRE scheme is said to be indistinguishable against adaptively chosen an identity and chosen plaintext attacks (IND-PrID-CPA) if for an arbitrary polynomial time adversary* $\mathcal{A}$, *the probability* $|\Pr[\kappa' = \kappa] - 1/2|$ *is negligible.*

Remark 2. The distinction of our security models from that for IB-PRE schemes with the semi-trust assumption (e.g., GA07 [10]) is reflected by the restrictions on queries. In GA07's IND-PrID-CPA security model [10], when $\mathcal{A}$ concludes phase 1 to give a challenge, he/she is restricted from choosing such id^* that "trivial" decryption is possible using keys extracted during this phase (e.g., by using re-encryption keys to translate from id^* to identity id for which $\mathcal{A}$ holds a decryption key). Moreover, during Phase 2, not only the query on challenge $\langle id^*, C^* \rangle$, but also any combination of queries resulting into the decryption of C^* is prohibited. In other words, if $rk_{id^* \to id_1}$, $rk_{id_1 \to id_2}$, ..., $rk_{id_{k-1} \to id_k}$ have been queried, then secret key queries on id_i for any $i = 1, ..., k$ are restricted. Thus, besides reflecting a multi-hop scenario, this restriction captures the characteristic that the proxy does not collude with delegatees and other proxies. In our defined IND-PrID-CPA security model, only condition $R_{id^*}^T s = 0$ is requested. That is,

(1) the secret key query on id^* is restricted and
(2) asking both the secret key query on id and the proxy re-encryption key query on $id^* \to id$ is restricted.

This restriction reflects the removal of the semi-trust assumption, because a re-encryption key query on $id \to id'$ and the secret key query on id' are allowed to be requested by $\mathcal{A}$.

Theorem 1. *If a single-hop IB-PRE scheme achieves the IND-PrID-CPA security defined above, then it is unidirectional, collusion-safe, and non-transitive.*

Proof. First, according to the definition of the IND-PrID-CPA security, the adversary $\mathcal{A}$ is allowed to ask for secret key sk_{id} and re-encryption key $rk_{id \to id^*}$ for arbitrary identity $id \neq id^*$ in Phase 2 of Game CPA; i.e., after receiving ciphertext C^*.

- Unidirectionality. Suppose that the discussed IB-PRE scheme is not unidirectional; i.e., $\mathcal{A}$ could perform reverse re-encryption. Then $\mathcal{A}$ can convert C^* that is encrypted by using target identity id^* into a valid ciphertext $\widetilde{C^*}$ for the user with identity id. Thus, $\mathcal{A}$ can break the IND-PrID-CPA security by using id's secret key sk_{id}. This is contradictory. Therefore the proposed IND-PrID-CPA security implies unidirectionality.

In addition, according to the definition of IND-PrID-CPA security, the adversary $\mathcal{A}$ is also allowed to ask for re-encryption keys $rk_{id^* \to id}$ and $rk_{id \to id'}$ as well as secret key $sk_{id'}$ in either Phase 1 or 2 of Game CPA. But $\mathcal{A}$'s query on secret key sk_{id} will be rejected. Here, we assume that all these three identities, id^*, id, and id', are distinct.

- Non-transitivity. Suppose that the discussed IB-PRE scheme fails to capture the property of non-transitivity, then $\mathcal{A}$ could derive re-encryption key $rk_{id^* \to id'}$ from re-encryption keys $rk_{id^* \to id}$ and $rk_{id \to id'}$. As a result, $\mathcal{A}$ can convert C^* into a valid ciphertext for user id'. Apparently, $\mathcal{A}$ can in turn use $sk_{id'}$ to break the IND-PrID-CPA security. This is also contradictory. Thus the proposed IND-PrID-CPA security implies Non-transitivity.
- Collusion "safeness". Suppose that the discussed IB-PRE scheme fails to capture the property of collusion "safeness", then $\mathcal{A}$ could derive secret key sk_{id} from $rk_{id \to id'}$ and $sk_{id'}$. Additionally, using $rk_{id^* \to id}$, $\mathcal{A}$ can convert C^* into

a valid ciphertext for user id. Apparently, $\mathcal{A}$ can in turn use sk_{id} to break the IND-PrID-CPA security. This is contradictory. Therefore the proposed IND-PrID-CPA security implies collusion "safeness". In fact, the property of collusion "safeness" is also implied by the property of non-transitivity in the single-hop scenarios (see Remark 1 for detailed arguments).   □

4   Proposed Schemes

4.1   A CPA-Secure and Non-transferable IB-PRE Scheme

Our basic IB-PRE scheme consists of the following seven algorithms:

1. **Setup (setup):** The actions performed by the trusted authority, PKG, are as follows:
   - Run the BDH parameter generator IG with an input 1^k to generate a prime p, two multiplicative groups $\mathbb{G}_1$ and $\mathbb{G}_2$ of order p, and an admissible bilinear map $\hat{e} : \mathbb{G}_1 \times \mathbb{G}_1 \to \mathbb{G}_2$.
   - Select $\alpha \in \mathbb{Z}_p^*$ and compute $g_1 = g^\alpha \in \mathbb{G}_1$, where g is a generator of $\mathbb{G}_1$; choose two elements $g_2, \eta \in \mathbb{G}_1$ at random, where η is called the re-encryption parameter.
   - Let $H : \{0,1\}^l \to \mathbb{G}_1$ (where l is a fixed positive integer) be a cryptographic hash function.

   The system parameters are given by

   $$params = (\mathbb{G}_1, \mathbb{G}_2, p, \hat{e}, g, g_1, g_2, \eta, H),$$

   and the master key is expressed as $msk = g_2^\alpha$.

2. **Extract (Ext):** The PKG generates a secret key for the user with identity $id \in \{0,1\}^l$ as follows:
   - Select $u \in \mathbb{Z}_p^*$ and compute $sk_{id} = (d_0, \ d_1) = (g_2^\alpha H(id)^u, \ g^u)$. Note that this u will be used later for issuing a proxy re-encryption key. But the PKG need not store each user's u. Instead, each u can be derived by the formula $u = h_{msk}(id)$ whenever it is required, where $h_{msk}(\cdot)$ is a keyed hash function.
   - Send sk_{id} to user id via a secure channel. The user can validate his/her secret key by checking whether

   $$\hat{e}(d_0, \ g) \overset{?}{=} \hat{e}(g_1, g_2)\hat{e}(H(id), \ d_1)$$

   holds.

3. **ReKeyGen (RKG):** There are two steps:
   - User id sends ("RK", id, id') to the PKG which returns the seed of re-encryption key

   $$\widetilde{rk}_{id \to id'} = \left(\frac{H(id)}{H(id')} \right)^{u'}$$

   to user id via a secure channel, where u' was selected by the PKG to generate secret key for id'.

- User id picks $\delta \in \mathbb{Z}_p^*$ at random and then computes the proxy re-encryption key as follows:

$$rk_{id \to id'} = (rk_1, rk_2) = (\eta^\delta \left(\frac{H(id)}{H(id')} \right)^{u'}, g^\delta).$$

Finally, user id sends $rk_{id \to id'}$ to his/her proxy via a secure channel.

4. **Encryption (Enc):** To encrypt message $m \in \mathcal{M}$ ($\subseteq \mathbb{G}_2$) for user id, the encryptor selects $r \in \mathbb{Z}_p^*$ and sends C to the decryptor, where

$$C = (C_1, C_2, C_3, C_4) = (m \cdot \hat{e}(g_1, g_2)^r, g^r, H(id)^r, \eta^r).$$

Note that if the encryptor does not want the ciphertext to be re-encrypted, he/she can omit C_4, or directly pick $C_4 \in \mathbb{G}_1$ at random.

5. **Re-Encryption (REnc):** Upon receiving $\langle C = (C_1, C_2, C_3, C_4), id, \ id' \rangle$, the proxy first checks whether $\hat{e}(C_2, \eta) = \hat{e}(C_4, g)$ holds: If not, outputs $\perp$ which means that the ciphertext is not allowed to be re-encrypted; otherwise, computes $C' = (C_1', C_2, C_3)$ and sends it to user id', where

$$C_1' = C_1 \cdot \frac{\hat{e}(C_4, rk_2)}{\hat{e}(C_2, rk_1)}.$$

6. **Decryption (Dec):** With the input $(C_1, C_2, C_3, -)$, the original decryptor id can decrypt the ciphertext by computing

$$m = C_1 \cdot \frac{\hat{e}(C_3, d_1)}{\hat{e}(C_2, d_0)}.$$

7. **Re-Decryption (RDec):** With the re-encrypted ciphertext (C_1', C_2, C_3) as the input, user id' performs the decryption by computing

$$m = C_1' \cdot \frac{\hat{e}(C_3, d_1')}{\hat{e}(C_2, d_0')},$$

where $(d_0', \ d_1')$ is user id''s secret key. Clearly, the algorithm RDec is exactly the same as the algorithm Dec. So, from the viewpoint of the decryptor, he/she need not know whether the ciphertext is re-encrypted.

Remark 3. Though the inputs of the decryption algorithm and the re-decryption algorithm are different, they are in fact indistinguishable in the sense that the decryptor cannot decide whether or not the received ciphertext has been re-encrypted, since the encryptor could send a triple if he/she does not want the message to be read by others apart from the original decryptor.

Theorem 2 (Consistency of IB-PRE). *The above IB-PRE scheme is consistent.*

Proof. The consistency can be checked by the following computations:

(1) Consistency between encryption and decryption.

$$C_1 \cdot \frac{\hat{e}(C_3, d_1)}{\hat{e}(d_0, C_2)} = m \cdot \hat{e}(g_1, g_2)^r \cdot \frac{\hat{e}(H(id)^r, g^u)}{\hat{e}(g_2^\alpha H(id)^u, g^r)}$$

$$= m \cdot \hat{e}(g_1, g_2)^r \cdot \frac{\hat{e}(H(id)^r, g^u)}{\hat{e}(g_2^\alpha, g^r) \cdot \hat{e}(H(id)^u, g^r)}$$

$$= m.$$

(2) Consistency among encryption, proxy re-encryption and decryption.

$$C_1' \cdot \frac{\hat{e}(C_3, d_1')}{\hat{e}(d_0', C_2)}$$

$$= \frac{C_1}{\hat{e}(C_2, rk_{id \to id'})} \cdot \frac{\hat{e}(C_3, d_1')}{\hat{e}(d_0', C_2)}$$

$$= \frac{m \cdot \hat{e}(g_1, g_2)^r}{\hat{e}(g^r, (H(id)/H(id'))^{u'})} \cdot \frac{\hat{e}(H(id)^r, g^{u'})}{\hat{e}(g_2^\alpha H(id')^{u'}, g^r)}$$

$$= m \cdot \hat{e}(g_1, g_2)^r \cdot \frac{\hat{e}(g^r, H(id)^{u'})}{\hat{e}(g^r, H(id)^{u'})} \cdot \frac{\hat{e}(H(id')^{u'}, g^r)}{\hat{e}(g_2, g_1^r) \cdot \hat{e}(H(id')^{u'}, g^r)}$$

$$= m.$$

$\square$

Theorem 3 (IND-PrID-CPA of IB-PRE). *Suppose that H is a random oracle. Then, the above IB-PRE scheme is indistinguishable against an adaptively chosen identity and plaintext attacks (IND-PrID-CPA) under the bilinear decisional Diffie-Hellman (BDDH) assumption. More precisely, if there is an adversary that can break the IND-PrID-CPA security of the above IB-PRE scheme with the probability at least ϵ within time t, then there is an algorithm that can solves the BDDH problem in $\mathbb{G}_2$ with the probability at least ϵ' within time t', such that*

$$\epsilon' \geq \frac{\epsilon}{e \cdot (q_s + 2q_r)} \quad \text{and} \quad t' = t + t_Q,$$

where e is the base of natural logarithm and t_Q denotes the time required for answering all queries, while q_s and q_r are the numbers of secret key queries and re-encryption key queries, respectively.

Proof. See Appendix A.

Corollary 1. *The proposed IB-PRE scheme is collusion-"safe" and non-transitive without the semi-trust assumption.*

Remark 4. Though it is not known whether the definition of the IND-PrID-CPA security guarantees the property of non-transferability (which is not proven in Theorem 1), the following informal argument indicates that our proposed IB-PRE scheme indeed achieves this desirable property.

Given re-encryption key $rk_{A \to B} = (\eta^\delta \left(\frac{H(A)}{H(B)} \right)^{u_B}, g^\delta)$ and secret key $sk_B = (g^{ab} H(B)^{u_B}, g^{u_B})$, under the aforementioned cryptographic assumption, we have not seen any method to retrieve the re-encryption key

$$rk_{A \to C} = (\eta^{\delta_c} \left(\frac{H(A)}{H(C)} \right)^{u_C}, g^{\delta_c})$$

without knowing u_C, which is computable only by the PKG. An equivalent way to obtain $rk_{A \to C}$ is to calculate $\widetilde{rk}_{A \to C} = \left(\frac{H(A)}{H(C)} \right)^{u_C}$; i.e., the seed of the re-encryption key. However, the following facts suggest the difficulty in working out $\widetilde{rk}_{A \to C}$:

- First, it is difficult to separate factors g^{ab} and $H(B)^{u_B}$ from the pair $(g^{ab} H(B)^{u_B}, g^{u_B})$ with given $g, g^a, g^b \in \mathbb{G}_1$ for unknown $a, b, u_B \in \mathbb{Z}_p^*$.
- Second, without loss of generality, let us view $H(A)$, $H(B)$ and $H(C)$ as g^x, g^y and $g^z \in \mathbb{G}_1$ for some unknown $x, y, z \in \mathbb{Z}_p^*$, respectively. Then, computing $\widetilde{rk}_{A \to C}$ is equivalent to computing $g^{(x-z)u_C} \in \mathbb{G}_1$ from the given $g, g^x, g^y, g^z, g^{(x-y)u_B}, g^{u_B}$ and $g^{u_C} \in \mathbb{G}_1$.

Both of the above cases are apparently infeasible under the assumption of the intractability of the CDH problem in $\mathbb{G}_1$.

4.2   CCA-Secure Construction

Let $H_1 : \mathbb{G}_2 \times \mathbb{G}_2 \to \mathbb{Z}_p^*$, $H_2 : \mathbb{G}_2 \to \{0,1\}^{\log |\mathcal{M}|}$ (where $\mathcal{M}$ is the message space) and $H_3 : \{0,1\}^* \to \mathbb{G}_1$ be three collision-resistant hash functions. In contrast to the CPA-secure IB-PRE scheme in section 4.1, only the following four algorithms must be modified:

Encryption (Enc): To encrypt the message $m \in \mathcal{M} = \{0,1\}^n$ for user id, the encryptor picks a random number $\sigma \in \mathbb{G}_2$, sets $r = H_1(m, \sigma) \in \mathbb{Z}_p^*$ and computes

$$(C_0, C_1, C_2, C_3, C_4) = (m \oplus H_2(\sigma), \sigma \cdot \hat{e}(g_1, g_2)^r, g^r, H(id)^r, \eta^r).$$

Finally, it sets $h = H_3(id||C_0||C_1||C_2||C_3||C_4)$ and $S = h^r$ and sends $\langle id, S, C \rangle$ to the decryptor, where $C = (C_0, C_1, C_2, C_3, C_4)$.

Re-Encryption (REnc): Upon receiving $\langle id, S, C = (C_0, C_1, C_2, C_3, C_4) \rangle$, the proxy who possesses the re-encryption key (rk_1, rk_2) first computes

$$h = H_3(id||C_0||C_1||C_2||C_3||C_4),$$

then checks whether

$$\hat{e}(g, S) = \hat{e}(h, C_2) \text{ and } \hat{e}(C_2, \eta) = \hat{e}(C_4, g)$$

hold simultaneously: If not, it outputs $\bot$ which means the ciphertext is invalid or not allowed to be re-encrypted; otherwise, it computes

$$C_1' = \frac{C_1 \cdot \hat{e}(C_4, rk_2)}{\hat{e}(C_2, rk_1)}$$

and sends $\langle id', C' \rangle$ to user id', where $C' = (C_0, C_1', C_2, C_3)$.

Decryption (Dec): With the input $\langle id, S, C = (C_0, C_1, C_2, C_3, -) \rangle$, the original decryptor id first computes $h = H_3(id||C_0||C_1||C_2||C_3||-)$, then checks whether $\hat{e}(g, S) = \hat{e}(h, C_2)$. If not, it outputs $\bot$; otherwise, it decrypts the ciphertext by computing

$$\sigma' = C_1 \cdot \frac{\hat{e}(C_3, d_1)}{\hat{e}(d_0, C_2)} \quad \text{and} \quad m' = C_0 \oplus H_2(\sigma'),$$

then outputs $m = m'$ if $C_2 = g^{H_1(m', \sigma')}$, or $\bot$ otherwise.

Re-Decryption (RDec): With a re-encrypted ciphertext $\langle id', C' = (C_0, C_1', C_2, C_3) \rangle$ as the input, user id' first decrypts the ciphertext by computing

$$\sigma' = C_1' \cdot \frac{\hat{e}(C_3, d_1')}{\hat{e}(d_0', C_2)} \quad \text{and} \quad m' = C_0 \oplus H_2(\sigma'),$$

and then outputs $m = m'$ if $C_2 = g^{H_1(m', \sigma')}$, or $\bot$ otherwise. Here (d_0', d_1') is the user id''s secret key.

Theorem 4 (IND-PrID-CCA). *Suppose that H, H_1, H_2 and H_3 are random oracles. Then, the above enhanced IB-PRE scheme is indistinguishable against an adaptively chosen identity and ciphertext attacks (IND-PrID-CCA) under the bilinear decisional Diffie-Hellman (BDDH) assumption.*

Proof. Considering the space limitation, this proof will be given in a full version that will soon be available.

Remark 5. Though the above transformation makes use of the well-known CHK technique [6] that was originally invented to derive a CCA-secure PKE from a CPA-secure IBE in the standard model, the above theorem is merely formulated in the random oracle model. This situation is very similar to what was done by Green and Ateniese [10]. At the present, we do not know whether the ROM assumption in the above theorem and in [10] can be removed.

5   Concluding Remarks

We have proposed two identity-based proxy re-encryption (IB-PRE) schemes that can prevent collusion attacks. The most important feature of our schemes is that we no longer need semi-trust assumption on the proxy. Our first construction has no ciphertext expansion during the re-encryption and is proven IND-PrID-CPA-secure in the random oracle model. It is also converted into an IND-PrID-CCA-secure IB-PRE scheme. In both schemes, the security is based on the bilinear decisional Diffie-Hellman assumption and the PKG takes part in generating the re-encryption keys. It is still open as to whether our proposal can be proved IND-PrID-CCA-secure in the standard model and the PKG's involvement in the re-encryption key generation can be removed.

Acknowledgments. The authors are grateful to the anonymous referees for their valuable comments. The work is supported by the Japan NICT International Exchange Program (No. 2009-002), and the second author is also partially supported by the China National Natural Science Foundation Programs (Nos. 90718001 and 60973159).

References

1. Ateniese, G., Benson, K., Hohenberger, S.: Key-private proxy re-encryption. In: Fischlin, M. (ed.) CT-RSA 2009. LNCS, vol. 5473, pp. 279–294. Springer, Heidelberg (2009)
2. Ateniese, G., Fu, K., Green, M., Hohenberger, S.: Improved proxy re-encryption schemes with applications to secure distributed storage. In: Internet Society (ISOC): NDSS 2005, pp. 29–43 (2005)
3. Blaze, M., Bleumer, G., Strauss, M.: Divertible protocols and atomic proxy cryptography. In: Nyberg, K. (ed.) EUROCRYPT 1998. LNCS, vol. 1403, pp. 127–144. Springer, Heidelberg (1998)
4. Boneh, D., Boyen, X.: Secure identity based encryption without random oracles. In: Franklin, M. (ed.) CRYPTO 2004. LNCS, vol. 3152, pp. 443–459. Springer, Heidelberg (2004)
5. Boneh, D., Franklin, M.: Identity-based encryption from the Weil pairing. In: Kilian, J. (ed.) CRYPTO 2001. LNCS, vol. 2139, pp. 213–229. Springer, Heidelberg (2001)
6. Canetti, R., Halevi, S., Katz, J.: Chosen-ciphertext security from identity-based encryption. In: Cachin, C., Camenisch, J.L. (eds.) EUROCRYPT 2004. LNCS, vol. 3027, pp. 207–222. Springer, Heidelberg (2004)
7. Canetti, R., Hohenberger, S.: Chosen-ciphertext secure proxy re-encryption. In: CCS 2007, pp. 185–194. ACM, New York (2007)
8. Chu, C., Tzeng, W.: Identity-based proxy re-encryption without random oracles. In: Garay, J.A., Lenstra, A.K., Mambo, M., Peralta, R. (eds.) ISC 2007. LNCS, vol. 4779, pp. 189–202. Springer, Heidelberg (2007)
9. Deng, R.H., Weng, J., Liu, S., Chen, K.: Chosen-ciphertext secure proxy re-encryption without pairings. In: Franklin, M.K., Hui, L.C.K., Wong, D.S. (eds.) CANS 2008. LNCS, vol. 5339, pp. 1–17. Springer, Heidelberg (2008)
10. Green, M., Ateniese, G.: Identity-based proxy re-encryption. In: Katz, J., Yung, M. (eds.) ACNS 2007. LNCS, vol. 4521, pp. 288–306. Springer, Heidelberg (2007)
11. Kamara, S., Lauter, K.: Cryptographic cloud storage. In: Sion, R., Curtmola, R., Dietrich, S., Kiayias, A., Miret, J.M., Sako, K., Sebé, F. (eds.) RLCPS, WECSR, and WLC 2010. LNCS, vol. 6054, pp. 136–149. Springer, Heidelberg (2010)
12. Libert, B., Vergnaud, D.: Unidirectional chosen-ciphertext secure proxy re-encryption. In: Cramer, R. (ed.) PKC 2008. LNCS, vol. 4939, pp. 360–379. Springer, Heidelberg (2008)
13. Libert, B., Vergnaud, D.: Tracing malicious proxies in proxy re-encryption. In: Galbraith, S.D., Paterson, K.G. (eds.) Pairing 2008. LNCS, vol. 5209, pp. 332–353. Springer, Heidelberg (2008)
14. Matsuo, T.: Proxy re-encryption systems for identity-based encryption. In: Takagi, T., Okamoto, T., Okamoto, E., Okamoto, T. (eds.) Pairing 2007. LNCS, vol. 4575, pp. 247–267. Springer, Heidelberg (2007)

15. Sakai, R., Ohgishi, K., Kasahara, M.: Cryptosystems based on pairing. In: SCIS 2000-C20 (2000)
16. Shamir, A.: Identity-based cryptosystems and signature schemes. In: Blakely, G.R., Chaum, D. (eds.) CRYPTO 1984. LNCS, vol. 196, pp. 47–53. Springer, Heidelberg (1985)
17. Shao, J., Cao, Z.: CCA-secure proxy re-encryption without pairings. In: Jarecki, S., Tsudik, G. (eds.) PKC 2009. LNCS, vol. 5443, pp. 357–376. Springer, Heidelberg (2009)
18. Tang, Q., Hartel, P.H., Jonker, W.: Inter-domain identity-based proxy re-encryption. In: Yung, M., Liu, P., Lin, D. (eds.) INSCRYPT 2008. LNCS, vol. 5487, pp. 332–347. Springer, Heidelberg (2008)
19. Yu, S., Wang, C., Ren, K., Lou, W.: Achieving secure, scalable, and fine-grained data access control in cloud computing. In: InfoCom 2010, pp. 534–542. IEEE, Los Alamitos (2010)

A   Proof of Theorem 3

Proof. Our proof technique is basically inherited from the Boneh-Boyen IBE scheme [4], which can be proven secure in the standard model. Unfortunately, in resorting merely to the Boneh-Boyen hash technique, we have failed to simulate the related re-encryption keys, though all involved secret keys can be effectively simulated. For the sake of simulating re-encryption keys, we make concessions by combining Boneh-Boyen hash technique and the Boneh-Franklin coin-tossing technique that is essentially a random oracle model [5].

Suppose that the adversary $\mathcal{A}$ has advantage ϵ in attacking IND-PrID-CPA security of IB-PRE. We now build an algorithm $\mathcal{C}$ that solves BDDH in $\mathbb{G}_2$. $\mathcal{C}$'s goal is to decide whether $T = \hat{e}(g,g)^{abc}$ for a given input (g, g^a, g^b, g^c, T). Let $g_1 = g^a, g_2 = g^b, g_3 = g^c$. Then, $\mathcal{C}$ performs the following steps:

Initialization. First, $\mathcal{C}$ initializes an identity re-encryption key matrix $R = (r_{ij}) = I$ and a zero secret key vector $\mathbf{s} = (s_1, ..., s_n) = 0$, where n is an estimation of the maximum number of users. Next, $\mathcal{C}$ picks a random number $x \in \mathbb{Z}_p^*$ and sets three empty tables with the following structures:

- Hash Table (KT): $(v, \ coin, \ h)$
- Secret Key Table (SKT): $(id, \ r, \ d_0, \ d_1)$
- Proxy Re-encryption Key Table (PRKT): $(id, \ id', \ rk^{(1)}_{id \to id'}, \ rk^{(2)}_{id \to id'})$
- Semantic agreement on table operation. We view each above table as an object-oriented database and possessing the so-called public method *lookup*: For a given table TBL with t-fields, the operation

$$\text{TBL.lookup}(\underline{v_1}, v_2, \cdots, v_{i-1}, \underline{v_i}, v_{i+1}, \cdots, v_t)$$

will return 0 if there is no tuple in TBL such that the first field is v_1 and the i-th field is v_i, and return 1 otherwise. Here, we assume that there are no repeated identical tuples in a table and the retrieval conditions are specified by the underlined values. We also agree that: (1) Whenever *lookup* returns

1, each $v_i (i = 1, \cdots, t)$ is assigned with the values recorded in i-th field of the corresponding item in TBL; (2) Whenever *lookup* returns 0, except for underlined inputs (e.g., v_1 and v_i in above example), other input parameters become undefined.

– Identity table settings. Without loss of generality, suppose there is an identity table, denoted by IDT, with a structure (i, id_i) that is automatically maintained by the system such that whenever a new user with identity id enrolls in the system, a new item is added into this table accordingly. Therefore, for any given legitimate identity id, we can determine an index $i \in \{1, \cdots, n\}$ such that $id = id_i$.

Setup. $\mathcal{C}$ sets $params = (\mathbb{G}_1, \mathbb{G}_2, p, \hat{e}; g, g_1, g_2, \eta)$ and sends it to $\mathcal{A}$, where re-encryption parameter $\eta \in \mathbb{G}_1$ is picked at random.

Phase 1. When $\mathcal{A}$ invokes the following queries, $\mathcal{C}$ works accordingly.

– Hash query on input v.
  1. If KT.lookup(v, *coin*, h) = 1 then answers $\mathcal{A}$ with h;
  2. Otherwise, $\mathcal{C}$ generates a random $coin \in \{0, 1\}$ so that $\Pr[coin = 0] = \xi$ for some ξ that will be determined later, and then sets

$$h = \begin{cases} g_1^x g^v, & \text{If } coin = 0; \\ g^v, & \text{Otherwise.} \end{cases}$$

  Then, answers $\mathcal{A}$ with h and adds $(v, coin, h)$ into KT.

– Secret key query on input id.
  1. Without loss of generality, suppose $\mathcal{A}$ has made a hash query on id. (Otherwise, $\mathcal{C}$ can make this query on behalf of $\mathcal{A}$.) This suggests there exists an item $(id, coin, h)$ in the hash table KT. Now, $\mathcal{C}$ checks whether $coin = 1$ in this item. If so, aborts; Otherwise, continues.
  2. If SKT.lookup($\underline{id}$, r, d_0, d_1) = 1 then $\mathcal{C}$ answers $\mathcal{A}$ with (d_0, d_1) and goes to Step 4; Otherwise, $\mathcal{C}$ continues.
  3. $\mathcal{C}$ picks $r \in \mathbb{Z}_p$ at random and sets

$$d_0 = g_2^{\frac{-id}{x}} H(id)^r, \quad d_1 = g_2^{\frac{-1}{x}} g^r.$$

  Then, $\mathcal{C}$ answers $\mathcal{A}$ with (d_0, d_1) and adds (id, r, d_0, d_1) into SKT. Here, we claim that the secret key for id is well-formed. To see this, let $\tilde{r} = r - \frac{b}{x}$. Then, we have that

$$d_0 = g_2^{\frac{-id}{x}} (g_1^x g^{id})^r$$
$$= g_2^a (g_1^x g^{id})^{\frac{-b}{x}} (g_1^x g^{id})^r$$
$$= g_2^a H(id)^{\tilde{r}},$$
$$d_1 = g^{\frac{-b}{x}} g^r$$
$$= g^{\tilde{r}}.$$

  4. $\mathcal{C}$ updates $s_i = 1$ for some $i \in \{1, \cdots, n\}$ such that $id_i = id$.

– Proxy re-encryption key query on input (id, id').

1. If PRKT.lookup(id, id', $rk^{(1)}_{id \to id'}$, $rk^{(2)}_{id \to id'}$) $= 1$, $\mathcal{C}$ answers $\mathcal{A}$ with $(rk^{(1)}_{id \to id'}, rk^{(2)}_{id \to id'})$ and goes to Step 5; otherwise, $\mathcal{C}$ continues.
2. Similarly, we can assume that $\mathcal{A}$ has made hash queries on id and id'. (Otherwise, $\mathcal{C}$ can make these queries on behalf of $\mathcal{A}$.) This suggests there exist two items $(id, coin, h)$ and $(id', coin', h')$ in the hash table KT. Now, $\mathcal{C}$ checks whether $coin = coin' = 0$. If not, it aborts; Otherwise, it continues.
3. If SKT.lookup($\underline{id'}$, r', d'_0, d'_1) $= 1$ then $\mathcal{C}$ goes to the step 4 directly; Otherwise, $\mathcal{C}$ picks $r' \in_R \mathbb{Z}^*_p$ and sets

$$d'_0 = g_2^{\frac{-id'}{x}} H(id)^{r'}, \quad d'_1 = g_2^{\frac{-1}{x}} g^{r'},$$

and then adds (id', r', d'_0, d'_1) into SKT.
4. $\mathcal{C}$ picks $\delta \in \mathbb{Z}_p$ at random, and answers $\mathcal{A}$ with

$$(rk^{(1)}_{id \to id'}, rk^{(2)}_{id \to id'}) = (\eta^\delta (g^{r'} g_2^{\frac{-1}{x}})^{id - id'}, g^\delta)$$

and adds $(id, id', rk^{(1)}_{id \to id'}, rk^{(2)}_{id \to id'})$ into PRKT. Here, we claim that $(rk^{(1)}_{id \to id'}, rk^{(2)}_{id \to id'})$ is well-formed. To see this, let $\widetilde{r}' = r' - \frac{b}{x}$. Then, we have that

$$rk^{(1)}_{id \to id'} = \eta^\delta \cdot (g^{r'} g_2^{\frac{-1}{x}})^{id - id'}$$
$$= \eta^\delta \cdot (g^{r' - \frac{b}{x}})^{id - id'}$$
$$= \eta^\delta \cdot (g^{id - id'})^{r' - \frac{b}{x}}$$
$$= \eta^\delta \cdot (\frac{g_1^x g^{id}}{g_1^x g^{id'}})^{r' - \frac{b}{x}}$$
$$= \eta^\delta \cdot \left(\frac{H(id)}{H(id')}\right)^{\widetilde{r}'},$$
$$rk^{(2)}_{id \to id'} = g^\delta.$$

5. $\mathcal{C}$ updates $r_{ij} = 1$ for some $i, j \in \{1, \cdots, n\}$ such that $id_i = id$ and $id_j = id'$.

Challenge. When the adversary $\mathcal{A}$ decides to end phase 1,

- The $\mathcal{A}$ submits a target identity id^* and messages (m_0, m_1) of equal length.
- The challenger $\mathcal{C}$ responds as follows:
  - At first, computes id^*'s relation vector R_{id^*} in the relation matrix, then rejects $\mathcal{A}$'s challenge targets and ends the simulation, if $R^T_{id^*} s \neq 0$; otherwise, continues.
  - Next, $\mathcal{C}$ checks whether $coin^* = 1$ in this item. If not, it aborts; Otherwise, it continues. Here, we suppose that $\mathcal{A}$ has made a hash query

on id^* and there exists an item $(id^*, coin^*, h^*)$ in the hash table KT; Otherwise, $\mathcal{C}$ can make this query on behalf of $\mathcal{A}$ before this step.

- Then, $\mathcal{C}$ flips a fair binary coin $\kappa \in \{0, 1\}$, and constructs a challenge ciphertext $C^* = (m_\kappa \cdot T, \ g_3, \ g_3^{id^*})$ for the message m_κ.
- Finally, $\mathcal{C}$ sends C^* to $\mathcal{A}$.

* Since $g_3 = g^c$ holds and $H(id^*) = g^{id^*}$ when $coin^* = 1$. Thus, we have that

$$C^* = (m_\kappa \cdot T, \ g^c, \ H(id^*)^c).$$

Hence, if $T = \hat{e}(g, g)^{abc} = \hat{e}(g_1, g_2)^c$ then C^* is a valid ciphertext of m_κ under the public key id^*. Otherwise, C^* is independent of κ in the adversary's view.

Phase 2. Phase 1 is repeated with the following restrictions: Whenever the adversary $\mathcal{A}$ makes a new secret key query on some id_i or a new proxy re-encryption key query on (id_i, id_j), $\mathcal{C}$ performs the following steps:

- First makes a virtual response (i.e., without sending the response to $\mathcal{A}$) as in Phase 1, except for the following case[3]:
  - If $\mathcal{A}$ invokes a proxy re-encryption key query on (id, id^*), the response could be arbitrary random pair $(rk^{(1)}_{id \to id^*}, \ rk^{(2)}_{id \to id^*}) \in \mathbb{G}_1^2$.
- Checks whether $R^T_{id^*} s \neq 0$ holds. If it does, $\mathcal{C}$ withdraws all modifications that have just been made and rejects $\mathcal{A}$'s query. Otherwise, $\mathcal{C}$ makes this virtual response into a real response (i.e., sending the response to $\mathcal{A}$).

Guess. Finally, $\mathcal{A}$ outputs a guess $\kappa' \in \{0, 1\}$. $\mathcal{C}$ concludes its own game by outputting a guess as follows. If $\kappa = \kappa'$ then $\mathcal{C}$ outputs 1 meaning $T = \hat{e}(g, g)^{abc}$. Otherwise, it outputs 0 meaning $T \neq \hat{e}(g, g)^{abc}$.

If $T = \hat{e}(g, g)^{abc}$ then $\mathcal{A}'$ guess must satisfy $|\Pr[\kappa = \kappa'] - 1/2| > \epsilon$. However, $\Pr[\kappa = \kappa'] = 1/2$ when T is uniform in $\mathbb{G}_2$. Therefore if $\mathcal{C}$ has not aborted and a, b, c are uniform in $\mathbb{Z}_p^*$, while T is uniform in $\mathbb{G}_2$, we have that

$$|\Pr[\mathcal{C}(g, g^a, g^b, g^c, \hat{e}(g, g)^{abc}) = 0] - \Pr[\mathcal{C}(g, g^a, g^b, g^c, T) = 0]|$$
$$\geq |(\frac{1}{2} \pm \epsilon) - \frac{1}{2}| \cdot \Pr[\overline{abort}]$$
$$= \epsilon.$$

Now, let us provide an estimation of the probability that $\mathcal{C}$ has not aborted. Note that $\mathcal{C}$ will never abort on replying to a hash query. At first $\mathcal{C}$ will not abort in the challenge phase with probability exactly $(1 - \xi)$. Similar to [5], suppose there are in total q_s secret key queries and q_r re-encryption key queries. For each secret key query, $\mathcal{C}$ will not abort with probability of exactly ξ, while for each re-encryption key query, $\mathcal{C}$ will not abort with probability of exactly ξ^2. Considering that the event for aborting a secret key and the event for aborting

---

[3] The response for $\mathcal{A}$'s proxy re-encryption key query on (id^*, id) can be generated by the same procedure described in phase 1 because that in our scheme, user id^*'s secret key, though it cannot be simulated, is not used in the re-encryption key generation.

a re-encryption key might be dependent, the probability for $\mathcal{C}$ not aborting the whole simulation is at least

$$\Pr[\overline{abort}] \geq (1-\xi)\xi^{q_s}\xi^{2q_r}.$$

The last item achieves maximum $\frac{1}{e \cdot (q_s+2q_r)}$ when $\xi = 1 - \frac{1}{q_s+2q_r}$. This suggests that in total, $\mathcal{C}$ will solve the BDDH problem with an advantage over $1/2$ with at least $\frac{\epsilon}{e \cdot (q_s+2q_r)}$, where e is the base of the natural logarithm, while q_s and q_r are the numbers of secret key queries and re-encryption key queries, respectively. This completes the proof of the theorem. $\qquad\square$

Fully Secure Anonymous HIBE and Secret-Key Anonymous IBE with Short Ciphertexts

Angelo De Caro, Vincenzo Iovino*, and Giuseppe Persiano

Dipartimento di Informatica ed Applicazioni
Università di Salerno, 84084 Fisciano (SA), Italy
{decaro,iovino,giuper}@dia.unisa.it

Abstract. Lewko and Waters [Eurocrypt 2010] presented a fully secure HIBE with short ciphertexts. In this paper we show how to modify their construction to achieve anonymity. We prove the security of our scheme under static (and generically secure) assumptions formulated in composite order bilinear groups.

In addition, we present a fully secure Anonymous IBE in the secret-key setting. Secret-Key Anonymous IBE was implied by the work of [Shen-Shi-Waters - TCC 2009] which can be shown secure in the selective-id model. No previous fully secure construction of secret-key Anonymous IBE is known.

1 Introduction

Identity-Based Encryption (IBE) was introduced by [14] to simplify the public-key infrastructure. An IBE is a public-key encryption scheme in which the public-key can be set to any string interpreted as one's identity. A central authority that holds the master secret key can produce a secret key corresponding to a given identity. Anyone can then encrypt messages using the identity, and only the owner of the corresponding secret key can decrypt the messages. First realizations of IBE are due to [2] which makes use of bilinear groups and to [8] which uses quadratic residues. Later, [10] introduced the more general concept of Hierarchical Identity-Based Encryption (HIBE) issuing a partial solution to it. An HIBE system is an IBE that allows delegation of the keys in a hierarchical structure. To the top of the structure there is the central authority that holds the master secret key, then several sub-authorities (or individual users) that hold delegated keys which can be used to decrypt only the messages addressed to the organization which the sub-authority belongs.

In this paper we are interested in *Anonymous* HIBE that are a special type of HIBE with the property that ciphertexts hide the identity for which they were encrypted. Interest in Anonymous IBE and HIBE was spurred by the observation that it can be used to build Public-Key Encryption with Keyword Search [1]. As noticed by [4], the first construction of Anonymous IBE was implicit

---

* Work done while visiting the Department of Computer Science of The Johns Hopkins University.

M. Joye, A. Miyaji, and A. Otsuka (Eds.): Pairing 2010, LNCS 6487, pp. 347–366, 2010.
© Springer-Verlag Berlin Heidelberg 2010

in [2] whose security relied on the random oracle assumption. Boneh and Waters [5] constructed Anonymous HIBE in the selective-id model. Recently, Lewko and Waters [12] used the Dual System Encryption methodology introduced by [17] to construct the first fully secure HIBE system with short ciphertexts. The construction given by [12] seems inherently non-anonymous. Another construction of Anonymous HIBE was given by [13] but their security proof is in the selective-ID model.

We show that a slight modification of the HIBE of [12] gives the *first* fully secure Anonymous HIBE. Our construction has, like the non-anonymous one of [12], short ciphertexts; that is, a ciphertext consists of a constant (that is independent of the depth of the hierarchy) number of elements from the underlying bilinear group. The full security of our construction is based on static (that is, independent from the running time of the adversary and the size of hierarchy) and generically secure assumptions.

Recently, a fully-secure hierarchical predicate encryption system has been given by [11]. Anonymous HIBE can be obtained as special case of the construction of [11] and, even though the construction of [11] is based on prime order gropus, the ciphertexts of the resulting Anonymous HIBE consist of $O(\ell^2)$ group elements and keys have $O(\ell^3)$ group elements. In [7] the authors constructed an Anonymous HIBE scheme based on hard lattice problems; in this construction the size of a ciphertext depends on the depth of the hierarchy.

We also study *Secret-Key* Anonymous IBE and show that if our public key construction is used in the secret key setting (that is, the public key is kept secret) then the scheme enjoys the additional property of key secrecy; that is, decryption keys for different identities are indistinguishable. We stress that key secrecy cannot be obtained in the public key setting as an adversary can test if a secret key Sk corresponds to identity ID by creating a cipertext Ct for ID (by using the *public key*) and then trying to decrypt Ct by using Sk. We mention that the Secret-Key Predicate Encryption Scheme of [15] has Secret-Key Anonymous IBE as a special case but its security is in the selective-id model. To the best of our knowledge, the concept of Secret-Key Hierarchical IBE has not been defined before and we defer its study to future work.

Organization of the work. In Section 2, we present a brief introduction of bilinear groups and state the complexity assumptions used to prove the security of our schemes. In Section 3 we present definitions and our construction for Public-key Anonymous HIBE. Finally, in Section 4, we present definitions and our construction for Secret-key Anonymous IBE. Due to lack of space the proof of security of our assumptions in the generic group model is omitted and can be found in the extended version [6].

2   Composite Order Bilinear Groups and Complexity Assumptions

Composite order bilinear groups were first used in cryptographic construction in [3]. We use groups of order product of four primes and a generator $\mathcal{G}$ which takes as input security parameter λ and outputs a description $\mathcal{I} = (N = p_1p_2p_3p_4, \mathbb{G}, \mathbb{G}_T, \mathbf{e})$ where p_1, p_2, p_3, p_4 are distinct primes of $\Theta(\lambda)$ bits, $\mathbb{G}$ and $\mathbb{G}_T$ are cyclic groups of order N, and $\mathbf{e} : \mathbb{G} \times \mathbb{G} \to \mathbb{G}_T$ is a map with the following properties:

1. (Bilinearity) $\forall\, g, h \in \mathbb{G}, a, b \in \mathbb{Z}_N, \mathbf{e}(g^a, h^b) = \mathbf{e}(g, h)^{ab}$.
2. (Non-degeneracy) $\exists\, g \in \mathbb{G}$ such that $\mathbf{e}(g, g)$ has order N in $\mathbb{G}_T$.

We further require that the group operations in $\mathbb{G}$ and $\mathbb{G}_T$ as well the bilinear map $\mathbf{e}$ are computable in deterministic polynomial time with respect to λ. Also, we assume that the group descriptions of $\mathbb{G}$ and $\mathbb{G}_T$ include generators of the respective cyclic groups. Furthermore, for $a, b, c \in \{1, p_1, p_2, p_3, p_4\}$ we denote by $\mathbb{G}_{abc}$ the subgroup of order abc. From the fact that the group is cyclic it is simple to verify that if g and h are group elements of different order (and thus belonging to different subgroups), then $\mathbf{e}(g, h) = 1$. This is called the *orthogonality property* and is a crucial tool in our constructions. We now give our complexity assumptions.

2.1 Assumption 1

For a generator $\mathcal{G}$ returning bilinear settings of order N product of four primes, we define the following distribution. First pick a random bilinear setting $\mathcal{I} = (N = p_1p_2p_3p_4, \mathbb{G}, \mathbb{G}_T, \mathbf{e})$ by running $\mathcal{G}(1^\lambda)$ and then pick

$$g_1, A_1 \leftarrow \mathbb{G}_{p_1}, A_2, B_2 \leftarrow \mathbb{G}_{p_2}, g_3, B_3 \leftarrow \mathbb{G}_{p_3}, g_4 \leftarrow \mathbb{G}_{p_4}, T_1 \leftarrow \mathbb{G}_{p_1p_2p_3}, T_2 \leftarrow \mathbb{G}_{p_1p_3}$$

and set $D = (\mathcal{I}, g_1, g_3, g_4, A_1A_2, B_2B_3)$. We define the advantage of an algorithm $\mathcal{A}$ in breaking Assumption 1 to be:

$$\mathsf{Adv}_{A1}^{\mathcal{A}}(\lambda) = |\mathrm{Prob}[\mathcal{A}(D, T_1) = 1] - \mathrm{Prob}[\mathcal{A}(D, T_2) = 1]|.$$

Assumption 1. *We say that Assumption 1 holds for generator $\mathcal{G}$ if for all probabilistic polynomial-time algorithms $\mathcal{A}$ $\mathsf{Adv}_{A1}^{\mathcal{A}}(\lambda)$ is a negligible function of λ.*

2.2 Assumption 2

For a generator $\mathcal{G}$ returning bilinear settings of order N product of four primes, we define the following distribution. First pick a random bilinear setting $\mathcal{I} = (N = p_1p_2p_3p_4, \mathbb{G}, \mathbb{G}_T, \mathbf{e})$ by running $\mathcal{G}(1^\lambda)$ and then pick

$$\alpha, s, r \leftarrow \mathbb{Z}_N, \ g_1 \leftarrow \mathbb{G}_{p_1}, \ g_2, A_2, B_2 \leftarrow \mathbb{G}_{p_2}, \ g_3 \leftarrow \mathbb{G}_{p_3}, \ g_4 \leftarrow \mathbb{G}_{p_4}, \ T_2 \leftarrow \mathbb{G}_T$$

and set $T_1 = \mathbf{e}(g_1, g_1)^{\alpha s}$ and $D = (\mathcal{I}, g_1, g_2, g_3, g_4, g_1^\alpha A_2, g_1^s B_2, g_2^r, A_2^r)$. We define the advantage of an algorithm $\mathcal{A}$ in breaking Assumption 2 to be:

$$\mathsf{Adv}_{A2}^{\mathcal{A}}(\lambda) = |\mathrm{Prob}[\mathcal{A}(D, T_1) = 1] - \mathrm{Prob}[\mathcal{A}(D, T_2) = 1]|.$$

Assumption 2. *We say that Assumption 2 holds for generator $\mathcal{G}$ if for all probabilistic polynomial time algorithm $\mathcal{A}$ $\mathsf{Adv}_{A2}^{\mathcal{A}}(\lambda)$ is a negligible function of λ.*

2.3   Assumption 3

For a generator $\mathcal{G}$ returning bilinear settings of order N product of four primes, we define the following distribution. First pick a random bilinear setting $\mathcal{I} = (N = p_1 p_2 p_3 p_4, \mathbb{G}, \mathbb{G}_T, \mathbf{e})$ by running $\mathcal{G}(1^\lambda)$ and then pick

$$\hat{r}, s \leftarrow \mathbb{Z}_N, \quad g_1, U, A_1 \leftarrow \mathbb{G}_{p_1}, \quad g_2, A_2, B_2, D_2, F_2 \leftarrow \mathbb{G}_{p_2}, \quad g_3 \leftarrow \mathbb{G}_{p_3},$$

$$g_4, A_4, B_4, D_4 \leftarrow \mathbb{G}_{p_4}, \quad A_{24}, B_{24}, D_{24} \leftarrow \mathbb{G}_{p_2 p_4}, \quad T_2 \leftarrow \mathbb{G}_{p_1 p_2 p_4}$$

and set $T_1 = A_1^s D_{24}$ and $D = (\mathcal{I}, g_1, g_2, g_3, g_4, U, U^s A_{24}, U^{\hat{r}}, A_1 A_4, A_1^{\hat{r}} A_2, g_1^{\hat{r}} B_2, g_1^s B_{24})$. We define the advantage of an algorithm $\mathcal{A}$ in breaking Assumption 3 to be:

$$\mathsf{Adv}_{A3}^{\mathcal{A}}(\lambda) = |\mathrm{Prob}[\mathcal{A}(D, T_1) = 1] - \mathrm{Prob}[\mathcal{A}(D, T_2) = 1]|.$$

Assumption 3. *We say that Assumption 3 holds for generator $\mathcal{G}$ if for all probabilistic polynomial time algorithm $\mathcal{A}$ $\mathsf{Adv}_{A3}^{\mathcal{A}}(\lambda)$ is a negligible function of λ.*

3   Public-Key Anonymous HIBE

3.1   Hierarchical Identity Based Encryption

A Hierarchical Identity Based Encryption scheme (henceforth abbreviated in HIBE) over an alphabet Σ is a tuple of five efficient and probabilistic algorithms: (Setup, Encrypt, KeyGen, Decrypt, Delegate).

Setup($1^\lambda, 1^\ell$): takes as input security parameter λ and maximum depth of an identity vector ℓ and outputs public parameters Pk and master secret key Msk.

KeyGen(Msk, ID $= (\mathsf{ID}_1, \ldots, \mathsf{ID}_j)$): takes as input master secret key Msk, identity vector $ID \in \Sigma^j$ with $j \leq \ell$ and outputs a private key $\mathsf{Sk}_{\mathsf{ID}}$.

Delegate(Pk, ID, $\mathsf{Sk}_{\mathsf{ID}}, \mathsf{ID}_{j+1}$): takes as input public parameters Pk, secret key for identity vector $\mathsf{ID} = (\mathsf{ID}_1, \ldots, \mathsf{ID}_j)$ of depth $j < \ell$, $\mathsf{ID}_{j+1} \in \Sigma$ and outputs a secret key for the depth $j + 1$ identity vector $(\mathsf{ID}_1, \ldots, \mathsf{ID}_j, \mathsf{ID}_{j+1})$.

Encrypt(Pk, M, ID): takes as input public parameters Pk, message M and identity vector ID and outputs a ciphertext Ct.

Decrypt(Pk, Ct, Sk): takes as input public parameters Pk, ciphertext Ct and secret key Sk and outputs the message M. We make the following obvious consistency requirement. Suppose ciphertext Ct is obtained by running the Encrypt algorithm on public parameters Pk, message M and identity ID and that Sk is a secret key for identity ID obtained through a sequence of KeyGen and Delegate calls using the same public parameters Pk. Then Decrypt, on input Pk, Ct and Sk, returns M except with negligible probability.

3.2   Security Definition

We give complete form of the security definition following [16]. Our security definition captures semantic security and ciphertext anonymity by means of the following game between an adversary $\mathcal{A}$ and a challenger $\mathcal{C}$.

Setup. The challenger $\mathcal{C}$ runs the Setup algorithm to generate public parameters Pk which it gives to the adversary $\mathcal{A}$. We let S denote the set of private keys that the challenger has created but not yet given to the adversary. At this point, $S = \emptyset$.

Phase 1. $\mathcal{A}$ makes Create, Delegate, and Reveal key queries. To make a Create query, $\mathcal{A}$ specifies an identity vector ID of depth j. In response, $\mathcal{C}$ creates a key for this vector by calling the key generation algorithm, and places this key in the set S. $\mathcal{C}$ only gives $\mathcal{A}$ a reference to this key, not the key itself. To make a Delegate query, $\mathcal{A}$ specifies a key $\mathsf{Sk_{ID}}$ in the set S and $\mathsf{ID}_{j+1} \in \Sigma$. In response, $\mathcal{C}$ appends ID_{j+1} to ID and makes a key for this new identity by running the delegation algorithm on ID, $\mathsf{Sk_{ID}}$ and ID_{j+1}. $\mathcal{C}$ adds this key to the set S and again gives $\mathcal{A}$ only a reference to it, not the actual key. To make a Reveal query, $\mathcal{A}$ specifies an element of the set S. $\mathcal{C}$ gives this key to $\mathcal{A}$ and removes it from the set S. We note that $\mathcal{A}$ needs no longer make any delegation queries for this key because it can run delegation algorithm on the revealed key for itself.

Challenge. $\mathcal{A}$ gives two pairs of message and identity $(M_0, \mathsf{ID}_0^\star)$ and $(M_1, \mathsf{ID}_1^\star)$ to $\mathcal{C}$. We require that no revealed identity in Phase 1 is a prefix of either $\mathsf{ID}_0^\star$ or $\mathsf{ID}_1^\star$. $\mathcal{C}$ chooses random $\beta \in \{0, 1\}$, encrypts M_β under $\mathsf{ID}_\beta^\star$ and sends the resulting ciphertext to $\mathcal{A}$.

Phase 2. This is the same as Phase 1 with the added restriction that any revealed identity vector must not be a prefix of either $\mathsf{ID}_0^\star$ or $\mathsf{ID}_1^\star$.

Guess. $\mathcal{A}$ must output a guess β' for β. The advantage of $\mathcal{A}$ is defined to be $\mathrm{Prob}[\beta' = \beta] - \frac{1}{2}$.

Definition 1. *An Anonymous Hierarchical Identity Based Encryption scheme is secure if all polynomial time adversaries achieve at most a negligible (in λ) advantage in the previous security game.*

3.3   Our Construction

In this section we describe our construction for an Anonymous HIBE scheme.

Setup($1^\lambda, 1^\ell$): The setup algorithm chooses random description $\mathcal{I} = (N = p_1 p_2 p_3 p_4, \mathbb{G}, \mathbb{G}_T, \mathbf{e})$ and random $Y_1, X_1, u_1, \ldots, u_\ell \in \mathbb{G}_{p_1}, Y_3 \in \mathbb{G}_{p_3}, X_4, Y_4 \in \mathbb{G}_{p_4}$ and $\alpha \in \mathbb{Z}_N$. The public parameters are published as:

$$\mathsf{Pk} = (N, Y_1, Y_3, Y_4, t = X_1 X_4, u_1, \ldots, u_\ell, \Omega = \mathbf{e}(Y_1, Y_1)^\alpha).$$

The master secret key is $\mathsf{Msk} = (X_1, \alpha)$.

KeyGen(Msk, ID $= (\mathsf{ID}_1, \ldots, \mathsf{ID}_j)$): The key generation algorithm chooses random $r_1, r_2 \in \mathbb{Z}_N$ and, for $i \in \{1, 2\}$, random $R_{i,1}, R_{i,2}, R_{i,j+1}, \ldots, R_{i,\ell} \in \mathbb{G}_{p_3}$. The secret key $\mathsf{Sk}_{\mathsf{ID}} = (K_{i,1}, K_{i,2}, E_{i,j+1}, \ldots, E_{i,\ell})$ is computed as

$$K_{1,1} = Y_1^{r_1} R_{1,1}, \quad K_{1,2} = Y_1^{\alpha} \left(u_1^{\mathsf{ID}_1} \cdots u_j^{\mathsf{ID}_j} X_1 \right)^{r_1} R_{1,2}$$

$$E_{1,j+1} = u_{j+1}^{r_1} R_{1,j+1}, \quad \ldots, \quad E_{1,\ell} = u_{\ell}^{r_1} R_{1,\ell},$$

$$K_{2,1} = Y_1^{r_2} R_{2,1}, \quad K_{2,2} = \left(u_1^{\mathsf{ID}_1} \cdots u_j^{\mathsf{ID}_j} X_1 \right)^{r_2} R_{2,2}$$

$$E_{2,j+1} = u_{j+1}^{r_2} R_{2,j+1}, \quad \ldots, \quad E_{1,\ell} = u_{\ell}^{r_2} R_{2,\ell}.$$

Notice that, $\mathsf{Sk}_{\mathsf{ID}}$ is composed by two sub-keys. The first sub-key, $(K_{1,1}, K_{1,2}, E_{1,j+1}, \ldots, E_{1,\ell})$, is used by the decryption algorithm to compute the blinding factor, the second, $(K_{2,1}, K_{2,2}, E_{2,j+1}, \ldots, E_{2,\ell})$, is used by the delegation algorithm and can be used also to verify that the identity vector of a given ciphertext matches the identity vector of the key.

Delegate(Pk, ID, $\mathsf{Sk}_{\mathsf{ID}}, \mathsf{ID}_{j+1}$): Given a key $\mathsf{Sk}_{\mathsf{ID}} = (K'_{i,1}, K'_{i,2}, E'_{i,j+1}, \ldots, E'_{i,\ell})$ for ID $= (\mathsf{ID}_1, \ldots, \mathsf{ID}_j)$, the delegation algorithm creates a key for $(\mathsf{ID}_1, \ldots, \mathsf{ID}_j, \mathsf{ID}_{j+1})$ as follows. It chooses random $\tilde{r}_1, \tilde{r}_2 \in \mathbb{Z}_N$ and, for $i \in \{1, 2\}$, random $R_{i,1}, R_{i,2}, R_{i,j+2}, \ldots, R_{i,\ell} \in \mathbb{G}_{p_3}$. The secret key $(K_{i,1}, K_{i,2}, E_{i,j+2}, \ldots, E_{i,\ell})$ is computed as

$$K_{1,1} = K'_{1,1} (K'_{2,1})^{\tilde{r}_1} R_{1,1}, \; K_{1,2} = K'_{1,2} (K'_{2,2})^{\tilde{r}_1} (E'_{1,j+1})^{\mathsf{ID}_{j+1}} (E'_{2,j+1})^{\tilde{r}_1 \mathsf{ID}_{j+1}} R_{1,2},$$

$$E_{1,j+2} = E'_{1,j+2} \cdot (E'_{2,j+2})^{\tilde{r}_1} R_{1,j+2}, \ldots, E_{1,\ell} = E'_{1,\ell} \cdot (E'_{2,\ell})^{\tilde{r}_1} R_{1,\ell}.$$

$$K_{2,1} = (K'_{2,1})^{\tilde{r}_2} R_{2,1}, \quad K_{2,2} = (K'_{2,2})^{\tilde{r}_2} \cdot (E'_{2,j+1})^{\tilde{r}_2 \mathsf{ID}_{j+1}} R_{2,2},$$

$$E_{2,j+2} = (E'_{2,j+2})^{\tilde{r}_2} R_{2,j+2}, \ldots, E_{2,\ell} = (E'_{2,\ell})^{\tilde{r}_2} R_{2,\ell}.$$

We observe that the new key has the same distributions as the key computed by the KeyGen algorithm on $(\mathsf{ID}_1, \ldots, \mathsf{ID}_j, \mathsf{ID}_{j+1})$ with randomness $r_1 = r'_1 + (r'_2 \cdot \tilde{r}_1)$ and $r_2 = r'_2 \cdot \tilde{r}_2$.

Encrypt(Pk, M, ID $= (\mathsf{ID}_1, \ldots, \mathsf{ID}_j)$): The encryption algorithm chooses random $s \in \mathbb{Z}_N$ and random $Z, Z' \in \mathbb{G}_{p_4}$. The ciphertext (C_0, C_1, C_2) for the message $M \in \mathbb{G}_T$ is computed as

$$C_0 = M \cdot e(Y_1, Y_1)^{\alpha s}, \quad C_1 = \left(u_1^{\mathsf{ID}_1} \cdots u_j^{\mathsf{ID}_j} t \right)^s Z, \quad C_2 = Y_1^s Z'.$$

Decrypt(Pk, Ct, Sk): The decryption algorithm assumes that the key and ciphertext both correspond to the same identity $(\mathsf{ID}_1, \ldots, \mathsf{ID}_j)$. If the key identity is a prefix of this instead, then the decryption algorithm starts by running the key delegation algorithm to create a key with identity matching the ciphertext identity exactly. The decryption algorithm then computes the blinding factor as:

$$\frac{e(K_{1,2}, C_2)}{e(K_{1,1}, C_1)} = \frac{e(Y_1, Y_1)^{\alpha s} e\left(u_1^{\mathsf{ID}_1} \cdots u_j^{\mathsf{ID}_j} X_1, Y_1 \right)^{r_1 s}}{e\left(Y_1, u_1^{\mathsf{ID}_1} \cdots u_j^{\mathsf{ID}_j} X_1 \right)^{r_1 s}} = e(Y_1, Y_1)^{\alpha s}.$$

By comparing our construction with the one of [12], we notice that component t of the public key and components C_1 and C_2 of the ciphertext have a $\mathbb{G}_{p_4}$ part. This addition makes the system anonymous. Indeed, if we remove from our construction the $\mathbb{G}_{p_4}$ parts of t and C_1 and C_2 (and thus obtain the scheme of [12]) then it is possible to test if ciphertext (C_0, C_1, C_2) is relative to identity $(\mathsf{ID}_1, \ldots, \mathsf{ID}_j)$ for public key $(N, Y_1, Y_3, Y_4, t, u_1, \ldots, u_\ell, \Omega)$ by testing $\mathbf{e}(C_2, t \cdot (u_1^{\mathsf{ID}_1} \cdots u_\ell^{\mathsf{ID}_\ell}))$ and $\mathbf{e}(C_1, Y_1)$ for equality.

3.4 Security

Following Lewko and Waters [12], we define two additional structures: *semi-functional ciphertexts* and *semi-functional keys*. These will not be used in the real scheme, but we need them in our proofs.

Semi-functional Ciphertext. We let g_2 denote a generator of $\mathbb{G}_{p_2}$. A semi-functional ciphertext is created as follows: first, we use the encryption algorithm to form a normal ciphertext (C'_0, C'_1, C'_2). We choose random exponents $x, z_c \in \mathbb{Z}_N$. We set:
$$C_0 = C'_0, \quad C_1 = C'_1 g_2^{xz_c}, \quad C_2 = C'_2 g_2^x.$$

Semi-functional Keys. To create a semi-functional key, we first create a normal key $(K'_{i,1}, K'_{i,2}, E'_{i,j+1}, \ldots, E'_{i,\ell})$ using the key generation algorithm. We choose random exponents $z, \gamma, z_k \in \mathbb{Z}_N$ and, for $i \in \{1,2\}$, random exponents $z_{i,j+1}, \ldots, z_{i,\ell} \in \mathbb{Z}_N$. We set:

$$K_{1,1} = K'_{1,1} \cdot g_2^\gamma, K_{1,2} = K'_{1,2} \cdot g_2^{\gamma z_k}, (E_{1,i} = E'_{1,i} \cdot g_2^{\gamma z_{1,i}})_{i=j+1}^\ell,$$

$$K_{2,1} = K'_{2,1} \cdot g_2^{z\gamma}, K_{2,2} = K'_{2,2} \cdot g_2^{z\gamma z_k}, (E_{2,i} = E'_{2,i} \cdot g_2^{z\gamma z_{2,i}})_{i=j+1}^\ell.$$

We note that when the first sub-key of a semi-functional key is used to decrypt a semi-functional ciphertext, the decryption algorithm will compute the blinding factor multiplied by the additional term $\mathbf{e}(g_2, g_2)^{x\gamma(z_k - z_c)}$. If $z_c = z_k$, decryption will still work. In this case, we say that the key is nominally semi-functional. If the second sub-key is used to test the identity vector of the ciphertext, then the decryption algorithm computes $\mathbf{e}(g_2, g_2)^{xz\gamma(z_k - z_c)}$ and if $z_c = z_k$, the test will still work.

To prove security of our Anonymous HIBE scheme, we rely on the static Assumptions $1, 2$ and 3. For a probabilistic polynomial-time adversary $\mathcal{A}$ which makes q key queries, our proof of security will consist of the following sequence of $q + 5$ games between $\mathcal{A}$ and a challenger $\mathcal{C}$.

Game$_{\mathsf{Real}}$: is the real Anonymous HIBE security game.

Game$_{\mathsf{Real'}}$: is the same as the real game except that all key queries will be answered by fresh calls to the key generation algorithm, ($\mathcal{C}$ will not be asked to delegate keys in a particular way).

Game$_{\mathsf{Restricted}}$: is the same as Game$_{\mathsf{Real'}}$ except that $\mathcal{A}$ cannot ask for keys for identities which are prefixes of one of the challenge identities modulo p_2. We will retain this restriction in all subsequent games.

Game_k: for k from 0 to q, we define Game_k like $\mathsf{Game}_{\mathsf{Restricted}}$ except that the ciphertext given to $\mathcal{A}$ is semi-functional and the first k keys are semi-functional. The rest of the keys are normal.

$\mathsf{Game}_{\mathsf{Final}_0}$: is the same as Game_q, except that the challenge ciphertext is a semi-functional encryption with C_0 random in $\mathbb{G}_T$ (thus the ciphertext is independent from the messages provided by $\mathcal{A}$).

$\mathsf{Game}_{\mathsf{Final}_1}$: is the same as $\mathsf{Game}_{\mathsf{Final}_0}$, except that the challenge ciphertext is a semi-functional encryption with C_1 random in $\mathbb{G}_{p_1 p_2 p_4}$ (thus the ciphertext is independent from the identity vectors provided by $\mathcal{A}$). It is clear that in this last game, no adversary can have advantage greater than 0.

We will show these games are indistinguishable in the following lemmata.

Indistinguishability of $\mathsf{Game}_{\mathsf{Real}}$ and $\mathsf{Game}_{\mathsf{Real}'}$

Lemma 1. *For any algorithm $\mathcal{A}$, $\mathsf{Adv}^{\mathcal{A}}_{\mathsf{Game}_{\mathsf{Real}}} = \mathsf{Adv}^{\mathcal{A}}_{\mathsf{Game}_{\mathsf{Real}'}}$.*

Proof. We note that the keys are identically distributed whether they are produced by the key delegation algorithm from a previous key or from a fresh call to the key generation algorithm. Thus, in the attacker's view, there is no difference between these games. □

Indistinguishability of $\mathsf{Game}_{\mathsf{Real}'}$ and $\mathsf{Game}_{\mathsf{Restricted}}$

Lemma 2. *Suppose that there exists a PPT algorithm $\mathcal{A}$ such that $\mathsf{Adv}^{\mathcal{A}}_{\mathsf{Game}_{\mathsf{Real}'}} - \mathsf{Adv}^{\mathcal{A}}_{\mathsf{Game}_{\mathsf{Restricted}}} = \epsilon$. Then there exists a PPT algorithm $\mathcal{B}$ with advantage $\geq \frac{\epsilon}{3}$ in breaking Assumption 1.*

Proof. Suppose that $\mathcal{A}$ has probability ϵ of producing an identity vector $\mathsf{ID} = (\mathsf{ID}_1, \ldots, \mathsf{ID}_k)$, that is a prefix of one of the challenge identities $\mathsf{ID}^\star = (\mathsf{ID}_1^\star, \ldots, \mathsf{ID}_j^\star)$ modulo p_2. That is, there exists i and $j \in \{0, 1\}$ such that that $\mathsf{ID}_i \neq \mathsf{ID}_{j,i}^\star$ modulo N and that p_2 divides $\mathsf{ID}_i - \mathsf{ID}_{j,i}^\star$ and thus $a = \gcd(\mathsf{ID}_i - \mathsf{ID}_{j,i}^\star, N)$ is a nontrivial factor of N. We notice that p_2 divides a and set $b = \frac{N}{a}$. The following three cases are exhaustive and at least one occurs with probability at least $\epsilon/3$.

1. $\mathrm{ord}(Y_1) \mid b$.
2. $\mathrm{ord}(Y_1) \nmid b$ and $\mathrm{ord}(Y_4) \mid b$.
3. $\mathrm{ord}(Y_1) \nmid b$, $\mathrm{ord}(Y_4) \nmid b$ and $\mathrm{ord}(Y_3) \mid b$.

Suppose case 1 has probability at least $\epsilon/3$. We describe algorithm $\mathcal{B}$ that breaks Assumption 1. $\mathcal{B}$ receives $(\mathcal{I}, g_1, g_3, g_4, A_1 A_2, B_2 B_3)$ and T and constructs Pk by running the Setup algorithm with the only exception that $\mathcal{B}$ sets $Y_1 = g_1, Y_3 = g_3$, and $Y_4 = g_4$. Notice that $\mathcal{B}$ has the master secret key Msk associated with Pk. Then $\mathcal{B}$ runs $\mathcal{A}$ on input Pk and uses knowledge of Msk to answer $\mathcal{A}$'s queries. At the end of the game, for all IDs for which $\mathcal{A}$ has asked for the key and for $\mathsf{ID}^\star \in \{\mathsf{ID}_0^\star, \mathsf{ID}_1^\star\}$, $\mathcal{B}$ computes $a = \gcd(\mathsf{ID}_i - \mathsf{ID}_i^\star, N)$. Then, if $\mathbf{e}\left((A_1 A_2)^a, B_2 B_3\right)$ is the identity element of $\mathbb{G}_T$ then $\mathcal{B}$ tests if $\mathbf{e}(T^b, A_1 A_2)$ is the identity element of $\mathbb{G}_T$. If this second test is successful, then $\mathcal{B}$ declares $T \in \mathbb{G}_{p_1 p_3}$. If it is not,

$\mathcal{B}$ declares $T \in \mathbb{G}_{p_1 p_2 p_3}$. It is easy to see that if p_2 divides a and $p_1 = \text{ord}(Y_1)$ divides b, then $\mathcal{B}$'s output is correct.

The other two cases are similar. Specifically, in case 2, $\mathcal{B}$ breaks Assumption 1 in the same way except that Pk is constructed by setting $Y_1 = g_4, Y_3 = g_3$, and $Y_4 = g_1$ (this has the effect of exchanging the roles of p_1 and p_4). Instead in case 3, $\mathcal{B}$ constructs Pk by setting $Y_1 = g_3, Y_3 = g_1$, and $Y_4 = g_4$ (this has the effect of exchanging the roles of p_1 and p_3). $\qquad\square$

Indistinguishability of Game$_{\text{Restricted}}$ and Game$_0$

Lemma 3. *Suppose that there exists a PPT algorithm $\mathcal{A}$ such that* $\text{Adv}^{\mathcal{A}}_{\text{Game}_{\text{Restricted}}} - \text{Adv}^{\mathcal{A}}_{\text{Game}_0} = \epsilon$. *Then there exists a PPT algorithm $\mathcal{B}$ with advantage ϵ in breaking Assumption 1.*

Proof. $\mathcal{B}$ receives $(\mathcal{I}, g_1, g_3, g_4, A_1 A_2, B_2 B_3)$ and T and simulates Game$_{\text{Restricted}}$ or Game$_0$ with $\mathcal{A}$ depending on whether $T \in \mathbb{G}_{p_1 p_3}$ or $T \in \mathbb{G}_{p_1 p_2 p_3}$.

$\mathcal{B}$ sets the public parameters as follows. $\mathcal{B}$ chooses random exponents $\alpha, a_1, \ldots, a_\ell, b, c \in \mathbb{Z}_N$ and sets $Y_1 = g_1, Y_3 = g_4, Y_4 = g_3$ $X_4 = Y_4^c, X_1 = Y_1^b$ and $u_i = Y_1^{a_i}$ for $i \in [\ell]$. $\mathcal{B}$ sends Pk $= (N, Y_1, Y_3, Y_4, t = X_1 X_4, u_1, \ldots, u_\ell, \Omega = \mathbf{e}(Y_1, Y_1)^\alpha)$ to $\mathcal{A}$. Notice that $\mathcal{B}$ knows the master secret key Msk $= (X_1, \alpha)$ associated with Pk and thus can answer all $\mathcal{A}$'s queries.

At some point, $\mathcal{A}$ sends $\mathcal{B}$ two pairs, $(M_0, \text{ID}_0^\star = (\text{ID}_{0,1}^\star, \ldots, \text{ID}_{0,j}^\star))$ and $(M_1, \text{ID}_1^\star = (\text{ID}_{1,1}^\star, \ldots, \text{ID}_{1,j}^\star))$. $\mathcal{B}$ chooses random $\beta \in \{0, 1\}$ and computes the challenge ciphertext as follows:

$$C_0 = M_\beta \cdot \mathbf{e}(T, Y_1)^\alpha, \quad C_1 = T^{a_1 \text{ID}_{\beta,1}^\star + \cdots + a_j \text{ID}_{\beta,j}^\star + b}, \quad C_2 = T.$$

We complete the proof with the following two observations. If $T \in \mathbb{G}_{p_1 p_3}$, then T can be written as $Y_1^{s_1} Y_3^{s_3}$. In this case (C_0, C_1, C_2) is a normal ciphertext with randomness $s = s_1, Z = Y_3^{s_3 a_1 \text{ID}_{\beta,1}^\star + \cdots + a_j \text{ID}_{\beta,j}^\star + b}$ and $Z' = Y_3^{s_3}$. If $T \in \mathbb{G}_{p_1 p_2 p_3}$, then T can be written as $Y_1^{s_1} g_2^{s_2} Y_3^{s_3}$ and this case (C_0, C_1, C_2) is a semi-functional ciphertext with randomness $s = s_1, Z = Y_3^{s_3 a_1 \text{ID}_{\beta,1}^\star + \cdots + a_j \text{ID}_{\beta,j}^\star + b}$, $Z' = Y_3^{s_3}$, $\gamma = s_2$ and $z_c = a_1 \text{ID}_{\beta,1}^\star + \cdots + a_j \text{ID}_{\beta,j}^\star + b$. $\qquad\square$

Indistinguishability of Game$_{k-1}$ and Game$_k$

Lemma 4. *Suppose there exists a PPT algorithm $\mathcal{A}$ such that* $\text{Adv}^{\mathcal{A}}_{\text{Game}_{k-1}} - \text{Adv}^{\mathcal{A}}_{\text{Game}_k} = \epsilon$. *Then, there exists a PPT algorithm $\mathcal{B}$ with advantage ϵ in breaking Assumption 1.*

Proof. $\mathcal{B}$ receives $(\mathcal{I}, g_1, g_3, g_4, A_1 A_2, B_2 B_3)$ and T and simulates Game$_{k-1}$ or Game$_k$ with $\mathcal{A}$ depending on whether $T \in \mathbb{G}_{p_1 p_3}$ or $T \in \mathbb{G}_{p_1 p_2 p_3}$.

$\mathcal{B}$ sets the public parameters by choosing random exponents $\alpha, a_1, \ldots, a_\ell, b, c \in \mathbb{Z}_N$ and setting $Y_1 = g_1, Y_3 = g_3, Y_4 = g_4, X_4 = Y_4^c, X_1 = Y_1^b$ and $u_i = Y_1^{a_i}$ for $i \in [\ell]$. $\mathcal{B}$ sends the public parameters Pk $= (N, Y_1, Y_3, Y_4, t = X_1 X_4, u_1, \ldots, u_\ell, \Omega = \mathbf{e}(Y_1, Y_1)^\alpha)$ to $\mathcal{A}$. Notice that $\mathcal{B}$ knows the master secret key Msk $= (X_1, \alpha)$

associated with Pk. Let us now explain how $\mathcal{B}$ answers the i-th key query for identity $(\mathsf{ID}_{i,1}, \ldots, \mathsf{ID}_{i,j})$.

For $i < k$, $\mathcal{B}$ creates a semi-functional key by choosing random exponents $r_1, r_2, f, z, w \in \mathbb{Z}_N$ and, for $i \in \{1,2\}$, random $w_{i,2}, w_{i,j+1}, \ldots, w_{i,\ell} \in \mathbb{Z}_N$ and setting:

$$K_{1,1} = Y_1^{r_1} \cdot (B_2 B_3)^f, \quad K_{1,2} = Y_1^{\alpha} \cdot (B_2 B_3)^w \left(u_1^{\mathsf{ID}_{i,1}} \cdots u_j^{\mathsf{ID}_{i,j}} X_1 \right)^{r_1} Y_3^{w_{1,2}},$$

$$E_{1,j+1} = u_{j+1}^{r_1} \cdot (B_2 B_3)^{w_{1,j+1}}, \ldots, E_{1,\ell} = u_{\ell}^{r_1} \cdot (B_2 B_3)^{w_{1,\ell}}.$$

and

$$K_{2,1} = Y_1^{r_2} \cdot (B_2 B_3)^{zf}, \quad K_{2,2} = (B_2 B_3)^{zw} \left(u_1^{\mathsf{ID}_{i,1}} \cdots u_j^{\mathsf{ID}_{i,j}} X_1 \right)^{r_2} Y_3^{w_{2,2}},$$

$$E_{2,j+1} = u_{j+1}^{r_2} \cdot (B_2 B_3)^{w_{2,j+1}}, \ldots, E_{2,\ell} = u_{\ell}^{r_2} \cdot (B_2 B_3)^{w_{2,\ell}}.$$

By writing B_2 as g_2^{ϕ}, we have that this is a properly distributed semi-functional key with $\gamma = \phi \cdot f$ and $\gamma \cdot z_k = \phi \cdot w$.

For $i > k$, $\mathcal{B}$ runs the KeyGen algorithm using the master secret key $\mathsf{Msk} = (X_1, \alpha)$.

To answer the k-th key query for $\mathsf{ID}_k = (\mathsf{ID}_{k,1}, \ldots, \mathsf{ID}_{k,j})$, $\mathcal{B}$ sets $z_k = a_1 \mathsf{ID}_{k,1} + \cdots + a_j \mathsf{ID}_{k,j} + b$, chooses random exponents $r_2' \in \mathbb{Z}_N$ and, for $i \in \{1,2\}$, random $w_{i,2}, w_{i,j+1}, \ldots, w_{i,\ell} \in \mathbb{Z}_N$, and sets:

$$K_{1,1} = T, \quad K_{1,2} = Y_1^{\alpha} \cdot T^{z_k} Y_3^{w_{1,2}}, \quad (E_{1,m} = T^{a_m} Y_3^{w_{1,m}})_{m=j+1}^{\ell}.$$

and

$$K_{2,1} = T^{r_2'}, \quad K_{2,2} = T^{r_2' \cdot z_k} Y_3^{w_{2,2}}, \quad (E_{2,m} = T^{r_2' \cdot a_m} Y_3^{w_{2,m}})_{m=j+1}^{\ell}.$$

We have the following two observations. If $T \in \mathbb{G}_{p_1 p_3}$, then T can be written as $Y_1^{r_1'} Y_3^{r_3}$ and $(K_{i,1}, K_{i,2}, E_{i,j+1}, \ldots, E_{i,\ell})$ is a normal key with randomness $r_1 = r_1'$, $r_2 = r_1' \cdot r_2'$. If $T \in \mathbb{G}_{p_1 p_2 p_3}$, then T can be written as $Y_1^{r_1'} g_2^{s_2} Y_3^{r_3}$. In this case the key is a semi-functional key with randomness $r_1 = r_1'$, $r_2 = r_1' \cdot r_2'$, $\gamma = s_2$ and $z = r_2'$.

At some point, $\mathcal{A}$ sends $\mathcal{B}$ two pairs, $(M_0, \mathsf{ID}_0^{\star} = (\mathsf{ID}_{0,1}^{\star}, \ldots, \mathsf{ID}_{0,j}^{\star}))$ and $(M_1, \mathsf{ID}_1^{\star} = (\mathsf{ID}_{1,1}^{\star}, \ldots, \mathsf{ID}_{1,j}^{\star}))$. $\mathcal{B}$ chooses random $\beta \in \{0,1\}$ and random $z, z' \in \mathbb{Z}_N$ and computes the challenge ciphertext as follows:

$$C_0 = M_{\beta} \cdot e(A_1 A_2, Y_1)^{\alpha}, \quad C_1 = (A_1 A_2)^{a_1 \mathsf{ID}_{\beta,1}^{\star} + \cdots + a_j \mathsf{ID}_{\beta,j}^{\star} + b} Y_4^z, \quad C_2 = A_1 A_2 Y_4^{z'}.$$

This implicitly sets $Y_1^s = A_1$ and $z_c = a_1 \mathsf{ID}_{\beta,1}^{\star} + \cdots + a_j \mathsf{ID}_{\beta,j}^{\star} + b \pmod{p_2}$. Since ID_k is not a prefix of $\mathsf{ID}_{\beta}^{\star}$ modulo p_2, we have that z_k and z_c are independent and randomly distributed. We observe that, if $\mathcal{B}$ attempts to test whether the k-th key is semi-functional by using the above procedure to create a semi-functional ciphertext for ID_k, then we will have that $z_k = z_c$ and thus decryption always works (independently of T).

We can thus conclude that, if $T \in \mathbb{G}_{p_1 p_3}$ then $\mathcal{B}$ has properly simulated Game_{k-1}. If $T \in \mathbb{G}_{p_1 p_2 p_3}$, then $\mathcal{B}$ has properly simulated Game_k. $\qquad\square$

Indistinguishability of Game$_q$ and Game$_{\mathsf{Final}_0}$

Lemma 5. *Suppose that there exists a PPT algorithm $\mathcal{A}$ such that* $\mathsf{Adv}^{\mathcal{A}}_{\mathsf{Game}_q}$ − $\mathsf{Adv}^{\mathcal{A}}_{\mathsf{Game}_{\mathsf{Final}_0}} = \epsilon$. *Then there exists a PPT algorithm $\mathcal{B}$ with advantage ϵ in breaking Assumption 2.*

Proof. $\mathcal{B}$ receives $(\mathcal{I}, g_1, g_2, g_3, g_4, g_1^{\alpha} A_2, g_1^s B_2, g_2^r, A_2^r)$ and T and simulates Game$_q$ or Game$_{\mathsf{Final}_0}$ with $\mathcal{A}$ depending on whether $T = \mathbf{e}(g_1, g_1)^{\alpha s}$ or T is a random element of $\mathbb{G}_T$.

$\mathcal{B}$ sets the public parameters as follows. $\mathcal{B}$ chooses random exponents $a_1, \ldots, a_\ell$, $b, c \in \mathbb{Z}_N$ and sets $Y_1 = g_1, Y_3 = g_3, Y_4 = g_4, X_4 = Y_4^c, X_1 = Y_1^b$, and $u_i = Y_1^{a_i}$ for $i \in [\ell]$. $\mathcal{B}$ computes $\Omega = \mathbf{e}(g_1^{\alpha} A_2, Y_1) = \mathbf{e}(Y_1^{\alpha}, Y_1)$ and send public parameters $\mathsf{Pk} = (N, Y_1, Y_2, Y_3, t = X_1 X_4, u_1, \ldots, u_\ell, \Omega)$ to $\mathcal{A}$.

Each time $\mathcal{B}$ is asked to provide a key for an identity $(\mathsf{ID}_1, \ldots, \mathsf{ID}_j)$, $\mathcal{B}$ creates a semi-functional key choosing random exponents $r_1, r_2, z, z' \in \mathbb{Z}_N$ and, for $i \in \{1, 2\}$, random $z_{i,j+1}, \ldots, z_{i,\ell}, w_{i,1}, w_{i,2}, w_{i,j+1}, \ldots, w_{i,\ell} \in \mathbb{Z}_N$ and setting:

$$K_{1,1} = Y_1^{r_1} \cdot g_2^z \cdot Y_3^{w_{1,1}}, \quad K_{1,2} = (g_1^{\alpha} A_2) \cdot g_2^{z'} \cdot \left(u_1^{\mathsf{ID}_1} \cdots u_j^{\mathsf{ID}_j} X_1 \right)^{r_1} \cdot Y_3^{w_{1,2}},$$

$$E_{1,j+1} = u_{j+1}^{r_1} \cdot g_2^{z_{1,j+1}} \cdot Y_3^{w_{1,j+1}}, \quad \ldots, \quad E_{1,\ell} = u_\ell^{r_1} \cdot g_2^{z_{1,\ell}} \cdot Y_3^{w_{1,\ell}}.$$

and

$$K_{2,1} = Y_1^{r_2} \cdot (g_2^r)^z \cdot Y_3^{w_{2,1}}, \quad K_{2,2} = A_2^r \cdot (g_2^r)^{z'} \cdot \left(u_1^{\mathsf{ID}_1} \cdots u_j^{\mathsf{ID}_j} X_1 \right)^{r_2} \cdot Y_3^{w_{2,2}},$$

$$E_{2,j+1} = u_{j+1}^{r_2} \cdot g_2^{z_{2,j+1}} \cdot Y_3^{w_{2,j+1}}, \quad \ldots, \quad E_{2,\ell} = u_\ell^{r_2} \cdot g_2^{z_{2,\ell}} \cdot Y_3^{w_{2,\ell}}.$$

At some point, $\mathcal{A}$ sends $\mathcal{B}$ two pairs, $(M_0, \mathsf{ID}_0^{\star} = (\mathsf{ID}_{0,1}^{\star}, \ldots, \mathsf{ID}_{0,j}^{\star}))$ and $(M_1, \mathsf{ID}_1^{\star} = (\mathsf{ID}_{1,1}^{\star}, \ldots, \mathsf{ID}_{1,j}^{\star}))$. $\mathcal{B}$ chooses random $\beta \in \{0, 1\}$ and random $z, z' \in \mathbb{Z}_N$ and computes the challenge ciphertext as follows:

$$C_0 = M_\beta \cdot T, \quad C_1 = (g_1^s B_2)^{a_1 \mathsf{ID}_{\beta,1}^{\star} + \cdots + a_j \mathsf{ID}_{\beta,j}^{\star} + b} \cdot Y_4^z, \quad C_2 = g_1^s B_2 \cdot Y_4^{z'}.$$

This implicitly sets $z_c = (a_1 \mathsf{ID}_{\beta,1}^{\star} + \cdots + a_j \mathsf{ID}_{\beta,j}^{\star} + b) \bmod p_2$. We note that $u_i = Y_1^{a_i \bmod p_1}$ and $X_1 = Y_1^{b \bmod p_1}$ are elements of $\mathbb{G}_{p_1}$, so when $a_1, \cdots, a_\ell$ and b are randomly chosen from $\mathbb{Z}_N$, their value modulo p_1 and modulo p_2 are random and independent.

We finish by observing that, if $T = \mathbf{e}(g, g)^{\alpha s}$, then the ciphertext constructed is a properly distributed semi-functional ciphertext with message M_β. If T instead is a random element of $\mathbb{G}_T$, then the ciphertext is a semi-functional ciphertext with a random message. $\qquad\square$

Indistinguishability of Game$_{\mathsf{Final}_0}$ and Game$_{\mathsf{Final}_1}$

Lemma 6. *Suppose that there exists a PPT algorithm $\mathcal{A}$ such that* $\mathsf{Adv}^{\mathcal{A}}_{\mathsf{Game}_{\mathsf{Final}_0}}$ − $\mathsf{Adv}^{\mathcal{A}}_{\mathsf{Game}_{\mathsf{Final}_1}} = \epsilon$. *Then there exists a PPT algorithm $\mathcal{B}$ with advantage ϵ in breaking Assumption 3.*

Proof. First, notice that if exists an adversary $\mathcal{A}'$ which distinguishes an encryption for an identity vector $\mathsf{ID}_0^\star$ from an encryption for an identity vector $\mathsf{ID}_1^\star$, where $\mathsf{ID}_0^\star$ and $\mathsf{ID}_1^\star$ are chosen by $\mathcal{A}'$, then there exists an adversary $\mathcal{A}$ which distinguishes an encryption for an identity $\mathsf{ID}^\star$ chosen by $\mathcal{A}$ from an encryption for a random identity vector. Hence, we suppose that we are simulating the games for a such adversary.

$\mathcal{B}$ receives $(\mathcal{I}, g_1, g_2, g_3, g_4, U, U^s A_{24}, U^{\hat{r}}, A_1 A_4, A_1^{\hat{r}} A_2, g_1^{\hat{r}} B_2, g_1^s B_{24})$ and T and simulates $\mathsf{Game}_{\mathsf{Final}_0}$ or $\mathsf{Game}_{\mathsf{Final}_1}$ with $\mathcal{A}$ depending on whether $T = A_1^s D_{24}$ or T is random in $\mathbb{G}_{p_1 p_2 p_4}$.

$\mathcal{B}$ sets the public parameters as follows. $\mathcal{B}$ chooses random exponents $\alpha, a_1, \ldots,$ $a_\ell \in \mathbb{Z}_N$ and sets $Y_1 = g_1, Y_3 = g_3, Y_4 = g_4, t = A_1 A_4, u_i = U^{a_i}$ for $i \in [\ell]$, and $\Omega = \mathbf{e}(Y_1, Y_1)^\alpha$. $\mathcal{B}$ sends the public parameters $\mathsf{Pk} = (N, Y_1, Y_2, Y_3, t, u_1, \ldots, u_\ell, \Omega)$ to $\mathcal{A}$.

Each time $\mathcal{B}$ is asked to provide a key for an identity $(\mathsf{ID}_1, \ldots, \mathsf{ID}_j)$, $\mathcal{B}$ creates a semi-functional key choosing random exponents $r_1', r_2' \in \mathbb{Z}_N$ and, for $\in \{1, 2\}$, random $z_{i,j+1}, \ldots, z_{i,\ell}, w_{i,1}, w_{i,2}, w_{i,j+1}, \ldots, w_{i,\ell} \in \mathbb{Z}_N$ and setting:

$$K_{1,1} = (g_1^{\hat{r}} B_2)^{r_1'} Y_3^{w_{1,1}}, \quad K_{1,2} = Y_1^\alpha \left(\left(U^{\hat{r}} \right)^{a_1 \mathsf{ID}_1 + \cdots + a_j \mathsf{ID}_j} \left(A_1^{\hat{r}} A_2 \right) \right)^{r_1'} Y_3^{w_{1,2}},$$

$$E_{1,j+1} = \left(U^{\hat{r}} \right)^{r_1' a_{j+1}} Y_2^{z_{1,j+1}} Y_3^{w_{1,j+1}}, \ldots, E_{1,\ell} = \left(U^{\hat{r}} \right)^{r_1' a_\ell} Y_2^{z_{1,\ell}} Y_3^{w_{1,\ell}}.$$

and

$$K_{2,1} = (g_1^{\hat{r}} B_2)^{r_2'} Y_3^{w_{2,1}}, \quad K_{2,2} = \left(\left(U^{\hat{r}} \right)^{a_1 \mathsf{ID}_1 + \cdots + a_j \mathsf{ID}_j} \left(A_1^{\hat{r}} A_2 \right) \right)^{r_2'} Y_3^{w_{2,2}},$$

$$E_{2,j+1} = \left(U^{\hat{r}} \right)^{r_2' a_{j+1}} Y_2^{z_{2,j+1}} Y_3^{w_{2,j+1}}, \ldots, E_{2,\ell} = \left(U^{\hat{r}} \right)^{r_2' a_\ell} Y_2^{z_{2,\ell}} Y_3^{w_{2,\ell}}.$$

This implicitly sets the randomness $r_1 = \hat{r} r_1'$ and $r_2 = \hat{r} r_2'$. At some point, $\mathcal{A}$ sends $\mathcal{B}$ two pairs, $(M_0, \mathsf{ID}^\star = (\mathsf{ID}_1^\star, \ldots, \mathsf{ID}_j^\star))$ and $(M_1, \mathsf{ID}^\star = (\mathsf{ID}_1^\star, \ldots, \mathsf{ID}_j^\star))$. $\mathcal{B}$ chooses random $C_0 \in \mathbb{G}_T$ and computes the challenge ciphertext as follows:

$$C_0, \quad C_1 = T \left(U^s A_{24} \right)^{a_1 \mathsf{ID}_1^\star + \cdots + a_j \mathsf{ID}_j^\star}, \quad C_2 = g_1^s B_{24}.$$

This implicitly sets x and z_c to random values.

If $T = A_1^s D_{24}$, then this is properly distributed semi-functional ciphertext with C_0 random and for identity vector $\mathsf{ID}^\star$. If T is a random element of $\mathbb{G}_{p_1 p_2 p_4}$, then this is a semi-functional ciphertext with C_0 random in $\mathbb{G}_T$ and C_1 and C_2 random in $\mathbb{G}_{p_1 p_2 p_4}$.

Hence, $\mathcal{B}$ can use the output of $\mathcal{A}$ to distinguish between these possibilities for T. $\qquad\square$

$\mathsf{Game}_{\mathsf{Final}_1}$ gives no advantage

Theorem 1. *If Assumptions $1, 2$ and 3 hold then our Anonymous HIBE scheme is secure.*

Proof. If the assumptions hold then we have proved by the previous lemmata that the real security game is indistinguishable from $\mathsf{Game}_{\mathsf{Final}_1}$, in which the value of β is information-theoretically hidden from the attacker. Hence the attacker can obtain no advantage in breaking the Anonymous HIBE scheme. □

4 Secret-Key Anonymous IBE

4.1 Secret Key Identity Based Encryption

A Secret-Key Identity Based Encryption scheme (IBE) is a tuple of four efficient and probabilistic algorithms: (Setup, Encrypt, KeyGen, Decrypt).

Setup(1^λ): Takes as input a security parameter λ and outputs the public parameters Pk and a master secret key Msk.

KeyGen(Msk, ID): Takes as input of the master secret key Msk, and an identity ID, and outputs a private key $\mathsf{Sk}_{\mathsf{ID}}$.

Encrypt(Msk, M, ID): Takes as input the master secret key Msk, a message M, and an identity ID and outputs a ciphertext Ct.

Decrypt(Ct, Sk): Takes as input a ciphertext Ct and a secret key Sk and outputs the message M, if the ciphertext was an encryption to an identity ID and the secret key is for the same identity.

4.2 Security Definitions

We present the security of an Anonymous IBE scheme in secret key model. In this model, we have two definition of security: *ciphertext security* and *key security*.

Ciphertext Security definition Security is defined through the following game, played by a challenger $\mathcal{C}$ and an adversary $\mathcal{A}$.

Setup. $\mathcal{C}$ runs the Setup algorithm to generate master secret key Msk which is kept secret.

Phase 1. $\mathcal{A}$ can make queries to the oracle Encrypt. To make a such query, $\mathcal{A}$ specifies a pair (M, ID) and receives an encryption of this pair computed using the Encrypt algorithm with Msk. $\mathcal{A}$ can make queries to the oracle KeyGen. To make a such query, $\mathcal{A}$ specifies an identity ID and receives a key of this identity computed using the KeyGen algorithm with Msk.

Challenge. $\mathcal{A}$ gives to $\mathcal{C}$ two pair message-identity (M_0, ID_0) and (M_1, ID_1). The identities must satisfy the property that no revealed identity in Phase 1 was either ID_0 or ID_1. $\mathcal{C}$ sets $\beta \in \{0, 1\}$ randomly and encrypts M_β under ID_β. $\mathcal{C}$ sends the ciphertext to the adversary.

Phase 2. This is the same as Phase 1 with the added restriction that any revealed identity must not be either ID_0 or ID_1.

Guess. $\mathcal{A}$ must output a guess β' for β. The advantage of $\mathcal{A}$ is defined to be $\mathrm{Prob}[\beta' = \beta] - \frac{1}{2}$.

Definition 2. *An Anonymous Identity Based Encryption scheme is ciphertext-secure if all polynomial time adversaries achieve at most a negligible (in λ) advantage in the previous security game.*

Key Security definition Security is defined through the following game, played by a challenger $\mathcal{C}$ and an attacker $\mathcal{A}$.

Setup. $\mathcal{C}$ runs the Setup algorithm to generate master secret key Msk which is kept secret.

Phase 1. $\mathcal{A}$ can make queries to the oracle Encrypt. To make a such query, $\mathcal{A}$ specifies a pair (M, ID) and receives an encryption of this pair computed using the Encrypt algorithm with the master secret key Msk. $\mathcal{A}$ can make queries to the oracle KeyGen. To make a such query, $\mathcal{A}$ specifies an identity ID and receives a key of this identity computed using the KeyGen algorithm with the master secret key Msk.

Challenge. $\mathcal{A}$ gives to $\mathcal{C}$ two identities ID_0 and ID_1. If in Phase 1 $\mathcal{A}$ did make a query (M, ID) to the oracle Encrypt such that ID was either ID_0 or ID_1, then the experiment fails. $\mathcal{C}$ sets $\beta \in \{0,1\}$ randomly and compute the secret key for ID_β. $\mathcal{C}$ sends the secret key to the adversary.

Phase 2. This is the same as Phase 1 with the added restriction that if $\mathcal{A}$ did make a query (M, ID) to the oracle Encrypt such that ID was either ID_0 or ID_1, then the experiment fails.

Guess. $\mathcal{A}$ must output a guess β' for β. The advantage $\mathcal{A}$ is defined to be $\text{Prob}[\beta' = \beta] - \frac{1}{2}$.

Definition 3. *A Secret-Key Anonymous Identity Based Encryption scheme is key-secure if all polynomial time adversaries achieve at most a negligible (in λ) advantage in the previous security game.*

Notice that no scheme with a deterministic KeyGen procedure can be key-secure.

4.3   Our Construction

In this section we describe our construction for a Secret-key Anonymous IBE scheme which is similar to its public key version from the previous sections.

Setup($1^\lambda, 1^\ell$): The setup algorithm chooses random description $\mathcal{I} = (N = p_1 p_2 p_3 p_4, \mathbb{G}, \mathbb{G}_T, \mathbf{e})$ and random $Y_1, X_1, u \in \mathbb{G}_{p_1}, Y_3 \in \mathbb{G}_{p_3}, X_4, Y_4 \in \mathbb{G}_{p_4}$ and $\alpha \in \mathbb{Z}_N$. The *fictitious* public parameters are:

$$\mathsf{Pk} = (N, Y_1, Y_3, Y_4, t = X_1 X_4, u, \Omega = \mathbf{e}(Y_1, Y_1)^\alpha).$$

The master secret key is $\mathsf{Msk} = (\mathsf{Pk}, X_1, \alpha)$.

KeyGen(Msk, ID): The key generation algorithm chooses random $r \in \mathbb{Z}_N$ and also random elements $R_1, R_2 \in \mathbb{G}_{p_3}$ The secret key $\mathsf{Sk}_{\mathsf{ID}} = (K_1, K_2)$ is computed as

$$K_1 = Y_1^r R_1, \quad K_2 = Y_1^\alpha (u^{\mathsf{ID}} X_1)^r R_2.$$

Encrypt(Msk, M, ID): The encryption algorithm chooses random $s \in \mathbb{Z}_N$ and random $Z, Z' \in \mathbb{G}_{p_4}$ The ciphertext (C_0, C_1, C_2) for the message $M \in \mathbb{G}_T$ is computed as

$$C_0 = M \cdot \mathbf{e}(Y_1, Y_1)^{\alpha s}, \quad C_1 = (u^{\mathsf{ID}} t)^s Z, \quad C_2 = Y_1^s Z'.$$

Decrypt(Msk, Ct, Sk): The decryption algorithm assumes that the key and ciphertext both correspond to the same identity ID. The decryption algorithm then computes the blinding factor similarly to the decryption procedure of the public-key version. Specifically,

$$\frac{\mathbf{e}(K_2, C_2)}{\mathbf{e}(K_1, C_1)} = \frac{\mathbf{e}(Y_1, Y_1)^{\alpha s}\mathbf{e}\left(u^{\mathsf{ID}}X_1, Y_1\right)^{rs}}{\mathbf{e}\left(Y_1, u^{\mathsf{ID}}X_1\right)^{rs}} = \mathbf{e}(Y_1, Y_1)^{\alpha s}.$$

4.4   Ciphertext Security

To prove ciphertext security of the Anonymous IBE scheme, we rely on the Assumptions $1, 2$ and 3 used in the proof of the public-key scheme.

We make the following considerations. If we instantiate the previous scheme as a public-key scheme by using the fictitious public-key parameter, it is identical to our public-key Anonymous IBE scheme (i.e., it is used in the non-hierarchical version). Thus, it is immediate to verify that from Assumptions $1, 2$ and 3 the security proof follows nearly identically. Generally, if a public-key IBE encryption scheme is semantically secure, its secret-key version is also semantically secure because we can simulate the encryption oracle by using the public-key. Therefore, we have the following theorem.

Theorem 2. *If Assumptions $1, 2$ and 3 hold, then our Secret-Key Anonymous IBE scheme is ciphertext-secure.*

4.5   Key Security

We will use *semi-functional ciphertexts* and *semi-functional keys* like defined previously. These will not be used in the real scheme, but we need them in our proofs. We include them for completeness.

Semi-functional Ciphertext. We let g_2 denote a generator of $\mathbb{G}_{p_2}$. A semi-functional ciphertext is created as follows: first, we use the encryption algorithm to form a normal ciphertext (C_0', C_1', C_2'). We choose random exponents $x, z_c \in \mathbb{Z}_N$. We set:

$$C_0 = C_0', \quad C_1 = C_1'g_2^{xz_c}, \quad C_2 = C_2'g_2^x.$$

Semi-functional Keys. To create a semi-functional key, we first create a normal key (K_1', K_2') using the key generation algorithm. We choose random exponents $\gamma, z_k \in \mathbb{Z}_N$. We set:

$$K_1 = K_1'g_2^{\gamma}, \quad K_2 = K_2'g_2^{\gamma z_k}.$$

We note that when a semi-functional key is used to decrypt a semi-functional ciphertext, the decryption algorithm will compute the blinding factor multiplied by the additional term $\mathbf{e}(g_2, g_2)^{x\gamma(z_k - z_c)}$. If $z_c = z_k$, decryption will still work. In this case, we say that the key is nominally semi-functional.

To prove the security of our scheme we rely on static Assumptions 1,2 and 3. For a PPT adversary $\mathcal{A}$ which makes q ciphertext queries, our proof of security will consist of the following $q + 3$ games between $\mathcal{A}$ and a challenger $\mathcal{C}$.

Game$_{\text{Real}}$: Is the real key security game.

Game$_{\text{Restricted}}$: Is the same as Game$_{\text{Real}}$ except that $\mathcal{A}$ cannot ask for keys for identities which are equal to one of the challenge identities modulo p_2. We will retain this restriction in all subsequent games.

Game$_k$: For k from 0 to q, Game$_k$ is like Game$_{\text{Restricted}}$, except that the key given to $\mathcal{A}$ is semi-functional and the first k ciphertexts are semi-functional. The rest of the ciphertexts are normal.

Game$_{\text{Final}}$: Is the same as Game$_q$, except that the challenge key is semi-functional with K_2 random in $\mathbb{G}_{p_1 p_2 p_4}$ (thus the key is independent from the identities provided by $\mathcal{A}$). It is clear that in this last game, no adversary can have advantage greater than 0.

We will show these games are indistinguishable in the following lemmata.

Indistinguishability of Game$_{\text{Real}}$ and Game$_{\text{Restricted}}$

Lemma 7. *Suppose that there exists a PPT algorithm $\mathcal{A}$ such that* $\text{Adv}^{\mathcal{A}}_{\text{Game}_{\text{Real}}} - \text{Adv}^{\mathcal{A}}_{\text{Game}_{\text{Restricted}}} = \epsilon$. *Then there exists a PPT algorithm $\mathcal{B}$ with advantage $\geq \frac{\epsilon}{3}$ in breaking Assumption 1.*

Proof. The proof is identical to that given in lemma 2. $\qquad\square$

Indistinguishability of Game$_{\text{Restricted}}$ and Game$_0$

Lemma 8. *Suppose that there exists a PPT algorithm $\mathcal{A}$ such that* $\text{Adv}^{\mathcal{A}}_{\text{Game}_{\text{Restricted}}} - \text{Adv}^{\mathcal{A}}_{\text{Game}_0} = \epsilon$. *Then there exists a PPT algorithm $\mathcal{B}$ with advantage ϵ in breaking Assumption 1.*

Proof. $\mathcal{B}$ receives $(\mathcal{I}, g_1, g_3, g_4, A_1 A_2, B_2 B_3)$ and T and simulates Game$_{\text{Restricted}}$ or Game$_0$ with $\mathcal{A}$ depending on whether $T \in \mathbb{G}_{p_1 p_3}$ or $T \in \mathbb{G}_{p_1 p_2 p_3}$.

$\mathcal{B}$ sets the fictitious public parameters as follows. $\mathcal{B}$ chooses random exponents $\alpha, a, b, c \in \mathbb{Z}_N$ and sets $Y_1 = g_1, Y_3 = g_3, Y_4 = g_4$ $X_4 = Y_4^c, X_1 = Y_1^b$ and $u = Y_1^a$. $\mathcal{B}$ uses $\text{Pk} = (N, Y_1, Y_3, Y_4, t = X_1 X_4, u, \Omega = \mathbf{e}(Y_1, Y_1)^{\alpha})$ to respond to the ciphertext queries issued by $\mathcal{A}$. Notice that $\mathcal{B}$ knows also the master secret key $\text{Msk} = (\text{Pk}, X_1, \alpha)$ and thus can simulate all $\mathcal{A}$'s key queries.

At some point, $\mathcal{A}$ sends $\mathcal{B}$ two identities, $\text{ID}_0^{\star}$ and $\text{ID}_1^{\star}$. $\mathcal{B}$ chooses random $\beta \in \{0, 1\}$ and computes the challenge key as follows:

$$K_1 = T, \quad K_2 = Y_1^{\alpha} T^{a\text{ID}_{\beta}^{\star} + b}.$$

We complete the proof with the following two observations. If $T \in \mathbb{G}_{p_1 p_3}$, then T can be written as $Y_1^{s_1} Y_3^{s_3}$. In this case (K_1, K_2) is a normal key with randomness $r = s_1, R_1 = Y_3^{s_3}, R_2 = (Y_3^{s_3})^{a\text{ID}_{\beta}^{\star} + b}$. If $T \in \mathbb{G}_{p_1 p_2 p_3}$, then T can be written as $Y_1^{s_1} g_2^{s_2} Y_3^{s_3}$ and this case (K_1, K_2) is a semi-functional key with randomness $r = s_1, R_1 = Y_3^{s_3}, R_2 = (Y_3^{s_3})^{a\text{ID}_{\beta}^{\star} + b}, \gamma = s_2$ and $z_c = a\text{ID}_{\beta}^{\star} + b$. Thus, in the former case we have properly simulated Game$_{\text{Restricted}}$, and in the latter case we have simulated Game$_0$. $\qquad\square$

Indistinguishability of Game$_{k-1}$ and Game$_k$

Lemma 9. *Suppose there exists a PPT algorithm $\mathcal{A}$ such that $\mathsf{Adv}^{\mathcal{A}}_{\mathsf{Game}_{k-1}} - \mathsf{Adv}^{\mathcal{A}}_{\mathsf{Game}_k} = \epsilon$. Then, there exists a PPT algorithm $\mathcal{B}$ with advantage ϵ in breaking Assumption 1.*

Proof. $\mathcal{B}$ receives $(\mathcal{I}, g_1, g_3, g_4, A_1 A_2, B_2 B_3)$ and T and simulates Game$_{k-1}$ or Game$_k$ with $\mathcal{A}$ depending on whether $T \in \mathbb{G}_{p_1 p_3}$ or $T \in \mathbb{G}_{p_1 p_2 p_3}$.

$\mathcal{B}$ sets the fictitious public parameters by choosing random exponents $\alpha, a, b, c \in \mathbb{Z}_N$ and setting $Y_1 = g_1, Y_3 = g_4, Y_4 = g_3, X_4 = Y_4^c, X_1 = Y_1^b$ and $u = Y_1^a$. Notice that $\mathcal{B}$ knows the master secret key $\mathsf{Msk} = (\mathsf{Pk}, X_1, \alpha)$ with $\mathsf{Pk} = (N, Y_1, Y_3, Y_4, t = X_1 X_4, u, \Omega = \mathbf{e}(Y_1, Y_1)^\alpha)$ and thus can respond to all $\mathcal{A}$'s key queries. Let us now explain how $\mathcal{B}$ answers the i-th ciphertext query for pair (M, ID).

For $i < k$, $\mathcal{B}$ creates a semi-functional ciphertext by choosing random exponents $s, w_1, w_2 \in \mathbb{Z}_N$ and setting:

$$C_0 = M\mathbf{e}(Y_1, Y_1)^{\alpha s}, \quad C_1 = (u^{\mathsf{ID}} X_1)^s (B_2 B_3)^{w_1}, \quad C_2 = Y_1^s Y_4^{w_2}$$

By writing B_2 as g_2^ϕ, we have that this is a properly distributed semi-functional ciphertext with $x = \phi$ and $z_c = w_1$.

For $i > k$, $\mathcal{B}$ runs the Encrypt algorithm using the master secret key $\mathsf{Msk} = (\mathsf{Pk}, X_1, \alpha)$.

To answer the k-th ciphertext query for (M_k, ID_k), $\mathcal{B}$ sets $z_c = a\mathsf{ID}_k + b$, chooses random exponent $w_1, w_2 \in \mathbb{Z}_N$, and sets:

$$C_0 = M_k \mathbf{e}(T, Y_1)^\alpha, \quad C_1 = T^{z_c} Y_4^{w_1}, \quad C_2 = T Y_4^{w_2}$$

We have the following two observations. If $T \in \mathbb{G}_{p_1 p_3}$, then T can be written as $Y_1^{r_1} Y_4^{r_4}$ In this case this is a properly distributed normal ciphertext with $s = r_1$. If $T \in \mathbb{G}_{p_1 p_2 p_3}$, then T can be written as $Y_1^{r_1} g_2^{r_2} Y_4^{r_4}$ and in this case it is a properly distributed semi-functional ciphertext with $x = r_2$.

At some point, $\mathcal{A}$ sends $\mathcal{B}$ two identities, $\mathsf{ID}_0^\star$ and $\mathsf{ID}_1^\star$. $\mathcal{B}$ chooses random $\beta \in \{0, 1\}$ and random $z, z' \in \mathbb{Z}_N$ and computes the challenge key as follows:

$$K_1 = (A_1 A_2) Y_3^z, \quad K_2 = Y_1^\alpha (A_1 A_2)^{a\mathsf{ID}_\beta^\star + b} Y_3^{z'}$$

This implicitly sets $Y_1^r = A_1$ and $z_k = a\mathsf{ID}_\beta^\star + b \mod p_2$. Since ID_k is not equal to $\mathsf{ID}_\beta^\star$ modulo p_2, we have that z_k and z_c are independent and randomly distributed.

We can thus conclude that, if $T \in \mathbb{G}_{p_1 p_3}$ then $\mathcal{B}$ has properly simulated Game$_{k-1}$. If $T \in \mathbb{G}_{p_1 p_2 p_3}$, then $\mathcal{B}$ has properly simulated Game$_k$. $\qquad\square$

Indistinguishability of Game$_q$ and Game$_{\mathsf{Final}}$

Lemma 10. *Suppose that there exists a PPT algorithm $\mathcal{A}$ such that $\mathsf{Adv}^{\mathcal{A}}_{\mathsf{Game}_q} - \mathsf{Adv}^{\mathcal{A}}_{\mathsf{Game}_{\mathsf{Final}}} = \epsilon$. Then there exists a PPT algorithm $\mathcal{B}$ with advantage ϵ in breaking Assumption 3.*

Proof. First, notice that if exists an adversary $\mathcal{A}'$ which distinguishes an encryption for an identity $\mathsf{ID}_0^\star$ from an encryption for an identity $\mathsf{ID}_1^\star$, where $\mathsf{ID}_0^\star$ and $\mathsf{ID}_1^\star$ are chosen by $\mathcal{A}'$, then there exists an adversary $\mathcal{A}$ which distinguishes an encryption for an identity $\mathsf{ID}^\star$ chosen by $\mathcal{A}$ from an encryption for a random identity. Hence, we suppose that we are simulating the games for a such adversary.

$\mathcal{B}$ receives $(\mathcal{I}, g_1, g_2, g_3, g_4, U, U^s A_{24}, U^{\hat{r}}, A_1 A_4, A_1^{\hat{r}} A_2, g_1^{\hat{r}} B_2, g_1^s B_{24})$ and T and simulates Game_q or $\mathsf{Game}_{\mathsf{Final}}$ with $\mathcal{A}$ depending on whether $T = A_1^s D_{24}$ or T is random in $\mathbb{G}_{p_1 p_2 p_4}$.

$\mathcal{B}$ chooses random exponents $\alpha \in \mathbb{Z}_N$ and sets $Y_1 = g_1, Y_3 = g_4, Y_4 = g_3$.

Each time $\mathcal{B}$ is asked to provide a ciphertext for an identity ID, $\mathcal{B}$ creates a semi-functional ciphertext choosing random exponents $r, w_1, w_2 \in \mathbb{Z}_N$ and sets

$$C_0 = M \cdot \mathbf{e}(g_1^{\hat{r}} B_2, Y_1)^{\alpha s}, \ C_1 = (A_1^{\hat{r}} A_2)^{r \mathsf{ID}} (U^{\hat{r}})^r Y_4^{w_1}, \ C_2 = (g_1^{\hat{r}} B_2)^r Y_4^{w_2}$$

This implicitly sets the randomness of the ciphertext to $\hat{r}r$, $u = A_1$ and $X_1 = U$.

Each time $\mathcal{B}$ is asked to provide a key for an identity ID, $\mathcal{B}$ creates a semi-functional key choosing random exponents $r, w_1, w_2 \in \mathbb{Z}_N$ and setting:

$$K_1 = (g_1^{\hat{r}} B_2)^r Y_3^{w_1}, \quad K_2 = Y_1^\alpha (A_1^{\hat{r}} A_2)^{r \mathsf{ID}} (U^{\hat{r}})^r Y_3^{w_2}.$$

This implicitly sets the randomness of the secret key to $\hat{r}r$.

At some point, $\mathcal{A}$ sends $\mathcal{B}$ two identities, $\mathsf{ID}_0^\star$ and $\mathsf{ID}_1^\star$. $\mathcal{B}$ chooses random $w_1, w_2 \in \mathbb{Z}_N$ and computes the challenge secret key as follows:

$$K_1 = (g_1^s B_{24}) Y_3^{w_1}, \quad K_2 = Y_1^\alpha T^{\mathsf{ID}_\beta^\star} (U^s A_{24}) Y_3^{w_2}.$$

This implicitly sets γ and z_k to random values.

If $T = A_1^s D_{24}$, then this is properly distributed semi-functional key for identity $\mathsf{ID}_\beta^\star$. If T is a random element of $\mathbb{G}_{p_1 p_2 p_4}$, then this is a semi-functional key with K_2 random in $\mathbb{G}_{p_1 p_2 p_4}$.

Hence, $\mathcal{B}$ can use the output of $\mathcal{A}$ to distinguish between these possibilities for T. □

$\mathsf{Game}_{\mathsf{Final}}$ gives no advantage

Theorem 3. *If Assumptions 1, 2 and 3 hold then our Anonymous IBE scheme is both ciphertext and key secure.*

Proof. If the assumptions hold then we have proved by the previous lemmata that the real security game is indistinguishable from $\mathsf{Game}_{\mathsf{Final}}$, in which the value of β is information-theoretically hidden from the attacker. Hence the attacker can obtain no non-negligible advantage in breaking the key security of the Secret-key Anonymous IBE scheme. We have showed previously that it is also ciphertext-secure. □

5   Conclusions and Open Problems

We constructed the first Fully Secure Anonymous HIBE system with short ciphertexts in the public key model and the first fully secure IBE in the secret key model and proved their security in the standard model from simple and non-interactive assumptions generically secure. A drawback of our construction is that it uses bilinear groups of composite order. An open problem is to build such a scheme in symmetric bilinear groups of prime order. The general technique of Freeman [9] does not seem to apply to our scheme.

We also stress that our decryption algorithm works if the key and the ciphertext correspond to the same identity. It would be interesting to construct an anonymous HIBE in which the decryption algorithm works provided that the identity of the key is a prefix of the identity of the ciphertext.

To the best of our knowledge, Secret-Key Hierarchical IBE has not been studied before and we defer it to future work.

Acknowledgements

The work of the authors has been supported in part by the European Commission through the EU ICT program under Contract ICT-2007-216646 ECRYPT II and in part by the Italian Ministry of University and Research Project PRIN 2008 *PEPPER: Privacy and Protection of Personal Data* (prot. 2008SY2PH4).

References

1. Boneh, D., Di Crescenzo, G., Ostrovsky, R., Persiano, G.: Public key encryption with keyword search. In: Cachin, C., Camenisch, J. (eds.) EUROCRYPT 2004. LNCS, vol. 3027, pp. 506–522. Springer, Heidelberg (2004)
2. Boneh, D., Franklin, M.K.: Identity based encryption from the Weil pairing. SIAM Journal on Computing 32(3), 586–615 (2003)
3. Boneh, D., Goh, E.-J., Nissim, K.: Evaluating 2-DNF formulas on ciphertexts. In: Kilian, J. (ed.) TCC 2005. LNCS, vol. 3378, pp. 325–341. Springer, Heidelberg (2005)
4. Boyen, X.: Multipurpose identity-based signcryption (a swiss army knife for identity-based cryptography). In: Boneh, D. (ed.) CRYPTO 2003. LNCS, vol. 2729, pp. 383–399. Springer, Heidelberg (2003)
5. Boyen, X., Waters, B.: Anonymous Hierarchical Identity-Based Encryption (Without Random Oracles). In: Dwork, C. (ed.) CRYPTO 2006. LNCS, vol. 4117, pp. 290–307. Springer, Heidelberg (2006)
6. De Caro, A., Iovino, V., Persiano, G.: Fully Secure Anonymous HIBE and Secret-Key Anonymous IBE with Short Ciphertexts. Cryptology ePrint Archive, Report 2010/197 (2010), http://eprint.iacr.org/
7. Cash, D., Hofheinz, D., Kiltz, E., Peikert, C.: Bonsai trees, or how to delegate a lattice basis. In: Gilbert, H. (ed.) EUROCRYPT 2010, pp. 523–552 (2010)
8. Cocks, C.: An identity based encryption scheme based on quadratic residues. In: Honary, B. (ed.) Cryptography and Coding 2001. LNCS, vol. 2260, pp. 360–363. Springer, Heidelberg (2001)

9. Freeman, D.M.: Converting pairing-based cryptosystems from composite-order groups to prime-order groups. In: Gilbert, H. (ed.) EUROCRYPT 2010. LNCS, vol. 6110, pp. 44–61. Springer, Heidelberg (2010)
10. Horwitz, J., Lynn, B.: Toward hierarchical identity-based encryption. In: Knudsen, L.R. (ed.) EUROCRYPT 2002. LNCS, vol. 2332, pp. 466–481. Springer, Heidelberg (2002)
11. Lewko, A., Okamoto, T., Sahai, A., Takashima, K., Waters, B.: Fully secure functional encryption: Attribute-based encryption and (hierarchical) inner product encryption. In: Gilbert, H. (ed.) EUROCRYPT 2010. LNCS, vol. 6110, pp. 62–91. Springer, Heidelberg (2010)
12. Lewko, A.B., Waters, B.: New techniques for dual system encryption and fully secure hibe with short ciphertexts. In: Micciancio, D. (ed.) TCC 2010. LNCS, vol. 5978, pp. 455–479. Springer, Heidelberg (2010)
13. Seo, J.H., Kobayashi, T., Ohkubo, M., Suzuki, K.: Anonymous hierarchical identity-based encryption with constant size ciphertexts. In: Jarecki, S., Tsudik, G. (eds.) PKC 2009. LNCS, vol. 5443, pp. 215–234. Springer, Heidelberg (2009)
14. Shamir, A.: Identity-based cryptosystems and signature schemes. In: Blakely, G.R., Chaum, D. (eds.) CRYPTO 1984. LNCS, vol. 196, pp. 47–53. Springer, Heidelberg (1985)
15. Shen, E., Shi, E., Waters, B.: Predicate privacy in encryption systems. In: Reingold, O. (ed.) TCC 2009. LNCS, vol. 5444, pp. 457–473. Springer, Heidelberg (2009)
16. Shi, E., Waters, B.: Delegating capabilities in predicate encryption systems. In: Aceto, L., Damgård, I., Goldberg, L.A., Halldórsson, M.M., Ingólfsdóttir, A., Walukiewicz, I. (eds.) ICALP 2008. LNCS, vol. 5126, pp. 560–578. Springer, Heidelberg (2008)
17. Waters, B.: Dual system encryption: Realizing fully secure IBE and HIBE under simple assumptions. In: Halevi, S. (ed.) CRYPTO 2009. LNCS, vol. 5677, pp. 619–636. Springer, Heidelberg (2009)

Chosen-Ciphertext Secure Identity-Based Encryption from Computational Bilinear Diffie-Hellman

David Galindo

University of Luxembourg
david.galindo@uni.lu

Abstract. We extend a technique by Hanaoka and Kurosawa that provides efficient chosen-ciphertext secure public key encryption based on the Computational Diffie-Hellman assumption to the identity-based encryption setting. Our main result is an efficient chosen-ciphertext secure identity-based encryption scheme with constant-size ciphertexts under the Computational Bilinear Diffie-Hellman assumption in the standard model.

Keywords: standard model, identity-based encryption, computational bilinear Diffie-Hellman assumption, hardcore bits.

1   Introduction

Designing efficient public key encryption schemes with chosen-ciphertext security under widely accepted hardness assumptions has been a challenging research direction in modern cryptography. The first breakthrough in this area was the scheme by Cramer and Shoup [5], which had security based on the Decisional Diffie-Hellman assumption in the standard model. It is not until very recently that schemes with similar properties based on the Computational Diffie-Hellman assumption have been proposed by Cash, Kiltz and Shoup [4], Hanaoka and Kurosawa [9] and Haralambiev, Jager, Kiltz and Shoup [10].

Identity-based encryption (IBE) [11,3] provides a public key encryption mechanism where public keys are arbitrary strings id such as an email address or any other distinguished user identifier. In this work we extend the technique by Hanaoka and Kurosawa to the identity-based setting and provide an efficient chosen-ciphertext secure identity-based key encapsulation mechanism with constant-size ciphertexts under the Computational Bilinear Diffie-Hellman assumption in the standard model.

2   Preliminaries

We introduce some basic notation. If S is a set then $s_1, \ldots, s_n \xleftarrow{\$} S$ denotes the operation of picking n elements s_i of S independently and uniformly at random.

M. Joye, A. Miyaji, and A. Otsuka (Eds.): Pairing 2010, LNCS 6487, pp. 367–376, 2010.

We write $\mathcal{A}(x, y, \ldots)$ to indicate that $\mathcal{A}$ is an algorithm with inputs $x, y, \ldots$ and by $z \leftarrow \mathcal{A}(x, y, \ldots)$ we denote the operation of running $\mathcal{A}$ with inputs $(x, y, \ldots)$ and letting z be the output. Throughout this paper we use the term "algorithm" as equivalent to "probabilistic polynomial-time algorithm". If st_1, st_2 are strings, then $st_1 \| st_2$ denotes the concatenation.

2.1 Identity-Based Encryption and Identity-Based Key Encapsulation

A IBE scheme Π for identities $id \in \mathbb{Z}_p^*$ is specified by four algorithms (Setup, KeyGen, Encrypt, Decrypt) [3]:

- Setup is a randomized algorithm which takes as input security parameter 1^k and returns a master public key PK and a master secret key SK. The master public key includes the description of a set of admissible messages and ciphertexts $\mathcal{M}_{PK}, \mathcal{C}_{PK}$ respectively, and a prime integer p. SK is kept secret by the trusted authority, while PK is publicly available and we consider it to be an implicit input to the rest of the algorithms.
- KeyGen takes as input SK and an identity $id \in \mathbb{Z}_p^*$. It outputs a user secret key $sk[id]$.
- Encrypt takes as input an identity $id \in \mathbb{Z}_p^*$ and $M \in \mathcal{M}_{PK}$. It returns a ciphertext C.
- Decrypt takes as inputs a private key $sk[id]$ and a ciphertext C, and it returns $M \in \mathcal{M}_{PK}$ or the special symbol $\bot$ indicating a decryption failure. In particular $\bot$ is returned if $C \notin \mathcal{C}_{PK}$.

These algorithms must satisfy a natural consistency constraint, namely that for any security parameter 1^k, identity $id \in \mathbb{Z}_p^*$ and message $M \in \mathcal{M}_{PK}$ it holds that $M \leftarrow \mathsf{Decrypt}(sk[id], \mathsf{Encrypt}(id, M))$ where $sk[id] \leftarrow \mathsf{KeyGen}(SK, id)$ and $(PK, SK) \leftarrow \mathsf{Setup}(1^k)$.

An IBE scheme can be obtained by combining an identity-based key encapsulation mechanism (IB-KEM) and a symmetric encryption scheme [6,1]. The IB-KEM is run to produce a symmetric encryption key that is later used to encrypt a message with the given symmetric encryption scheme. Formally, an IB-KEM for identities $id \in \mathbb{Z}_p^*$ is specified by four algorithms (KEM.Setup, KEM.KeyGen, KEM.Encap, KEM.Decap):

- KEM.Setup is a randomized algorithm that takes as inputs a security parameter 1^k and a positive integer κ. It works almost exactly as the Setup algorithm of an IBE scheme, except that no set of admissible plaintexts is output. The integer κ denotes the bit-length of the symmetric encryption keys output by the IB-KEM and is returned as part of PK.
- KEM.KeyGen takes as input SK and an identity $id \in \mathbb{Z}_p^*$. It outputs a user secret key $sk[id]$.
- KEM.Encap takes as input an identity $id \in \mathbb{Z}_p^*$ and outputs a symmetric key $K \in \{0, 1\}^\kappa$ and a ciphertext C.

– KEM.Decap takes as inputs a private key $sk[id]$ and a ciphertext C, and it returns K or the special symbol $\perp$ indicating a decryption failure. In particular $\perp$ is returned if $C \notin \mathcal{C}_{PK}$.

Similarly to IBE, any IB-KEM must satisfy natural consistency constraints, namely that for any security parameter 1^k, integer κ and identity $id \in \mathbb{Z}_p^*$ it holds that $K \leftarrow$ KEM.Decap$(sk[id], C)$ where

$$(K, C) \leftarrow \text{KEM.Encap}(id, M), \quad sk[id] \leftarrow \text{KEM.KeyGen}(SK, id)$$

and $(PK, SK) \leftarrow$ KEM.Setup$(1^k, \kappa)$.

Selective-identity chosen-ciphertext security. [3,1] Let us consider the following game:

Initialization. The adversary outputs a security parameter 1^k, a positive integer κ and an identity id^* it wants to attack. κ must be polynomially-bounded in k.

Setup. The challenger runs $(PK, SK) \leftarrow$ KEM.Setup$(1^k, \kappa)$. The challenger sets $(K^*, C^*) \leftarrow$ KEM.Encap(PK, id^*). It picks $\beta \xleftarrow{\$} \{0, 1\}$ and sends C^* to the adversary, together with K^* if $\beta = 1$ or a fresh key $K^\dagger \xleftarrow{\$} \{0, 1\}^\kappa$ if $\beta = 0$. It gives PK to the adversary and keeps SK to itself.

Find. The adversary makes a polynomial number of queries of the following types:
  – *User secret key.* The adversary asks the challenger to run and deliver $sk[id] \leftarrow$ KEM.KeyGen(SK, id) for adversarial input $id \neq id^*$.
  – *Decryption query.* The adversary asks the challenger to output the result of KEM.Decap$(sk[id], C)$ for adversarial input $(id, C) \neq (id^*, C^*)$.

Guess. The adversary outputs a guess $\beta' \in \{0, 1\}$.

The advantage of such an adversary $\mathcal{A}$ is defined as

$$\text{Adv}_{\text{IBKEM}, \mathcal{A}}^{\text{sID–CCA}}(1^k) = \left| \Pr[\mathcal{A}(K^*, C^*) = 1] - \Pr[\mathcal{A}(K^\dagger, C^*) = 1] \right|.$$

Definition 1. *An identity-based key encapsulation mechanism* IBKEM *is secure under selective-identity and chosen-ciphertext attacks if for any* IND-sID-CCA *adversary $\mathcal{A}$ the function* $\text{Adv}_{\text{IBKEM}, \mathcal{A}}^{\text{IND–sID–CCA}}(1^k)$ *is negligible.*

2.2 Diffie-Hellman Assumptions on Pairing Groups

Let $\mathbb{G}_1 = \langle g_1 \rangle$ and $\mathbb{G}_2 = \langle g_2 \rangle$ be (cyclic) groups of order p prime. A map $e : \mathbb{G}_1 \times \mathbb{G}_2 \to \mathbb{G}_3$ to a group $\mathbb{G}_3$ is called a *bilinear* map, if it satisfies the following two properties:

Bilinearity: $e(g_1^a, g_2^b) = e(g_1, g_2)^{ab}$ for all integers a, b
Non-Degenerate: $e(g_1, g_2)$ has order p in $\mathbb{G}_3$.

We assume there exists an efficient bilinear pairing instance generator algorithm $\mathcal{IG}$ that on input a security parameter 1^k outputs the description of $\langle e, \mathbb{G}_1, \mathbb{G}_2, \mathbb{G}_3, g_1, g_2, p \rangle$, with p a k-bit length prime.

Definition 2 (BDH assumption). *Let* $\langle\, e, \mathbb{G}_1, \mathbb{G}_2, \mathbb{G}_3, p, g_1, g_2\,\rangle \leftarrow \mathcal{IG}(1^k)$. *Let us define*

$$\mathbf{Z} \leftarrow \left(\, e, \mathbb{G}_1, \mathbb{G}_2, \mathbb{G}_3, p, g_1, g_2, g_1^a, g_2^a, g_1^b, g_2^b, g_1^c\,\right)$$

where $a, b, c \xleftarrow{\$} \mathbb{Z}_p$. *We say that* $\mathcal{IG}$ *satisfies the* Computational Bilinear Diffie-Hellman assumption *if*

$$\mathsf{Adv}_{\mathcal{IG},\mathcal{A}}^{\mathrm{BDH}}(k) := \Pr[\mathcal{A}(\mathbf{Z}) = e(g_1, g_2)^{abc}]$$

is negligible in k. *The probabilities are computed over the internal random coins of* $\mathcal{A}$, $\mathcal{IG}$ *and the random coins of the inputs.*

Our definition of bilinear pairings and BDH assumption encompasses all pairing-type categories arising from elliptic curves as classified by Galbraith, Paterson and Smart [8], and hence it is as general as possible.

Definition 3 (BDH hardcore predicate). *Let* $\langle\, e, \mathbb{G}_1, \mathbb{G}_2, \mathbb{G}_3, p, g_1, g_2\,\rangle \leftarrow \mathcal{IG}(1^k)$. *Let us define*

$$\mathbf{Z} \leftarrow \left(\, e, \mathbb{G}_1, \mathbb{G}_2, \mathbb{G}_3, p, g_1, g_2, g_1^a, g_2^a, g_1^b, g_2^b, g_1^c\,\right)$$

where $a, b, c \xleftarrow{\$} \mathbb{Z}_p$. *Let* $h : \mathbb{G}_3 \to \{0, 1\}$ *be a function and consider*

$$\mathsf{Adv}_{\mathcal{IG},\mathcal{A}}^{h}(k) := \left|\Pr\left[\mathcal{A}\left(\,\mathbf{Z}, h, h(e(g_1, g_2)^{abc})\,\right) = 1\right] - \Pr\left[\mathcal{A}\left(\mathbf{Z}, h, \beta\right) = 1\right]\right|$$

where $\beta \xleftarrow{\$} \{0, 1\}$. *We say that* h *is a* BDH hardcore predicate *if the the BDH assumption for* $\mathcal{IG}$ *implies that* $\mathsf{Adv}_{\mathcal{IG},\mathcal{A}}^{h}(k)$ *is negligible. The probabilities are computed over the internal random coins of* $\mathcal{A}$, $\mathcal{IG}$ *and the random coins of the inputs.*

2.3 Lagrange Interpolation

Let $f(x) = \sum_{0 \leq l \leq t} b_l x^l$ be a polynomial over $\mathbb{Z}_p$ with degree t and $\big(\,(x_0, f(x_0)),$ $\ldots, (x_t, f(x_t))\big)$ be $t+1$ distinct points where $f(x)$ has been evaluated over $\mathbb{Z}_p^*$. Then one can recover $f(x)$ as $f(x) = f(x_0)\lambda_{x_0}(x) + \ldots + f(x_0)\lambda_{x_0}(x)$, where $\lambda_{x_j}(x) \in \mathbb{Z}_p[x]$ for $0 \leq l \leq t$ are called *Lagrange coefficients* and are defined as

$$\lambda_{x_l}(x) = \frac{(x - x_0)(x - x_1)\cdots(x - x_{l-1})(x - x_{l+1})\cdots(x - x_t)}{(x_l - x_0)(x_l - x_1)\cdots(x_l - x_{l-1})(x_l - x_{l+1})\cdots(x_j - x_t)}.$$

It can be seen that given $\big(g_1, (g_1^{x_0}, g_1^{f(x_0)}), \ldots, (g_1^{x_t}, g_1^{f(x_t)})\big)$ it is possible to compute any $g_1^{b_l}$ for $0 \leq l \leq t$ thanks to the Lagrange coefficients, where $\mathbb{G}_1 = \langle g_1 \rangle$ has prime order p. Similarly, given

$$\big(g_1, g_1^{b_0}, \ldots, g_1^{b_{j-1}}, (x_0, f(x_0)), \ldots, (x_{t-j}, f(x_{t-j}))\big)$$

it is possible to reconstruct any $g_1^{b_l}$ for $j \leq l \leq t$. These facts are used in our scheme and in its security reduction.

3 Chosen-Ciphertext Secure IB-KEM from the BDH Assumption in the Standard Model

In this section we describe a new IB-KEM which is obtained by extending the techniques that Hanaoka and Kurosawa [9] applied to ElGamal encryption scheme [7]. The new IB-KEM is the result of applying these extended techniques to Boneh and Boyen's identity-based encryption scheme [2] and it has security based on the Computational BDH assumption. We assume the existence of global pairing parameters $\langle\, e, \mathbb{G}_1, \mathbb{G}_2, \mathbb{G}_3, p, g_1, g_2 \,\rangle \leftarrow \mathcal{IG}(1^k)$ known to all the parties. Our IB-KEM is defined as follows:

- KEM.Setup($1^k, \kappa$) chooses $a, \gamma \xleftarrow{\$} \mathbb{Z}_p^*$ and sets $u_0 \leftarrow g_1^a, v_0 \leftarrow g_2^a, u_{-1} \leftarrow g_1^\gamma, v_{-1} \leftarrow g_2^\gamma$. Next, it randomly picks $b_0, b_1, \ldots, b_{\kappa+2} \xleftarrow{\$} \mathbb{Z}_p^*$ and defines the polynomial $f(x) = b_0 + b_1 x^1 + \ldots + b_{\kappa+2} x^{\kappa+2} \in \mathbb{Z}_p[x]$. It computes $y_l = g_1^{b_l}$, $Y_l = e(u_0, g_2^{b_l})$ for $l = 0, \ldots, \kappa + 2$. It chooses a target collision resistant hash function $\mathsf{TCR} : \mathbb{G}_1 \times \{0,1\} \to \mathbb{Z}_p^*$, as well as a BDH hardcore predicate $h : \mathbb{G}_3 \to \{0,1\}$. It defines the functions $H_1 : \mathbb{Z}_p^* \to \mathbb{G}_1$ that maps $id \mapsto u_0^{id} u_{-1}$ and $H_2 : \mathbb{Z}_p^* \to \mathbb{G}_2$ that maps $id \mapsto v_0^{id} v_{-1}$. Finally, let $\mathcal{C}_{PK}$ be $\mathbb{G}_1^4$ and

$$PK \leftarrow \langle\, e, \mathbb{G}_1, \mathbb{G}_2, \mathbb{G}_3, p, g_1, g_2, u_0, v_0, u_{-1}, v_{-1}, y_0, \ldots, y_{\kappa+2},$$
$$Y_0, \ldots, Y_{\kappa-1}, H_1 \,\rangle \text{ and}$$
$$SK \leftarrow \langle\, a, b_0, \ldots, b_{\kappa+2}, H_2 \,\rangle.$$

- KEM.KeyGen(SK, id) outputs $(sk_0[id], \ldots, sk_{\kappa-1}[id])$, where

$$sk_l[id] \leftarrow \left(g_2^{ab_l} H_2(id)^{r_l}, g_2^{r_l}\right) \in \mathbb{G}_2^2 \text{ and } r_l \xleftarrow{\$} \mathbb{Z}_p^* \text{ for } 0 \le l \le \kappa - 1$$

- KEM.Encap(PK, id) computes

$$C \leftarrow \left(g_1^r, g_1^{r \cdot f(\underline{t})}, g_1^{r \cdot f(\overline{t})}, H_1(id)^r\right) \in \mathbb{G}_1^4,$$

$$K \leftarrow h(Y_0^r)\|\ldots\|h(Y_{\kappa-1}^r) \in \{0,1\}^\kappa,$$

where $\underline{t} = \mathsf{TCR}(g_1^r, 0)$ and $\overline{t} = \mathsf{TCR}(g_1^r, 1)$. It outputs (K, C).

- KEM.Decap takes as inputs a user key $sk[id] = (sk_0[id], \ldots, sk_{\kappa-1}[id])$, and a ciphertext $C = (C_0, C_1, C_2, C_3)$. It first checks if $e(C_0, g_1^{f(\underline{t})}) = e(g_1, C_1)$ and $e(C_0, g_1^{f(\overline{t})}) = e(g_1, C_2)$. If not it returns $\perp$. Otherwise it parses $sk_l[id]$ as (A_l, B_l) for $l = 0, \ldots, \kappa - 1$ and it returns

$$K \leftarrow h\left(\frac{e(C_0, A_0)}{e(C_3, B_0)}\right)\|\ldots\|h\left(\frac{e(C_0, A_{\kappa-1})}{e(C_3, B_{\kappa-1})}\right)$$

The above scheme is consistent since for a honestly generated ciphertext, we have

$$\frac{e(C_0, A_l)}{e(C_3, B_l)} = \frac{e\big(g_1^r, g_2^{ab_l} H_2(id)^{r_l}\big)}{e\big(H_1(id)^r, g_2^{r_l}\big)} = e(g_1^r, g_2^{ab_l}) \cdot \frac{e\big(g_1^r, H_2(id)^{r_l}\big)}{e\big(H_1(id)^r, g_2^{r_l}\big)} =$$

$$= Y_l^r \cdot \frac{e\big(g_1^r, g_2^{r_l(a \cdot id + \gamma)}\big)}{e\big(g_1^{r(a \cdot id + \gamma)}, g_2^{r_l}\big)} = Y_l^r \text{ for } l = 0, \ldots, \kappa - 1$$

Theorem 1. *Let h be a BDH hardcore predicate and TCR be a target collision-resistant hash function. Then the above IB-KEM scheme is secure against selective-identity and chosen-ciphertext attacks if $\mathcal{IG}$ is an instance generator algorithm for which the Bilinear Diffie-Hellman assumption holds.*

Proof. An adversary starts by outputting a security parameter 1^k, a key length κ and a target identity $id^\star \in \mathbb{Z}_p^*$. Given a BDH instance

$$\mathbf{Z} \leftarrow \big(e, \mathbb{G}_1, \mathbb{G}_2, \mathbb{G}_3, g_1, g_2, p, g_1^a, g_2^a, g_1^b, g_2^b, g_1^c \big)$$

and a successful adversary $\mathcal{A}$ against the IND-sID-CCA security of our IB-KEM, we construct an algorithm $\mathcal{B}$ distinguishing $h\big(e(g_1, g_2)^{abc}\big)$ from random with non-negligible advantage.

To do so we need to apply a hybrid argument that we explain next. Assume that for challenge ciphertext $C^\star = \big(g_1^c, g_1^{c \cdot f(\underline{t}^\star)}, g_1^{c \cdot f(\overline{t}^\star)}, H_1(id^\star)^c\big)$, where $\underline{t}^\star = \mathsf{TCR}(g_1^c, 0)$ and $\overline{t}^\star = \mathsf{TCR}(g_1^c, 1)$, there exists an adversary $\mathcal{A}$ which distinguishes $h(Y_0^c)|| \ldots ||h(Y_{\kappa-1}^c)$ from random (thus breaking the IB-KEM security). Then, there exists another adversary $\mathcal{A}'$ that for some j such that $0 \leq j \leq \kappa - 1$ distinguishes $h(Y_0^c)|| \ldots ||h(Y_j^c)||rand_{\kappa-j-1}$ from $h(Y_0^c)|| \ldots ||h(Y_{j-1}^c)||rand_{\kappa-j}$, where $rand_l$ denotes a l-bit uniformly random string.

Therefore we can assume the existence of such an adversary $\mathcal{A}'$. We use it to construct an algorithm $\mathcal{B}$ that given $(g_1, g_2, g_1^a, g_2^a, g_1^b, g_2^b, g_1^c)$ distinguishes $h\big(e(g_1, g_2)^{abc}\big)$ from random. $\mathcal{B}$ is defined as follows:

Generate system parameters. $\mathcal{B}$ starts by setting $\langle e, \mathbb{G}_1, \mathbb{G}_2, \mathbb{G}_3, p, g_1, g_2 \rangle$ to be the global parameters of the system. It continues by simulating the master public key of the IB-KEM and the challenge encapsulation to $\mathcal{A}'$:

1. It sets $u_0 \leftarrow g_1^a, v_0 \leftarrow g_2^a$ and $u_{-1} \leftarrow u_0^{-id^\star} g_1^\delta$, $v_{-1} \leftarrow v_0^{-id^\star} g_2^\delta$ for random $\delta \xleftarrow{\$} \mathbb{Z}_p^*$.
2. It sets $\underline{t}^\star = \mathsf{TCR}(g_1^c, 0)$ and $\overline{t}^\star = \mathsf{TCR}(g_1^c, 1)$.
3. It sets $y_j \leftarrow g_1^b \in \mathbb{G}_1$ and $z_j \leftarrow g_2^b \in \mathbb{G}_2$ and picks random

$$rnd_j, \ldots, rnd_{\kappa-1} \xleftarrow{\$} \mathbb{Z}_p^* \backslash \{\underline{t}^\star, \overline{t}^\star\}$$

   Additionally it chooses

$$u_{\underline{t}^\star}, u_{\overline{t}^\star}, b_0, \ldots, b_{j-1}, u_j, \ldots, u_{\kappa-1} \xleftarrow{\$} \mathbb{Z}_p^*$$

4. It sets $y_l = g_1^{b_l} \in \mathbb{G}_1$ and $z_l = g_2^{b_l} \in \mathbb{G}_2$ for $0 \le l \le j-1$.
5. By using Lagrange interpolation, it computes $y_{j+1}, \ldots, y_{\kappa+2}$ such that for a function $F_1(x) = \prod_{0 \le l \le \kappa+2} y_l^{x^l}$ it holds that $F_1(\underline{t}^\star) = g_1^{u_{\underline{t}^\star}}$, $F_1(\overline{t}^\star) = g_1^{u_{\overline{t}^\star}}$, and $F_1(rnd_j) = g_1^{u_j}, \ldots, F_1(rnd_{\kappa-1}) = g_1^{u_{\kappa-1}}$.
6. By using Lagrange interpolation, it computes $z_{j+1}, \ldots, z_{\kappa+2}$ such that for a function $F_2(x) = \prod_{0 \le l \le \kappa+2} z_l^{x^l}$ it holds that $F_2(\underline{t}^\star) = g_2^{u_{\underline{t}^\star}}$, $F_2(\overline{t}^\star) = g_2^{u_{\overline{t}^\star}}$, and $F_2(rnd_j) = g_2^{u_j}, \ldots, F_2(rnd_{\kappa-1}) = g_2^{u_{\kappa-1}}$.
7. It sets $Y_l = e(g_1^a, z_l)$ for $l = 0, \ldots, \kappa-1$.
8. Let $\mathcal{C}_{PK}$ be $\mathbb{G}_1^4$.
9. $\mathcal{B}$ sets the master public key to be

$$PK \leftarrow \langle\, e, \mathbb{G}_1, \mathbb{G}_2, \mathbb{G}_3, p, g_1, g_2, u_0, v_0, u_{-1}, v_{-1}, y_0, \ldots, y_{\kappa+2},$$
$$Y_0, \ldots, Y_{\kappa-1}, H_1 \,\rangle,$$

the challenge ciphertext

$$C^\star \leftarrow (g_1^c, (g_1^c)^{u_{\underline{t}^\star}}, (g_1^c)^{u_{\overline{t}^\star}}, (g_1^c)^\delta)$$

and the challenge key

$$K^\star = \big(h(e(g_1^c, g_2^a)^{b_0})||h(e(g_1^c, g_2^a)^{b_1})|| \ldots ||h(e(g_1^c, g_2^a)^{b_{j-1}})||\beta||rand_{k-j-1}\big),$$

where $\beta \overset{\$}{\leftarrow} \{0,1\}$.

Finally $\mathcal{B}$ initializes $\mathcal{A}'$ with PK and $(K^\star, C^\star)$. Notice that, because of the properties of Lagrange interpolation, the distribution of the resulting master public key is statistically-close to the distribution of a honestly generated key.

$\mathcal{B}$ answers the queries by $\mathcal{A}'$ in the following way:

Create user secret key queries. For a secret key query with $id \ne id^\star$, it computes $sk_l[id] \leftarrow \big(z_l^{\frac{-\delta}{id-id^\star}} H_2(id)^{r_l}, z_l^{\frac{-1}{id-id^\star}} g_2^{r_l}\big)$, where $r_l \overset{\$}{\leftarrow} \mathbb{Z}_p^*$ for $l = 0, \ldots, \kappa-1$. To see that this is a valid random user secret key, let us write $z_l = g_2^{b_l}$ for $b_l \in \mathbb{Z}_p^*$ and let $s_l = r_l - b_l/(id - id^\star)$. Notice that b_l is unknown to $\mathcal{B}$ for $j \le l \le \kappa+2$, and thus s_l is not defined explicitly but implicitly. It turns out that

$$z_l^{\frac{-\delta}{id-id^\star}} H_2(id)^{r_l} = g_2^{\frac{-\delta b_l}{id-id^\star}} \big(g_2^{a(id-id^\star)} g_2^\delta\big)^{r_l} =$$
$$= g_2^{ab_l} \big(g_2^{a(id-id^\star)} g_2^\delta\big)^{r_l - \frac{b_l}{id-id^\star}} = g_2^{ab_l} H_2(id)^{s_l}$$

and $z_l^{\frac{-1}{id-id^\star}} g_2^{r_l} = g_2^{s_j}$.

Decryption queries. For decryption queries of the form (id, C) for $id \ne id^\star$, it first checks whether $C \in \mathcal{C}_{PK}$. If not, it answers $\bot$. Otherwise it obtains $sk[id]$ by running the user key generation algorithm and returns $\mathsf{KEM.Decap}(sk[id], C)$.

For decryption queries $(id^\star, C)$, it parses C as (C_0, C_1, C_2, C_3) and it proceeds as follows:

1. If $C_0 = g_1^c$ it answers $\perp$.
2. If $C_0 \neq g_1^c$ and the intersection

$$\{\mathsf{TCR}(C_0, 0), \mathsf{TCR}(C_0, 1)\} \cap \{\underline{t}^\star, \overline{t}^\star, rnd_j, \ldots, rnd_{\kappa-1}\}$$

   is non-empty then $\mathcal{B}$ aborts and outputs a random bit β'.
3. If $C_0 \neq g_1^c$ and the intersection

$$\{\mathsf{TCR}(C_0, 0), \mathsf{TCR}(C_0, 1)\} \cap \{\underline{t}^\star, \overline{t}^\star, rnd_j, \ldots, rnd_{\kappa-1}\}$$

   is empty, it first checks if $e(C_0, g_1^{f(\underline{t})}) = e(g_1, C_1)$ and $e(C_0, g_1^{f(\overline{t})}) = e(g_1, C_2)$, where $\underline{t} = \mathsf{TCR}(C_0, 0)$, $\overline{t} = \mathsf{TCR}(C_0, 1)$. If not $\mathcal{B}$ outputs $\perp$ indicating decryption failure. Otherwise $\mathcal{B}$ computes $C_0^{u_{\underline{t}^\star}}, C_0^{u_{\overline{t}^\star}}, C_0^{u_j}, \ldots, \ C_0^{u_{\kappa-1}}$. Let $f_1 \in \mathbb{Z}_p[x]$ be a polynomial of degree $\kappa + 2$ whose coefficients for the terms x^l are b_l for $0 \leq l \leq j - 1$. Additionally, f_1 satisfies that

$$\begin{aligned}&\left(f_1(\underline{t}), f_1(\overline{t}), f_1(\underline{t}^\star), f_1(\overline{t}^\star), f_1(rnd_{j+1}), \ldots, f_1(rnd_{\kappa-1})\right) = \\ &\left(\log_{C_0} C_1, \log_{C_0} C_2, u_{\underline{t}^\star}, u_{\overline{t}^\star}, u_{j+1}, \ldots, u_{\kappa-1}\right)\end{aligned}$$

Next by using Lagrange interpolation it computes $C_0^{b_{1,l}}$ for $j \leq l \leq \kappa - 1$, where $b_{1,l} \in \mathbb{Z}_p$ denote the coefficients of the x_l term of the polynomial f_1 respectively. Then it answers the decryption query (id, C) with

$$h(e(C_0^{b_0}, g_2^a)) || \ldots || h(e(C_0^{b_{j-1}}, g_2^a)) || h(e(C_0^{b_{1,j}}, g_2^a)) || \ldots || h(e(C_0^{b_{1,\kappa-1}}, g_2^a)).$$

Guess. At some point $\mathcal{A}'$ outputs a guess $\beta' \in \{0, 1\}$ and $\mathcal{B}$ outputs the same guess for $h(e(g_1, g_2)^{abc})$.

Success analysis of algorithm $\mathcal{B}$. Let us define a series of events:

- Win denotes the event that $\mathcal{A}'$ correctly distinguishes

$$h(Y_0^c) || \ldots || h(Y_j^c) || rand_{\kappa-j-1} \text{ from } h(Y_0^c) || \ldots || h(Y_{j-1}^c) || rand_{\kappa-j}$$

- Abort is the event that $\mathcal{A}'$ makes a decryption query $(id^\star, (C_0, C_1, C_2, C_3))$ such that $C_0 \neq g_1^c$ and the intersection $\{\mathsf{TCR}(C_0, 0), \mathsf{TCR}(C_0, 1)\} \cap \{\underline{t}^\star, \overline{t}^\star, rnd_j, \ldots, rnd_{\kappa-1}\}$ is non-empty
- Invalid denotes the event that $\mathcal{A}'$ makes a decryption query of the form $(id^\star, (C_0, C_1, C_2, C_3))$ such that $e(C_0, g_1^{f(\underline{t})}) = e(g_1, C_1)$ and $e(C_0, g_1^{f(\overline{t})}) = e(g_1, C_2)$ but $\mathcal{B}$'s decryption answer is incorrect

Then, analogously to [9], $\mathcal{B}$'s advantage in distinguishing $h(e(g_1, g_2)^{abc})$ from a random bit β is bounded as follows

$$\mathsf{Adv}^h_{\mathcal{IG},\mathcal{B}}(k) = \left| \Pr\left[\mathcal{B}\left(\mathbf{Z}, h, h(e(g_1, g_2)^{abc}) \right) = 1 \right] - \Pr\left[\mathcal{B}\left(\mathbf{Z}, h, \beta \right) = 1 \right] \right|$$
$$\geq \left| \Pr[\mathsf{Win}|\overline{\mathsf{Abort}} \wedge \overline{\mathsf{Invalid}}]\Pr[\overline{\mathsf{Abort}} \wedge \overline{\mathsf{Invalid}}] - 1/2 \right|$$
$$\geq \left| \Pr[\mathsf{Win}] - \Pr[\mathsf{Abort}] - \Pr[\mathsf{Invalid}] \right|$$

Lemmas 3 and 4 in [9] imply that the probabilities $\Pr[\mathsf{Abort}]$ and $\Pr[\mathsf{Invalid}]$ are negligible. Finally $\Pr[\mathsf{Win}] = \frac{1}{\kappa} \cdot \mathsf{Adv}^{\mathsf{sID-CCA}}_{\mathsf{IBKEM},\mathcal{A}}$ due to the hybrid argument.   □

4 Extensions

The IB-KEM presented in the last section admits several extensions.

Identity-based encryption. Coupling our IB-KEM with a chosen ciphertext secure symmetric key encryption scheme yields an IBE scheme with IND-sID-CCA security based on the BDH assumption. The resulting IBE scheme is fairly efficient and has constant size ciphertexts.

Identity-based encryption with adaptive-identity security. It is straightforward to upgrade our IBE scheme to have adaptive-identity security by replacing the function H_1 (and H_2 accordingly) by the function used by Waters [12] to attain adaptive-identity security.

Hierarchical identity-based encryption. Our modification to the IBE scheme by Boneh and Boyen can also be applied to the hierarchical IBE scheme presented by the same authors in [2], and results in an hierarchical IBE scheme with IND-sID-CCA security based on the BDH assumption. For example, the resulting hierarchical IB-KEM scheme has a ciphertext for a ℓ-level hierarchical identity $(id_1, \ldots, id_\ell)$ with the form

$$\left(g_1^r, g_1^{r \cdot f(\underline{t})}, g_1^{r \cdot f(\overline{t})}, H_1^1(id)^r, \ldots, H_1^\ell(id)^r \right) \in \mathbb{G}_1^{3+\ell},$$

where $H_1^1, \ldots, H_1^\ell$ are ℓ independent instances of the function H_1 and $\underline{t} = \mathsf{TCR}(g_1^r, 0)$, $\overline{t} = \mathsf{TCR}(g_1^r, 1)$. The key $K \leftarrow h(Y_0^r)\|\ldots\|h(Y_{\kappa-1}^r) \in \{0,1\}^\kappa$ remains unchanged.

References

1. Bentahar, K., Farshim, P., Malone-Lee, J., Smart, N.P.: Generic Constructions of Identity-Based and Certificateless KEMs. J. Cryptology 21(2), 178–199 (2008)
2. Boneh, D., Boyen, X.: Efficient Selective-ID Secure Identity Based Encryption Without Random Oracles. In: Cachin, C., Camenisch, J.L. (eds.) EUROCRYPT 2004. LNCS, vol. 3027, pp. 223–238. Springer, Heidelberg (2004)
3. Boneh, D., Franklin, M.K.: Identity-Based Encryption From The Weil Pairing. In: Kilian, J. (ed.) CRYPTO 2001. LNCS, vol. 2139, pp. 213–229. Springer, Heidelberg (2001)

4. Cash, D., Kiltz, E., Shoup, V.: The Twin Diffie-Hellman Problem and Applications. In: Smart, N.P. (ed.) EUROCRYPT 2008. LNCS, vol. 4965, pp. 127–145. Springer, Heidelberg (2008)
5. Cramer, R., Shoup, V.: A Practical Public Key Cryptosystem Provably Secure Against Adaptive Chosen Ciphertext Attack. In: Krawczyk, H. (ed.) CRYPTO 1998. LNCS, vol. 1462, pp. 13–25. Springer, Heidelberg (1998)
6. Cramer, R., Shoup, V.: Design and analysis of practical public-key encryption schemes secure against adaptive chosen ciphertext attack. SIAM Journal of Computing 33(1), 167–226 (2004)
7. El Gamal, T.: A Public Key Cryptosystem and a Signature Scheme Based on Discrete Logarithms. In: Blakely, G.R., Chaum, D. (eds.) CRYPTO 1984. LNCS, vol. 196, pp. 10–18. Springer, Heidelberg (1985)
8. Galbraith, S.D., Paterson, K.G., Smart, N.P.: Pairings for cryptographers. Discrete Applied Mathematics 156(16), 3113–3121 (2008)
9. Hanaoka, G., Kurosawa, K.: Efficient Chosen Ciphertext Secure Public Key Encryption under the Computational Diffie-Hellman Assumption. In: Pieprzyk, J. (ed.) ASIACRYPT 2008. LNCS, vol. 5350, pp. 308–325. Springer, Heidelberg (2008)
10. Haralambiev, K., Jager, T., Kiltz, E., Shoup, V.: Simple and Efficient Public-Key Encryption from Computational Diffie-Hellman in the Standard Model. In: Nguyen, P.Q., Pointcheval, D. (eds.) PKC 2010. LNCS, vol. 6056, pp. 1–18. Springer, Heidelberg (2010)
11. Shamir, A.: Identity-based cryptosystems and signature schemes. In: Blakely, G.R., Chaum, D. (eds.) CRYPTO 1984. LNCS, vol. 196, pp. 47–53. Springer, Heidelberg (1985)
12. Waters, B.: Efficient Identity-Based Encryption Without Random Oracles. In: Cramer, R. (ed.) EUROCRYPT 2005. LNCS, vol. 3494, pp. 114–127. Springer, Heidelberg (2005)

A Survey of Local and Global Pairings on Elliptic Curves and Abelian Varieties

Joseph H. Silverman*

Mathematics Department, Brown University, Providence, RI 02912 USA
jhs@math.brown.edu

Abstract. There are many bilinear pairings that naturally appear when one studies elliptic curves, abelian varieties, and related groups. Some of these pairings, notably the Weil and Lichtenbaum–Tate pairings, can be defined over finite fields and have important applications in cryptography. Others, such as the Néron–Tate canonical height pairing and the Cassels–Tate pairing on the Shafarevich–Tate group, are of fundamental importance in number theory and arithmetic geometry, but have seen limited use in cryptography. In this article I will present a survey of some of the pairings that are used to study elliptic curves and abelian varieties. I will attempt to show why they are natural pairings and how they fit into a wider framework.

Keywords and Phrases: elliptic curve, abelian variety, cryptography, pairings.

The *elliptic curve discrete logarithm problem* (ECDLP) has attracted considerable attention since Neal Koblitz [9] and Victor Millier [14] independently proposed its use as the basis for crytography. To date, the fastest general algorithms for ECDLP are exponential, as opposed to the subexponential algorithms known for the classical discrete logarithm problem (DLP) and the integer factorization problem (IFP). This is one reason why elliptic curve cryptography (ECC) has gained in popularity in recent years. Another reason is the existence of certain functorial pairings that exist on elliptic curves, and more generally on abelian varieties of all dimensions. The first cryptographic application of pairings was the reduction of the ECDLP to the DLP [13], which in some cases led to a significant decrease in security. Subsequently, pairings were found to have useful positive applications in cryptography, for example to tripartite Diffie–Hellman key exchange [8] and to the construction of identity-based cryptosystems [3].

There are many natural pairings on elliptic curves, abelian varieties, and associated groups, including the Weil pairing on m-torsion, the Lichtenbaum–Tate pairing over finite fields, the Tate pairing over local fields, the Cassels–Tate pairing on the Shafarevich–Tate group, and the Néron–Tate canonical height pairing

---

* Research supported by DMS-0650017 and DMS-0854755. This article is an expanded version of a talk at Pairing 2010, Yamanaka Onsen, Japan, December 13–15, 2010.
2010 *MSC*: Primary 14G50, Secondary 94A60.

on algebraic points.[1] In this note I will describe these pairings and their inter-relationships. Although at present only a few of these pairing are being directly used in cryptography, all of them are of fundamental importance in studying the arithmetic properties of abelian varieties, and it seems a safe bet that more of them will find cryptographic applications in the years ahead.

1   Pairings in the Abstract

As a good starting point to understanding pairings on abelian varieties, we consider the following two abstract questions:

1. **What is a pairing?**
2. **What makes a pairing functorial, i.e., natural?**

Abstractly, to create a pairing, one starts with a ring R and an R-module M. By definition, a pairing on M is an R-bilinear map

$$\langle \,\cdot\, , \,\cdot\, \rangle : M \times M \longrightarrow R,$$

or equivalently, an R-module homomorphism

$$\phi : M \longrightarrow \mathrm{Hom}_R(M, R).$$

The equivalence of these definitions is clear via the identification

$$\langle a, b \rangle = \phi(a)(b).$$

If M is a finitely generated free R-module, i.e., if it has a finite R-basis, then M is isomorphic to R^d and there are isomorphisms

$$\mathrm{Hom}_R(M, R) \cong \mathrm{Hom}_R(R^d, R) = \mathrm{Hom}_R(R, R)^d = R^d \cong M.$$

The isomorphism $\mathrm{Hom}_R(M, R) \cong M$ defines a non-degenerate pairing on M.

However, and this is very important, the isomorphism $\mathrm{Hom}_R(M, R) \cong M$ and the pairing on M *depend on the choice of a basis for M*. Once we choose a basis, we get a pairing. But there are many reasons why it is not desirable to have to choose a basis.

For example, from a cryptographic perspective, defining a pairing using a basis means that computing the pairing requires solving a (multidimensional) discrete logarithm problem. Explicitly suppose that we fix an R-basis $\mathbf{e}_1, \ldots, \mathbf{e}_d$ for M. Then any choice of elements $c_{ij} \in R$ defines a pairing

$$\left\langle \sum_{i=1}^d u_i \mathbf{e}_i, \sum_{i=1}^d v_i \mathbf{e}_i \right\rangle = \sum_{i=1}^d \sum_{j=1}^d u_i v_j c_{ij}.$$

---

[1] The alert reader may have noted that John Tate's name appears frequently in this list. As Marcus du Sautoy said in his remarks [5] when Tate was awarded the Abel Prize in 2010, "If one measured the influence of a mathematician by the number of mathematical ideas that bear their name, then John Tate would be a clear winner."

Suppose now that we are given two elements $\mathbf{u}, \mathbf{v} \in M$. In order to compute the value of the pairing $\langle \mathbf{u}, \mathbf{v} \rangle$, we need to write $\mathbf{u}$ and $\mathbf{v}$ in terms of the given basis, which in general is a very difficult problem.

Thus the task of defining computationally useful pairings is closely tied to finding a "good", i.e., functorial, isomorphism

$$M \xrightarrow{\ \sim\ } \mathrm{Hom}_R(M, R)$$

whose definition does not depend on choosing a basis for M.

2 Hom and Ext

Let

$$0 \longrightarrow M \longrightarrow N \longrightarrow P \longrightarrow 0$$

be an exact sequence of R-modules. If we look at the homomorphisms from M, N, and P to some other R-module Q, then the arrows are reversed and we get an exact sequence

$$0 \longrightarrow \mathrm{Hom}_R(P, Q) \longrightarrow \mathrm{Hom}_R(N, Q) \longrightarrow \mathrm{Hom}_R(M, Q), \quad (1)$$

but the right-hand map need not be surjective. In order to extend the sequence, we use the Ext_R construction. The elements of $\mathrm{Ext}_R(M, Q)$ are represented by maps

$$f : M \times M \longrightarrow Q$$

that satisfy the relation

$$f(x, y) - f(x, y + z) + f(x + y, z) - f(y, z) = 0 \qquad \text{for all } x, y, z \in M. \quad (2)$$

Two such maps f_1 and f_2 are considered to be the same if there is a map $g : M \to Q$ such that the difference $f_1 - f_2$ has the form

$$f_1(x, y) - f_2(x, y) = g(x + y) - g(x) - g(y).$$

The reason that Ext_R modules are useful is because they allow us to extend the exact sequence (1).

Proposition 1. *Let*

$$0 \longrightarrow M \longrightarrow N \longrightarrow P \longrightarrow 0$$

be an exact sequence of R-modules. Then there is an exact sequence

$$0 \longrightarrow \mathrm{Hom}_R(P, Q) \longrightarrow \mathrm{Hom}_R(N, Q) \longrightarrow \mathrm{Hom}_R(M, Q)$$

$$\longrightarrow \mathrm{Ext}_R(P, Q) \longrightarrow \mathrm{Ext}_R(N, Q) \longrightarrow \mathrm{Ext}_R(M, Q).$$

The map from $\mathrm{Hom}_R(M, Q)$ to $\mathrm{Ext}_R(P, Q)$ is easy to describe. We are given that the map $N \to P$ is surjective, so for each $x \in P$ we can choose some $\alpha(x) \in N$ that maps to x. (In other words, $\alpha : P \to N$ is a section to the given map $N \to P$.) Then for any $x, y \in P$, the element

$$\gamma_{x,y} := \alpha(x + y) - \alpha(x) - \alpha(y)$$

is in the kernel of $N \to P$, because the map $N \to P$ is a homomorphism. Thus $\gamma_{x,y}$ comes from a (unique) element of M, which we also denote $\gamma_{x,y}$. Now let $F \in \mathrm{Hom}_R(M, Q)$. We use F to create an element of $\mathrm{Ext}_R(P, Q)$ via the formula

$$P \times P \longrightarrow Q, \qquad (x, y) \longmapsto F(\gamma_{x,y}).$$

Finally, we mention an alternative description of $\mathrm{Ext}_R(M, Q)$ that is sometime useful. An element of $\mathrm{Ext}_R(M, Q)$ gives an *extension of M by Q*, that is, it can be used to define an R-module L that fits into the exact sequence

$$0 \longrightarrow Q \longrightarrow L \longrightarrow M \longrightarrow 0. \tag{3}$$

To do this, let $f : M \times M \to Q$ be an element of $\mathrm{Ext}_R(M, Q)$. We use f to define a twisted R-module structure on the set $M \times Q$ via the rule

$$(x_1, z_1) \oplus_f (x_2, z_2) = \big(x_1 + x_2, z_1 + z_2 + f(x_1, x_2)\big).$$

This twisted R-module fits into the exact sequence (3).

Conversely, if we are given an exact sequence (3), then we can produce an element of $\mathrm{Ext}_R(M, Q)$ as follows. First, for each $x \in M$, choose some $\beta(x) \in L$ that maps to x. Then we get an element of $\mathrm{Ext}_R(M, Q)$ from the formula

$$M \times M \longrightarrow Q, \qquad (x, y) \longmapsto \beta(x + y) - \beta(x) - \beta(y).$$

3   The Dual Abelian Variety and the Weil Pairing

The prototypical example of a functorial pairing in algebraic geometry is the Weil pairing, which on elliptic curves has the form

$$E[m] \times E[m] \longrightarrow \boldsymbol{\mu}_m.$$

This is a pairing of $\mathbb{Z}/m\mathbb{Z}$ modules, but the usual definition for elliptic curves is unenlightening and somewhat misleading. In general for an abelian variety A, the Weil pairing is a pairing

$$A[m] \times \hat{A}[m] \longrightarrow \boldsymbol{\mu}_m,$$

where $\hat{A}$ is the *dual abelian variety* to A. There are several useful ways to characterize $\hat{A}$. We start with an abstract definition,

$$\hat{A} = \mathrm{Ext}(A, \mathbb{G}_m).$$

Using this definition, we consider the exact sequence

$$0 \longrightarrow A[m] \longrightarrow A \overset{m}{\longrightarrow} A \longrightarrow 0$$

and apply $\mathrm{Hom}(\,\cdot\,, \mathbb{G}_m)$. This reverses the direction of the arrows and (using Proposition 1) gives an exact sequence

$$
\begin{array}{ccccccc}
0 & \longrightarrow & \mathrm{Hom}(A[m], \mathbb{G}_m) & \overset{\delta}{\longrightarrow} & \mathrm{Ext}(A, \mathbb{G}_m) & \overset{m}{\longrightarrow} & \mathrm{Ext}(A, \mathbb{G}_m) \\
& & \| & & \| & & \| \\
0 & \longrightarrow & \mathrm{Hom}(A[m], \boldsymbol{\mu}_m) & \overset{\delta}{\longrightarrow} & \hat{A} & \overset{m}{\longrightarrow} & \hat{A}
\end{array}
\tag{4}
$$

(We are also using the fact that $\mathrm{Hom}(A, \mathbb{G}_m) = 0$, since there are no algebraic maps from the compact variety A to the affine variety $\mathbb{G}_m$, and the fact that any map from $A[m]$ to $\mathbb{G}_m$ must have image in $\boldsymbol{\mu}_m$.) The exact sequence (4) gives a natural isomorphism

$$\hat{A}[m] \cong \mathrm{Hom}\big(A[m], \boldsymbol{\mu}_m\big), \tag{5}$$

and (5) is exactly what we need to define the *Weil pairing*

$$e_m \,:\, A[m] \times \hat{A}[m] \cong A[m] \times \mathrm{Hom}\big(A[m], \boldsymbol{\mu}_m\big) \longrightarrow \boldsymbol{\mu}_m.$$

Thus the existence of the Weil pairing is a natural consequence of the definition of the dual abelian variety.

In order to compute the Weil pairing, we use an alternative characterization of the dual abelian variety as the group of divisor classes

$$\mathrm{Pic}^0(A) = \frac{\{\text{Divisors algebraically equivalent to } 0\}}{\{\text{Divisors linearly equivalent to } 0\}}.$$

The identification of $\mathrm{Ext}(A, \mathbb{G}_m)$ with $\mathrm{Pic}^0(A)$ is reasonably straightforward. By definition, an element of $\mathrm{Pic}^0(A)$ is represented by a divisor D that is algebraically equivalent to 0. (If A is an elliptic curve, this just means that D is a divisor of degree 0.) For any point $P \in A$, we let

$$\tau_P : A \longrightarrow A, \qquad Q \longmapsto Q + P, \tag{6}$$

be the translation-by-P map. Then one can show that $\tau_P^* D - D$ is the divisor of a function,

$$\tau_P^* D - D = \mathrm{div}(F_{D,P}).$$

Recall that elements of $\mathrm{Ext}(A, \mathbb{G}_m)$ are represented by maps

$$A \times A \longrightarrow \mathbb{G}_m$$

satisfying condition (2). We map a divisor class $[D] \in \mathrm{Pic}^0(A)$ to an element of $\mathrm{Ext}(A, \mathbb{G}_m)$ via

$$
\begin{array}{ccc}
\mathrm{Pic}^0(A) & \longrightarrow & \mathrm{Ext}(A, \mathbb{G}_m) \\[4pt]
[D] & \longmapsto & \left((P, Q) \mapsto \dfrac{F_{D,P}(Q)}{F_{D,P}(O)} \right).
\end{array}
$$

(If $F_{D,P}$ has a zero or a pole at Q, then some adjustment needs to be made.) If you trace through all of these definitions, you'll end up with the usual formula for the Weil pairing in terms of functions and divisors.

4   Principal Polarizations and the Weil Pairing

The Weil pairing and many other pairings on abelian varieties are most naturally defined as pairings between an abelian variety A and its dual abelian variety $\hat{A}$. For applications, one generally wants a pairing on A itself. This can be accomplished by using a natural map from A to $\hat{A}$.

Recall that one description of $\hat{A}$ is as the group of divisor classes $\mathrm{Pic}^0(A)$. Let τ_P be the be the translation-by-P map (6). Then for any divisor $D \in \mathrm{Div}(A)$, the divisor $\tau_P^* D - D$ is algebraically equivalent to zero, so we get a map

$$\phi_D : A \longrightarrow \mathrm{Pic}^0(A), \qquad P \longmapsto [\tau_P^* D - D].$$

Example 1. On an elliptic curve E, we can take $D = (O)$, and then the map ϕ_D is simply

$$\phi_D(P) = \big[(-P) - (O)\big]. \tag{7}$$

The map ϕ_D is a *principal polarization* if it defines an isomorphism $A \cong \hat{A}$. For example, the map (7) on elliptic curves is a principal polarization. More generally, the Jacobian variety of any (nonsingular) curve has a natural principal polarization, so Jacobians are self-dual.

More precisely, let C/K be a nonsingular curve of genus $g \geq 1$ and let $P_0 \in C(K)$. The Jacobian variety of C is an abelian variety J_C that is isomorphic, as a group, to $\mathrm{Pic}^0(C)$. One can show that the map

$$C^g \longrightarrow J_C = \mathrm{Pic}^0(C), \qquad (P_1, \ldots, P_g) \longmapsto \big[(P_1) + (P_2) + \cdots + (P_g) - g(P_0)\big]$$

is almost everywhere finite-to-one, so J_C has dimension g. The image of the map

$$C^{g-1} \longrightarrow J_C = \mathrm{Pic}^0(C), \qquad (P_1, \ldots, P_{g-1}) \longmapsto \big[(P_1) + \cdots + (P_{g-1}) - (g-1)(P_0)\big]$$

is a subvariety of codimension one in J_C. It is called the *theta divisor* and denoted Θ. Then the map

$$\phi_\Theta : J_C \longrightarrow \mathrm{Pic}^0(J_C) = \hat{J}_C$$

is a principal polarization that identifies J_C with its dual.

If A comes equipped with a principal polarization $\phi_D : A \to \hat{A}$, then we can define the Weil pairing

$$e_m : A[m] \times A[m] \longrightarrow \boldsymbol{\mu}_m, \qquad (P, Q) \longmapsto e_m\big(P, \phi_D(Q)\big).$$

For example, using the principal polarization (7) on an elliptic curve leads to the familiar definition of the Weil pairing found in standard textbooks.

5   A Primer on Galois Cohomology

The Weil pairing is a fundamental tool used to define other pairings on abelian varieties. In order to define these new pairings in a functorial way, we need to use a little bit of group cohomology, about which we briefly recall some basic definitions and facts. (For more on group cohomology and Galois cohomology, see for example [1,19].)

Let M be an abelian group on which the Galois group $G_K = \mathrm{Gal}(\bar{K}/K)$ acts.[2] The cohomology group $H^0(G_K, M)$ is the group of elements fixed by G_K,

$$H^0(G_K, M) = \{a \in M : a^\sigma = a \text{ for all } \sigma \in G_K\}.$$

A 1-*cocycle* is a map[3]

$$\xi : G_K \longrightarrow M \quad \text{satisfying} \quad \xi(\sigma\tau) = \xi(\tau)^\sigma + \xi(\sigma) \quad \text{for all } \sigma, \tau \in G_K.$$

A 1-*coboundary* is a map of the form

$$\eta : G_K \longrightarrow M, \quad \eta(\sigma) = a^\sigma - a,$$

for some $a \in M$. The cohomology group $H^1(G_K, M)$ is the group

$$H^1(G_K, M) = \frac{(\text{group of 1-cocyles})}{(\text{group of 1-coboundaries})}.$$

There are higher cohomology groups that are defined in a similar manner using "cocyles modulo coboundaries". In particular, the 2-*cocycles* that are used to form $H^2(G_K, M)$ are maps

$$\xi : G_K \times G_K \longrightarrow M$$

satisfying the 2-cocycle condition

$$\xi(\tau, \mu)^\sigma - \xi(\sigma\tau, \mu) + \xi(\sigma, \tau\mu) - \xi(\sigma, \tau) = 0 \qquad \text{for all } \sigma, \tau, \mu \in G_K.$$

The 2-*coboundaries* are maps of the form

$$\eta(\sigma, \tau) = f(\tau)^\sigma - f(\sigma\tau) + f(\sigma) \quad \text{for some map } f : G_K \to M.$$

Elements of cohomology groups can be multiplied using the *cup product*

$$\cup : H^i(G_K, M) \times H^j(G_K, N) \longrightarrow H^{i+j}(G_K, M \otimes N).$$

---

[2] Here $\bar{K}$ is an algebraic closure of K. We will assume that our fields K are perfect, so for example K could be a finite field or a field of characteristic zero.

[3] We are cheating a little bit, because we should really take only cocycles that are continuous with respect to the profinite topology on G_K. For ease of exposition, we will ignore this subtlety.

The general definition of the cup product is somewhat complicated, but we'll only need $(i, j) = (1, 0)$ and $(1, 1)$. The former is easy, it is simply

$$H^1(G_K, M) \times H^0(G_K, N) \longrightarrow H^1(G_K, M \otimes N), \quad (\xi \cup a)(\sigma) = \xi(\sigma) \otimes a.$$

The latter is similar, but with a twist,[4]

$$H^1(G_K, M) \times H^1(G_K, N) \longrightarrow H^2(G_K, M \otimes N), \quad (\xi \cup \eta)(\sigma, \tau) = \xi(\sigma) \otimes \eta(\tau)^\sigma.$$

(Note that we are using a left action, so $(a^\sigma)^\tau = a^{\tau\sigma}$.)

The group $H^0(G_K, M)$ occurs naturally in many contexts, since we often want to know which elements are fixed by a group. For example, if A/K is an abelian variety, then $H^0(G_K, A) = A(K)$ is the group of rational points defined over K. (For ease of notation, we write $H^i(G_K, A)$ instead of $H^i(G_K, A(\bar{K}))$.) Higher cohomology groups occur naturally when we apply the functor $H^0(G_k, \cdot)$ to an exact sequence.

Proposition 2. *Let*

$$0 \longrightarrow M \longrightarrow N \longrightarrow P \longrightarrow 0$$

be an exact sequence of G_K-modules. Then there is a long exact sequence

$$0 \longrightarrow H^0(G_K, M) \longrightarrow H^0(G_K, N) \longrightarrow H^0(G_K, P)$$
$$\xrightarrow{\delta} H^1(G_K, M) \longrightarrow H^1(G_K, N) \longrightarrow H^1(G_K, P)$$
$$\xrightarrow{\delta} H^2(G_K, M) \longrightarrow H^2(G_K, N) \longrightarrow H^2(G_K, P)$$

(The maps denoted δ are called connecting homomorphisms.*)*

A particularly imporant example of this theorem is associated to the exact sequence

$$0 \longrightarrow M[m] \longrightarrow M \xrightarrow{m} M \longrightarrow M/mM \longrightarrow 0.$$

Here m is an integer, and the middle map is the *multiplication-by-m map*

$$a \longmapsto ma.$$

---

[4] We illustrate typical cohomological calculations by verifying that if ξ and η are 1-cocycles, then $\xi \cup \eta$ is a 2-cocycle.

$(\xi \cup \eta)(\tau, \mu)^\sigma - (\xi \cup \eta)(\sigma\tau, \mu) + (\xi \cup \eta)(\sigma, \tau\mu) - (\xi \cup \eta)(\sigma, \tau)$

$= (\xi(\tau) \otimes \eta(\mu)^\tau)^\sigma - \xi(\sigma\tau) \otimes \eta(\mu)^{\sigma\tau} + \xi(\sigma) \otimes \eta(\tau\mu)^\sigma - \xi(\sigma) \otimes \eta(\tau)^\sigma$

$= \xi(\tau)^\sigma \otimes \eta(\mu)^{\sigma\tau} - \xi(\sigma\tau) \otimes \eta(\mu)^{\sigma\tau} + \xi(\sigma) \otimes \eta(\tau\mu)^\sigma - \xi(\sigma) \otimes \eta(\tau)^\sigma$

$= [\xi(\sigma\tau) - \xi(\sigma)] \otimes \eta(\mu)^{\sigma\tau} - \xi(\sigma\tau) \otimes \eta(\mu)^{\sigma\tau} + \xi(\sigma) \otimes [\eta(\mu)^\tau + \eta(\tau)]^\sigma - \xi(\sigma) \otimes \eta(\tau)^\sigma$

$= \xi(\sigma\tau) \otimes \eta(\mu)^{\sigma\tau} - \xi(\sigma) \otimes \eta(\mu)^{\sigma\tau} - \xi(\sigma\tau) \otimes \eta(\mu)^{\sigma\tau} + \xi(\sigma) \otimes \eta(\mu)^{\sigma\tau} + \xi(\sigma) \otimes \eta(\tau)^\sigma - \xi(\sigma) \otimes \eta(\tau)^\sigma$

$= 0.$

In our applications we will have $M/mM = 0$, so taking cohomology gives the *Kummer sequence*

$$0 \to H^0(G_K, M)/mH^0(G_K, M) \xrightarrow{\delta} H^1(G_K, M[m]) \to H^1(G_K, M)[m] \to 0.$$

For our purposes, the most important G_K-modules will be[5]

1. the multiplicative group $\bar{K}^*$,
2. the group $\boldsymbol{\mu}_m$ of mth-roots of unity in $\bar{K}^*$,
3. an abelian variety $A(\bar{K})$,
4. the group $A[m]$ of m-torsion points in $A(\bar{K})$.

We state two basic, but very important, cohomological properties of the multiplicative group.

Theorem 1. (Hilbert's Theorem 90)

(a) $H^1(G_K, \bar{K}^*) = 0$.
(b) $H^1(G_K, \boldsymbol{\mu}_m) \cong K^*/K^{*m}$.

The isomorphism in (b) can be made quite explicit. Let $a \in K^*$ and take any $\alpha \in \bar{K}^*$ satisfying $\alpha^m = a$. Then the corresponding element of $H^1(G_K, \boldsymbol{\mu}_m)$ is the cohomology class of the cocycle $\sigma \longrightarrow \alpha^\sigma/\alpha$.

6   The Lichtenbaum–Tate Pairing

The Lichtenbaum–Tate pairing is an example of a pairing whose functorial definition uses the Weil pairing in an essential way. A point $P \in A(K)$ defines a class $[P] \in A(K)/mA(K)$. The Kummer sequence (note that $A(\bar{K})/mA(\bar{K}) = 0$) then gives us a 1-cocycle via

$$A(K)/mA(K) = H^0(G_K, A)/mH^0(G_K, A) \xrightarrow{\delta} H^1(G_K, A[m]), \quad [P] \longmapsto F_P.$$

Using this map and the Weil pairing, we can define a map

$$A(K)/mA(K) \times \hat{A}[m](K) \longrightarrow H^1(G_K, \boldsymbol{\mu}_m), \quad (P, Q) \longmapsto [\sigma \mapsto e_m(F_P(\sigma), Q)].$$

(Note that we need Q to be in $\hat{A}(K)$, since otherwise the map $\sigma \mapsto e_m(F_P(\sigma), Q)$ is not a 1-cocycle.)

The second part of Hilbert's Theorem 90 says that $H^1(G_K, \boldsymbol{\mu}_m) \cong K^*/K^{*m}$. The resulting map

$$\langle \cdot, \cdot \rangle_{\text{LT}} : A(K)/mA(K) \times \hat{A}[m](K) \longrightarrow K^*/K^{*m}$$

is the *Lichtenbaum–Tate pairing*.

---

[5] If K has positive characteristic p, we will always assume that p does not divide m.

More precisely, the Lichtenbaum–Tate pairing is constructed out of connecting homomorphisms, the cup product, and the Weil pairing. Thus

$$
\begin{aligned}
A(K) \times \hat{A}[m](K) \xrightarrow{\ =\ } & H^0(G_K, A) \times H^0(G_K, \hat{A}[m]) \\
\xrightarrow{\ \delta \times 1\ } & H^1(G_K, A[m]) \times H^0(G_K, \hat{A}[m]) \\
\xrightarrow{\ \cup\ } & H^1(G_K, A[m] \otimes \hat{A}[m]) \\
\xrightarrow{\ e_m\ } & H^1(G_K, \boldsymbol{\mu}_m) \\
\xrightarrow{\ \delta^{-1}\ } & K^*/K^{*m}.
\end{aligned}
$$

If A is an elliptic curve, or more generally if we have a principal polarization that allows us to identify $\hat{A}$ with A, then we obtain a pairing on A itself,

$$
\langle \cdot, \cdot \rangle_{\mathrm{LT}} : A(K)/mA(K) \times A[m](K) \longrightarrow K^*/K^{*m}.
$$

In cryptography, the Lichtenbaum–Tate pairing is used when K is a finite field, say $K = \mathbb{F}_q$. One chooses the abelian variety A and integer m such that $A(\mathbb{F}_q)/mA(\mathbb{F}_q)$ and $A[m](\mathbb{F}_q)$ are cyclic groups of order m and such that the natural map

$$
A[m](\mathbb{F}_q) \longrightarrow A(\mathbb{F}_q)/mA(\mathbb{F}_q)
$$

is an isomorphism. Then one can show that the Lichtenbaum–Tate pairing

$$
\langle \cdot, \cdot \rangle_{\mathrm{LT}} : A[m](\mathbb{F}_q) \times A[m](\mathbb{F}_q) \longrightarrow \mathbb{F}_q^*/\mathbb{F}_q^{*m}
$$

is non-degenerate and symmetric. In other words,

$$
\langle P, Q \rangle_{\mathrm{LT}} = \langle Q, P \rangle_{\mathrm{LT}}
$$

and

$$
\langle P, Q \rangle_{\mathrm{LT}} = 0 \iff P = O \text{ or } Q = O.
$$

In particular, if $P \in A[m](\mathbb{F}_q)$ has order m, then $\langle P, P \rangle_{\mathrm{LT}}$ is an element of order m in $\mathbb{F}_q^*/\mathbb{F}_q^{*m}$.

Finally, it is often convenient to raise the value of the Lichtenbaum–Tate pairing to the $(q-1)/m$ power, so that its values are specific m^{th}-roots of unity, rather than equivalence classes modulo m^{th}-powers. In other words, we use the isomorphism

$$
\mathbb{F}_q^*/\mathbb{F}_q^{*m} \xrightarrow{\ \sim\ } \boldsymbol{\mu}_m(\mathbb{F}_q), \qquad z \longrightarrow z^{(q-1)/m},
$$

to obtain a modified Lichtenbaum–Tate pairing

$$
\langle \cdot, \cdot \rangle_{\mathrm{LT}}' : A[m](\mathbb{F}_q) \times A[m](\mathbb{F}_q) \longrightarrow \boldsymbol{\mu}_m(\mathbb{F}_q), \qquad \langle P, Q \rangle_{\mathrm{LT}}' = \langle P, Q \rangle_{\mathrm{LT}}^{(q-1)/m}.
$$

7   Homogeneous Spaces and the Weil–Châtelet Group

Let A/K be an abelian variety. A *homogeneous space for A/K* is an algebraic variety B/K together with a group action of A on B. In other words, there is a map (defined over K)

$$\mu : A \times B \longrightarrow B$$

that satisfies

$$\mu\big(P, \mu(Q,T)\big) = \mu(P + Q, T).$$

We further require that the action be *principal*. This means that for any fixed $T \in B$, the map

$$\mu(\,\cdot\,, T) : A \longrightarrow B, \qquad P \longmapsto \mu(P,T),$$

is an isomorphism. Note, however, that this isomorphism is only defined over $\bar{K}$, so B might not be K-isomorphic to A. However, if $B(K) \neq \emptyset$, we can take $T \in B(K)$ to get a K-isomorphism $A \xrightarrow{\sim} B$.

Example 2. The curve

$$C_{abc} : ax^3 + by^3 + cz^3 = 0$$

is a homogeneous space for the elliptic curve

$$E_{abc} : x^3 + y^3 + abcz^3 = 0.$$

Notice that C_{abc} is isomorphic to the curve E_{abc} if we work over the field $K(\sqrt[3]{a}, \sqrt[3]{b}, \sqrt[3]{c})$, but they need not be isomorphic over K. To define μ, one takes a $\bar{K}$ isomorphism $\phi : C_{abc} \to E_{abc}$ and sets $\mu(P,T) = \phi^{-1}\big(P + \phi(T)\big)$.

We say that two homogeneous spaces B and B' are *K-equivalent* if there is an isomorphism $B \cong B'$ defined over K that respects the action of A. The *Weil–Châtelet group of A/K* is

$$\mathrm{WC}(A/K) = \frac{\{\text{homogeneous spaces for } A/K\}}{K\text{-equivalence}}.$$

We say that a homogeneous space B/K is *trivial* if it is K-equivalent to A/K. One can show that B/K is trivial if and only if $B(K)$ is non-empty. Of course, we know that $A(K)$ is non-empty, since $O \in A(K)$. Homogeneous spaces arise naturally when one studies the number theoretic properties of an abelian variety A/K.

As its name suggests, the set of homogeneous spaces modulo K-equivalence can be given the structure of a group. Let $B/K \in \mathrm{WC}(A/K)$ and choose some point $T \in B(\bar{K})$. Then one can show that for each $\sigma \in G_K$ there is a point $\xi_T(\sigma) \in A(\bar{K})$ such that

$$\mu(P, T^\sigma) = \mu(P + \xi_T(\sigma), T) \qquad \text{for all } P \in A(\bar{K}).$$

The map

$$G_K \longrightarrow A(\bar{K}), \qquad \sigma \longrightarrow \xi_T(\sigma),$$

is a 1-cocycle, so it represents an element of $H^1(G_K, A)$. Further, choosing a different point $T \in B(\bar{K})$ has the effect of changing $\xi_T(\sigma)$ by a 1-coboundary, so B itself determines an element $[\xi_B]$ of $H^1(G_K, A)$. The map

$$\mathrm{WC}(A/K) \longrightarrow H^1(G_K, A), \qquad [B] \longrightarrow [\xi_B],$$

is a bijection of sets, and we know that $H^1(G_K, A)$ is a group, so this makes $\mathrm{WC}(A/K)$ into a group.

Why is this useful? The Kummer sequence associated to A is the short exact sequence (note that $H^0(G_K, A) = A(K)$)

$$0 \to A(K)/mA(K) \to H^1(G_K, A[m]) \to H^1(G_K, A)[m] \to 0.$$

Thus $A(K)/mA(K)$ is a subgroup of $H^1(G_K, A[m])$. In practice, it is often possible to do explicit computations and exactly determine the group $H^1(G_K, A[m])$. So calculating $A(K)/mA(K)$ comes down to understanding $H^1(G_K, A)$. This last group can be studied using geometry because it is isomorphic to $\mathrm{WC}(A/K)$.

8   The Tate Pairing for Local Fields

The *Brauer group* of a field K classifies the finite rank central simple algebras over K. We won't have to worry about what that means, because the Brauer group is isomorphic to the cohomology group $H^2(G_K, \bar{K}^*)$. Since Hilbert's Theorem 90 says that $H^1(G_K, \bar{K}^*) = 0$, the Kummer sequence for $\bar{K}^*$ gives

$$0 \longrightarrow H^2(G_K, \boldsymbol{\mu}_m) \longrightarrow H^2(G_K, \bar{K}^*) \xrightarrow{\eta \to \eta^m} H^2(G_K, \bar{K}^*),$$

so

$$H^2(G_K, \boldsymbol{\mu}_m) \cong H^2(G_K, \bar{K}^*)[m].$$

If $K/\mathbb{Q}_p$ is a p-adic field, i.e., a finite extension of $\mathbb{Q}_p$, then an important theorem from (local) class field theory (see [18]) says that

$$H^2(G_K, \bar{K}^*) \cong \mathbb{Q}/\mathbb{Z} \qquad \text{and} \qquad H^2(G_K, \boldsymbol{\mu}_m) \cong \mathbb{Z}/m\mathbb{Z}.$$

This makes the Brauer group a nice target for pairings.

The starting point for the functorial definition of the Tate pairing is the Kummer sequence for the abelian variety A,

$$0 \longrightarrow A(K)/mA(K) \longrightarrow H^1(G_K, A[m]) \longrightarrow H^1(G_K, A)[m] \longrightarrow 0,$$

and the analogous sequence for its dual,

$$0 \longrightarrow \hat{A}(K)/m\hat{A}(K) \longrightarrow H^1(G_K, \hat{A}[m]) \longrightarrow H^1(G_K, \hat{A})[m] \longrightarrow 0.$$

Given an point $P \in A(K)/mA(K)$, we map it to a 1-cocycle $F_P : G_K \to A[m]$. Similarly, given an element $\xi \in H^1(G_K, \hat{A})[m]$, we can lift it to $H^1(G_K, \hat{A}[m])$

and get a 1-cocycle $F_\xi : G_K \longrightarrow \hat{A}[m]$. This and the Weil pairing let us define a map

$$T_{P,\xi} : G_K \times G_K \longrightarrow \boldsymbol{\mu}_m, \qquad T_{P,\xi}(\sigma, \tau) = e_m\big(F_P(\sigma), F_\xi(\tau)^\sigma\big).$$

(In fancier language, the 2-cocycle $T_{P,\xi}$ may be defined by composing the Weil pairing with the cup product of F_P and F_ξ, i.e., $T_{P,\xi} = e_m \circ (F_P \cup F_\xi)$.) The map $T_{P,\xi}$ is a 2-cocycle, so it gives an element $[T_{P,\xi}] \in H^2(G_K, \boldsymbol{\mu}_m)$. We have defined a pairing

$$A(K)/mA(K) \times H^1(G_K, \hat{A})[m] \longrightarrow H^2(G_K, \boldsymbol{\mu}_m), \qquad (P, \xi) \longrightarrow [T_{P,\xi}].$$

Up to here, the construction works for any field K. But if we now make the assumption that $K/\mathbb{Q}_p$ is a p-adic field, then $H^2(G_K, \boldsymbol{\mu}_m)$ is isomorphic to $\mathbb{Z}/m\mathbb{Z}$, so we get a pairing

$$\langle \cdot\,, \cdot \rangle_{\mathrm{Tate}} : A(K)/mA(K) \times H^1(G_K, \hat{A})[m] \longrightarrow \mathbb{Z}/m\mathbb{Z}.$$

Tate [23] proved that this pairing is nondegenerate. Further, by taking an appropriate limit as $m \to \infty$ and using the structure of $A(K)$ when K is a local field, he showed that one obtains a nondegenerate pairing

$$\langle \cdot\,, \cdot \rangle_{\mathrm{Tate}} : A(K) \times H^1(G_K, \hat{A}) \longrightarrow \mathbb{Q}/\mathbb{Z}.$$

This is somewhat surprising, since it says that if K is a p-adic field, then the set of K-rational points of A is dual to the collection of homogeneous spaces of $\hat{A}$.

9 The Shafarevich–Tate Group

An important, but mysterious, group that appears when one studies elliptic curves or abelian varieties over number fields is the *Shafarevich–Tate group* III. The elements of $\mathrm{III}(A/K)$ are homogeneous spaces B/K for A with the property that $B(K_v)$ is non-empty for every completion K_v of K. In other words, $\mathrm{III}(A/K)$ is the subgroup of $\mathrm{WC}(A/K)$ defined to be the kernel of the map

$$\mathrm{WC}(A/K) \longrightarrow \prod_v \mathrm{WC}(A/K_v).$$

The Shafarevich–Tate group appears as a natural obstruction when trying to compute the group of rational points $A(K)$. More precisely, for each integer $m \geq 2$ there is an exact sequence

$$0 \longrightarrow A(K)/mA(K) \longrightarrow S^{(m)}(A/K) \longrightarrow \mathrm{III}(A/K)[m] \longrightarrow 0,$$

where the Selmer group $S^{(m)}(A/K)$ is the kernel of the map

$$H^1(G_K, A[m]) \longrightarrow \prod_v H^1(G_{K_v}, A) = \prod_v \mathrm{WC}(A/K_v).$$

In theory, and often in practice, it is possible to compute the Selmer group $S^{(m)}(A/K)$, so computing generators for $A(K)/mA(K)$ is reduced to determining the elements of order m in Ш.

It is conjectured that the Shafarevich–Tate group is always finite, although this is not known in general. The first proofs that $Ш(A/K)$ is finite for certain types of abelian varieties were given by Rubin [17] and Kolyvagin [10] in the 1990s. However, it is known that if $Ш(A/K)$ is finite for a principally polarized A, then its order is a perfect square. The way that this is proven is via a pairing, which is the subject of the next section.

10   The Cassels–Tate Pairing on the Shafarevich–Tate Group

The *Cassels–Tate* pairing on Ш is a bilinear pairing

$$\langle\,\cdot\,,\,\cdot\,\rangle_{\mathrm{CT}} : Ш(A/K) \times Ш(\hat{A}/K) \longrightarrow \mathbb{Q}/\mathbb{Z}.$$

If A is principally polarized, so $\hat{A} \cong A$, then the pairing is alternating, and it is an amusing exercise to show that if a *finite* group admits an alternating bilinear pairing, then the order of the group must be a perfect square.

There are several equivalent definitions of the Cassels–Tate pairing, but they are all somewhat complicated; see [4,15,25]. We briefly describe one definition [24] in order to illustrate the connection with the Weil pairing and the similarity with Tate's local pairing.

Every element of Ш has finite order, so it suffices to give a pairing on $Ш(A/K)[m] \times Ш(\hat{A}/K)[m]$. Let $f : G_K \to A(\bar{K})$ and $f' : G_K \to \hat{A}(\bar{K})$ be 1-cocycles representing elements of $Ш(A/K)[m]$ and $Ш(\hat{A}/K)[m]$, respectively. Choose some map

$$g : G_K \longrightarrow A[m^2] \quad \text{satisfying} \quad f(\sigma) = mg(\sigma) \quad \text{for all } \sigma \in G_K.$$

Then there is a unique map

$$
\begin{aligned}
&d : G_K \times G_K \longrightarrow A[m] \\
&\quad \text{satisfying} \quad d(\sigma,\tau) = g(\tau)^\sigma - g(\sigma\tau) + g(\sigma) \quad \text{for all } \sigma,\tau \in G_K. \quad (8)
\end{aligned}
$$

The map d is a 2-cocyle as a map from G_K to $A[m]$, but need not be a 2-coboundary, although it is a 2-coboundary if we treat it as a map from G_K to $A[m^2]$.

The assumption that f represents an element of $Ш(A/K)$ means that it becomes a coboundary over every completion K_v, so not only can we find a map

$$g_v : G_{K_v} \longrightarrow A[m^2] \quad \text{satisfying} \quad f(\sigma) = mg_v(\sigma) \quad \text{for all } \sigma \in G_{K_v},$$

we can choose g_v to be a 1-cocycle. We further note that the image of the difference $g - g_v$ lies in $A[m]$.

To summarize, we have the following maps:

f a 1-cocycle $G_K \to A[m]$.
f' a 1-cocycle $G_K \to \hat{A}[m]$.
g a map $G_k \to A[m^2]$ satisfying $f = mg$.
g_v a 1-cocycle $G_{K_v} \to A[m^2]$ satisfying $f = mg_v$.
d a 2-cocycle $G_K \times G_K \to A[m]$ related to g by (8).

The cup product $d \cup f'$ represents an element of $H^3(G_K, A[m] \otimes \hat{A}[m])$, so composing with the Weil pairing $e_m : A[m] \times \hat{A}[m] \to \boldsymbol{\mu}_m$, we obtain a map

$$e_m \circ (\delta \cup f') : G_K \times G_K \times G_K \longrightarrow \bar{K}^* \tag{9}$$

representing an element of $H^3(G_K, \bar{K}^*)$. However, a theorem of Tate (see [15]) says that $H^3(G_K, \bar{K}^*) = 0$, so (9) is a coboundary. In other words, there is a map $h : G_K \times G_K \to \bar{K}^*$ such that

$$\big(e_m \circ (\delta \cup f')\big)(\sigma, \tau, \rho) = h(\tau, \mu)^\sigma - h(\sigma\tau, \mu) + h(\sigma, \tau\mu) - h(\sigma, \tau) \text{ for all } \sigma, \tau, \mu \in G_K.$$

We use g, g_v, f', h, and the Weil pairing to define a map

$$\ell_v : G_{K_v} \times G_{K_v} \longrightarrow \bar{K}_v^*, \qquad \ell_v = \frac{e_m \circ \big((g - g_v) \cup f'\big)}{h}.$$

One can check that ℓ_v is in fact a 2-cocycle. (The map $e_m \circ \big((g - g_v) \cup f'\big)$ is not a cocycle, but the map h acts as a sort of correction factor that makes it into a cocycle.) Thus ℓ_v represents an element of $H^2(G_{K_v}, \bar{K}_v^*)$. We recall from Section 8 that for local fields there is an isomorphism

$$\mathrm{inv}_v : H^2(G_{K_v}, \bar{K}_v^*) \overset{\sim}{\longrightarrow} \mathbb{Q}/\mathbb{Z}.$$

(If $K_v = \mathbb{R}$ or $\mathbb{C}$, the map has image $\frac{1}{2}\mathbb{Z}/\mathbb{Z}$ or 0, respectively.) The Cassels–Tate pairing

$$\langle \cdot, \cdot \rangle_{\mathrm{CT}} : \text{Ш}(A/K)[m] \times \text{Ш}(\hat{A}/K)[m] \longrightarrow \mathbb{Q}/\mathbb{Z}$$

is then defined by the formula

$$\langle f, f' \rangle_{\mathrm{CT}} = \sum_v \mathrm{inv}_v(\ell_v).$$

Of course, there are many things to check, for example that the choices we made do not matter and that only finitely many of the terms in the sum are nonzero. We refer the reader to the references listed above for further details.

11   The Néron–Tate Canonical Height Pairing

Another important pairing on elliptic curves and abelian varieties is the canonical height pairing originally defined by Néron [16] and Tate [12]. This pairing on the group of points $A(K)$ defined over a number field K is closely related to the information-theoretic content of the points.

We start with the height of an algebraic number, and rather than giving the precise definition, which is unenlightening and somewhat technical, we instead describe what the height represents. (For the exact definition, and for additional information about heights, both canonical and non-canonical, see for example [2,6,11,21].) We define the height of an element $\alpha \in K$ to be (roughly)

$$h(\alpha) = \# \text{ of bits needed to describe the number } \alpha.$$

For example, if $K = \mathbb{Q}$ and $\alpha = a/b$ is written in lowest terms, then we can take

$$h(\alpha) = \log_2 |a| + \log_2 |b|$$

to be the number of bits required to store the two numbers a and b. In general, if $V \subset \mathbb{P}^n$ is an algebraic variety embedded in projective space and $P \in V(K)$, we define the height of P to be

$$h(P) = \text{the sum of the heights of the coordinates of } P.$$

So, roughly, it takes $h(P)$ bits to specify the coordinates of the point P.

Different embeddings of V into projective space will give different height functions. For an elliptic curve, one typically takes a Weierstrass equation

$$Y^2 + a_1 XY + a_3 Y = X^3 + a_2 X^2 + a_4 X + a_6,$$

and then the height of a point $P = (x, y)$ would be $h(P) = h(x) + h(y)$, or we could just take $h(P) = h(x)$.

Now let A/K be an abelian variety, which for the moment we assume is embedded in projective space, i.e., we are given a specific set of coordinates and equations that define A. Then it turns out that the height function

$$h : A(K) \longrightarrow \mathbb{R}$$

interacts with the group law on A in a very interesting way. For example, if we take a point $P \in A(K)$ and add it to itself m times, then the height $h(mP)$ is, more-or-less, equal to $m^2 h(P)$. This leads to the Néron–Tate construction.

Definition 1. *Let A/K be an abelian variety defined over a number field K, together with an embedding $A \subset \mathbb{P}^n$. The canonical height (relative to the embedding) is the function*

$$\hat{h} : A(K) \longrightarrow \mathbb{R}, \qquad \hat{h}(P) = \lim_{m \to \infty} \frac{1}{m^2} h(mP).$$

The canonical (or Néron–Tate) height pairing is the map

$$\langle \cdot, \cdot \rangle_{\text{NT}} : A(K) \times A(K) \longrightarrow \mathbb{R}, \qquad \langle P, Q \rangle_{\text{NT}} = \hat{h}(P + Q) - \hat{h}(P) - \hat{h}(Q).$$

The canonical height and its associated pairing have a number of remarkable properties, of which we mention only three.

Theorem 2. (a) *The canonical height satisfies*

$$\frac{1}{2}\langle P, P\rangle_{\mathrm{NT}} = \hat{h}(P) = h(P) + O(1).$$

Thus $\hat{h}(P)$ encodes the information-theoretic content of the point P.
(b) *The height pairing $\langle\,\cdot\,,\,\cdot\,\rangle_{\mathrm{NT}}$ is bilinear.*
(c) *The extension of $\langle\,\cdot\,,\,\cdot\,\rangle_{\mathrm{NT}}$ to the vector space $A(K) \otimes \mathbb{R}$ is non-degenerate, i.e., $\hat{h}$ is a positive definite quadratic form on the finite-dimensional real vector space $A(K) \otimes \mathbb{R}$.*

The Mordell–Weil theorem says that $A(K)$ is a finitely generated group. Let $P_1, \ldots, P_r \in A(K)$ be generators for $A(K)$ modulo its torsion subgroup. The *elliptic regulator*

$$\mathrm{Reg}(A/K) = \det\big(\langle P_i, P_j\rangle_{\mathrm{NT}}\big)_{1 \le i,j \le r}$$

plays an important role in the arithmetic of A, similar to the role played by the classical regulator in studying the group of units in a number field. In particular, it appears in the Birch–Swinnerton-Dyer conjectural formula for the leading coefficient of the L-series $L(E/K, s)$ at $s = 1$.

Canonical heights have been used in cryptography to show that certain potential attacks on the elliptic curve discrete logarithm problem (ECDLP) are not feasible. Recall that the fastest algorithm to solve the classical discrete logarithm problem on $\mathbb{F}_q^*$ is the *index calculus*, which has subexponential running time. Victor Miller, in his original paper [14], said that he did not see how to formulate an index calculus attack on the ECDLP. The author and Suzuki [22] used canonical heights and other tools to show that a direct index calculus attack on ECDLP is highly unlikely. Further, a reverse index attack [20] was also shown to be infeasible [7] using canonical heights. Indeed, Neal Koblitz, one of the inventors of elliptic curve cryptography, gave a talk at ECC 2000 with the provocative title

Miracles of the Height Function—A Golden Shield Protecting ECC.

Our description of the canonical height may seem somewhat *ad hoc*, so we now explain why it is a "natural pairing" in the same sense that the Weil and Tate pairings are natural pairings. To do this, we need to go back to the observation that the canonical height, as we have defined it, depends on the coordinates we use to describe the points of A, or equivalently, on the embedding of A into $\mathbb{P}^n$. So we first briefly recall how projective embeddings of varieties are related to divisors and divisor classes.

If a nonsingular variety V is embedded in $\mathbb{P}^n$, then the intersection of V with a hyperplane $H \subset \mathbb{P}^n$ gives a divisor $V \cap H$ on V. If we change to another hyperplane H', we get a divisor $V \cap H'$ that is linearly equivalent to $V \cap H$. Conversely, let D be a (positive) divisor on V and let $f_0, \ldots, f_n$ be a basis for the set of algebraic functions on V whose poles are no worse than D. Then we get a map $\phi_D = [f_0, \ldots, f_n]$ of V into $\mathbb{P}^n$. The divisor D is said to be *very ample* if the map ϕ_D is an embedding. A general theorem says that every divisor can be written as the difference of two very ample divisors. Finally, we say that two divisors D_1 and D_2 are *linearly equivalent* if their difference $D_1 - D_2$ is the divisor of an algebraic function on V. We write $D_1 \sim D_2$ to denote linear equivalence. The group of divisors is denoted $\mathrm{Div}(V)$, and the group of divisor classes modulo linear equivalence is denoted $\mathrm{Pic}(V)$ and is called the *Picard group of* V.

We have already (more-or-less) defined the height of points in projective space. Let V be a nonsingular variety. For each divisor $D \in \mathrm{Div}(V)$, we write $D = D_1 - D_2$ as a difference of very ample divisors and define a height function (relative to D) by

$$h_{V,D} : V(K) \longrightarrow \mathbb{R}, \qquad h_{V,D}(P) = h\big(\phi_{D_1}(P)\big) - h\big(\phi_{D_2}(P)\big).$$

There are a number of choices in this definition, but one can show that making different choices leads to a function $h'_{V,D}$ that differs from $h_{V,D}$ by a bounded amount, i.e.,

$$h_{V,D}(P) = h'_{V,D}(P) + O(1) \qquad \text{for all } P \in V(K).$$

These height functions are not yet canonical, but they have many useful properties, of which we state two.

Proposition 3. *Let V be a nonsingular variety defined over a number field K.*

(a) $h_{V,D_1+D_2} = h_{V,D_1} + h_{V,D_2} + O(1)$.

(b) *If $D_1, D_2 \in \mathrm{Div}(V)$ are linearly equivalent, then $h_{V,D_1} = h_{V,D_2} + O(1)$. In particular, if $D \in \mathrm{Div}(V)$ is linearly equivalent to 0, then $h_{V,D}$ is a bounded function.*

A fundamental principle in algebraic geometry is that divisor class relations on an algebraic variety determine many of the varieties geometric properties. On the other hand, height functons measure number theoretic properties of points. The content of Proposition 3 is that heights transform geometry, in the form of divisor class relations, into number theory, in the form of height relations.

Since an abelian variety A is a variety that is also a group, it is desirable to describe meaningful relationships between its geometry and its group law. The next theorem gives one such result in the form of an important divisor relation.

Theorem 3. (Theorem of the Cube) *Let A be an abelian variety, let $D \in \mathrm{Div}(A)$, let $m \in \mathbb{Z}$, and let $[m] : A \to A$ denote the multiplication-by-m map. Then*

$$[m]^* D \sim \left(\frac{m^2 + m}{2}\right) D + \left(\frac{m^2 - m}{2}\right) [-1]^* D.$$

In particular, if D is symmetric, i.e., if $[-1]^* D \sim D$, then $[m]^* D \sim m^2 D$.

The *canonical height relative to the divisor* D is the function

$$\hat{h}_{A,D} : A(K) \longrightarrow \mathbb{R}, \qquad \hat{h}_{A,D}(P) = \lim_{m \to \infty} \frac{1}{m^2} h_{A,D}(mP).$$

Using Proposition 3 and the Theorem of the Cube, one can show that the limit exists and depends only on the linear equivalence class of D. The canonical height is a quadratic form on $A(K) \otimes \mathbb{R}$, and if D is very ample, then it is a positive definite form.

The definition of $\hat{h}_{A,D}$ is very nice, but it depends on choosing a divisor D. As our earlier experience suggests, the functorial canonical height pairing will not be a bilinear map on $A(K) \times A(K)$, but instead it will be a bilinear map on $A(K) \times \hat{A}(K)$. This height is associated to a natural divisor class that lives on $A \times \hat{A}$.

Proposition 4. *We identify $\hat{A}$ with $\mathrm{Pic}^0(A)$, which is a subgroup of $\mathrm{Pic}(A)$. There is a divisor $\mathcal{P}$ on $A \times \hat{A}$, called the Poincaré divisor, satisfying:*

1. *Let $\xi \in \hat{A}$, so we can identify ξ with the divisor class of a divisor $D \in \mathrm{Div}(A)$. Then*
$$\mathcal{P} \cap (A \times \{\xi\}) \sim D.$$

2. *Let $P \in A$. Then*
$$\mathcal{P} \cap (\{P\} \times \hat{A}) \sim 0.$$

The Poincaré divisor is unique up to linear equivalence.

Theorem 4. *Let A be an abelian variety defined over a number field K. The canonical height on $A \times \hat{A}$ relative to the Poincaré divisor,*
$$\hat{h}_{A \times \hat{A}, \mathcal{P}} : A(K) \times \hat{A}(K) \longrightarrow \mathbb{R},$$

is a bilinear form, and it is non-degenerate in the sense that
$$\hat{h}_{A \times \hat{A}, \mathcal{P}}(P, Q) = 0 \text{ for all } P \in A(K) \quad \Longleftrightarrow \quad Q \in \hat{A}(K)_{\mathrm{tors}},$$
$$\hat{h}_{A \times \hat{A}, \mathcal{P}}(P, Q) = 0 \text{ for all } Q \in \hat{A}(K) \quad \Longleftrightarrow \quad P \in A(K)_{\mathrm{tors}}.$$

Acknowledgements. The author would like to thank Chris Rasmussen for his helpful suggestions.

References

1. Atiyah, M.F., Wall, C.T.C.: Cohomology of groups. In: Algebraic Number Theory (Proc. Instructional Conf., Brighton, 1965), pp. 94–115. Thompson, Washington (1967)
2. Bombieri, E., Gubler, W.: Heights in Diophantine geometry. New Mathematical Monographs, vol. 4. Cambridge University Press, Cambridge (2006)
3. Boneh, D., Franklin, M.: Identity-based encryption from the Weil pairing. In: Kilian, J. (ed.) CRYPTO 2001. LNCS, vol. 2139, pp. 213–229. Springer, Heidelberg (2001)
4. Cassels, J.W.S.: Arithmetic on curves of genus 1. IV. Proof of the Hauptvermutung. J. Reine. Angew. Math. 211, 95–112 (1962)
5. du Sautoy, M.: John Tate wins the Abel Prize (2010), http://www.abelprisen.no/en/prisvinnere/2010/marcus/

6. Hindry, M., Silverman, J.H.: Diophantine Geometry: An Introduction. Graduate Texts in Mathematics, vol. 201. Springer, New York (2000)

7. Jacobson, M.J., Koblitz, N., Silverman, J.H., Stein, A., Teske, E.: Analysis of the xedni calculus attack. Des. Codes Cryptogr. 20(1), 41–64 (2000)

8. Joux, A.: A one round protocol for tripartite Diffie-Hellman. In: Bosma, W. (ed.) ANTS 2000. LNCS, vol. 1838, pp. 385–393. Springer, Heidelberg (2000)

9. Koblitz, N.: Elliptic curve cryptosystems. Math. Comp. 48(177), 203–209 (1987)

10. Kolyvagin, V.A.: Finiteness of $E(\mathbb{Q})$ and $Ш(E,\mathbb{Q})$ for a subclass of Weil curves. Izv. Akad. Nauk SSSR Ser. Mat. 52(3), 522–540, 670–671 (1988)

11. Lang, S.: Fundamentals of Diophantine Geometry. Springer, New York (1983)

12. Lang, S.: Les formes bilinéaires de Néron et Tate. In: Séminaire Bourbaki, Soc. Math. France, Paris, vol. 8, Exp. No. 274, pp. 435–445 (1995)

13. Menezes, A.J., Okamoto, T., Vanstone, S.A.: Reducing elliptic curve logarithms to logarithms in a finite field. IEEE Trans. Inform. Theory 39(5), 1639–1646 (1993)

14. Miller, V.S.: Use of elliptic curves in cryptography. In: Williams, H.C. (ed.) CRYPTO 1985. LNCS, vol. 218, pp. 417–426. Springer, Heidelberg (1986)

15. Milne, J.S.: Arithmetic Duality Theorems. Perspectives in Mathematics, vol. 1. Academic Press Inc., Boston (1986)

16. Néron, A.: Quasi-fonctions et hauteurs sur les variétés abéliennes. Ann. of Math. 82(2), 249–331 (1965)

17. Rubin, K.: Tate-Shafarevich groups and L-functions of elliptic curves with complex multiplication. Invent. Math. 89(3), 527–559 (1987)

18. Serre, J.-P.: Local fields. Graduate Texts in Mathematics, vol. 67. Springer, New York (1979); Translated from the French by Marvin Jay Greenberg

19. Serre, J.-P.: Galois cohomology, English edn. Springer Monographs in Mathematics. Springer, Berlin (2002); Translated from the French by Patrick Ion and revised by the author

20. Silverman, J.H.: The xedni calculus and the elliptic curve discrete logarithm problem. Des. Codes Cryptogr. 20(1), 5–40 (2000)

21. Silverman, J.H.: The Arithmetic of Elliptic Curves, 2nd edn. Graduate Texts in Mathematics, vol. 106. Springer, Dordrecht (2009)

22. Silverman, J.H., Suzuki, J.: Elliptic curve discrete logarithms and the index calculus. In: Ohta, K., Pei, D. (eds.) ASIACRYPT 1998. LNCS, vol. 1514, pp. 110–125. Springer, Heidelberg (1998)

23. Tate, J.: WC-groups over p-adic fields. In: Textes des Conférences; Exposés 152 à 168; 2e éd. corrigée, Exposé 156. Séminaire Bourbaki; 10e année: 1957/1958, vol. 13, Secrétariat mathématique, Paris (1958)

24. Tate, J.: Letter to J.W.S. Cassels (August 1, 1962) (unpublished)

25. Tate, J.: Duality theorems in Galois cohomology over number fields. In: Proc. Internat. Congr. Mathematicians (Stockholm, 1962), pp. 288–295. Inst. Mittag-Leffler, Djursholm (1963)

Compact Hardware for Computing the Tate Pairing over 128-Bit-Security Supersingular Curves

Nicolas Estibals

Équipe-projet CARAMEL, LORIA, Nancy Université / CNRS / INRIA
Campus Scientifique, BP 239 54506 Vandoeuvre-lès-Nancy Cedex, France

Abstract. This paper presents a novel method for designing compact yet efficient hardware implementations of the Tate pairing over supersingular curves in small characteristic. Since such curves are usually restricted to lower levels of security because of their bounded embedding degree, aiming for the recommended security of 128 bits implies considering them over very large finite fields. We however manage to mitigate this effect by considering curves over field extensions of moderately-composite degree, hence taking advantage of a much easier tower field arithmetic. This technique of course lowers the security on the curves, which are then vulnerable to Weil descent attacks, but a careful analysis allows us to maintain their security above the 128-bit threshold.

As a proof of concept of the proposed method, we detail an FPGA accelerator for computing the Tate pairing on a supersingular curve over $\mathbb{F}_{3^{5 \cdot 97}}$, which satisfies the 128-bit security target. On a mid-range Xilinx Virtex-4 FPGA, this accelerator computes the pairing in 2.2 ms while requiring no more than 4755 slices.

Keywords: Tate pairing, supersingular elliptic curves, FPGA implementation.

1 Introduction

Pairings were first introduced in cryptography in 1993 by Menezes, Okamoto, & Vanstone [36] and Frey & Rück [24] as an attack against the elliptic curve discrete logarithm problem (ECDLP) for some families of curves over finite fields. Since then, constructive properties of pairings have also been discovered and exploited in several cryptographic protocols: starting independently in 2000 with Joux's one-round tripartite Diffie–Hellman key agreement [31] and Sakai–Ohgishi–Kasahara cryptosystem [46], many others have followed, such as Mitsunari–Sakai–Kasahara broadcast encryption scheme [39], Boneh–Franklin identity-based encryption [12] or Boneh–Lynn–Shacham short signature [13] for instance. Pairings nowadays being the cornerstone of various protocols, their efficient implementation on a wide range of targets became a great challenge, especially on low-resource environments.

M. Joye, A. Miyaji, and A. Otsuka (Eds.): Pairing 2010, LNCS 6487, pp. 397–416, 2010.

Although many FPGA implementations of pairing accelerators have been proposed [34,43,30,2,6,9,7,47], none of them allows to reach the AES-128 security level. However, recent ASIC implementations of pairings over Barreto–Naehrig (BN) [4] curves with 128 bits of security have been published [22,33]. The main difficulty for computing a pairing at the 128-bit security level is to implement an efficient arithmetic over a quite large finite field.

In contrast with the ASIC implementation, we chose to implement pairings over supersingular elliptic curves over small-characteristic finite fields so as to benefit from the many optimizations available in the literature. As a drawback, since supersingular curves are restricted to low embedding degrees, this implies considering unbalanced settings, where the curve offers potentially much more security than the required 128 bits. Nonetheless we took advantage of this excess of security and defined our curves over finite fields of composite extension degree: on the one hand, the curves might be weaker because of, for instance, the Gaudry–Hess–Smart attack [27,26,17]; on the other hand, the arithmetic algorithm can really benefit from this tower field structure. This article is devoted to the demonstration that this compromise is very effective in the context of a low-resources hardware implementation.

After a reminder on the Tate pairing and its security in a general context (Section 2), we present the consequences on security of defining an elliptic curve over a composite-extension field (Section 3). We then detail the algorithms for computing the Tate pairing over such curves in Section 4 and present a low-area FPGA accelerator implementing these algorithms for a test-case curve in Section 5. Finally we report our performance results and compare them against other implementations from the literature (Section 6) and conclude in Section 7.

2   Definition and Security of the Tate Pairing

Given an elliptic curve E defined over a finite field $\mathbb{F}_q$, take ℓ a prime number dividing the cardinal of the curve $\#E(\mathbb{F}_q)$. The embedding degree k of E is then defined as the smallest integer such that $\ell \mid q^k - 1$, that is to say such that the group of ℓ-th roots of unity $\mu_\ell = \{x \in \overline{\mathbb{F}_q} \mid x^\ell = 1\}$ is in $\mathbb{F}_{q^k}^*$. Assuming further that $k > 1$ and that there are no points of order ℓ^2 in $E(\mathbb{F}_{q^k})$, we can then define the Tate pairing over E as the map:

$$e : E(\mathbb{F}_q)[\ell] \times E(\mathbb{F}_{q^k})[\ell] \to \mathbb{F}_{q^k}^* / \left(\mathbb{F}_{q^k}^*\right)^\ell \cong \mu_\ell,$$

where $E(\mathbb{F}_q)[\ell] = \{P \in E(\mathbb{F}_q) \mid [\ell]P = \mathcal{O}\}$ denotes the $\mathbb{F}_q$-rational ℓ-torsion subgroup. The embedding degree k, also called security multiplier in this context, acts as a cursor to adjust the size of the multiplicative group $\mathbb{F}_{q^k}^*$ with respect to that of $\mathbb{F}_q$, which directly constrains $\#E(\mathbb{F}_q)$ to Hasse's bounds, therefore limiting the achievable values of ℓ. Given that the discrete logarithm problem (DLP) is exponential in the subgroup $E(\mathbb{F}_q)[\ell]$ but subexponential in the finite field $\mathbb{F}_{q^k}^* \supset \mu_\ell$ (*cf.* Section 2.2), one might want to choose a curve giving a security multiplier k that balances the security on both the input and the output of the Tate pairing.

As we are targeting the AES-128 security level, elliptic curves with an embedding degree between 12 and 15 seem to be a good choice. Barreto–Naehrig (BN) curves are a family of such curves with prime cardinal $\ell = \#E(\mathbb{F}_q)$ and embedding degree $k = 12$ [4]; as a result BN curves perfectly balance the security between the ℓ-torsion and μ_ℓ at the 128-bit level. However, since BN curves are defined over prime fields, computing a pairing over them requires expensive modular arithmetic, which is far less better-suited to hardware implementation than arithmetic over small-characteristic finite fields. Last but not least, BN curves are ordinary curves: point doubling and tripling formulae are not as efficient as in the supersingular case in characteristic 2 and 3 respectively.

As a consequence, we chose to consider supersingular elliptic curves even if their embedding degree is bounded by 6 [3]. Due to this bound, the security on the curve will be too high with respect to the security on μ_ℓ. We however decided to take advantage of this: using finite fields with composite extension degree will decrease the security on the curves but make the field arithmetic better suited to low-resource hardware implementations. Those points will be detailed and quantified in the next two sections.

We now detail the definition, security and computation of the Tate pairing over the considered supersingular elliptic curves.

2.1   Pairing over Supersingular Elliptic Curves

Our study focuses on pairings on supersingular curves over finite fields $\mathbb{F}_q$ with $q = p^m$ and $p = 2$ or 3. We thus define the two following families [3]:

$$E_{2,b}/\mathbb{F}_2 : y^2 + y = x^3 + x + b, \quad \text{where } b \in \{0,1\}; \text{ and}$$
$$E_{3,b}/\mathbb{F}_3 : \quad y^2 = x^3 - x + b, \quad \text{where } b = \pm 1.$$

When m is coprime to 2 and 6 in characteristic 2 and 3 respectively, the cardinal of those curves reaches the Hasse bounds:

$$\#E_{2,b}(\mathbb{F}_q) = 2^m \pm 2^{\frac{m+1}{2}} + 1,$$
$$\#E_{3,b}(\mathbb{F}_q) = 3^m \pm 3^{\frac{m+1}{2}} + 1.$$

Moreover, their embedding degree is 4 and 6 in characteristic 2 and 3, respectively. Thanks to their supersingularity, there exists a distortion map over those elliptic curves, mapping the $\mathbb{F}_q$-rational ℓ-torsion group to another subgroup of $E(\mathbb{F}_{q^k})[\ell]$:

$$\delta : E(\mathbb{F}_q)[\ell] \to E(\mathbb{F}_{q^k})[\ell],$$

which is used to define the modified Tate pairing as:

$$\hat{e} : \begin{cases} E(\mathbb{F}_q)[\ell] \times E(\mathbb{F}_q)[\ell] & \to \mathbb{F}_{q^k}^* / \left(\mathbb{F}_{q^k}^*\right)^\ell \cong \mu_\ell \\ (P, Q) & \mapsto e(P, \delta(Q)) \end{cases}.$$

One can furthermore show that $\hat{e}$ is not degenerate. We refer the reader to [3] and [9, Table I] for the mathematical details of pairing construction over supersingular curves.

2.2   Attacks against Pairings over Supersingular Curves

The security of the pairing is determined by the difficulty of the discrete logarithm problem (DLP) on the input curve and on the output multiplicative group.

Since ℓ is a prime, the best known algorithm to attack the DLP on the ℓ-torsion is Pollard's ρ method [42], which requires an average of $\sqrt{\pi\ell/2}$ group operations. As Duursma $et\ al.$ showed in [25,19], we should take into account the group of automorphisms on the curve, which has order 24 and 12 in characteristic 2 and 3, respectively, [48, Chap. III, §10] as well as the m iterated Frobenius endomorphisms $(x,y) \mapsto (x^{p^i}, y^{p^i})$, for $0 \leqslant i < m$ as they allow to speed up Pollard's ρ method by a $\sqrt{m \cdot \#\mathrm{Aut}(E)}$ factor. All in all the average cost of Pollard's method on $E(\mathbb{F}_q)[\ell]$ is:

$$\begin{cases} \sqrt{\dfrac{\pi\cdot\ell}{48m}} & \text{if } p = 2, \text{ and} \\[2mm] \sqrt{\dfrac{\pi\cdot\ell}{24m}} & \text{if } p = 3. \end{cases}$$

Additionally, one may attack the DLP on $\mu_\ell \subset \mathbb{F}_{q^k}^*$; this is the fundamental idea behind the attacks of Menezes, Okamoto, & Vanstone [36] and Frey & Rück [24]. Since the ℓ-th roots of unity are defined in the multiplicative group of a finite field, the DLP may be attacked by sieving algorithms. In our case, where the characteristic p is 2 or 3, one can use the function field sieve (FFS) [1]; the complexity of this attack is subexponential:

$$\exp\left(\left(\frac{32}{9} + o(1)\right)^{\frac{1}{3}} \cdot \log^{\frac{1}{3}} q^k \cdot \log\log^{\frac{2}{3}} q^k\right).$$

If we consider our 128-bit security level target, we need to take m between 1100 and 1200 in characteristic 2 and around 500 in characteristic 3.

3   Elliptic Curves over Composite-Extension Fields

We examine, in this section, the consequences on security of defining supersingular elliptic curves over a finite field of the form $\mathbb{F}_{q^n}$, where $q = p^m$, n is a small integer and m a prime. This corresponds to substituting q^n for q and $m \cdot n$ for m in the previous section.

It is important to remark that such elliptic curves defined over composite-extension fields have already been described for cryptographic use under the name Trace-Zero Variety (TZV) [23]. Applying the Weil descent to $E(\mathbb{F}_{q^n})$, we obtain an isomorphic variety $W_E(\mathbb{F}_q)$ which is also isomorphic to the product $E(\mathbb{F}_q) \times B(\mathbb{F}_q)$ where $B(\mathbb{F}_q)$ is the TZV. It is a variety defined over the base field $\mathbb{F}_q$ which might also be represented as the quotient $E(\mathbb{F}_{q^n})/E(\mathbb{F}_q)$. As we consider in this work an ℓ-torsion subgroup of $E(\mathbb{F}_{q^n})$ which is not contained in $E(\mathbb{F}_q)$, this ℓ-torsion is a subgroup of the corresponding TZV. In the context of pairings, TZVs have also been studied, chiefly for point compression [44, 45, 16].

3.1   The Gaudry–Hess–Smart Attack

As soon as one defines a curve on a field of composite extension degree, one should also consider other attacks: the Weil descent can indeed be applied on those curves and have some "destructive facets." The Weil descent allows one to map an elliptic curve defined over $\mathbb{F}_{q^n}$ to the Jacobian of a curve of genus at least n over $\mathbb{F}_q$.

Thus the discrete logarithm problem on the elliptic curve defined over $\mathbb{F}_{q^n}$ might be transported to the DLP on the Jacobian of a genus-n curve over $\mathbb{F}_q$. This last DLP can then be solved using an index calculus algorithm. Gaudry, Hess, & Smart have shown that this attack (GHS) runs in $\tilde{O}(q^{2-\frac{2}{n}})$ in some cases (Weil restrictions) [27]. More generally Gaudry [26] and Diem [17] showed that this also holds in the general case, but with a very bad dependency in n (hidden in the big-O notation).

3.2   The Static Diffie–Hellman Problem

Recent studies [28,32] showed that defining a curve over a finite field of composite extension degree makes it weaker regarding the static Diffie–Hellman problem (SDH). The SDH problem on a curve consists in: given two points $P, [d]P \in E(\mathbb{F}_q)$ (where d is a secret integer) and an oracle $Q \mapsto [d]Q$, compute $[d]R$ where R is randomly chosen point.

The cryptographic consequence of solving SDH problem is breaking the Diffie–Hellman key exchange protocol when one participant never changes his private key, as it occurs in the El Gamal encryption scheme for instance [20].

Granger discovered the best known algorithm that solves the SDH problem on elliptic curves defined over a field of composite extension degree $\mathbb{F}_{q^n}$ with $O(q^{1-\frac{1}{n+1}})$ calls to the oracle and in $\tilde{O}(q^{1-\frac{1}{n+1}})$ time [28].

One should notice that the attacker not only needs a great computational power but also a great number of calls to the oracle: a simple but efficient protection against this attack is revoking a key after a certain amount of use.

3.3   Finding Curves with 128-Bit Security Level

To the best of our knowledge, the literature does not mention any other attack on curves over fields of composite extension degree.

In order to find suitable curves for our method, we enumerated all the supersingular curves of characteristic 2 and 3 on fields with moderately-composite extension degrees $m \cdot n$ ($n < 15$) large enough for the 128-bit security level. We then evaluated an approximation (constants hidden in big-O are not taken into

Table 1. Different curves and their security in bits against the different known attacks. A security of N bits means that approximately 2^N operations are required to perform the attack.

q	n	b	$\log_2 \ell$	Cost of the attacks (bits)			
				Pollard's ρ	FFS	GHS	SDH
2^{1117}	1	1	1076	531	128	–	–
2^{367}	3	1	698	342	128	489	275
2^{227}	5	1	733	359	129	363	189
2^{163}	7	1	753	370	129	279	142
2^{127}	9	1	487	236	130	225	114
2^{103}	11	1	922	454	129	187	94
2^{89}	13	0	1044	515	164	130	82
2^{73}	15	0	492	239	136	127	68
3^{503}	1	1	697	342	132	–	–
3^{97}	**5**	**−1**	**338**	**163**	**130**	**245**	**128**
3^{67}	7	−1	612	300	129	182	92
3^{53}	11	−1	672	330	140	152	77
3^{43}	13	1	764	376	138	125	63

account) of the computation time of each of the attacks mentioned in the paper: Pollard's ρ, FFS, GHS and SDH. A selection of curves reaching the 128-bit level of security is given in Table 1; since that is not necessarily a security issue for all protocols, we also present curves that are not resistant to Granger's SDH attack.

The main difficulty in computing Table 1 is to factor the cardinal of the different curves because they contains more than 350 digits in characteristic 2 and 240 in characteristic 3. Luckily those cardinals are the Aurifeuillean factors of Cunningham numbers and many of them are referenced in the factor tables maintained by Wagstaff [49] and Leyland [35].

The security estimations given in Table 1 confirm the intuition: the more composite the extension degree of the field of definition, the more effective the attacks using Weil descent, until they become the best attack on the curves.

As a proof of concept, we finally chose to implement the pairing over the supersingular curve $E_{3,-1}$ over $\mathbb{F}_{3^{5\cdot97}}$, as this curve has an embedding degree equal to 6 and is resistant to all the attacks, even for the SDH problem.

4   Computation of the Tate Pairing over Composite-Extension Fields

As we have identified some curves that allow us to reach the 128-bit level of security, we now focus on the algorithms for computing the pairing over such curves.

4.1   Algorithms for Computing the Tate Pairing

The computation of the Tate pairing is split into two parts: Miller's loop [37,38] and a final exponentiation in the multiplicative group $\mathbb{F}_{q^{k \cdot n}}^{*}$.

Many improvements of Miller's algorithm have been published since its discovery. Duursma & Lee adapted it to exploit the simple point-tripling formulae in characteristic 3 by turning the double-and-add into a triple-and-add algorithm [18]. Furthermore Barreto *et al.* put forward the η_T approach which divides by two the length of the loop by exploiting the action of the Verschiebung on the ℓ-torsion [5].[1] Those improvements and a careful implementation of the arithmetic of the extension over $\mathbb{F}_{q^{k \cdot n}}$ leads to the algorithms presented by Beuchat *et al.* in [8,6].

To implement the pairing of our test case, we chose the unrolled loop algorithm in [8, Algorithm 5] because it minimizes the number of multiplications on the field of definition $\mathbb{F}_{q^n}$ which represents the major cost on a field large enough to reach the AES-128 security level. Moreover this algorithm requires only additions, multiplications and cubings over $\mathbb{F}_{q^n}$ but not any cube rooting; therefore it represents a substantial saving in hardware resources requirements.

We have now determined the sequence of operations in $\mathbb{F}_{q^n}$ to compute the η_T pairing over $\mathbb{F}_{q^n}$. Nonetheless we want to design compact hardware to execute them: the datapath of a circuit directly handling elements of $\mathbb{F}_{q^n}$ would be very large. Therefore we take advantage of the composite extension degree of our field of definition and implement the pairing as sequence of operations over $\mathbb{F}_q$: the datapath of a coprocessor dealing with elements of $\mathbb{F}_q$ only will be much smaller. Thus we have to express the arithmetic of $\mathbb{F}_{q^n}$ in terms of operation over $\mathbb{F}_q$ in an efficient way.

4.2   Representation and Computation over the Extension

Pairing computation requires a large number of multiplications. Using normal basis would thus be very harmful. As a consequence $\mathbb{F}_{q^n}$ is represented using a polynomial basis: $\mathbb{F}_{q^n} \cong \mathbb{F}_q[X]/(f(X))$ where f is a degree-n irreducible polynomial over $\mathbb{F}_q$. Hence an element of $\mathbb{F}_{q^n}$ is represented as a polynomial of degree at most $n - 1$ over $\mathbb{F}_q$, and operations over $\mathbb{F}_{q^n}$ are mapped to operations over $\mathbb{F}_q[X]$ followed if necessary by a reduction modulo f.

The irreducible polynomial f could be taken among all irreducible polynomials of degree n over $\mathbb{F}_q$ but we restricted this choice to polynomials over $\mathbb{F}_p$ in order to avoid multiplications over $\mathbb{F}_q$ during the different reductions modulo f. This is possible because n is coprime to m. We also chose f to have a low Hamming weight, *i.e.* a trinomial or a pentanomial, so as to further reduce the cost of the reductions.

Frobenius automorphism over $\mathbb{F}_{q^n}$. During the pairing computation, many iterated applications of the Frobenius, *i.e.* p^i-th powering, are required. By linearity of this operation, we have:

---

[1] The η_T pairing is in fact a power of the actual Tate pairing but the conversion between the two is free [6].

$$(a_0 + a_1 X + \cdots + a_{n-1} X^{n-1})^{p^i} = a_0^{p^i} + a_1^{p^i} X^{p^i} + \cdots + a_{n-1} X^{(n-1) \cdot p^i}.$$

Moreover we have that $X^{p^n} \equiv 1 \pmod{f}$ because f is defined over $\mathbb{F}_p$. Therefore computing the i-th iterated Frobenius over $\mathbb{F}_{q^n}$ is tantamount to computing the i-th iterated Frobenius over all coefficients and then applying a linear combination on them that only depends on the value of $i \bmod n$.

Multiplications over $\mathbb{F}_{q^n}$. Multiplication is the most expensive operation and it can be greatly optimized by using subquadratic multiplication schemes. Choosing the best algorithm to compute the products of two degree-$(n - 1)$ polynomials depends on many criteria and we studied how different solutions fit our case.

Many subquadratic multiplication algorithms can be used: Karatsuba, Montgomery's Karatsuba-like formulae [40,21], or CRT-based algorithms [14,15]. The common point between those algorithms is that they can all be expressed as the linear combination of a set of products of linear combinations of the coefficients of the operands.

The Toom–Cook algorithm and its variants cannot be used easily in the case of polynomials over low-characteristic fields, as it is based on an evaluate–interpolate scheme. To be efficient, evaluation points, their inverse, and their successive powers should have a small representation. However, we cannot find enough "simple elements" in low-characteristic fields: taking interpolation points in $\mathbb{F}_q$ instead of $\mathbb{F}_p$ will increase the number of multiplications and defeat the whole point of the method.

Furthermore, as we will see in Section 5.1, additions do not have a negligible cost when compared to multiplications as it is often assumed in estimations of multiplication complexity. Thus we have to express the formulae given by the different algorithms and count the total number of operations of each type.

Inversion over $\mathbb{F}_{q^n}$. During the final exponentiation step of the pairing computation, an inversion over $\mathbb{F}_{q^n}$ has to be carried out. Because there is only one inversion in the whole pairing computation, there is no gain to dedicate specific hardware resources to speed up its computation. However, thanks to the Itoh–Tsujii algorithm [29] which consists in applying Fermat's little theorem, the inversion over $\mathbb{F}_{q^n}$ is computed with $(n - 1) \cdot m$ applications of the Frobenius in $\mathbb{F}_{q^n}$, some multiplications over $\mathbb{F}_{q^n}$ and one inversion over $\mathbb{F}_q$. We also used another Itoh–Tsujii's algorithm to compute this last inversion over $\mathbb{F}_q$ and then do not need any other inversion since inversion over $\mathbb{F}_p$ is the identity when $p = 2$ or 3.

4.3   Our Test Case: $\mathbb{F}_{3^{5 \cdot 97}}$

We chose to construct the extension for our test case as $\mathbb{F}_{3^{5 \cdot 97}} \cong \mathbb{F}_{3^{97}}[X]/(X^5 - X + 1)$, $\mathbb{F}_{3^{97}}$ itself being represented as $\mathbb{F}_3[t]/(t^{97} + t^{16} - 1)$. Thus we evaluated multiplication over the extension cost thanks to different algorithms (*cf.* Table 2):

- the quadratic and so-called schoolbook method;
- one-level Karatsuba, where the sub-products are computed using the schoolbook method;
- recursive Karatsuba, where the sub-products are also computed thanks to Karatsuba algorithm;
- Montgomery's Karatsuba-like formulae [40];
- algorithm based on the Chinese Remainder Theorem (CRT) by Cenk & Özbudak [14] (*cf.* Section A for detailed algorithm).

Since $n = 5$ is odd, Montgomery's trick [40, Section 2.3] for applying the Karatsuba formulae can be used and saves one extra sub-product.

As we have now expressed a variety of algorithms for multiplication over $\mathbb{F}_{3^{5 \cdot 97}}$, choosing one of them is a matter of algorithm–architecture co-design. Indeed, timing for each algorithm heavily depends on:

- the cost of multiplication on $\mathbb{F}_{3^{97}}$ compared to the addition,
- the data dependencies, and
- the scheduling of the operations in regards to the memory architecture.

Table 2. Cost of different multiplication algorithms over $\mathbb{F}_{3^{5 \cdot 97}}$

Algorithm	Multiplications over $\mathbb{F}_{3^{97}}$	Additions over $\mathbb{F}_{3^{97}}$	Add./Mul. Ratio
Schoolbook	25	8	0.32
One-level Karatsuba (Montgomery's trick)	21	29	1.38
Recursive Karatsuba	15	39	2.60
Recursive Karatsuba (Montgomery's trick)	14	43	3.07
Montgomery's Karatsuba-like formulae [40]	13	54	4.153
Cenk & Özbudak [14]	**12**	**53**	**4.42**

Finally, it turned out that the algorithm by Cenk & Özbudak [14] best fitted our arithmetic coprocessor (*cf.* Section 5). In conclusion, the overall cost of the arithmetic over the extension field $\mathbb{F}_{3^{5 \cdot 97}}$ is presented in Table 3. Table 4 summarizes the number of operations over the field $\mathbb{F}_{3^{97}}$ and its extension $\mathbb{F}_{3^{5 \cdot 97}}$ needed to perform Miller's loop and the final exponentiation from [8].

5   Hardware Accelerator for Computing the Tate Pairing

5.1   An Arithmetic Coprocessor over $\mathbb{F}_q$

As we have now reduced the pairing computation to a sequence of operations over $\mathbb{F}_q$ with $q = p^m$, we need a coprocessor able to perform additions, multiplications

Table 3. Cost of the arithmetic over $\mathbb{F}_{3^{5\cdot97}}$ in terms of operations over $\mathbb{F}_{3^{97}}$

	$\times$	$+$	$(.)^3$
Addition	$-$	5	$-$
Multiplication	12	53	$-$
Iterated Frobenius, $(.)^{3^i}$ where $i \equiv 0 \pmod 5$	$-$	$-$	$5i$
where $i \equiv 1 \pmod 5$	$-$	5	$5i$
where $i \equiv 2 \pmod 5$	$-$	6	$5i$
where $i \equiv 3 \pmod 5$	$-$	8	$5i$
where $i \equiv 4 \pmod 5$	$-$	7	$5i$
Inverse	41	129	484

Table 4. Count of operations for full-pairing computation over $\mathbb{F}_{3^{5\cdot97}}$, and the corresponding cost over $\mathbb{F}_{3^{97}}$

	$\times$	$+$	$(.)^3$	$1/.$
$\mathbb{F}_{3^{5\cdot97}}$	3104	13127	4123	1
$\mathbb{F}_{3^{97}}$	37289	253314	21099	$-$

and Frobenius (squarings and cubings) over this field. To this intent, we chose the coprocessor that Beuchat *et al.* developed for the final exponentiation in [10].

The architecture of this coprocessor is reproduced in Fig. 1 and is composed of three units running in parallel: a register file implemented by means of a dual-ported RAM, a unit performing additions and Frobenius applications, and a parallel-serial multiplier. Several direct feedback paths exist between the inputs and outputs of the units, for instance allowing a product to be used in an addition without having to go through the register file: this allows us to save time while decreasing the pressure on the memory, which is a major bottleneck of the architecture.

Frobenius computations are quite scarce in the overall pairing algorithm (*cf.* Table 4) but long sequences of iterated squarings or cubings occur several times. The coprocessor is designed to fit this observation: the addition unit shares most of its datapath with a Frobenius unit which can carry out both single and double applications of the Frobenius in one clock cycle. One should also notice that there is a direct feedback loop from its output to one of its inputs so as to further speed up sequences of Frobenius.

Products are processed in a parallel-serial fashion: at each cycle the first operand is multiplied by D coefficients of the second operand. The complete multiplication over $\mathbb{F}_{p^m}$ is then computed in $\lceil \frac{m}{D} \rceil$ clock cycles. D is a parameter of the processor and is chosen as trade-off between computation time of the multiplication and the operating frequency (a large value of D lengthens the critical path and this deteriorates the frequency).

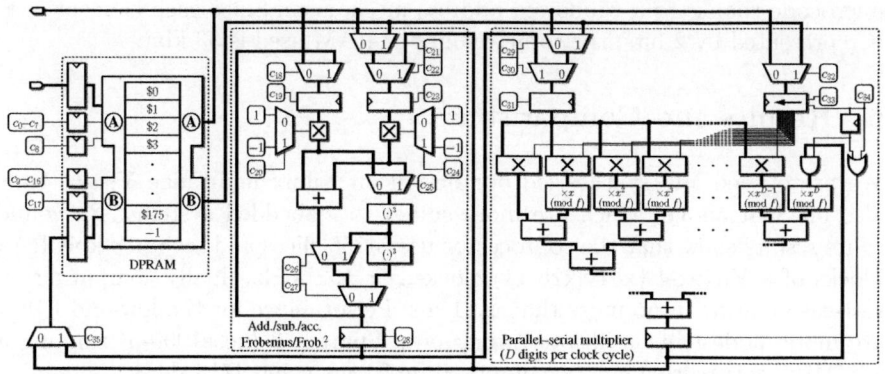

Fig. 1. Finite field coprocessor over $\mathbb{F}_q$

In our case of computing the Tate pairing over $\mathbb{F}_{3^{5 \cdot 97}}$, we chose $D = 14$. The product on $\mathbb{F}_{3^{97}}$ then takes 7 clock cycles, *i.e.* 7 times longer than an addition. Given this cost ratio between multiplications and additions, the multiplication algorithm over $\mathbb{F}_{3^{5 \cdot 97}}$ by Cenk & Özbudak fit best the coprocessor, that is to say we managed to find a scheduling of the algorithm that hides all the additions behind the 12 multiplications over $\mathbb{F}_{3^{97}}$. A multiplication algorithm with less sub-products and more additions would not yield a better execution time since the bottleneck would be in the memory access. Indeed memory ports are near to be saturated in our scheduling of Cenk & Özbudak's algorithm.

5.2 Micro- and Macrocode

Considering the total number of multiplications over $\mathbb{F}_q$ (*cf.* Table 4) and their cost, the pairing needs a minimum of 260 000 clock cycles to be calculated. During those cycles, the 36 control bits (the c_i's in Fig. 1) should be set: this represents a total amount of 10 Mbit of memory for the pairing program. Thus we cannot store those control bits directly in an instruction memory: it would use up much more resources than the coprocessor itself.

In order to reduce instruction memory requirements, we implemented two levels of code. In the lower one, the microcode, we implemented the arithmetic over the extension $\mathbb{F}_{3^{5 \cdot 97}}$. These operations are called in a macro-program that computes the actual pairing. Given that the non-reduced pairing is computed thanks to Miller's loop, we also constructed a loop mechanism on the macrocode.

Finally the implementation of the Tate pairing over $E(\mathbb{F}_{3^{5 \cdot 97}})$ is a sequence of 464 macro-operations which takes 428 853 clock cycles to be executed. Although microcoding implies a loss of parallelism, it allows us to drastically reduce the size of the instruction memory, which now fits in 24 kbit.

The register file is split into two parts: the first one contains 32 macro-variables (elements of $\mathbb{F}_{3^{5 \cdot 97}}$) and the second serves as a scratch space of 16 temporary variables (elements of $\mathbb{F}_{3^{97}}$) for use inside the microcode. Macro-variables are blocks of 5 consecutive addresses in the register file that are accessed in the microcode thanks to a windowed address mechanism. Since each element of $\mathbb{F}_3$ is represented by 2 bits, the total amount of RAM used is 33 kbit.

6   Results and Comparisons

We prototyped and synthesized our design on Xilinx mid-range Virtex 4 and also on Spartan-3's, which are more suited to embedded systems. Place-and-route results show that the coprocessor uses 4755 slices and seven 18 kbit RAM blocks of a Virtex-4 (xc4vlx25-11) clocked at 192 MHz, finally computing our test-case pairing in no more than 2.11 ms. Performance for the low-end FPGA are more modest but still interesting: on a Spartan 3 (xc3s1000-5) running at 104 MHz, this pairing can be computed in 4.1 ms using 4713 slices.

To the best of our knowledge, this design is the first FPGA implementation of a pairing reaching 128 bits of security; thus we compared our design to FPGA implementations of less secure pairings (Table 5), along with ASIC (Table 6) and software (Table 7) implementations of 128-bit security pairings.

Table 5. Tate pairing computation on FPGA

	Curve	Sec. (bits)	FPGA	Area (slices)	Freq. (MHz)	Time (μs)	Area×time (slices.s)
Barenghi et al. [2]	$E(\mathbb{F}_{p_{512}})$	87	xc2v8000-5	33857	135	1610	54.5
Shu et al. [47]	$E(\mathbb{F}_{2^{457}})$	88	xc4vlx200-10	58956	100	100.8	5.94
Beuchat et al. [9]	$E(\mathbb{F}_{2^{457}})$	88	xc4vlx100-11	44223	215	7.52	0.33
Beuchat et al. [6]	$E(\mathbb{F}_{2^{459}})$	89	xc2vp20-6	8153	115	327	2.66
Beuchat et al. [6]	$E(\mathbb{F}_{3^{193}})$	89	xc2vp20-6	8266	90	298	2.46
Beuchat et al. [9]	$E(\mathbb{F}_{3^{193}})$	89	xc4vlx200-11	47260	179	9.33	0.44
Shu et al. [47]	$E(\mathbb{F}_{2^{557}})$	96	xc4vlx200-10	37931	66	675.5	25.62
Beuchat et al. [9]	$E(\mathbb{F}_{2^{557}})$	96	xc4vlx200-11	55156	139	13.2	0.73
Beuchat et al. [9]	$E(\mathbb{F}_{3^{239}})$	97	xc4vlx200-11	66631	179	11.5	0.77
Beuchat et al. [9]	$E(\mathbb{F}_{2^{613}})$	100	xc4vlx200-11	62418	143	15.1	0.95
Beuchat et al. [9]	$E(\mathbb{F}_{2^{691}})$	105	xc4vlx200-11	78874	130	18.8	1.48
Beuchat et al. [9]	$E(\mathbb{F}_{3^{313}})$	109	xc4vlx200-11	97105	159	16.9	1.64
This paper	$E(\mathbb{F}_{3^{5 \cdot 97}})$	**128**	**xc4vlx25-11**	**4755**	**192**	**2227**	**10.59**
			xc3s1000-5	**4713**	**104**	**4113**	**19.38**

The literature about pairing computation on FPGAs only focuses on low-security pairings because they already reach the limit of the available FPGA resources. Indeed the designs presented in [2, 6, 7, 9, 47] have a datapath that handles the field of definition of their respective curves and thus increasing the security means increasing the designs' area. In contrast our approach allows us to "split" elements of the field of definition into smaller parts and thus achieve a smaller area: the coprocessor is very compact compared to the other published architectures. However we have to pay the price of security in terms of computation time: computing a pairing over $E(\mathbb{F}_{3^{5 \cdot 97}})$ (128 bits of security) with our processor is 130 times slower than computing one over $E(\mathbb{F}_{3^{313}})$ (109 bits of security) with Beuchat et al.'s hardware [9]. It is however 20 times smaller.

Table 6. Performance of some ASIC accelerators for pairings over BN-curves of AES-128 security level

	Platform	Area without RAM	RAM	Frequency	Time
Fan et al. [22]	130 nm ASIC	113 kGates	32 kbits	204 MHz	2.91 ms
Kammler et al. [33]	130 nm ASIC	97 kGates	64 kbits	308 MHz	15.8 ms

Table 7. Computation of pairing at the AES-128 security level in software

	Curve	Processor	Time (ms)
Beuchat et al. [11]	Supersingular over $\mathbb{F}_{2^{1223}}$	2.4 GHz Intel Core2	11.9
	Supersingular over $\mathbb{F}_{3^{509}}$	2.4 GHz Intel Core2	7.59
Naehrig et al. [41]	257 bit BN curve	2.4 GHz Intel Core2	1.87

The first ASIC implementations of pairings with 128 bits of security were presented in [33, 22]. The two implementations use BN-curves so as to exploit their optimal embedding degree $k = 12$ while targeting 128 bits of security. Although we did not synthesize our design on ASIC, a very rough and pessimistic estimation places our coprocessor around the 100-kGate mark, not counting the register file. That is to say roughly the same area as required by the two accelerators presented in Table 6. We also use 33 kbit of dual-ported RAM: a bit more than Fan et al. and half of the amount used by Kammler et al. As a result, our architecture seems to be very comparable with the ones from [33, 22] in terms of area, and its performance is also very closed to the ASICs' one.

Finally we compared our results against single-core software implementations of 128-bits pairings over supersingular curves [11] and BN curves [41]. Even though, we targeted our implementation to embedded systems and low-resource hardware, our timings are very comparable to that of the software implementations: specific hardware for small-characteristic finite field arithmetic proves to be very efficient when compared to software implementations.

7 Conclusion

We presented a compact hardware implementation of a pairing reaching 128 bits of security, which is perfectly suited for embedded systems. To this end, we showed that the Tate pairing on supersingular curves over composite-extension field is a pertinent solution, even though their embedding degree k could be deemed too small at first glance. This also demonstrates that the efficiency of the underlying arithmetic plays a key role in pairing computation, and should be taken into account, right along with the size of the base field and the embedding degree, when designing pairing-based cryptosystems.

Furthermore, the idea to use curves defined over finite fields $\mathbb{F}_{q^n}$ of moderately-composed extension degree might be exploited in other areas of cryptography. While targeting the AES-128 level of security, the attacks based on Weil descent do not introduce extra weaknesses on the curve as long as n is kept small enough. This is an interesting result in itself: expanding the fauna of pairing-friendly curves suited to the 128-bit security level is indeed very relevant for cryptography. Moreover, computations on such curves can be carried out in a more efficient and parallel way, which yields better overall performances.

An interesting development of this work is to implement this idea in characteristic 2. Indeed, arithmetic over binary fields is simpler than in characteristic 3; as a consequence, characteristic 2 might also be a good choice, even though the embedding degree is even lower. We are planning to explore this direction in the near future.

Implementing the pairing on all the supersingular elliptic curves shown in Table 1 would also give a better coverage of the area–time trade-off for computing pairings with 128 bits of security: the more composite the extension degree, the smaller the base field $\mathbb{F}_q$ and thus the coprocessor. Additionally, in our approach, products over $\mathbb{F}_q$ are performed thanks to a quadratic scheme but the algorithms used for multiplications over $\mathbb{F}_{q^n}$ are subquadratic; therefore using a larger n for a same size of the field $\mathbb{F}_{q^n}$ might lead to a more efficient multiplication.

Furthermore, Cesena has noticed that the extra structure in curves defined over a composite-degree extension field—or TZVs—leads to a natural parallelization of Miller's algorithm [16]. It might be of interest to design a more parallel accelerator exploiting this fact. Such a circuit might achieve a lower latency for computing the Tate pairing with 128 bits of security at the cost of a larger silicon footprint.

Last but not least, the method presented in this article might scale to higher levels of security. For instance, the curve $E_{3,1}(\mathbb{F}_{3^{17 \cdot 67}})$ reaches 192 bits of security, while keeping the hardware requirements to a minimum. Finding other such curves and comparing them against higher-embedding-degree ordinary curves might help finding the crossover point between the two and assessing the actual relevance of supersingular elliptic curves in the context of low-resource pairing-based cryptography.

Acknowledgements

We sincerely thank Jérémie Detrey and Pierrick Gaudry for their insightful discussions and accurate comments that helped us to improve this work.

We would like to thank the anonymous reviewer for pointing out the following references [23, 44, 45, 16].

The author also wants to thank Paul Leyland and Sam W. Wagstaff for maintaining the factor tables of Cunningham numbers.

References

1. Adleman, L.M.: The function field sieve. In: Huang, M.-D.A., Adleman, L.M. (eds.) ANTS 1994. LNCS, vol. 877, pp. 108–121. Springer, Heidelberg (1994)
2. Barenghi, A., Bertoni, G., Breveglieri, L., Pelosi, G.: A FPGA coprocessor for the cryptographic Tate pairing over $\mathbb{F}_p$. In: Proceedings of the Fourth International Conference on Information Technology: New Generations (ITNG 2008). IEEE Computer Society, Los Alamitos (2008)
3. Barreto, P.S.L.M., Kim, H.Y., Lynn, B., Scott, M.: Efficient algorithms for pairing-based cryptosystems. In: Yung, M. (ed.) CRYPTO 2002. LNCS, vol. 2442, pp. 354–369. Springer, Heidelberg (2002)
4. Barreto, P.S.L.M., Naehrig, M.: Pairing-friendly elliptic curves of prime order. In: Preneel, B., Tavares, S. (eds.) SAC 2005. LNCS, vol. 3897, pp. 319–331. Springer, Heidelberg (2005)
5. Barreto, P., Galbraith, S., hÉigeartaigh, C.Ó., Scott, M.: Efficient pairing computation on supersingular abelian varieties. Designs, Codes and Cryptography 42(3), 239–271 (2007)
6. Beuchat, J.L., Brisebarre, N., Detrey, J., Okamoto, E., Rodríguez-Henríquez, F.: A comparison between hardware accelerators for the modified Tate pairing over $\mathbb{F}_{2^m}$ and $\mathbb{F}_{3^m}$. In: Galbraith, S., Paterson, K. (eds.) Pairing 2008. LNCS, vol. 5209, pp. 297–315. Springer, Heidelberg (2008)
7. Beuchat, J.L., Doi, H., Fujita, K., Inomata, A., Ith, P., Kanaoka, A., Katouno, M., Mambo, M., Okamoto, E., Okamoto, T., Shiga, T., Shirase, M., Soga, R., Takagi, T., Vithanage, A., Yamamoto, H.: FPGA and ASIC implementations of the η_T pairing in characteristic three. Computers and Electrical Engineering 36(1), 73–87 (2010)
8. Beuchat, J.L., Brisebarre, N., Detrey, J., Okamoto, E., Shirase, M., Takagi, T.: Algorithms and arithmetic operators for computing the η_t pairing in characteristic three. IEEE Transactions on Computers – Special Section on Special-Purpose Hardware for Cryptography and Cryptanalysis 57(11), 1454–1468 (2008)
9. Beuchat, J.L., Detrey, J., Estibals, N., Okamoto, E., Rodríguez-Henríquez, F.: Fast architectures for the η_T pairing over small-characteristic supersingular elliptic curves (2009), cryptology ePrint Archive, Report 2009/398
10. Beuchat, J.L., Detrey, J., Estibals, N., Okamoto, E., Rodríguez-Henríquez, F.: Hardware accelerator for the Tate pairing in characteristic three based on Karatsuba-Ofman multipliers. In: Clavier, C., Gaj, K. (eds.) CHES 2009. LNCS, vol. 5747, pp. 225–239. Springer, Heidelberg (2009)
11. Beuchat, J.L., López-Trejo, E., Martínez-Ramos, L., Mitsunari, S., Rodríguez-Henríquez, F.: Multi-core implementation of the Tate pairing over supersingular elliptic curves. In: Garay, J., Miyaji, A., Otsuka, A. (eds.) CANS 2009. LNCS, vol. 5888, pp. 413–432. Springer, Heidelberg (2009)
12. Boneh, D., Franklin, M.K.: Identity-based encryption from the Weil pairing. In: Kilian, J. (ed.) CRYPTO 2001. LNCS, vol. 2139, pp. 213–229. Springer, Heidelberg (2001)
13. Boneh, D., Lynn, B., Shacham, H.: Short Signatures from the Weil Pairing. Journal of Cryptology 17(4), 297–319 (2004)

14. Cenk, M., Özbudak, F.: Efficient multiplication in $\mathbb{F}_{3^{\ell m}}, m \geqslant 1$ and $5 \leqslant \ell \leqslant 18$. In: Vaudenay, S. (ed.) AFRICACRYPT 2008. LNCS, vol. 5023, pp. 406–414. Springer, Heidelberg (2008)

15. Cenk, M., Özbudak, F.: On multiplication in finite fields. Journal of Complexity 26(2), 172–186 (2010)

16. Cesena, E.: Pairing with supersingular Trace Zero Varieties revisited (2008), cryptology ePrint Archive, Report 2008/404

17. Diem, C.: On the discrete logarithm problem in class groups of curves. Mathemathics of Computation (to appear)

18. Duursma, I., Lee, H.S.: Tate pairing implementation for hyperelliptic curves $y^2 = x^p - x + d$. In: Laih, C. (ed.) ASIACRYPT 2003. LNCS, vol. 2894, pp. 111–123. Springer, Heidelberg (2003)

19. Duursma, I.M., Gaudry, P., Morain, F.: Speeding up the discrete log computation on curves with automorphisms. In: Lam, K.Y., Okamoto, E., Xing, C. (eds.) ASIACRYPT 1999. LNCS, vol. 1716, pp. 103–121. Springer, Heidelberg (1999)

20. El-Gamal, T.: A public key cryptosystem and a signature scheme based on discrete logarithms. In: Blakely, G.R., Chaum, D. (eds.) CRYPTO 1985. LNCS, vol. 196, pp. 10–18. Springer, Heidelberg (1984)

21. Fan, H., Hasan, M.A.: Comments on Montgomery's "Five, Six, and Seven-Term Karatsuba-Like Formulae". IEEE Transactions on Computers 56(5), 716–717 (2007)

22. Fan, J., Vercauteren, F., Verbauwhede, I.: Faster $\mathbb{F}_p$-arithmetic for Cryptographic Pairings on Barreto-Naehrig Curves. In: Clavier, C., Gaj, K. (eds.) CHES 2009. LNCS, vol. 5747, pp. 240–253. Springer, Heidelberg (2009)

23. Frey, G.: Applications of arithmetical geometry to cryptographic constructions. In: Proceedings of the Fifth International Conference on Finite Fields and Applications (Augsburg, 1999), pp. 128–161. Springer, Heidelberg (2001)

24. Frey, G., Rück, H.G.: A remark concerning m-divisibility and the discrete logarithm in the divisor class group of curves. Mathematics of Computation 62(206), 865–874 (1994)

25. Gallant, R.P., Lambert, R.J., Vanstone, S.A.: Improving the parallelized Pollard lambda search on anomalous binary curves. Math. Comput. 69(232), 1699–1705 (2000)

26. Gaudry, P.: Index calculus for abelian varieties and the elliptic curve discrete logarithm problem. Journal of Symbolic Compution 44(12), 1690–1702 (2009)

27. Gaudry, P., Hess, F., Smart, N.: Constructive and destructive facets of Weil descent on elliptic curves. Journal of Cryptology 15(1), 19–46 (2002)

28. Granger, R.: On the static Diffie–Hellman problem on elliptic curves over extension fields. In: ASIACRYPT. LNCS, Springer, Heidelberg (2010) (to appear)

29. Itoh, T., Tsujii, S.: A fast algorithm for computing multiplicative inverses in $GF(2^m)$ using normal bases. Information and Computation 78(3), 171–177 (1988)

30. Jiang, J.: Bilinear pairing (Eta_T Pairing) IP core. Tech. rep., City University of Hong Kong – Department of Computer Science (May 2007)

31. Joux, A.: A one round protocol for tripartite Diffie-Hellman. In: Bosma, W. (ed.) ANTS 2000. LNCS, vol. 1838, pp. 385–394. Springer, Heidelberg (2000)

32. Joux, A., Vitse, V.: Elliptic curve discrete logarithm problem over small degree extension fields. Application to the static Diffie–Hellman problem on $E(\mathbb{F}_{q^5})$ (2010), cryptology ePrint Archive, Report 2010/157

33. Kammler, D., Zhang, D., Schwabe, P., Scharwaechter, H., Langenberg, M., Auras, D., Ascheid, G., Mathar, R.: Designing an ASIP for cryptographic pairings over Barreto-Naehrig curves. In: Clavier, C., Gaj, K. (eds.) CHES 2009. LNCS, vol. 5747, pp. 254–271. Springer, Heidelberg (2009)
34. Kerins, T., Marnane, W., Popovici, E., Barreto, P.: Efficient hardware for the Tate pairing calculation in characteristic three. In: Rao, J., Sunar, B. (eds.) CHES 2005. LNCS, vol. 3659, pp. 412–426. Springer, Heidelberg (2005)
35. Leyland, P.: Cunningham numbers,
    http://www.leyland.vispa.com/numth/factorization/cunningham/main.htm
36. Menezes, A.J., Okamoto, T., Vanstone, S.A.: Reducing elliptic curve logarithms to logarithms in a finite field. IEEE Transactions on Information Theory 39(5), 1639–1646 (1993)
37. Miller, V.S.: Short programs for functions on curves. IBM, Thomas J. Watson Research Center (1986)
38. Miller, V.S.: The Weil pairing, and its efficient calculation. J. Cryptology 17(4), 235–261 (2004)
39. Mitsunari, S., Sakai, R., Kasahara, M.: A new traitor tracing. IEICE Trans. Fundamentals E85-A(2), 481–484 (2002)
40. Montgomery, P.L.: Five, six, and seven-term Karatsuba-like formulae. IEEE Transactions on Computers 54(3), 362–369 (2005)
41. Naehrig, M., Niederhagen, R., Schwabe, P.: New software speed records for cryptographic pairings (2010), cryptology ePrint Archive, Report 2010/186
42. Pollard, J.: Monte Carlo methods for index computation (mod p). Mathematics of Computation, 918–924 (1978)
43. Ronan, R., hÉigeartaigh, C.Ó., Murphy, C., Scott, M., Kerins, T.: Hardware acceleration of the Tate pairing on a genus 2 hyperelliptic curve. Journal of Systems Architecture 53, 85–98 (2007)
44. Rubin, K., Silverberg, A.: Supersingular abelian varieties in cryptology. In: Yung, M. (ed.) CRYPTO 2002. LNCS, vol. 2442, pp. 336–353. Springer, Heidelberg (2002)
45. Rubin, K., Silverberg, A.: Using abelian varieties to improve pairing-based cryptography. Journal of Cryptology 22(3), 330–364 (2009)
46. Sakai, R., Ohgishi, K., Kasahara, M.: Cryptosystems based on pairing. In: 2000 Symposium on Cryptography and Information Security (SCIS 2000), Okinawa, Japan, pp. 26–28 (January 2000)
47. Shu, C., Kwon, S., Gaj, K.: Reconfigurable computing approach for Tate pairing cryptosystems over binary fields. IEEE Transactions on Computers 58(9), 1221–1237 (2009)
48. Silverman, J.H.: The Arithmetic of Elliptic Curves. Graduate Texts in Mathematics, vol. 106. Springer, Heidelberg (1986)
49. Wagstaff, S.: The Cunningham Project,
    http://homes.cerias.purdue.edu/~ssw/cun/index.html

A   Multiplication Algorithm over $\mathbb{F}_{3^{5m}}$

Given a characteristic-3 finite field $\mathbb{F}_{3^m}$ with $5 \nmid m$, its degree-5 extension $\mathbb{F}_{3^{5m}}$, and $\alpha \in \mathbb{F}_{3^{5m}}$ such that $\alpha^5 = \alpha - 1$, we present explicit formulae for multiplication over $\mathbb{F}_{3^{5m}}$ in Algorithm 1. These formulae come from the CRT-based algorithm presented by Cenk & Özbudak in [14].

Algorithm 1. Multiplication in $\mathbb{F}_{3^{5m}}$

Input: $A = a_0 + a_1\alpha + a_2\alpha^2 + a_3\alpha^3 + a_4\alpha^4$ and $B = b_0 + b_1\alpha + b_2\alpha^2 + b_3\alpha^3 + b_4\alpha^4$
where $a_i, b_i \in \mathbb{F}_{3^m}$.

Output: $C = A \cdot B$

1. $a_5 \leftarrow a_0 + a_1$; $b_5 \leftarrow b_0 + b_1$; $a_6 \leftarrow a_2 + a_3$; $b_6 \leftarrow b_2 + b_3$
2. $a_7 \leftarrow a_2 - a_3$; $b_7 \leftarrow b_2 - b_3$; $a_8 \leftarrow a_0 + a_4$; $b_8 \leftarrow b_0 + b_4$
3. $a_9 \leftarrow a_4 + a_5$; $b_9 \leftarrow b_4 + b_5$
4. $p_0 \leftarrow a_0 \cdot b_0$; $p_1 \leftarrow a_1 \cdot b_1$; $p_2 \leftarrow a_4 \cdot b_4$; $p_3 \leftarrow a_5 \cdot b_5$
5. $p_4 \leftarrow (a_1 - a_3) \cdot (b_1 - b_3)$; $p_5 \leftarrow (a_1 - a_6) \cdot (b_1 - b_6)$
6. $p_6 \leftarrow (a_2 - a_8) \cdot (b_2 - b_8)$; $p_7 \leftarrow (a_6 + a_9) \cdot (b_6 + b_9)$
7. $p_8 \leftarrow (a_6 - a_9) \cdot (b_6 - b_9)$; $p_9 \leftarrow (a_0 - a_4 + a_7) \cdot (b_0 - b_4 + b_7)$
8. $p_{10} \leftarrow (a_1 - a_7 - a_8) \cdot (b_1 - b_7 - b_9)$; $p_{11} \leftarrow (a_3 - a_4 + a_5) \cdot (b_3 - b_4 + b_5)$
9. $t_0 \leftarrow p_1 - p_3$; $t_1 \leftarrow p_7 + p_{10}$; $t_2 \leftarrow p_7 - p_{10}$
10. $t_3 \leftarrow p_2 + p_4$; $t_4 \leftarrow p_8 - p_6$; $t_5 \leftarrow p_{11} - t_0$
11. $c_0 \leftarrow t_4 - t_0 - t_2 - p_4$; $c_1 \leftarrow p_2 - p_0 - p_9 - t_1 - t_5$
12. $c_2 \leftarrow p_5 - p_8 - t_3 - t_5$; $c_3 \leftarrow t_2 - t_3 + t_4$
13. $c_4 \leftarrow p_4 - p_0 - p_6 + t_1$
14. **return** $C = c_0 + c_1\alpha + c_2\alpha^2 + c_3\alpha^3 + c_4\alpha^4$

B ℓ-torsion of Presented Curves

We provide in this appendix the largest factor ℓ of the cardinals of the curves used in Table 1 and the one with 192-bit security level mentioned in the conclusion.

B.1 Characteristic 2

- $E_{2,1}(\mathbb{F}_{2^{1117}})$:
  $\ell = 619074192321273307277438691119233058790820634893360057193377\backslash$
  $122275541424570658263412019435765493074195820297417376747932\backslash$
  $248094264569966237629582261870883925353443145570692187335548\backslash$
  $683837515601099459860669193973764482753436531478745981766945\backslash$
  $411504253650899252459234448440440995323058733022537477753547\backslash$
  54333152682491155152733 (1076 bits)
- $E_{2,1}(\mathbb{F}_{2^{3 \cdot 367}})$:
  $\ell = 969479603278186730289503541042886691666123364558568701313813\backslash$
  $359745290668184127412771736661726167918225502828978334069274\backslash$
  $912792960814346728226407067755389529068277515573173949951347\backslash$
  $909708753667984651964976566357$ (698 bits)
- $E_{2,1}(\mathbb{F}_{2^{5 \cdot 227}})$:
  $\ell = 387807566127257652326614188973820517854886880596071365340579\backslash$
  $081203820130327954203519023617159911950745211425564080964614\backslash$
  $422638824727652925391132558155138126471653299809062407679361\backslash$
  $06177017769993501924867181334979908226181$ (733 bits)

- $E_{2,1}(\mathbb{F}_{2^{7\cdot163}})$:

  $\ell = 52733342365707641873577119849074213590259398884049828283$ $5698\backslash$
  $22546908155123401741300939392434961539105788302597014340$ $9298\backslash$
  $82552537701109573887517388556155980296275926022847976526$ $2681\backslash$
  $1637983037702889357614490116553202067374175$ 6101 (753 bits)

- $E_{2,1}(\mathbb{F}_{2^{9\cdot127}})$:

  $\ell = 34542876351533745532051334659583845756215258693728250218$ $8431\backslash$
  $30874656232121949901762883256301982063468282701977079908$ $9846\backslash$
  90637173458395849296993 3493 (487 bits)

- $E_{2,1}(\mathbb{F}_{2^{11\cdot103}})$:

  $\ell = 45509321813123175395060361962904749784797443088158411434$ $2263\backslash$
  $23353958009277072667876691739270258562662583597889984307$ $5414\backslash$
  $30519887805045990553583565505084709320278329943002107535$ $6835\backslash$
  $01474553542164657008510080303855519722927476592523127872$ $8563\backslash$
  $73610510150628859847392052958851491$ 521 (922 bits)

- $E_{2,0}(\mathbb{F}_{2^{13\cdot89}})$:

  $\ell = 21180233225268233482694475848150309365517775211089551565$ $6983\backslash$
  $70775392175515775846644799547412757706093592438545570301$ $6958\backslash$
  $75813798319427995249914672065928536456368400930285025966$ $6276\backslash$
  $19564185584076451841932717943171855425417458132108398294$ $7322\backslash$
  $98152880401123094762543205186697053592109913518433241299$ $5164\backslash$
  09337087016 7293 (1044 bits)

- $E_{2,0}(\mathbb{F}_{2^{15\cdot73}})$:

  $\ell = 98005790863005484959098240780258945758899979356253262817$ $6670\backslash$
  $09778512984540986418865971216565813297866385024242045378$ $3649\backslash$
  646693497098428583250379 8101 (492 bits)

B.2   Characteristic 3

- $E_{3,1}(\mathbb{F}_{3^{503}})$:

  $\ell = 54552365767611244726090456357891273837330786721968621584$ $9632\backslash$
  $46980147111242687893977672522229043765371847396273376087$ $4627\backslash$
  $31593093312658124846589965112048106611183908157516496458$ $9811\backslash$
  $98588571901721493851456380$ 4313 (697 bits)

- $E_{3,-1}(\mathbb{F}_{3^{5\cdot97}})$:

  $\ell = 58873245301175350801369450309110049026192845915751464730$ $9296\backslash$
  $941697666832507096172127852923090672$ 844101 (338 bits)

- $E_{3,-1}(\mathbb{F}_{3^{7\cdot67}})$:

  $\ell = 19887717320605289481263507429615793274133203442415703865$ $0419\backslash$
  $39585411607577163706616759949915010995822199996762235890$ $0215\backslash$
  $86006727622355473185242460494206772736456553494378139587$ $6039\backslash$
  31 839 (612 bits)

- $E_{3,-1}(\mathbb{F}_{3^{11\cdot53}})$:

  $\ell = 21148181949374608678249357006334414413702540085806470705398$\\
  $97000729752368024704685306862541134701003798735457009614028$\\
  $13116723324601418928273763475905777298379241455002360540104$\\
  4286639185637504527958 (672 bits)

- $E_{3,1}(\mathbb{F}_{3^{13\cdot43}})$:

  $\ell = 7808051312160133221359091175319630305251045334543792929999$\\
  $28517660502894770417816131776562968915427140891407141754078$\\
  $06506539184480137644034449742235329868025905007719019241492$\\
  $27880097531140775099199845730350794443332650516869$ (764 bits)

- $E_{3,1}(\mathbb{F}_{3^{17\cdot67}})$:

  $\ell = 5808072514435807490151578921038797742632036461481865457337$\\
  $30132796400506185678101366120502447887732601930123554923819$\\
  $69412766420336309764610144880018466688688553400561834490675$\\
  $47047267951415460826580941969794736857658127539041949502719$\\
  $91750229880954255856215951978882932420615958213429340785946$\\
  $86490019106602344957487333330342781841117126897138950731648$\\
  $83669741340526840114895744180591754787695685413862673603698$\\
  $30619628696210257143075238078897118661671487526012459979702$\\
  43146595818195017 (1650 bits)

A Variant of Miller's Formula and Algorithm

John Boxall[1], Nadia El Mrabet[2,*], Fabien Laguillaumie[3], and Duc-Phong Le[4,*]

[1] LMNO – Université de Caen Basse-Normandie, France
john.boxall@unicaen.fr
[2] LIASD – Université Paris 8, France
nelmrabe@mime.univ-paris8.fr
[3] GREYC – Université de Caen Basse-Normandie, France
fabien.laguillaumie@unicaen.fr
[4] Faculty of Information Technology
College of Technology,
Vietnam National University, Vietnam
phongld@vnu.edu.vn

Abstract. Miller's algorithm is at the heart of all pairing-based cryptosystems since it is used in the computation of pairing such as that of Weil or Tate and their variants. Most of the optimizations of this algorithm involve elliptic curves of particular forms, or curves with even embedding degree, or having an equation of a special form. Other improvements involve a reduction of the number of iterations.

In this article, we propose a variant of Miller's formula which gives rise to a *generically* faster algorithm for any pairing friendly curve. Concretely, it provides an improvement in cases little studied until now, in particular when denominator elimination is not available. It allows for instance the use of elliptic curve with embedding degree not of the form $2^i 3^j$, and is suitable for the computation of *optimal pairings*. We also present a version with denominator elimination for even embedding degree. In our implementations, our variant saves between 10% and 40% in running time in comparison with the usual version of Miller's algorithm without any optimization.

1 Introduction

Pairings were first introduced into cryptography in Joux' seminal paper describing a tripartite (bilinear) Diffie-Hellman key exchange [20]. Since then, the use of cryptosystems based on bilinear maps has had a huge success with some notable breakthroughs such as the first identity-based encryption scheme [10]. Nevertheless, pairing-based cryptography has a reputation of being inefficient, because it is computationally more expensive than cryptography based on modular arithmetic. On the other hand, the use of pairings seems to be essential in the definition of protocols with specific security properties and also allows one to reduce bandwidth in certain protocols.

---

* This work was done while these two authors held post-doc positions at the Université de Caen Basse-Normandie.

M. Joye, A. Miyaji, and A. Otsuka (Eds.): Pairing 2010, LNCS 6487, pp. 417–434, 2010.

Ever since it was first described, Miller's algorithm [24] has been the central ingredient in the calculation of pairings on elliptic curves. Many papers are devoted to improvements in its efficiency. For example, it can run faster when the elliptic curves are chosen to belong to specific families (see for example [4,6,12]), or different coordinate systems (see for example [18,13,7]). Another standard method of improving the algorithm is to reduce the number of iterations by introducing pairings of special type, for example particular *optimal pairings* [26,17,16] or using addition chains (see for example [8]).

In this paper we study a variant of Miller's algorithm for elliptic curves which is *generically* faster than the usual version. Instead of using the formula $f_{s+t} = f_s f_t \frac{\ell_{s,t}}{v_{s+t}}$ (see Subsection 2.2 for notation and Lemma 1) on which the usual Miller algorithm is based, our variant is inspired by the formula $f_{s+t} = \frac{1}{f_{-s} f_{-t} \ell_{-s,-t}}$ (see Lemma 2 for a proof). An important feature is that the only vertical line that appears is f_{-1}, in other words the vertical line passing through P, and even this does not appear explicitly except at initialization. We shall see in § 3.1 why it does not appear in the addition step. Our algorithm is of particular interest to compute the Ate-style pairings [3,17] on elliptic curves with small embedding degrees k, and in situations where denominator elimination using a twist is not possible (for example on curves with embedding degree prime to 6). A typical example is the case of *optimal pairings* [26], which by definition only require about $\log_2(r)/\varphi(k)$ (where r is the group order) iterations of the basic loop. If k is prime, then $\varphi(k \pm 1) \leq \frac{k+1}{2}$ which is roughly $\frac{\varphi(k)}{2} = \frac{k-1}{2}$, so that at least twice as many iterations are necessary if curves with embedding degrees $k \pm 1$ are used instead of curves of embedding degree k.

The paper is organized as follows. In Section 2 we recall some background on pairings, and recall the usual Miller algorithm (Figure 1). In Section 3 we explain and analyze generically our version of Miller's algorithm, which is resumed by the pseudocode in Figure 2 when the elliptic curve is given in Jacobian coordinates. Section 4 discusses a variant without denominators applicable when k is even (see Figure 5). Section 5 describes some numerical experiments and running times; in an example with $k = 18$ and r having 192 bits, the algorithm of Figure 5 is roughly 40% faster than the usual Miller algorithm, and about as fast as the algorithm of [4].

Further work is needed to see whether many of the recent ideas used to improve the usual Miller algorithm can be adapted to the variant presented here. We believe that doing so should lead to further optimizations.

2   Background on Pairings

2.1   Basics on Pairings

We briefly recall the basic definitions and some examples of pairings used in cryptography. For further information, see for example [9,11].

We let $r \geq 2$ denote an integer which, unless otherwise stated, is supposed to be prime. We let $(\mathbb{G}_1, +)$, $(\mathbb{G}_2, +)$ and $(\mathbb{G}_T, \cdot)$ denote three finite abelian groups, which are supposed to be of order r unless otherwise indicated. A *pairing* is a map $e : \mathbb{G}_1 \times \mathbb{G}_2 \to \mathbb{G}_T$ such that $e(P_1 + P_2, Q) = e(P_1, Q)e(P_2, Q)$ and $e(P, Q_1 + Q_2) = e(P, Q_1)e(P, Q_2)$ for all P, P_1, $P_2 \in \mathbb{G}_1$ and for all Q, Q_1, $Q_2 \in \mathbb{G}_2$. We say that the pairing e is *left non degenerate* if, given $P \in \mathbb{G}_1$ with $P \neq 1$, there exists $Q \in \mathbb{G}_2$ with $e(P, Q) \neq 1$. The notion of a right degenerate pairing is defined similarly and e is said to be *non degenerate* if it is both left and right non degenerate

We recall briefly one of the most frequent choices for the groups $\mathbb{G}_1$, $\mathbb{G}_2$ and $\mathbb{G}_T$ in pairing-based cryptography. Here, $\mathbb{G}_1$ is the group generated by a point P of order r on an elliptic curve E defined over a finite field $\mathbb{F}_q$ of characteristic different to r. Thus, $\mathbb{G}_1 \subseteq E(\mathbb{F}_q)$ is cyclic of order r but, in general, the whole group $E[r]$ of points of order dividing r of E is not rational over $E(\mathbb{F}_q)$. Recall that the *embedding degree* of E (with respect to r) is the smallest integer $k \geq 1$ such that r divides $q^k - 1$. A result of Balasubramanian and Koblitz [2] asserts that, when $k > 1$, all the points of $E[r]$ are rational over the extension $\mathbb{F}_{q^k}$ of degree k of $\mathbb{F}_q$. The group $\mathbb{G}_2$ is chosen as another subgroup of $E[r]$ of order r. Finally, $\mathbb{G}_T$ is the subgroup of order r in $\mathbb{F}_{q^k}^\times$; it exists and is unique, since r divides $q^k - 1$ and $\mathbb{F}_{q^k}^\times$ is a cyclic group.

Let $P \in E(\mathbb{F}_q)$ be an r-torsion point, let D_P be a degree zero divisor with $D_P \sim [P] - [O_E]$, and let f_{r,D_P} be such that $\text{div} f_{r,D_P} = rD_P$. Let Q be a point of $E(\mathbb{F}_{q^k})$ (not necessarily r-torsion) and $D_Q \sim [Q] - [O_E]$ of support disjoint with D_P. Consider

$$e_r^T(P, Q) = f_{r,D_P}(D_Q). \tag{1}$$

Weil reciprocity shows that if D_Q is replaced by $D_Q' = D_Q + \text{div } h \sim D_Q$, then (1) is multiplied by $h(D_P)^r$. So the value is only defined up to r-th powers. Replacing D_P by $D_P' = D_P + \text{div } h$ changes f_{r,D_P} to $f_{r,D_P'} = f_{r,D_P}h^r$, and the value is well-defined modulo multiplication by r-th powers. If then Q is replaced by $Q + rR$, the value changes again by an r-th power. This leads to adapting the range and domain of e_r^T as follows.

Theorem 1. *The Tate pairing is a map*

$$e_r^T : E(\mathbb{F}_q)[r] \times E(\mathbb{F}_{q^k})/rE(\mathbb{F}_{q^k}) \to \mathbb{F}_{q^k}^\times / (\mathbb{F}_{q^k}^\times)^r$$

satisfying the following properties:

1. *Bilinearity,*
2. *Non-degeneracy,*
3. *Compatibility with isogenies.*

The *reduced* Tate pairing computes the unique rth root of unity belonging to the class of $f_{r,D_P}(D_Q)$ modulo $(\mathbb{F}_{q^k}^\times)^r$ as $f_{r,D_P}(D_Q)^{(q^k-1)/r}$. In practice, we take Q to lie in some subgroup $\mathbb{G}_2$ of order r of $E(\mathbb{F}_{q^k})$ that injects into $E(\mathbb{F}_{q^k})/rE(\mathbb{F}_{q^k})$ *via* the canonical map. The more popular Ate pairing [3] and its variants (see [23] for instance) are optimized versions of the Tate pairing when restricted to Frobenius eigenspaces. Besides its use in cryptographic protocols, the Tate pairing is also useful in other applications, such as walking on isogeny volcanoes [19], which can be used in the computation of endomorphism rings of elliptic curves.

However, in this article we concentrate on the computation of $f_{n,D_P}(D_Q)$ (which we write as $f_{n,P}(Q)$ in the sequel). This is done using Miller's algorithm described in the next subsection.

2.2   Computation of Pairings and Miller's Algorithm

In order to emphasize that our improvement can be applied in a very general context, we explain briefly in this subsection how pairings are computed. In what follows, $\mathbb{F}$ denotes a field (not necessarily finite), E an elliptic curve over $\mathbb{F}$ and r an integer not divisible by the characteristic of $\mathbb{F}$. We suppose that the group $E(\mathbb{F})$ of $\mathbb{F}$-rational points of E contains a point P of order r. Since r is prime to the characteristic of $\mathbb{F}$, the group $E[r]$ of points of order r of E is isomorphic to a direct sum of two cyclic groups of order r. In general, a point $Q \in E[r]$ that is not a multiple of P will be defined over some extension $\mathbb{F}'$ of $\mathbb{F}$ of finite degree.

If P, P' are two points in $E(\mathbb{F})$, we denote by $\ell_{P,P'}$ a function with divisor $[P] + [P'] + [-(P + P')] - 3[O_E]$ and by v_P a function with divisor $[P]+[-P]-2[O_E]$. Clearly these functions are only defined up to a multiplicative constant; we recall at the end of this section how to normalize functions so that they are uniquely determined by their divisor. Note that v_P is just the same as $\ell_{P,-P}$.

If s and t are two integers, we denote by $f_{s,P}$ (or simply f_s if there is no possibility of confusion) a function whose divisor is $s[P] - [sP] - (s-1)[O_E]$. We abbreviate $\ell_{sP,tP}$ to $\ell_{s,t}$ and v_{sP} to v_s. As understood in this paper, the purpose of Miller's algorithm is to calculate $f_{s,P}(Q)$ when $Q \in E[r]$ is not a multiple of P. All pairings can be expressed in terms of these functions for appropriate values of s.

Miller's algorithm is based on the following Lemma describing the so-called *Miller's formula*, which is proved by considering divisors.

Lemma 1. *For s and t two integers, up to a multiplicative constant, we have* $f_{s+t} = f_s f_t \frac{\ell_{s,t}}{v_{s+t}}.$

The usual Miller algorithm makes use of Lemma 1 with $t = s$ in a doubling step and $t = 1$ in an addition step. It is described by the pseudocode in Figure 1, which presents the algorithm updating numerators and denominators separately, so that just one inversion is needed at the end. We write the functions ℓ and v as quotients $(N\ell)/(D\ell)$ and $(Nv)/(Dv)$, where each of the terms $(N\ell)$, $(D\ell)$, (Nv), (Dv) is computed using only additions and multiplications, and no inversions. Here the precise definitions of $(N\ell)$, $(D\ell)$, (Nv), (Dv) will depend on the representations that are used; in Section 3.2 we indicate one such choice when short Weierstrass coordinates and the associated Jacobian coordinates are used. In the algorithm, T is always a multiple of P, so that the hypothesis that Q is not a multiple of P implies that at the functions $\ell_{T,T}$, $\ell_{T,P}$, v_{2T} and v_{T+P}

Algorithm 1. Miller(P, Q, s) usual

Data: $s = \sum_{i=0}^{l-1} s_i 2^i$ (radix 2), $s_i \in \{0, 1\}$, $Q \in E(\mathbb{F}')$ not a multiple of P.
Result: $f_{s,P}(Q)$.
$T \leftarrow P$, $f \leftarrow 1$, $g \leftarrow 1$,
for $i = l - 2$ **to** 0 **do**

> $f \leftarrow f^2(N\ell)_{T,T}(Dv)_{2T}$,
> $g \leftarrow g^2(D\ell)_{T,T}(Nv)_{2T}$,
> $T \leftarrow 2T$
> **if** $s_i = 1$ **then**
>> $f \leftarrow f(N\ell)_{T,P}(Dv)_{T+P}$,
>> $g \leftarrow g(D\ell)_{T,P}(Nv)_{T+P}$,
>> $T \leftarrow T + P$
>
> **end**

end
return f/g

Fig. 1. The usual Miller algorithm

cannot vanish at Q. It follows that f and g never vanish at Q so that the final quotient f/g is well-defined and non-zero.

Obviously in any implementation it is essential for the functions appearing in the programs to be uniquely determined, and so we end this section by recalling briefly how this can be done in our context. If w is a uniformizer at O_E, we say that the non-zero rational function f on E is normalized (or monic) with respect to w if the Laurent expansion of f at O_E is of the form

$$f = w^n + c_{n+1}w^{n+1} + c_{n+2}w^{n+2} + \cdots, \quad c_i \in \mathbb{F},$$

(i. e. if the first non-zero coefficient is 1). Any non-zero rational function on E can be written in a unique way as a product of a constant and a normalized function, and the normalized rational functions form a group under multiplication that is isomorphic to the group of principal divisors.

As a typical example, when E is in short Weierstrass form

$$y^2 = x^3 + ax + b, \ a, b \in \mathbb{F} \tag{2}$$

one can take $w = \frac{x}{y}$. Any rational function on E can be written as a quotient of two functions whose polar divisors are supported at the origin O_E of E. If f is a function whose only pole is at O_E, then there exist two polynomials $U(x)$ and $V(x)$ such that $f = U(x) + V(x)y$: here U and V are uniquely determined by f. Furthermore, if the order of the pole of f is n, then $n \geq 2$ and $U(x)$ is of degree $\frac{n}{2}$ and $V(x)$ of degree at most $\frac{n-4}{2}$ if n is even and $V(x)$ is of degree $\frac{n-3}{2}$ and $U(x)$ of degree at most $\frac{n-1}{2}$ when n is odd. Then f is normalized if and only if, when f is written in the form $U(x) + V(x)y$, $U(x)$ is monic when n is even and $V(x)$ is monic when n is odd. In general, a rational function on E is normalized if and only if it is a quotient of two normalized functions whose polar divisors are supported by the origin.

For example, when E is given by the equation (2), the functions

$$\ell_{T,P}(x,y) = \begin{cases} y - y_P - \frac{y_T - y_P}{x_T - x_P}(x - x_P), & T \neq O_E, \pm P, \\ y - y_P - \frac{3x_P^2 + a}{2y_P}(x - x_P), & T = P, 2P \neq O_E. \end{cases}$$

and

$$v_P(x,y) = x - x_P$$

are normalized, so that if they are used in implementations of the algorithms given in Figures 1 and 2, then these algorithms output $f_{s,P}(Q)$ with $f_{s,P}$ the normalized function with divisor $s[P] - [sP] - (s-1)[O_E]$.

3   Our Variant of Miller's Algorithm

In this section, we describe our variant of Miller's formula and algorithm and analyze the cost of the latter in terms of basic operations.

3.1   The Algorithm

The main improvement comes from the following Lemma.

Lemma 2. *For s and t two integers, up to a multiplicative constant, we have*

$$f_{s+t} = \frac{1}{f_{-s}f_{-t}\ell_{-s,-t}}.$$

Proof. This lemma is again proved by considering divisors. Indeed,

$$\begin{aligned} \mathrm{div}(f_{-s}f_{-t}\ell_{-s,-t}) &= (-s)[P] - [(-s)P] + (s+1)[O_E] \\ &\quad + (-t)[P] - [(-t)P] + (t+1)[O_E] \\ &\quad + [-sP] + [-tP] + [(s+t)P] - 3[O_E] \\ &= -(s+t)[P] + [(s+t)P] + (s+t-1)[O_E] \\ &= -\mathrm{div}(f_{s+t}), \end{aligned}$$

which concludes the proof.   □

We shall seek to exploit the fact that here the right hand member has only three terms whereas that of Lemma 1 has four.

Our variant of Miller's algorithm is described by the pseudocode in Figure 2. It was inspired by the idea of applying Lemma 2 with $t = s$ or $t \in \{\pm 1\}$. However, the scalar input is given in binary representation. It updates numerators and denominators separately, so that only one final inversion appears at the end. As in Figure 1, the hypothesis that Q is not a multiple of P implies that at no stage do f and g vanish, so that the final quotient f/g makes sense. Note that T is always a *positive* multiple of P. We use the notation $\ell'_{-T,-P}$ for the function $f_{-1}\ell_{-T,-P}$, since in many situations it can be computed faster than simply by

---

Algorithm 2. Miller(P, Q, r) modified

Data: $s = \sum_{i=0}^{l-1} s_i 2^i$, $s_i \in \{0, 1\}$, $s_{l-1} = 1$, h Hamming weight of s, $Q \in E(\mathbb{F}')$
not a multiple of P

Result: $f_{s,P}(Q)$;

$f \leftarrow 1, T \leftarrow P,$

if $l + h$ *is odd* **then**

 | $\delta \leftarrow 1, g \leftarrow f_{-1}$

end

else

 | $\delta \leftarrow 0, g \leftarrow 1$

end

for $i = l - 2$ **to** 0 **do**

1   **if** $\delta = 0$ **then**

     $f \leftarrow f^2(N\ell)_{T,T},$

     $g \leftarrow g^2(D\ell)_{T,T},$

     $T \leftarrow 2T, \delta \leftarrow 1$

2   **if** $s_i = 1$ **then**

      $g \leftarrow g(N\ell')_{-T,-P},$

      $f \leftarrow f(D\ell')_{-T,-P},$

      $T \leftarrow T + P, \delta \leftarrow 0$

    end

   end

3   **else**

     $g \leftarrow g^2(N\ell)_{-T,-T},$

     $f \leftarrow f^2(D\ell)_{-T,-T},$

     $T \leftarrow 2T, \delta \leftarrow 0$

4   **if** $s_i = 1$ **then**

      $f \leftarrow f(N\ell)_{T,P},$

      $g \leftarrow g(D\ell)_{T,P},$

      $T \leftarrow T + P, \delta \leftarrow 1$

    end

   end

end

return f/g

Fig. 2. Our modified Miller algorithm

computing f_{-1} and $\ell_{-T,-P}$ and taking the product. For example, when E is given in short Weierstrass coordinates by the equation $y^2 = x^3 + ax + b$, a, $b \in \mathbb{F}$, we have

$$\ell'_{-T,-P} = f_{-1}\ell_{-T,-P} = \frac{1}{x_Q - x_P}(y_Q + y_P + \lambda(x_Q - x_P)) = \frac{y_Q + y_P}{x_Q - x_P} + \lambda, \quad (3)$$

where $\frac{y_Q + y_P}{x_Q - x_P}$ can be precomputed (at the cost of one inversion and one multiplication in the big field) and λ denotes the slope of the line joining T to P.

 The tables in Figures 3 and 4 show that our variant is more efficient than the classical Miller's algorithm as we save a product in the big field at each doubling and each addition step. We also save some multiplications and squarings in $\mathbb{F}$. The following subsection discusses all this in more detail. In Section 4 we describe a version without denominators that works for elliptic curves with even embedding degree.

3.2 Generic Analysis

In this subsection, we compare the number of operations needed to compute $f_{s,P}(Q)$ using the algorithms in Figures 1 and 2. In order to fix ideas, we make our counts using Jacobian coordinates (X, Y, Z) associated to a short Weierstrass model $y^2 = x^3 + ax + b$, a, $b \in \mathbb{F}$, so that $x = X/Z^2$ and $y = Y/Z^3$. We suppose that the Jacobian coordinates of P lie in $\mathbb{F}$ and that those of Q lie in some extension $\mathbb{F}'$ of $\mathbb{F}$ of whose degree is denoted by k. We denote by $\mathbf{m}_a$ the multiplication by the curve coefficient a and we denote respectively by $\mathbf{m}$ and $\mathbf{s}$ multiplications and squares in $\mathbb{F}$, while the same operations in $\mathbb{F}'$ are denoted respectively by $\mathbf{M}_k$ and $\mathbf{S}_k$ if k is the degree of the extension $\mathbb{F}'$. We assume that $\mathbb{F}'$ is given by a basis as a $\mathbb{F}$-vector space one of whose elements is 1, so that multiplication of an element of $\mathbb{F}'$ by an element of $\mathbb{F}$ counts as k multiplications in $\mathbb{F}$. We ignore additions and multiplications by small integers.

If S is any point of E, then X_S, Y_S and Z_S denote the Jacobian coordinates of S, so that when $S \neq O_E$, the Weierstrass coordinates of S are $x_S = X_S/Z_S^2$ and $y_S = Y_S/Z_S^3$. As before, T is a multiple of P, so that X_T, Y_T and Z_T all lie in $\mathbb{F}$. Since P and Q are part of the input, we assume they are given in Weierstrass coordinates and that $Z_P = Z_Q = 1$.

We need to define the numerators and the denominators of the quantities appearing in the algorithms of Figures 1 and 2. The cost of computing these quantities and the total cost of these algorithms are analyzed in Figures 3 and 4.

The doubling step. We first deal with doubling. Suppose that $2T \neq O_E$, which will always be the case if r is odd. Then $y_T \neq 0$ and the slope of the tangent to E at T is

$$\mu_T = \frac{3x_T^2 + a}{2y_T} = \frac{(N\mu)_T}{(D\mu)_T},$$

where $(N\mu)_T = 3X_T^2 + aZ_T^4$ and $(D\mu)_T = 2Y_TZ_T = (Y_T + Z_T)^2 - Y_T^2 - Z_T^2$. Hence the value of $\ell_{T,T}$ at Q can be written

$$\ell_{T,T}(Q) = y_Q - y_T - \mu_T(x_Q - x_T) = \frac{(N\ell)_{T,T}}{(D\ell)_{T,T}},$$

where now

$$(N\ell)_{T,T} = (D\mu)_T Z_T^2 y_Q - (N\mu)_T Z_T^2 x_Q - 2Y_T^2 - (N\mu)_T X_T$$

and

$$(D\ell)_{T,T} = (D\mu)_T Z_T^2.$$

The coordinates X_{2T}, Y_{2T} and Z_{2T} of $2T$ are given by

$$\begin{cases} X_{2T} = (N\mu)_T^2 - 8X_T Y_T^2, \\ Y_{2T} = (N\mu)_T(4X_T Y_T^2 - X_{2T}) - 8Y_T^4, \\ Z_{2T} = (D\mu)_T. \end{cases}$$

Hence the value of v_{2T} at Q can be calculated as

$$v_{2T}(Q) = x_Q - x_{2T} = \frac{(Nv)_{2T}}{(Dv)_{2T}},$$

where $(Dv)_{2T} = Z_{2T}^2$ and $(Nv)_{2T} = Z_{2T}^2 x_Q - X_{2T}$.

In the modified algorithm, we also need to compute $\ell_{-T,-T}$ at Q. But the coordinates of $-T$ are $(x_T, -y_T)$, so that we can write

$$\ell_{-T,-T}(Q) = y_Q + y_T + \mu_T(x_Q - x_T) = \frac{(N\ell)_{-T,-T}}{(D\ell)_{-T,-T}},$$

with

$$(N\ell)_{-T,-T} = (D\mu)_T Z_T^2 y_Q + (N\mu)_T Z_T^2 x_Q + 2Y_T^2 - (N\mu)_T X_T$$

and

$$(D\ell)_{-T,-T} = (D\mu)_T Z_T^2.$$

All the operations needed in the doubling steps of the algorithms in Figures 1 and 2 are shown in detail in Figure 3. The quantities are to be computed in the order shown in the table, a blank entry indicating that the corresponding quantity need not be computed in the case indicated at the top of the corresponding column. The costs of the computations are calculated assuming that the results of intermediate steps are kept in memory. An entry 0 indicated that the quantity has already been calculated at a previous stage.

The addition step. Next we deal with addition. When $T \neq \pm P$, the slope of the line joining T and P is

$$\lambda_{T,P} = \frac{y_T - y_P}{x_T - x_P} = \frac{(N\lambda)_{T,P}}{(D\lambda)_{T,P}},$$

where $(N\lambda)_{T,P} = Y_T - y_P Z_T^3$ and $(D\lambda)_{T,P} = X_T Z_T - x_P Z_T^3$.

It follows that the value of $\ell_{T,P}$ at the point Q is given by

$$y_Q - y_P - \lambda(x_Q - x_P) = \frac{(N\ell)_{T,P}}{(D\ell)_{T,P}},$$

where we precompute $y_Q - y_P$ and $x_Q - x_P$, and the numerator and denominator are given by

$$(N\ell)_{T,P} = (D\lambda)_{T,P}(y_Q - y_P) - (N\lambda)_{T,P}(x_Q - x_P)$$

Quantity	Formula	Classical Miller (Fig. 1)	Modified Miller (Fig. 2, loop 1)	Modified Miller (Fig. 2, loop 3)
$(N\mu)_T$	$3X_T^2 + aZ_T^4$	$m_a + 3s$	$m_a + 3s$	$m_a + 3s$
$(D\mu)_T$	$2Y_TZ_T = (Y_T + Z_T)^2 - Y_T^2 - Z_T^2$	2s	2s	2s
$(N\ell)_{T,T}$	$(D\mu)_TZ_T^2y_Q - (N\mu)_TZ_T^2x_Q$ $-2Y_T^2 - (N\mu)_TX_T$	$(3+2k)\mathbf{m}$	$(3+2k)\mathbf{m}$	
$(D\ell)_{T,T}$	$(D\mu)_TZ_T^2$	0	0	
X_{2T}	$(N\mu)_T^2 - 8X_TY_T^2$	$s+m$	$s+m$	$s+m$
Y_{2T}	$(N\mu)_T(4X_TY_T^2 - X_{2T}) - 8Y_T^4$	$s+m$	$s+m$	$s+m$
Z_{2T}	$(D\mu)_T$	0	0	0
$(Dv)_{2T}$	Z_{2T}^2	s		
$(Nv)_{2T}$	$Z_{2T}^2x_Q - X_{2T}$	km		
$(N\ell)_{-T,-T}$	$(D\mu)_TZ_T^2y_Q + (N\mu)_TZ_T^2x_Q$ $+2Y_T^2 - (N\mu)_TX_T$			$(3+2k)\mathbf{m}$
$(D\ell)_{-T,-T}$	$(D\mu)_TZ_T^2$			0
f	$\leftarrow f^2(N\ell)_{T,T}(Dv)_{T,T}$	$km + \mathbf{S}_k + \mathbf{M}_k$		
g	$\leftarrow g^2(D\ell)_{T,T}(Nv)_{T,T}$	$km + \mathbf{S}_k + \mathbf{M}_k$		
f	$\leftarrow f^2(N\ell)_{T,T}$		$\mathbf{S}_k + \mathbf{M}_k$	
g	$\leftarrow g^2(D\ell)_{T,T}$		$km + \mathbf{S}_k$	
f	$\leftarrow f^2(D\ell)_{-T,-T}$			$km + \mathbf{S}_k$
g	$\leftarrow g^2(N\ell)_{-T,-T}$			$\mathbf{S}_k + \mathbf{M}_k$
TOTAL		$m_a + 8s$ $+(5+5k)\mathbf{m}$ $+2\mathbf{S}_k + 2\mathbf{M}_k$	$m_a + 7s$ $+(5+3k)\mathbf{m}$ $+2\mathbf{S}_k + \mathbf{M}_k$	$m_a + 7s$ $+(5+3k)\mathbf{m}$ $+2\mathbf{S}_k + \mathbf{M}_k$

Fig. 3. Analysis of the cost of generic doubling

and

$$(D\ell)_{T,P} = (D\lambda)_{T,P}.$$

The coordinates X_{T+P}, Y_{T+P} and Z_{T+P} of $T + P$ are then given by

$$\begin{cases} X_{T+P} = (N\lambda)_{T,P}^2 - (X_T + x_PZ_T^2)(X_T - x_PZ_T^2)^2, \\ Y_{T+P} = -(D\lambda)_{T,P}^3y_P + (N\lambda)_{T,P}(x_P(D\lambda)_{T,P}^2 - X_{T+P}), \\ Z_{T+P} = (D\lambda)_{T,P}. \end{cases}$$

It follows that we can write the value of v_{T+P} at Q as

$$v_{T+P}(Q) = x_Q - x_{T+P} = \frac{(Nv)_{T+P}}{(Dv)_{T+P}},$$

with $(Nv)_{T+P} = x_QZ_{T+P}^2 - X_{T+P}$ and $(Dv)_{T+P} = Z_{T+P}^2$.

When the loop beginning with line **2** of Figure 2 is used, we need to calculate $\ell'_{-T,-P}$ at Q. In fact, using equation (3), we can write

$$\ell'_{-T,-P}(Q) = \frac{y_Q + y_P}{x_Q - x_P} + \frac{(N\lambda)_{T,P}}{(D\lambda)_{T,P}}.$$

If $\frac{y_Q+y_P}{x_Q-x_P}$ has been precomputed and has value $\alpha_{Q,P}$, we can write

$$\ell'_{-T,-P}(Q) = \frac{(N\ell')_{-T,-P}}{(D\ell')_{-T,-P}},$$

with $(N\ell')_{-T,-P} = (D\lambda)_{T,P}\alpha_{Q,P} + (N\lambda)_{T,P}$ and $(D\ell')_{-T,-P} = (D\lambda)_{T,P}$.

All the operations needed in the addition steps of the algorithms in Figures 1 and 2 are shown in detail in Figure 4. As with the doubling step, the quantities are to be computed in the order shown in the table, a blank entry indicating that the corresponding quantity need not be computed in the case indicated at the top of the corresponding column. The costs of the computations are calculated assuming that the results of intermediate steps are kept in memory. An entry 0 indicated that the quantity has already been calculated at a previous stage.

Quantity	Formula	Classical Miller (Fig. 1)	Modified Miller (Fig. 2, loop 2)	Modified Miller (Fig. 2, loop 4)
$(D\lambda)_{T,P}$	$(X_T - x_P Z_T^2)Z_T$	$s + 2m$	$s + 2m$	$s + 2m$
$(N\lambda)_{T,P}$	$Y_T - y_P Z_T^3$	$2m$	$2m$	$2m$
$(N\ell)_{T,P}$	$(D\lambda)_{T,P}(y_Q - y_P)$ $-(N\lambda)_{T,P}(x_Q - x_P)$	$2km$		$2km$
$(D\ell)_{T,P}$	$(D\lambda)_{T,P}$	0		0
X_{T+P}	$(N\lambda)_{T,P}^2 - X_T(X_T - x_P Z_T^2)^2$ $-x_P Z_T^2(X_T - x_P Z_T^2)^2$	$2s + 2m$	$2s + 2m$	$2s + 2m$
Y_{T+P}	$-y_P Z_T^3(X_T - x_P Z_T^2)^2$ $-x_P Z_T^2(X_T - x_P Z_T^2)^2)$ $+(N\lambda)_{T,P}(x_P Z_T^2(X_T - x_P Z_T^2)^2$ $-X_{T+P})$	$2m$	$2m$	$2m$
Z_{T+P}	$(D\lambda)_{T,P}$	0	0	0
$(Nv)_{T+P}$	$x_Q Z_{T+P}^2 - X_{T+P}$	$s + km$		
$(Dv)_{T+P}$	Z_{T+P}^2	0		
$(N\ell')_{-T,-P}$	$\alpha_{Q,P}(D\lambda)_{T,P} + (N\lambda)_{T,P}$		km	
$(D\ell')_{-T,-P}$	$(D\lambda)_{T,P}$		0	
f	$\leftarrow f(N\ell)_{T,P}(Dv)_{T+P}$	$km + \mathbf{M}_k$		
g	$\leftarrow g(D\ell)_{T,P}(Nv)_{T+P}$	$km + \mathbf{M}_k$		
f	$\leftarrow f(D\ell')_{-T,-P}$		km	
g	$\leftarrow g(N\ell')_{-T,-P}$		$\mathbf{M}_k$	
f	$\leftarrow f(N\ell)_{T,P}$			$\mathbf{M}_k$
g	$\leftarrow g(D\ell)_{T,P}$			km
TOTAL		$4s + (8 + 5k)m$ $+2\mathbf{M}_k$	$3s + (8 + 2k)m$ $+\mathbf{M}_k$	$3s + (8 + 3k)m$ $+\mathbf{M}_k$

Fig. 4. Analysis of the cost of generic addition

3.3   The Main Result

The following theorem recapitulates the number of operations in our variant of Miller's algorithm.

Theorem 2. *Suppose E is given in short Weierstrass form $y^2 = x^3 + ax + b$ with coefficients $a, b \in \mathbb{F}$. Let $P \in E(\mathbb{F})$ be a point of odd order $r \geq 2$ and let Q be a point of E of order r with coordinates in an extension field $\mathbb{F}'$ of $\mathbb{F}$ of degree k. We assume P and Q given in Weierstrass coordinates (x_P, y_P) and (x_Q, y_Q).*

1. *Using the associated Jacobian coordinates, the algorithms of Figures 1 and 2 can be implemented in such a way that all the denominators $(D\ell)_{T,T}$, $(D\ell)_{T,P}$, $(Dv)_{2T}$, $(Dv)_{T+P}$ and $(D\ell')_{-T,-P}$ belong to $\mathbb{F}$.*
2. *When this is the case:*
    (a) *Each doubling step of the generic usual Miller algorithm takes $\mathbf{m}_a + 8\mathbf{s} + (5 + 5k)\mathbf{m} + 2\mathbf{S}_k + 2\mathbf{M}_k$ operations while in the generic modified Miller algorithm it requires only $\mathbf{m}_a + 7\mathbf{s} + (5 + 3k)\mathbf{m} + 2\mathbf{S}_k + \mathbf{M}_k$ operations.*
    (b) *Each addition step of the generic usual Miller algorithm takes $4\mathbf{s} + (8 + 5k)\mathbf{m} + 2\mathbf{M}_k$ operations. On the other hand, the generic modified Miller algorithm requires only $3\mathbf{s} + (8 + 2k)\mathbf{m} + \mathbf{M}_k$ operations when line 2 is needed and $3\mathbf{s} + (8 + 3k)\mathbf{m} + \mathbf{M}_k$ operations when line 4 is needed.*

We have made no serious attempt to minimize the number of operations, for example by using formulas similar to those in [1].

Since the first part of Theorem 2 implies that, when $\mathbb{F}$ is a finite field with q elements, we have

$$(D\ell)_{T,T}^{q-1} = (D\ell)_{T,P}^{q-1} = (Dv)_{2T}^{q-1} = (Dv)_{T+P}^{q-1} = (D\ell')_{-T,-P}^{q-1} = 1,$$

denominator elimination is possible when we only need to calculate $f_{s,P}(Q)$ to some power divisible by $q - 1$. Such an algorithm saves at least $k\mathbf{m}$ operations both in the classical version and our variant.

Our new version of Miller's algorithm works perfectly well for arbitrary embedding degree. For example, using Theorem 2 of [16], it should be possible to find an elliptic curve with a prime embedding degree minimizing the number of iterations. *Optimal pairings* [26] involve in their computation a product $\prod_{i=0}^{\ell} f_{c_i,Q}^{q^i}(P)$ whose terms can be computed with our algorithm. Note that switching P and Q will lead to more computations in the extension field, but it is shown in [23] that optimized versions of the Ate and the twisted Ate pairing can be computed at least as fast as the Tate pairing. Note that Heß [16] §5, also mentions pairings of potential interest when k is odd and the elliptic curve has discriminant -4 and when k is not divisible by 3 and the elliptic curve has discriminant -3.

4   Curves with Even Embedding Degree

Currently, most implementations (where $\mathbb{F}$ is a finite field) are adapted to curves with embedding degree $2^i 3^j$, since the usual version of Miller's algorithm can be implemented more efficiently. Indeed, such curves admit an even twist which allows denominator elimination [4,5]. In the case of a cubic twist, denominator elimination is also possible [22]. Another advantage of embedding degrees of the form $2^i 3^j$ is that the corresponding extensions of $\mathbb{F}$ can be written as composite extensions of degree 2 or 3, which allows faster basic arithmetic operations [21].

In what follows, we discuss a version with denominator elimination of our variant adapted to even embedding degrees. Similar ideas have been used before, for example in [22] or [13]. We suppose that $\mathbb{F}$ is a finite field of odd cardinality

q which we denote by $\mathbb{F}_q$. We consider an elliptic curve E with short Weierstrass model $y^2 = x^3 + ax + b$ with a, $b \in \mathbb{F}_q$. We suppose that $E(\mathbb{F}_q)$ contains a point P with affine coordinates (x_P, y_P) of order an odd prime r and embedding degree k. If $n \geq 1$ is an integer, we denote by $\mathbb{F}_{q^n}$ the extension of degree n of $\mathbb{F}_q$ in a fixed algebraic closure of $\mathbb{F}_q$.

We suppose for the rest of this section that k is even. Let γ be a non-square element of $\mathbb{F}_{q^{k/2}}$ and fix a square root β in $\mathbb{F}_{q^k}$ of γ, so that every element of $\mathbb{F}_{q^k}$ can be written in a unique way in the form $x + y\beta$ with x, $y \in \mathbb{F}_{q^{k/2}}$. Then one knows that $E[r]$ contains non zero points $Q = (x_Q, y_Q)$ that satisfy x_Q, $y_Q/\beta \in \mathbb{F}_{q^{k/2}}$. In fact, if π denotes the Frobenius endomorphism of E over $\mathbb{F}_q$ given by $\pi(x, y) = (x^q, y^q)$, then π restricts to an endomorphism of $E[r]$ viewed as a vector space over the field with r elements, and a point $Q \in E[r] \setminus \{O_E\}$ has the desired property if and only if Q belongs to the eigenspace V of π with eigenvector q. We can easily construct such points Q as follows: if $R \in E[r]$, then the point

$$Q := [k](R) - \sum_{i=0}^{k-1} \pi^i(R)$$

lies in V and, when r is large, the probability that it is non zero when R is selected at random is overwhelming.

As is well-known, this leads to speed-ups in the calculation of $f_{s,P}(Q)$ whenever $Q \in V$. For example, the calculation of $(N\ell)_{T,P}$ (see Figure 4) takes only $(2\frac{k}{2})\mathbf{m} = k\mathbf{m}$ operations. On the other hand, in general neither $\alpha_{Q,P}$ nor $\alpha_{Q,P}/\beta$ belongs to $\mathbb{F}_{q^{k/2}}$, so that the calculation of $(N\ell')_{-T,-P}$ also requires $k\mathbf{m}$ operations. So, the improvement obtained in the generic case is lost.

When $P \in E[r](\mathbb{F}_q)$ and $Q \in V$, it is well-known that the Tate pairing $e_r^T(P, Q)$ is given by $e_r^T(P, Q) = f_{r,P}(Q)^{\frac{q^k-1}{r}}$. Denominator elimination is possible since $q^{k/2} - 1$ divides $\frac{q^k-1}{r}$, as follows from the fact that r is prime and the definition of the embedding degree k.

Let $v = x + y\beta$ with x, $y \in \mathbb{F}_{q^{k/2}}$ be an element of $\mathbb{F}_{q^k}$. The conjugate of v over $\mathbb{F}_{q^{k/2}}$ is then $\bar{v} = x - y\beta$. It follows that, if $v \neq 0$, then

$$\frac{1}{v} = \frac{\bar{v}}{x^2 - \gamma y^2} \tag{4}$$

where $x^2 - \gamma y^2 \in \mathbb{F}_{q^{k/2}}$. Thus, in a situation where elements of $\mathbb{F}_{q^{k/2}}$ can be ignored, $\frac{1}{v}$ can be replaced by $\bar{v}$, thereby saving an inversion in $\mathbb{F}_{q^k}$.

We exploit all this in the algorithm in Figure 5, where we use the same notation as in Figure 2. It outputs an element f of $\mathbb{F}_{q^k}$ such that $f^{q^{\frac{k}{2}}-1} = f_{s,P}(Q)^{q^{\frac{k}{2}}-1}$. Thus, when $s = r$, we find, since $q^{\frac{k}{2}} - 1$ divides $\frac{q^k-1}{r}$, that $e_r^T(P, Q) = f^{\frac{q^k-1}{r}}$.

We replace the denominators $\ell_{-T,-T}$ and $\ell_{-T,-P}$ (updated in the function g) by their conjugates $\overline{\ell_{-T,-T}}$ and $\overline{\ell_{-T,-P}}$. In Jacobian coordinates, one has

$$\overline{(N\ell')_{-T,-P}} = \overline{\alpha_{Q,P}}(D\lambda)_{T,P} + (N\lambda)_{T,P}, \qquad \text{and} \qquad (5)$$

$$\overline{(N\ell)_{-T,-T}} = 2Y_T(-y_Q Z_T^3 + Y_T) + (N\mu)_T(x_Q Z_T^2 - X_T). \qquad (6)$$

Cost analysis of the doubling and addition steps in this algorithm can be found in Figures 6 and 7. We summarize our conclusions in the following result.

Theorem 3. *Let β be a non-square element of $\mathbb{F}_{q^k}$ whose square lies in $\mathbb{F}_{q^{k/2}}$. Suppose E is given in short Weierstrass form $y^2 = x^3 + ax + b$ with coefficients a, $b \in \mathbb{F}_q$. Let $P \in E(\mathbb{F}_q)$ be a point of odd prime order r with even embedding degree k. Let Q be a point of E of order r with coordinates in the extension field $\mathbb{F}_{q^k}$ of $\mathbb{F}_q$ of degree k. We assume P and Q given in Weierstrass coordinates (x_P, y_P) and (x_Q, y_Q) with $x_Q, y_Q/\beta \in \mathbb{F}_{q^{k/2}}$. Using the associated Jacobian coordinates, every doubling step of the algorithm of Figure 5 requires $\mathbf{m}_a + 7\mathbf{s} + (5+k)\mathbf{m} + \mathbf{S}_k + \mathbf{M}_k$ operations and every addition step requires $3\mathbf{s} + (8+k)\mathbf{m} + \mathbf{M}_k$.*

As before, we have made no serious attempt to minimize the number of operations, for example by using formulas similar to those in [1]. Moreover, we believe that there is scope for further improvement in the case of curves in special families or curves with efficient arithmetic (see for example [14]).

5   Experiments

We ran some experiments comparing usual Miller (Figure 1) with the variant of Figure 2 when $k = 17$ and $k = 19$. When $k = 18$, we compared the performance of the algorithms of Figures 1 and 5 and also the algorithm proposed in 2003 by [4]. In each case, the group order r has 192 bits and the rho-value $\rho = \frac{\log q}{\log r}$ is a little under 1.95, q being the cardinality of the base field. In the example with $k = 17$, the big field $\mathbb{F}_{q^k}$ was generated as $\mathbb{F}_q[x]/(x^{17} + x + 12)$ while when $k = 18$ and $k = 19$, $\mathbb{F}_{q^k}$ was generated as $\mathbb{F}_q[x]/(x^k + 2)$. Our curves were constructed using the Cocks-Pinch method (see [15]).

– For $k = 17$, the curve is $E : y^2 = x^3 + 6$ and

$r = 6277101735386680763835789423207666416102355444464039939857$
$q = 22052206043753515622219125759263352673620079371332192473307$
$84762783175923330153479572768090060536491226188438277$1

– For $k = 18$, the curve is $E : y^2 = x^3 + 3$ and

$r = 6277101735386680763835789423207666416102355444464046918739$
$q = 3528684539263022042925513517751522924824248454977423175709$
$6903373137251646468704115267717074091166525440296813891$

Algorithm 3. Miller(P, Q, s) modified with even embedding degree

---

Data: $s = \sum_{i=0}^{l-1} s_i 2^i$, $s_i \in \{0,1\}$, $s_{l-1} = 1$, h Hamming weight of s, $Q \in E[r]$ a non-zero element with x_Q, $y_Q/\beta \in \mathbb{F}_{q^{k/2}}$.

Result: An element f of $\mathbb{F}_{q^k}$ satisfying $f^{q^{k/2}-1} = f_{s,P}(Q)^{q^{k/2}-1}$

$f \leftarrow 1, T \leftarrow P,$

if $l + h$ is odd then
  | $\delta \leftarrow 1$
end
else
  | $\delta \leftarrow 0$
end
for $i = l - 2$ to 0 do
 1   if $\delta = 0$ then
       | $f \leftarrow f^2(N\ell)_{T,T}, \; T \leftarrow 2T, \; \delta \leftarrow 1$
 2 | if $s_i = 1$ then
         | $f \leftarrow f\overline{(N\ell')}_{-T,-P}, \; T \leftarrow T + P, \; \delta \leftarrow 0$
       | end
     end
 3   else
       | $f \leftarrow f^2\overline{(N\ell)}_{-T,-T}, \; T \leftarrow 2T, \; \delta \leftarrow 0$
 4 | if $s_i = 1$ then
         | $f \leftarrow f(N\ell)_{T,P}, \; T \leftarrow T + P, \; \delta \leftarrow 1$
       | end
     end
end
return f

Fig. 5. The modified Miller algorithm for even embedding degree

Quantity	Formula	Modified Miller (Fig. 5, loop 1)	Modified Miller (Fig. 5, loop 3)
$(N\mu)_T$	$3X_T^2 + aZ_T^4$	$\mathbf{m}_a + 3\mathbf{s}$	$\mathbf{m}_a + 3\mathbf{s}$
$(D\mu)_T$	$2Y_T Z_T = (Y_T + Z_T)^2 - Y_T^2 - Z_T^2$	$2\mathbf{s}$	$2\mathbf{s}$
$(N\ell)_{T,T}$	$(D\mu)_T Z_T^2 y_Q - (N\mu)_T Z_T^2 x_Q - 2Y_T^2 - (N\mu)_T X_T$	$(3 + 2(k/2))\mathbf{m}$	
X_{2T}	$(N\mu)_T^2 - 8X_T Y_T^2$	$\mathbf{s} + \mathbf{m}$	$\mathbf{s} + \mathbf{m}$
Y_{2T}	$(N\mu)_T(4X_T Y_T^2 - X_{2T}) - 8Y_T^4$	$\mathbf{s} + \mathbf{m}$	$\mathbf{s} + \mathbf{m}$
Z_{2T}	$(D\mu)_T$	0	0
$(N\ell)_{-T,-T}$	$(D\mu)_T Z_T^2 y_Q + (N\mu)_T Z_T^2 x_Q + 2Y_T^2 - (N\mu)_T X_T$		$(3 + 2(k/2))\mathbf{m}$
f	$\leftarrow f^2(N\ell)_{T,T}$	$\mathbf{S}_k + \mathbf{M}_k$	
f	$\leftarrow f^2\overline{(N\ell)}_{-T,-T}$		$\mathbf{S}_k + \mathbf{M}_k$
TOTAL		$\mathbf{m}_a + 7\mathbf{s}$ $+(5 + k)\mathbf{m}$ $+\mathbf{S}_k + \mathbf{M}_k$	$\mathbf{m}_a + 7\mathbf{s}$ $+(5 + k)\mathbf{m}$ $+\mathbf{S}_k + \mathbf{M}_k$

Fig. 6. Analysis of doubling for even embedding degree

Quantity	Formula	Modified Miller (Fig. 5, loop **2**)	Modified Miller (Fig. 5, loop **4**)
$(D\lambda)_{T,P}$	$(X_T - x_P Z_T^2)Z_T$	$s + 2m$	$s + 2m$
$(N\lambda)_{T,P}$	$Y_T - y_P Z_T^3$	$2m$	$2m$
$(N\ell)_{T,P}$	$(D\lambda)_{T,P}(y_Q - y_P) - (N\lambda)_{T,P}(x_Q - x_P)$		$2(k/2)m$
X_{T+P}	$(N\lambda)_{T,P}^2 - X_T(X_T - x_P Z_T^2)^2$ $\qquad -x_P Z_T^2(X_T - x_P Z_T^2)^2$	$2s + 2m$	$2s + 2m$
Y_{T+P}	$-y_P Z_T^3(X_T(X_T - x_P Z_T^2)^2$ $\qquad -x_P Z_T^2(X_T - x_P Z_T^2)^2)$ $\quad +(N\lambda)_{T,P}(x_P Z_T^2(X_T - x_P Z_T^2)^2$ $\qquad -X_{T+P})$	$2m$	$2m$
Z_{T+P}	$(D\lambda)_{T,P}$	0	0
$(N\ell')_{-T,-P}$	$\alpha_{Q,P}(D\lambda)_{T,P} + (N\lambda)_{T,P}$	km	
f	$\leftarrow f(N\ell')_{-T,-P}$	$\mathbf{M}_k$	
f	$\leftarrow f(N\ell)_{T,P}$		$\mathbf{M}_k$
TOTAL		$3s + (8+k)m$ $+\mathbf{M}_k$	$3s + (8+k)m$ $+\mathbf{M}_k$

Fig. 7. Cost analysis of addition for even embedding degree

– For $k = 19$, the curve is $E : y^2 = x^3 + 2$ and

$$r = 6277101735386680763835789423207666416102355444464038231927$$
$$q = 328331730495825080256136908360300125975534969742697656 7975$$
$$265167703768467341628884414224762338965233382661234563 7$$

For the computations, we used the NTL library [25] and implemented the algorithms without any optimization on an Intel(R) Core(TM)2 Duo CPU E8500 @
3.16Ghz using Ubuntu Operating System 9.04. The computations of the Miller
function (without any final exponentiation) were executed on 100 random inputs.
The experimental average results are summarized in Figure 8.

k	Usual Miller (Fig. 1)	Our variant (Fig. 2)	Our variant with k even (Fig. 5)	Miller without denominators [4]
17	$0.0664s$	$0.0499s$	–	–
18	$0.0709s$	–	$0.0392s$	$0.0393s$
19	$0.0769s$	$0.0683s$	–	–

Fig. 8. Timings

6   Conclusion

In this paper we presented a variant of Miller's formula and algorithm. Generically, it is more efficient than the usual Miller algorithm as in Figure 1, calculation suggest that it can lead to a real improvement in cases where denominator
elimination is not available. Consequently, we believe it will have applications in

pairing-based cryptography using elliptic curves with embedding degree not being on the form $2^i 3^j$, for example when the optimal Ate or Twisted Ate pairing is used. Further work is needed to clarify such questions.

Acknowledgments

This work was supported by Project ANR-07-TCOM-013-04 PACE financed by the Agence National de Recherche (France). We would like to thank Andreas Enge for several helpful discussions and the anonymous reviewers for their numerous suggestions and remarks which have enables us to substantially improve the paper.

References

1. Arène, C., Lange, T., Naehrig, M., Ritzenthaler, C.: Faster computation of the Tate pairing. Preprint, Cryptology ePrint Archive: Report 2009/155 (2009), http://eprint.iacr.org/2009/155
2. Balasubramanian, R., Koblitz, N.: The improbability that an elliptic curve has subexponential discrete log problem under the Menezes–Okamoto–Vanstone algorithm. J. Cryptology 11(2), 141–145 (1998)
3. Barreto, P.S.L.M., Galbraith, S.D., O'Eigeartaigh, C., Scott, M.: Efficient pairing computation on supersingular abelian varieties. Des. Codes Cryptography 42(3), 239–271 (2007)
4. Barreto, P.S.L.M., Lynn, B., Scott, M.: On the selection of pairing-friendly groups. In: Matsui, M., Zuccherato, R.J. (eds.) SAC 2003. LNCS, vol. 3006, pp. 17–25. Springer, Heidelberg (2003)
5. Barreto, P.S.L.M., Lynn, B., Scott, M.: Efficient implementation of pairing-based cryptosystems. J. Cryptology 17(4), 321–334 (2004)
6. Barreto, P.S.L.M., Naehrig, M.: Pairing-friendly elliptic curves of prime order. In: Preneel, B., Tavares, S. (eds.) SAC 2005. LNCS, vol. 3897, pp. 319–331. Springer, Heidelberg (2006)
7. Bernstein, D., Lange, T.: Explicit-formulas database (2010), http://www.hyperelliptic.org/EFD/
8. Blake, I.F., Kumar Murty, V., Xu, G.: Refinements of Miller's algorithm for computing the Weil/Tate pairing. J. Algorithms 58(2), 134–149 (2006)
9. Blake, I.F., Seroussi, G., Smart, N.P.: Advances in Elliptic Curves Cryptography. London Mathematical Society Lecture Note Series, vol. 317. Cambridge University Press, Cambridge (2005)
10. Boneh, D., Franklin, M.K.: Identity-based encryption from the Weil pairing. SIAM J. Comput. 32(3), 586–615 (2003); Extended abstract in proc. of Crypto 2001
11. Cohen, H., Frey, G., Avanzi, R., Doche, C., Lange, T., Nguyen, K., Vercauteren, F.: Handbook of Elliptic and Hyperelliptic Curve Cryptography. Discrete mathematics and its applications. Chapman & Hall, Boca Raton (2006)
12. Costello, C., Hisil, H., Boyd, C., Nieto, J.M.G., Wong, K.K.-H.: Faster pairings on special Weierstrass curves. In: Shacham, H., Waters, B. (eds.) Pairing 2009. LNCS, vol. 5671, pp. 89–101. Springer, Heidelberg (2009)
13. Costello, C., Lange, T., Naehrig, M.: Faster pairing computations on curves with high-degree twists. In: Nguyen, P.Q., Pointcheval, D. (eds.) PKC 2010. LNCS, vol. 6056, pp. 224–242. Springer, Heidelberg (2010)

14. El Mrabet, N., Nègre, C.: Finite field multiplication combining AMNS and DFT approach for pairing cryptography. In: Boyd, C., González Nieto, J. (eds.) ACISP 2009. LNCS, vol. 5594, pp. 422–436. Springer, Heidelberg (2009)
15. Freeman, D., Scott, M., Teske, E.: A taxonomy of pairing-friendly elliptic curves. J. Cryptology 23(2), 224–280 (2010)
16. Heß, F.: Pairing lattices. In: Galbraith, S.D., Paterson, K.G. (eds.) Pairing 2008. LNCS, vol. 5209, pp. 18–38. Springer, Heidelberg (2008)
17. Heß, F., Smart, N.P., Vercauteren, F.: The Eta pairing revisited. IEEE Transactions on Information Theory 52(10), 4595–4602 (2006)
18. Ionica, S., Joux, A.: Another approach to pairing computation in Edwards coordinates. In: Chowdhury, D.R., Rijmen, V., Das, A. (eds.) INDOCRYPT 2008. LNCS, vol. 5365, pp. 400–413. Springer, Heidelberg (2008)
19. Ionica, S., Joux, A.: Pairing the volcano. In: Hanrot, G., Morain, F., Thomé, E. (eds.) ANTS-IX. LNCS, vol. 6197, pp. 201–218. Springer, Heidelberg (2010)
20. Joux, A.: A one round protocol for tripartite Diffie-Hellman. J. Cryptology 17(4), 263–276 (2000); Extended abstract in proc. of ANTS 2000
21. Koblitz, N., Menezes, A.: Pairing-based cryptography at high security levels. In: Smart, N.P. (ed.) Cryptography and Coding 2005. LNCS, vol. 3796, pp. 13–36. Springer, Heidelberg (2005)
22. Lin, X., Zhao, C., Zhang, F., Wang, Y.: Computing the Ate pairing on elliptic curves with embedding degree $k = 9$. IEICE Transactions 91-A(9), 2387–2393 (2008)
23. Matsuda, S., Kanayama, N., Heß, F., Okamoto, E.: Optimised versions of the Ate and twisted Ate pairings. IEICE Transactions 92-A(7), 1660–1667 (2009)
24. Miller, V.S.: The Weil pairing, and its efficient calculation. J. Cryptology 17(4), 235–261 (2004)
25. Shoup, V.: NTL: a library for doing number theory (2009), http://www.shoup.net/ntl/
26. Vercauteren, F.: Optimal pairings. IEEE Transactions of Information Theory 56, 455–461 (2009)

Pairing Computation on Elliptic Curves with Efficiently Computable Endomorphism and Small Embedding Degree

Sorina Ionica[2] and Antoine Joux[1,2]

[1] Université de Versailles Saint-Quentin-en-Yvelines, 45 avenue des États-Unis, 78035 Versailles CEDEX, France
[2] DGA
{sorina.ionica,antoine.joux}@m4x.org

Abstract. Scott uses an efficiently computable isomorphism in order to optimize pairing computation on a particular class of curves with embedding degree 2. He points out that pairing implementation becomes thus faster on these curves than on their supersingular equivalent, originally recommended by Boneh and Franklin for Identity Based Encryption. We extend Scott's method to other classes of curves with small embedding degree and efficiently computable endomorphism.

1 Introduction

Pairings were first used in cryptography for attacking the discrete logarithm problem on the elliptic curve [21], but nowadays they are also used as bricks for building new cryptographic protocols such as the tripartite Diffie-Hellman protocol [15], identity-based encryption [5], short signatures [6], and others.

A cryptographic pairing is a bilinear map $e : \mathbb{G}_1 \times \mathbb{G}_2 \to \mathbb{G}_3$, where $\mathbb{G}_1$, $\mathbb{G}_2$ and $\mathbb{G}_3$ are groups of large prime order r. Known pairings on elliptic curves, i.e. the Weil, Tate pairings, map to the multiplicative group of the minimal extension of the ground field $\mathbb{F}_q$ containing the r-th roots of unity. The degree of this extension, denoted usually by k, is called the embedding degree with respect to r. The basic algorithm used in pairing computation was given by Miller and is an extension of the double-and-add method for finding a point multiple. The cost of the computation heavily depends on costs of operations in $\mathbb{F}_{q^k}$. Consequently, in practice we need curves with small embedding degree.

The reduction of the loop length in Miller's algorithm is one of the main directions taken by research in pairing computation during the past few years. These results concern only pairings on $\mathbb{G}_1 \times \mathbb{G}_2$ or $\mathbb{G}_2 \times \mathbb{G}_1$, where subgroups $\mathbb{G}_1$ and $\mathbb{G}_2$ are given by

$$\mathbb{G}_1 = E[r] \cap \mathrm{Ker}(\pi - [1]) \quad \text{and} \quad \mathbb{G}_2 = E[r] \cap \mathrm{Ker}(\pi - [q]),$$

where π is the Frobenius morphism of E, i.e. $\pi : E \to E$, $(x,y) \mapsto (x^q, y^q)$. The pairings computed by the new algorithms [2,13] are actually powers of the

M. Joye, A. Miyaji, and A. Otsuka (Eds.): Pairing 2010, LNCS 6487, pp. 435–449, 2010.
© Springer-Verlag Berlin Heidelberg 2010

Tate pairing and are called in the literature the Eta ($\mathbb{G}_1 \times \mathbb{G}_2$, $\mathbb{G}_2 \times \mathbb{G}_1$), Ate ($\mathbb{G}_2 \times \mathbb{G}_1$) and twisted Ate pairing ($\mathbb{G}_1 \times \mathbb{G}_2$).

Furthermore, Hess and, independently, Vercauteren [12,27] showed that on some families of curves with small Frobenius trace, the complexity of Miller's algorithm is $\mathcal{O}(\frac{1}{\varphi(k)} \log r)$, where φ is the Euler totient function.

In this paper, we propose the use of efficiently computable endomorphisms, other than the Frobenius map, to optimize pairing computation. Our method, which works on curves having a small embedding degree, improves pairing computation on curves constructed by the Cocks-Pinch method.

The remainder of this paper is organized as follows. Section 2 presents background on pairings and the Cocks-Pinch method for constructing curves with complex multiplication. Section 3 presents our results which make use of endomorphisms to compute pairings. Section 4 presents an evaluation of costs of an implementation of our algorithm and compares performances to those of the Tate pairing computation. In Appendix 5 we give examples of curves constructed using the Cocks-Pinch method, with small embedding degree and endomorphism of small degree.

2   Background on Pairings

The definition of the Tate pairing and of Miller's algorithm [20] used in pairing computations. This algorithm heavily relies on the double-and-add method for finding a point multiple. Let E be an elliptic curve given by a Weierstrass equation:

$$y^2 = x^3 + a_4 x + a_6, \tag{1}$$

defined over a finite field $\mathbb{F}_q$, with $\text{char}(\mathbb{F}_q) \neq 2, 3$. Let P_∞ denote the neutral element on the elliptic curve. Consider r a large prime dividing $\#E(\mathbb{F}_q)$ and k the embedding degree with respect to r.

Let P be an r-torsion point and for any integer i, denote by $f_{i,P}$ a function with divisor $\text{div}\,(f_{i,P}) = i(P) - (iP) - (i-1)(P_\infty)$ (see [24] for an introduction to divisors). Note that $f_{r,P}$ is such that $\text{div}\,(f_{r,P}) = r(P) - r(P_\infty)$.

In order to define the Tate pairing we take Q a point in $E(\mathbb{F}_{q^k})$ representing an element of $E(\mathbb{F}_{q^k})/rE(\mathbb{F}_{q^k})$. Let T be a point on the curve such that the support of the divisor $D = (Q + T) - (T)$ is disjoint from the one of $f_{r,P}$. We then define the Tate pairing as

$$t_r(P, Q) = f_{r,P}(D). \tag{2}$$

This value is a representative of an element of $\mathbb{F}_{q^k}^*/(\mathbb{F}_{q^k}^*)^r$. However for cryptographic protocols it is essential to have a unique representative so we will raise it to the $((q^k - 1)/r)$-th power, obtaining an r-th root of unity. We call the resulting value the *reduced* Tate pairing

$$T_r(P, Q) = t_r(P, Q)^{\frac{q^k - 1}{r}}.$$

As stated in [10], if the function $f_{r,P}$ is normalized, i.e. $(u_{P_\infty}^r f_{r,P})(P_\infty) = 1$ for some $\mathbb{F}_q$-rational uniformizer u_{P_∞} at P_∞, then one can ignore the point T and compute the pairing as

$$T_r(P,Q) = f_{r,P}(Q)^{(q^k-1)/r}.$$

In the sequel of this paper we only consider normalized functions. Before going into the details of Miller's algorithm, we recall the standard addition law on an elliptic curve in Weierstrass form. Suppose we want to compute the sum of iP and jP for $i, j \geq 1$. Let l be the line through iP and jP. Then l intersects the cubic curve E at one further point R according to Bezout's theorem (see [11]). We take v the line between R and P_∞ (which is a vertical line when R is not P_∞). Then v intersects E at one more point and we define the sum of iP and jP to be this point.

The lines l and v are functions on the curve E and the corresponding divisors are

$$\mathrm{div}\,(l) = (iP) + (jP) + (R) - 3(P_\infty),$$
$$\mathrm{div}\,(v) = (R) + ((i+j)P) - 2(P_\infty).$$

One can then easily check the following relation:

$$f_{i+j,P} = f_{i,P} f_{j,P} \frac{l}{v}. \tag{3}$$

Turning back to Miller's algorithm, suppose we want to compute $f_{r,P}(Q)$. We compute at each step of the algorithm on one side mP, where m is the integer with binary expansion given by the i topmost bits of the binary expansion of r, and on the other side $f_{m,P}$ evaluated at Q, by exploiting the formula above. We call the set of operations executed for each bit i of r a *Miller operation*.

Algorithm 1. Miller's algorithm

INPUT: An elliptic curve E defined over a finite field $\mathbb{F}_q$, P an r-torsion point on the curve and $Q \in E(\mathbb{F}_{q^k})$.

OUTPUT: the Tate pairing $t_r(P,Q)$.

1: Let $i = \lceil \log_2(r) \rceil$, $K \leftarrow P, f \leftarrow 1$
2: **while** $i \geq 1$ **do**
3:    Compute the equation of l arising in the doubling of K
4:    $K \leftarrow 2K$ and $f \leftarrow f^2 l(Q)$
5:    **if** the i-th bit of r is 1 **then**
6:       Compute the equation of l arising in the addition of K and P
7:       $K \leftarrow P + K$ and $f \leftarrow fl(Q)$
8:    **end if**
9:    Let $i \leftarrow i - 1$.
10: **end while**
11: **return** f.

Table 1. Cost of one step in Miller's algorithm for even embedding degree

	Doubling	Mixed addition
$\mathcal{J}$ [1,14]	$(1+k)\mathbf{m} + 11\mathbf{s} + 1\mathbf{M} + 1\mathbf{S}$	$(6+k)\mathbf{m} + 6\mathbf{s} + 1\mathbf{M}$
$\mathcal{J}, y^2 = x^3 + b$ $d = 2, 6$ [7]	$(2k/d+2)\mathbf{m} + 7\mathbf{s} + 1\mathbf{M} + 1\mathbf{S}$	$(2k/d+9)\mathbf{m} + 2\mathbf{s} + 1\mathbf{M}$
$\mathcal{J}, y^2 = x^3 + ax$ $d = 2, 4$ [7]	$(2k/d+2)\mathbf{m} + 8\mathbf{s} + 1\mathbf{M} + 1\mathbf{S}$	$(2k/d+12)\mathbf{m} + 4\mathbf{s} + 1\mathbf{M}$

Implementing pairings. In implementations, we usually prefer curves with even embedding degree. On these curves, thanks to the existence of twists, most computations in a Miller operation are done in proper subfields of $\mathbb{F}_{q^k}$. Moreover, thanks to the final exponentiation, terms contained in proper subfields of $\mathbb{F}_{q^k}$ can be ignored (see [18]). Algorithm 1 gives the pseudocode of Miller's algorithm for curves with even embedding degree. In Table 1 we give costs for the doubling and the addition step in Algorithm 1 for an implementation on $\mathbb{G}_1 \times \mathbb{G}_2$, on curves with twists[1] of degree d. We denote by $\mathbf{m}, \mathbf{s}$ the costs of multiplication and squaring in $\mathbb{F}_q$ and $\mathbf{M}, \mathbf{S}$ the costs of multiplication and squaring in $\mathbb{F}_{q^k}$.

Security issues. A secure pairing-based cryptosystem needs to be implemented on elliptic curve subgroups $\mathbb{G}_1$ and $\mathbb{G}_2$ such that the discrete logarithm problem is computationally difficult in $\mathbb{G}_1$, $\mathbb{G}_2$ and in $\mathbb{F}_{q^k}^*$. The best known algorithm for computing discrete logarithms on elliptic curves is the Pollard-rho method [25,22], which has complexity $O(\sqrt{r})$, where r is the order of the groups $\mathbb{G}_1$ and $\mathbb{G}_2$. Meanwhile, the best known algorithm for solving the discrete logarithm problem in the multiplicative group of a finite field is the index calculus algorithm, which has sub-exponential running time [17,16]. Consequently, in order to achieve the same level of security in both the elliptic curve subgroups and in the finite field subgroup, we need to choose a q^k which is significantly larger than r. It is therefore interesting to consider the ratio of these sizes

$$\frac{k \log q}{\log r}.$$

As the efficiency of the implementation will depend critically on the so-called ρ-value

$$\rho = \frac{\log q}{\log r},$$

it is preferable to keep this value as small as possible.

The Cocks-Pinch method for constructing pairing friendly curves. Let E be an ordinary curve defined over a finite field $\mathbb{F}_q$. We denote by π the Frobenius morphism and by t its trace. Given the fact that curve must have a subgroup of

---

[1] We briefly recall that a twist E' of degree d of an elliptic curve E defined over $\mathbb{F}_q$ is a curve isomorphic to E, such that the isomorphism between the two curves is minimally defined over $\mathbb{F}_{q^d}$. The reader should check [24] for more details on twists.

large order r and that the number of points on the curve is $\#E(\mathbb{F}_q) = q + 1 - t$ we write

$$q + 1 - t = hr.$$

Furthermore, the fact that the Frobenius is an element of an order in a quadratic imaginary field $\mathbb{Q}(\sqrt{-D})$ gives

$$Dy^2 = 4q - t^2 = 4hr - (t - 2)^2.$$

To sum up, in order to generate a pairing friendly curve, we are looking for q, r, k, D, t and y satisfying the following conditions

$$r \mid Dy^2 + (t - 2)^2,$$
$$r \mid q^k - 1,$$
$$t^2 + Dy^2 = 4q.$$

Cocks and Pinch gave an algorithm (which is presented in [4]) which finds, given r and a small k, parameters q prime and t satisfying the equations above.

Algorithm 2. The Cocks-Pinch algorithm

INPUT: k, r a prime number, D and $k | (r - 1)$.
OUTPUT: q, t such that there is a curve with CM by $\sqrt{-D}$ over $\mathbb{F}_q$ with $q + 1 - t$
   points where $r | (q + 1 - t)$ and $r | (q^k - 1)$.
1: Choose a primitive k-th root of unity g in $\mathbb{F}_r$.
2: Choose an integer $t \leftarrow g + 1 (\mathrm{mod}\ r)$.
3: **if** $\gcd(t, D) \neq 1$ **then**
4:    exit (or choose another g).
5: **end if**
   Choose an integer $y_0 = \pm(t - 2)/\sqrt{-D}(\mathrm{mod}\ r)$.
   $j \rightarrow 0$
6: **repeat**
7:    $q \leftarrow (t^2 + D(y_0 + jr)^2)/4$
8:    $j \leftarrow j + 1$
9: **until** q is prime
10: **return** q and t

This method produces ordinary curves with a ρ-value approximatively 2, which is less preferred in practice. However, Vercauteren [8, Prop. 7.1] showed that for certain embedding degrees and certain values of the CM discriminant $-D$, there are no ordinary curves with smaller ρ-value.

Proposition 1. *Let E be an elliptic curve over $\mathbb{F}_q$ with a subgroup of prime order $r > 3$ and embedding degree $k > 1$ with respect to r. If E has a twist $E'/\mathbb{F}_q$ of degree k and $r \geq 4\sqrt{q}$, then E is supersingular.*

It follows that in some cases, like $k = 2$ or $k = 4$ and[2] discriminant -4, the curves produced by the Cocks-Pinch algorithm have optimal ρ-value. Moreover, with this method we may choose the value of r from the very beginning. This is an advantage, because we may choose r with low Hamming weight. On curves with such r, in Algorithm 1, we perform mostly doublings and very few additions.

The Eta pairing and its variants. To our knowledge, isogenies were proposed to speed up pairing computation for the first time by Barreto et al. [2], who introduced the Eta pairing. This idea was extended by Hess et al. [13]. We present here the main result in [13], without giving the proof.

Theorem 1. *Let E be an elliptic curve defined over $\mathbb{F}_q$, r a large prime with $r \mid \#E(\mathbb{F}_q)$ and k the embedding degree with respect to r. Assume that $k > 1$ and denote by t the trace of the Frobenius.*

(a) For $T = t - 1$, $Q \in \mathbb{G}_2 = E[r] \cap Ker(\pi - [q])$, $P \in \mathbb{G}_1 = E[r] \cap Ker(\pi - [1])$ we have

(i) $f_{T,Q}(P)$ defines a bilinear pairing, which we call the Ate pairing;

(ii) Let $N = gcd(T^k - 1, q^k - 1)$ and $T^k - 1 = LN$, with k the embedding degree, then

$$t_r(Q, P)^L = f_{T,Q}(P)^{c(q^k - 1)/N}$$

where $c = \sum_{i=0}^{k-1} T^{k-1-i} q^i \equiv kq^{k-1} \mod r$. For $r \nmid L$, the Ate pairing is non-degenerate.

(b) Assume E has a twist of degree d and set $m = gcd(k, d)$ and $e = k/m$. We denote by $c = \sum_{i=0}^{m-1} T^{e(m-1-i)} q^{ei} \equiv mq^{e(m-1)} \mod r$. We have

(i) $f_{T^e, P}(Q)$ defines a bilinear pairing, which we call the twisted Eta pairing;

(ii) $t_r(P, Q)^L = f_{T^e, P}(Q)^{c(q^k - 1)/N}$ and the twisted Eta pairing is non-degenerate if $r \nmid L$.

The Ate and twisted Eta pairing can be computed using Miller's algorithm with a loop length of $\log T$ and $\log T^e$, respectively. Consequently, if the trace t is smaller than r, these pairings may be significantly faster than the Tate pairing.

3   Speeding Up Pairing Computation Using Endomorphisms of Small Degree

The following result was given by Verheul [28], whose purpose was to investigate the existence of distortion maps for points of order r, i.e. maps ϕ such that for a point P, $\phi(P) \notin \langle P \rangle$. We will make use of this result to improve pairing computation.

Theorem 2. *Let E be an ordinary elliptic curve defined over $\mathbb{F}_q$ and let P be a point over $\mathbb{F}_q$ of E, whose order is a prime integer $r \neq q$. Suppose the embedding degree k is greater than 1 and denote by Q a point defined over $\mathbb{F}_{q^k}$, such that $\pi(Q) = qQ$. Then P and Q are eigenvectors of any other endomorphism of E.*

---

[2] Only curves with discriminant -4 have twists of degree 4.

Proof. Let ϕ be an endomorphism of E. For the point P we have

$$\phi(\pi(P)) = \pi(\phi(P)) \text{ and } \phi(\pi(P)) = \phi(P). \tag{4}$$

The first equality comes from the fact that the ring $\text{End}(E)$ is commutative, the second one is due to the fact that $P \in E(\mathbb{F}_q)$. It follows that $\pi(\phi(P)) = \phi(P)$, so $\phi(P)$ is an eigenvector for the eigenvalue 1 of π. This means that $\phi(P) \in \langle P \rangle$. The proof for Q is similar.

Notation 1. *In the sequel we denote the correction of two points R_1 and R_2 as follows:*

$$corr_{R_1,R_2} = \frac{l_{R_1,R_2}}{v_{R_1+R_2}},$$

where l_{R_1,R_2} is the line passing through R_1 and R_2 and $v_{R_1+R_2}$ is the vertical line through $R_1 + R_2$.

Our starting idea is a method to exploit efficiently computable endomorphisms in pairing computation suggested by Scott [23], for a family of curves called NSS. These curves are defined over $\mathbb{F}_q$ with $q \equiv 1 \mod 3$ and given by an equation of the form $y^2 = x^3 + B$. Since they have $k = 2$ and $\rho \sim 2$, the Eta and Ate pairings will not bring any improvement to pairing computation. However, these curves admit an endomorphism $\phi : (x,y) \to (\beta x, y)$, where β is a non-trivial cube root of unity. Its characteristic equation is $\phi^2 + \phi + 1 = 0$. If P is an eigenvalue of ϕ such that $\phi(P) = \lambda P$, then λ verifies the equation

$$\lambda^2 + \lambda + 1 = cr.$$

We obtain

$$f^c_{r,P}(Q) = f_{\lambda^2 + \lambda, P}(Q) = f_{\lambda(\lambda+1),P}(Q) = f^{\lambda+1}_{\lambda,P}(Q) \cdot f_{\lambda+1,[\lambda]P}(Q) \cdot \frac{l_{[\lambda]P,P}}{v_{[\lambda+1]P}}.$$

Since for $P = (x,y)$, λP is given by $(\beta x, y)$, we can easily compute $f_{\lambda,\lambda P}(Q)$ and $f_{\lambda,P}(Q)$ at the same time when running Miller's algorithm, by replacing x with βx when computing doublings, additions and line equations. Note that pairing computation on these curves has been recently improved by Zhao and al. [29].

We apply similar techniques to curves with endomorphisms that verify a characteristic equation $x^2 + ax + b = 0$, with a, b small. In all cases, we use the Cocks-Pinch method to construct curves such that there is a $\lambda \sim \sqrt{r}$ which verifies $\lambda^2 + a\lambda + b = cr$. This can be done by exhaustive search on λ. Thanks to the density of prime numbers, we are able to produce couples (λ, r) within seconds with MAGMA.

We obtain a new algorithm for pairing computation, whose loop is shorter than that of the algorithm computing the Tate pairing.

Lemma 1. *Let E be an elliptic curve defined over a finite field $\mathbb{F}_q$ and ϕ an endomorphism of E whose degree is b. Let P, Q be two points on the curve E. Then for any integer λ the following equality is true up to a constant:*

$$f_{\lambda,\phi(P)}(\phi(Q)) = f^b_{\lambda,P}(Q) \left(\prod_{K \in Ker\,\phi \backslash \{P_\infty\}} corr_{P,K}(Q) \right)^\lambda \left(\prod_{K \in Ker\,\phi \backslash \{P_\infty\}} corr_{\lambda P,K}(Q) \right)^{-1}$$

Proof. We have

$$
\phi^*(f_{\lambda,\phi(P)}) = \lambda \sum_{K \in \mathrm{Ker}\phi} (P+K) - \sum_{K \in \mathrm{Ker}\phi} (\lambda P + K) - (\lambda - 1) \sum_{K \in \mathrm{Ker}\phi} (K)
$$

$$
= \lambda \sum_{K \in \mathrm{Ker}\phi} ((P+K) - (K)) - \sum_{K \in \mathrm{Ker}\phi} ((\lambda P + K) - (K))
$$

$$
= \lambda \sum_{K \in \mathrm{Ker}\phi} ((P) - (O)) - \sum_{K \in \mathrm{Ker}\phi} (\lambda P) - (O) + \mathrm{div}\left(\left(\prod_{K \in \mathrm{Ker}\,\phi} \frac{l_{K,P}}{v_{K+P}} \right)^{\lambda} \right)
$$

$$
- \mathrm{div}\left(\prod_{K \in \mathrm{Ker}\,\phi} \frac{l_{K,\lambda P}}{v_{K+\lambda P}} \right) = \mathrm{div}(f_{\lambda,P}) + \mathrm{div}\left(\prod_{K \in \mathrm{Ker}\,\phi \backslash \{P_\infty\}} corr_{\lambda P,K} \right)
$$

$$
- \mathrm{div}\left(\prod_{K \in \mathrm{Ker}\,\phi \backslash \{P_\infty\}} corr_{\lambda P,K} \right).
$$

Using the fact that $\phi^*(f_{\lambda,\phi(P)}) = f_{\lambda,\phi(P)} \circ \phi$, we obtain the equality we have announced.

In the sequel, we make use of the following relation which holds for all $m, n \in \mathbb{Z}$ and any point P on the elliptic curve

$$
f_{mn,P} = f_{m,P}^n \cdot f_{n,mP}. \tag{5}
$$

This equality can be easily checked using divisors. In the sequel, we denote by $\hat{\phi}$ the dual of an isogeny ϕ. The reader is referred to [24] for the definition of the dual.

Theorem 3. *Let E be an elliptic curve defined over a finite field $\mathbb{F}_q$, r a prime number such that $r | \#E(\mathbb{F}_q)$ and k the embedding degree with respect to r. Let ϕ be an efficiently computable separable endomorphism of E, whose characteristic equation is $X^2 + aX + b = 0$. Let $\mathbb{G}_1$ and $\mathbb{G}_2$ be the the subgroups of order r whose elements are eigenvectors of ϕ defined over $\mathbb{F}_q$ and $\mathbb{F}_{q^k}$, respectively. Let λ be the eigenvalue of ϕ on $\mathbb{G}_1$, verifying $\lambda^2 + a\lambda + b = cr$, with $r \nmid bc$. Then the map $a_\phi(\cdot, \cdot) : \mathbb{G}_1 \times \mathbb{G}_2 \to \mathbb{F}_{q^k}^* / (\mathbb{F}_{q^k}^*)^r$ given by*

$$
a_\phi(P, Q) = f_{\lambda,P}^{\lambda+a}(bQ) f_{\lambda,P}^b(\hat{\phi}(Q)) f_{a,\lambda P}(bQ) f_{b,P}(bQ) \left(\prod_{K \in Ker\phi \backslash \{P_\infty\}} corr_{P,K}(\hat{\phi}(Q)) \right)^{\lambda}
$$

$$
\cdot \left(\prod_{K \in Ker\phi \backslash \{P_\infty\}} corr_{\lambda P,K}(\hat{\phi}(Q)) \right)^{-1} corr_{\lambda^2 P, a\lambda P}(bQ) \, l_{\lambda^2 P + a\lambda P, bP}(bQ)
$$

is a bilinear non-degenerate pairing.

Proof. The following equality is obtained by repeatedly applying the equality at (5)

$$f_{\lambda^2+a\lambda+b,P} = (f_{\lambda,P}^\lambda) \cdot (f_{\lambda,\lambda P}) \cdot (f_{\lambda,P}^a) \cdot (f_{a,\lambda P}) \cdot (f_{b,P})$$
$$\cdot corr_{\lambda^2 P,a\lambda P} \cdot l_{\lambda^2 P+a\lambda P,bP}. \tag{6}$$

By applying Lemma 1, we obtain

$$f_{\lambda,\lambda P}(bQ) = f_{\lambda,P}^b(\hat{\phi}(Q)) \left(\prod_{K\in\mathrm{Ker}\phi\setminus\{P_\infty\}} corr_{P,K}(\hat{\phi}(Q)) \right)^\lambda$$
$$\cdot \left(\prod_{K\in\mathrm{Ker}\phi\setminus\{P_\infty\}} corr_{\lambda P,K}(\hat{\phi}(Q)) \right)^{-1}$$

By replacing this term in equation (6), we derive that $a_\phi(P,Q)$ is a power of $t_r(P,Q)$. Since $(bc,r) = 1$, we conclude that a_ϕ defines a non-degenerate pairing on $\mathbb{G}_1 \times \mathbb{G}_2$.

If the value of λ is close to $\sqrt{r}$ and a and b are small, Theorem 3 gives an efficient algorithm to compute the Tate pairing (actually a small power of the Tate pairing). This is Algorithm 3. The complexity of the new algorithm is $O(\log ab\lambda)$.

Algorithm 3. Our algorithm for pairing computation for curves with an efficiently computable endomorphism

---

INPUT: An elliptic curve E, P, Q points on E and ϕ such that $\phi(P) = \lambda P$, $Q' = \hat{\phi}(Q)$.
OUTPUT: A power of the Tate pairing $T_r(P,Q)$.
1: Let $i = \lceil\log_2(\lambda)\rceil$, $K \leftarrow P$, $f \leftarrow 1$, $g \leftarrow 1$
2: **while** $i \geq 1$ **do**
3: Compute equation of l arising in the doubling of K
4: $K \leftarrow 2K$ and $f \leftarrow f^2 l(bQ)$ and $g \leftarrow g^2 l(Q')$
5: **if** the i-th bit of λ is 1 **then**
6:   Compute equation of l arising in the addition of K and P
7:   $K \leftarrow P + K$ and $f \leftarrow fl(bQ)$ and $g \leftarrow gl(Q')$
8: **end if**
9: Let $i \leftarrow i - 1$
10: **end while**
11: Compute $A \leftarrow f^{\lambda+a}$
12: Compute $B \leftarrow g^b$
13: Compute $C \leftarrow \left(\prod_{K\in\mathrm{Ker}\phi\setminus\{P_\infty\}} corr_{P,K}(Q')\right)^\lambda \left(\prod_{K\in\mathrm{Ker}\phi\setminus\{P_\infty\}} corr_{\lambda P,K}(Q')\right)^{-1}$
14: Compute $D \leftarrow f_{a,\lambda P}(bQ)f_{b,P}(bQ)$
15: $F \leftarrow corr_{\lambda^2 P,a\lambda P}(bQ)l_{\lambda^2 P+a\lambda P,bP}(bQ)$
16: Return $A \cdot B \cdot C \cdot D \cdot F$

---

4   Computational Costs

Suppose we use an endomorphism ϕ whose characteristic equation is

$$\phi^2 + a\phi + b = 0,$$

with a and b small. We also neglect the cost of computing the dual of ϕ at Q, $\hat{\phi}(Q)$, because $\hat{\phi}$ can be precomputed by Vélu's formulae [26] and is given by polynomials of small degree. Note that in some protocols Q is a fixed point, so all the precomputations on this point may be done before the computation of the pairing.

We also note that the endomorphism is defined over $\mathbb{F}_q$, because the curve E is ordinary. Moreover, the points in $\operatorname{Ker}\phi$ are eigenvectors for the Frobenius endomorphism. Indeed, since $\operatorname{End}(E)$ is a commutative ring, we have $\phi(\pi(K)) = \pi(\phi(K)) = O$, for all $K \in \operatorname{Ker}\phi$. It follows that $\pi(K) \in \operatorname{Ker}\phi$. Thus the points of $\operatorname{Ker}\phi$ are defined over an extension field of $\mathbb{F}_q$ of degree smaller than b. Furthermore, if $\operatorname{Ker}\phi$ is cyclic, we have

$$\left(\prod_{K \in \operatorname{Ker}\phi \setminus \{P_\infty\}} corr_{P,K}(\hat{\phi}(Q)) \right) \in \mathbb{F}_{q^k}.$$

Consequently, given that the degree of ϕ is small, we assume that the number of operations needed to compute the correction $\prod_{K \in \operatorname{Ker}\phi} corr_{P,K}(\hat{\phi}(Q))$ is negligible. Since a and b are small, we also assume that the costs of the exponentiation at line 12 and that of the computation of functions at line 14 of Algorithm 3 are negligible.

Since in practice we usually consider curves with even embedding degree, we present only results for these curves. We assume that the curves have an efficiently computable endomorphism and eigenvalues of size $\sqrt{r}$. In our evaluation, we only counted the number of operations performed in the doubling part of Miller's algorithm, because we suppose that λ and r have low Hamming weight (which is possible if the curve is constructed with the Cocks-Pinch method).
For operations in extension fields of degree 2, we use tower fields. For example, to construct an extension field of degree 4 we have

$$\mathbb{F}_q \subset \mathbb{F}_{q^2} \subset \mathbb{F}_{q^4}.$$

With Karatsuba's method the cost of an operation in the extension field of degree 2 is three times the cost of the same operation in the base field, while with Toom-Cook a multiplication in an extension field of degree 3 costs 5 multiplications in the base field. Using the formulas in Table 1 the total cost of the doubling step in Algorithm 3 and of the exponentiation at line 11 is

$$(11\mathbf{s} + (1 + 2k)\mathbf{m} + 2\mathbf{M} + 2\mathbf{S}) \log \lambda + \log \lambda \mathbf{M} \text{ if } D \neq 3, 4.$$

Our computations showed that our method gives better performances than the Tate pairing for some families of ordinary curves with embedding degree 2, 3 and

4. Indeed, using the complexity estimations above and making the assumption that $s \approx m$, our algorithm is faster than the Tate pairing if and only if

$$(12 + 2k)m + 5M > 2((12 + k)m + 2M).$$

A simple computation shows that this is true if and only if $k \leq 4$. In Table 2, we compare the performances of our method to those of Miller's algorithm, for curves with embedding degree 2 and 4 constructed via the Cocks-Pinch method. Note that for $k = 2$, the Eta pairing algorithm (and its variants) is not faster than the Tate pairing algorithm, because $t \approx r$. We assume that for curves with embedding degree 4 the CM discriminant is not -4, because for such curves the Tate and the twisted Ate pairing have comparable costs. Note that for $D = -4$ the twisted Ate pairing computation has complexity $O(\log t)$ and is thus faster than the Tate pairing and also faster than our method if t is careffully chosen of small size.

Table 2. Our method versus the Tate pairing

bit length of r	$k = 2$		$k \geq 4$ and $D \neq 4$	
	Tate pairing	This work	Tate pairing	This work
160 bits	3040	2400	5120	4880

As explained in [8], curves with small embedding degrees are preferred in implementations at low security levels (80 bits). Thus, if we want to set up a pairing-based cryptosystem with a 160-bit elliptic curve subgroup, we may choose a MNT curve with embedding degree 6 and ρ-value close to 1 or we may take a curve with embedding degree $k \in \{2, 3, 4\}$ and ρ-value close to 2. Table 3 presents a list of families of curves that have been proposed for pairing-based cryptography at 80 bits security level. Note that we may use an ordinary curve with embedding degree 2 and ρ-value approximatively 3 constructed by the Cocks-Pinch method or a supersingular curve with $k = 2$ and $\rho \sim 3$ (see Algorithm 3.3 in [8]). Table 6 presents the operation count in $\mathbb{F}_q$ for pairing computation on curves with different embedding degrees. We assume that ordinary curves with embedding degrees 2 are constructed via the Cocks-Pinch method and we evaluate the cost of the computation performed in Algorithm 3 for these curves, as explained above. For supersingular curves and MNT curves with embedding degree 6 we estimate the cost of the doubling part in Algorithm 1 computing the Tate pairing using formulae in Table 1. Since on these curves, the parameter r does not necessarily have low Hamming weight, we also count the number of operations performed in the mixed addition part of the Miller operation, if a NAF representation of r is used. For $k = 4$, the Cocks-Pinch method allows constructing curves with small trace of the Frobenius (by choosing a small g and then taking r a divisor of $\Phi_4(g)$ in Algorithm 2). In this case, if the curve has a twist o degree 4, i.e. $D = 4$, then the twisted Ate pairing is optimal. Otherwise, the Tate pairing is optimal. Finally, we also give a family with $k = 4$ and $D = 3$ from [8]. On curves from this family, the optimal pairing can be computed using lattice reduction in $\frac{\log r}{2}$ Miller iterations (see [12] for details).

Table 3. Bit sizes of curve parameters for pairing-based cryptography at 80 bits security level

bit length of r	bit length of q	bit length of q^k	value of k and ρ
160	480	960	supersingular curves $k = 2$ and $\rho \sim 3$
160	480	960	ordinary curves $k = 2$ and $\rho \sim 3$
160	160	960	MNT curves $k = 6$ and $\rho \sim 1$
160	320	1280	ordinary curves $k = 4$ and $\rho \sim 2$
160	220	960	ordinary curves $k = 4$ and $\rho \sim 1.5$

Table 4. Pairing computation at 80 bits security level

value of k and ρ	doubling step (operations in $\mathbb{F}_q$)	mixed addition (operations in $\mathbb{F}_q$)
supersingular curves $k = 2$ and $\rho \sim 3$	3040	-
ordinary curves $k = 2$ and $\rho \sim 3$	2400	-
MNT curves $k = 6$ and $\rho \sim 1$	7680	1760
ordinary curves $k = 4$ and $\rho \sim 2$ and $D = 4$	2400	-
ordinary curves $k = 4$ and $\rho \sim 1.5$ and $D = 3$	2480	-

Table 5. Pairing computation at 80 bits security level

value of k and ρ	Miller loop	final exponentiation	total cost
supersingular curves $k = 2$ and $\rho \sim 3$	13680	5760	19350
ordinary curves $k = 2$ and $\rho \sim 3$	10800	5760	16470
MNT curves $k = 6$ and $\rho \sim 1$	9440	6400	15840
$k = 4$ and $\rho \sim 2$ and $D = 4$	5880	14138	20018
$k = 4$ and $\rho \sim 1.5$ and $D = 3$	3100	4980	8080

Table 5 presents total costs for the Miller loop and for the final exponentiation, in terms of number of operations in $\mathbb{F}_q$, for different types of curves and embedding degrees. In the final exponentiation, we assume that applying the Frobenius operator can be done for free and we estimate only the cost of $\Phi_r(q)/r$. The last column presents global costs of pairing computation. Note that the size of q is different for these families of curves. We have therefore taken into account costs of integer multiplication for different bit lengths (see [3] for GMP benchmarks).

We have realised a simple implementation under MAGMA 2.15-1.5. We give the time in seconds for the computation of 100 pairings on a 2.6 GHz Intel Core 2 Duo processor. Note that in theory an implementation of pairings on MNT curves is expected to be faster.

We conclude by giving in Appendix 5 examples of curves constructed with the Cocks-Pinch method, with endomorphism verifying an equation of the form $X^2 + aX + b = 0$ and roots $\lambda \sim \sqrt{r}$. We also note that on these curves, the GLV method [9] can be used to speed up scalar multiplication in the implementation of a pairing-based protocol. We therefore believe that these curves offer a good choice for pairing-based cryptography at 80 bits security level.

Table 6. Execution time on a 2.6 GHz Intel Core 2 Duo processor

value of k and ρ	bit length of r	This work	Eta/Tate
ordinary curves $k = 2$ and $\rho \sim 3$	160	1.02	1.42
MNT curves $k = 6$ and $\rho \sim 1$	172	-	1.96
ordinary curves $k = 4$ and $\rho \sim 1.5$ and $D = 3$	160	-	0.41

5 Conclusion

We have given a new algorithm for pairing computation on curves with endomorphisms of small degree. Our pairing on curves constructed with the Cocks-Pinch method is more efficient than than known pairings for some curves with embedding degree 2 and 4.

Acknowledgements

The authors thank anonymous reviewers for helpful comments.

References

1. Arène, C., Lange, T., Naehrig, M., Ritzenthaler, C.: Faster computation of the Tate pairing, http://eprint.iacr.org/2009/155
2. Barreto, P., Galbraith, S., Héigeartaigh, C., Scott, M.: Efficient Pairing Computation on Supersingular Abelian Varieties. Des. Codes Cryptography 42(3), 239–271 (2007)
3. Bernstein, D.: Integer multiplication benchmarks, http://cr.yp.to/speed/mult/gmp.html
4. Blake, I.F., Seroussi, G., Smart, N.P.: Advances in Elliptic Curve Cryptography. London Mathematical Society Lecture Note Series, vol. 317. Cambridge University Press, Cambridge (2005)
5. Boneh, D., Franklin, M.K.: Identity-based encryption from the Weil pairing. In: Kilian, J. (ed.) CRYPTO 2001. LNCS, vol. 2139, pp. 213–229. Springer, Heidelberg (2001)
6. Boneh, D., Lynn, B., Shacham, H.: Short signatures from the Weil pairing. In: Boyd, C. (ed.) ASIACRYPT 2001. LNCS, vol. 2248, pp. 514–532. Springer, Heidelberg (2001)
7. Costello, C., Lange, T., Naehrig, M.: Faster Pairing Computations on Curves with High-Degree Twists. In: Nguyen, P.Q., Pointcheval, D. (eds.) PKC 2010. LNCS, vol. 6056, pp. 224–242. Springer, Heidelberg (2010)
8. Freeman, D., Scott, M., Teske, E.: A taxonomy of pairing-friendly elliptic curves. Journal of Cryptology 23, 224–280 (2010)
9. Gallant, R.P., Lambert, R.J., Vanstone, S.A.: Faster point multiplication on elliptic curves with efficient endomorphisms. In: Kilian, J. (ed.) CRYPTO 2001. LNCS, vol. 2139, pp. 190–200. Springer, Heidelberg (2001)

10. Granger, R., Hess, F., Oyono, R., Thériault, N., Vercauteren, F.: Ate pairing on hyperelliptic curves. In: Naor, M. (ed.) EUROCRYPT 2007. LNCS, vol. 4515, pp. 430–447. Springer, Heidelberg (2007)
11. Hartshorne, R.: Algebraic geometry. Graduate Texts in Mathematics, vol. 52. Springer, Heidelberg (1977)
12. Hess, F.: A note on the Tate pairing of curves over finite fields. Arch. Math. 82, 28–32 (2004)
13. Hess, F., Smart, N.P., Vercauteren, F.: The Eta pairing revisited. IEEE Transactions on Information Theory 52, 4595–4602 (2006)
14. Ionica, S., Joux, A.: Another approach to pairing computation in Edwards coordinates. In: Chowdhury, D.R., Rijmen, V., Das, A. (eds.) INDOCRYPT 2008. LNCS, vol. 5365, pp. 400–413. Springer, Heidelberg (2008)
15. Joux, A.: A one round protocol for tripartite Diffie-Hellman. Journal of Cryptology 17(4), 263–276 (2004)
16. Joux, A., Lercier, R.: The function field sieve in the medium prime case. In: Vaudenay, S. (ed.) EUROCRYPT 2006. LNCS, vol. 4004, pp. 254–270. Springer, Heidelberg (2006)
17. Joux, A., Lercier, R., Smart, N., Vercauteren, F.: The number field sieve in the medium prime case. In: Dwork, C. (ed.) CRYPTO 2006. LNCS, vol. 4117, pp. 326–344. Springer, Heidelberg (2006)
18. Koblitz, N., Menezes, A.: Pairing-based cryptography at high security levels. In: Smart, N.P. (ed.) Cryptography and Coding 2005. LNCS, vol. 3796, pp. 13–36. Springer, Heidelberg (2005)
19. MAGMA Computational Algebra System. MAGMA version V2.16-5 (2010), http://magma.maths.usyd.edu.au/magma
20. Miller, V.: The Weil pairing, and its efficient calculation. Journal of Cryptology 17(4), 235–261 (2004)
21. Okamoto, T., Menezes, A., Vanstone, S.A.: Reducing elliptic curve logarithms to logarithms in the finite field. In: Proceedings 23rd Annual ACM Symposium on Theory of Computing (STOC), pp. 80–89. ACM Press, New York (1991)
22. Pollard, J.: Monte Carlo methods for index computation (mod p). Mathematics of Computation (32), 918–924 (1978)
23. Scott, M.: Faster pairings using an elliptic curve with an efficient endomorphism. In: Maitra, S., Veni Madhavan, C.E., Venkatesan, R. (eds.) INDOCRYPT 2005. LNCS, vol. 3797, pp. 258–269. Springer, Heidelberg (2005)
24. Silverman, J.H.: The Arithmetic of Elliptic Curves. Graduate Texts in Mathematics, vol. 106. Springer, Heidelberg (1986)
25. van Oorschot, P.C., Wiener, M.J.: Parallel collision search with cryptanalytic applications. Journal of Cryptology (12), 1–18 (1999)
26. Vélu, J.: Isogenies entre courbes elliptiques. Comptes Rendus De Academie Des Sciences Paris, Serie I-Mathematique, Serie A 273, 238–241 (1971)
27. Vercauteren, F.: Optimal pairings. IEEE Transactions on Information Theory (2009) (to appear)
28. Verheul, E.R.: Evidence that XTR is more secure than supersingular elliptic curve cryptosystems. In: Pfitzmann, B. (ed.) EUROCRYPT 2001. LNCS, vol. 2045, pp. 195–201. Springer, Heidelberg (2001)
29. Zhao, C., Xie, D., Zhang, F., Zhang, J., Chen, B.: Computing the Bilinear Pairings on Elliptic Curves with Automorphisms. Designes, Codes and Cryptography (to appear)

Appendix 1

In order to display the equations of the endomorphism easily, we give first a small example.

Example 1. A toy example

We take $D = -4 \cdot 2$ and we want a curve with an endomorphism whose characteristic equation will be

$$X^2 + 2 = 0.$$

We choose $\lambda = 66543$ verifying the equation $\lambda^2 + 2 = r$, with

$$r = 4427970851.$$

Our implementation of the Cocks-Pinch method in MAGMA [19] found the following curve

$$y^2 = x^3 + 49768875163241226962283x + 22119500007255165642796$$

over the prime field $\mathbb{F}_q$, with

$$q = 14930662548972368088859.$$

This curve has $k = 2$ with respect to r. The endomorphism corresponding to $\alpha = \sqrt{-2}$ in $\mathbb{Z}[\sqrt{-2}]$ is

$$[\alpha](x, y) = \left(7465331274486184044429 \frac{x^2 + 49768875163241226962285x + 2}{x + 49768875163241226962285}, \right.$$
$$\left. 11197940817690300409659 \frac{x^2 + 9953775032648245392570x + 82948125272206871160477}{(x + 49768875163241226962285)^2} y \right).$$

As observed in [9], computing this endomorphism is slightly harder than doubling. The equations of the dual of α are similar.

Example 2. Consider $D = 3$ and an endomorphism with characteristic equation given by $X^2 + 2X + 4 = 0$. We found $\lambda = 2^{40} + 2^{29} + 1$ verifying $\lambda^2 + 2\lambda + 4 = r$ where r is given by $r = 1210106699470122931716103$. We have

$$q = 126422926680861157408034773550955195230739769633357.$$

The curve E given by the equation $y^2 = x^3 + 1$ has embedding 2 with respect to r.

Example 3. Consider $D = 4$ and an endomorphism with characteristic equation given by $X^2 + 2X + 2 = 0$. We found $\lambda = 2^{40} + 2^{25} + 1$ verifying $\lambda^2 + 2\lambda + 2 = r$ where r is given by $r = 1208999607721222100484101$. We have

$$q = 198386524985776646431182137712777804938134063565449557.$$

The curve E given by the equation $y^2 = x^3 + x$ has embedding 4 with respect to r.

High Speed Flexible Pairing Cryptoprocessor on FPGA Platform

Santosh Ghosh, Debdeep Mukhopadhyay, and Dipanwita Roychowdhury

Department of Computer Science and Engineering,
Indian Institute of Technology,
Kharagpur, India
{santosh,debdeep,drc}@cse.iitkgp.ernet.in

Abstract. This paper presents a Pairing Crypto Processor (PCP) over Barreto-Naehrig curves (BN curves). The proposed architecture is specifically designed for field programmable gate array (FPGA) platforms. The design of PCP utilizes the efficient implementation of the underlying finite field primitives. The techniques proposed maximize the utilization of in-built features of an FPGA device which significantly improves the performance of the primitives.

Extensive parallelism techniques have been proposed to realize a PCP which requires lesser clock cycles than the existing designs. The proposed design is the first reported result on an FPGA platform for 128-bit security. The PCP provides flexibility to choose the curve parameters for pairing computations.

The cryptoprocessor needs 1730 k, 1206 k, and 821 k cycles for the computation of Tate, ate, and R-ate pairings, respectively. On a Virtex-4 FPGA device it consumes 52 kSlices at 50MHz and computes the Tate, ate, and R-ate pairings in 34.6 ms, 24.2 ms, and 16.4 ms, respectively, which is comparable to known CMOS implementations.

Keywords: $\mathbb{F}_{p^k}$-arithmetic, FPGA, Barreto-Naehrig curves, elliptic-curve cryptography (ECC), pairing-based cryptography.

1   Introduction

CRYPTOGRAPHIC PAIRING [24] is a bilinear map $\mathbb{G}_1 \times \mathbb{G}_2 \to \mathbb{G}_3$ where $\mathbb{G}_1$ and $\mathbb{G}_2$ are typically additive groups and $\mathbb{G}_3$ is a multiplicative group. Many cryptographic pairings such as the Tate pairing [27], ate pairing [20], and R-ate pairing [8] choose $\mathbb{G}_1$ and $\mathbb{G}_2$ to be specific cyclic subgroups of $E(\mathbb{F}_{p^k})$, and $\mathbb{G}_3$ to be a subgroup of $\mathbb{F}_{p^k}$. Selection of such groups as well as field types have a strong impact on the security and computation cost of pairing. Barreto-Naehrig curves (BN curves) [19] are a type of elliptic curves which support the computation of cryptographic pairings with a 128-bit security level. It is defined over a 256-bit prime field having embedding degree $k = 12$.

Related works. The software implementation results of pairings over BN curve have been shown in [1], [15], [13], and [16]. The highly optimized software

M. Joye, A. Miyaji, and A. Otsuka (Eds.): Pairing 2010, LNCS 6487, pp. 450–466, 2010.

codes run on a 64-bit core2 processor which computes a R-ate pairing in only $10,000,000$ cycles. The software implementation of [1] gives the speed record for the computation of Optimal-ate pairing on BN curves, which is computed by $4,470,408$ cycles on a Intel Core 2 Quad Q6600 processor.

An application specific instruction-set processor (ASIP) has been proposed in [5]. It is designed by extending a RISC core with additional scalable functional units. It requires a special programming environment in order to execute pairings. Therefore, the authors have developed a special C compiler. Implementation result shows that the ASIP can compute an Optimal-ate pairing in 15.8 ms over a 256-bit BN curve at 338 MHz with a 130 nm CMOS library.

A pairing processor specially for BN curves has been proposed in [6]. It exploits the characteristic of the field defined by BN curves and choose curve parameters such that the underlying $\mathbb{F}_p$ multiplication becomes more efficient. It shows a 5.4 times speedup of a pairing computation compared to the ASIP proposed in [5]. However, the main limitation of the pairing processor [6] is that it is useful only for computing pairings over a fixed BN curve.

Contribution. This paper proposes a flexible cryptoprocessor for the computation of pairings over BN curves. Field programmable gate array (FPGA) is one of the suitable platforms for implementing cryptographic algorithms. In this paper, we propose new implementation techniques of addition and multiplication on FPGAs. The in-built features available inside an FPGA device have been utilized to develop a high speed 256-bit adder circuit. We show that when utilizing such adder circuits and adopting a parallelism technique, the multiplication in $\mathbb{F}_p$ can be substantially improved. Based on such $\mathbb{F}_p$ arithmetic cores, we develop a parallel configurable hardware for computing addition, subtraction, and multiplication on $\mathbb{F}_p$ and $\mathbb{F}_{p^2}$. Existing techniques to speed up arithmetic in extension fields (see [16,21]) for fast computation in $\mathbb{F}_{p^6}$ and $\mathbb{F}_{p^{12}}$ are used on top of it. The major contributions of the paper are highlighted here.

- The paper implements underlying primitives for $\mathbb{F}_p$ arithmetic on FPGA platforms, which provides a significant speedup from existing platform-independent techniques.
- It proposes a pairing hardware that is flexible for curve parameters.
- Parallelism techniques are adopted in different levels including underlying finite field operations which drastically reduces the overall cycle count of pairing computation.
- The proposed FPGA design achieves a comparable speed with the existing CMOS design.

The proposed configurable $\mathbb{F}_{p^k}$ arithmetic cores and parallel computation result in a significant improvement on the performance of Tate, ate, and R-ate pairing over BN curves.

Organization of the paper. Section 2 of the paper gives a brief description of cryptographic pairings and BN curves. Efficient design of finite field primitives on FPGA platforms are described in section 3. Section 4 describes the pairing cryptoprocessor. Section 5 shows the experimental results based on BN curves and provides comparative studies with existing contemporary designs. The paper is concluded in section 6.

2   Background of Pairings

The name bilinear pairing indicates that it takes a pair of vectors as input and returns a number, and it performs linear transformation on each of its variables. For example, the dot product of vectors is a bilinear pairing [11]. Similarly, for cryptographic applications the bilinear pairing (or pairing) operations are defined on elliptic or hyperelliptic Jacobian curves. Pairing is a mapping $\mathbb{G}_1 \times \mathbb{G}_2 \rightarrow \mathbb{G}_3$, where $\mathbb{G}_1$ is a additive subgroup of $E(\mathbb{F}_q)$ on some elliptic or hyperelliptic Jacobian curve defined over a finite field $\mathbb{F}_q$, $\mathbb{G}_2$ is an another similar kind of subgroup of $E(\mathbb{F}_{q^k})$ over the lowest extension field $\mathbb{F}_{q^k}$, and $\mathbb{G}_3$ is a subgroup of the multiplicative group of $\mathbb{F}_{q^k}$. Here q is the characteristic field representative. It is normally 2^m, 3^m, or a large prime p. The parameter k corresponds to the embedding degree, often referred to as security multiplier in pairing computation, i.e. the smallest positive integer such that r divides $q^k - 1$ and r is a large odd prime which divides the order of the curve group $(\#E(\mathbb{F}_q))$. If the point P be a r-torsion point then the Tate pairing of order r is a map: $e_r : E(\mathbb{F}_q)[r] \times E(\mathbb{F}_{q^k})[r] \rightarrow \mathbb{F}_{q^k}^*/(\mathbb{F}_{q^k}^*)^r$, where $E(\mathbb{F}_q)[r]$ denote the subgroup of $E(\mathbb{F}_q)$ of all points of order dividing r, and similarly for $\mathbb{F}_{q^k}$. Tate pairing of order r satisfies the following properties:

- **Non-degeneracy** : For each $P \neq \mathcal{O}$ there exist $Q \in E(\mathbb{F}_{q^k})[r]$ such that $e_r(P, Q) \neq r$.
- **Bilinearity** : For any integer n, $e_r([n]P, Q) = e_r(P, [n]Q) = e_r(P, Q)^n$ for all $P \in E(\mathbb{F}_q)[r]$ and $Q \in E(\mathbb{F}_{q^k})[r]$.
- **Computability** : There exists an efficient algorithm to compute $e_r(P, Q)$ given P and Q.

The value e_r is representative of an element of the quotient group $\mathbb{F}_{q^k}^*/(\mathbb{F}_{q^k}^*)^r$. However for cryptographic protocols it is essential to have a unique element so it is raised to the $((q^k - 1)/r)$-th power, obtaining an rth root of unity. The resulting value is called reduced Tate pairing.

2.1   Choice of Elliptic Curve

The most important parameters for cryptographic pairings are the underlying finite field, the order of the curve, the embedding degree, and the order of $\mathbb{G}_1, \mathbb{G}_2$ and $\mathbb{G}_3$. These parameters should be chosen such that the best exponential time algorithms to solve the discrete logarithm problem (DLP) in $\mathbb{G}_1$ and $\mathbb{G}_2$ and the sub-exponential time algorithms to solve the DLP in $\mathbb{G}_3$ take longer than a chosen security level. This paper uses the 128-bit symmetric key security level. For the 128-bit security level, the National Institute of Standards and Technology (NIST) recommends a prime group order of 256 bits for $E(\mathbb{F}_p)$ and of 3072 bits for the finite field $\mathbb{F}_{p^k}$ [17].

Barreto-Naehrig curves, introduced in [19], are elliptic curves over fields of prime order p with embedding degree $k = 12$. The BN curve is represented as :

$$E_{\mathbb{F}_p} : Y^2 = X^3 + 3$$

with BN parameter $z = 6000000000001F2D$ (in hexadecimal) [16]. It forms the group $E(\mathbb{F}_p)$ with order $\#E(\mathbb{F}_p) = r = 36z^4 + 36z^3 + 18z^2 + 6z + 1$, which is a 256-bit prime of Hamming weight 91. The field characteristic $p = 36z^4 + 36z^3 + 24z^2 + 6z + 1$ is a 256-bit prime of Hamming weight 87, and $t - 1 = p - r = 6z^2 + 1$ is a 128-bit integer of Hamming weight 28. Here $t = p + 1 - r$ is the trace of $E_{\mathbb{F}_p}$. The prime $p \equiv 7 \pmod 8$ (so -2 is a quadratic non-residue, we represent it by β) and $p \equiv 1 \pmod 6$.

2.2   Pairing Computation

Pairing computation consists of two major steps : the computation of Miller function and the final exponentiation. Algorithm 1 shows computation of Tate pairing. The first part is computed by one of the optimized version of Miller algorithm [25]. Several optimizations of this algorithm have been presented in [27]. The resulting algorithm is called BKLS algorithm.

Algorithm 1. Computing the Tate pairing

Input: $P \in G_1$ and $Q \in G_2$.
Output: $e_r(P, Q)$.

Write r in binary : $r = \sum_{i=0}^{L-1} r_i 2^i$.
$T \leftarrow P, f \leftarrow 1$.
for i from $L - 2$ *downto* 0 **do**
$\quad T \leftarrow 2T$.
$\quad f \leftarrow f^2 \cdot l_{T,T}(Q)$.
$\quad$ **if** $r_i = 1$ *and* $i \neq 0$ **then**
$\quad\quad T \leftarrow T + P$.
$\quad\quad f \leftarrow f \cdot l_{T,P}(Q)$.
$\quad$ **end**
end
return $f^{(q^k - 1)/r}$.

The BN curves also admits a sextic twist [15], which means that the point Q is mapped on a point Q' defined over $\mathbb{F}_{p^2}$. Thus the line functions $l_{T,T}(Q)$ and $l_{T,P}(Q)$ is computed over $\mathbb{F}_{p^2}$ instead of $\mathbb{F}_{p^{12}}$. Value of the line functions are represented as : $l_0 + l_1 W^2 + l_2 W^3$, with $l_0 \in \mathbb{F}_p$, $l_1, l_2 \in \mathbb{F}_{p^2}$, and a quadratic non-residue W over $\mathbb{F}_{p^2}$. The Miller function f is computed over $\mathbb{F}_{p^{12}}$, which is represented as : $f_0 + f_1 W + f_2 W^2 + f_3 W^3 + f_4 W^4 + f_5 W^5$, with $f_i \in \mathbb{F}_{p^2}$. So in the Tate pairing computation f^2, $f \cdot l_{T,T}(Q)$, and $f \cdot l_{T,P}(Q)$ are performed on $\mathbb{F}_{p^{12}}$. Whereas all other computations are performed on $\mathbb{F}_p$ and $\mathbb{F}_{p^2}$.

The detailed procedure of pairing computation including the final exponentiation on BN curve is described in [15] and [16]. Another efficient way of computing final exponentiation is described in [7]. This paper follows the descriptions that are given in [15] for computing the Tate, ate, and R-ate pairings. We use Jacobian coordinate systems for performing elliptic curve operations, where a point (X, Y, Z) corresponds to the point (x, y) in affine coordinates with $x = X/Z^2$ and $y = Y/Z^3$. Let (m, s, i) denote the cost of multiplication, squaring, inversion in $\mathbb{F}_p$. Using Jacobian coordinate system the Miller function of Tate pairing on BN curve requires $27934m$ and the final exponentiation requires $7246m + i$ [15]. Thus the total cost for Tate pairing on BN curve is $35180m + i$. Similarly, the cost of ate pairing is $23047m + i$ and the cost of R-ate pairing is $15093m + 2i$.

3   Implementing $\mathbb{F}_p$ Primitives on FPGA

In 1983 Blakley introduced an interesting algorithm to perform modular multiplication of two integers A and B modulo an integer M [30,31]. It is an iterative binary double-and-add algorithm. The main idea of the algorithm is that it keeps the intermediate result after each iteration below the modulus value. Thus it avoids final division. Let A be the multiplicand, B be the multiplier, and $R = (A \cdot B) \bmod M$, where A, B, M, R are represented in two's complement number system. The binary representation of $B = \sum_{i=0}^{l-1} b_i 2^i$, and R is initialized by A. The algorithm first computes $R = 2R \bmod M$, and if $b_i = 1$ then it computes $R = (R + A) \bmod M$. The mod M operation are performed by single subtraction in both cases.

In the context of $\mathbb{F}_p$ multiplication the modulus M corresponds to p. All arithmetics in $\mathbb{F}_p$ are performed in two's complement number system. Therefore, all values are kept in two's complement number system throughout the whole pairing computation, which avoids input and output conversions like existing implementations [5,6].

3.1   Fast Carry Chains for $\mathbb{F}_p$ Primitives

The main difficulty of the Blakley algorithm is that the computation of addition on large operands. The modified Blakley algorithm for large operands are shown in [22,26]. The use of carry save adder (CSA) helps to speed up the repeated additions on large operands. However these modified versions require at least one final addition on large carry chain. Also they use some pre-computed values which require additional time and storage area.

This paper exploits the features available in an FPGA device for efficient computation of Blakley algorithm on large operands. The specific features that are available in an FPGA device are efficiently utilized for developing arithmetic primitives in $\mathbb{F}_{2^m}$ fields in [14]. However, to the best of the authors knowledge no work is existing in the literature for the same in case of $\mathbb{F}_p$ field. The modern FPGA consists of 16 slices (or 32 LUTs) within a single row which are connected through an in-built fast carry chain (FCC). The FCC can perform addition on two 32-bit operands most efficiently compared to any other adder structures [3,29]. It is experimentally shown that on a Virtex-4 FPGA device the latency of a 32-bit addition using a fast carry chain takes only 5.8 ns; whereas, the same using a carry lookahead structure takes 8.7 ns. Hence, fast carry chain is 1.5 times faster than carry lookahead structure for computing addition of two 32-bit operands on an FPGA platform. In order to compute an addition of two operands longer than 32 bits, the FPGA will utilize more than one row which requires additional routing delay. For example, the addition $(A+B)$ of two 64-bit operands (A, B) using a single 64-bit carry chain is slower than the same using three 32-bit FCC and a 2:1 multiplexor [3].

We develop an efficient 256-bit adder using 32-bit fast carry chains. The repeated Karatsuba decomposition is applied on 256-bit operands. An operand is decomposed upto a depth of three for converting it into eight pieces of 32-bit operands. A 64-bit addition is performed by using three 32-bit fast carry chains with a carry select structure. Let, $A = A_1 2^{32} + A_0$, $B = B_1 2^{32} + B_0$, and $C = A + B$, where A_i, B_i are 32-bit integers. We compute $A_0 + B_0$, $A_1 + B_1 + 0$, and $A_1 + B_1 + 1$ in parallel on three FCC. Then the carry out of the least significant addition $(A_0 + B_0)$ is used to multiplex the results of the most significant additions. Thus the latency of a 64-bit adder is 1 FCC + 1 MUX, where MUX corresponds to a 2:1 multiplexer. Similarly, a 128-bit adder is developed by three 64-bit adders, and a 256-bit adder is developed by three 128-bit adders. Therefore, a 256-bit adder is developed hierarchically from 32-bit adder. At every level of hierarchy it adds one additional MUX in the critical path. Thus the latency of a 256-bit adder is 1 FCC + 3 MUX delay, which is $9.9ns$ on a Virtex-4 FPGA. Whereas, the latency of a 256-bit carry lookahead adder on the same platform is $16.7ns$, which is 1.7 times slower than the above technique.

3.2 Programmable $\mathbb{F}_p$ Primitive

The 256-bit high speed adder circuit that is designed by utilizing fast carry chains and Karatsuba decomposition technique is exploited to develop a programmable $\mathbb{F}_p$ primitive. Figure 1 shows the overall resulting structure of the $\mathbb{F}_p$ adder/subtractor/multiplier unit. We use the same structure for all three $\mathbb{F}_p$ operations for pairing computation. The configuration of the design to perform addition and subtraction can be easily formed. The input parameters (A, B) are added by the right most adder unit and then reduced (if necessary) by adding 2's complement of p. The select lines c_1 and c_2 of mx blocks (multiplexors) are generated from the carry out of 256-bit adders, which decide whether the resultant values are greater than the modulus p. The unit performs addition and subtraction in $\mathbb{F}_p$ in one clock cycle.

Proposed hardware follows the parallelism technique of Montgomery ladder [28] for computing Blakley multiplication algorithm in $\mathbb{F}_p$. The choice of this algorithm is due to its lower hardware cost. The algorithm also provides the scope of adaptability with Montgomery ladder for parallelism. The Montgomery ladder computes two intermediate results (R_0, R_1) in each iteration. It computes independent modular doubling and modular addition in each iteration, which are

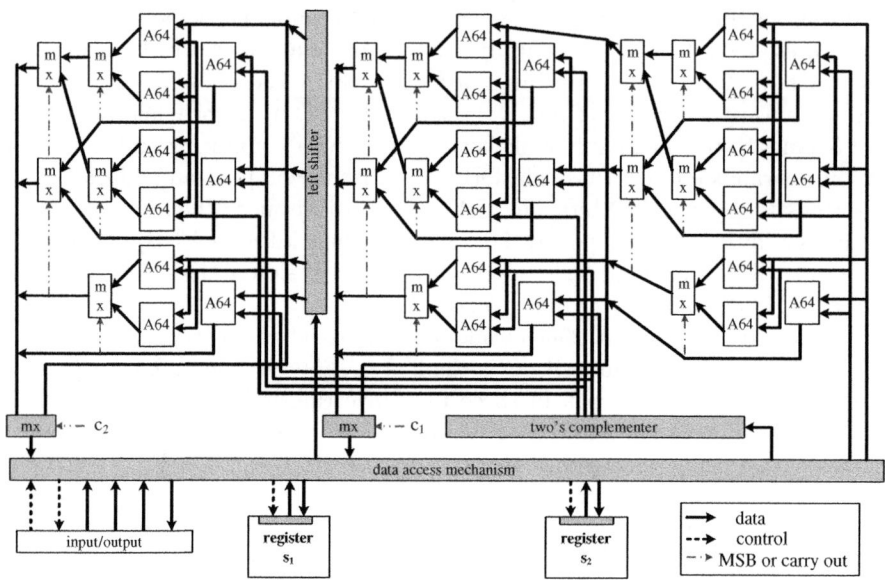

Fig. 1. The architecture of $\mathbb{F}_p$ adder/subtractor/multiplier unit

performed in parallel on the proposed unit. The data transfer in the registers for computing $(A \cdot B) \bmod p$ is as follows :

- The *register* s_1 is initialized by zero, and *register* s_2 is initialized by A, which are used to hold the intermediate result R_0 and R_1, respectively.
- Iterative execution starts from $i = 255$ and goes down to zero, where it is considered that the operands are the members of $\mathbb{F}_p$ with a 256-bit p.
- The right part of *left-shifter* block performs $(R_0 + R_1) \bmod p$. This modular addition is performed by two 256-bit adder units. The first adder performs $R_0 + R_1$, whereas the second one makes the reduction by a 2's complement subtraction. The final result is multiplexed by the mx_1 block. The *left-shifter* performs $2R_{b_i}$ and its left part computes the reduction $(2R_{b_i}) \bmod p$. The final result is multiplexed by mx_2 block. In the architecture, an $A64$ block corresponds to a 64-bit adder unit.
- The *data access mechanism* restores the intermediate results $R_{\bar{b}_i} \leftarrow (R_0 + R_1) \bmod p$ and $R_{b_i} \leftarrow (2R_{b_i}) \bmod p$, where $\bar{b}_i$ represents the complement of b_i.
- After final iteration, the *register* s_1 holds the result of $(A \cdot B) \bmod p$.

The procedures runs iteratively for 256 times. Thus, the above unit performs one 256-bit $\mathbb{F}_p$ multiplication in only 256 cycles.

4   The Pairing Cryptoprocessor (PCP)

This section describes the proposed architecture of the cryptoprocessor. The architecture is based on the efficient utilization of FPGA features. The parallelism techniques are adopted in different level of computations for overall speed up. First, we explain the top level of the design followed by the internal parts.

4.1   The Datapath Design

The major operations for pairing computations are point doubling (PD), point addition (PA), line computation $(l(Q))$, f^2, and $f \cdot l(Q)$. In case of Tate pairing on BN curve, the PA and PD are performed on $E(\mathbb{F}_p)$. Hence, the underline operations are performed in $\mathbb{F}_p$. Similarly, the operation $l(Q)$ is performed in $\mathbb{F}_{p^2}$ while the other two operations are performed in $\mathbb{F}_{p^{12}}$. In case of ate and R-ate pairings, the PA, PD, $l(Q)$ are performed in $\mathbb{F}_{p^2}$, and f^2, $f \cdot l(Q)$ are performed in $\mathbb{F}_{p^{12}}$. However, each of the above computations are well defined and constitute a sequence of $\mathbb{F}_p$ operations which provides a scope of parallelism. The proposed datapath exploits such properties to speed up pairing computations on FPGA platforms.

Figure 2 shows the overall resulting structure of the datapath. Two Configurable $\mathbb{F}_{p^k}$ Arithmetic Units (CAU) are included which perform arithmetic in $\mathbb{F}_p$ and $\mathbb{F}_{p^2}$ depending on their mode of configurations. The instructions that are decoded to configure the CAUs are stored into a small instruction memory segment. There is a special instruction fetch and decode (IFD) unit which reads the respective instructions and converts them to proper configuration signals for both the CAUs. The input data to the CAUs come in parallel from respective registers. The mechanism and regularity of data access for computing above operations are fairly simple. The distribution of access to the registers and resolution of access conflicts are handled efficiently at the runtime by a dedicated hardware block, the data access unit (DAU) that distributes the data access to the CAUs from the correct register and vice versa.

Each CAU performs atmost three $\mathbb{F}_p$ operations in parallel. Thus, overall twelve independent operands along with modulus p and six outputs are accessed in either directions between memory elements and the CAUs. This on-demand concurrent data requests result in multiple independent read or write connections between CAUs and DAU. The DAU takes care of granting accesses. Therefore, a simple multiplexing protocol is used between CAUs and registers, which is able to confirm a request within the same cycle in order not to cause any delay cycles when trying to access data in parallel. The data accesses and instruction sequences are hard coded into the *sequence control* of the architecture which avoids the additional software development costs.

The data access conflicts have been resolved prior to design the DAU protocol. The proposed one is a custom hardware for pairing computations which executes a fixed set of operations. The dependency of the instructions are predefined and thus the access conflicts are known. The priority of the data processing and the respective execution is rearranged accordingly which achieves maximum utilization of CAUs.

Figure 3 depicts the functionality of data access unit. The major connections between registers and CAUs are shown while the DAU performs as a mediator. Due to the demand of parallel access, the proposed pairing hardware stores all

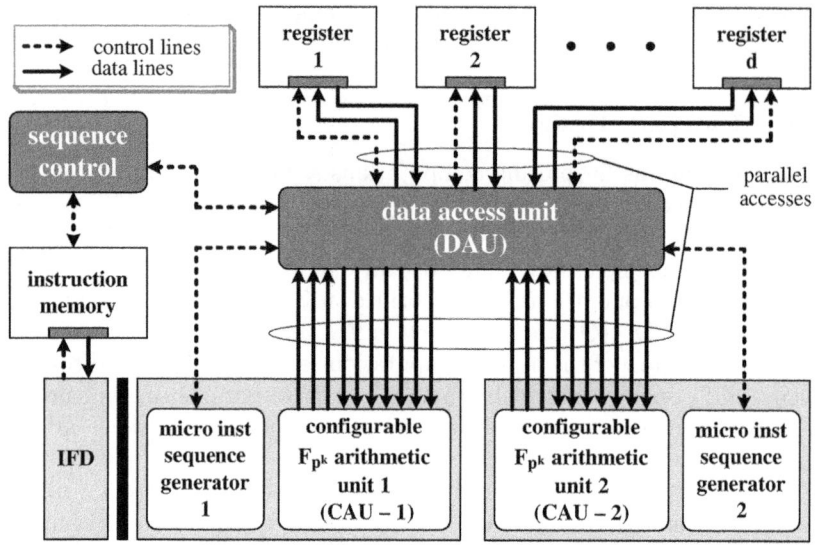

Fig. 2. The datapath of the pairing cryptoprocessor

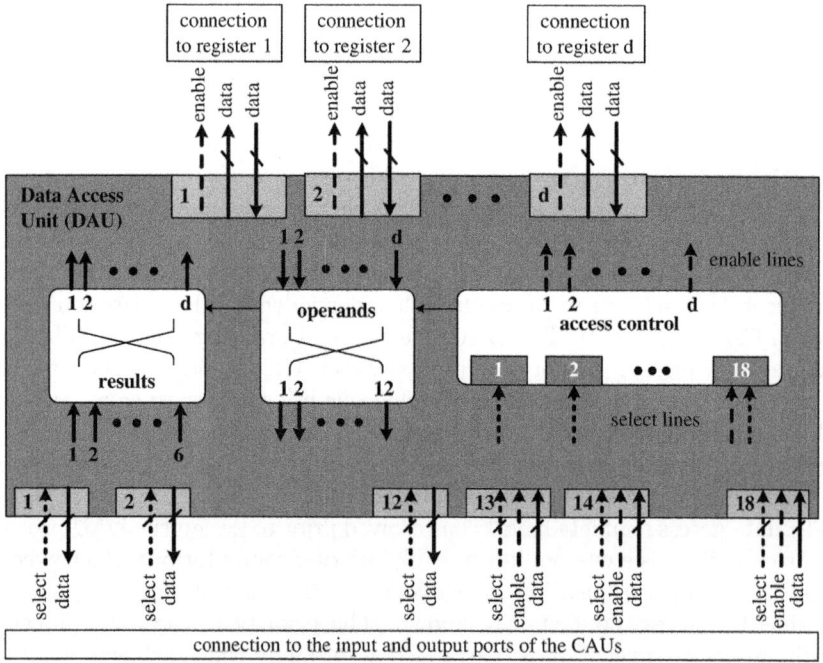

Fig. 3. The interconnect of data access unit (DAU)

intermediate results in its active registers. Fifty 256-bit registers ($d = 50$) are incorporated. Each of the register consists of data-in, data-out, and enable lines. It gets updated by data-in lines when the respective enable signal is invoked. The crossbar switch (*results*) redirects the outputs of each operation to registers. Similarly, the *operands* are redirected from registers to the input ports of the CAUs. The respective select signals are generated prior to the above two redirection procedures. The *access control* block synchronizes the select lines of the multiplexors for operands and results. It also synchronizes the enable signals of registers for restoring the intermediate results.

The *sequence control* runs synchronously with the IFD unit. It generates the select and enable signals with respect to each of the operand and result port which are going inside the DAU. In brief, the *sequence control* generates the signals for controlling input/output to the CAUs and registers while IFD is responsible to generate respective signals for configuring each of the CAUs.

A Configurable $\mathbb{F}_{p^k}$ Arithmetic Unit (CAU). It is observed that the major operations for pairing computations over BN curves are performed either on $\mathbb{F}_p$ or on $\mathbb{F}_{p^2}$. Thus we design a configurable architecture for performing arithmetics in $\mathbb{F}_p$ and $\mathbb{F}_{p^2}$. Figure 4 shows the data processing inside the proposed CAU. It consists of three $\mathbb{F}_p$ arithmetic units, which are individually capable to perform addition, subtraction, and multiplication in $\mathbb{F}_p$. Thus a CAU performs atmost three parallel $\mathbb{F}_p$ operations, which demand six operands and a modulus p in parallel. The final outputs are stored in first three registers, respectively.

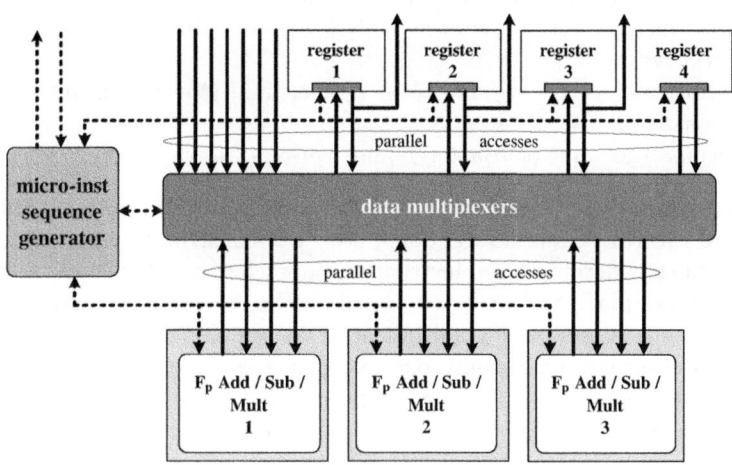

Fig. 4. The architecture of Configurable $\mathbb{F}_{p^k}$ Arithmetic Unit (CAU)

The major $\mathbb{F}_{p^2}$-operations are multiplication and squaring. Let an element $\alpha \in \mathbb{F}_{p^2}$ be represented as $\alpha_0 + \alpha_1 X$, where $\alpha_0, \alpha_1 \in \mathbb{F}_p$ and X is an indeterminate. The formula of Karatsuba multiplication $c = ab$ on $\mathbb{F}_{p^2}$ is :

$$v_0 = a_0 b_0, \; v_1 \; = \; a_1 b_1,$$
$$c_0 = v_0 + \beta v_1,$$
$$c_1 = (a_0 + a_1)(b_0 + b_1) - v_0 - v_1,$$

where $v_0, v_1, c_0, c_1, a_0, a_1, b_0, b_1 \in \mathbb{F}_p$. Here β is a quadratic non-residue in $\mathbb{F}_p$ which is -2 in case of BN curve. Similarly, the squaring $c = a^2$ on $\mathbb{F}_{p^2}$ using Complex method is computed by :

$$v_0 = a_0 a_1,$$
$$c_0 = (a_0 + a_1)((a_0 + \beta a_1)) - v_0 - \beta v_0$$
$$c_1 = 2v_0.$$

Hence the costs of multiplication and squaring in $\mathbb{F}_{p^2}$ are $3m$ and $2m$ [21]. However, due to the parallel independent structure, the cost of both operations on a CAU is only m.

The *micro inst sequence generator* finds the current operation type and generates the respective micro instructions. For example, let the current operation is a $\mathbb{F}_{p^2}$ multiplication. The sequence of operations for computing $c = ab$ in $\mathbb{F}_{p^2}$ on a CAU is as follows :

1. It computes $t_3 \leftarrow a_0 + a_1$ and $t_4 \leftarrow b_0 + b_1$ and stores them into *register-3* and *register-4*, respectively.
2. It performs $t_1 \leftarrow a_0 b_0$, $t_2 \leftarrow a_1 b_1$, and $t_3 \leftarrow t_3 t_4$ in parallel, and stores the results into *register-1*, *register-2*, and *register-3*, respectively.
3. It performs $t_1 \leftarrow t_1 - t_2$ and $t_4 \leftarrow t_1 + t_2$, and stores them into *register-1* and *register-4*.
4. It performs $t_1 \leftarrow t_1 - t_2$ and $t_2 \leftarrow t_3 - t_4$, and stores them into *register-1* and *register-2*.
5. It outputs the value of *register-1* and *register-2* as c_0 and c_1, respectively.

The *micro inst sequence generator* generates micro instruction like register enable, CAU reset, operand select, done, etc. It is constructed as a typical state machine which generates micro instructions at each state. Its deterministic state transition takes place at every cycle based on the current state and overall status of the CAU. In case of $\mathbb{F}_p$ multiplication, it remains in a same state for 256 cycles. Whereas, it remains only one cycle in a same state for computing $\mathbb{F}_p$ addition and subtraction. Thus the computation of $c = ab$ in $\mathbb{F}_{p^2}$ takes only 260 cycles which costs approximately m.

4.2 Computation of the Pairings on PCP

In this paper we follow the formula and algorithms for the computation of cryptographic pairings that are given in [15]. However, due to the multiple functional units the operations are performed in parallel by the proposed PCP. In Jacobian coordinates the formulae for doubling a point $T = (X, Y, Z)$ are $2T = (X_3, Y_3, Z_3)$ where $X_3 = 9X^4 - 8XY^2$, $Y_3 = (3X^2)(4XY^2 - X_3) - 8Y^4$

and $Z_3 = 2YZ$. The computation of $2T$, $l_{T,T}(Q)$, and f^2 are performed in parallel. The tangent line at T, after clearing denominators, is $l(x, y) = 3X^3 - 2Y^2 - 3X^2Z^2x + Z_3Z^2y$ [23].

In case of Tate pairing computation on BN curves $X, Y, Z, X_3, Y_3, Z_3 \in \mathbb{F}_p$ and $x, y \in \mathbb{F}_{p^2}$. The above operations are performed by one of the CAUs which costs $6m$. At the same time another CAU starts the computation of f^2. We represent the Miller function f as $(f_{0,0} + f_{0,1}v + f_{0,2}v^2) + (f_{0,0} + f_{0,1}v + f_{0,2}v^2)w$, where $f_{i,j} \in \mathbb{F}_{p^2}$. The equivalent representations of f are :

$$f = f_0 + f_1 w, \text{ where} f_0, f_1 \in \mathbb{F}_{p^6}$$
$$= (f_{0,0} + f_{0,1}v + f_{0,2}v^2) + (f_{1,0} + f_{1,1}v + f_{1,2}v^2)w, \text{ where} f_{i,j} \in \mathbb{F}_{p^2}$$
$$= f_{0,0} + f_{1,0}W + f_{0,1}W^2 + f_{1,1}W^3 + f_{0,2}W^4 + f_{1,2}W^5$$

Thus, the squaring (f^2) is performed in $\mathbb{F}_{p^{12}}$ which requires two $\mathbb{F}_{p^6}$ multiplications using Complex method. Now, one $\mathbb{F}_{p^6}$ multiplication requires six multiplications in $\mathbb{F}_{p^2}$. The first $\mathbb{F}_{p^6}$ multiplication is computed in parallel with $2T$, $l_{T,T}(Q)$ by another CAU which costs $6m$. The second $\mathbb{F}_{p^6}$ multiplication is performed by both the CAUs, which costs only $3m$ in the PCP. Therefore, the total cost of computing $2T$, $l_{T,T}(Q)$, and f^2 by the PCP is $9m$.

The l_Q is represented as : $(l_0 + l_1 v) + (l_2 v)w$, where $l_0 \in \mathbb{F}_p$, $l_1, l_2 \in \mathbb{F}_{p^2}$, which is equivalent to $l_0 + l_1 W + l_2 W^3$. The computation of $f \cdot l(Q)$ is done by three $\mathbb{F}_{p^6}$ multiplications using Karatsuba multiplication procedure. Due to the sparse representation of l_Q the costs are lesser than the actual costs. In total the computation of $f \cdot l(Q)$ requires 37 $\mathbb{F}_p$ multiplications, which costs only $7m$ in PCP.

The formulae for mixed Jacobian-affine addition are the following: if $P = (X_1, Y_1, Z_1)$ is in Jacobian coordinates and $Q = (X_2, Y_2)$ is in affine coordinates, then $P+Q = (X_3, Y_3, Z_3)$ where $X_3 = (Y_2Z_1^3 - Y_1)^2 - (X_2Z_1^2 - X_1)^2(X_1 + X_2Z_1^2)$, $Y_3 = (Y_2Z_1^3 - Y_1)[X_1(X_2Z_1^2 - X_1)^2 - X_3] - Y_1(X_2Z_1^2 - X_1)^3$, $Z_3 = (X_2Z_1^2 - X_1)Z_1$. The line through T and P is $l(x, y) = [(Y_2Z_1^3 - Y_1)X_2 - Y_2Z_3] - (Y_2Z_1^3 - Y_1)x + Z_3y$. The cost of computing $T + P$, $l_{T,P}(Q)$ is $5m$ in the PCP.

Therefore, the cost of computing doubling step of the Miller algorithm is $16m$ and the cost for evaluating addition step is $12m$ in the PCP. The total cost for evaluating iterative Miller function of the Tate pairing computation over BN curves is $5176m$. The final exponentiation is computed by the exact procedures that are described in [15]. It requires one inversion in $\mathbb{F}_p$ which we perform by a^{p-2}. The cost for computing the final exponentiation is $1477m$ in our PCP. Hence, the total cost for computing a Tate pairing over BN curves by our cryptoprocessor is $6653m$, which takes $1,729,780$ cycles.

The ate pairing interchanges the input points of Tate pairing and it runs a smaller number of iterations. It uses $t - 1$ (instead of r) to determine the number of iterations in the Miller operation [15]. In case of BN curve $t \approx \sqrt{r}$, which makes the number of iterations halved. The computation costs $3165m$, $1477m$, and $4642m$ for the Miller operation, the final exponentiation, and the ate pairing, respectively on our proposed hardware. Hence, the number of cycles required to compute an ate pairing by the PCP is $1,206,902$.

The R-ate pairing follows the same procedures of ate pairing but it uses $a = 6z + 2$ (instead of $t - 1$) to determine the number of iterations in the Miller operation. Since, $a \approx \sqrt{t}$ on BN curves, the Miller operation in R-ate pairing has half as many iterations as in ate pairing computation. In case of R-ate pairing the Miller operation consists an additional step which requires an inversion in $\mathbb{F}_p$. The costs $1681m$, $1477m$, and $3158m$ for the Miller operation, the final exponentiation, and the R-ate pairing, respectively. The total number of cycles required to compute an R-ate pairing is $821,080$ by the proposed hardware.

5   Results

The whole design has been done in Verilog (HDL). All results have been obtained from the place-and-route report of Xilinx ISE Design Suit [10] using a Virtex-4 xc4vlx200-12ff1531 FPGA device with a supply voltage of $1.2V$. The design can run at a maximum frequency of $50MHz$. The pairing hardware uses around $52k$ logic slices including controllers and data access unit. It uses $27k$ flip flops for registers. It finishes one Tate, ate, and R-ate pairing computations in $34.6ms$, $24.2ms$, and $16.4ms$, respectively. Table 1 shows the implementation results.

Table 1. Implementation results of proposed hardware on xc4vlx200 device

Operation	Slice	LUT	FF	Frequency [MHz]	Cycles	Security [bit]	Times [ms]
Tate					1730 k		34.6
ate	52 k	101 k	27 k	50	1207 k	128	24.2
R-ate					821 k		16.4

The critical path of the design goes through the data access mechanism, then through two 256-bit adders, the multiplexer mx_1, and back through data access mechanism. In § 3.1 it is shown that the latency of a 256-bit adder circuit is $9.9ns$. However, this addition latency consists of input buffer delay of $1.3ns$, addition logic delay of $6.2ns$, and output buffer delay of $2.4ns$. The individual delays of the addition logic includes input and output buffer delays. In our architecture the critical path is within two internal registers which includes neither the input buffer nor the output buffer. Therefore, the total latency of the critical path of the design is calculated as $3.8ns + 2 \times 6.2ns + 1.6ns + 2.2ns = 20ns$.

5.1   Comparison with Pairing Implementations

This section provides the performance comparison with related pairing implementations over BN curves. Performances are compared with actual implementations of cryptographic pairings on software and dedicated hardware achieving a 128-bit security level. Hardware implementations of η_T pairing over binary and cubic curves are shown in [9,18]. These designs are for lower security levels

Table 2. Hardware and software implementations of pairing over BN-curves

Reference designs	Platform	Pairing	Frequency [MHz]	Area	Cycles	Times [ms]
our design	Virtex-4	Tate	50	52 kSlices	1730 k	34.6
		ate			1206 k	24.2
		R-ate			821 k	16.4
Kammler et al. [5]	130 nm CMOS	Tate	338	97 kGates	11 627 k	34.4
		ate			7706 k	22.8
		R-ate			5340 k	15.8
Fan et al. [6] §	130 nm CMOS	ate	204	183 kGates	862 k	4.2
		R-ate			593 k	2.9
Naehrig et al. [1]	core2 Q6600	Opt.-ate	-	-	4 470 k	-
Beuchat et al. [4]	core i7 2:8GHz	Opt.-ate	-	-	2 630 k	-
Hankerson et al. [15]	Pentium-4	ate	2400	-	81 000 k	-
	64-bit core2	R-ate			10 000 k	-
Grabher et al. [13]	64-bit core2	ate	2400	-	14 640 k	-
Devegili et al. [16]	Pentium-4	Tate	3400	-	156 740 k	-
		ate			133 620 k	-

§ implementation specifically for BN-curves with fixed parameters.

(72-bit) and hence it shall be unfair to compare with the present work. Table 2 gives a performance comparison of related hardware and software implementations.

Due to the parallel structure our PCP computes six $\mathbb{F}_p$ multiplications in parallel which are completed in 256 cycles. The main features that strengthen the proposed PCP for pairing computations are as follows :

- The proposed cryptoprocessor is the first FPGA results for pairing computation with 128-bit security.
- Our adopted parallelism and efficient use of two $\mathbb{F}_{p^k}$ arithmetic cores reduce the total number of cycles drastically.
- Due to the inherent properties the frequency of a design in FPGA is much lower than that in ASIC (CMOS standard cell). However, the speed achieved of the PCP is comparable to the CMOS standard cell design.
- The PCP is flexible to configure for different curve parameters.

The underlying platform plays a crucial role in determining the performance of a design. Thus, existing designs on different platforms does not lead to a fair comparison. We try to find out the platform independent features of existing designs and compare them with our proposed one. The cycles required to compute pairings on different designs may be considered such a parameter.

Kammler et al. [5] reported the first hardware implementation of cryptographic pairings achieving a 128-bit security. In [5] the proposed hardware is not only a cryptoprocessor, but an actual ASIP : it is in fact a general purpose processor, augmented with finite field arithmetic units in order to compute pairings. It uses the same z that we have considered to generate a 256-bit BN curve. The Montgomery algorithm is used for $\mathbb{F}_p$ multiplication. The platform of the design is 130 nm CMOS standard cell library, whereas our design is on Virtex-4 FPGA. The main feature of the design [5] is the fast modular multiplication in $\mathbb{F}_p$ which takes only 68 cycles. The average cycle count of our PCP for one $\mathbb{F}_p$ multiplication is only 43 which is 1.6 times faster than [5]. With respect to the Tate pairing computation, the design of [5] takes 11 627 k cycles, whereas our design takes only 1730 k cycles, which is much less (0.15 times) compared to [5].

Fan *et al.* [6] proposed a processor for cryptographic pairing over BN curves. They designed a fast modular multiplier in $\mathbb{F}_p$ only for BN parameters which takes only 23 cycles. The 130 nm ASIC design of [6] provides the best known performance which takes only $2.9ms$ for computing a R-ate pairing over BN-curve. This design also attain smaller area-latency product than that in [5]. But the main drawback of the design proposed in [6] is that it does not provide the flexibility to compute pairings on chosen parameters. Whereas, our design provides the above flexibility in all aspects which indeed requires more cycles.

The results of software implementations [15,13] are quite impressive. On an Intel 64-bit core2 processor, R-ate pairing requires only $10,000,000$ cycles. The advantages of Intel core2 is that it has a fast multiplier (two full 64-bit multiplications in 8 cycles) and relatively high clock frequency. It takes 13 times more clock cycles than our cryptoprocessor. In a very recent work by Naehrig et al. [1] shows that the Optimal-ate pairing on BN curves can be computed by $4,470,408$ cycles on an Intel Core 2 Quad Q6600 processor. The software implementation of same pairing on a different curve is described in [4]. It takes only 2.63 million clock cycles on a Intel Core i7 : 2.8 GHz processor. However, the exact time required to compute pairings by executing softwares on a Desktop or Server systems are not predictable. It depends on so many other factors like available cache memory, context switching, bus speed of the system, etc.

6   Conclusion

In this paper we explored the inherent FPGA features for designing efficient $\mathbb{F}_p$ primitives. The parallel scheduling has been applied to speed up multiplication in $\mathbb{F}_{p^2}$ and $\mathbb{F}_{p^{12}}$. The proposed pairing cryptoprocessor can be programmed for any curve parameters. It provides a comparable speed with the existing ASIC designs. The overall clock cycles required to compute pairings over BN curves are less than existing designs. To the best of our knowledge it is the first FPGA result for high security (128-bit) cryptographic pairings.

The recent work by Granger and Scott [2] shows that a 1-2-4-12 towering is to be preferred to a 1-2-6-12 towering as implemented in this work. Therefore, in future it could be considered for further speedup of pairing computations.

Acknowledgement

The authors would like to thank the anonymous reviewers for their critical suggestions that greatly improved the quality of this paper.

References

1. Naehrig, M., Niederhagen, R., Schwabe, P.: New software speed records for cryptographic pairings. Cryptology ePrint Archive, Report 2010/186, http://eprint.iacr.org/
2. Granger, R., Scott, M.: Faster Squaring in the Cyclotomic Subgroup of Sixth Degree Extensions. In: Nguyen, P.Q., Pointcheval, D. (eds.) PKC 2010. LNCS, vol. 6056, pp. 209–223. Springer, Heidelberg (2010)
3. Ghosh, S., Mukhopadhyay, D., Roychowdhury, D.: High Speed F_p Multipliers and Adders on FPGA Platform. In: DASIP 2010, Scotland (2010)
4. Beuchat, J.L., Díaz, J.E.G., Mitsunari, S., Okamoto, E., Henríquez, F.R., Teruya, T.: High-Speed Software Implementation of the Optimal Ate Pairing over Barreto-Naehrig Curves. Cryptology ePrint Archive, Report 2010/354, http://eprint.iacr.org/
5. Kammler, D., Zhang, D., Schwabe, P., Scharwaechter, H., Langenberg, M., Auras, D., Ascheid, G., Mathar, R.: Designing an ASIP for cryptographic pairings over Barreto-Naehrig curves. In: Clavier, C., Gaj, K. (eds.) CHES 2009. LNCS, vol. 5747, pp. 254–271. Springer, Heidelberg (2009)
6. Fan, J., Vercauteren, F., Verbauwhede, I.: Faster F_p-arithmetic for cryptographic pairings on Barreto-Naehrig curves. In: Clavier, C., Gaj, K. (eds.) CHES 2009. LNCS, vol. 5747, pp. 240–253. Springer, Heidelberg (2009)
7. Scott, M., Benger, N., Charlemagne, M., Perez, L.J.D., Kachisa, E.J.: On the Final Exponentiation for Calculating Pairings on Ordinary Elliptic Curves. In: Shacham, H., Waters, B. (eds.) Pairing 2009. LNCS, vol. 5671, pp. 78–88. Springer, Heidelberg (2009)
8. Lee, E., Lee, H.S., Park, C.M.: Efficient and generalized pairing computation on abelian varieties. Cryptology ePrint Archive, Report 2009/040, http://eprint.iacr.org/
9. Beuchat, J., Detrey, J., Estibals, N., Okamoto, E., Rodríguez-Henríquez, F.: Hardware accelerator for the Tate pairing in characteristic three based on Karatsuba-Ofman multipliers. Cryptology ePrint Archive, Report 2009/122 (2009)
10. Xilinx ISE Design Suit (2009), http://www.xilinx.com/tools/designtools.htm
11. Hoffstein, J., Pipher, J., Silverman, J.H.: An introduction to mathmatical cryptography. Springer, Heidelberg (2008)
12. Barenghi, A., Bertoni, G., Breveglieri, L., Pelosi, G.: A FPGA coprocessor for the cryptographic Tate pairing over F_p. In: Proc. Fifth Intl. Conf. Information Technology: New Generations - ITNG 2008, pp. 112–119 (2008)
13. Grabher, P., Großschädl, J., Page, D.: On software parallel implementation of cryptographic pairings. In: Avanzi, R.M., Keliher, L., Sica, F. (eds.) SAC 2008. LNCS, vol. 5381, pp. 35–50. Springer, Heidelberg (2009)
14. Rebeiro, C., Mukhopadhyay, D.: High speed compact elliptic curve cryptoprocessor for FPGA platforms. In: Chowdhury, D.R., Rijmen, V., Das, A. (eds.) INDOCRYPT 2008. LNCS, vol. 5365, pp. 376–388. Springer, Heidelberg (2008)
15. Hankerson, D., Menezes, A., Scott, M.: Software implementation of pairings. In: Joye, M., Neven, G. (eds.) Identity-Based Cryptography (2008)
16. Devegili, A.J., Scott, M., Dahab, R.: Implementing cryptographic pairings over Barreto-Naehrig curves. In: Takagi, T., Okamoto, T., Okamoto, E., Okamoto, T. (eds.) Pairing 2007. LNCS, vol. 4575, pp. 197–207. Springer, Heidelberg (2007)

17. Barke, E., Barker, W., Burr, W., Polk, W., Smid, M.: Recommendation for key management - part 1: General (revised). National Institute of Standards and Technology, NIST Special Publication 800-57 (2007)
18. Shu, C., Kwon, S., Gaj, K.: FPGA accelerated Tate pairing based cryptosystems over binary fields. In: FPT 2006, pp. 173–180 (2006)
19. Barreto, P.S.L.M., Naehrig, M.: Pairing-friendly elliptic curves of prime order. In: Preneel, B., Tavares, S. (eds.) SAC 2005. LNCS, vol. 3897, pp. 319–331. Springer, Heidelberg (2006)
20. Hess, F., Smart, N.P., Vercauteren, F.: The Eta pairing revisited. IEEE Transactions on Information Theory 52(10), 4595–4602 (2006)
21. Devegili, A., ÓhÉigeartaigh, C., Scott, M., Dahab, R.: Multiplication and squaring on pairing-friendly fields. Cryptology ePrint Archive, Report 2006/471 (2006)
22. Amanor, D.N., Paar, C., Pelzl, J., Bunimov, V., Schimmler, M.: Efficient hardware architectures for modular multiplication on FPGAs. In: International Conference on Field Programmable Logic and Applications 2005, pp. 539–542 (2005)
23. Chatterjee, S., Sarkar, P., Barua, R.: Efficient computation of Tate pairing in projective coordinate over general characteristic fields. In: Park, C.-s., Chee, S. (eds.) ICISC 2004. LNCS, vol. 3506, pp. 168–181. Springer, Heidelberg (2005)
24. Galbraith, S.: Pairings. In: Blake, I.F., Seroussi, G., Smart, N.P. (eds.) Advances in Elliptic Curve Cryptography. London Mathematical Society Lecture Note Series, ch. IX, Cambridge University Press, Cambridge (2005)
25. Miller, V.S.: The Weil pairing, and its efficient calculation. Journal of Cryptology 17, 235–261 (2004)
26. Bunimov, V., Schimmler, M.: Area and time efficient modular multiplication of large integers. In: ASAP 2003. IEEE Computer Society, Los Alamitos (2003)
27. Barreto, P.S.L.M., Kim, H.Y., Lynn, B., Scott, M.: Efficient algorithms for pairing based cryptosystems. In: Yung, M. (ed.) CRYPTO 2002. LNCS, vol. 2442, pp. 354–368. Springer, Heidelberg (2002)
28. Joye, M., Yen, S.M.: The Montgomery powering ladder. In: Kaliski Jr., B.S., Koç, Ç.K., Paar, C. (eds.) CHES 2002. LNCS, vol. 2523, pp. 291–302. Springer, Heidelberg (2003)
29. Hauck, S., Hosler, M.M., Fry, T.W.: High-performance carry chains for FPGAs. In: FPGA 1998, pp. 223–233 (1998)
30. Blakley, G.R.: A computer algorithm for calculating the product A*B modulo M. IEEE Transactions on Computers C-32(5), 497–500 (1983)
31. Sloan, K.R.: Comments on a computer algorithm for calculating the product A*B modulo M. IEEE Transactions on Computers C-34(3), 290–292 (1985)

Author Index

GPSR Compliance

The European Union's (EU) General Product Safety Regulation (GPSR) is a set of rules that requires consumer products to be safe and our obligations to ensure this.

If you have any concerns about our products, you can contact us on ProductSafety@springernature.com

In case Publisher is established outside the EU, the EU authorized representative is:

Springer Nature Customer Service Center GmbH
Europaplatz 3
69115 Heidelberg, Germany

Batch number: 09478804

Printed by Printforce, the Netherlands